Quantum Statistical Properties of Radiation

WILEY SERIES IN PURE AND APPLIED OPTICS

Advisory Editor

Stanley S. Ballard, University of Florida

BLOOM • *Gas Lasers*

CAULFIELD AND LU • *The Applications of Holography*

FRANCON AND MALLICK • *Polarization Interferometers*

GREEN • *The Middle Ultraviolet; Its Science and Technology*

HEARD • *Laser Parameter Measurements Handbook*

HUDSON • *Infrared System Engineering*

LENGYEL • *Introduction to Laser Physics*

LENGYEL • *Lasers*, Second Edition

LEVI • *Applied Optics, A Guide to Optical System Design*, Volume I

LOUISELL • *Quantum Statistical Properties of Radiation*

MOLLER AND ROTHSCHILD • *Far-Infrared Spectroscopy*

PRATT • *Laser Communication Systems*

ROSS • *Laser Receivers*

SHULMAN • *Optical Data Processing*

STEELE • *Optical Lasers in Electronics*

WILLIAMS AND BECKLUND • *Optics*

ZERNIKE AND MIDWINTER • *Applied Nonlinear Optics*

Quantum Statistical Properties of Radiation

WILLIAM H. LOUISELL
Professor of Physics and Electrical Engineering
University of Southern California

John Wiley & Sons, New York / London / Sydney / Toronto

Library of Congress Cataloging in Publication Data

Louisell, William Henry, 1924–
Quantum statistical properties of radiation.

"A Wiley-Interscience publication."
Includes bibliographical references.
1. Quantum electrodynamics. 2. Quantum statistics.
I. Title.

QC680.L65 537.6 73-547
ISBN 0-471-54785-9

Printed in the United States of America

10 9 8 7 6 5 4 3 2 1

Preface

The invention of the laser was directly responsible for a tremendous development in the field of nonequilibrium quantum statistical mechanics. Although many people have made important contributions, W. E. Lamb, Jr., of Yale University, H. Haken of the *Technische Hochschule* in Stuttgart, Germany, and M. Lax of Bell Telephone Laboratories and City College of New York and their collaborators have been the trail blazers. One of the purposes of this book is to present some of the developments in the theory of the quantum statistical properties of radiation in the hope they may be useful in other areas of physics. This book is not intended to review the field, since only principles and techniques are stressed. References have been chosen only to give credit to the source or to illustrate the fundamental techniques.

Emphasis has been placed on the work of Lax,* since I have worked closely with him. Furthermore, Haken† has written an excellent article which presents his work while Lamb‡ in collaboration with M. O. Scully and M. Sargent is in the process of writing a book giving the Lamb school viewpoint. As a result, I feel that these two monumental works adequately cover the field and no duplication here is attempted.

Little mention is made in the book of the tremendous advances of R. J. Glauber of Harvard and L. Mandel and E. Wolf of the University of Rochester on the coherence properties of laser light. The reader should consult these authors. Furthermore, no attempt has been made to cover the field of nonlinear optics§ led by Bloembergen of Harvard and Franken of Michigan to mention but a few of the major contributors.

For the sake of presenting a cohesive account of the quantum statistical properties of radiation from first principles, I have borrowed heavily from material from my book *Radiation and Noise in Quantum Electronics,*

* M. Lax, *Brandeis University Summer Institute of Theoretical Physics, 1966, Statistical Physics*, Vol. 2, M. Cretien, E. P. Gross, and S. Leser (Eds.), Gordon and Breach, 1968.

† H. Haken, *Handbuch der Physik*, XXV 12C, *Light and Matter*, L. Genzel (Ed.).

‡ W. E. Lamb, Jr., M. O. Scully, and M. Sargent, to be published.

§ N. Bloembergen, *Nonlinear Optics*, New York: W. A. Benjamin, 1965.

(McGraw-Hill, 1964). The publisher has very generously allowed the use of this material in the present work.

The book is intended for physics and electrical engineering graduate students interested in the field of quantum electronics, although it should have interest to those in other areas such as solid-state and low temperature where quantum stochastic processes are of importance. Some background in quantum mechanics and statistical mechanics is necessary.

In Chapter 1, I present the Dirac formulation of nonrelativistic quantum mechanics which many readers may be able to omit. Sections 1.20 and 1.21 introduce the density operator and the reduced density operator, respectively, which play a central role in the book.

Chapter 2 discusses some elementary quantum systems. In Part I the quantum theory of a simple harmonic oscillator is presented and the boson creation and annihilation operators are introduced which play a central role in the quantum theory of radiation. The coherent state whose properties have been studied extensively by Glauber, Klauder, and many others is introduced.

In Part II of Chapter 2, orbital angular momentum and Pauli spin operators are studied. In Part III a brief discussion of the interaction of a nonrelativistic electron with an electric and magnetic field is given.

Chapter 3 gives a rather extensive development of operator algebra. Much of this material has never appeared in book form before. No effort has been made at mathematical rigor. Rather the intent is to present the techniques in a form which will make them useful. The generalization of ordering techniques to arbitrary quantum operators introduced by Lax should prove useful in other areas of physics rather than just in the field of laser physics. By their use one may avoid annoying quantum commutation questions and yet retain all the quantum features of a problem. Characteristic functions, the Wigner distribution function, Wick's theorem, and the generalized Wick theorem for bosons are treated. Several applications of the techniques are presented as well as the principle of maximum entropy introduced by Shannon in Communication theory and developed further by Jaynes and co-workers in statistical mechanics applications. Some readers may justifiably find criticism of this chapter due to the great amount of formalism, but it is felt that familiarity with the techniques presented may be useful in the future in simplifying other physical problems.

The electromagnetic field is quantized in Chapter 4 in a standard presentation and the density operator as it applies to a radiation field is discussed.

Chapter 5 studies the interaction of radiation with matter. Various topics include the absorption and emission of radiation by an atom, the Wigner–Weisskopf theory of natural linewidth and the Lamb shift, the Kramers–Heisenberg scattering cross-section with applications to Thomson and

Raman scattering as well as resonance fluorescence, the Doppler effect, and such.

A rather complete discussion of the quantum theory of damping using the density operator is presented in Chapter 6. A model for a loss mechanism is introduced which has applications in many areas of physics. The Markoff approximation is given in both the Schrödinger and Heisenberg pictures. Fokker–Planck equations for a damped harmonic oscillator and for atoms with linewidth are derived and discussed. The rotating wave van der Pol oscillator which arises in the theory of laser linewidth is treated.

The Langevin approach to the quantum theory of damping is the subject of Chapter 7. The fluctuating random forces present in classical Langevin theory become operators in the quantum treatment which causes unique problems which are discussed. The use of generalized associated distribution functions introduced in Chapter 3 aids materially in the clarification of such problems.

To illustrate the principles of the first seven Chapters, I consider in Chapter 9 the statistical properties of laser radiation. For completeness and by way of introduction, I present a brief account of Lamb's semiclassical theory of a laser in Chapter 8. The theory which was developed earlier has also proved useful in studying the statistical properties of optical parametric devices, and M. J. Stephens and M. O. Scully have used these techniques very profitably to study the statistical properties of superconducting Josephson junctions.

Many people over the years have been very helpful in making this book possible. In particular it is a pleasure to acknowledge help and encouragement from M. Lax, J. P. Gordon, and L. R. Walker of Bell Telephone Laboratories, as well as H. Heffner of Stanford University. Also special thanks are due to Beatrice Shube and James Gaughan of John Wiley & Sons, Inc., for their help on the manuscript.

WILLIAM H. LOUISELL

University of Southern California
April 1973

Contents

PART II BOSON CREATION AND ANNIHILATION OPERATORS

Quantum Statistical Properties of Radiation

1

Dirac Formulation of Quantum Mechanics

The failure of classical mechanics to account for many experimental results such as the stability of atoms and matter, blackbody radiation, specific heat of solids, wave-particle duality of light and material particles, and such, led physicists to the realization that classical concepts were inherently inadequate to describe the physical behavior of events on an atomic scale. To explain these phenomena, a fundamental departure from classical mechanics was necessary. This departure took the form of postulating, as a fundamental law of nature, that there is a limit to the accuracy with which a measurement (or observation) on a physical system can be made. That is, the actual measurement itself disturbs the system being measured in an uncontrollable way, regardless of the care, skill, or ingenuity of the experimenter. The disturbance produced by the measurement in turn requires modification of the classical concept of causality, since, in the classical sense, there is a causal connection between the system and the measurement. This leads to a theory in which one can predict only the probability of obtaining a certain result when a measurement is made on a system rather than an exact value, as in the classical case. This probability interpretation of the theory is fundamentally different from classical statistical theory. In the latter case, probability is necessary for the practical reason that one cannot measure, for example, the coordinates and momenta of 10^{23} gas molecules in a container, although in principle it would be possible to do so with complete precision. In quantum mechanics, the precise measurement of both coordinates and momenta is not possible even in principle because of the disturbance caused by a measurement.

If, on the other hand, no measurements are made for a certain time, the system follows a causal law of development during this interval. That is, when no intervening measurement is made, the state of a system at time t develops from the state at an earlier time t_0 in a perfectly predictable and therefore causal way.

The lack of causality and the probability interpretation of quantum mechanics require for its description a type of mathematics different in many ways from the mathematics used to describe classical mechanics. The presentation of this mathematics will form a large part of this chapter.

Classical mechanics must be contained as a limiting case in quantum mechanics because, if the disturbance caused by an observation may be neglected, classical mechanics is valid. The quantum description of a system must shift to a classical description in this limit, provided the quantum system has a classical analog. This is called the *correspondence principle* and restricts the possible forms that a quantum theory may have.

In this chapter we give a simplified treatment of the Dirac formulation of nonrelativistic quantum mechanics. We restrict ourselves to one-dimensional problems, for the most part, since the extension to three dimensions is fairly straightforward. For simplicity, the problem of degeneracy is also discussed only briefly. In the latter part of the chapter we show how the Schrödinger formulation may be obtained as a special case of the more general Dirac formulation.

No effort is made to be mathematically rigorous, and for simplicity many of the subtle and more difficult points are omitted. The postulates of the theory are not complete but should be sufficient to give the reader a working acquaintance with the mathematical methods and the physical concepts involved in quantum mechanics. In short, this chapter is intended only as an introduction to the Dirac formulation of quantum mechanics; for a deeper insight, the reader is referred to any of a number of excellent books on quantum mechanics [1–7].

The Dirac formulation involves the concept of vectors (and operators) in a space that may have a finite or an infinite number of dimensions. Let us give a simple illustration of the way in which such vectors arise in the theory. We shall consider a particle of mass m constrained to move in one dimension in a potential $V(q)$, where q is the coordinate of the particle which may have any value from $-\infty$ to $+\infty$; that is, the particle may be anywhere in the one-dimensional space. According to the Schrödinger formulation of wave mechanics [4], the state of the particle at time t is described by a wave function in the position representation, $\psi(q, t)$. If no intervening measurements are made, this state develops in a completely causal way from the state at time t_0, $\psi(q, t_0)$, according to the postulated Schrödinger wave equation

$$\left[-\frac{\hbar^2}{2m}\frac{\partial^2}{\partial q^2} + V(q)\right]\psi(q, t) = i\hbar\frac{\partial}{\partial t}\psi(q, t),$$

where $\hbar$ is Planck's constant divided by 2π. The probability interpretation (necessary when a measurement is made to determine the position of the particle) of $\psi(q, t)$ is as follows: $|\psi(q, t)|^2\, dq$ gives the probability of finding the particle between q and $q + dq$ at time t when a measurement of position is made.

We may take the Fourier transform of $\psi(q, t)$ to obtain another wave function*

$$\varphi(p, t) = \frac{1}{\sqrt{2\pi\hbar}} \int_{-\infty}^{\infty} \psi(q, t) \exp\left(-\frac{ipq}{\hbar}\right) dq.$$

This is called the wave function in the momentum representation, where p represents the momentum of the particle. It is completely determined by $\psi(q, t)$, which represents the state of the system at time t. It is therefore reasonable to say that $\varphi(p, t)$ represents the *same* dynamical state as $\psi(q, t)$. It is just another way of describing the same state. For the momentum wave function the probability interpretation is that $|\varphi(p, t)|^2\, dp$ gives the probability that a measurement of the momentum will yield a value between p and $p + dp$.

The theory can be developed in an entirely equivalent way in either the position or the momentum representation. In fact, the representation plays a role analogous to a coordinate system in geometry. Since, in ordinary geometry, problems may be solved by means of vectors, without the use of a coordinate system (and in more generality), it is interesting to ask if quantum mechanics may be formulated without the use of a particular representation. The results would be independent of any particular representation then. The obvious advantages of using a representation in such a formulation would not be lost, however. A convenient representation could always be used to carry out a calculation just as a coordinate system may be chosen when vectors are used. This is the goal of the Dirac formulation of quantum mechanics: to develop the theory independent of any specific representation.

To see how to go about this program, let us attempt† to give a geometrical interpretation to the wave function $\psi(q)$ at time t to take advantage of the concept of vectors. The coordinate q can have any value from $-\infty$ to $+\infty$, as noted earlier. For each specific value, say $q_1, q_2, q_3, \ldots$, the wave function has a value $\psi(q_1)$, $\psi(q_2)$, $\psi(q_3), \ldots$. We may imagine that an infinite-dimensional space has a set of mutually perpendicular axes each labeled by one of the values of q $(q_1, q_2, \ldots)$, and that $\psi(q_1)$ is the projection of some vector on the q_1 axis, $\psi(q_2)$ is the projection of the same vector on the q_2

* The Fourier transform exists if $\int_{-\infty}^{\infty} |\psi(q)|^2\, dq$ exists; that is, $\psi(q)$ [and also $\varphi(p)$] must be square integrable.

† The geometrical interpretation presented here is heuristic at best and is actually not correct. However, it may help to give the reader some intuitive feeling for the vector space, called a Hilbert space, which is defined as the space of square integrable functions in configuration space.

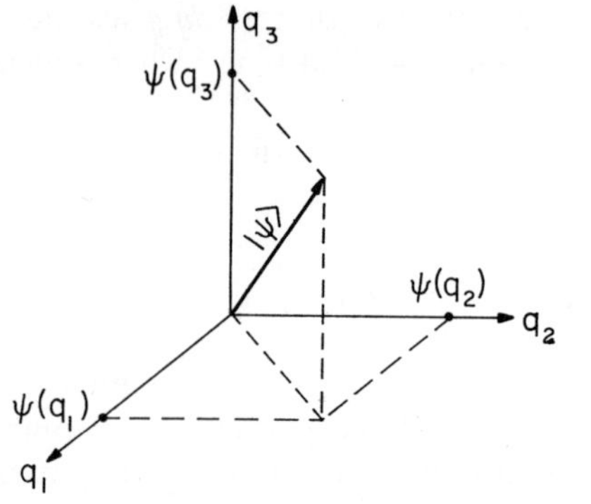

Figure 1.1 Pictorial diagram of a ket vector and three of its coordinate representatives.

axis, and so on. The vector then represents the state of the system just as its components do. This vector is not an ordinary vector since it has a complex character, and we must have a special notation to designate it, just as we do for an ordinary vector. Dirac uses the symbol $|\ \rangle$ to designate a vector of this type and calls it a ket vector, or simply a ket, to distinguish it from ordinary vectors. The particular vector whose components are $\psi(q_1)$, $\psi(q_2), \ldots$ is called ket ψ and written $|\psi\rangle$. Figure 1.1 shows a very diagrammatic sketch of the vector $|\psi\rangle$ and its "components" along the mutually perpendicular axes described above. Unfortunately, only three of these axes can be shown.

By way of analogy, if $\mathbf{A}$ is an ordinary vector and (x, y, z) represent a cartesian coordinate system, $\mathbf{A}$ may be specified by giving its components along these axes: $\mathbf{A} = (A_x, A_y, A_z)$; that is, $\mathbf{A}$ can be *represented* by its components. Similarly, $|\psi\rangle$ may be specified by its components along the orthogonal q axes: $|\psi\rangle = [\psi(q_1), \psi(q_2), \ldots]$. Thus $\mathbf{A}$ represents the vector equally as well as its components along certain axes, and $|\psi\rangle$ represents the state of the system just as well as its components. The vector in this case is said to be given in the position representation.

The vector $\mathbf{A}$ may also be specified by giving its components along another cartesian coordinate system (x', y', z') rotated with respect to (x, y, z): $\mathbf{A} = (A_{x'}, A_{y'}, A_{z}')$. So too $|\psi\rangle$ may be expressed in another representation: $|\psi\rangle = [\varphi(p_1), \varphi(p_2), \varphi(p_3), \ldots]$. This is called the momentum representation and is visualized roughly as the components of the same vector on a rotated orthogonal set of axes; this is shown in Figure 1.2. The relation between the q and p axes is given by the Fourier transform.

It should be clear that there must exist an infinite number of other equivalent representations that might not have been so obvious without the introduction of the concept of vectors into the theory. We must now specify the

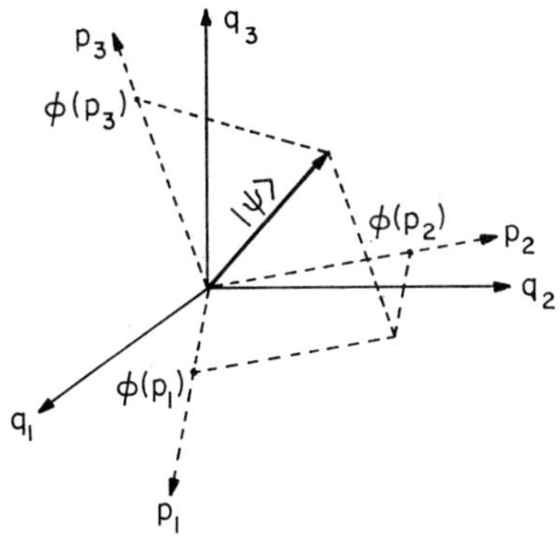

Figure 1.2 Pictorial diagram of a ket vector and three of its momentum representatives.

properties of ket vectors in more precise fashion to develop the theory further.

1.1 KET VECTORS

As noted above, Dirac calls vectors designated by the symbol $|a\rangle$, $|x\rangle$, and such ket vectors. A general ket is denoted by $|\ \rangle$, and the labels inside designate particular kets.

From the discussion above, we associate a ket vector with each state of the dynamical system under study. Since we shall postulate that a linear superposition of states of the system is also a state of the system, the ket vector space must be a linear vector space [8]. A vector space is said to be linear in the following sense. If c_1 and c_2 are complex numbers and $|a\rangle$ and $|b\rangle$ are two kets, the linear combination

$$|u\rangle = c_1|a\rangle + c_2|b\rangle \tag{1.1.1a}$$

is also a ket vector, since a linear combination of two states associated with $|a\rangle$ and $|b\rangle$ is also a state of the system. If a ket depends on a parameter q', which may take on any value in a certain range, $q_1' \leq q' \leq q_2'$, we may generalize (1.1.1a) to read

$$|v\rangle = \int_{q_1'}^{q_2'} c(q')|q'\rangle\, dq', \tag{1.1.1b}$$

where $c(q')$ is an ordinary (complex) function of q' and the vector $|v\rangle$ is in ket space. Kets such as $|u\rangle$ (and $|v\rangle$) above are said to be linearly dependent on $|a\rangle$ and $|b\rangle$ (or $|q'\rangle$). If, in a certain set of ket vectors (two or more), none of them can be expressed as a linear combination of the others, the vectors are said to be linearly independent.

Although the classical and quantum superposition principles are different, as we shall see below, it may be stated by way of analogy that, if $\hat{\mathbf{i}}$, $\hat{\mathbf{j}}$, and $\hat{\mathbf{k}}$ are three mutually perpendicular unit vectors in ordinary space, any other vector may be written as a linear combination of these three; that is, any other constant vector $\mathbf{A}$ may be written as $\mathbf{A} = c_1\hat{\mathbf{i}} + c_2\hat{\mathbf{j}} + c_3\hat{\mathbf{k}}$. On the other hand, $\hat{\mathbf{i}}$ cannot be expressed as a linear combination of $\hat{\mathbf{j}}$ and $\hat{\mathbf{k}}$ and is said to be linearly independent of $\hat{\mathbf{j}}$ and $\hat{\mathbf{k}}$.

Another assumption in the theory is that if a state is superimposed with itself, there results not a new state vector but only the original state again; that is, when $c_1|a\rangle$ and $c_2|a\rangle$ are added, where c_1 and c_2 are arbitrary complex numbers, the result is

$$c_1|a\rangle + c_2|a\rangle = (c_1 + c_2)|a\rangle,$$

and the kets $c_1|a\rangle$, $c_2|a\rangle$, $(c_1 + c_2)|a\rangle$ all represent the *same* state of the system, with the exception of the case $c_1 + c_2 = 0$, which corresponds to no state at all. Thus a state is specified entirely by the *direction* of the ket vector. It may be concluded that $+|a\rangle$ and $-|a\rangle$ represent the same state. Therefore, there is a one-to-one correspondence between a state of a system and a *direction* in ket vector space. This assumption is a departure from classical mechanics and shows that the classical and quantum superposition principles are different.

The ket vector space may have a finite or an infinite number of dimensions. The dimensionality is determined by the number of linearly independent kets in the space. Since independent states of a quantum system are represented by independent kets, the dimensionality is determined by the number of independent states of the quantum system.

1.2 SCALAR PRODUCT; BRA VECTORS

We have introduced ket vectors in an abstract linear vector space by saying that their projection on a given set of orthogonal axes in an infinite-dimensional space gives the values of the wave function $\psi(q, t)$ in the position representation at time t. This is only a pedantic crutch, but it helps to visualize the procedure. The essential definition of kets is that a direction in ket space and every state of the system are in one-to-one correspondence.

In the study of ordinary vector analysis, we cannot proceed very far before defining the scalar product of two vectors. We may define the scalar product of $\mathbf{A}$ and $\mathbf{B}$ as follows: with every two vectors $\mathbf{A}$ and $\mathbf{B}$ in the space, there is associated a real number f, which is written

$$f = \mathbf{A} \cdot \mathbf{B}.$$

The scalar product of any two vectors is then defined, since the number to associate with any pair of them is known. This definition may seem strange at first but a little reflection shows that it is a more general definition than any formulas we might give for finding the number f, having been given **A** and **B**. One such formula is $f = |\mathbf{A}|\,|\mathbf{B}| \cos\theta$, where the first two factors are the magnitudes of **A** and **B**, and θ is the angle between them. But the length itself is defined only in terms of the scalar product of the vector with itself, and so the formula does not serve as an effective definition of a scalar product, although it is very useful in practice.

More generally, the scalar product of a particular vector **B** with all other vectors **A** in the space may be regarded as a way of defining **B**. If the set of numbers $f(\mathbf{B})$ for all **A**'s is given, **B** is determined. For three-dimensional space, it is sufficient to take for **A** the three unit vectors $\hat{\mathbf{i}}$, $\hat{\mathbf{j}}$, and $\hat{\mathbf{k}}$, which are linearly independent, and define **B** by giving its scalar product with each. Thus

$$B_x = \mathbf{B}\cdot\hat{\mathbf{i}} \qquad B_y = \mathbf{B}\cdot\hat{\mathbf{j}} \qquad B_z = \mathbf{B}\cdot\hat{\mathbf{k}},$$

and the three numbers B_x, B_y, and B_z define **B**.

It is a postulate of the theory of ordinary vectors that the function $f(\mathbf{B})$ a linear function of **B**. This means that, if $\mathbf{B}_1$ and $\mathbf{B}_2$ are two vectors,

$$\mathbf{A}\cdot(\mathbf{B}_1 + \mathbf{B}_2) = \mathbf{A}\cdot\mathbf{B}_1 + \mathbf{A}\cdot\mathbf{B}_2$$
$$\mathbf{A}\cdot(c\mathbf{B}) = c(\mathbf{A}\cdot\mathbf{B}),$$

where c is a number. It is clear that the numbers $f(\mathbf{B})$ may be considered a function of **B** since for every **A** there is a number, $f(\mathbf{B})$. This is what is meant by the expression a function $\varphi(x)$ of a continuous variable x: with each x is associated a number $\varphi(x)$.

The scalar products defined above all involve vectors in the same space. There is an interesting example in crystallography in which the scalar product is used to define vectors in another space. We let **a**, **b**, **c** be the primitive translation vectors in a crystal lattice in ordinary space. We may define the primitive translations **a***, **b***, **c*** in the reciprocal lattice space by giving the scalar products of **a***, **b***, and **c*** with **a**, **b**, **c**. We have

$$\begin{array}{lll} \mathbf{a}^*\cdot\mathbf{a} = 1 & \mathbf{b}^*\cdot\mathbf{a} = 0 & \mathbf{c}^*\cdot\mathbf{a} = 0 \\ \mathbf{a}^*\cdot\mathbf{b} = 0 & \mathbf{b}^*\cdot\mathbf{b} = 1 & \mathbf{c}^*\cdot\mathbf{b} = 0 \\ \mathbf{a}^*\cdot\mathbf{c} = 0 & \mathbf{b}^*\cdot\mathbf{c} = 0 & \mathbf{c}^*\cdot\mathbf{c} = 1. \end{array}$$

Therefore, by defining the scalar product of **a*** with three independent vectors in ordinary space, we have defined a vector **a*** in another space.

After this lengthy introduction, we now define scalar products of ket vectors in the following way. With each ket $|a\rangle$ is associated a *complex* number f. (In the examples above the numbers were real but ket vectors are more general vectors than those in ordinary space.) The set of numbers associated with different $|a\rangle$'s is a function of $|a\rangle$. This function must be a linear function,* which means that if $|a_1\rangle$ and $|a_2\rangle$ are two kets, the number associated with $|a_1\rangle$ and $|a_2\rangle$ is the sum of the numbers associated with $|a_1\rangle$ and $|a_2\rangle$ separately, and the number associated with $c|a\rangle$, where c is a complex number, is c times the number associated with $|a\rangle$, that is,

$$\begin{aligned} f(|a_1\rangle + |a_2\rangle) &= f(|a_1\rangle) + f(|a_2\rangle) \\ f(c|a_1\rangle) &= cf(|a_1\rangle). \end{aligned} \tag{1.2.1}$$

Just as in the case of the reciprocal-lattice example above, we may visualize the numbers f associated with all the kets in ket space as defining a vector in another space, designated by the symbol $\langle f|$. Dirac calls the vectors denoted by the symbol $\langle\ |$ bra vectors. We may write the scalar product of $\langle f|$ and $|a\rangle$ as

$$f(|a\rangle) = \langle f|a\rangle. \tag{1.2.2}$$

If we give all the numbers f for each ket $|a\rangle$, we have defined $\langle f|$. The space of bra vectors is different from the space of ket vectors, just as the reciprocal lattice space was different from the original crystal space. The definition here is more general, however, because f may be a complex number in (1.2.2) whereas it was real in the crystal example.

When we use the scalar product notation of (1.2.2), we may rewrite (1.2.1) as

$$\begin{aligned} \langle f|(|a_1\rangle + |a_2\rangle) &= \langle f|a_1\rangle + \langle f|a_2\rangle \\ \langle f|(c|a\rangle) &= c\langle f|a\rangle. \end{aligned} \tag{1.2.3}$$

Since a bra is defined by its scalar product with a ket, $\langle b| = 0$ if $\langle b|a\rangle = 0$ for every ket $|a\rangle$. Similarly, $\langle b_1| = \langle b_2|$ if $\langle b_1|a\rangle = \langle b_2|a\rangle$ for every $|a\rangle$.

The sum of two bras is defined by its scalar product with $|a\rangle$. Thus

$$\begin{aligned} (\langle b_1| + \langle b_2|)|a\rangle &= \langle b_1|a\rangle + \langle b_2|a\rangle \\ (c\langle b|)|a\rangle &= c\langle b|a\rangle. \end{aligned} \tag{1.2.4}$$

Thus far we have defined bras only in terms of their scalar products with kets, and there is no definite relation between them. To give a connection, we make the following *assumption:* each ket may be associated with a single bra in a unique way; that is, a one-to-one correspondence between kets and bras is assumed. It is therefore reasonable to give the bra the same label as the ket with which it is associated. Thus $\langle a|$ is the bra associated with $|a\rangle$. Similarly, with the ket

$$|u\rangle = |a\rangle + |b\rangle, \tag{1.2.5a}$$

* To be mathematically precise, one whould say that f is a linear functional on the vector space.

there is associated the bra

$$\langle u| = \langle a| + \langle b|, \tag{1.2.5b}$$

and with the ket

$$|v\rangle = c|a\rangle, \tag{1.2.6a}$$

where c is a complex number, there is associated the bra

$$\langle v| = c^*\langle a|, \tag{1.2.6b}$$

where c^* is the complex conjugate of c. We shall not go into the reason for taking c^* instead of c but just accept it as a new assumption for simplicity [1]. It is therefore reasonable to call the bra associated with a ket its hermitian adjoint, and vice versa, and write

$$\langle u| = (|u\rangle)^\dagger \qquad |u\rangle = (\langle u|)^\dagger, \tag{1.2.7}$$

where the dagger means that the bra is changed to its associated ket (and vice versa) and the complex conjugate of any numbers involved is taken, as in (1.2.6).

Since by assumption there is a unique correspondence between bras and kets, the direction of a bra vector may represent the state of a quantum system equally as well as does the direction of a ket. They are said to be duals of one another.

As yet we have not defined the length of a bra or ket. We shall consider two kets $|a\rangle$ and $|b\rangle$ and the associated bras $\langle a|$ and $\langle b|$. From these vectors we may form four numbers $\langle a|b\rangle$, $\langle b|a\rangle$, $\langle a|a\rangle$, and $\langle b|b\rangle$. In general, $\langle a|b\rangle$ and $\langle b|a\rangle$ will be complex, and we make the additional *assumption* that they are related by

$$\langle a|b\rangle = \langle b|a\rangle^*, \tag{1.2.8}$$

where the asterisk means complex conjugate in the future. With this assumption, if we let $|b\rangle = |a\rangle$, we conclude that $\langle a|a\rangle$ is real. We define the length, or norm, of $|a\rangle$ as $\langle a|a\rangle$, and so assumption (1.2.8) is necessary if we want the vectors to have a real norm. We make the further *assumption* that the length of a vector must be positive or zero, that is,

$$\langle a|a\rangle \geq 0. \tag{1.2.9}$$

The equality holds only if $|a\rangle = 0$.

The assumptions (1.2.8) and (1.2.9) may be given motivation from a consideration of the wave function $\psi(q, t)$ and its complex conjugate $\psi^*(q, t)$. We visualized $\psi(q, t)$ as components of $|\psi\rangle$ in ket space. Likewise we may visualize $\psi^*(q, t)$ as the components of $\langle\psi|$ in bra space. We then know from wave mechanics that the complex numbers $\psi^*(q, t)\chi(q, t)$ and $\chi^*(q, t)\psi(q, t)$

are related by

$$\psi^*(q)\chi(q) = [\chi^*(q)\psi(q)]^*,$$

and

$$\int_{-\infty}^{\infty} |\psi(q)|^2 \, dq \geq 0.$$

Similar relations should hold for bras and kets since they can be intimately related to wave functions. This motivated the assumptions (1.2.8) and (1.2.9).

The concept of orthogonality is also important where vectors are concerned. In the case of bras and kets, if the scalar product $\langle a|b\rangle = 0$, the vectors are orthogonal. In wave mechanics, $\psi^*(q)$ and $\chi(q)$ are orthogonal if $\int \psi^*(q)\chi(q)\, dq = 0$. The orthogonality involved here is different from the orthogonality of two ordinary vectors **A** and **B**. If $\mathbf{A} \cdot \mathbf{B} = 0$, **A** and **B** are at right angles to one another. But **A** and **B** are in the same vector space. In the present case, $\langle a|$ and $|b\rangle$ are in different vector spaces. (See the crystal-lattice example treated earlier.) Nevertheless, if $\langle a|b\rangle = 0$, it may be said that $|a\rangle$ and $|b\rangle$ are orthogonal as well as $\langle a|$ and $\langle b|$. When $\langle a|b\rangle = 0$, it may also be said that the associated quantum states of the system that they represent are orthogonal.

If the norm of all vectors in the space is finite, the space is called *Hilbert* space [8]. The theory must include vectors of infinite norm, as we shall see later. The space of these vectors forms an even more general vector space which is called ket or bra space. Including vectors of infinite norm requires the introduction of the Dirac δ function at a later stage.

1.3 LINEAR OPERATORS

The concept of linear operators is already familiar to the reader. For example, if $f(t)$ is a square integrable function of a continuous variable t, the function belongs to Hilbert space [8]. We may then define the linear operator d/dt in this space by associating another function $g(t)$ with $f(t)$ and write

$$g(t) = \frac{d}{dt} f(t).$$

If, with every $f(t)$ in the space, we associate another $g(t)$, we have defined the operator d/dt. If, furthermore, we require that

$$\frac{d}{dt} [f_1(t) + f_2(t)] = g_1(t) + g_2(t)$$

$$\frac{d}{dt} cf(t) = cg(t),$$

where g_1, g_2, and g are the three functions associated with f_1, f_2, and f respectively, and c is a number, then d/dt is a linear operator.

We may similarly define other linear operators such as integration, multiplication by a constant, and many others and build up a whole scheme of linear operators. Clearly, such operators are needed also in vector space to extend its range of applicability.

We must now introduce linear operators in the space of ket and bra vectors. If with each ket $|a\rangle$ in the space we associate another ket $|b\rangle$, the association may be used to define an operator D which we may write in the form

$$|b\rangle = D|a\rangle, \tag{1.3.1}$$

where D might mean differentiation, integration, or something else. Note the convention that an operator appears to the left of the ket on which it operates.

We are interested only in linear operators; this means that if $|a_1\rangle$, $|a_2\rangle$, and $|a\rangle$ are any three kets and c is a number, D must satisfy the relations

$$\begin{aligned} D(|a_1\rangle + |a_2\rangle) &= D|a_1\rangle + D|a_2\rangle \\ D(c|a\rangle) &= c(D|a\rangle). \end{aligned} \tag{1.3.2}$$

Since an operator is completely defined when its effect on every ket in the space is known, two operators D_1 and D_2 are equal if $D_1|a\rangle = D_2|a\rangle$ for every $|a\rangle$. The null operator, $D = 0$, is defined by $D|a\rangle = 0$ for every $|a\rangle$. The identity operator, $D = I$, is defined by $D|a\rangle = |a\rangle$ for every $|a\rangle$.

At this stage we may build up an algebra of linear operators. We may define the sum of two operators $D_1 + D_2$ by their action on $|a\rangle$:

$$(D_1 + D_2)|a\rangle = D_1|a\rangle + D_2|a\rangle, \tag{1.3.3}$$

a product

$$(D_1 D_2)|a\rangle = D_1(D_2|a\rangle). \tag{1.3.4}$$

From this, if $D_1 = D_2$, we can define powers of operators, and so on.

We also have, for example,

$$\begin{aligned} (D_1 + D_2)|a\rangle &= (D_2 + D_1)|a\rangle \\ [(D_1 + D_2) + D_3]|a\rangle &= [D_1 + (D_2 + D_3)]|a\rangle \\ [D_1(D_2 + D_3)]|a\rangle &= D_1 D_2|a\rangle + D_1 D_3|a\rangle. \end{aligned} \tag{1.3.5}$$

The algebra of N-dimensional square matrices is the same as the algebra of linear operators as the reader should recognize.

The commutator of two operators D_1 and D_2 is written $[D_1, D_2]$ and is defined by

$$[D_1, D_2] \equiv D_1 D_2 - D_2 D_1. \tag{1.3.6}$$

In general, $D_1D_2 \neq D_2D_1$, which is a property held in common with matrices. The algebra of quantum mechanics is a noncommutative algebra. Two familiar linear operators that do not commute are $D_1 = x$ (multiplication by x) and $D_2 = d/dx$ (differentiation). It is easily verified that, if $f(x)$ is a continuous function of x,

$$\left[x, \frac{d}{dx}\right] f(x) \equiv \left(x \frac{d}{dx} - \frac{d}{dx} x\right) f(x) = -f(x),$$

so that noncommutating operators are already familiar.

Multiplication by a constant is a linear operation. A constant operator commutes with all linear operators.

If two operators D_1 and D_2 satisfy the equations

$$D_1D_2 = D_2D_1 = I, \tag{1.3.7}$$

where I is the identity operator, then D_2 is the inverse of D_1, and D_1 is the inverse of D_2, if the inverse exists. This is written as

$$D_2 = D_1^{-1} \qquad D_1 = D_2^{-1}. \tag{1.3.8}$$

The inverse of a product of operators is

$$(D_1D_2D_3)^{-1} = D_3^{-1}D_2^{-1}D_1^{-1}. \tag{1.3.9}$$

As noted earlier, these properties of operators are common to finite square matrices. In fact, later we shall *represent* operators by matrices.

We have defined the action of linear operators on kets; we must now give meaning to their operation on a bra. We shall consider the ket

$$|b\rangle = D|a\rangle.$$

We may take the scalar product of this ket with any bra, say $\langle c|$; this scalar product $\langle c|b\rangle = \langle c|(D|a\rangle)$ depends linearly on $|a\rangle$ since D is linear. From the definition of a bra, the scalar product $\langle c|b\rangle$ may be considered as the scalar product of $|a\rangle$ with some bra, say $\langle d|$. Then for each $\langle c|$ there corresponds a bra $\langle d|$. The bra $\langle d|$ depends linearly on $\langle c|$ so that $\langle d|$ is obtained from $\langle c|$ by the application of a linear operator to $\langle c|$. Since this operator is uniquely determined by D, we may reasonably write

$$\langle d| = \langle c|D.$$

We adopt the convention that operators always appear to the right of bras and summarize the definition above by the relation

$$\langle c|(D|a\rangle) = (\langle c|D)|a\rangle. \tag{1.3.10}$$

It therefore is unnecessary to use the parentheses, and either side may be written $\langle c|D|a\rangle$. Therefore, D may first operate on $\langle c|$ and the result applied to $|a\rangle$, or vice versa. The operator properties given in (1.3.2) to (1.3.5) are equally valid whether they are applied to bras or kets. Note also that $\langle c|D|a\rangle$ is a closed-bracket expression and is therefore a complex number in general.

A simple example of a linear operator that occurs frequently in the quantum theory is $|a\rangle\langle b| = P$. We see that P may operate on a ket to give

$$P|c\rangle = |a\rangle\langle b|c\rangle,$$

which is a ket $|a\rangle$ multiplied by the number $\langle b|c\rangle$, and

$$\langle c|P = \langle c|a\rangle\langle b|$$

is a bra $\langle b|$ multiplied by the number $\langle c|a\rangle$. It is left as an exercise to show that P satisfies the requirements of a linear operator. An example in ordinary vector analysis that corresponds approximately to an operator such as P is the dyadic $\hat{\mathbf{i}}\hat{\mathbf{j}}$. In this case, $\hat{\mathbf{i}}\hat{\mathbf{j}} \cdot \hat{\mathbf{k}} = 0$, $\hat{\mathbf{i}} \cdot \hat{\mathbf{i}}\hat{\mathbf{j}} = \hat{\mathbf{j}}$, and so on.

Linear operators play a central role in the physical interpretation of the theory. Following Dirac, we make the *assumption* that each quantity that can be measured for a physical system (which is called a dynamical variable) can be represented by a particular kind of linear operator, to be described in the following section. Examples of dynamical variables associated with linear operators are position (q), momentum (p), angular momentum $\mathbf{L}$, energy (H), and such which occur in classical mechanics, as well as spin angular momentum ($\boldsymbol{\sigma}$) which has no classical analog. Classically these variables commute with each other, but quantum-mechanically it may be postulated that some of these operators do not commute. The commutation relations determine the type of algebra the operators obey and mark the departure of quantum mechanics from classical mechanics.

1.4 HERMITIAN OPERATORS

Linear operators are, in general, complex quantities; if we let them correspond to dynamical variables, they would be complex. However, physically, quantities such as momentum, position, and the like give real numbers when they are measured. Therefore, the linear operators that represent dynamical variables must be restricted to real linear operators. Such operators are said to be hermitian and are defined as follows:

The bra associated with the ket $|q\rangle = L|p\rangle$, where L is a linear operator, is written

$$\langle q| = \langle p|L^\dagger = (L|p\rangle)^\dagger = (|q\rangle)^\dagger.$$

The symbol $L^\dagger$ is called the hermitian adjoint of L; that is, the bra $\langle q|$, which is the hermitian adjoint of $|q\rangle$, may be considered the result of some linear operator acting on $\langle p|$, which is designated by $L^\dagger$ [1].

We next show that

$$L^{\dagger\dagger} = L. \tag{1.4.1}$$

We let

$$|b\rangle = L|p\rangle, \tag{1.4.2}$$

where $|p\rangle$ is an arbitrary ket. Its adjoint (associated bra) is

$$\langle b| = \langle p|L^{\dagger}. \tag{1.4.3}$$

If we take the adjoint again, we obtain

$$|b\rangle = L^{\dagger\dagger}|p\rangle. \tag{1.4.4}$$

If we take the scalar product of an arbitrary bra $\langle a|$ with both (1.4.2) and (1.4.4) we have

$$\langle a|b\rangle = \langle a|L|p\rangle = \langle a|L^{\dagger\dagger}|p\rangle. \tag{1.4.5}$$

Since $\langle a|$ and $|p\rangle$ are arbitrary, (1.4.1) follows.

If in (1.2.8), we let $\langle a| = \langle p|L^{\dagger}$ and $|a\rangle = L|p\rangle$ we have

$$\langle p|L^{\dagger}|b\rangle = \langle b|L|p\rangle^*. \tag{1.4.6}$$

If a linear operator is self-adjoint,

$$L = L^{\dagger}, \tag{1.4.7}$$

the operator is said to be hermitian. From (1.4.6), if L is hermitian, it must satisfy the equation

$$\langle p|L|b\rangle = \langle b|L|p\rangle^*, \tag{1.4.8}$$

for any $|b\rangle$ and $|p\rangle$. Therefore, any operator that satisfies (1.4.8) is hermitian.

The following properties may be proved for any linear operator:

$$(cL|a\rangle)^{\dagger} = c^*\langle a|L^{\dagger},$$

where c is a constant,

$$\begin{aligned} [(L_1 + L_2)|a\rangle]^{\dagger} &= \langle a|(L_1^{\dagger} + L_2^{\dagger}) \\ (L_1L_2|a\rangle)^{\dagger} &= \langle a|L_2^{\dagger}L_1^{\dagger} \\ (\langle a|L_1L_2L_3)^{\dagger} &= L_3^{\dagger}L_2^{\dagger}L_1^{\dagger}|a\rangle \\ \langle a|L_1L_2|b\rangle^* &= \langle b|L_2^{\dagger}L_1^{\dagger}|a\rangle \\ (|a\rangle\langle b|)^{\dagger} &= |b\rangle\langle a|. \end{aligned} \tag{1.4.9}$$

The algebra of adjoints of operators is the same as for finite square matrices.

1.5 THE EIGENVALUE PROBLEM

Bras and ket vectors, or rather directions of bras and kets, are associated with states of a system, and linear hermitian operators are associated with dynamical variables that describe the system. In the next section we show how to relate these mathematical concepts to physical measurements made on the system. Before this, we must introduce the concept of eigenvalues of hermitian operators.

An eigenvalue problem is a familiar one in classical mathematics as well as in classical physics. One of the simplest examples involves the solution of the equation

$$Lu(x) = \lambda u(x),$$

where L is known to be $-d^2/dx^2$ and $u(x)$ and λ are unknown. If we add the boundary conditions that $u(0) = u(l) = 0$, we find that λ can take on only a certain discrete set of *eigenvalues* given by $\lambda_n = \pi^2 n^2/l^2$, where $n = 0, \pm 1, \pm 2, \ldots$. The associated *eigenfunctions* $u_n(x)$ are $u_n(x) = \sin(\pi n x/l)$. Note that the effect of an operator L on an eigenfunction $u_n(x)$ is to reproduce $u_n(x)$. If L operates on an arbitrary function $u(x)$, it will not, in general, reproduce $u(x)$ times a number.

We may similarly formulate an eigenvalue problem for operators in ket (and bra) space. We let L be a linear operator and $|a\rangle$ a ket. If L operates on $|a\rangle$ and gives $|a\rangle$ multiplied by a number l, then $|a\rangle$ is an eigenket of L and l is the associated eigenvalue. This may be written

$$L|a\rangle = l|a\rangle.$$

This is an eigenvalue problem: L is a known operator, and l and $|a\rangle$ are unknown, and we are asked to find kets which, when acted on by L, reproduce the same ket times a number subject to a set of boundary conditions. It is customary to label an eigenket with its eigenvalue; with this convention, we may rewrite the eigenvalue equation as

$$L|l\rangle = l|l\rangle. \tag{1.5.1}$$

The eigenvalue problem may equally well be formulated in terms of bras:

$$\langle d|D = d\langle d|. \tag{1.5.2}$$

For simplicity, in this book we shall usually consider cases in which there is only one eigenvalue for each eigenvector. If more than one independent eigenvector can be associated with a given eigenvalue, the system is said to be degenerate. Degeneracy can be treated easily, but it complicates the formulas needlessly.

If $|l\rangle$ is an eigenket of L, then, by (1.5.1), any constant c times $|l\rangle$ is also an eigenket with the same eigenvalue. In line with earlier assumptions, the states represented by $|l\rangle$ and $c|l\rangle$ are the same state.

We shall be interested usually in the solution of the eigenvalue problem for linear hermitian operators for reasons that should become clear in the next section. Before attempting the solutions of any specific eigenvalue problem, we shall prove two very important theorems valid for all linear hermitian operators.

THEOREM 1

The eigenvalues of a linear hermitian operator are real.

PROOF

We let L be a linear hermitian operator. The eigenvalues of L satisfy the equation

$$L|l\rangle = l|l\rangle.$$

If we form the scalar product of both sides with $\langle l|$, we have

$$\langle l|L|l\rangle = l\langle l|l\rangle. \tag{1.5.3}$$

If we take the complex conjugate of both sides, we obtain, by means of (1.4.6),

$$\langle l|L|l\rangle^* = \langle l|L^\dagger|l\rangle = l^*\langle l|l\rangle. \tag{1.5.4}$$

But since $L^\dagger = L$ and $\langle l|l\rangle \neq 0$, comparison of (1.5.3) and (1.5.4) shows that $l = l^*$, and the theorem is proved. We see that $\langle l|l\rangle = 0$ only in the trivial case, in which $|l\rangle = 0$. Note that the norm $\langle l|l\rangle$ is real.

THEOREM 2

Two eigenvectors of a linear hermitian operator L belonging to different eigenvalues are orthogonal.

PROOF

We let l' and l'' be two eigenvalues of L and $|l'\rangle$ and $|l''\rangle$ be the associated eigenkets. Then ($L = L^\dagger$; l' and l'' are real)

$$L|l'\rangle = l'|l'\rangle \tag{1.5.5}$$

$$\langle l''|L = l''\langle l''|. \tag{1.5.6}$$

If we form the scalar product of (1.5.5) with $\langle l''|$, the scalar product of (1.5.6) with $|l'\rangle$, and subtract, we find that

$$(l' - l'')\langle l'|l''\rangle = 0.$$

Since $l' \neq l''$ by assumption, then $\langle l'|l''\rangle = 0$, and the theorem is proved. From (1.5.5) and (1.5.6) we see that the eigenvalues associated with eigenkets are the same as those associated with eigenbras.

The solution of an eigenvalue problem in many cases is complicated. We shall now solve a particularly simple one to illustrate the method. In a later chapter we discuss a physical system that may be described by this example but for the moment we shall consider it merely as a mathematical example.

We suppose a linear hermitian operator σ_z that satisfies an auxiliary condition

$$\sigma_z{}^2 = I, \tag{1.5.7}$$

where I is the identity operator, and we wish to solve the eigenvalue problem

$$\sigma_z|s\rangle = s|s\rangle. \tag{1.5.8}$$

From Theorem 1, we know that s is real, and from Theorem 2 we know that $\langle s'|s''\rangle = 0$ if $s' \neq s''$.

To solve for the eigenvalues and eigenvectors, we multiply both sides of (1.5.8) from the left by σ_z, use (1.5.7) and (1.5.8), and obtain

$$\sigma_z{}^2|s\rangle = |s\rangle = s\sigma_z|s\rangle = s^2|s\rangle,$$

or

$$(s^2 - 1)|s\rangle = 0.$$

If we form the scalar product of this with $\langle s|$, we see that, since $\langle s|s\rangle$ is positive and not zero, the eigenvalues of σ_z are given by

$$s = \pm 1.$$

Since by assumption there can be no degeneracy (two eigenvalues the same), there are only two eigenvalues, and so we may rewrite (1.5.8) as

$$\sigma_z|+1\rangle = +1|+1\rangle \qquad \sigma_z|-1\rangle = -1|-1\rangle. \tag{1.5.9}$$

By Theorem 2 we know that

$$\langle +1|-1\rangle = 0 = \langle -1|+1\rangle. \tag{1.5.10}$$

These are the *orthogonality* relations obeyed by eigenvectors belonging to different eigenvalues.

As we know, any eigenket multiplied by a constant is also an eigenket belonging to the same eigenvalue. We may therefore choose a constant so that the norm of the eigenvectors is unity as long as the norm is finite and write

$$\langle +1|+1\rangle = \langle -1|-1\rangle = 1. \tag{1.5.11}$$

These are the *normalization* conditions. Normalization does not specify the vector uniquely; we may still multiply $|+1\rangle$ by exp $(i\alpha)$ since $\langle +1|$ will be multiplied by exp $(-i\alpha)$, where α is real, and (1.5.11) will be left unchanged. Such a phase shift is of no physical significance in the theory, and we shall usually choose $\alpha = 0$.

For any eigenvalue problem in which the norm of the vectors is finite, the eigenvectors may always be normalized and (1.5.10) and (1.5.11) combined into the *orthogonality relations*

$$\langle l'|l''\rangle = \delta_{l'l''}, \tag{1.5.12}$$

where δ_{ij} is the Kronecker δ defined by

$$\delta_{ij} = \begin{cases} 1 & \text{if } \; i = j \\ 0 & \text{if } \; i \neq j. \end{cases} \tag{1.5.13}$$

When the vectors have an infinite norm, these results have to be generalized, as we discuss later.

Anticipating future work, we shall now show that σ_z may be represented by a 2×2 matrix given by

$$\sigma_z = \begin{bmatrix} 1 & 0 \\ 0 & -1 \end{bmatrix} \tag{1.5.14}$$

To show this, we form the scalar products of both equations (1.5.9) with $\langle +1|$ and $\langle -1|$, respectively. If we use (1.5.10) and (1.5.11), we obtain the so-called "matrix elements" of σ_z given by

$$\begin{aligned} \langle +1|\sigma_z|+1\rangle &= +1 \qquad & \langle +1|\sigma_z|-1\rangle &= 0 \\ \langle -1|\sigma_z|+1\rangle &= 0 \qquad & \langle -1|\sigma_z|-1\rangle &= -1. \end{aligned}$$

We then may group these results into a matrix such as (1.5.14), with the convention that the rows are labeled by the eigenbras and the columns by the eigenkets.

Any ket in the space may be expressed in terms of the eigenkets $|+1\rangle$ and $|-1\rangle$. When this can be done, it is said that the eigenkets form a complete set by definition. Again we are anticipating the results of the next section.

To show that any ket $|P\rangle$ in the space may be expanded in terms of $|+1\rangle$ and $|-1\rangle$, we write the identity

$$\begin{aligned} |P\rangle \equiv I|P\rangle &= \tfrac{1}{2}(I + \sigma_z + I - \sigma_z)|P\rangle \\ &= \tfrac{1}{2}(I + \sigma_z)|P\rangle + \tfrac{1}{2}(I - \sigma_z)|P\rangle. \end{aligned} \tag{1.5.15}$$

We consider each factor separately. On using (1.5.7), we have

$$\sigma_z[\tfrac{1}{2}(I + \sigma_z)|P\rangle] = 1\,[\tfrac{1}{2}(\sigma_z + I)|P\rangle],$$

so that $\frac{1}{2}(I + \sigma_z)|P\rangle$ is an eigenket of σ_z with eigenvalue $+1$. It may therefore differ from $|+1\rangle$ only by a constant, and we may write

$$\tfrac{1}{2}(I + \sigma_z)|P\rangle = c_1|+1\rangle, \tag{1.5.16}$$

where c_1 is a constant. Similarly, we see that the last term in (1.5.15) is given by

$$\sigma_z[\tfrac{1}{2}(I - \sigma_z)|P\rangle] = -1\,[\tfrac{1}{2}(I - \sigma_z)]|P\rangle,$$

so that we may write

$$\tfrac{1}{2}(I - \sigma_z)|P\rangle = c_2|-1\rangle, \tag{1.5.17}$$

where c_2 is another constant. Thus (1.5.15) may be written by using (1.5.16) and (1.5.17) as

$$|P\rangle = c_1|+1\rangle + c_2|-1\rangle, \tag{1.5.18}$$

as originally stated. Any ket is therefore linearly dependent on the kets $|+1\rangle$ and $|-1\rangle$, and we have proved that the set $\{|+1\rangle, |-1\rangle\}$ is complete.

We may also derive the so-called *completeness* relation in this simple example. We multiply (1.5.18) from the left alternatively by $\langle +1|$ and $\langle -1|$, use the orthonormality relations (1.5.10) and (1.5.11), and see that

$$c_1 = \langle +1|P\rangle \qquad c_2 = \langle -1|P\rangle. \tag{1.5.19}$$

If we substitute these into (1.5.18), we obtain

$$|P\rangle = (|+1\rangle\langle +1| + |-1\rangle\langle -1|)|P\rangle.$$

Since $|P\rangle$ is arbitrary, this equation will be satisfied if

$$|+1\rangle\langle +1| + |-1\rangle\langle -1| \equiv I, \tag{1.5.20}$$

which is the *completeness* or *closure* relation. We discuss the completeness relation for general hermitian operators in the next section.

The Hilbert space in this example is two-dimensional because we considered only nondegenerate eigenvalues.

If we substitute (1.5.19) in (1.5.16) and (1.5.17), we have the results

$$\begin{aligned} \tfrac{1}{2}(I + \sigma_z) &= |+1\rangle\langle +1| \\ \tfrac{1}{2}(I - \sigma_z) &= |-1\rangle\langle -1|. \end{aligned} \tag{1.5.21}$$

We may subtract these equations to obtain σ_z:

$$\sigma_z = |+1\rangle\langle +1| - |-1\rangle\langle -1|, \tag{1.5.22}$$

so that we have expressed σ_z in terms of operators of the type $|a\rangle\langle a|$ mentioned near the end of Section 1.3. Eigenvalues of an operator are sometimes referred to as its spectrum.

1.6 OBSERVABLES, COMPLETENESS, EXPANSION IN EIGENKETS; DIRAC δ FUNCTION

In the preceding section we solved a very simple eigenvalue problem by finding the eigenvalues and eigenkets of the hermitian operator σ_z; we showed that the set of eigenkets was complete in that any ket in the space could be expanded in terms of the eigenkets of σ_z. In this section we give a physical interpretation to the eigenvalues. We also discuss the expansion of an arbitrary ket in terms of eigenkets of a hermitian operator and show that the orthonormality relation (1.5.12) must be generalized when the eigenvalues of a hermitian operator have a continuous range of values.

With every dynamical variable of a system there is associated a hermitian operator. When a measurement of the variable is made, a real number is obtained. It is therefore reasonable to make the following physical *assumption* in the theory. If the quantum system is in a particular eigenstate of L, say $|l\rangle$, then if we measure L, we obtain the value l. We also assume that, if we measure L of a system and always obtain l with certainty, the system is in eigenstate $|l\rangle$; that is, if we measure L for a large number of systems, each prepared in an identical way, and always get l, then the system is in state $|l\rangle$.

Furthermore, when a single measurement of L is made on a system in an arbitrary state, one of the eigenvalues of L is obtained. When the measurement is made on a system in an arbitrary state, the act of measurement [1] disturbs the system and causes it to jump into one of the eigenstates of the measured quantity. If L is measured immediately a second time, the same eigenvalue (with certainty) obtained on the first measurement is obtained here.

It is further assumed that any state of the system is linearly dependent on the eigenstates of L; that is, the eigenstates of L form a complete set. Those hermitian operators having a complete set of eigenvectors are called *observables*.

Proof of completeness and therefore than an operator is an observable is, in general, impossible.* The example of σ_z in the preceding section is one simple case where completeness could be proved. We therefore always assume that, if a quantity can be measured, its eigenkets form a complete set.

The completeness assumption allows expansion of an arbitrary state of the system in terms of the eigenkets of L. We have seen this in the special case given in (1.5.18). An even more familiar example of completeness involves the expansion of a certain class of periodic functions in a Fourier series of sine and cosine functions. The set of sines and cosines form a complete set with respect to this class of functions. We work out this example later. In the case of discrete eigenvalues of an observable, the set $\{|l\rangle\}$ is complete, and we may expand any ket $|\psi\rangle$ as the linear combination

$$|\psi\rangle = \sum_l c_l |l\rangle, \tag{1.6.1}$$

where the sum extends over the entire range of values that l may have. This range may be finite, as in the σ_z example of (1.5.18), or infinite, as in the

* There is, however, a spectral decomposition theorem which says that every "reasonable" hermitian operator has a decomposition into projections such as (1.5.22) (see Ref. 2, pp. 260ff).

case of a harmonic oscillator to be considered in Chapter 2. We may obtain the value of the expansion coefficients c_l (which may be complex numbers, in general) by the orthonormality relations (1.5.12). To do this, we multiply both sides of (1.6.1) from the left by the eigenket $\langle l'|$, use (1.5.12), and have

$$\langle l'|\psi\rangle = \sum_l c_l \langle l'|l\rangle = \sum_l c_l \delta_{l,l'} = c_{l'}. \tag{1.6.2}$$

When we put this back into (1.6.1), we obtain

$$|\psi\rangle = \sum_l |l\rangle\langle l|\psi\rangle \tag{1.6.3}$$

as the expansion of an arbitrary $|\psi\rangle$ in terms of eigenkets of L. Since $|\psi\rangle$ is arbitrary, (1.6.3) is satisfied if

$$\sum_l |l\rangle\langle l| = I. \tag{1.6.4}$$

This is the *completeness* or closure relation for discrete eigenvalues. An example of this relation was given for the operator σ_z in the preceding section [see (1.5.20)].

To illustrate the orthogonality relations (1.5.12) and the expansion postulate (1.6.1) and (1.6.3), as well as to introduce the Dirac δ function which will be needed shortly, let us consider the problem of expanding a continuous $f(x)$ in terms of the complete set of "eigenfunctions"

$$u_n(x) = \frac{1}{\sqrt{x_0}} e^{2\pi i n x/x_0} \qquad n = 0, \pm 1, \pm 2, \ldots. \tag{1.6.5}$$

If any continuous function could be expanded in the set $\{u_n(x)\}$, the set would be said to be complete *with respect to* the class of continuous functions. We see immediately that $u_n(x + x_0) = u_n(x)$ so that each member of the set is periodic with period x_0. Therefore, a continuous function $f(x)$ may be expanded only if it too is periodic with period x_0. The set is complete with respect to the class of continuous periodic functions with period x_0 if any such function may be expanded in the form

$$f(x) = \sum_{n=-\infty}^{\infty} c_n u_n(x) \tag{1.6.6}$$

(A more general class of functions may be expanded in this way but we are not being rigorous or exhaustive.) This expansion is the analog of (1.6.1).

The orthonormality relations are given by

$$\int_0^{x_0} u_{n'}^*(x) u_n(x)\, dx = \delta_{n'n}, \tag{1.6.7}$$

which are the analog of (1.5.12) when we use (1.6.5).

To obtain the expansion coefficients c_n in (1.6.6), we multiply both sides by $u_{n'}^*(x)$ and integrate from $x = 0$ to $x = x_0$. We find

$$\int_0^{x_0} f(x)u_{n'}^*(x)\,dx = \sum_{n=-\infty}^{\infty} c_n \int_0^{x_0} u_{n'}^*(x)u_n(x)\,dx, \qquad (1.6.8)$$
$$= c_{n'},$$

where we used (1.6.7). This is the analog of (1.6.2). [We are assuming that the integral for $c_{n'}$ exists and that it is legal to interchange the order of summation and integration in (1.6.8).] If we substitute (1.6.8) into (1.6.6) and again interchange the order of summation and integration (it is not "legal" but we do it anyway), we have

$$f(x) = \int_0^{x_0} dx'\, f(x') \sum_{n=-\infty}^{\infty} u_n^*(x')u_n(x). \qquad (1.6.9)$$

This is the analog of (1.6.3). To obtain the analog of the completeness relation (1.6.4) requires the introduction of the Dirac δ function.

Dirac defined an improper function $\delta(x - x')$ which has the property that

$$\delta(x - x') = \begin{cases} 0 & \text{if } x \neq x' \\ \infty & \text{if } x = x'. \end{cases} \qquad (1.6.10)$$

Clearly, this is not a function in the ordinary sense. However, it is very useful, and its use can be rigorously justified by the theory of distributions.† We shall not be rigorous and merely define it by its integral properties, namely,

$$\int_{x'-\epsilon}^{x'+\epsilon} \delta(x - x')\,dx = 1, \qquad (1.6.11)$$

where ϵ is arbitrary in size, but positive, and

$$\int_{x'-\epsilon}^{x'+\epsilon} f(x)\,\delta(x - x')\,dx = f(x'). \qquad (1.6.12)$$

Also it is symmetric:

$$\delta(x) = \delta(-x). \qquad (1.6.13)$$

There are many interesting representations of the δ function. One that is quite useful is

$$\delta(x) = \lim_{a\to\infty} \frac{\sin ax}{\pi x}. \qquad (1.6.14)$$

† See Ref. 2, Appendix A.

It has the property that, at $x = 0$, $\delta(0) = \infty$, and its integral from $-\infty$ to $+\infty$ is unity. As a result, we see that

$$\lim_{a\to\infty} \int_{-a}^{a} e^{ikx}\, dx = \lim_{a\to\infty} \frac{2 \sin ka}{k} = 2\pi\, \delta(k). \tag{1.6.15}$$

The reader is referred to standard quantum mechanics books [1–5] for further discussion of this interesting "function."

Since both sides of (1.6.9) must represent $f(x)$, comparison with (1.6.12) shows that we must have

$$\sum_{n=-\infty}^{\infty} u_n^*(x')u_n(x) = \delta(x' - x).$$

This is the completeness relation analogous to (1.6.4).

Let us return to the case in which the eigenvalues of the observable L are continuous. If $l' \neq l''$, we still have the orthogonality relation $\langle l'|l''\rangle = 0$. We now try to generalize (1.6.1) (since by hypothesis the eigenvectors form a complete set) and expand two vectors $|C\rangle$ and $|D\rangle$ as

$$\begin{aligned} |C\rangle &= \int c(l')|l'\rangle\, dl' \\ |D\rangle &= \int d(l'')|l''\rangle\, dl''. \end{aligned} \tag{1.6.16}$$

The scalar product of these is

$$\langle C|D\rangle = \int dl'\, c^*(l') \int d(l'')\langle l'|l''\rangle\, dl''. \tag{1.6.17}$$

Since $\langle l'|l''\rangle = 0$ if $l'' \neq l'$, the integral

$$\int d(l'')\langle l'|l''\rangle\, dl''$$

is zero if $\langle l'|l'\rangle$ is finite. This is true since the integrand will be zero everywhere except at the one point $l'' = l'$ and such an integral is zero. Since, in general, $\langle C|D\rangle \neq 0$, we can only conclude that $\langle l'|l'\rangle = \infty$. But the Dirac δ function comes to the rescue. With its use the orthonormality relations (1.5.12) for continuous eigenvalues may be generalized to

$$\langle l'|l''\rangle = \delta(l' - l''). \tag{1.6.18}$$

From the property (1.6.11) for the δ function, we have normalized the eigenvectors so that

$$\int \langle l'|l''\rangle\, dl'' = 1.$$

If we use (1.6.18) in (1.6.17) and let $|C\rangle = |D\rangle$, we see that

$$\langle C|C\rangle = \int dl'\; c^*(l') \quad c(l'')\, \delta(l' - l'')\, dl''$$

$$= \int |c(l')|^2\, dl' \geq 0 \tag{1.6.19}$$

by assumption. The integrals are over the range of eigenvalues of L.

From (1.6.16) and (1.6.18) we may obtain the completeness relation as follows: We form the scalar product

$$\langle l''|C\rangle = \int c(l')\langle l''|l'\rangle\, dl'$$

$$= \int c(l')\, \delta(l' - l'')\, dl' = c(l''). \tag{1.6.20}$$

This is the analog of (1.6.2) in the discrete case. If we substitute this into the expansion (1.6.16), we obtain

$$|C\rangle = \int |l'\rangle\, dl'\, \langle l'|C\rangle. \tag{1.6.21}$$

Since $|C\rangle$ is arbitrary, we have the *completeness* relation

$$\int |l'\rangle\, dl'\, \langle l'| = I. \tag{1.6.22}$$

The δ function is also useful in Fourier transforms. The expansion of $\psi(q)$ in a Fourier integral is given by

$$\psi(q) = \frac{1}{\sqrt{2\pi\hbar}} \int_{-\infty}^{\infty} \varphi(p) \exp\left(+\frac{ipq}{\hbar}\right) dp, \tag{1.6.23}$$

and the inverse transform is

$$\varphi(p) = \frac{1}{\sqrt{2\pi\hbar}} \int_{-\infty}^{\infty} \psi(q') \exp\left(-\frac{ipq'}{\hbar}\right) dq'. \tag{1.6.24}$$

If we substitute $\varphi(p)$ into the integral for $\psi(q)$ and interchange the order of integration with respect to p and q', we have

$$\psi(q) = \int_{-\infty}^{\infty} dq'\, \psi(q') \left\{\frac{1}{2\pi} \int_{-\infty}^{\infty} \frac{dp}{\hbar} \exp\left[\frac{ip(q - q')}{\hbar}\right]\right\}. \tag{1.6.25}$$

This equation yields the identity $\psi(q) = \psi(q)$ if

$$\frac{1}{2\pi} \int_{-\infty}^{\infty} \frac{dp}{\hbar} \exp\left[\frac{ip(q' - q'')}{\hbar}\right] = \delta(q' - q''). \tag{1.6.26}$$

This is the representation of the δ function given in (1.6.15), with suitable change in notation.

Some observables have both a discrete and a continuous spectrum. The extension of the orthonormality and completeness relations in this case is straightforward.

Before closing this section, we discuss very briefly a function of an observable, $f(L)$. If $f(L)$ can be expanded in a power series, by repeated application of the eigenvalue equation,

$$L|l\rangle = l|l\rangle, \tag{1.6.27}$$

we see that

$$f(L)|l\rangle = f(l)|l\rangle. \tag{1.6.28a}$$

For example [1], if $f(L) = L^2$, then $L^2|l\rangle = l^2|l\rangle$. However, we shall postulate that (1.6.28a) holds even if the function cannot be expanded in a power series. We consider the function $f(L) = L^{-1}$. Then

$$L^{-1}|l\rangle = l^{-1}|l\rangle$$

provided none of the eigenvalues $l = 0$. But this is just a requirement that L^{-1} exist. Another simple example is the function $f(L) = L^{1/2}$. In this case,

$$L^{1/2}|l\rangle = \pm l^{1/2}|l\rangle,$$

and there is an ambiguity in sign. The operator $L^{1/2}$ exists and its eigenvalues are real if the eigenvalues of L are positive. The ambiguity in sign may be removed by choosing a sign for each eigenvalue. Usually, in practice, one selects the positive square root.

Finally, the adjoint of (1.6.28a) is

$$\langle l|f^\dagger(L) = f^*(l)\langle l| \tag{1.6.28b}$$

or if G is a function of L,

$$\langle l|G(L) = G(l)\langle l|. \tag{1.6.28c}$$

1.7 MATRICES

In the following section, we give a matrix representation for ket and bra vectors as well as linear operators in a space. Although a knowledge of finite matrices is assumed, we discuss here a few of the less familiar properties as well as extend the ideas of finite matrices to infinite matrices heuristically.

The trace of a square finite matrix is defined as the sum of the diagonal elements. Thus if A is a square finite matrix, then

$$\text{Tr}\,(A) \equiv \sum_i A_{ii}, \tag{1.7.1}$$

where Tr is an abbreviation for the trace, and A_{ii} is the ith diagonal element. Also, the trace of a product of finite square matrices is invariant under cyclic permutations, that is,

$$\text{Tr}\,(ABC) = \text{Tr}\,(BCA) = \text{Tr}\,(CAB) \tag{1.7.2}$$

as may be proved easily provided the traces exist.

The hermitian adjoint of a matrix A, written $A^\dagger$, is obtained by interchanging rows and columns and taking the complex conjugate of each element. If

$$A = A^\dagger \tag{1.7.3}$$

the matrix is said to be *hermitian;* that is, $(A^\dagger)_{ij} = A^*_{ji} = A_{ij}$.

A matrix A is *unitary* if

$$AA^\dagger = A^\dagger A = I \rightarrow A^\dagger = A^{-1} \tag{1.7.4}$$

where A^{-1} is the inverse. The inverse exists if the determinant of A is not zero.

These and all the more familiar properties may be used for matrices with an infinite number of rows and columns. The rows and columns may be labeled by discrete indices, or by a set of continuous indices that extend over some range of values, or a combination of both. For example, if q and q' can have any value from $-\infty$ to $+\infty$, we may write a matrix element of A with labels $A_{q;q'}$ or, equivalently, $A(q;q')$. When we multiply this by another matrix B with elements $B_{q;q'}$, by analogy with the rule of matrix multiplication for finite matrices, we write

$$C_{q;q'} = \int_{-\infty}^{\infty} A(q;q'')B(q'';q')\,dq''. \tag{1.7.5}$$

When the integrals converge, all is well. Similarly, the trace of A is

$$\text{Tr}\,(A) = \int A(q';q')\,dq'.$$

This assumes that the integral exists. A diagonal matrix is written

$$A(q';q'') = A(q';q')\,\delta(q'-q'').$$

The extension to infinite matrices is straightforward.

1.8 MATRIX REPRESENTATION OF KETS, BRAS, AND OPERATORS

It has previously been pointed out that representations play the role of coordinates in ordinary vector analysis. In this section we give a more precise meaning to these intuitive concepts. We show how to find a representation and how to express operators and eigenvectors in this representation. In particular, we show that arbitrary kets and bras in the vector space may be *represented* in terms of column and row vectors (more generally called matrices) and that operators may be represented in terms of matrices. We have already seen a simple example of this in (1.5.14). The advantage of using a particular representation in solving quantum problems is the same as the advantage of using a particular coordinate system in ordinary geometry.

We develop the theory of representations in terms of an observable L which has a discrete spectrum in parallel with an observable q with a continuous spectrum.

We begin by considering an observable L (or q) that satisfies the eigenvalue equations

$$L|l\rangle = l|l\rangle \qquad q|q'\rangle = q'|q'\rangle, \tag{1.8.1}$$

where the eigenvalues l are discrete (and q' continuous). Since L (and q) are observables, the eigenkets $\{|l\rangle\}$ (and $\{|q'\rangle\}$) form a complete set. By (1.5.12) and (1.6.18) these eigenkets satisfy the orthonormality conditions

$$\langle l'|l''\rangle = \delta_{l'l''} \qquad \langle q'|q''\rangle = \delta(q' - q''), \tag{1.8.2}$$

and, by (1.6.4) [and (1.6.22)], the completeness relations

$$\sum_{l'} |l'\rangle\langle l'| = I \qquad \int |q'\rangle\, dq'\, \langle q'| = I. \tag{1.8.3}$$

By virtue of (1.8.3), an arbitrary ket $|\psi\rangle$ may be written

$$|\psi\rangle = \sum_{l'} |l'\rangle\langle l'|\psi\rangle \qquad |\psi\rangle = \int |q'\rangle\langle q'|\psi\rangle\, dq'. \tag{1.8.4}$$

By way of analogy, if $\hat{\mathbf{i}}$, $\hat{\mathbf{j}}$, and $\hat{\mathbf{k}}$ are three unit orthogonal vectors in ordinary space, we may write an arbitrary vector $\mathbf{A}$ as

$$\mathbf{A} = A_x\hat{\mathbf{i}} + A_y\hat{\mathbf{j}} + A_z\hat{\mathbf{k}}. \tag{1.8.5}$$

This is the analog of (1.8.4). The analog of (1.8.2) is

$$\hat{\mathbf{i}}\cdot\hat{\mathbf{j}} = \hat{\mathbf{j}}\cdot\hat{\mathbf{k}} = \cdots = 0;$$

$\hat{\mathbf{i}}\cdot\hat{\mathbf{i}} = \hat{\mathbf{j}}\cdot\hat{\mathbf{j}} = \hat{\mathbf{k}}\cdot\hat{\mathbf{k}} = 1$, and the set $\{\hat{\mathbf{i}}, \hat{\mathbf{j}}, \hat{\mathbf{k}}\}$ is complete since any vector $\mathbf{A}$ may be expanded as in (1.8.5). It is said that $\mathbf{A}$ is expanded in the $\hat{\mathbf{i}}$, $\hat{\mathbf{j}}$, $\hat{\mathbf{k}}$ representation. These vectors may be called basis vectors for the space.

Analogously, the set of vectors $\{|l\rangle\}$ (and $\{|q'\rangle\}$) may be regarded as a particular set of orthogonal unit basis vectors in the sense of (1.8.2), and it is said that (1.8.4) gives the expansion of $|\psi\rangle$ in the L (or q) representation. The numbers $\langle l'|\psi\rangle$ ($\langle q'|\psi\rangle$) are the "components" or, as Dirac calls them, the *representatives* of $|\psi\rangle$ in the L (or q) representation.

Just as we may select a complete set of basis vectors other than $\hat{\mathbf{i}}$, $\hat{\mathbf{j}}$, $\hat{\mathbf{k}}$ to represent $\mathbf{A}$, we may choose eigenvectors of observables other than L (or q) to use as basis vectors to represent $|\psi\rangle$. We discuss this situation in the next section.

From the original definition of scalar products (see Section 1.2) we know that $\langle l'|\psi\rangle$ ($\langle q'|\psi\rangle$) may be considered as a function of l' (or q') which may be written $\psi_{l'}$ [or $\psi(q')$] since with each l' (or q') there is associated a number $\psi_{l'}$ [or $\psi(q')$]. These numbers, or functions, determine $|\psi\rangle$ uniquely (if the eigenkets have been specified) just as the numbers A_x, A_y, and A_z specify $\mathbf{A}$ uniquely (when $\hat{\mathbf{i}}$, $\hat{\mathbf{j}}$, and $\hat{\mathbf{k}}$ have been chosen).

We may write the vector $\mathbf{A}$ as a column vector (or matrix) in the form

$$\mathbf{A} = \begin{bmatrix} A_x \\ A_y \\ A_z \end{bmatrix} = A_x\hat{\mathbf{i}} + A_y\hat{\mathbf{j}} + A_z\hat{\mathbf{k}}, \tag{1.8.6}$$

where we label the rows by x, y, z. Similarly, we may write $|\psi\rangle$ in (1.8.4) as a matrix of one column in which we label the rows by l' (or q'):

$$|\psi\rangle = \sum_{l'} \langle l'|\psi\rangle|l'\rangle \qquad |\psi\rangle = \int dq' \, \langle q'|\psi\rangle|q'\rangle. \tag{1.8.7}$$

Thus there is a way of *representing* an arbitrary ket as a column vector.* The number of rows is determined by the number of eigenvalues that l' (or q') may have. The representatives $\langle l'|\psi\rangle$ (or $\langle q'|\psi\rangle$) are usually complex.

We may illustrate (1.8.7) with the simple example in Section 1.5. From (1.5.18) and (1.5.19) we have

$$|P\rangle = \begin{bmatrix} \langle 1|P\rangle \\ \langle -1|P\rangle \end{bmatrix} = \begin{bmatrix} c_1 \\ c_2 \end{bmatrix} \equiv c_1\begin{bmatrix} 1 \\ 0 \end{bmatrix} + c_2\begin{bmatrix} 0 \\ 1 \end{bmatrix} \tag{1.8.8}$$

as the vector representing an arbitrary ket in the space. From this, we see that the basis vectors themselves may be represented by

$$|+1\rangle = \begin{bmatrix} \langle +1|+1\rangle \\ \langle -1|+1\rangle \end{bmatrix} = \begin{bmatrix} 1 \\ 0 \end{bmatrix}$$

$$|-1\rangle = \begin{bmatrix} \langle +1|-1\rangle \\ \langle -1|-1\rangle \end{bmatrix} = \begin{bmatrix} 0 \\ 1 \end{bmatrix}. \tag{1.8.9}$$

* Representing $\langle q'|\psi\rangle$ as a column vector labeled by the continuous variable q' is not well defined. However, there is no harm in visualizing this by analogy with the discrete case $\langle l'|\psi\rangle$.

We have already pointed out earlier in the introduction that the representatives $\langle q'|\psi\rangle = \psi(q')$ are the Schrödinger wave functions in the coordinate representation when q is the observable associated with the coordinate of a particle constrained to move in one dimension. We go into these matters more fully later in the chapter.

In complete analogy with the above, an arbitrary bra $\langle\psi|$ may be represented by a matrix with one row and the columns labeled by l' (or q'). Thus

$$\langle\psi| = \sum_{l'} \langle\psi|l'\rangle\langle l'| \qquad \langle\psi| = \int \langle\psi|q'\rangle\, dq'\, \langle q'|. \tag{1.8.10}$$

From the general property (1.2.8), we have

$$\langle\psi|l'\rangle = \langle l'|\psi\rangle^* \equiv \psi_{l'}^* \qquad \langle\psi|q'\rangle = \langle q'|\psi\rangle^* \equiv \psi^*(q'). \tag{1.8.11}$$

Therefore, the hermitian adjoint of a column vector is a row vector whose corresponding elements are the complex conjugates of the row vector.

We next consider the problem of expressing any linear operator A as a matrix in the L (or q) representation. To do this, we apply the completeness relations (1.8.3) twice and write the identities

$$\begin{aligned} A &= \sum_{l',l''} |l'\rangle\langle l'|A|l''\rangle\langle l''| \\ A &= \iint dq'\, dq''\, |q'\rangle\langle q'|A|q''\rangle\langle q''|. \end{aligned} \tag{1.8.12}$$

The numbers $\langle l'|A|l''\rangle$ ($\langle q'|A|q''\rangle$) are functions of l' and l'' (q' and q'') (see the definition of a linear operator in Section 1.3), and we may write

$$\langle l'|A|l''\rangle \equiv A_{l';l''} \tag{1.8.13}$$

as the matrix elements of A in the L representation and

$$\langle q'|A|q''\rangle = A(q';q'') \tag{1.8.14}$$

in the (continuous) q representation. In particular, if $A = L$ (or q), the matrix elements reduce by means of (1.8.2) and (1.8.1) to

$$\langle l'|L|l''\rangle = l'\delta_{l'l''} \qquad \langle q'|q|q''\rangle = q'\,\delta(q' - q'') \tag{1.8.15}$$

so that L (or q) is diagonal in its own representation. Thus (1.8.13) gives A in the representation in which L is diagonal and (1.8.14) gives A in the representation in which q is diagonal. If we return once more to the example of Section 1.5, we see that (1.5.14) is a special case of (1.8.15); that is,

$$\sigma_z = \begin{bmatrix} \langle +1|\sigma_z|+1\rangle & \langle +1|\sigma_z|-1\rangle \\ \langle -1|\sigma_z|+1\rangle & \langle -1|\sigma_z|-1\rangle \end{bmatrix} \equiv \begin{bmatrix} 1 & 0 \\ 0 & -1 \end{bmatrix}, \tag{1.8.16}$$

which is diagonal. This is called the representation in which σ_z is diagonal, or simply the σ_z representation.

It has been pointed out several times previously that the algebra of operators is the same as the algebra of matrices; thus it is not surprising that operators may be represented by matrices. Similarly, the algebra of ket and bra vectors is the same as the algebra of one-column or one-row matrices (or vectors).

We have introduced a set of basis vectors as eigenvectors of a particular observable L (or q). This is not essential, and a set of basis vectors may be introduced quite arbitrarily; every basis chosen gives rise to a representation of vectors by column matrices and operators by square matrices. The basis vectors may always be chosen in such a way that they are orthonormal in the sense of (1.8.2). In the next section we show how to change from one set of basis vectors to another.

1.9 TRANSFORMATION FUNCTIONS; CHANGE OF REPRESENTATION; DIAGONALIZATION

The choice of the observable used to represent state vectors and operators is not unique. As an example, if the system is a particle constrained to move in one dimension, either the momentum p or the coordinate q would serve as a suitable observable in representing state vectors and operators in matrix form. For some calculations, one representation may be more convenient than the other. In this section we consider the effect on state vectors and operators of changing from one representation to another. In the next section we work out a particular example.

This problem is completely analogous to a coordinate rotation in ordinary geometry. Let us illustrate this for a two-dimensional rotation. If x, y is a rectangular system and if x', y' is a system rotated counterclockwise through angle θ, we have the transformation equations

$$\begin{aligned} x' &= \cos\theta x + \sin\theta y \\ y' &= -\sin\theta x + \cos\theta y, \end{aligned} \tag{1.9.1}$$

or, in matrix notation,

$$\begin{bmatrix} x' \\ y' \end{bmatrix} = \begin{bmatrix} \cos\theta & \sin\theta \\ -\sin\theta & \cos\theta \end{bmatrix} \begin{bmatrix} x \\ y \end{bmatrix}. \tag{1.9.2}$$

The transformation matrix

$$K = \begin{bmatrix} \cos\theta & \sin\theta \\ -\sin\theta & \cos\theta \end{bmatrix} \tag{1.9.3}$$

satisfies the conditions that the determinant of K, written det K, equals 1 and

$$K\tilde{K} = K\tilde{K} = I \equiv \begin{bmatrix} 1 & 0 \\ 0 & 1 \end{bmatrix}, \tag{1.9.4}$$

where $\tilde{K}$ is the transpose of K. From this we conclude that

$$\tilde{K} = K^{-1}, \tag{1.9.5}$$

where K^{-1} is the inverse. When $\tilde{K} = K^{-1}$ and det $K = 1$, the transformation between $x'y'$ and xy is a rotation. We may use (1.9.5) to invert (1.9.2), and obtain

$$\begin{bmatrix} x \\ y \end{bmatrix} = \tilde{K}\begin{bmatrix} x' \\ y' \end{bmatrix} = \begin{bmatrix} \cos\theta & -\sin\theta \\ \sin\theta & \cos\theta \end{bmatrix}\begin{bmatrix} x' \\ y' \end{bmatrix}. \tag{1.9.6}$$

Similarly, the components of an arbitrary vector $\mathbf{A}$ in the x', y' frame are related to the components in the x, y frame by

$$\begin{bmatrix} A_{x'} \\ A_{y'} \end{bmatrix} = K\begin{bmatrix} A_x \\ A_y \end{bmatrix}, \tag{1.9.7}$$

or

$$\begin{bmatrix} A_x \\ A_y \end{bmatrix} = \tilde{K}\begin{bmatrix} A_{x'} \\ A_{y'} \end{bmatrix}. \tag{1.9.8}$$

These components are shown in Figure 1.3.

Transforming from one representation to another is no more complicated in principle than the simple example above. Let us assume that we have two observables L and M that satisfy the eigenvalue equations

$$L|l'\rangle = l'|l'\rangle \qquad M|m'\rangle = m'|m'\rangle. \tag{1.9.9}$$

Both l' and m' may be discrete, both continuous, one discrete and the other continuous, or l' as well as m' may have some discrete values and some continuous. For simplicity, we assume that both l' and m' are discrete although

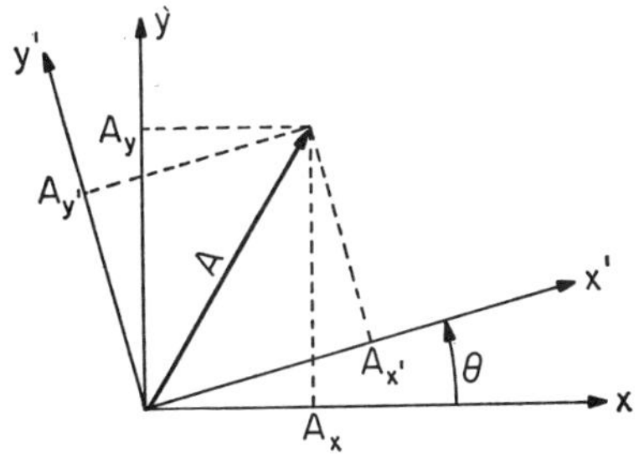

Fi gure 1.3 Components of a vector in two rotated coordinate systems.

any of the possibilities above may be treated in analogous fashion. Since L and M are observables,

$$\langle l'|l''\rangle = \delta_{l'l''} \qquad \langle m'|m''\rangle = \delta_{m'm''}, \tag{1.9.10}$$

and both sets are complete so that

$$\sum_{l'} |l'\rangle\langle l'| = I \qquad \sum_{m'} |m'\rangle\langle m'| = I. \tag{1.9.11}$$

We may therefore expand any $|l'\rangle$ in terms of the $\{|m'\rangle\}$ or vice versa and obtain

$$|l'\rangle = \sum_{m'} |m'\rangle\langle m'|l'\rangle \qquad |m'\rangle = \sum_{l'} |l'\rangle\langle l'|m'\rangle. \tag{1.9.12}$$

The numbers $\langle m'|l'\rangle$ and $\langle l'|m'\rangle$, which by (1.2.8) are related by

$$\langle m'|l'\rangle = \langle l'|m'\rangle^*, \tag{1.9.13}$$

are called the transformation function between the L and M representations. Because of the completeness and orthogonality relations

$$\begin{aligned} \sum_{m'} \langle l'|m'\rangle\langle m'|l''\rangle &= \delta_{l'l''} \\ \sum_{l'} \langle m'|l'\rangle\langle l'|m''\rangle &= \delta_{m'm''}. \end{aligned} \tag{1.9.14}$$

Because of (1.9.13) and (1.9.14), the transformations (1.9.12) are called unitary.

We may arrange the numbers $\langle m'|l'\rangle$ as a matrix with the eigenvalues m' labeling the columns and l' labeling the rows.

We may also use either the L or M representation to represent any operator A. Thus $A_{l'l''} \equiv \langle l'|A|l''\rangle$ are the matrix elements of A in the L-representation while $A_{m'm''} \equiv \langle m'|A|m''\rangle$ are the matrix elements in the M-representation. If we use the completeness relations, we see that

$$\begin{aligned} \langle l'|A|l''\rangle &= \sum_{m',m''} \langle l'|m'\rangle\langle m'|A|m''\rangle\langle m''|l''\rangle \\ \langle m'|A|m''\rangle &= \sum_{l',l''} \langle m'|l'\rangle\langle l'|A|l''\rangle\langle l''|m''\rangle \end{aligned} \tag{1.9.15}$$

Therefore, the transformation functions are needed to obtain the connection between the matrix elements in the two representations. When we transform from one representation to the other under a unitary transformation, the operators, according to (1.9.15), are said to transform under a similarity transformation.

We therefore must obtain the transformation functions. Explicit examples are given later.

Suppose that A is an observable and that we wish to solve the eigenvalue problem

$$A|a'\rangle = a'|a'\rangle, \tag{1.9.16}$$

where the $\{|a'\rangle\}$ are orthogonal and complete. Assume in addition that we know the matrix elements of A in the L-representation. If we multiply both sides of (1.9.16) by $\langle l'|$, we obtain

$$\langle l'|A|a'\rangle = a'\langle l'|a'\rangle. \tag{1.9.17}$$

If we insert the completeness relation, we have

$$\sum_{l''} \langle l'|A|l''\rangle\langle l''|a'\rangle = a'\langle l'|a'\rangle. \tag{1.9.18}$$

This may be rewritten as

$$\sum_{l''} \{\langle l'|A|l''\rangle - a'\delta_{l'',l'}\}\langle l''|a'\rangle = 0. \tag{1.9.19}$$

This represents a set of homogeneous algebraic equations to determine the transformation function $\langle l'|a'\rangle$ from the L-representation to the representation in which A is diagonal. To have nontrivial solutions, it is necessary and sufficient that the determinant of the coefficients vanish, namely,

$$\|A_{l'l''} - a'\delta_{l'l''}\| = 0. \tag{1.9.20}$$

The values of a' that satisfy this are the eigenvalues of A. Corresponding to each eigenvalue a', we can solve (1.9.19) for the transformation function $\langle l''|a'\rangle$. Thus diagonalizing the matrix representing A is equivalent to solving the eigenvalue problem.

We next prove the following theorem. Two matrices may be simultaneously diagonalized by the *same* transformation if and only if they commute.

PROOF

Assume A and B are two operators which are diagonal in the M representation. Then

$$\begin{aligned} \langle m'|A|m''\rangle &= A_{m'}\langle m'|m''\rangle \\ \langle m'|B|m''\rangle &= B_{m'}\langle m'|m''\rangle. \end{aligned} \tag{1.9.21}$$

We then have

$$\begin{aligned} \langle m'|AB|m''\rangle &= \sum_{m'''} \langle m'|A|m'''\rangle\langle m'''|B|m''\rangle \\ &= \sum_{m'''} A_{m'}B_{m''}\langle m'|m'''\rangle\langle m'''|m''\rangle \\ &= A_{m'}B_{m'}\langle m'|m''\rangle. \end{aligned} \tag{1.9.22}$$

Also,

$$\langle m'|BA|m''\rangle = B_{m'}A_{m'}\langle m'|m''\rangle. \tag{1.9.23}$$

Therefore, we see that

$$AB = BA.$$

The converse is easily proved.

We have therefore developed a complete scheme for transforming vectors and operators between two representations. We illustrate this by a simple example in the next section.

1.10 QUANTIZATION; EXAMPLE OF CONTINUOUS SPECTRUM

If a single measurement of an observable is made, one of its eigenvalues is obtained. The ability to solve eigenvalue problems is therefore essential to relate the theory to experiment. Thus far we have solved only one such problem, in Section 1.5, where the observable satisfied the algebraic equation $\sigma_z^2 = 1$. In that case the Hilbert space consisted of only two eigenvectors, and the eigenvalue spectrum had only the two discrete values $+1$ and -1. In this section we solve an eigenvalue problem in which the eigenvalue spectrum is continuous. This simple example demonstrates how to treat quantum-mechanically a system that has a classical analog.

Again we consider a single particle of mass m constrained to move in one dimension in a field of force. Classically, this system may be described completely by a position coordinate q and a momentum p. If we specify both these quantities at a certain time, we have specified the classical state of the system at this time.

To treat this system quantum-mechanically, according to the theory we have developed thus far, we associate with each of these dynamical variables (since they are observable) a linear hermitian operator which we shall call q and p. As operators, they satisfy

$$p = p^\dagger \qquad q = q^\dagger. \tag{1.10.1}$$

The classical hamiltonian for the system is also an observable, and we may associate with it the hermitian operator,

$$H = \frac{1}{2m}p^2 + V(q) = H^\dagger \tag{1.10.2}$$

which is expressible in terms of p and q.

After the operators needed to describe the physical system are enumerated, the next step in setting up the quantum problem is to specify the algebra that the operators must obey. This requires an additional *postulate* for the theory; it is given in terms of the commutation relations for p and q, namely,

$$[q, q] = 0 \qquad [p, p] = 0$$
$$[q, p] \equiv (qp - pq) = i\hbar, \tag{1.10.3}$$

where $\hbar$ is Planck's constant divided by 2π; that is, we postulate that p and q satisfy the commutation relations above. Classically, q and p commute so that, to the extent they do not commute, the quantum and classical systems differ. Accordingly, the classical system is quantized when the observables q and p satisfy (1.10.3). The justification for the quantum postulate is the remarkable agreement between theory and experiment. It is possibly the most profound and fundamental postulate in the theory.

If $\hbar \to 0$, q and p will commute so that classical mechanics should be contained in the quantum formulation in the limit as $\hbar \to 0$. This is just the correspondence principle.

It is said that q and p obey a noncommutative algebra. Before proceeding, let us develop a few useful algebraic relations. If l is an integer, we may prove by mathematical induction from (1.10.3) that (see Problem 1.1)

$$[q, p^l] = i\hbar l p^{l-1} \equiv i\hbar \frac{\partial}{\partial p} p^l$$
$$[p, q^l] = -i\hbar l q^{l-1} \equiv -i\hbar \frac{\partial}{\partial q} q^l. \tag{1.10.4}$$

From these commutator relations, it follows directly that, if $F(p)$ and $G(q)$ are functions that may be expanded in a power series in p and q, respectively, then

$$[q, F(p)] \equiv i\hbar \frac{\partial F}{\partial p} \tag{1.10.5a}$$

$$[p, G(q)] \equiv -i\hbar \frac{\partial G}{\partial q}. \tag{1.10.5b}$$

We postulate that these are also true even if F and G cannot be expanded in a power series.

These relations may be generalized even further to the case where $F(p, q)$ is a function of p and q:

$$[q, F(p, q)] \equiv i\hbar \frac{\partial F}{\partial p} \tag{1.10.6a}$$

$$[p, F(p, q)] \equiv -i\hbar \frac{\partial F}{\partial q}. \tag{1.10.6b}$$

Extreme care must be exercised in the use of (1.10.6) since p and q do not commute. For example, if $F(p, q) = pqp$, this is *not* equal to p^2q, and in this case the application of (1.10.6) gives

$$[q, pqp] = i\hbar \frac{\partial}{\partial p} pqp = i\hbar(pq + qp).$$

In other words, the order of the factors in $F(p, q)$ must be preserved when using (1.10.6).

We are now ready to solve the two eigenvalue problems

$$p|p'\rangle = p'|p'\rangle \qquad q|q'\rangle = q'|q'\rangle \tag{1.10.7}$$

for the momentum and coordinate. This involves finding the eigenvalues and eigenvectors. To do this, we shall introduce two Dirac operators, called translation operators for reasons that will become apparent shortly.

We consider the operator [2]

$$Q(\xi) = \exp\left(-\frac{i\xi p}{\hbar}\right), \tag{1.10.8}$$

where ξ is an arbitrary real parameter. Since p is hermitian, we see that

$$Q^\dagger(\xi) = \exp\left(\frac{i\xi p}{\hbar}\right) = Q^{-1}(\xi) \tag{1.10.9}$$

so that Q is a unitary operator if ξ is real.*

If we use (1.10.5a) we see that

$$[q, Q] = i\hbar \frac{\partial Q}{\partial p} = \xi Q$$

so that

$$qQ = Qq + \xi Q. \tag{1.10.10}$$

If we now multiply both sides of this equation by the eigenket $|q'\rangle$ of q with eigenvalue q', by (1.10.7) we have

$$q\{Q|q'\rangle\} = (q' + \xi)\{Q|q'\rangle\}. \tag{1.10.11}$$

This says that, if $|q'\rangle$ is an eigenket with eigenvalue q', then $Q|q'\rangle$ is also an eigenket with eigenvalue $q' + \xi$. Now by Theorem 1, Section 1.5, the eigenvalues of q must be real since q is hermitian; thus ξ must be real, which we assumed in (1.10.9), but otherwise it is completely arbitrary. Therefore, the eigenvalue $q' + \xi$ may be any value from $-\infty$ to $+\infty$. The eigenvalues of q' have a continuous spectrum.

This is the justification for calling Q a translation operator since it translates the eigenvalue from q' to $q' + \xi$. If we multiply (1.10.10) from the left

* Eigenvalue of q must be real; therefore, ξ must be real.

by Q^{-1}, we obtain

$$Q^{-1}qQ = q + \xi, \tag{1.10.12}$$

which shows that Q translates the operator q by an amount ξ.

Since q' is continuous, the eigenkets may be normalized in such a way that they satisfy the "orthonormality" relation (1.8.2),

$$\langle q'|q''\rangle = \delta(q' - q''). \tag{1.10.13}$$

Also, since q is an observable, we have the completeness relation

$$\int_{-\infty}^{\infty} |q'\rangle\, dq'\, \langle q'| = I. \tag{1.10.14}$$

Now we have shown that $Q(\xi)|q'\rangle$ is an eigenket of q with eigenvalue $q' + \xi$. It may therefore differ from the eigenket $|q' + \xi\rangle$ by a multiplicative constant so that we may therefore let

$$Q(\xi)|q'\rangle = c(q', \xi)|q' + \xi\rangle. \tag{1.10.15}$$

It also follows directly from this that

$$\langle q''|Q^\dagger(\xi) = c^*(q'', \xi)\langle q'' + \xi|. \tag{1.10.16}$$

If we multiply these together and use (1.10.9), we obtain

$$\langle q''|Q^\dagger Q|q'\rangle = \langle q''|q'\rangle = c^*(q'', \xi)c(q', \xi)\langle q'' + \xi|q' + \xi\rangle.$$

If we use (1.10.13) and integrate both sides over all dq'', we see that

$$|c(q', \xi)|^2 = 1$$

independent of q' and ξ. Aside from a trivial phase, we may let $c = 1$ so that

$$\exp\left(-\frac{i\xi p}{\hbar}\right)|q'\rangle \equiv Q(\xi)|q'\rangle = |q' + \xi\rangle. \tag{1.10.17}$$

In particular, if we take $q' = 0$ we have

$$Q(\xi)|0\rangle_q = |\xi\rangle_q,$$

where the subscript q denotes that $|0\rangle_q$ is an eigenket of q with eigenvalue zero. If we next let ξ be some arbitrary eigenvalue, say q', then this becomes

$$\exp\left(-\frac{iq'p}{\hbar}\right)|0\rangle_q = Q(q')|0\rangle_q = |q'\rangle. \tag{1.10.18}$$

The state with eigenvalue q' can be generated from the state with eigenvalue $q' = 0$ by the translation operator $Q(q')$.

We may take the adjoint of (1.10.17) to obtain

$$\langle q'|Q^\dagger(\xi) = \langle q'| \exp\left(\frac{ip\xi}{\hbar}\right) = \langle q' + \xi| \tag{1.10.19}$$

so that

$$_q\langle 0|Q^\dagger(q') = \langle q'|. \tag{1.10.20}$$

Therefore, all eigenkets (and eigenbras) may be generated from one eigenket (or eigenbra). This is sufficient to determine all the properties of eigenvectors needed for the theory.

We may analogously introduce another unitary translation operator

$$P(\xi) = \exp\left(\frac{i\xi q}{\hbar}\right), \tag{1.10.21}$$

and show that the eigenvalues of p may have any value from $-\infty$ to $+\infty$ and that

$$P(\xi)|p'\rangle = |p' + \xi\rangle \tag{1.10.22}$$

$$P(p')|0\rangle_p \equiv \exp\left(\frac{ip'q}{\hbar}\right)|0\rangle_p = |p'\rangle \tag{1.10.23}$$

or

$$_p\langle 0|P^\dagger(p') = {}_p\langle 0| \exp\left(-\frac{ip'q}{\hbar}\right) = \langle p'|. \tag{1.10.24}$$

We also have

$$P^{-1}pP = p + \xi. \tag{1.10.25}$$

The eigenkets satisfy the orthonormality and completeness relations

$$\langle p'|p''\rangle = \delta(p' - p'') \qquad \int_{-\infty}^{\infty} |p'\rangle\, dp'\, \langle p'| = I. \tag{1.10.26}$$

We have therefore solved the eigenvalue problem for both q and p. The set $|q'\rangle$ and the set $|p'\rangle$ are both complete and either may be used as a set of basis vectors to represent arbitrary state vectors and operators. From the discussion in the preceding section, we need to know the transformation function

$$S(p'; q') = \langle p'|q'\rangle = \langle q'|p'\rangle^* \tag{1.10.27}$$

to transform from one representation to the other. This function may be calculated quite simply by means of the translation operators Q and P. By (1.10.18) we have

$$\langle p'|q'\rangle = \langle p'| \exp\left(-\frac{iq'p}{\hbar}\right)|0\rangle_q. \tag{1.10.28}$$

But [see (1.6.28)]

$$\langle p'|F(p) = F(p')\langle p'| \tag{1.10.29}$$

so that

$$\langle p'|q'\rangle = \exp\left(-\frac{iq'p'}{\hbar}\right)\langle p'|0\rangle_q. \tag{1.10.30}$$

By (1.10.24) and the fact that [see (1.6.28)]

$$F(q)|q'\rangle = F(q')|q'\rangle, \tag{1.10.31}$$

(1.10.30) may be written

$$\langle p'|q'\rangle = \exp\left(-\frac{iq'p'}{\hbar}\right) {}_p\langle 0|\exp\left(-\frac{ip'q}{\hbar}\right)|0\rangle_q$$

$$= \exp\left(-\frac{iq'p'}{\hbar}\right) {}_p\langle 0|0\rangle_q \equiv \langle q'|p'\rangle^*. \tag{1.10.32}$$

But ${}_p\langle 0|0\rangle_q$ is just a constant that may be evaluated as follows: by means of the orthonormality relation for $|p'\rangle$ (Eqs. 1.10.26) and the completeness relation for $|q'\rangle$ (Eq. 1.10.14) we have the result

$$\int_{-\infty}^{\infty} \langle p'|q'\rangle\, dq'\, \langle q'|p''\rangle = \delta(p' - p''). \tag{1.10.33}$$

If we substitute (1.10.32) into (1.10.33), we find that

$$|{}_p\langle 0|0\rangle_q|^2 \int_{-\infty}^{\infty} \exp\left[\frac{iq'(p'' - p')}{\hbar}\right] dq' = \delta(p' - p'').$$

With suitable change in notation, we see by (1.6.26) that the integral equals $2\pi\hbar\delta(p' - p'')$ so that the constant is

$$|{}_p\langle 0|0\rangle_q|^2 = \frac{1}{2\pi\hbar}. \tag{1.10.34}$$

We may choose the phase of ${}_p\langle 0|0\rangle_q$ so that

$$S(p'; q') = \langle p'|q'\rangle = \frac{1}{\sqrt{2\pi\hbar}} \exp\left(-\frac{ip'q'}{\hbar}\right) = \langle q'|p'\rangle^*. \tag{1.10.35}$$

This is the transformation function discussed in Section 1.9.

We saw in the preceding section that an operator in one representation transforms to another representation by means of (1.9.15). To apply this result to an operator A in the p and q representatives, we let the operator L in (1.9.15) correspond to q and the operator M correspond to p. Since the eigenvalues are continuous in this case, the sums in (1.9.15) must be replaced by integrals, and we have

$$\langle q'|A|q''\rangle = \int_{-\infty}^{\infty} dp' \int_{-\infty}^{\infty} dp''\, \langle q'|p'\rangle\langle p'|A|p''\rangle\langle p''|q''\rangle.$$

If we use (1.10.35), this becomes

$$\langle q'|A|q''\rangle = \frac{1}{2\pi\hbar}\int_{-\infty}^{\infty} dp' \int_{-\infty}^{\infty} dp'' \exp\left[\frac{i(p'q' - p''q'')}{\hbar}\right]\langle p'|A|p''\rangle. \quad (1.10.36)$$

We now consider the special case of $A = p$. By (1.10.26) and (1.10.7), we have

$$\langle p'|p|p''\rangle = p'\,\delta(p' - p''). \quad (1.10.37)$$

If we put (1.10.37) into (1.10.36) and use (1.6.12) to carry out the integral over p'', we find that

$$\begin{aligned}\langle q'|p|q''\rangle &= \frac{1}{2\pi\hbar}\int_{-\infty}^{\infty} dp'\, p' \exp\left[\frac{ip'(q' - q'')}{\hbar}\right] \\ &= \frac{\hbar}{i}\frac{\partial}{\partial q'}\frac{1}{2\pi\hbar}\int_{-\infty}^{\infty} \exp\left[\frac{ip'(q' - q'')}{\hbar}\right] dp' \\ &= -\frac{\hbar}{i}\frac{\partial}{\partial q''}\frac{1}{2\pi\hbar}\int_{-\infty}^{\infty} \exp\left[\frac{ip'(q' - q'')}{\hbar}\right] dp', \quad (1.10.38)\end{aligned}$$

where we may differentiate the integrals with respect to q' or q'' inside the integral sign. By (1.6.26) we see that (1.10.38) may be written as

$$\begin{aligned}\langle q'|p|q''\rangle &= \frac{\hbar}{i}\frac{\partial}{\partial q'}\,\delta(q' - q'') \\ &= -\frac{\hbar}{i}\frac{\partial}{\partial q''}\,\delta(q' - q''). \quad (1.10.39)\end{aligned}$$

But, by (1.10.13), this may be written in still another way:

$$\begin{aligned}\langle q'|p|q''\rangle &= \frac{\hbar}{i}\frac{\partial}{\partial q'}\langle q'|q''\rangle \\ &= -\frac{\hbar}{i}\frac{\partial}{\partial q''}\langle q'|q''\rangle. \quad (1.10.40)\end{aligned}$$

This shows how the operator p transforms from the p to the q-representation.

We may generalize this result easily to show that, if F is a function of p,

$$\langle q'|F(p)|q''\rangle = F\left(\frac{\hbar}{i}\frac{\partial}{\partial q'}\right)\langle q'|q''\rangle \quad (1.10.41)$$

while if V is a function of q, we have, by (1.10.31),

$$\langle q'|V(q)|q''\rangle = V(q')\langle q'|q''\rangle. \quad (1.10.42)$$

We may similarly show how q transforms to the p-representation. We find that

$$\langle p'|q|p''\rangle = -\frac{\hbar}{i}\frac{\partial}{\partial p'}\langle p'|p''\rangle$$
$$= +\frac{\hbar}{i}\frac{\partial}{\partial p''}\langle p'|p''\rangle. \qquad (1.10.43)$$

As another application of the theory given in the preceding section, let us show how the representatives of an arbitrary state vector $|\psi\rangle$ are related in the two representations. By the completeness relations, we have

$$\psi(q') \equiv \langle q'|\psi\rangle = \int_{-\infty}^{\infty} \langle q'|p'\rangle\, dp'\, \langle p'|\psi\rangle$$
$$= \frac{1}{\sqrt{2\pi\hbar}}\int_{-\infty}^{\infty} \exp\left(\frac{ip'q'}{\hbar}\right)\langle p'|\psi\rangle\, dp' \qquad (1.10.44)$$

and similarly

$$\varphi(p) \equiv \langle p'|\psi\rangle = \frac{1}{\sqrt{2\pi\hbar}}\int_{-\infty}^{\infty} \exp\left(-\frac{ip'q'}{\hbar}\right)\langle q'|\psi\rangle\, dq', \qquad (1.10.45)$$

where we have used the transformation function (1.10.35). We know that $\langle q'|\psi\rangle = \psi(q')$ is a function of q' and $\langle p'|\psi\rangle = \varphi(p')$ is another function of p'. From (1.10.44) and (1.10.45) we see that $\psi(q')$ and $\varphi(p')$ are Fourier transforms of each other. They express the same state vector $|\psi\rangle$ in two different representations, in line with earlier discussions. They are called the Schrödinger wave functions.

We may generalize (1.10.41) and (1.10.42) as follows: let us multiply (1.10.41) from the right by $\langle q''|\psi\rangle$ and integrate over dq'' from $-\infty$ to $+\infty$. If we use the completeness relation, (1.10.41) reduces to

$$\langle q'|F(p)|\psi\rangle = F\left(\frac{\hbar}{i}\frac{\partial}{\partial q'}\right)\langle q'|\psi\rangle$$
$$= F\left(\frac{\hbar}{i}\frac{\partial}{\partial q'}\right)\psi(q'), \qquad (1.10.46)$$

where $|\psi\rangle$ is an arbitrary state vector. A similar procedure for (1.10.42) shows that

$$\langle q'|V(q)|\psi\rangle = V(q')\langle q'|\psi\rangle$$
$$= V(q')\psi(q'). \qquad (1.10.47)$$

These are very useful results.

There is still another important observable for this system, namely, the hamiltonian (1.10.2), whose eigenvalues we do not yet know. It will satisfy the eigenvalue equation

$$H|E\rangle = E|E\rangle = \left[\frac{p^2}{2m} + V(q)\right]|E\rangle. \tag{1.10.48}$$

Obviously we cannot solve this problem until we specify the potential $V(q)$. [Note that the eigenvectors and eigenvalues of p and q are independent of $V(q)$.] Chapter 2 is devoted to a study of (1.10.48) when $V(q) = \frac{1}{2}kq^2$, where k is a constant. This potential corresponds to a simple harmonic oscillator. A simpler example is a free particle, in which case $V(q) = 0$. The eigenvalue problem for H in this case is extremely simple but we shall postpone its solution until Section 1.19.

We may express the energy-eigenvalue problem (1.10.48) in the coordinate representation. (The eigenkets $|E\rangle$ are the energy representation.) To do this, we take the scalar product of both sides of (1.10.48) with an eigenbra of q, $\langle q'|$. We then have

$$\langle q'|\left[\frac{p^2}{2m} + V(q)\right]|E\rangle = E\langle q'|E\rangle.$$

We may now use (1.10.46) and (1.10.47) with $|\psi\rangle = |E\rangle$ (since $|\psi\rangle$ was arbitrary) and write the equation above as

$$\left[-\frac{\hbar^2}{2m}\frac{d^2}{dq'^2} + V(q')\right]\langle q'|E\rangle = E\langle q'|E\rangle. \tag{1.10.49}$$

The solution of this equation gives the transformation function $\langle q'|E\rangle$ between the energy and coordinate representations. It is also the energy eigenfunction in the q representation or the eigenfunction of q in the energy representation. Equation (1.10.49) is called the time-independent Schrödinger wave equation. The term $\langle q'|E\rangle$ is the Schrödinger wave function, which is written as

$$\psi_E(q') \equiv \langle q'|E\rangle.$$

Equation (1.10.49) is a derived result from the postulates given previously. It does not require a separate postulate in the theory. All the analysis up to this point assumes that we are looking at the system at a particular time; we shall need another postulate to tell how the system develops in time when it is left undisturbed (see Section 1.13).

Since the hamiltonian is an observable, we conclude that the eigenkets $|E\rangle$ will form a complete set and satisfy the orthonormality conditions. We cannot say in advance of specifying $V(q)$ whether the eigenvalues will be discrete, continuous, or some of both, or whether there will be any degeneracy. Before solving (1.10.49) for a particular $V(q')$, we interrupt the discussion to give a physical interpretation of states that are not eigenstates of an observable. This involves a probability interpretation of the theory alluded to in the introduction.

1.11 MEASUREMENT OF OBSERVABLES; PROBABILITY INTERPRETATION

We have given a physical interpretation to eigenstates of observables. We have made the assumption that if the system is in the eigenstate $|l\rangle$ of an observable L, a measurement of L will yield the value l. We have also assumed that, when a single measurement of L is made for a system in an arbitrary state $|\psi\rangle$, the result will be one of the eigenvalues of L. The disturbance involved in the measurement will cause the system to jump into one of its eigenstates in an uncontrollable way. We cannot predict which of the eigenvalues will be obtained, since the disturbance caused by the interaction of the system with the observation mechanism will destroy the causal connection between the measured value and the state of the system before a measurement was made. We shall now make an *assumption* that makes it possible to find the *probability* of obtaining a given eigenvalue of L when it is measured on a system in an arbitrary state $|\psi\rangle$.

If we measure an observable L a large number (actually an infinite number) of times, each time with the system in the same state $|\psi\rangle$, and average all these measurements, we shall *assume* that the average is

$$\langle L\rangle = \frac{\langle\psi|L|\psi\rangle}{\langle\psi|\psi\rangle}\,. \tag{1.11.1}$$

All physically realizable states $|\psi\rangle$ are represented by vectors of finite norm.

We may generalize the assumption (1.11.1) to the case where f is a function of L and say that the average of $f(L)$ is

$$\langle f(L)\rangle = \frac{\langle\psi|f(L)|\psi\rangle}{\langle\psi|\psi\rangle}\,. \tag{1.11.2}$$

The quantum averages in (1.11.1) and (1.11.2) are ensemble averages; that is, it is assumed that there are an infinite number of identical quantum systems, each prepared in an identical way, with no interactions between them. Each system is called an element of the ensemble. Then L, or $f(L)$, is measured on each element of the ensemble and the results are averaged. The averages are those defined by (1.11.1) and (1.11.2). It should never be overlooked that a quantum average is an ensemble average with every element of the ensemble in state $|\psi\rangle$. If the eigenkets of L form a complete orthonormal set, we have

$$I = \sum_l |l\rangle\langle l|$$

if the eigenvalues are discrete. If we multiply both sides by a function of L, then

$$f(L) = \sum_l f(l)|l\rangle\langle l|. \tag{1.11.3}$$

The operator $|l\rangle\langle l|$ is called a projection operator and (1.11.3) allows any $f(L)$ to be expressed as a weighted sum of projection operators.

If we use the completeness relation twice, we may write any operator A as

$$A = \sum_{l,m} |l\rangle\langle l|A|m\rangle\langle m| = \sum_{l,m} A_{lm}|l\rangle\langle m|. \tag{1.11.4}$$

If we choose the function $f(L)$ such that $f(l) = \delta_{l,l'}$, then (1.11.3) reduces to

$$f(L) = |l'\rangle\langle l'|. \tag{1.11.5}$$

Equation (1.5.22) is an example of the expansion (1.11.5).

According to (1.11.2), the average value of $f(L)$ given in (1.11.5) is the probability $P_{l'}$ of L "having the value l'" when a measurement is made on the system in state $|\psi\rangle$. Therefore, by (1.11.5) and (1.2.8), we have

$$P_l = \langle\psi|f(L)|\psi\rangle = \langle\psi|l'\rangle\langle l'|\psi\rangle = |\langle l'|\psi\rangle|^2, \tag{1.11.6}$$

where we tacitly assumed $\langle\psi|\psi\rangle = 1$. The wave function $\langle l'|\psi\rangle = \psi(l')$ in the L-representation is called the probability amplitude, and $|\psi(l')|^2$ is the probability of obtaining the value l' when a single measurement of L is made on the system in state $|\psi\rangle$.

We now consider the analog of the example above when the operator $L = q$ and has a continuous spectrum from $-\infty$ to $+\infty$. If $q|q'\rangle = q'|q'\rangle$ and

$$\langle q'|q''\rangle = \delta(q' - q'') \tag{1.11.7a}$$

$$I = \int_{-\infty}^{\infty} |q''\rangle\, dq''\, \langle q''|, \tag{1.11.7b}$$

we may multiply both sides of (1.11.7b) by $f(q)$ and obtain

$$f(q) = \int_{-\infty}^{\infty} f(q'')|q''\rangle\, dq''\, \langle q''| \tag{1.11.8}$$

as the analog of (1.11.3). As a special case, we define the function $f(q'')$ to be

$$f(q'') = \begin{cases} 1 & \text{if } \; q' < q'' < q' + dq' \\ 0 & \text{otherwise} \end{cases}. \tag{1.11.9}$$

For this special case, the operator $f(q)$ in (1.11.8) reduces to

$$f(q) = \int_{q'}^{q'+dq'} |q''\rangle\, dq''\, \langle q''|. \tag{1.11.10}$$

The probability that a measurement of q will give a value between q' and $q' + dq'$ when the system is in state $|\psi\rangle$ is the average value of $f(q)$ in (1.11.10) and is

$$P_{q'}\, dq' = \langle\psi|f(q)|\psi\rangle = \int_{q'}^{q'+dq'} \langle\psi|q''\rangle\, dq''\, \langle q''|\psi\rangle$$

$$= |\langle q'|\psi\rangle|^2\, dq'. \tag{1.11.11}$$

Again the wave function $\langle q'|\psi\rangle = \psi(q')$ in the q representation is the probability amplitude, and $|\psi(q')|^2\, dq'$ is the probability that a measurement of q on the system in state $|\psi\rangle$ will give a value between q' and $q' + dq'$.

Two probability interpretations may be given to the transformation function $\langle l'|m'\rangle = \langle m'|l'\rangle^*$ from the L to the M-representation. We assume that l' and m' are discrete. First, we may say that $|\langle l'|m'\rangle|^2$ gives the probability of obtaining the value l' when we measure L for a system in state $|m'\rangle$. Alternatively, we may say that it gives the probability of obtaining the value m' when we measure M for the system in state $|l'\rangle$. These two probabilities are equal. The term $\langle l'|m'\rangle$ may be called the wave function for the observable M in the L-representation and it may also be called the wave function for L in the M-representation.

The theory of measurement is important in applications of quantum mechanics to measuring electromagnetic fields or detecting signals in communication problems. We discuss this in more detail in later chapters and here refer the reader to the books of von Neumann [9] and Bohm [5] for a very thorough treatment of the theory of measurements. These questions are intimately connected with the uncertainty principle of Heisenberg which we derive in the next section.

1.12 THE HEISENBERG UNCERTAINTY PRINCIPLE

Let us assume an ensemble of identical noninteracting quantum systems, each in state $|\psi\rangle$. On half the ensemble, we measure one observable A; on the other half of the ensemble, we measure another observable B. Each half of the ensemble has an infinite number of elements. The measurements of A and B are called simultaneous since the state of each element of the ensemble is the same when A and B are measured.

A measurement of A on one element of the ensemble gives one of the eigenvalues of A, say a, and after that measurement, the element of the ensemble jumps from state $|\psi\rangle$ to state $|a\rangle$. Similarly, a measurement of B on another element of the ensemble gives an eigenvalue of B, say b, and that ensemble element jumps to state $|b\rangle$. The probability that a single measurement of A will give the value a is $|\langle a|\psi\rangle|^2$, and the probability that a single measurement of B will give the value b is $|\langle b|\psi\rangle|^2$, by the previous section.

The quantum- or ensemble-average value of A and B for all these many measurements is

$$\langle A\rangle = \langle\psi|A|\psi\rangle \qquad \langle B\rangle = \langle\psi|B|\psi\rangle \tag{1.12.1}$$

where $\langle\psi|\psi\rangle = 1$.

The measurements of A and B, in general, have fluctuations about the average value of $\langle A\rangle$ and $\langle B\rangle$. These fluctuations are not to be thought of as ordinary fluctuations because the measuring instruments are not perfect. The latter type of error is assumed to be nonexistent. If we let

$$\langle A^2\rangle = \langle\psi|A^2|\psi\rangle \qquad \langle B^2\rangle = \langle\psi|B^2|\psi\rangle \tag{1.12.2}$$

the mean-square deviations or fluctuations in A and B that are of quantum origin are

$$(\Delta A)^2 = \langle A^2\rangle - \langle A\rangle^2 \qquad (\Delta B)^2 = \langle B^2\rangle - \langle B\rangle^2. \tag{1.12.3}$$

These fluctuations will be zero if and only if the state is an eigenstate of either A, B, or both. In fact, this is the way an eigenstate of an observable is defined: every measurement always gives the same eigenvalue with no fluctuations.

We now suppose that the observables A and B do not commute but satisfy the commutation relation

$$[A, B] = iC, \tag{1.12.4}$$

where C is a constant or another observable. We shall show that, in this case, both variables cannot be simultaneously measured with complete precision (i.e., with no fluctuations) and that their mean-square deviations satisfy the inequality

$$(\Delta A)^2(\Delta B)^2 \geq \tfrac{1}{4}|\langle C\rangle|^2, \tag{1.12.5}$$

where

$$\langle C\rangle = \langle\psi|C|\psi\rangle. \tag{1.12.6}$$

This is called the Heisenberg uncertainty relation.

Before proving this relation, let us discuss its significance briefly. We know that $|\psi\rangle$ must be an eigenstate of A if we are to obtain a precise value (i.e., with no fluctuation) when we measure A. Similarly, $|\psi\rangle$ must be an eigenstate of B if we are to obtain a precise value when we measure B. That means that if we measure A and B simultaneously and obtain precise values for both (i.e., eigenvalues for both), the state would have to be a *simultaneous* eigenstate for both A and B.* This implies that $\Delta A = \Delta B = 0$, which, by

* There may be some states $|\psi\rangle$ in which $\langle\psi|C|\psi\rangle = 0$, in which case A and B may be measured simultaneously for the state $|\psi\rangle$ although $[A, B] = iC \neq 0$. An example arises in the case of angular momentum, to be considered later. If L_x, L_y, and L_z are the components of angular momentum, they satisfy $[L_x, L_y] = i\hbar L_z$ while all three components commute with L^2, the total angular momentum. The state for which $L_z|\psi\rangle = 0$ gives $L_x|\psi\rangle$ and $L_y|\psi\rangle$ different from zero; they may be measured simultaneously.

the uncertainty principle, can be true only if $|\langle C\rangle| = 0$. This is possible if $C = 0$, in which case A and B commute, by (1.12.4). Two observables may be measured simultaneously with complete precision if they commute.

If A and B do not commute but satisfy (1.12.4), the mean-square fluctuations, sometimes called the uncertainty in the measurement, satisfy (1.12.5).

To prove (1.12.5), let us define two new variables, α and β, by

$$\alpha = A - \langle A\rangle \qquad \beta = B - \langle B\rangle. \tag{1.12.7}$$

Since $\langle A\rangle$ and $\langle B\rangle$ are numbers, it follows from (1.12.4) that α and β satisfy the commutation relation

$$[\alpha, \beta] = iC. \tag{1.12.8}$$

Since $\langle\alpha\rangle = \langle\beta\rangle = 0$, by (1.12.7), we see that

$$(\Delta\alpha)^2 = (\Delta A)^2 = \langle\alpha^2\rangle \qquad (\Delta\beta)^2 = (\Delta B)^2 = \langle\beta^2\rangle. \tag{1.12.9}$$

The product of $(\Delta\alpha)^2(\Delta\beta)^2$ is

$$(\Delta\alpha)^2(\Delta\beta)^2 = \langle\psi|\alpha^2|\psi\rangle\langle\psi|\beta^2|\psi\rangle. \tag{1.12.10}$$

We may now use the Schwarz inequality: if $|\varphi\rangle$ and $|\chi\rangle$ are any two kets, then

$$|\langle\varphi|\chi\rangle|^2 \leq \langle\varphi|\varphi\rangle\langle\chi|\chi\rangle, \tag{1.12.11}$$

where the equality holds if and only if

$$|\varphi\rangle = c|\chi\rangle, \tag{1.12.12}$$

where c is a constant (see Problem 1.4). Since A and B are observables, it follows that $\alpha = \alpha^\dagger$ and $\beta = \beta^\dagger$. If we let $|\chi\rangle = \beta|\psi\rangle$ and $|\varphi\rangle = \alpha|\psi\rangle$ in (1.12.11), we see that (1.12.10) satisfies the inequality

$$(\Delta\alpha)^2(\Delta\beta)^2 \geq |\langle\psi|\alpha\beta|\psi\rangle|^2. \tag{1.12.13}$$

We may always write the identity

$$\alpha\beta = \tfrac{1}{2}(\alpha\beta + \beta\alpha) + \tfrac{1}{2}(\alpha\beta - \beta\alpha) = \tfrac{1}{2}(\alpha\beta + \beta\alpha) + \frac{i}{2}C,$$

where we have used (1.12.8). If we put this into (1.12.13), then

$$(\Delta\alpha)^2(\Delta\beta)^2 \geq \tfrac{1}{4}|\langle\psi|(\alpha\beta + \beta\alpha)|\psi\rangle + i\langle\psi|C|\psi\rangle|^2. \tag{1.12.14}$$

Since $\alpha\beta + \beta\alpha$ as well as C are hermitian, the numbers $\langle\psi|(\alpha\beta + \beta\alpha)|\psi\rangle$ and $\langle C\rangle = \langle\psi|C|\psi\rangle$ are real. Accordingly, (1.12.14) may be written

$$(\Delta\alpha)^2(\Delta\beta)^2 = (\Delta A)^2(\Delta B)^2 \geq \tfrac{1}{4}|\langle C\rangle|^2,$$

where (1.12.9) was used. This is the uncertainty principle (1.12.5).

For the equality to hold, by (1.12.12) and (1.12.14), where $|\varphi\rangle = \alpha|\psi\rangle$ and $|\chi\rangle = \beta|\psi\rangle$

$$\alpha|\psi\rangle = c\beta|\psi\rangle \tag{1.12.15a}$$

$$\langle\psi|(\alpha\beta + \beta\alpha)|\psi\rangle = 0, \tag{1.12.15b}$$

where c is a constant. If the state $|\psi\rangle$ satisfies these relations, the uncertainty product $\Delta A\, \Delta B = \frac{1}{2}|\langle C\rangle|$, its minimum possible value.

As a special application, we take the operator $A = q$ and $B = p$ for a particle in one dimension. By (1.10.3), $[q, p] = i\hbar$, and so (1.12.5) becomes in this special case

$$\Delta p\, \Delta q \geq \tfrac{1}{2}\hbar. \tag{1.12.16}$$

This indicates that if, for example, we measure p when the system is in an eigenstate of p, say $|\psi\rangle = |p'\rangle$, then $\langle p^2\rangle = \langle p\rangle^2$ and $(\Delta p)^2 = 0$. By (1.12.16) we conclude that $(\Delta q)^2$, the mean-square fluctuation for a simultaneous measurement of q, is infinite. That is, if we measure p repeatedly and always obtain p', we know nothing about q on a simultaneous measurement. The same argument applies if $|\psi\rangle = |q'\rangle$, an eigenstate of q. Thus $\Delta q = 0$ and so $\Delta p = \infty$.

These limiting cases in which $|\psi\rangle$ is an eigenstate of either p or q are in agreement with the probability interpretation of the theory given in the preceding section. The probability that a measurement of q will yield a value between q' and $q' + dq'$ when it is known with certainty that a measurement of p will give the value p' is

$$P_{q'}\, dq' = |\langle q'|p'\rangle|^2\, dq' = \frac{dq'}{2\pi\hbar}, \tag{1.12.17}$$

where the transformation function (1.10.35) was used. This probability is independent of q'; thus it is equally probable that the particle will be found anywhere from $-\infty$ to $+\infty$. That is, if $\Delta p = 0$, $\Delta q = \infty$, in agreement with the uncertainty principle.

Similarly, the probability that a measurement of the momentum will yield a value between p' and $p' + dp'$ when it is known with certainty that a measurement of q will give the value q' is

$$P_{p'}\, dp' = |\langle q'|p'\rangle|^2\, dp' = \frac{dp'}{2\pi\hbar} \tag{1.12.18}$$

by (1.10.35). Again, if we know q, we cannot say what the momentum is. The measurement of q with complete precision causes such a profound disturbance on the system that nothing can be known about the momentum.

To summarize, if the system is in a state $|\psi\rangle$ and a single measurement of p is made, the result is one of the eigenvalues, p', and the measurement forces the system into state $|p'\rangle$. This represents a way of "preparing" the system in state $|p'\rangle$. The probability of obtaining the value p' when one makes a measurement of p is $|\langle p'|\psi\rangle|^2\, dp'$ when the system is in state $|\psi\rangle$. In general, the system will not be in an eigenstate of p or q; thus Δp and Δq will be finite and nonzero. It is interesting to inquire what state of the system will give the minimum uncertainty product for $\Delta p\, \Delta q$. This state corresponds to localizing the particle as precisely as possible in momentum space when it is located in the region Δq in coordinate space. To find this state, we must solve (1.12.15).

If we put (1.12.15a) and its adjoint into (1.12.15b), we find that

$$(c + c^*)\langle\psi|\beta^2|\psi\rangle = 0.$$

Since $\langle\beta^2\rangle = (\Delta p)^2 \neq 0$ ($|\psi\rangle$ is not an eigenstate of p), we conclude that c is pure imaginary. We write it $c = -i\xi$, where ξ is real.

For $\alpha = q - \langle q\rangle$ and $\beta = p - \langle p\rangle$, (1.12.15a) becomes

$$(q - \langle q\rangle)|\psi\rangle = -i\xi(p - \langle p\rangle)|\psi\rangle.$$

If we take the scalar product of both sides with $\langle q'|$, an eigenbra of q, and use (1.10.46) and (1.10.47), we may write this as

$$\frac{i}{\xi}(q' - \langle q\rangle)\psi(q') = \left(\frac{\hbar}{i}\frac{d}{dq'} - \langle p\rangle\right)\psi(q'),$$

where $\psi(q') = \langle q'|\psi\rangle$. This is a simple ordinary first-order differential equation for the wave function $\psi(q')$ whose solution is

$$\psi(q') = c_2 \exp\left[\frac{i}{\hbar}\langle p\rangle q' - \frac{1}{2\hbar\xi}(q' - \langle q\rangle)^2\right], \qquad (1.12.19)$$

where c_2 is a constant of integration. We must still determine ξ and c_2. We obtain these by requiring that

$$\langle\psi|\psi\rangle = \int_{-\infty}^{\infty}\langle\psi|q'\rangle\, dq'\,\langle q'|\psi\rangle = \int_{-\infty}^{\infty}|\psi(q')|^2\, dq' = 1, \qquad (1.12.20)$$

where we used the completeness relation. The second requirement is that the prescribed mean-square fluctuation in q be

$$\begin{aligned}(\Delta q)^2 &= \langle(q - \langle q\rangle)^2\rangle \\ &= \int_{-\infty}^{\infty}|\psi(q')|^2(q' - \langle q\rangle)^2\, dq'.\end{aligned} \qquad (1.12.21)$$

If we use the integrals

$$\int_{-\infty}^{\infty} e^{-sx^2}\,dx = \sqrt{\frac{\pi}{s}}$$
$$\int_{-\infty}^{\infty} x^2 e^{-sx^2}\,dx = \frac{1}{2}\frac{\sqrt{\pi}}{s^{3/2}}, \tag{1.12.22}$$

then, when we use (1.12.19) for $\psi(q')$ in (1.12.20) and (1.12.21), we find

$$|c_2|^2\sqrt{\pi\hbar\xi} = 1$$
$$|c_2|^2\sqrt{\pi}(\hbar\xi)^{3/2} = 2(\Delta q)^2.$$

From these we find that $\hbar\xi = 2(\Delta q)^2$ and $|c_2|^2 = [2\pi(\Delta q)^2]^{-1/2}$. We choose the phase of c_2 so that it is real; thus (1.12.19) becomes

$$\psi(q') = \langle q'|\psi\rangle = \frac{1}{[2\pi(\Delta q)^2]^{1/4}} \exp\left[\frac{i\langle p\rangle q'}{\hbar} - \frac{(q' - \langle q\rangle)^2}{4(\Delta q)^2}\right]. \tag{1.12.23}$$

This is called a minimum uncertainty wave function in the coordinate representation. Repeated measurements of q for the system in this state give the average $\langle q\rangle$ with mean-square fluctuations $(\Delta q)^2$, and repeated measurements of p give an average of $\langle p\rangle$. However, the mean-square fluctuation of p is given by $(\Delta p)^2 = \hbar^2/4(\Delta q)^2$. Since $\langle p\rangle$, $\langle q\rangle$, and $(\Delta q)^2$ are arbitrary, there is a triple infinity of minimum uncertainty states. These minimum uncertainty states will play a unique role in our later considerations of the measurement of the electromagnetic field.

From the probability interpretation of the theory, the probability that the particle will be located between q' and $q' + dq'$ when a measurement of q is made is, from (1.12.23),

$$|\psi(q')|^2\,dq' = \frac{dq'}{\sqrt{2\pi(\Delta q)^2}} \exp\left[-\frac{(q' - \langle q\rangle)^2}{2(\Delta q)^2}\right]. \tag{1.12.24}$$

This is a familiar gaussian probability distribution function centered at $q' = \langle q\rangle$ which has a standard deviation of Δq.

Equation (1.12.23) gives the minimum uncertainty state in the coordinate representation. We may use (1.10.45) to express it in the momentum representation as

$$\varphi(p') = \langle p'|\psi\rangle = \frac{1}{[2\pi(\Delta p)^2]^{1/4}} \exp\left[-\frac{i}{\hbar}\langle q\rangle(p' - \langle p\rangle) - \frac{(p' - \langle p\rangle)^2}{4(\Delta p)^2}\right], \tag{1.12.25}$$

which is the momentum representation of the same minimum uncertainty state. Again, $|\varphi(p')|^2$ is a gaussian probability distribution function in momentum space centered at $p' = \langle p \rangle$ with standard deviation $\Delta p = \hbar/2(\Delta q)$. It is uniquely determined by $\psi(q')$, by (1.10.45).

By (1.10.44) we may visualize $\psi(q')$ as a superposition of plane waves $\exp(ip'q'/\hbar)$ of wavelength

$$\lambda = \frac{\hbar 2\pi}{p'} = \frac{h}{p'}. \tag{1.12.26}$$

This is the de Broglie wavelength to be associated with a particle of mass m and explains the wave nature of particles in diffraction experiments.

In (1.10.44) $\varphi(p')$ was visualized as the amplitude of each superposed plane wave; for the minimum uncertainty state, these waves interfere constructively in a region Δp in momentum space to give $|\psi(q')|^2$ a large value in a range Δq whereas they interfere destructively outside this range to make $|\psi(q')|^2$ small. Thus $\psi(q')$ represents a wave packet, and (1.12.23) is a minimum uncertainty wave packet at a fixed time. Use of a wave packet makes it possible to localize the particle in a limited region of coordinate and momentum space so that waves exhibit particle-like character.

It should be noted that the results given in this section are independent of the field of force in which a particle is located since the motion of the system is not yet involved.

1.13 DYNAMICAL BEHAVIOR OF A QUANTUM SYSTEM

As yet, the formulation of quantum mechanics is not complete since we have not specified how the system behaves dynamically; that is, we have not shown how the state of the system changes in time. The theory up to this point has been developed for some fixed time.

When a quantum system is unperturbed by any measurements, the system develops in time in a completely causal manner. It is only the disturbance caused by the interaction of the measuring device with the system that makes the behavior cease to be strictly causal.

To give the time development of a quantum system, we *postulate* the existence of a hamiltonian H for the system and *require* that the state vector for the system $|\psi(t)\rangle$ change in accordance with the Schrödinger equation

$$i\hbar \frac{\partial}{\partial t} |\psi(t)\rangle = H|\psi(t)\rangle, \tag{1.13.1}$$

where H is to be treated as an observable of the system and must therefore be hermitian.

Two cases may arise. In the first case, the system is conservative and H is explicitly independent of time. In that case we may formally integrate

(1.13.1) and obtain

$$|\psi(t)\rangle = U(t, t_0)|\psi(t_0)\rangle, \tag{1.13.2}$$

where

$$U(t, t_0) = \exp\left[-\frac{iH(t - t_0)}{\hbar}\right], \tag{1.13.3}$$

and $|\psi(t_0)\rangle$ is the state of the system at time t_0. This solution may be verified by differentiation and substitution back into (1.13.1). [The derivative of the operator $U(t, t_0)$ with respect to t is defined exactly like the differentiation of ordinary functions.]

From (1.13.3) it follows that U satisfies the equation

$$i\hbar\frac{dU}{dt} = HU, \tag{1.13.4}$$

while from (1.13.2), at $t = t_0$, U must satisfy the initial condition

$$U(t_0, t_0) = I. \tag{1.13.5}$$

Since H is hermitian, it follows from (1.13.3) that

$$U^\dagger(t, t_0) = \exp\left[\frac{iH(t - t_0)}{\hbar}\right] = U^{-1}(t, t_0), \tag{1.13.6}$$

which shows that U is a unitary operator. Therefore, it may be said that the state of the system at time t develops in a completely causal way from the state at time t_0 by a unitary transformation. Based on a geometric picture of state vectors, we may visualize (1.13.2) as a continuous generalized "rotation" of the state vector in ket space from an initial direction at t_0 to a final direction at t. Since $U^{-1} = U^\dagger$, the norm of $|\psi(t)\rangle$ is

$$\langle\psi(t)|\psi(t)\rangle = \langle\psi(t_0)|U^\dagger(t, t_0)U(t, t_0)|\psi(t_0)\rangle = \langle\psi(t_0)|\psi(t_0)\rangle \tag{1.13.7}$$

and does not change. The direction changes but the norm is unchanged.

Since any two functions of H commute, it is easy to see from (1.13.3) that U satisfies the so-called group property

$$U(t, t_2) = U(t, t_1)U(t_1, t_2), \tag{1.13.8}$$

where $t > t_1 > t_2$.

The adjoint equations are

$$-i\hbar\frac{\partial}{\partial t}\langle\psi(t)| = \langle\psi(t)|H, \tag{1.13.9}$$

since $H = H^\dagger$, and

$$-i\hbar\frac{\partial U^\dagger}{\partial t} = U^\dagger H. \tag{1.13.10}$$

The second case arises when H depends explicitly on time. We shall discuss this case in more detail in Section 1.16, but we should note that the solution to (1.13.1) is *not* given by

$$|\psi(t)\rangle = \exp\left[-\frac{i}{\hbar}\int_{t_0}^{t} H(t')\,dt\right]|\psi(t_0)\rangle. \tag{1.13.11}$$

This follows since in general $\int_{t_0}^{t} H(t')\,dt'$ does *not* commute with $H(t)$ so that when we attempt to differentiate the exponential, the order of factors would be ambiguous. We should note that just because two operators may commute at one time does not ensure that they commute at two different times.

When H depends on time explicitly, we may still look for a solution of the form

$$|\psi(t)\rangle = U(t, t_0)|\psi(t_0)\rangle. \tag{1.13.12}$$

When we put this into (1.13.1), we see that U satisfies

$$i\hbar\frac{\partial U}{\partial t}|\psi(t_0)\rangle = H(t)U|\psi(t_0)\rangle. \tag{1.13.13}$$

Since $|\psi(t_0)\rangle$ is completely arbitrary, it follows that U must satisfy the equation

$$i\hbar\frac{\partial U}{\partial t} = H(t)U \tag{1.13.14}$$

subject to the initial condition

$$U(t_0, t_0) = 1; \tag{1.13.15}$$

$H(t)$ must still be hermitian. It can be shown that U must be unitary and satisfy the group property (1.13.8) even when H is time dependent.

1.14 THE SCHRÖDINGER PICTURE OF QUANTUM MECHANICS

The development of quantum mechanics up to this point has been in the so-called Schrödinger picture. Let us review briefly what this implies for a particle in one dimension. The observables (p, q, and H) were taken as hermitian operators and were time-independent. A subscript S indicates that operators and vectors are in the Schrödinger picture. Thus we write p_S, q_S, H_S. The eigenvectors of these operators are written $|p'\rangle_S$, $|q'\rangle_S$, $|E\rangle_S$; any of these may be taken as stationary (time-independent) basis vectors to represent operators or state vectors. In fact, in the Schrödinger picture state vectors are stationary and act like a fixed coordinate system in ordinary geometry.

At a fixed instant of time, any state is represented by a linear superposition of a set of stationary basis vectors.

The operators obey the communication relations $[q_S, p_S] = i\hbar$, and $[q_S, q_S] = [p_S, p_S] = 0$ in the Schrödinger picture.

The state vector describing the dynamical behavior of the system as a function of time is $|\psi_S(t)\rangle$. In the Schrödinger picture it moves continuously according to the Schrödinger equation from an initial direction $|\psi(t_0)\rangle$ to a final direction $|\psi_S(t)\rangle$ at time t.

If $f(p_S, q_S)$ is a function of p_S and q_S, then the quantum (ensemble) average at time t is

$$\langle f(p_S, q_S)\rangle = \langle \psi_S(t)| f(p_S, q_S)|\psi_S(t)\rangle. \tag{1.14.1}$$

If the system is in state $|\psi(t_0)\rangle$ at time t_0, the state at time t, by (1.13.2), is $|\psi_S(t)\rangle = U(t, t_0)|\psi(t_0)\rangle$, and the probability that the system will be in some fixed state $\langle u|$ at time t_1 is

$$|\langle u|\psi_S(t_1)\rangle|^2 = |\langle u|U(t_1, t_0)|\psi(t_0)\rangle|^2. \tag{1.14.2}$$

This description in which the basis vectors are stationary and the dynamical state vector $|\psi_S(t)\rangle$ moves is called the Schrödinger picture of quantum mechanics and is the picture used up to now.

1.15 THE HEISENBERG PICTURE

In the Schrödinger picture the basis vectors (any eigenvectors of observables) are visualized as a fixed set of vectors and the state vector as moving. The same system can be described equally well by letting the basis vectors move and the state vector remain stationary, as in classical mechanics. This mode of formulating quantum mechanics is physically equivalent to the Schrödinger picture and is called the Heisenberg picture. It is clear that, if operators are stationary (time-independent) in the Schrödinger picture, they must be time-dependent in the Heisenberg picture in order that the two descriptions be physically equivalent.

The state vectors in the two pictures are related by definition by (1.13.2)

$$|\psi_S(t)\rangle = U(t, t_0)|\psi_H(t_0)\rangle, \tag{1.15.1}$$

where the subscript H designates the Heisenberg picture. The vector $|\psi_H(t_0)\rangle$ is stationary while $|\psi_S(t)\rangle$ is moving. Since $U(t_0, t_0) = 1$, the state vectors coincide at $t = t_0$ in the two pictures.

The average value of an operator A_S is [compare with (1.14.1)]

$$\begin{aligned}\langle A\rangle &= \langle \psi_S(t)|A_S|\psi_S(t)\rangle \\ &= \langle \psi_H(t_0)|U^\dagger A_S U|\psi_H(t_0)\rangle;\end{aligned} \tag{1.15.2}$$

where we have used (1.15.1) and its adjoint. We define the operator in the Heisenberg picture by

$$A_H(t) = U^\dagger(t, t_0)A_S U(t, t_0). \tag{1.15.3}$$

From this definition, we see that operators that are stationary in the Schrödinger picture usually depend on time in the Heisenberg picture. The transformation (1.15.3) is called a similarity transformation when $U^\dagger = U^{-1}$. With this definition, the average values of A at time t may be written

$$\langle A\rangle = \langle\psi_H(t_0)|A_H(t)|\psi_H(t_0)\rangle. \tag{1.15.4}$$

The transformation law (1.15.3) is necessary in order that the average of A be the same in both pictures, making the two physically equivalent.

From the transformation law (1.15.3), which is valid even if A_S has an explicit time dependence, the equation of motion for an observable in the Heisenberg picture may be obtained. If we differentiate both sides of (1.15.3) with respect to t and use (1.13.14), its adjoint, and the fact that $U^\dagger U = UU^\dagger = 1$, we obtain

$$\begin{aligned} i\hbar\frac{dA_H}{dt} &= U^\dagger A_S H U - U^\dagger H A_S U + i\hbar U^\dagger \frac{\partial A_S}{\partial t} U \\ &= U^\dagger A_S U U^\dagger H U - U^\dagger H U U^\dagger A_S U + i\hbar U^\dagger \frac{\partial A_S}{\partial t} U \\ &\equiv [A_H, H_H] + i\hbar U^\dagger \frac{\partial A_S}{\partial t} U, \end{aligned} \tag{1.15.5}$$

where we used (1.15.3) and define

$$H_H(t) = U^\dagger(t, t_0)H_S(t)U(t, t_0), \tag{1.15.6}$$

which is the hamiltonian in the Heisenberg picture. Equation (1.15.5) is called the Heisenberg equation of motion for the observable A. If $dA_H/dt = 0$, then A_H is a constant of the motion.

As a special case of (1.15.5), we let $A_S = H_S$. If the system is conservative, $\partial H_S/\partial t = 0$, and $U(t, t_0)$ is given by (1.13.3). Since in this case $[H_S, U] = 0$, we see by (1.15.6) that $H_H = H_S$. By (1.15.5), we then have

$$\frac{dH_H}{dt} = 0, \tag{1.15.7}$$

which shows that H is a constant of the motion.

If A_S has no explicit time dependence and the system is conservative, (1.15.5) reduces to

$$\frac{dA_H}{dt} = \frac{1}{i\hbar}[A_H, H]. \tag{1.15.8}$$

In this case, if A_H commutes with H, A_H is a constant of the motion.

Another important theorem is that commutation relations have the same *form* in the two pictures. This again is obviously necessary if the two pictures are to be physically equivalent. As an example, we let A_S, B_S, and C_S be three observables in the Schrödinger picture that satisfy

$$[A_S, B_S] = iC_S \tag{1.15.9}$$

If we multiply both sides from the left by $U^\dagger$ and both sides from the right by U, we have

$$U^\dagger A_S B_S U - U^\dagger B_S A_S U = iU^\dagger C_S U,$$

or, inserting $UU^\dagger = 1$ between the A_S and B_S, we have

$$(U^\dagger A_S U)(U^\dagger B_S U) - (U^\dagger B_S U)(U^\dagger A_S U) = iU^\dagger C_S U,$$

which, by (1.15.3), becomes

$$[A_H(t), B_H(t)] = iC_H(t). \tag{1.15.10}$$

This is the same form as (1.15.9). In short, the Heisenberg and Schrödinger pictures are physically equivalent.

If A_H is a constant of the motion, it must commute with the hamiltonian. Its eigenvalues are the same in both the Schrödinger and Heisenberg pictures and are then said to be *good quantum numbers*.

The Heisenberg picture is extremely useful in demonstrating the formal analogy between a quantum system and its classical analog. We shall consider the example of a particle in one dimension. We shall show that in this case the Heisenberg-operator equations of motion are identical in form with the classical hamiltonian equations of motion. For any system having a classical analog, this correspondence permits a check on the validity of the theory.

Let us consider the hamiltonian (1.10.2), which is a function of p and q. By (1.15.10) we have shown that (1.10.6) are valid in the Heisenberg picture, and so the equations of motion (1.15.5) for $A_H = q_H$ or p_H are, by (1.10.6),

$$\begin{aligned} \frac{dq_H}{dt} &= \frac{1}{i\hbar}[q_H, H_H(p_H, q_H)] = \frac{\partial H_H}{\partial p_H} \\ \frac{dp_H}{dt} &= \frac{1}{i\hbar}[p_H, H_H(p_H, q_H)] = -\frac{\partial H_H}{\partial q_H}, \end{aligned} \tag{1.15.11}$$

since $\partial q_S/\partial t = \partial p_S/\partial t = 0$.

These are identical in form with the classical equations of motion in hamiltonian form:

$$\frac{dq}{dt} = \frac{\partial H}{\partial p}$$
$$\frac{dp}{dt} = -\frac{\partial H}{\partial q}. \tag{1.15.12}$$

Aside from questions of ordering of operators in (1.15.11), the classical equations correspond to the quantum equations.

In classical mechanics, if A is a function of p, q, and t, then A satisfies the equation of motion

$$\begin{aligned}\frac{dA}{dt} &= \frac{\partial A}{\partial q}\frac{dq}{dt} + \frac{\partial A}{\partial p}\frac{dp}{dt} + \frac{\partial A}{\partial t} \\ &= \frac{\partial A}{\partial q}\frac{\partial H}{\partial p} - \frac{\partial A}{\partial p}\frac{\partial H}{\partial q} + \frac{\partial A}{\partial t},\end{aligned} \tag{1.15.13}$$

where we used (1.15.12). The classical Poisson bracket is defined by

$$\{A, H\} = \frac{\partial A}{\partial q}\frac{\partial H}{\partial p} - \frac{\partial A}{\partial p}\frac{\partial H}{\partial q} \tag{1.15.14}$$

so that (1.15.13) may be written

$$\frac{dA}{dt} = \{A, H\} + \frac{\partial A}{\partial t}. \tag{1.15.15}$$

If we compare (1.15.15) with (1.15.5), we see that we may shift from classical mechanics to quantum mechanics by replacing the Poisson bracket by $(i\hbar)^{-1}$ times the commutator bracket, that is

$$\{A, B\} \to \frac{1}{i\hbar}[A, B]. \tag{1.15.16}$$

It may be shown that the algebra of commutators and Poisson brackets is the same (see Problem 1.8). This gives a clue to the motivation for the quantization postulate (1.10.3). If a system has no quantum analog, the quantization rules practically are a matter of pure intuition. The only check is comparison with experiment (as it always is).

1.16 THE INTERACTION PICTURE; TIME-DEPENDENT PERTURBATION THEORY, DYSON TIME ORDERING OPERATOR

There is another picture besides the two discussed in the previous sections which is especially useful when the hamiltonian may be written as the sum of two terms of the form

$$H^S = H_0{}^S + V^S. \tag{1.16.1}$$

We shall assume $H_0{}^S$ is independent of time but V^S may depend explicitly on time although it need not. We call this the *interaction* picture which is defined by the unitary transformation $U_0(t, t_0)$ from the Schrödinger picture by means of

$$|\psi_S(t)\rangle = U_0(t, t_0)|\psi_I(t)\rangle, \tag{1.16.2}$$

where the subscript I refers to the interaction picture and U_0 satisfies the equation

$$i\hbar \frac{\partial U_0}{\partial t} = H_0{}^S U_0 \tag{1.16.3}$$

with

$$U_0(t, t_0) = \exp\left[-\frac{i}{\hbar} H_0{}^S(t - t_0)\right] \tag{1.16.4}$$

$$U_0^\dagger = U_0{}^{-1} \tag{1.16.5}$$

$$U_0(t_0, t_0) = 1. \tag{1.16.6}$$

We are assuming that the Schrödinger equation when $V^S = 0$ may be solved explicitly.

The Schrödinger equation is

$$i\hbar \frac{\partial|\psi_S\rangle}{\partial t} = [H_0{}^S + V^S]|\psi_S\rangle. \tag{1.16.7}$$

If we use (1.16.2), this becomes

$$i\hbar \frac{\partial U_0}{\partial t}|\psi_I\rangle + i\hbar U_0 \frac{\partial|\psi_I\rangle}{\partial t} = [H_0{}^S + V^S]U_0|\psi_I(t)\rangle.$$

If we use (1.16.3), the first terms on the left and right cancel. If we then multiply both sides from the left by $U_0^\dagger = U_0{}^{-1}$, we obtain

$$i\hbar \frac{\partial|\psi_I(t)\rangle}{\partial t} = V_I(t)|\psi_I(t)\rangle \tag{1.16.8}$$

where we have let

$$V_I(t) = U_0^\dagger V_S U_0. \tag{1.16.9}$$

Equation (1.16.8) is the Schrödinger equation for the state vector in the interaction picture (IP). Even if V^S is not explicitly time dependent, we see by (1.16.9) that $V_I(t)$ will usually depend on time.

Before proceeding to obtain a formal solution of (1.16.8) let us see how operators transform under the unitary transformation (1.16.2). The average value of A^S is

$$\begin{aligned}\langle A\rangle &= \langle\psi_S(t)|A_S|\psi_S(t)\rangle \\ &= \langle\psi_I(t)|U_0^\dagger A_S U_0|\psi_I(t)\rangle,\end{aligned} \tag{1.16.10}$$

where we used (1.16.2) and its adjoint.

We therefore see that it is natural to define the operator in the IP as

$$A_I(t) = U_0^\dagger A_S U_0. \tag{1.16.11}$$

Thus (1.16.9) gives the interaction energy V_S in the IP. Thus

$$\langle A\rangle = \langle\psi_I(t)|A_I(t)|\psi_I(t)\rangle. \tag{1.16.12}$$

We may obtain the equation of motion for $A_I(t)$ if we differentiate both sides of (1.16.11) with respect to t and use (1.16.3) and its adjoint. We have

$$\begin{aligned}i\hbar\frac{dA_I}{dt} &= U_0^\dagger A_S i\hbar\frac{\partial U_0}{\partial t} + i\hbar\frac{\partial U_0^\dagger}{\partial t}A_S U_0 + U_0^\dagger i\hbar\frac{\partial A_S}{\partial t}U_0 \\ &= U_0^\dagger A_S H_0{}^S U_0 - U_0^\dagger H_0{}^S A_S U_0 + U_0^\dagger i\hbar\frac{\partial A_S}{\partial t}U_0.\end{aligned} \tag{1.16.13}$$

But since $H_0{}^S$ is time independent,

$$[H_0{}^S, U_0] = 0 \rightarrow H_0{}^S = H_0{}^I, \tag{1.16.14}$$

thus when we use (1.16.11), (1.16.13) becomes

$$\begin{aligned}i\hbar\frac{dA_I}{dt} &= [A_I, H_0{}^I] + U_0^\dagger i\hbar\frac{\partial A_S}{\partial t}U_0 \\ &= [A_I, H_0{}^S] + U_0^\dagger i\hbar\frac{\partial A_S}{\partial t}U_0.\end{aligned} \tag{1.16.15}$$

Note that if $V_S = 0$, (1.16.15) becomes simply the Heisenberg equation of motion.

Let us return to (1.16.8) and look for a solution of the form

$$|\psi_I(t)\rangle = U(t, t_0)|\psi_S(t_0)\rangle, \tag{1.16.16}$$

where by (1.16.2) and (1.16.6)

$$|\psi_I(t_0)\rangle = |\psi_S(t_0)\rangle. \tag{1.16.17}$$

That is, the two pictures coincide at $t = t_0$. If we substitute (1.16.16) into (1.16.8) and note that $|\psi_S(t_0)\rangle$ is arbitrary, U satisfies

$$i\hbar\frac{\partial U}{\partial t} = V_I(t)U, \tag{1.16.18}$$

or on integrating both sides we have

$$U(t, t_0) = 1 + \frac{1}{i\hbar}\int_{t_0}^{t} V_I(t')U(t', t_0)\,dt', \tag{1.16.19}$$

where

$$U(t_0, t_0) = 1. \tag{1.16.20}$$

If we let $t = t'$ and the dummy integration variable $t' \to t''$, we may rewrite this as

$$U(t', t_0) = 1 + \frac{1}{i\hbar}\int_{t_0}^{t'} V_I(t'')U(t'', t_0)\,dt''. \tag{1.16.21}$$

If we substitute this into the integrand in (1.16.19), we obtain

$$U(t, t_0) = 1 + \frac{1}{i\hbar}\int_{r_0}^{t} V_I(t_1)\,dt_1 + \left(\frac{1}{i\hbar}\right)^2\int_{t_0}^{t} dt_1\, V_I(t_1)\int_{t_0}^{t_1} dt_2\, V_I(t_2)U(t_2, t_0). \tag{1.16.22}$$

We may proceed indefinitely with iteration and obtain

$$U(t, t_0) = 1 + \sum_{n=1}^{\infty}\left(\frac{1}{i\hbar}\right)^n\int_{t_0}^{t} dt_1\int_{t_0}^{t_1} dt_2 \cdots \int_{t_0}^{tn-1} dt_n\, V_I(t_1)V_I(t_2)\cdots V_I(t_n). \tag{1.16.23}$$

If the interaction energy V_I is small compared with H_0 this series converges rapidly and we have a power series solution in the perturbation V_I. By (1.16.2) and (1.16.16) we have

$$|\psi_S(t)\rangle = U_0(t, t_0)U(t, t_0)|\psi(t_0)\rangle. \tag{1.16.24}$$

Let us assume that in the absence of the perturbation ($V_S = 0$), we may solve the energy eigenvalue problem

$$H_0{}^S|n\rangle = E_n{}^0|n\rangle, \tag{1.16.25}$$

so that for any function of $H_0{}^S$

$$f(H_0{}^S)|n\rangle = f(E_n{}^0)|n\rangle, \tag{1.16.26}$$

where

$$\langle n|m\rangle = \delta_{nm} \qquad \sum_n |n\rangle\langle n| = 1. \tag{1.16.27}$$

We may then expand $|\psi_I(t)\rangle$ as

$$\begin{aligned} |\psi_I(t)\rangle &= \sum_n |n\rangle\langle n|\psi_I(t)\rangle \\ &\equiv \sum_n c_n(t)|n\rangle, \end{aligned} \tag{1.16.28}$$

where we have let

$$c_n(t) \equiv \langle n|\psi_I(t)\rangle \tag{1.16.29a}$$

Therefore, by (1.16.2), (1.16.4), and (1.16.26) we have

$$|\psi_S(t)\rangle = U_0|\psi_I(t)\rangle = \sum_n c_n(t) \exp\left[-\frac{i}{\hbar} E_n{}^0(t - t_0)\right]|n\rangle. \tag{1.16.29b}$$

From the orthogonality relations, we have

$$\langle m|\psi_S(t)\rangle = c_m(t) \exp\left[-\frac{i}{\hbar} E_m{}^0(t - t_0)\right], \tag{1.16.30}$$

while

$$|\langle m|\psi_S(t)\rangle|^2 = |c_m(t)|^2 = |\langle m|\psi_I(t)\rangle|^2 \tag{1.16.31}$$

is the probability of finding the system in state $|m\rangle$ at time t if we measure its energy. We used (1.16.29). If we use (1.16.16), we obtain

$$\begin{aligned} |c_m(t)|^2 &= |\langle m|U(t, t_0)|\psi(t_0)\rangle|^2 \\ &= \langle m|U(t, t_0)|\psi(t_0)\rangle\langle\psi(t_0)|U^\dagger(t, t_0)|m\rangle \end{aligned} \tag{1.16.32}$$

since by (1.4.6)

$$\langle l|A|r\rangle^* = \langle r|A^\dagger|l\rangle. \tag{1.16.33}$$

If we assume at t_0 we made a measurement and determined that the system was in the energy eigenstate $|\psi(t_0)\rangle = |i\rangle$, when we use (1.16.23), (1.16.32) becomes

$$\begin{aligned} |c(m, t|i, t_0)|^2 = \Bigg|\langle m|1 + \sum_{n=1}^{\infty}\left(\frac{1}{i\hbar}\right)^n \int_{t_0}^{t} dt_1 \int_{t_0}^{t_1} dt_2 \cdots \int_{t_0}^{t_{n-1}} dt_n \\ \times V_I(t_1)V_I(t_2)\cdots V_I(t_n)|i\rangle\Bigg|^2, \end{aligned} \tag{1.16.34}$$

which gives the probability of finding the system in state $|m\rangle$ at time t given that it was in the state $|i\rangle$ at time t_0.

In zeroth order in the interaction energy, we have

$$|c^{(0)}(m, t|i, t_0)|^2 = |\langle m|i\rangle|^2 = \delta_{mi}. \tag{1.16.35}$$

To this approximation the system remains in its initial state. In first order we have

$$\begin{aligned} |c^{(1)}(m, t|i, t_0)|^2 &= \frac{1}{\hbar^2}\left|\int_{t_0}^{t} dt_1 \langle m|V_I(t_1)|i\rangle\right|^2 \\ &= \frac{1}{\hbar^2}\left|\int_{t_0}^{t} dt_1\, \langle m|e^{(i/\hbar)H_0(t_1-t_0)}V_S e^{-(i/\hbar)H_0(t_1-t_0)}|i\rangle\right|^2 \\ &= \frac{1}{\hbar^2}\left|\int_{t_0}^{t} dt_1 e^{i\omega_{mi}(t_1-t_0)}\langle m|V_S|i\rangle\right|^2, \end{aligned} \tag{1.16.36}$$

where we used (1.16.9), (1.16.4), and (1.16.26) and have let

$$\omega_{mi} = \frac{E_m{}^0 - E_i^0}{\hbar}. \tag{1.16.37}$$

If V_S is explicitly time-independent, we may carry out the integral and (1.16.36) reduces to

$$|c^{(1)}(m, t|i, t_0)|^2 = \frac{|\langle m|V_S|i\rangle|^2}{\hbar^2} \frac{4\sin^2 \frac{1}{2}\omega_{mi}(t - t_0)}{\omega_{mi}{}^2}. \tag{1.16.38}$$

If V_S is sinusoidal

$$V_S = V_S^{(0)} \sin(\omega t - \varphi), \tag{1.16.39}$$

(1.16.36) becomes

$$|c^{(1)}(m, t|i, t_0)|^2 = \frac{|\langle m|V_S^{(0)}|i\rangle|^2}{4\hbar^2} \left| e^{i(\omega t_0 - \varphi)} \frac{[e^{i(\omega_{mi}+\omega)(t-t_0)} - 1]}{(\omega_{mi} + \omega)} - e^{-i(\omega t_0 - \varphi)} \frac{[e^{i(\omega_{mi}-\omega)(t-t_0)} - 1]}{(\omega_{mi} - \omega)} \right|^2. \tag{1.16.40}$$

The factor $4 \sin^2 \frac{1}{2}\omega_{mi}(t - t_0)/\omega_{mi}{}^2$ in (1.16.38) is a very strongly peaked function of ω_{mi} (see Figure 1.4). At $\omega_{mi} = 0$ its amplitude increases as $(t - t_0)^2$ and decreases to zero when $\omega_{mi} = 2\pi/(t - t_0)$. The probability that V_S induces the system to make a transition between state $|i\rangle$ and state $|m\rangle$ is thus very small unless energy is conserved between the initial and final states. Energy conservation was not put in the theory in an *ad hoc* manner but is a derived result.

If $\hbar\omega_{mi} = E_m^{(0)} - E_i^{(0)} > 0$, the second factor in (1.16.40) is large compared with the first when $\omega_{mi} \cong \omega$. Then we have

$$|c^{(1)}(m, t|i, t_0)|^2 \cong \frac{|\langle m|V_S^{(0)}|i\rangle|^2}{\hbar^2} \frac{\sin^2 \frac{1}{2}(\omega_{mi} - \omega)(t - t_0)}{(\omega_{mi} - \omega)^2}, \tag{1.16.41}$$

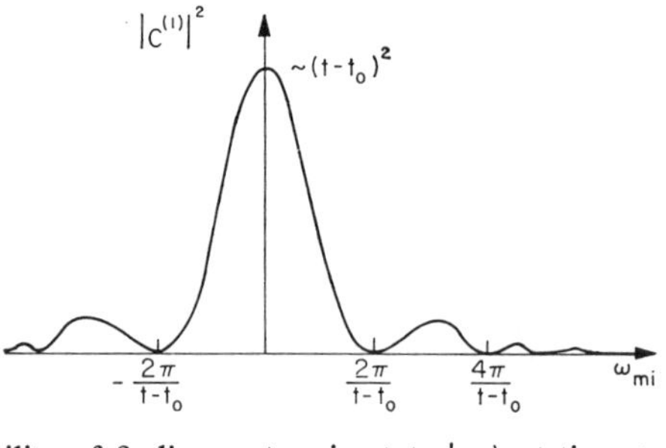

Figure 1.4 Probability of finding system in state $|m\rangle$ at time t given it is in state $|i\rangle$ initially as a function of $(E_m^{(0)} - E_i^{(0)})/\hbar$.

which again is peaked at $\hbar\omega_{mi} = \hbar\omega$. In this case the final energy is increased by $\hbar\omega$. If $\omega_{mi} < 0$, the first term is important and $\hbar\omega_{mi} = -\hbar\omega$, so that the final energy is reduced by $\hbar\omega$. We shall see in Chapter 5 that these may correspond to the emission and absorption of a photon of electromagnetic energy by an atom.

So far we have assumed that the energy levels of the system are infinitely sharp. We show in a later chapter that they must always have some linewidth. Alternately, in (1.16.39) we assumed the interaction was perfectly monochromatic whereas in practice there will always be some bandwidth associated with any sinusoidal perturbation. Accordingly, we shall assume that there are a number of closely spaced energy levels which are distributed with a density $g(E_m)\,dE_m$ between E_m and $E_m + dE_m$. We may obtain the total transition probability to any of these levels for which (1.16.38) is approximately valid by summing

$$\sum_m |c^{(1)}(m, t|i, t_0)|^2 = \sum_m \frac{|\langle m|V_S|i\rangle|^2}{\hbar^2} \frac{4\sin^2 \frac{1}{2}\omega_{mi}(t - t_0)}{\omega_{mi}^2}$$

$$\rightarrow \int g(E_m)\,dE_m \frac{|\langle m|V_S|i\rangle|^2}{\hbar^2} \frac{4\sin^2 \frac{1}{2}\omega_{mi}(t - t_0)}{\omega_{mi}^2}. \quad (1.16.42)$$

Because the levels are closely spaced, we have converted the sum to an integral. Because the integrand is strongly peaked at $\omega_{mi} \simeq 0$, we may remove the slowly varying factors $g(E_m)$ and $|\langle m|V_S|i\rangle|^2$ from the integral and evaluate them at $E_m = E_i$ and extend the limits from $-\infty$ to $+\infty$. Since

$$4\int_{-\infty}^{\infty} \frac{\sin^2 \frac{1}{2}ax}{x^2}\,dx = 2\pi a, \quad (1.16.43)$$

and $dE_m = \hbar\,d\omega_{mi}$, we have

$$\sum_m |c^{(1)}(m, t|i, t_0)|^2 \simeq \frac{2\pi}{\hbar} g(E_i)|\langle m|V_S|i\rangle|^2_{E_m = E_i}(t - t_0). \quad (1.16.44)$$

We may therefore replace the highly peaked factor $\sin^2 \frac{1}{2}ax/x^2$ by a δ-function in (1.16.41) and write the transition probability per second from state i to state m as

$$w_{mi} = \frac{d}{dt}|c^{(1)}(m, t|i, t_0)|^2 = \frac{2\pi}{\hbar}|\langle m|V_S|i\rangle|^2\,\delta(E_m - E_i). \quad (1.16.45)$$

When we sum over a range of final states, the transition probability per second is

$$w = \frac{d}{dt}\sum_m |c^{(1)}(m, t|i, t_0)|^2 = \frac{2\pi}{\hbar}|\langle m|V_S|i\rangle|^2 g(E_i), \quad (1.16.46)$$

which is called Fermi's golden rule. Only transitions in which energy is approximately conserved ($E_m \simeq E_i$) contribute. This is one of the most used results of time-dependent perturbation theory.

In some instances the matrix element between the initial and final states vanish but they do not vanish for some other state $|l\rangle$. In this case, the transition may take place in second order. If we insert a completeness relation in the $n = 2$ term in (1.16.34) we obtain

$$|c^{(2)}(m, t|i, t_0)|^2 = \left| \frac{1}{\hbar^2} \int_{t_0}^{t} dt_1 \int_{t_0}^{t_1} dt_2 \sum_l \langle m|V_S|l\rangle\langle l|V_S|i\rangle e^{i[\omega_{ml}(t_1-t_0)+\omega_{li}(t_2-t_0)]} \right|^2. \tag{1.16.47}$$

If V_S is time-independent, this reduces to

$$|c^{(2)}(m, t|i, t_0)|^2 = \left| \frac{1}{\hbar^2} \sum_l \frac{\langle m|V_S|l\rangle\langle l|V_S|i\rangle}{\omega_{li}} \left[\frac{e^{i\omega_{mi}(t-t_0)} - 1}{\omega_{mi}} - \frac{e^{i\omega_{ml}(t-t_0)} - 1}{\omega_{ml}} \right] \right|^2. \tag{1.16.48}$$

In this case the transition from $|i\rangle$ to $|m\rangle$ takes place through the intermediate states $|l\rangle$ which have nonzero matrix elements.

The first term is large in (1.16.48) when $\omega_{mi} = 0$, that is, when energy between initial and final states is conserved. The second term is large when $\omega_{ml} \simeq 0$ which need not conserve energy between initial and intermediate states. It arises from turning on the interaction suddenly at $t = t_0$ and is not usually important.

We next introduce the Dyson time-ordering operator which allows us to write the perturbation expansion for $U(t, t_0)$ in (1.16.23) in extremely compact form.

We notice first that in (1.16.23) that $t_0 < t_n < t_{n-1} < \cdots < t_2 < t_1 < t$. Furthermore, the interaction operators $V_I(t_1) \cdots V_I(t_n)$ are time-ordered in the sense that the operator at the earliest time is on the right, followed by the operator with the next later time, and so on. Consider now two non-commuting operators $A(t_1)$ and $B(t_2)$. The time-ordering operator when applied to AB is defined by

$$P\{A(t_1)B(t_2)\} = \begin{cases} A(t_1)B(t_2) & \text{if } t_2 < t_1 \\ B(t_2)A(t_1) & \text{if } t_1 < t_2 \end{cases}. \tag{1.16.49}$$

That is, when operators are written inside the P-operator, we treat them as c-numbers since

$$P\{A(t_1)B(t_2)\} = P\{B(t_2)A(t_1)\}. \tag{1.16.50}$$

We then positional order them so that the earlier time is on the right, and they again become operators. In Chapter 3 we gain more experience with the ordering concept and technique. We might put bars over the A and B *inside* the P-operator to remind us they commute:

$$P\{\bar{A}(t_1)\bar{B}(t_2)\} = \begin{cases} A(t_1)B(t_2) & t_2 < t_1 \\ B(t_2)A(t_1) & t_1 < t_2 \end{cases}. \tag{1.16.51}$$

If the operators are evaluated at the same time, the P-operator does nothing.

When many operators are involved, P is generalized so that the operators are positional ordered from right to left with the earliest operator on the right, the next later operator to its left and so on. In (1.16.23) since the operators are already time-ordered, we have

$$P\{\bar{V}_I(t_1)\bar{V}_I(t_2)\cdots\bar{V}_I(t_n)\} = V_I(t_1)V_I(t_2)\cdots V_I(t_n). \tag{1.16.52}$$

We next show that (1.16.23) may be written as

$$\begin{aligned} U(t, t_0) &= P\left\{\exp -\frac{i}{\hbar}\int_{t_0}^{t}\bar{V}_I(t')\,dt'\right\} \\ &\equiv P\left\{\sum_{n=0}^{\infty}\left(-\frac{i}{\hbar}\right)^n\frac{1}{n!}\left(\int_{t_0}^{t}\bar{V}_I(t')\,dt'\right)^n\right\} \\ &\equiv P\left\{1+\sum_{n=1}^{\infty}\left(-\frac{i}{\hbar}\right)^n\frac{1}{n!}\int_{t_0}^{t}\bar{V}_I(t_1)\,dt_1\int_{t_0}^{t}\bar{V}_I(t_2)\,dt_2\cdots\int_{t_0}^{t}dt_n\,\bar{V}_I(t_n)\right\} \\ &\equiv P\left\{1+\sum_{n=1}^{\infty}\left(-\frac{i}{\hbar}\right)^n\frac{1}{n!}\int_{t_0}^{t}dt_1\int_{t_0}^{t}dt_2\cdots\int_{t_0}^{t}dt_n\,\bar{V}_I(t_1)\cdots\bar{V}_I(t_n)\right\}. \end{aligned} \tag{1.16.53}$$

We shall proceed to show that this is identical with (1.16.23) up to the $n = 2$ term. The remaining can be shown by induction which we leave as an exercise for the reader.

The $n = 0$ term is obviously identical. For the $n = 1$ term we have

$$P\left(-\frac{i}{\hbar}\right)\int_{t_0}^{t}\{\bar{V}_I(t_1)\}\,dt_1 = -\frac{i}{\hbar}\int_{t_0}^{t}V_I(t_1)\,dt_1.$$

Since only one time is involved, the P-operator does nothing and this is the same as the $n = 1$ term in (1.16.23). For the $n = 2$ term we have

$$I \equiv \left(-\frac{i}{\hbar}\right)^2\frac{1}{2!}\int_{t_0}^{t}dt_1\int_{t_0}^{t}dt_2\,P\{\bar{V}_I(t_1)\bar{V}_I(t_2)\}, \tag{1.16.54}$$

where the region of integration is shaded in Figure 1.5. We may therefore break this up into two integrals over the two separate triangular areas so that

$$I = \left(-\frac{i}{\hbar}\right)^2\frac{1}{2!}\left[\int_{t_0}^{t}dt_1\int_{t_0}^{t_1}dt_2\,P\{\bar{V}_I(t_1)\bar{V}_I(t_2)\} + \int_{t_0}^{t}dt_2\int_{t_0}^{t_2}dt_1\,P\{\bar{V}_I(t_1)\bar{V}_I(t_2)\}\right],$$

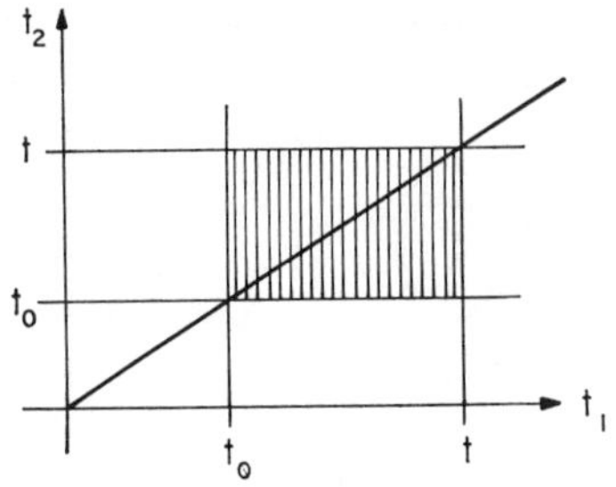

Figure 1.5 Region of integration in (1.16.54).

where in the first integral we integrated over t_2 first and then t_1 and did the reverse in the second. In the first integral we see that $t_2 < t_1$ so that

$$P\{\bar{V}_I(t_1)\bar{V}_I(t_2)\} = V_I(t_1)V_I(t_2),$$

while in the second integral $t_1 < t_2$ so that

$$P\{\bar{V}_I(t_1)\bar{V}_I(t_2)\} = V_I(t_2)V_I(t_1),$$

so that I becomes

$$I = \left(-\frac{i}{\hbar}\right)^2 \frac{1}{2}\left[\int_{t_0}^{t} dt_1 \int_{t_0}^{t_1} dt_2\, V_I(t_1)V_I(t_2) + \int_{t_0}^{t} dt_2 \int_{t_0}^{t_2} dt_1\, V_I(t_2)V_I(t_1)\right],$$

and the V's are now positional ordered and their operator character fully restored. We now note that in the second integral, we could change variables and call t_2 anything we want, say x or even t_1, and call t_1 anything else, say y or t_2, without changing the value of these definite integrals. Therefore,

$$I = \left(-\frac{i}{\hbar}\right)^2 \frac{1}{2}\left[\int_{t_0}^{t} dt_1 \int_{t_0}^{t_1} dt_2\, V_I(t_1)V_I(t_2) + \int_{t_0}^{t} dt_1 \int_{t_0}^{t_1} dt_2\, V_I(t_1)V_I(t_2)\right],$$

so that both integrals are equal and I is then identical with the $n = 2$ term in (1.16.23). We therefore see that the P-operator allows us to write U in the very compact form

$$U(t, t_0) = P\left\{\exp -\frac{i}{\hbar}\int_{t_0}^{t} \bar{V}_I(t')\, dt'\right\}. \tag{1.16.55}$$

We therefore have

$$|\psi_S(t)\rangle = \exp\left[-\frac{i}{\hbar}H_0(t - t_0)\right]P\left\{\exp -\frac{i}{\hbar}\int_{t_0}^{t} \bar{V}_I(t')\, dt'\right\}|\psi_H(t_0)\rangle. \tag{1.16.56}$$

Finally, for future reference we should like to obtain explicit equations of motion for the expansion coefficients $c_n(t)$ in (1.16.29). We have

$$
\begin{aligned}
i\hbar \frac{dc_n}{dt} &= \langle n| i\hbar \frac{\partial}{\partial t} |\psi_I(t)\rangle \\
&= \langle n| i\hbar \frac{\partial U}{\partial t} |\psi(t_0)\rangle \\
&= \langle n| V_I(t) U(t) |\psi(t_0)\rangle, \qquad (1.16.57)
\end{aligned}
$$

where we used (1.16.16) and (1.16.18). If we insert a completeness relation and use

$$\langle n| V_I(t) |l\rangle = e^{i\omega_{nl}(t-t_0)} \langle n| V_S |l\rangle, \qquad (1.16.58)$$

this becomes

$$
\begin{aligned}
i\hbar \frac{dc_n}{dt} &= \sum_l \langle n| V_S |l\rangle e^{i\omega_{nl}(t-t_0)} \langle l| U(t) |\psi(t_0)\rangle \\
&= \sum_l \langle n| V_S |l\rangle e^{i\omega_{nl}(t-t_0)} c_l(t), \qquad (1.16.59)
\end{aligned}
$$

where we again used (1.16.29) and (1.16.16). If we expand $c_l(t)$ in powers of V, we have

$$c_l(t) = c_l^{(0)}(t) + c_l^{(1)}(t) + c_l^{(2)}(t) + \cdots. \qquad (1.16.60)$$

If we use this on both sides of (1.16.59) and equate equal powers of V, we obtain

$$\frac{dc_n^{(0)}}{dt} = 0 \qquad (1.16.61a)$$

$$\frac{dc_n^{(1)}}{dt} = \frac{1}{i\hbar} \sum_l \langle n| V_S |l\rangle e^{i\omega_{nl}(t-t_0)} c_l^{(0)}(t) \qquad (1.16.61b)$$

$$\frac{dc_n^{(2)}}{dt} = \frac{1}{i\hbar} \sum_l \langle n| V_S |l\rangle e^{i\omega_{nl}(t-t_0)} c_l^{(1)}(t) \qquad (1.16.61c)$$

$$\cdots$$

Integrating (1.16.61a) we have

$$c_n^{(0)}(t) = \delta_{ni}. \qquad (1.16.62a)$$

If we put this into (1.16.61b), we have

$$c_n^{(1)}(t) = \frac{1}{i\hbar} \int_{t_0}^{t} \langle n| V_S |i\rangle e^{i\omega_{ni}(t_1-t_0)} \, dt_1. \qquad (1.16.62b)$$

If we put this in (1.16.61c), we have

$$c_n^{(2)}(t) = \left(\frac{1}{i\hbar}\right)^2 \sum_l \int_{t_0}^{t} dt_1 \langle n|V_S|l\rangle e^{i\omega_{nl}(t_1-t_0)} \int_{t_0}^{t_1} dt_2 \langle l|V_S|i\rangle e^{i\omega_{li}(t_2-t_0)}, \quad (1.16.62c)$$

and so on. These are seen to be in agreement with the prior results.

1.17 PERTURBATION THEORY FOR A HEISENBERG OPERATOR

If a system is described by a hamiltonian

$$H = H_0 + V, \quad (1.17.1)$$

and M is an arbitrary operator which is explicitly time-independent in the Schrödinger picture, then M obeys the Heisenberg equation of motion (1.15.5)

$$i\hbar \frac{dM^H}{dt} = [M^H, H^H] = [M^H, H_0^H + V^H], \quad (1.17.2)$$

where all operators are in the Heisenberg picture. If we consider only time-independent hamiltonians, it is evident that the total hamiltonian in the Heisenberg and Schrödinger pictures are equal. This is seen as follows. For H time-independent, if we let $M^H = H^H$, it follows from (1.17.2) that

$$\frac{dH^H(t)}{dt} = \frac{1}{i\hbar}[H^H(t), H^H(t)] = 0, \quad (1.17.3)$$

so that $H^H(t)$ is time-independent. Since at $t = t_0$, both the Heisenberg and Schrödinger pictures coincide, we have $H^H = H^S$. It is not necessarily true that $H_0^H(t)$ equals H_0^S since by (1.15.6)

$$H_0^H(t) = e^{(i/\hbar)(H_0+V)(t-t_0)} H_0^S e^{-(i/\hbar)(H_0+V)(t-t_0)}$$
$$\neq H_0^S$$

unless $[H_0^S, V^S] = 0$, an uninteresting case. Similarly, $V^H(t) \neq V^S$ but $(H_0 + V)^H = (H_0 + V)^S$. We may therefore write (1.17.2) as

$$\frac{dM^H(t)}{dt} = \frac{1}{i\hbar}[M^H(t), H_0^S + V^S]. \quad (1.17.4)$$

Next we make the change of variable

$$M^H(t) = U_0^\dagger(t, t_0)X(t)U_0(t, t_0), \quad (1.17.5)$$

where

$$U_0(t, t_0) = e^{-(i/\hbar)H_0^S(t-t_0)}. \quad (1.17.6)$$

If we differentiate (1.17.5) with respect to t and use (1.17.6) and its adjoint, we find

$$\frac{dM^H}{dt} = U_0^\dagger X \frac{dU_0}{dt} + \frac{dU_0^\dagger}{dt} X U_0 + U_0^\dagger \frac{dX}{dt} U_0$$

$$= \frac{1}{i\hbar} [U_0^\dagger X U_0, H_0{}^S] + U_0^\dagger \frac{dX}{dt} U_0$$

$$= \frac{1}{i\hbar} [M^H(t), H_0{}^S] + U_0^\dagger \frac{dX}{dt} U_0. \tag{1.17.7}$$

If we compare this with (1.17.4), we have on using (1.17.5)

$$U_0^\dagger \frac{dX}{dt} U_0 = \frac{1}{i\hbar} [M^H(t), V^S]$$

$$= \frac{1}{i\hbar} [U_0^\dagger X U_0, V^S]. \tag{1.17.8}$$

If we multiply from the left on both sides by U_0 and from the right by $U_0^\dagger$ and use the fact that $U_0 U_0^\dagger = U_0^\dagger U_0 = 1$, we have

$$\frac{dX}{dt} = \frac{1}{i\hbar} [X(t), U_0 V^S U_0^\dagger]. \tag{1.17.9}$$

Note from (1.16.9) that

$$V^I(t - t_0) = U_0^\dagger(t, t_0) V^S U_0(t, t_0), \tag{1.17.10}$$

when U_0 is given by (1.17.6). We easily see that

$$V^I(t_0 - t) = U_0(t, t_0) V^S U_0^\dagger(t, t_0). \tag{1.17.11}$$

With this notation we may integrate both sides of (1.17.9) to obtain

$$X(t) = X(t_0) + \frac{1}{i\hbar} \int_{t_0}^{t} [X(t'), V^I(t_0 - t')]\, dt', \tag{1.17.12}$$

where by (1.17.5)

$$X(t_0) = M^H(t_0) \equiv M^S. \tag{1.17.13}$$

We proceed to iterate (1.17.12) in the familiar way to obtain

$$X(t) = M^S + \frac{1}{i\hbar} \int_{t_0}^{t} [M^S, V^I(t_0 - t_1)]\, dt_1$$

$$+ \left(\frac{1}{i\hbar}\right)^2 \int_{t_0}^{t} dt_1 \int_{t_0}^{t} dt_2 \{[M^S, V^I(t_0 - t_2)], V^I(t_0 - t_1)\} + \cdots. \tag{1.17.14}$$

We next use (1.17.5) to transform X back to the Heisenberg operator:

$$M^H(t) = U_0^\dagger(t, t_0) M^S U_0(t, t_0) + \frac{1}{i\hbar}\int_{t_0}^{t} U_0^\dagger(t, t_0)[M^S, V^I(t_0 - t_1)]U_0(t, t_0)\, dt_1 + \cdots. \quad (1.17.15)$$

If we use (1.16.11) and (1.17.11) and insert $U_0 U_0^\dagger = U_0^\dagger U_0 = 1$ in appropriate places, this becomes

$$M^H(t) = M^I(t) + \frac{1}{i\hbar}\int_{t_0}^{t}[M^I(t), V^I(t - t_1)]\, dt_1 + \left(\frac{1}{i\hbar}\right)^2 \int_{t_0}^{t} dt_1 \int_{t_0}^{t} dt_2\, \{[M^I(t), V^I(t - t_2)], V^I(t - t_1)\} + \cdots. \quad (1.17.16)$$

which is the desired perturbation expansion of a Heisenberg operator.

1.18 WAVE MECHANICS

We have mentioned wave functions so often in the course of this development that it is redundant to say much about wave mechanics except to reemphasize the concepts. Wave mechanics is quantum mechanics formulated in the Schrödinger picture in the position or momentum representation. We shall summarize some of the results for a particle in one dimension.

The coordinate representation is determined by the eigenvalue equation (all in the Schrödinger picture)

$$q|q'\rangle = q'|q'\rangle. \quad (1.18.1)$$

The orthonormality relation

$$\langle q''|q'\rangle = \delta(q' - q'') \quad (1.18.2)$$

may be regarded as the coordinate wave function in the coordinate representation. It is also the representative of the eigenket $|q'\rangle$ in the q representation. The set $\{|q'\rangle\}$ is complete, and so

$$\int_{-\infty}^{\infty} |q'\rangle\, dq'\, \langle q'| = I. \quad (1.18.3)$$

The state vectors $|\psi(t)\rangle$ satisfy the equation of motion

$$i\hbar \frac{\partial}{\partial t}|\psi(t)\rangle = H|\psi(t)\rangle = \left[\frac{p^2}{2m} + V(q)\right]|\psi(t)\rangle. \quad (1.18.4)$$

The representative of $|\psi(t)\rangle$ in the q representation is

$$\psi(q', t) \equiv \langle q'|\psi(t)\rangle, \quad (1.18.5)$$

and is the Schrödinger wave function at time t for state $|\psi(t)\rangle$. If we take the scalar product of (1.18.4) with $\langle q'|$ and use (1.18.5), (1.10.46), and (1.10.47), we have the time-dependent Schrödinger wave equation

$$i\hbar \frac{\partial}{\partial t} \psi(q', t) = \left[-\frac{\hbar^2}{2m} \frac{\partial^2}{\partial q'^2} + V(q') \right] \psi(q', t). \tag{1.18.6}$$

We could take $\partial/\partial t$ from $\langle q'|\partial/\partial t|\psi(t)\rangle$ since $\langle q'|$ is stationary in the Schrödinger picture.

The average value of $F(p, q)$ at time t is, by (1.11.1),

$$\begin{aligned} \langle F(p, q)\rangle &= \langle \psi(t)|F(p, q)|\psi(t)\rangle \\ &= \iint_{-\infty}^{+\infty} \langle \psi(t)|q'\rangle \, dq' \, \langle q'|F(p, q)|q''\rangle \, dq'' \, \langle q''|\psi(t)\rangle \\ &= \int_{-\infty}^{+\infty} \psi^*(q', t) F\left(\frac{\hbar}{i} \frac{\partial}{\partial q'}, q'\right) \psi(q', t) \, dq', \end{aligned} \tag{1.18.7}$$

where we have used the completeness relation (1.18.3) twice, as well as (1.10.41), (1.10.42), and (1.18.2). The order of factors in $F(p, q)$ must be strictly preserved.

The eigenket $|p'\rangle$ of p in the q-representation is, by (1.10.35),

$$\langle q'|p'\rangle = \frac{1}{\sqrt{2\pi\hbar}} \exp\left(\frac{ip'q'}{\hbar}\right). \tag{1.18.8}$$

This is the momentum eigenfunction in the coordinate representation. By means of this transformation function, the Schrödinger picture may be expressed in the momentum representation. From the previous work this is straightforward.

1.19 THE FREE PARTICLE; CHANGE IN TIME OF MINIMUM UNCERTAINTY WAVE PACKET

In this section, we consider a free particle in which the potential $V(q) = 0$ and the hamiltonian becomes

$$H = \frac{p^2}{2m} = H^\dagger. \tag{1.19.1}$$

Since the system is conservative, H is a constant of the motion. Also,

$$[p, H] = 0, \tag{1.19.2}$$

and so p is a constant of the motion for a free particle.

The operator p satisfies the eigenvalue problem

$$p|p'\rangle = p'|p'\rangle, \tag{1.19.3}$$

where $-\infty < p' < +\infty$. If we operate on both sides of (1.19.3) with H, we have

$$Hp|p'\rangle = pH|p'\rangle = p'H|p'\rangle, \tag{1.19.4}$$

since H and p commute. We therefore see that $H|p'\rangle$ is an eigenket of p with eigenvalue p'. Thus $H|p'\rangle$ can differ from $|p'\rangle$ only by a constant factor which we call E, and so we may write

$$H|p'\rangle = E|p'\rangle. \tag{1.19.5}$$

This is just the energy-eigenvalue problem. Since p and H commute, the eigenket of p may be a simultaneous eigenket of H. Thus we may write (1.19.5) as

$$H|p', E\rangle = E|p', E\rangle. \tag{1.19.6}$$

By (1.19.1) and (1.19.3),

$$\frac{p^2}{2m}|p', E\rangle = \frac{p'^2}{2m}|p', E\rangle = E|p', E\rangle, \tag{1.19.7}$$

or the eigenvalues of H are

$$E = \frac{p'^2}{2m}, \tag{1.19.8}$$

which are continuous and have any value from 0 to $+\infty$. Therefore, the solution of the energy-eigenvalue problem is very simple for a free particle. We note, however, that the eigenvalues $+p'$ and $-p'$ give the same E; that is, the states $|+p', E\rangle$ and $|-p', E\rangle$, which correspond to a particle moving to the right and to the left, have the same energy. This is an example of degeneracy.

The solution of the Schrödinger equation

$$i\hbar\frac{\partial|\psi(t)\rangle}{\partial t} = \frac{p^2}{2m}|\psi(t)\rangle \tag{1.19.9}$$

is, by (1.13.3),

$$|\psi(t)\rangle = \exp\left(-\frac{ip^2t}{2m\hbar}\right)|\psi(0)\rangle. \tag{1.19.10}$$

In the coordinate representation we have

$$\psi(q', t) = \langle q'|\psi(t)\rangle = \langle q'|\exp\left(-\frac{ip^2t}{2m\hbar}\right)|\psi(0)\rangle \tag{1.19.11}$$

for the time development of the wave function. If we use the completeness relation for $|p'\rangle$, this may be written as

$$\begin{aligned}\psi(q', t) &= \int_{-\infty}^{\infty} \langle q'| \exp\left(-\frac{ip^2t}{2m\hbar}\right)|p'\rangle\, dp'\, \langle p'|\psi(0)\rangle \\ &= \int_{\infty}^{\infty} \exp\left(-\frac{ip'^2t}{2m\hbar}\right)\langle q'|p'\rangle\, dp'\, \varphi(p', 0) \\ &= \frac{1}{\sqrt{2\pi\hbar}}\int_{-\infty}^{\infty} \exp\left(-\frac{ip'^2t}{2m\hbar}\right) \exp\left(\frac{ip'q'}{\hbar}\right)\varphi(p', 0)\, dp', \end{aligned} \tag{1.19.12}$$

where we used (1.10.35) and let

$$\langle p'|\psi(0)\rangle = \varphi(p', 0). \tag{1.19.13}$$

This gives the wave function in the coordinate representation when $\varphi(p', 0)$ is the initial arbitrary wave function in the p-representation.

As a special case of (1.19.12), we let $\varphi(p', 0)$ be a minimum uncertainty wave packet given by (1.12.25). As time develops, the wave packet for a free particle is given by (1.19.12):

$$\begin{aligned}\psi(q', t) &= \frac{1}{\sqrt{2\pi\hbar}\,[2\pi(\Delta p)^2]^{1/4}} \\ &\times \int_{-\infty}^{\infty} \exp\left[-\frac{ip'^2t}{2m\hbar} + \frac{ip'q'}{\hbar} - \frac{i}{\hbar}\langle q\rangle(p' - \langle p\rangle) - \frac{(p' - \langle p\rangle)^2}{4(\Delta p)^2}\right] dp'.\end{aligned} \tag{1.19.14}$$

For simplicity, we let $\langle p\rangle_{t=0} = \langle q\rangle_{t=0} = 0$. If we replace $(\Delta p)_{t=0}$ by $\hbar/2(\Delta q)_{t=0}$, we may easily carry out the integral to obtain

$$\psi(q', t) = \frac{1}{(2\pi)^{1/4}(\Delta q + i\hbar t/2m\,\Delta q)^{1/2}} \exp\left[-\frac{q'^2}{4(\Delta q)^2 + 2i\hbar t/m}\right], \tag{1.19.15}$$

where

$$(\Delta q)^2 = \langle\psi(0)|q^2|\psi(0)\rangle = \int_{-\infty}^{\infty} q'^2|\psi(q', 0)|^2\, dq'.$$

If we calculate the average value of q at time t, we have, by (1.18.7) and (1.19.15),

$$\langle q\rangle_t = \int_{-\infty}^{\infty} |\psi(q', t)|^2 q'\, dq' = 0. \tag{1.19.16}$$

That is, the particle stays at $q' = 0$, on an average, during the course of time. However, at time t

$$\langle q^2 \rangle_t = \int_{-\infty}^{\infty} |\psi(q', t)|^2 q'^2 \, dq' = [\Delta q]_{t=0}^2 + \frac{\hbar^2 t^2}{4m^2 [\Delta q]_{t=0}^2}, \tag{1.19.17}$$

so that the mean-square deviation at time t is greater than its value at $t = 0$ by an amount proportional to t^2. As time goes on, the packet spreads in coordinate space. The smaller Δq is at $t = 0$, the faster the packet spreads.

It is also seen that the average of the momentum at time t is, by (1.18.7) and (1.19.15),

$$\langle p \rangle_t = \int_{-\infty}^{\infty} \psi^*(q', t) \frac{\hbar}{i} \frac{\partial \psi(q', t)}{\partial q'} \, dq' = 0, \tag{1.19.18}$$

while the mean-square deviation of p is found directly to be

$$\begin{aligned} \langle p^2 \rangle_t &= -\hbar^2 \int_{-\infty}^{\infty} \psi^*(q', t) \frac{\partial^2}{\partial q'^2} \psi(q', t) \, dq' \\ &= \frac{\hbar^2}{4[\Delta q]_{t=0}^2} = \langle p^2 \rangle_{t=0}. \end{aligned} \tag{1.19.19}$$

Therefore, the uncertainty product at time t is, by (1.19.17) and (1.19.19),

$$[\Delta p]_t^2 [\Delta q]_t^2 = \frac{\hbar^2}{4} + \left[\frac{\hbar^2}{4m[\Delta q]_{t=0}^2} \right]^2 t^2 \geq \frac{\hbar^2}{4}. \tag{1.19.20}$$

Thus it may be concluded that, as time goes on, the minimum uncertainty wave packet ceases to be a minimum uncertainty packet. When we study electromagnetic fields, we shall attribute this spread to the zero-point fluctuations of the field.

1.20 THE DENSITY OPERATOR [9–13]; PERTURBATION THEORY

Let M be an arbitrary operator in the Schrödinger picture (SP). If $|\psi_S(t)\rangle$ is the state vector at time t in the SP, then we have seen that the average value of M is given by

$$\langle M \rangle = \langle \psi_S(t) | M_S | \psi_S(t) \rangle. \tag{1.20.1a}$$

We have also seen that we may obtain the same mean if we work in the Heisenberg picture (HP), namely,

$$\langle M \rangle = \langle \psi_H(t_0) | M_H(t) | \psi_H(t_0) \rangle, \tag{1.20.1b}$$

where $|\psi_H(t_0)\rangle$ gives the state of the system at t_0 and

$$M_H(t) = e^{(i/\hbar)H_T(t-t_0)}M_S e^{-(i/\hbar)H_T(t-t_0)}, \tag{1.20.2}$$

where H_T is the total hamiltonian in the SP.

We next note that

$$\mathrm{Tr}\,(|u\rangle\langle v|) = \langle v|u\rangle. \tag{1.20.3}$$

This may be proved as follows. If $\{|n\rangle\}$ represents any complete set of state vectors, then in this representation, we may write the trace as the sum of the diagonal matrix elements

$$\mathrm{Tr}\,(|u\rangle\langle v|) = \sum_n \langle n|u\rangle\langle v|n\rangle. \tag{1.20.4}$$

Since $\langle n|u\rangle$ and $\langle v|n\rangle$ are simply numbers, we may rewrite this as

$$\mathrm{Tr}\,(|u\rangle\langle v|) = \sum_n \langle v|n\rangle\langle n|u\rangle.$$

From the completeness relation,

$$\sum_n |n\rangle\langle n| = 1,$$

the trace above reduces to (1.20.3).

If we next identify $M|\psi\rangle$ in (1.20.1) with $|u\rangle$ and $\langle\psi|$ with $\langle v|$, we may use (1.20.3) to rewrite (1.20.1) as

$$\begin{aligned}\langle M\rangle &= \mathrm{Tr}\,M_S|\psi_S(t)\rangle\langle\psi_S(t)| \\ &= \mathrm{Tr}\,M_H(t)|\psi_H(t_0)\rangle\langle\psi_H(t_0)|.\end{aligned} \tag{1.20.5}$$

Up to this point we have tacitly assumed that we have made sufficient measurements to determine that the system is in the state $|\psi_H(t_0)\rangle$ initially. In many cases it may be impractical to make enough measurements to do this. For example, if our system consists of 10^{23} gas molecules in a container, we could not hope to determine exactly $|\psi(t_0)\rangle$. Suppose we only know that it is in such a state with a probability p_ψ where

$$\sum_\psi p_\psi = 1; \qquad p_\psi \geq 0. \tag{1.20.6}$$

We should then average (1.20.5) over this probability distribution so that the mean value of M becomes

$$\begin{aligned}\langle\langle M\rangle\rangle &= \sum_\psi p_\psi\,\mathrm{Tr}\,M_S|\psi_S(t)\rangle\langle\psi_S(t)| \\ &= \sum_\psi p_\psi\,\mathrm{Tr}\,M_H(t)|\psi_H(t_0)\rangle\langle\psi_H(t_0)|.\end{aligned} \tag{1.20.7}$$

The first averaging was due to the statistical interpretation inherent in quantum mechanics (see Section 1.11) whereas the second averaging is of a classical nature and would be necessary if we were treating the system classically. In the event we can make enough measurements to determine that the system is in state $|\psi(t_0)\rangle$, then

$$p_\psi = \delta_{\psi,\psi_0}, \tag{1.20.8}$$

and (1.20.7) reduces to (1.20.5).

We may next define the density operator in the SP and HP, respectively, as

$$\begin{aligned} \rho_S(t) &= \sum_\psi p_\psi |\psi_S(t)\rangle\langle\psi_S(t)| \\ \rho_H(t_0) = \rho_S(t_0) &= \sum_\psi p_\psi |\psi_H(t_0)\rangle\langle\psi_H(t_0)|, \end{aligned} \tag{1.20.9}$$

so that (1.20.7) becomes

$$\begin{aligned} \langle\!\langle M \rangle\!\rangle &= \text{Tr } M_S \rho_S(t) \\ &= \text{Tr } M_H(t)\, \rho_H(t_0), \end{aligned} \tag{1.20.10}$$

since we may interchange the averaging (summing) over ψ and tracing (also summing) operations.

We may easily obtain the equation of motion for $\rho_S(t)$ from its definition and the Schrödinger equation

$$\begin{aligned} i\hbar \frac{\partial|\psi_S\rangle}{\partial t} &= H_S|\psi_S\rangle \\ -i\hbar \frac{\partial}{\partial t}\langle\psi_S| &= \langle\psi_S|H_S. \end{aligned} \tag{1.20.11}$$

For from (1.20.9) and (1.20.11), we have (p_ψ is time-independent)

$$\begin{aligned} i\hbar \frac{\partial \rho_S}{\partial t} &= \sum_\psi p_\psi \left\{ i\hbar \frac{\partial|\psi_S\rangle}{\partial t}\langle\psi_S| + |\psi_S\rangle i\hbar \frac{\partial\langle\psi_S|}{\partial t} \right\} \\ &= \sum_\psi p_\psi \{H_S|\psi_S\rangle\langle\psi_S| - |\psi_S\rangle\langle\psi_S|H_S\} \\ &= [H_S, \rho_S], \end{aligned} \tag{1.20.12}$$

since p_ψ is a number and therefore commutes with H.

From (1.20.9), we see that ρ is hermitian so that

$$\rho^\dagger = \rho. \tag{1.20.13}$$

In addition, we have that

$$\text{Tr } \rho = \sum_\psi p_\psi \langle\psi(t)|\psi(t)\rangle = \sum_\psi p_\psi = 1, \tag{1.20.14}$$

where we used (1.20.3) and (1.20.6).

We next note that if $|\chi\rangle$ is any ket then

$$\langle\chi|\rho|\chi\rangle = \sum_{\psi} p_{\psi}\langle\chi|\psi\rangle\langle\psi|\chi\rangle = \sum_{\psi} p_{\psi}|\langle\chi|\psi\rangle|^2 \geq 0. \tag{1.20.15}$$

That is, the diagonal matrix elements of ρ are all real and positive in any representation. Since the sum of the diagonal elements equals 1 by (1.20.14) in any representation, it follows that the diagonal elements always lie between 0 and 1:

$$0 \leq \langle\chi|\rho|\chi\rangle \leq 1. \tag{1.20.16}$$

If at t_0 we know the state of the system, then (1.20.8) is satisfied, and ρ reduces to $|\psi(t)\rangle\langle\psi(t)|$. We then have

$$\rho^2 = |\psi(t)\rangle\langle\psi(t)|\psi(t)\rangle\langle\psi(t)| = \rho,$$

since $\langle\psi|\psi\rangle = 1$; therefore,

$$\text{Tr}\ \rho^2 = 1. \tag{1.20.17}$$

When this condition is satisfied we say the ensemble is in a pure state.

For a mixed state, we do not know the state at t_0 precisely and we have

$$\text{Tr}\ \rho^2 < 1. \tag{1.20.18}$$

To show that $\text{Tr}\ \rho^2 \leq 1$, we note that because ρ is hermitian, it may be diagonalized by a unitary transformation

$$\rho' = S\rho S^{\dagger}, \tag{1.20.19}$$

where

$$S^{\dagger} = S^{-1}. \tag{1.20.20}$$

Since a trace is invariant under a unitary transformation we have that

$$\text{Tr}\ \rho^2 = \text{Tr}\ \rho'^2 = \sum_n (\rho_n')^2 \leq \left(\sum_n \rho_n'\right)^2 = 1. \tag{1.20.21}$$

This follows since ρ'^2 is diagonal when ρ' is and ρ_n' are the diagonal elements, which by (1.20.16) lie between 0 and 1. Since for a pure state $\text{Tr}\ \rho^2 = 1$, (1.20.18) corresponds to a mixed state.

We next rederive our time dependent perturbation theory results using the density operator. Assume our hamiltonian may be written as the sum

$$H = H_0 + V, \tag{1.20.22}$$

where we assume we have solved the energy eigenvalue problem

$$H_0|n\rangle = E_n^{(0)}|n\rangle, \tag{1.20.23}$$

when the set $\{|n\rangle\}$ forms a complete orthonormal set. Then (1.20.12) becomes

$$i\hbar \frac{\partial \rho_S}{\partial t} = [H_0{}^S + V_S, \rho_S(t)]. \tag{1.20.24}$$

We shall transform to the IP (interaction picture) if we let

$$\rho_S(t) = U_0(t, t_0)\rho_I(t)U_0^\dagger(t, t_0), \tag{1.20.25a}$$

where

$$U_0(t, t_0) = \exp\left[-\frac{i}{\hbar} H_0{}^S(t - t_0)\right], \tag{1.20.25b}$$

when $H_0{}^S$ is time-dependent.

If we next substitute (1.20.25) into (1.20.24) we obtain

$$i\hbar \frac{\partial U_0}{\partial t} \rho_I U_0^\dagger + U_0\rho_I i\hbar \frac{\partial U_0^\dagger}{\partial t} + U_0\left(i\hbar \frac{\partial \rho_I}{\partial t}\right) U_0^\dagger = [H_0{}^S + V_S, U_0\rho_I U_0^\dagger]. \tag{1.20.26}$$

If we use (1.20.25b) and its adjoint, we see this may be written as

$$[H_0{}^S, U_0\rho_I U_0^\dagger] + U_0\left(i\hbar \frac{\partial \rho_I}{\partial t}\right) U_0^\dagger = [H_0{}^S + V_S, U_0\rho_I U_0^\dagger],$$

and we see that the commutator on the left cancels the first one on the right. This leaves

$$U_0\left(i\hbar \frac{\partial \rho_I}{\partial t}\right) U_0^\dagger = V_S U_0\rho_I U_0^\dagger - U_0\rho_I U_0^\dagger V_S. \tag{1.20.27}$$

Since $U_0^\dagger = U^{-1}$, if we multiply both sides from the left by $U_0^\dagger$ and both sides from the right by U_0, we obtain the equation of motion for the density operator in the IP, namely,

$$i\hbar \frac{\partial \rho_I}{\partial t} = [V_I(t - t_0), \rho_I(t)], \tag{1.20.28}$$

where [cf. (1.16.11)]

$$V_I(t - t_0) = U_0^\dagger(t, t_0)V_S U_0(t, t_0) \tag{1.20.29}$$

is the interaction in the IP. We have not restricted V_S to be time independent in the SP.

Note that since $U_0(t_0, t_0) = 1$, it follows from (1.20.25) that

$$\rho_S(t_0) = \rho_I(t_0). \tag{1.20.30}$$

In general, (1.20.28) cannot be solved exactly. When the interaction is small, we may obtain an approximate solution in powers of V which will converge rapidly. On iteration we obtain

$$\rho_I(t) = \rho(t_0) + \frac{1}{i\hbar}\int_{t_0}^{t}[V_I(t_1 - t_0), \rho(t_0)]\,dt_1$$
$$+ \left(\frac{1}{i\hbar}\right)^2 \int_{t_0}^{t} dt_1 \int_{t_0}^{t_1} dt_2 [V_I(t_1 - t_0), [V_I(t_2 - t_0), \rho(t_0)]] + \cdots. \tag{1.20.31}$$

We now show that under special circumstances, we may obtain the results given in Section 1.16 of time-dependent perturbation theory, but the results here are more general than that presented in Section 1.16. [Why?]

From (1.16.2) we have that

$$|\psi_S(t)\rangle = U_0(t, t_0)|\psi_I(t)\rangle, \tag{1.20.32}$$

while by (1.20.25) we have

$$\rho_S(t) = U_0(t, t_0)\rho_I(t)U_0^\dagger(t, t_0), \tag{1.20.33}$$

and by (1.20.9), we have

$$\rho_S(t) = \sum_\psi p_\psi |\psi_S(t)\rangle\langle\psi_S(t)|. \tag{1.20.34}$$

If we use (1.20.32) and its adjoint, this may be written as

$$\rho_S(t) = U_0(t, t_0) \sum_\psi p_\psi |\psi_I(t)\rangle\langle\psi_I(t)| U_0^\dagger(t, t_0). \tag{1.20.35}$$

If we compare (1.20.33) with this we see that in the I.P.

$$\rho_I(t) = \sum_\psi p_\psi |\psi_I(t)\rangle\langle\psi_I(t)|, \tag{1.20.36}$$

where p_ψ is the probability the system is in state $|\psi(t_0)\rangle$ at time t_0.

If we next take a diagonal matrix element of both sides of (1.20.36) in the representation (1.20.23) we obtain

$$\begin{aligned}\langle m|\rho_I|m\rangle &= \sum_\psi p_\psi \langle m|\psi_I(t)\rangle\langle\psi_I(t)|m\rangle \\ &= \sum_\psi p_\psi |\langle m|\psi_I(t)\rangle|^2 \\ &= \sum_\psi p_\psi |c_m(t)|^2,\end{aligned} \tag{1.20.37}$$

where the last step follows from (1.16.31). If we use (1.20.31), we have

$$\sum_\psi p_\psi |c_m(t)|^2 = \langle m|\rho(t_0)|m\rangle + \frac{1}{i\hbar}\int_{t_0}^{t} \langle m|[V_I(t_1 - t_0), \rho(t_0)]|m\rangle\,dt_1$$
$$+ \left(\frac{1}{i\hbar}\right)^2 \int_{t_0}^{t} dt_1 \int_{t_0}^{t_1} dt_2\,\langle m|[V_I(t_1 - t_0), [V_I(t_2 - t_0), \rho(t_0)]]|m\rangle$$
$$+ \cdots. \tag{1.20.38}$$

If we next assume that at t_0 the system is in some unperturbed initial energy eigenstate $|i\rangle$ where $i \neq m$, then

$$p_\psi = \delta_{\psi,i}, \tag{1.20.39}$$

and

$$\rho(t_0) = |i\rangle\langle i|. \tag{1.20.40}$$

By the orthogonality relations, the first two terms on the right in (1.20.38) vanish, and we have

$$|c(m, t|i, t_0)|^2 = \left(\frac{1}{i\hbar}\right)^2 \int_{t_0}^{t} dt_1 \int_{t_0}^{t_1} dt_2$$
$$\times \langle m|[V_I(t_1 - t_0), [V_I(t_2 - t_0), |i\rangle\langle i|]]|m\rangle + \cdots. \tag{1.20.41}$$

If we use (1.20.29) and the fact that

$$U_0|m\rangle = \exp\left[-\frac{i}{\hbar} E_m^{(0)}(t - t_0)\right]|m\rangle, \tag{1.20.42}$$

(1.20.41) reduces to

$$|c(m, t|i, t_0)|^2 = +\frac{1}{\hbar^2} \int_{t_0}^{t} dt_1 \int_{t_0}^{t_1} dt_2$$
$$\times \{e^{i\omega_{mi}(t_1-t_2)}\langle m|V_S(t_1)|i\rangle\langle i|V_S(t_2)|m\rangle$$
$$+ e^{-i\omega_{mi}(t_1-t_2)}\langle m|V_S(t_2)|i\rangle\langle i|V_S(t_1)|m\rangle\} + \cdots, \tag{1.20.43}$$

where we not only have used (1.20.29), but we have also allowed V_S to be explicitly time-dependent and have defined ω_{im} by

$$\hbar\omega_{im} = E_i^{(0)} - E_m^{(0)} = -\hbar\omega_{mi}. \tag{1.20.44}$$

We next show that this is identical with (1.16.36). We have from (1.16.36)

$$|c^{(1)}(m, t|i, t_0)|^2 = \frac{1}{\hbar^2} \int_{t_0}^{t} dt_1\, e^{i\omega_{mi}(t_1-t_0)}\langle m|V_S(t_1)|i\rangle$$
$$\times \int_{t_0}^{t} dt_2\, e^{-i\omega_{mi}(t_2-t_0)}\langle i|V_S(t_2)|m\rangle, \tag{1.20.45}$$

where since $V_S^\dagger = V_S$

$$\langle m|V_S(t)|i\rangle^* = \langle i|V_S(t)|m\rangle. \tag{1.20.46}$$

We shall reduce (1.20.45) to (1.20.43) by using the same techniques used in Section 1.16 with the Dyson time-ordering operator. We rewrite (1.20.45) as

(see Figure 1.5)

$$
\begin{aligned}
|c^{(1)}|^2 &= \frac{1}{\hbar^2}\int_{t_0}^{t} dt_1 \int_{t_0}^{t} dt_2\, e^{i\omega_{mi}(t_1-t_2)}\langle m|V_S(t_1)|i\rangle\langle i|V_S(t_2)|m\rangle \\
&= \frac{1}{\hbar^2}\int_{t_0}^{t} dt_1 \int_{t_0}^{t_1} dt_2\, e^{i\omega_{mi}(t_1-t_2)}\langle m|V_S(t_1)|i\rangle\langle i|V_S(t_2)|m\rangle \\
&\quad + \frac{1}{\hbar^2}\int_{t_0}^{t} dt_2 \int_{t_0}^{t_2} dt_1\, e^{i\omega_{mi}(t_1-t_2)}\langle i|V_S(t_2)|m\rangle\langle m|V_S(t_1)|i\rangle \\
&= \frac{1}{\hbar^2}\int_{t_0}^{t} dt_1 \int_{t_0}^{t_1} dt_2\, e^{i\omega_{mi}(t_1-t_2)}\langle m|V_S(t_1)|i\rangle\langle i|V_S(t_2)|m\rangle \\
&\quad + \frac{1}{\hbar^2}\int_{t_0}^{t} dt_1 \int_{t_0}^{t_1} dt_2\, e^{-i\omega_{mi}(t_1-t_2)}\langle i|V_S(t_1)|m\rangle\langle m|V_S(t_2)|i\rangle. \qquad (1.20.47)
\end{aligned}
$$

Here we used the fact that the matrix elements are c-numbers and commute, and in the last integral we have interchanged the dummy variables of integration t_1 and t_2. This is seen to be in exact agreement with the first term in (1.20.43).

We therefore see that (1.20.37) reduces to the perturbation theory results when the system is known with certainty to be in a state $|i\rangle$ initially but includes the more general case when we know that it is in state $|\psi\rangle$ initially only with a certain probability p_ψ.

1.21 THE REDUCED DENSITY OPERATOR

Let us consider two independent systems A and B which are described by hamiltonians H_A and H_B, respectively. Assume we have solved the energy eigenvalue problems

$$
\begin{aligned}
H_A|A'\rangle &= E_{A'}|A'\rangle; \qquad & H_B|B'\rangle &= E_{B'}|B'\rangle \\
\langle A'|A''\rangle &= \delta_{A'A''}; & \langle B'|B''\rangle &= \delta_{B'B''} \\
\sum_{A'}|A'\rangle\langle A'| &= 1; & \sum_{B'}|B'\rangle\langle B'| &= 1.
\end{aligned} \qquad (1.21.1)
$$

If they are put into contact at time t_0, the total hamiltonian is

$$H_T = H_A + H_B + V_{AB} \equiv H_0 + V_{AB}. \qquad (1.21.2)$$

Note that operators of the A and B systems commute since they are independent. In particular

$$[H_A, H_B] = 0. \qquad (1.21.3)$$

The density operator for the two coupled systems in the SP satisfies the equation of motion

$$i\hbar \frac{\partial \rho_{AB}{}^S}{\partial t} = [H_T, \rho_{AB}{}^S]. \tag{1.21.4}$$

At $t = t_0$, the systems are uncoupled so that the density operator factors as

$$\rho(t_0) = \rho_A(t_0)\rho_B(t_0), \tag{1.21.5}$$

where

$$\begin{aligned} \mathrm{Tr}_A\, \rho_A(t_0) &= \sum_{A'} \langle A'|\rho_A(t_0)|A'\rangle = 1 \\ \mathrm{Tr}_B\, \rho_B(t_0) &= \sum_{B'} \langle B'|\rho_B(t_0)|B'\rangle = 1, \end{aligned} \tag{1.21.6}$$

when we evaluate the traces in the representations in which H_A and H_B are diagonal. Therefore, we have

$$\mathrm{Tr}_{AB}\, \rho(t_0) = \sum_{A',B'} \langle A', B'|\rho_A(t_0)\rho_B(t_0)|A', B'\rangle, \tag{1.21.7}$$

where

$$|A', B'\rangle \equiv |A'\rangle|B'\rangle. \tag{1.21.8}$$

Therefore,

$$\begin{aligned} \mathrm{Tr}_{AB}\, \rho(t_0) &= \sum_{A'} \langle A'|\rho_A(t_0)|A'\rangle \sum_{B'} \langle B'|\rho_B(t_0)|B'\rangle \\ &= \mathrm{Tr}_A\, \rho_A(t_0)\, \mathrm{Tr}_B\, \rho_B(t_0) = 1. \end{aligned} \tag{1.21.9}$$

In the IP, by the previous section, we have

$$\rho_{AB}{}^S(t) = U_0(t, t_0)\rho_{AB}{}^I(t)U_0^\dagger\,(t, t_0) \tag{1.21.10}$$

$$i\hbar \frac{\partial \rho_{AB}{}^I(t)}{\partial t} = [V_{AB}{}^I(t - t_0), \rho_{AB}{}^I(t)] \tag{1.21.11}$$

$$V_{AB}{}^I(t - t_0) = U_0^\dagger(t, t_0)V_{AB}{}^S U_0(t, t_0) \tag{1.21.12}$$

$$\begin{aligned} U_0(t, t_0) &= \exp -\frac{i}{\hbar}(H_A + H_B)(t - t_0) \\ &\equiv U_0{}^A(t, t_0)U_0{}^B(t, t_0), \end{aligned} \tag{1.21.13}$$

where the last step follows from (1.21.3).

Now suppose we are only interested in making measurements on system A. If M is any arbitrary function of operators of system A only in the SP, we have that

$$\begin{aligned} \langle\langle M\rangle\rangle &= \mathrm{Tr}_{AB}\, M^S\rho_{AB}{}^S(t) \\ &= \sum_{A',B'} \langle A', B'|M^S\rho_{AB}{}^S(t)|A', B'\rangle \\ &= \sum_{A'} \langle A'|\Big\{M^S \sum_{B'} \langle B'|\rho_{AB}{}^S(t)|B'\rangle\Big\}|A'\rangle. \end{aligned} \tag{1.21.14}$$

The last step is allowed because M is only a function of A-system operators. We may therefore write this

$$\langle\langle M\rangle\rangle = \mathrm{Tr}_A \{M^S[\mathrm{Tr}_B\, \rho_{AB}{}^S(t)]\}. \tag{1.21.15}$$

We therefore see that if we are interested in averages for system A only, we do not need the full density operator $\rho_{AB}{}^S(t)$ but only the simpler *reduced* density operator defined by

$$\rho_A{}^S(t) = \mathrm{Tr}_B\, \rho_{AB}{}^S(t). \tag{1.21.16}$$

If we use (1.21.10) and (1.21.13), we have

$$\begin{aligned}\rho_A{}^S(t) &= \mathrm{Tr}_B \{U_0{}^A U_0{}^B \rho_{AB}{}^I(t) U_0{}^{\dagger A} U_0{}^{\dagger B} \\ &= \sum_{B'} \langle B'|U_0{}^A U_0{}^B \rho_{AB}{}^I(t) U_0{}^{\dagger A} U_0{}^{\dagger B}|B'\rangle.\end{aligned} \tag{1.21.17}$$

But

$$\begin{aligned}U_0{}^{\dagger B}|B'\rangle &= \exp \frac{i}{\hbar} E_{B'}(t - t_0)|B'\rangle \\ \langle B'|U_0{}^B &= \exp -\frac{i}{\hbar} E_{B'}(t - t_0)\langle B'|.\end{aligned} \tag{1.21.18}$$

Therefore, (1.21.17) reduces to

$$\begin{aligned}\rho_A{}^S(t) &= U_0{}^A \sum_{B'} \langle B'|\rho_{AB}{}^I(t)|B'\rangle U_0{}^{\dagger A} \\ &= U_0{}^A\, \mathrm{Tr}_B\, \rho_{AB}{}^I(t) U_0{}^{\dagger A} \\ &\equiv U_0{}^A \rho_A{}^I(t) U_0{}^{\dagger A},\end{aligned} \tag{1.21.19}$$

where $\rho_A{}^I(t)$ is the reduced density operator in the IP.

If we now iterate (1.21.11) and trace over the B-system we have to second order in V

$$\begin{aligned}\rho_A{}^I(t) = \rho_A(t_0) &+ \frac{1}{i\hbar}\int_{t_0}^{t} dt_1\, \mathrm{Tr}_B\, [V_I(t_1 - t_0), \rho_A(t_0)\rho_B(t_0)] \\ &+ \left(\frac{1}{i\hbar}\right)^2 \int_{t_0}^{t} dt_1 \int_{t_0}^{t_1} dt_2\, \mathrm{Tr}_B\, [V_I(t_1 - t_0), [V_I(t_2 - t_0), \rho_A(t_0)\rho_B(t_0)]],\end{aligned} \tag{1.21.20}$$

where we used (1.21.5). The traces over B cannot be carried out until we specify the particular problem under discussion.

The probability of finding system A in the energy eigenstate $|A^f\rangle$ at time

t, when at $t = t_0$ it is described by $\rho_A(t_0)$, is

$$\langle A^f|\rho_A{}^I(t)|A^f\rangle$$
$$= \langle A^f|\rho_A(t_0)|A^f\rangle + \frac{1}{i\hbar}\int_{t_0}^{t} dt_1\langle A^f|\{\mathrm{Tr}_B\,[V_I(t_1), \rho_A(t_0)\rho_B(t_0)]\}|A^f\rangle$$
$$+ \left(\frac{1}{i\hbar}\right)^2\int_{t_0}^{t} dt_1\int_{t_0}^{t_1} dt_2\langle A^f|\{\mathrm{Tr}_B\,[V_I(t_1), [V_I(t_2), \rho_A(t_0)\rho_B(t_0)]]\}|A^f\rangle + \cdots, \tag{1.21.21}$$

to second order in V_I. If at t_0, system A is in state $|A^i\rangle$, then

$$\rho_A(t_0) = |A^i\rangle\langle A^i| \tag{1.21.22}$$

and (1.21.21) reduces to

$$\langle A^f|\rho_A{}^I(t)|A^f\rangle \equiv |c_A(f, t/i, t_0)|^2$$
$$= \frac{1}{\hbar^2}\int_{t_0}^{t} dt_1\int_{t_0}^{t_1} dt_2\,\mathrm{Tr}_B\,\{\langle A^f|V_I(t_1)|A^i\rangle\rho_B(t_0)\langle A^i|V_I(t_2)|A^f\rangle$$
$$+ \langle A^f|V_I(t_2)|A^i\rangle\rho_B(t_0)\langle A^i|V_I(t_1)|A^f\rangle\} + \cdots, \tag{1.21.23}$$

since $\langle A^f|A^i\rangle = 0$. This is the probability of finding system A in state $|A^f\rangle$ at time t when it was in state $|A^i\rangle$ at t_0.

In the second integral, we may interchange the order of integration as in the previous section and obtain

$$\frac{1}{\hbar^2}\int_{t_0}^{t} dt_2\int_{t_2}^{t} dt_1\,\mathrm{Tr}_B\,\langle A^f|V_I(t_2)|A^i\rangle\rho_B(t_0)\langle A^i|V_I(t_1)|A^f\rangle.$$

If we then interchange the dummy integration variables $t_1 \to t_2$, we may combine the two integrals in (1.21.23) to obtain

$$|c_A(f, t/i, t_0)|^2 = \frac{1}{\hbar^2}\int_{t_0}^{t} dt_1\int_{t_0}^{t} dt_2\,\mathrm{Tr}_B\,\langle A^f|V_I(t_1)|A^i\rangle\rho_B(t_0)\langle A^i|V_I(t_2)|A^f\rangle + \cdots. \tag{1.21.24}$$

If we next use (1.21.12), (1.21.13), and (1.21.18), this reduces to

$$|c_A^{(1)}(f, t/i, t_0)|^2 = \frac{1}{\hbar^2}\,\mathrm{Tr}_B\left\{\int_{t_0}^{t} dt_1\, e^{i\omega_{fi}{}^A t_1}U_0{}^{B^\dagger}(t_1)\langle A^f|V_{AB}{}^S|A^i\rangle U_0{}^B(t_1)\right.$$
$$\left.\times\ \rho_B(t_0) \times \int_{t_0}^{t} dt_2\, e^{-i\omega_{fi}{}^A t_2}U_0{}^{B^\dagger}(t_2)\langle A^i|V_{AB}{}^S|A^f\rangle U_0{}^B(t_2)\right\}, \tag{1.21.25}$$

where we have let

$$\hbar\omega_{fi}{}^A = E_{A^f} - E_{A^i} \tag{1.21.26}$$

If we use the cyclic property of traces and the fact that $V_{AB}{}^S$ is hermitian, we may rewrite (1.21.25) as

$$|c_A^{(1)}(f, t/i, t_0)|^2 = \frac{1}{\hbar^2} \operatorname{Tr}_B II^\dagger \rho_B(t_0), \tag{1.21.27a}$$

where

$$I = \int_{t_0}^{t} dt_2 \, e^{-i\omega_{fi}{}^A t_2} U_0{}^{B^\dagger}(t_2)\langle A^i | V_{AB}{}^S | A^f\rangle U_0{}^B(t_2) \tag{1.21.27b}$$

It should be emphasized that this result assumes that we have measured the energy of system A at t_0 and found it to be E_{A^i} and therefore it is in state $|A^i\rangle$. System B is described by $\rho_B(t_0)$. If enough measurements are made, then its state would be known completely also at t_0. However, we discuss in Chapter 3, Part IV, how we estimate $\rho_B(t_0)$ when an incomplete set of measurements are made. At any rate, (1.21.27) then gives the probability of finding system A in state $|A^f\rangle$ at time t if we again measure its energy, but *no* measurements are made on system B at time t.

PROBLEMS

1.1 Derive the commutation relations (1.10.4) by mathematical induction.

1.2 Evaluate the following commutators:

(*a*) $[q, \sin(p^2q)]$; (*b*) $[q, \sin(pqp)]$; (*c*) $[q, \sin(qp^2)]$, where p and q satisfy $[q, p] = i\hbar$.

1.3 Solve the eigenvalue problem $p|p'\rangle = p'|p'\rangle$ in the coordinate representation. *Hint:* Use (1.10.46).

1.4 Given any two kets $|\varphi\rangle$ and $|\chi\rangle$, show that they satisfy the Schwarz inequality

$$|\langle\varphi|\chi\rangle|^2 \leq \langle\varphi|\varphi\rangle\langle\chi|\chi\rangle.$$

Show that the equality holds if and only if $|\varphi\rangle = c|\chi\rangle$ where c is a constant.

1.5 Show that any function of the hermitian operator

$$H(t) = p \sin \omega t + q \cos \omega t,$$

where $[q, p] = i\hbar$ and q and p are in the Schrödinger picture, commutes with $H(t)$ but that $\left[\int_0^t H(t')\,dt',\ H(t)\right] \neq 0$. Can you solve the Schrödinger equation $i\hbar[\partial|\psi(t)\rangle/\partial t] = (p \sin \omega t + q \cos \omega t)|\psi(t)\rangle$?

1.6 If $H(t) = (p + q)f(t)$, where $f(t)$ is any continuous function of time and q and p are hermitian and satisfy $[q, p] = i\hbar$, show that $\left[\int_0^t H(t')\,dt', H(t)\right] = 0$. Contrast this result with Problem 1.5. Solve the Schrödinger equation $(p + q)f(t)|\psi(t)\rangle = i\hbar[\partial|\psi(t)\rangle/\partial t]$.

1.7 Let A_S, B_S, C_S, and D_S be observables in the Schrödinger picture that satisfy the equation

$$A_S = B_S C_S + D_S.$$

Transform this equation to the Heisenberg picture.

1.8 Prove that the following properties hold for Poisson brackets as well as commutator brackets.

$$\begin{aligned}
&(a)\ \{u, v\} = -\{v, u\}\\
&(b)\ \{u, c\} = 0\\
&(c)\ \{u_1 + u_2, v\} = \{u_1, v\} + \{u_2, v\}\\
&(d)\ \{u, v_1 + v_2\} = \{u, v_1\} + \{u, v_2\}\\
&(e)\ \{u_1 u_2, v\} = \{u_1, v\}u_2 + u_1\{u_2, v\}\\
&(f)\ \{u, v_1 v_2\} = \{u, v_1\}v_2 + v_1\{u, v_2\}\\
&(g)\ \{u, \{v, w\}\} + \{v, \{w, u\}\} + \{w, \{u, v\}\} = 0.
\end{aligned}$$

The u, v, and w are considered functions of p and q.

1.9 Prove that commutation relations in the Heisenberg and interaction pictures have the same form.

1.10 Consider two observables q and N that do not commute. The eigenvalues of q are continuous and may have any value from $-\infty$ to $+\infty$ while the eigenvalues of N are the positive integers and 0. We write $q|q'\rangle = q'|q'\rangle$ and $N|n\rangle = n|n\rangle$. Formally, expand a particular $|q'\rangle$ in terms of the set $\{|n\rangle\}$ and a particular $|n\rangle$ in terms of the set $\{|q'\rangle\}$. Show that the transformation function between these two representations is unitary.

1.11 If $|\varphi\rangle$ and $|\chi\rangle$ are two kets with finite norm, find the trace of the operators $|\varphi\rangle\langle\varphi|$ and $|\chi\rangle\langle\varphi|$. Show that the trace (O), where O is any operator, is independent of the representation used.

1.12 Let A be a hermitian operator that satisfies an equation of the form

$$f(A) = (A - a_1)(A - a_2)(A - a_3) = 0,$$

where a_1, a_2, and a_3 are real numbers and no two are equal. Show that A is an observable. Find the eigenvalues of A and express an arbitrary ket $|\psi\rangle$ as a linear combination of the eigenkets. [Proceed as in (1.5.15).] Express $|\psi\rangle$, A, and the eigenkets as matrices in the A-representation.

1.13 Consider an electron beam whose energy is 100 eV moving in the x direction. Its position in the y direction is measured by allowing it to pass through a slit 0.01 mm wide in the y direction. By virtue of the uncertainty principle, what will be the spread in the beam after it has traveled 1 m from the slit? 100 m?

1.14 If A and B are two noncommuting operators, show that Tr (AB) = Tr (BA), provided that the traces exist.

1.15 If the perturbation in (1.16.7) has a first- and second-order term where $V = H_1 + H_2$ and H_2 is of order H_1^2, show that the probability amplitudes (1.16.34) up to second order are given by

$$c^{(1)}(f, t|i, 0) = \frac{1}{i\hbar}\int_0^t dt_1\, e^{i\omega_{fi}t_1}\langle f|H_1|i\rangle$$

$$c^{(2)}(f, t|i, 0) = \frac{1}{i\hbar}\int_0^t dt_1\, e^{i\omega_{fi}t_1}\langle f|H_2|i\rangle$$

$$-\frac{1}{\hbar^2}\int_0^t dt_1\int_0^{t_1} dt_2 \sum_k \langle f|H_1|k\rangle\langle k|H_1|i\rangle\, e^{i(\omega_{fk}t_1+\omega_{ki}t_2)}.$$

REFERENCES

[1] P. A. M. Dirac, *The Principles of Quantum Mechanics*, 4th ed., Oxford: Clarendon, 1958, Chaps 1–5.

[2] A. Messiah, *Quantum Mechanics*, Vol. I., New York: Interscience, 1961, Chaps. 7 and 8.

[3] E. Merzbacher, *Quantum Mechanics*, 2nd ed., New York: Wiley, 1970.

[4] L. I. Schiff, *Quantum Mechanics*, 3rd ed., New York: McGraw-Hill, 1968, Chaps. 1–3, 6.

[5] D. Bohm, *Quantum Theory*, Englewood Cliffs, N.J.: Prentice-Hall, 1951.

[6] R. H. Dicke and J. P. Wittke, *Introduction to Quantum Mechanics*, Reading, Mass.: Addison-Wesley, 1960.

[7] L. Pauling and E. B. Wilson, *Introduction to Quantum Mechanics*, New York: McGraw-Hill, 1935.

[8] See, for example, M. H. Stone, *Linear Transformations in Hilbert Space*, New York: American Mathematical Society, 1932.

[9] J. von Neumann, *Mathematical Foundations of Quantum Mechanics*, Princeton, N.J.: Princeton University Press, 1955, Chaps. 5 and 6.

[10] J. von Neumann, *Gesellschaft der Wissenschaften zu Göttinger Math. Phys. Nachrichten*, 245–272 (1927). Also P. A. M. Dirac, *Proc. Camb. Phil. Soc.*, **25**, 62 (1929); **26**, 376 (1930); **27**, 240 (1930).

[11] R. C. Tolman, *The Principles of Statistical Mechanics*, Oxford: Clarendon Press, 1938.

[12] U. Fano, *Rev. Mod. Phys.*, **29**, 74 (1957).

[13] D. ter Haar, *Rept. Progr. Phys.*, **24**, 304 (1961).

2

Elementary Quantum Systems

The harmonic oscillator plays a central role in the quantum theory of electromagnetic fields. The quantum features of these fields have become increasingly important with the recent advent of amplifiers at optical frequencies (lasers). The oscillator is also important in the quantum theory of lattice vibrations in solids (phonons) as well as in quantum electrodynamics. Our primary concern is the electromagnetic field, although we use the quantum theory of lattice vibrations for a model of an attenuator. For these reasons and also because the oscillator offers a simple example of a dynamical system with a classical analog to illustrate the general theory given in the previous chapter, we devote the first part of this chapter to its study.

There is another simple dynamical system which has no classical analog, namely, the spin momentum of an electron; we study this system in Part II of this chapter. The electron spin is of fundamental importance in many areas of physics although our primary concern will be its use in a more or less phenomenological model for a laser. The spin also offers an illustration of how a quantum theory of a system with no classical analog is formulated. The rigorous treatment of electron spin is given by the relativistic formulation of quantum mechanics due to Dirac [1, 2]. However, we limit ourselves to the nonrelativistic theory of Pauli.

Part III is devoted to the interaction of an electron with an electromagnetic field. Chapter 5 will present a more detailed discussion of the interaction of radiation with atoms when the electromagnetic field is quantized.

We begin the study of the oscillator in the Heisenberg picture. The Heisenberg equations of motion are identical in form with the classical equations of motion, as noted in Section 1.15, so that the Heisenberg picture shows the formal analogy between the classical and quantized oscillator. It also offers an opportunity to introduce in a very simple way creation and annihilation operators, which are a great convenience in calculations. They are used in Section 2.2 to solve for the energy eigenvalues of the oscillator and the basis vectors for the energy representation and will play a very decisive role in the theory of the quantized radiation field later.

The following section makes an effort to give a physical interpretation of the creation and annihilation operators, although to do this correctly involves the study of symmetry properties of wave functions representing an assembly of oscillators and boson particles. In the interest of simplicity, we must be content with a simplified version and refer the reader to a more advanced text for a rigorous treatment [3, 4]. In the same section we introduce operators that describe fermions. The rigorous interpretation in this case involves the theory of second quantization [3, 4]. However, the operators that describe electron spin can be put into a form analogous to fermion operators, as we show in Part II of the chapter.

In Section 1.10 we solved for the eigenvalues and eigenvectors for the coordinate and momentum of a particle in one dimension; the analysis was independent of the potential $V(q)$. Therefore, the position and momentum eigenvectors for the oscillator are already known. In Section 2.4 of Part I, we give the representatives of the energy basis vectors obtained in Section 2.2 in the coordinate representation. In wave-mechanics language these are called the oscillator energy eigenfunctions. In the final section of Part I, we introduce the coherent states which have proved very useful in the study of lasers.

In Section 1.19 we considered the time development of a minimum uncertainty wave packet which described a free particle. In the following chapter we develop much more powerful mathematical methods for handling wave packets. Minimum uncertainty wave packets for an oscillator, as noted in Chapter 1, will play a very important part in our later considerations of the electromagnetic field.

In Part II we introduce orbital angular momentum operators and solve for their eigenvalues and eigenvectors. With an added constraint, we next study electron spin. The added constraint gives a model which has no classical analog and is justified since it predicts a splitting of spectral lines in agreement with experiment. In the Dirac relativistic formulation of the theory of an electron [1], the spin is predicted and is not put into the theory in an *ad hoc* manner as it is in the present nonrelativistic formulation. In Section 2.8 we introduce the Pauli spin operators and in Section 2.9 we give these operators in the HP when the electron is in a d-c magnetic field.

In Part III we give the nonrelativistic hamiltonian for an electron in an unquantized electromagnetic field.

PART I. THE HARMONIC OSCILLATOR

2.1 THE OSCILLATOR IN THE HEISENBERG PICTURE

Let us consider a classical harmonic oscillator of unit mass in one dimension described by a coordinate q and momentum p. The hamiltonian is

$$H = \tfrac{1}{2}(p^2 + \omega^2 q^2), \tag{2.1.1}$$

where ω^2 is a constant related to the restoring force on the particle. The classical hamiltonian equations of motion for q and p are, by (1.15.12) and (2.1.1),

$$\frac{dq}{dt} = \frac{\partial H}{\partial p} = p \tag{2.1.2a}$$

$$\frac{dp}{dt} = -\frac{\partial H}{\partial q} = -\omega^2 q. \tag{2.1.2b}$$

The usual way to solve such a set of coupled equations is to differentiate both sides of (2.1.2a) with respect to t and use (2.1.2b) to eliminate dp/dt. This procedure gives

$$\frac{d^2q}{dt^2} = -\omega^2 q, \tag{2.1.3}$$

and the solution is

$$q(t) = A \cos \omega t + B \sin \omega t, \tag{2.1.4a}$$

where A and B are constants. To find $p(t)$, we substitute this in (2.1.2a) and find that

$$p(t) = -\omega A \sin \omega t + \omega B \cos \omega t. \tag{2.1.4b}$$

If, at $t = 0$, q has the value $q(0)$ and p is $p(0)$, these terms may be used to express A and B so that (2.1.4) become

$$\begin{aligned} q(t) &= q(0) \cos \omega t + \frac{p(0)}{\omega} \sin \omega t \\ p(t) &= -\omega q(0) \sin \omega t + p(0) \cos \omega t. \end{aligned} \tag{2.1.5}$$

There is a commonly used alternative way [5] to decouple (2.1.2a) and (2.1.2b). If we multiply both sides of (2.1.2a) by $\sqrt{\omega/2}$ and both sides of (2.1.2b) by $\pm i/\sqrt{2\omega}$ and add both equations, we obtain the two decoupled equations

$$\frac{da}{dt} = -i\omega a \tag{2.1.6a}$$

$$\frac{da^*}{dt} = i\omega a^*. \tag{2.1.6b}$$

Here we define

$$\begin{aligned} a &= \frac{1}{\sqrt{2\omega}}(\omega q + ip) \\ a^* &= \frac{1}{\sqrt{2\omega}}(\omega q - ip), \end{aligned} \tag{2.1.7a}$$

where a^* is the complex conjugate of a. We may easily solve these equations for p and q and obtain

$$\begin{aligned} q &= \frac{1}{\sqrt{2\omega}}(a^* + a) \\ p &= i\sqrt{\frac{\omega}{2}}(a^* - a). \end{aligned} \tag{2.1.7b}$$

The solutions of (2.1.6a) and (2.1.6b) are given directly as

$$\begin{aligned} a(t) &= a(0)e^{-i\omega t} \equiv \frac{1}{\sqrt{2\omega}}[\omega q(0) + ip(0)]e^{-i\omega t} \\ a^*(t) &= a^*(0)e^{i\omega t} = \frac{1}{\sqrt{2\omega}}[\omega q(0) - ip(0)]e^{i\omega t}. \end{aligned} \tag{2.1.8}$$

The introduction of a and a^* has made the solution of the simple equations (2.1.2) even simpler.

We may express the hamiltonian (2.1.1) in terms of a and a^* by means of (2.1.7b). After minor algebra, we obtain

$$H = \omega a^* a, \tag{2.1.9}$$

which also looks simpler in these variables than in terms of p and q. In fact, we may formally obtain (2.1.6) directly from the hamiltonian (2.1.9) if we take as the hamiltonian equations of motion

$$\begin{aligned} i\frac{da}{dt} &= \frac{\partial H}{\partial a^*} = \omega a \\ i\frac{da^*}{dt} &= -\frac{\partial H}{\partial a} = -\omega a^*. \end{aligned} \tag{2.1.10}$$

The oscillator has one degree of freedom and one normal mode of oscillation. Pierce calls a (or a^*) the normal mode amplitude [5].

Let us turn now to the quantum treatment of the oscillator in the Heisenberg picture. From the general theory of Chapter 1, we associate hermitian operators with the observables q, p, and H and postulate that q and p satisfy the commutation relation (1.10.3)

$$[q, p] = i\hbar. \tag{2.1.11}$$

The hamiltonian is

$$H = \tfrac{1}{2}(p^2 + \omega^2 q^2) = H^\dagger. \tag{2.1.12}$$

All these operators are in the Schrödinger picture and are independent of time. The Schrödinger equation of motion (1.13.1) is

$$i\hbar \frac{\partial|\psi(t)\rangle}{\partial t} = H|\psi(t)\rangle, \tag{2.1.13}$$

and, by (1.13.2) and (1.13.3), the solution is

$$|\psi_S(t)\rangle = U(t, 0)|\psi_H(0)\rangle = \exp\left(-\frac{iHt}{\hbar}\right)|\psi_H(0)\rangle, \tag{2.1.14}$$

where U is unitary. This is the transformation law (1.15.1) between state vectors in the Schrödinger and Heisenberg pictures. Operators transform between the two pictures by the similarity transformation (1.15.3) so that

$$\begin{aligned} q_H(t) &= U^\dagger(t, 0) q_S U(t, 0) \\ p_H(t) &= U^\dagger(t, 0) p_S U(t, 0). \end{aligned} \tag{2.1.15}$$

For a conservative system, we showed in Section 1.15 that the hamiltonian in the two pictures is the same so that we may write

$$H_H = \tfrac{1}{2}[p_H{}^2(t) + \omega^2 q_H{}^2(t)]. \tag{2.1.16}$$

The Heisenberg equations of motion for $q_H(t)$ and $p_H(t)$ are, by (1.15.11),

$$\frac{dq_H}{dt} = \frac{\partial H_H}{\partial p_H} = p_H \tag{2.1.17a}$$

$$\frac{dp_H}{dt} = -\frac{\partial H_H}{\partial q_H} = -\omega^2 q_H. \tag{2.1.17b}$$

The only difference between these and the classical equations (2.1.2) is the operator character, where q_H and p_H satisfy

$$[q_H(t), p_H(t)] = i\hbar. \tag{2.1.18}$$

The solution of (2.1.17) is given by (2.1.5):

$$q_H(t) = U^\dagger(t, 0)q_S U(t, 0) = q_S \cos \omega t + \frac{p_s}{\omega} \sin \omega t$$
$$p_H(t) = U^\dagger(t, 0)p_S U(t, 0) = -\omega q_S \sin \omega t + p_S \cos \omega t, \qquad (2.1.19)$$

where (2.1.15) was used, and q_S and p_S are the operators in the Schrödinger picture at $t = 0$. They are not initial conditions as in the classical sense of (2.1.5). Since $U(t, 0) = \exp[-i(p_S^2 + \omega^2 q_S^2)t/2\hbar]$, we see by (2.1.19) the effect of commuting U through q_S and p_S. In the next chapter we show other techniques for obtaining the result of $U^\dagger qU$ and $U^\dagger pU$.

By analogy with (2.1.7), we may introduce two convenient nonhermitian operators a and $a^\dagger$ defined by

$$a = \frac{1}{\sqrt{2\hbar\omega}}(\omega q + ip)$$
$$a^\dagger = \frac{1}{\sqrt{2\hbar\omega}}(\omega q - ip) \qquad (2.1.20a)$$

or

$$q = \sqrt{\frac{\hbar}{2\omega}}(a^\dagger + a)$$
$$p = i\sqrt{\frac{\hbar\omega}{2}}(a^\dagger - a). \qquad (2.1.20b)$$

For reasons that will become apparent later, a is called an annihilation operator and $a^\dagger$ a creation operator.

The operators a and $a^\dagger$, like q and p, do not commute. If we substitute (2.1.20b) into (2.1.11) and use the fact that all operators commute with themselves, we find that a and $a^\dagger$ satisfy the commutation relation

$$[a, a^\dagger] = 1. \qquad (2.1.21)$$

If we use this and substitute (2.1.20b) into (2.1.12), the hamiltonian becomes

$$H = \frac{\hbar\omega}{2}(aa^\dagger + a^\dagger a) = \hbar\omega(a^\dagger a + \tfrac{1}{2}). \qquad (2.1.22)$$

This differs from (2.1.9) because a and $a^\dagger$ do not commute as they do classically. The term $\hbar\omega/2$ is called the zero-point energy of the oscillator.

We shall immediately need the commutation relations

$$[a, a^\dagger a] = a$$
$$[a^\dagger, a^\dagger a] = -a^\dagger, \qquad (2.1.23)$$

which may be proved directly from (2.1.21).

The Heisenberg equations (1.15.5) apply to nonhermitian as well as to hermitian operators. Therefore, the equations of motion for $a_H(t)$ and $a_H^\dagger(t)$ become

$$\begin{aligned}\frac{da_H}{dt} &= \frac{1}{i\hbar}[a_H, H_H] = -i\omega a_H \\ \frac{da_H^\dagger}{dt} &= \frac{1}{i\hbar}[a_H^\dagger, H_H] = +i\omega a_H^\dagger,\end{aligned} \tag{2.1.24}$$

where we used (2.1.22) and (2.1.23) which are equally valid in either the Schrödinger or Heisenberg picture. The solutions of (2.1.24) are

$$\begin{aligned}a_H(t) &= U^\dagger(t, 0)a_S U(t, 0) = a_S e^{-i\omega t} \\ a_H^\dagger &= U^\dagger(t, 0)a_S^\dagger U(t, 0) = a_S^\dagger e^{i\omega t},\end{aligned} \tag{2.1.25}$$

where we used (1.15.3). Also

$$U(t, 0) = \exp(-i\omega t a_S^\dagger a_S)\exp\left(-\frac{i\omega t}{2}\right). \tag{2.1.26}$$

Both a_S and $a_S^\dagger$ are in the Schrödinger picture.

In the future we usually designate an operator M in the Heisenberg picture by $M(t)$ and in the Schrödinger picture by M, instead of using H and S subscripts, if there is no likelihood of confusion.

The operator $a^\dagger a$, which is hermitian, occurs so frequently that we shall let

$$N = a^\dagger a = N^\dagger. \tag{2.1.27}$$

It is called the number operator for reasons that will become clear later. In terms of N, we may rewrite (2.1.23) as

$$Na = a(N - 1) \tag{2.1.28a}$$

$$Na^\dagger = a^\dagger(N + 1). \tag{2.1.28b}$$

The hamiltonian is related to N by

$$N = \frac{1}{\hbar\omega}H - \frac{1}{2}. \tag{2.1.29}$$

From these results, the close formal analogy between the classical and quantum treatments of an oscillator can be seen.

2.2 THE ENERGY-EIGENVALUE PROBLEM FOR THE OSCILLATOR

According to the physical interpretation of quantum mechanics, the eigenvalues of the energy are the only values obtainable by an experimental measurement of the energy. To compare theory and experiment, we must therefore solve the eigenvalue equation

$$H|E\rangle = E|E\rangle. \tag{2.2.1}$$

Because of the simple connection between N and H given by (2.1.29), the eigenvalue problem for N,

$$N|n'\rangle = n'|n'\rangle, \tag{2.2.2}$$

is entirely equivalent to (2.2.1). Since H and therefore N are observables, the eigenkets $\{|n'\rangle\}$ form a complete orthonormal set of basis vectors in the N-representation.

To solve (2.2.2), let us review briefly the solution of the eigenvalue problem for the coordinate and momentum given in Section 1.10. We found an operator which, when applied to a known eigenket, would generate another eigenket. In the present case, we show that the operators a and $a^\dagger$ generate new eigenkets from a given eigenket just as the translation operators perform this function for p and q. Both derivations are due to Dirac.

We assume that $|n'\rangle$ is a known eigenket of N with eigenvalue n' which satisfies (2.2.2). If we operate both sides of (2.1.28a) on $|n'\rangle$ and use (2.2.2), we see that

$$N\{a|n'\rangle\} = (n' - 1)\{a|n'\rangle\}.$$

Similarly, by (2.1.28b) we have

$$N\{a^\dagger|n'\rangle\} = (n' + 1)\{a^\dagger|n'\rangle\}.$$

This shows that, if $|n'\rangle$ is an eigenket of N with eigenvalue n', then $a|n'\rangle$ is an eigenket of N with eigenvalue $n' - 1$ and $a^\dagger|n'\rangle$ is an eigenket with eigenvalue $n' + 1$. From $|n'\rangle$ we have generated two more eigenkets.

We may repeat the process and apply both sides of (2.1.28a) to eigenket $a|n'\rangle$, use the result above, and generate eigenket $a^2|n'\rangle$ with eigenvalue $n' - 2$. Similarly, (2.1.28b) applied to $a^\dagger|n'\rangle$ will generate eigenket $a^{\dagger 2}|n'\rangle$ with eigenvalue $n' + 2$. Obviously this process may be continued indefinitely and an infinite set of eigenkets and eigenvalues generated from a known eigenket; they may be listed as

$$\begin{array}{cccc} |n'\rangle & a|n'\rangle & a^2|n'\rangle & \cdots \\ n' & n'-1 & n'-2 & \cdots \end{array} \tag{2.2.3a}$$

$$\begin{array}{cccc} |n'\rangle & a^\dagger|n'\rangle & a^{\dagger 2}|n'\rangle & \cdots \\ n' & n'+1 & n'+2 & \cdots . \end{array} \tag{2.2.3b}$$

Note that, in these series, successive eigenvalues differ by 1.

We next show that n' can only be zero or a positive integer. Since N is hermitian, from the general theory of Chapter 1, n' must be real and by *assumption* the norm of any vector must be greater than or equal to zero. If the norm is zero, the vector is zero. If $|n'\rangle$ is an eigenket, then

$$\langle n'|n'\rangle > 0,$$

since $|n'\rangle = 0$ is trivial. Let us form the scalar product of (2.2.2) with $\langle n'|$. We have

$$\langle n'|N|n'\rangle = \langle n'|a^\dagger a|n'\rangle = n'\langle n'|n'\rangle. \tag{2.2.4}$$

But this is the norm of the vector $a|n'\rangle$, which must be greater than or equal to zero. Since $\langle n'|n'\rangle > 0$ and $\langle n'|a^\dagger a|n'\rangle \geq 0$, we conclude by (2.2.4) that $n' \geq 0$. Therefore, the eigenvalues of N are real and nonnegative. If $n' = 0$, then

$$a|0\rangle = 0, \tag{2.2.5}$$

since the norm is zero by (2.2.4).

If $n' \neq 0$, the norm of $a|n'\rangle$ is given by (2.2.4). However, by (2.2.3a), if n' is not an integer, the sequence of eigenvalues would eventually become negative and the norms of the associated vectors would become negative, which is not allowed. The only way to prevent this is for n' to be a positive integer or zero. Therefore, the eigenvalues of N are the positive integers or zero, as stated.

The norm of the vector $a^\dagger|n\rangle$ is

$$\langle n|aa^\dagger|n\rangle = \langle n|(1 + a^\dagger a)|n\rangle = (1 + n)\langle n|n\rangle, \tag{2.2.6}$$

where we used (2.1.21). Since $n \geq 0$ and $\langle n|n\rangle > 0$, we see that $a^\dagger|n\rangle$ can never be zero.

The eigenkets generated by successive application of a and $a^\dagger$ are not normalized to unity as yet. They may be normalized as follows: since $a|n\rangle$ is an eigenket of N with eigenvalue $n - 1$, $a|n\rangle$ can differ from $|n - 1\rangle$ by a constant. We therefore write

$$a|n\rangle = c_n|n - 1\rangle. \tag{2.2.7}$$

The norm of this vector is, by (2.2.4),

$$\langle n|a^\dagger a|n\rangle = n\langle n|n\rangle = |c_n|^2\langle n - 1|n - 1\rangle.$$

If $\langle n - 1|n - 1\rangle$ is normalized to unity and if we choose $|c_n|$ to be $\sqrt{n}$, then $\langle n|n\rangle = 1$. The phase of c_n is arbitrary and we choose it zero; thus (2.2.7) becomes

$$a|n\rangle = \sqrt{n}\,|n - 1\rangle. \tag{2.2.8}$$

We restrict n to be greater than zero since state $|-1\rangle$ has no meaning. When $n = 0$, (2.2.8) reduces to (2.2.5).

Similarly, we may use (2.2.6) and write

$$a^\dagger|n\rangle = \sqrt{n+1}\,|n+1\rangle. \tag{2.2.9}$$

If $\langle 0|0\rangle$ is normalized, then all others will be normalized.

We may collect these important results:

$$\begin{aligned} N|n\rangle &= n|n\rangle \\ a|0\rangle &= 0 \\ a|n\rangle &= \sqrt{n}\,|n-1\rangle \\ a^\dagger|n\rangle &= \sqrt{n+1}\,|n+1\rangle. \end{aligned} \tag{2.2.10}$$

An extremely useful result is obtained if we use (2.2.9) and apply the operator $a^\dagger$ to the state $|0\rangle$ n times. From this we generate the state $|n\rangle$ given by

$$|n\rangle = \frac{a^{\dagger n}}{\sqrt{n!}}\,|0\rangle. \tag{2.2.11}$$

This formula will be used repeatedly.

From the general theory, the orthonormality relations are

$$\langle n'|n''\rangle = \delta_{n'n''}, \tag{2.2.12}$$

and the completeness relation is

$$\sum_{n=0}^{\infty} |n\rangle\langle n| = I. \tag{2.2.13}$$

Since the norm of these eigenvectors is finite, they form a complete set of basis vector for a Hilbert space.

From (2.2.10), their adjoints and the completeness relation it follows that

$$\begin{aligned} a^\dagger a &= \sum_0^\infty n|n\rangle\langle n| \\ a &= \sum_0^\infty \sqrt{n}|n-1\rangle\langle n| = \sum_0^\infty \sqrt{n+1}|n\rangle\langle n+1| \\ a^\dagger &= \sum_0^\infty \sqrt{n+1}|n+1\rangle\langle n| = \sum_0^\infty \sqrt{n}|n\rangle\langle n-1| \\ f(a^\dagger a) &= \sum_0^\infty f(n)|n\rangle\langle n|, \end{aligned} \tag{2.2.14}$$

where f is any function of $a^\dagger a$.

We may use (2.2.10) and (2.2.12) to obtain the matrix elements of a, $a^\dagger$, and

N in the N-representation. They are

$$\begin{aligned} \langle n'|a|0\rangle &= 0 \\ \langle n'|a|n''\rangle &= \sqrt{n''}\delta_{n',n''-1} \\ \langle n'|a^\dagger|n''\rangle &= \sqrt{n''+1}\delta_{n',n''+1} \\ \langle n'|N|n''\rangle &= n''\delta_{n'n''}. \end{aligned} \tag{2.2.15}$$

The energy eigenvalues are

$$E_n = \hbar\omega(n + \tfrac{1}{2}), \tag{2.2.16}$$

where $n = 0, 1, 2, \ldots, \infty$. Classically, any positive value of energy may be obtained when the energy is measured, but quantum-mechanically only discrete values may be obtained. In the limit of large n (n is called a quantum number), the discrete character of (2.2.16) is unnoticeable and the quantum result becomes the classical result. Since $\hbar \simeq 10^{-34}$ J-sec, $\hbar\omega$ is small up to optical frequencies where the quantum features become important.

2.3 PHYSICAL INTERPRETATION OF N, a, AND $a^\dagger$; BOSONS AND FERMIONS

The operators q, p, and H for an oscillator have the physical significance of position, momentum, and energy, respectively. The operators a, $a^\dagger$, and $a^\dagger a$ were defined in terms of q, p, and H, and we would like to give them a physical interpretation.

Figure 2.1 shows an energy-level diagram for a quantized oscillator. Along the vertical axis the energy as given by (2.2.16) is plotted. The horizontal axis has no significance. Beside each level is the corresponding eigenstate of the operator N. If the oscillator is in the energy eigenstate corresponding to $|n\rangle$, this is indicated by a dot on the line of energy $(n + \frac{1}{2})\hbar\omega$. The state of the oscillator in this case is indicated by the value of n and the separation $\hbar\omega$.

There is no inconsistency in another interpretation for the oscillator in state $|n\rangle$. We may assume that the hamiltonian describes a system of n identical noninteracting quanta, each of which is in the *same* dynamical

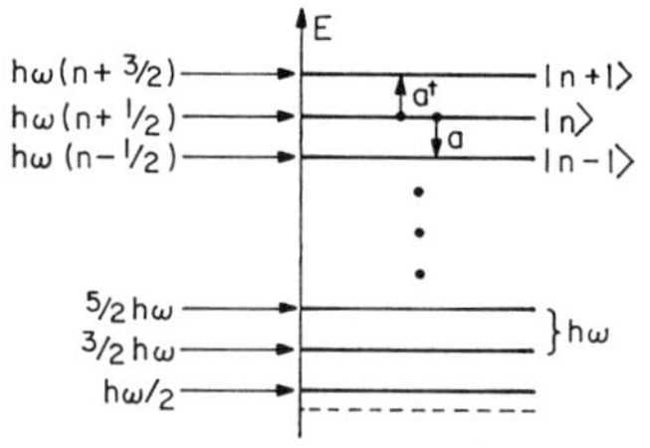

Figure 2.1 Energy-level diagram for quantized harmonic oscillator.

state with energy $\hbar\omega$. We may therefore interpret the state $|n\rangle$ as a state with n quanta while the state $|0\rangle$ has no quanta and is called the vacuum state.

The rigorous justification for this interpretation is given by Dirac [1], and we do not discuss it here. It is one of the most fundamental consequences of quantum theory since it allows a unification of the particle and wave properties of light. The quanta in that case are called photons, and we devote a large part of this book to a study of the quantum properties of light.

The operator N is called the number operator because a measurement of N yields one of the eigenvalues $0, 1, \ldots, \infty$ which is interpreted as the number of quanta in the state. According to this interpretation, any number of quanta may occupy the same dynamical state. Particles in nature having this property are called *bosons*. Light quanta (photons), elastic vibration quanta in crystal (phonons), and α particles, among others, are bosons.

It is now easy to see why $a^\dagger$ and a may be interpreted as creation and annihilation operators, respectively. From (2.2.9), if the oscillator is started in state $|n\rangle$ with n quanta and we operate with $a^\dagger$, we generate state $|n + 1\rangle$, with $n + 1$ quanta. This is shown in Figure 2.1. Therefore, $a^\dagger$ is a creation or a raising operator. Similarly, by (2.2.8), if a operates on state $|n\rangle$, it gives state $|n - 1\rangle$ with $n - 1$ quanta, and a is an annihilation or lowering operator.

The eigenvalues of N are a direct consequence of the commutation relation $[a, a^\dagger] = 1$. Therefore, particles that obey this commutation relation are bosons. There is another class of physical particles in nature, called *fermions*, which have the property that no two of them can occupy the same dynamical state. Electrons, protons, and neutrons are examples of fermions. The Pauli exclusion principle states that no two identical particles can occupy the same state; thus fermions obey the exclusion principle. The fundamental quantum postulate $[a, a^\dagger] = 1$ cannot therefore be a suitable postulate for fermions. Although the complete theory of fermions involves the theory of second quantization [1, 2], we may give an oversimplified version in which we formally indicate a different quantization procedure due to Jordan and Wigner [6]; this version avoids the difficulty of having a state multiply occupied.

In the theory of second quantization for a system with only one dynamical state, there are still two operators b and $b^\dagger$ which are interpreted as a fermion annihilation and creation operator, respectively. These have no connection with the momentum or coordinate of the fermion, as their boson brothers do. There is also an operator $N = b^\dagger b = N^\dagger$ which is interpreted as a number operator defined in terms of the hamiltonian by

$$H = Eb^\dagger b, \tag{2.3.1}$$

where E is the energy eigenvalue of the one dynamical state of the system.

Jordan and Wigner then postulate that b and $b^\dagger$ obey the *anticommutation* relations

$$\begin{aligned} \{b, b\}_+ &= 0 \qquad \{b^\dagger, b^\dagger\}_+ = 0 \\ \{b, b^\dagger\}_+ &= 1 \end{aligned} \tag{2.3.2}$$

rather than commutation relations, where the anticommutator of A and B is defined by

$$\{A, B\}_+ \equiv AB + BA. \tag{2.3.3}$$

From (2.3.2),

$$b^2 = b^{\dagger 2} = 0. \tag{2.3.4}$$

We next derive the eigenvalues of N. If we use (2.3.2) and (2.3.4), we see that

$$N^2 = b^\dagger b b^\dagger b = b^\dagger(1 - b^\dagger b)b = b^\dagger b = N. \tag{2.3.5}$$

This simple algebraic equation that N satisfies is sufficient to find the eigenvalues of N. For if

$$N|n\rangle = n|n\rangle, \tag{2.3.6}$$

then, by (2.3.5) and (2.3.6),

$$n^2|n\rangle = N^2|n\rangle = N|n\rangle = n|n\rangle. \tag{2.3.7}$$

Therefore, the eigenvalues of N are

$$n^2 = n \qquad \text{or} \qquad n = 1, 0. \tag{2.3.8}$$

Since we assume that there is no degeneracy, the Hilbert space will consist of only two vectors, designated as $|0\rangle$ and $|1\rangle$.

From the general theory, the eigenkets of N satisfy the orthonormality and completeness relations

$$\begin{aligned} \langle 1|1\rangle = \langle 0|0\rangle = 1 \qquad \langle 1|0\rangle = \langle 0|1\rangle = 0 \\ |1\rangle\langle 1| + |0\rangle\langle 0| = I, \end{aligned} \tag{2.3.9}$$

where

$$\begin{aligned} N|1\rangle &= |1\rangle \\ N|0\rangle &= 0. \end{aligned} \tag{2.3.10}$$

The matrix elements of N are, by (2.3.9) and (2.3.10),

$$N = \begin{bmatrix} \langle 1|N|1\rangle & \langle 1|N|0\rangle \\ \langle 0|N|1\rangle & \langle 0|N|0\rangle \end{bmatrix} = \begin{bmatrix} 1 & 0 \\ 0 & 0 \end{bmatrix}. \tag{2.3.11}$$

It also follows from (2.3.9) and (2.3.10) that

$$N = |1\rangle\langle 1|. \tag{2.3.12}$$

This is the analog of (2.2.14).

Since the eigenvalues of N are 1 and 0, the state can be occupied by only one particle or else be empty, in agreement with the exclusion principle. Therefore, the quantum postulate of Jordan and Wigner is sufficient to ensure that a dynamical state can be occupied by only one particle at a time.

Next we must show that b and $b^\dagger$ act as creation and annihilation operators. If we start with state $|0\rangle$, with no particle present, $b^\dagger|0\rangle$ gives a state with one particle present and is therefore a creation operator. Since by (2.3.4) $b^{\dagger 2} = 0$, we are prevented from generating a state $b^{\dagger 2}|0\rangle$ with two particles. Also, $b|1\rangle$ gives a state $|0\rangle$ with no particles present and therefore acts as an annihilation operator. Also, $b^2|1\rangle = 0$.

To show these results, we note the following *commutation* relations for b and $b^\dagger$ with $N = b^\dagger b$:

$$\begin{aligned} [b, N] &= b \\ [b^\dagger, N] &= -b^\dagger. \end{aligned} \tag{2.3.13}$$

These may be proved directly from (2.3.2) and (2.3.4). They are identical in form with the boson creation and annihilation operators in (2.1.28). We proceed exactly as in that case to show that

$$b|n\rangle = c_n|1 - n\rangle \tag{2.3.14a}$$

generates an eigenket of N with eigenvalue $1 - n$. In contrast to the boson case, however, the series stops because $b^2|n\rangle = 0$, by (2.3.4). The norm is

$$\langle n|b^\dagger b|n\rangle = n\langle n|n\rangle = |c_n|^2\langle 1 - n|1 - n\rangle.$$

We see that if $n = 0$,

$$b|0\rangle = 0, \tag{2.3.14b}$$

while if $n = 1$, since $\langle 0|0\rangle = 1 = \langle 1|1\rangle$,

$$|c_1|^2 = 1.$$

Therefore, (2.3.14a) for $n = 1$ is

$$b|1\rangle = |0\rangle, \tag{2.3.14c}$$

where we choose the phase of c_1 to be real. Similarly, we may show that

$$\begin{aligned} b^\dagger|0\rangle &= |1\rangle \\ b^\dagger|1\rangle &= 0. \end{aligned} \tag{2.3.15}$$

From (2.3.9), (2.3.14b), (2.3.14c), and (2.3.15), the matrix elements of b and

$b^\dagger$ in the N-representation are

$$b = \begin{bmatrix} \langle 1|b|1\rangle & \langle 1|b|0\rangle \\ \langle 0|b|1\rangle & \langle 0|b|0\rangle \end{bmatrix} = \begin{bmatrix} 0 & 0 \\ 1 & 0 \end{bmatrix} \tag{2.3.16a}$$

$$b^\dagger = \begin{bmatrix} \langle 1|b^\dagger|1\rangle & \langle 1|b^\dagger|0\rangle \\ \langle 0|b^\dagger|1\rangle & \langle 0|b^\dagger|0\rangle \end{bmatrix} = \begin{bmatrix} 0 & 1 \\ 0 & 0 \end{bmatrix}. \tag{2.3.16b}$$

The state vectors may be represented by the matrices

$$|0\rangle = \begin{bmatrix} \langle 1|0\rangle \\ \langle 0|0\rangle \end{bmatrix} = \begin{bmatrix} 0 \\ 1 \end{bmatrix} \qquad |1\rangle = \begin{bmatrix} \langle 1|1\rangle \\ \langle 0|1\rangle \end{bmatrix} = \begin{bmatrix} 1 \\ 0 \end{bmatrix}. \tag{2.3.17}$$

We show in Part II how spin operators can be put in a one-to-one correspondence with fermion annihilation and creation operators.

2.4 TRANSFORMATION FUNCTION FROM N TO q REPRESENTATION FOR OSCILLATOR

The energy-eigenvalue problem for an oscillator was solved in Section 2.2 in the N-representation. It is sometimes useful to work in the coordinate representation; this requires the transformation function $\langle q'|n\rangle$, which is the energy wave function for the oscillator. These are the representatives of the energy eigenkets $|n\rangle$ in the coordinate representation.

There are two ways of finding the transformation function

$$u_n(q') \equiv \langle q'|n\rangle. \tag{2.4.1}$$

First, the energy-eigenvalue problem may be written as

$$\tfrac{1}{2}(p^2 + \omega^2 q^2)|n\rangle = \hbar\omega(n + \tfrac{1}{2})|n\rangle. \tag{2.4.2}$$

If we take the scalar product with $\langle q'|$, an eigenbra of q, and use (1.10.46) and (1.10.47), then (2.4.2) reduces to

$$\left(-\frac{\hbar^2}{2}\frac{d^2}{dq'^2} + \frac{\omega^2}{2}q'^2\right)u_n(q') = \hbar\omega(n + \tfrac{1}{2})u_n(q'), \tag{2.4.3}$$

where we used (2.4.1). The transformation function is therefore the solution of (2.4.3), the Schrödinger equation, which is square integrable; that is, the solutions must be chosen so that $\int_{-\infty}^{\infty} |u_n(q')|^2\,dq'$ exists. The solution of (2.4.3) may be found in Schiff [7].

Another method of obtaining $u_n(q')$ is to start with (2.2.5)

$$a|0\rangle = 0,$$

and replace a in terms of p and q by (2.1.20a). This becomes

$$(\omega q + ip)|0\rangle = 0.$$

If we take the scalar product of both sides with $\langle q'|$ and use (1.10.46) and (1.10.47), we have

$$\left(\omega q' + \hbar \frac{d}{dq'}\right) u_0(q') = 0,$$

where we used (2.4.1). The solution of this equation normalized so that

$$\int_{-\infty}^{\infty} |u_0(q')|^2 \, dq' = 1 \tag{2.4.4}$$

is

$$u_0(q') \equiv \langle q'|0\rangle = \left(\frac{\omega}{\pi\hbar}\right)^{1/4} \exp\left(\frac{-\omega q'^2}{2\hbar}\right). \tag{2.4.5}$$

This is the coordinate representation of the vacuum state and is also called the oscillator ground-state wave function.

Next, by (2.2.9) and (2.1.20a), we see that

$$\langle q'|a^\dagger|0\rangle = \frac{1}{\sqrt{2\hbar\omega}} \langle q'|\omega q - ip|0\rangle = \langle q'|1\rangle.$$

If we use (1.10.46), (1.10.47), and (2.4.5), we have

$$u_1(q') \equiv \langle q'|1\rangle = \frac{1}{\sqrt{2\hbar\omega}} \left(\omega q' - \hbar \frac{d}{dq'}\right) u_0(q')$$

$$= \left[\left(\frac{2\omega}{\hbar}\right)^2 \frac{\omega}{\pi\hbar}\right]^{1/4} q' \exp\left(-\frac{\omega}{2\hbar} q'^2\right). \tag{2.4.6}$$

It is easy to verify that $\int_{-\infty}^{\infty} |u_1|^2 \, dq' = 1$.

In this manner we may successively generate $u_2, u_3, \ldots$. The result is summarized by

$$\langle q'|n\rangle = u_n(q') = \left(\frac{\delta}{\pi^{1/2} 2^n n!}\right)^{1/2} H_n(\delta q') e^{-\frac{1}{2}\delta^2 q'^2}, \tag{2.4.7}$$

where $\delta = \sqrt{\omega/\hbar}$, and $H_n(x)$ is the Hermite polynomial of order n. These are the eigenkets $|n\rangle$ in the coordinate representation. After we have developed more techniques in the manipulation of operators in the following chapter, we shall derive a generating function that will give $u_n(q')$ more neatly.

2.5 THE COHERENT STATES[8]

So far we have used the number representation $|n\rangle$ which satisfies the eigenvalue equation

$$a^\dagger a|n\rangle = n|n\rangle \qquad n = 0, 1, 2, \ldots . \tag{2.5.1}$$

These form a complete orthogonal set of basis vectors to describe a harmonic oscillator. We now introduce the coherent state which is extremely useful in dealing with radiation problems. Actually, we show that a coherent state is just a minimum uncertainty wave packet state.

We define the eigenvector of the nonhermitian operator a to be a coherent state. To find it we must solve the eigenvalue problem

$$a|\alpha\rangle = \alpha|\alpha\rangle. \tag{2.5.2}$$

Since a is nonhermitian, we cannot use our prior theorems which say that eigenvalues are real and that eigenvectors are orthogonal and complete.

To solve (2.5.2), we use the completeness relation for the N-representation to expand $|\alpha\rangle$ as

$$|\alpha\rangle = \sum_{n=0}^{\infty} |n\rangle\langle n|\alpha\rangle \equiv \sum_{n=0}^{\infty} c_n(\alpha)|n\rangle, \tag{2.5.3}$$

where $c_n(\alpha) = \langle n|\alpha\rangle$ is the transformation between the number and coherent state representations. The $|\langle n|\alpha\rangle|^2$ gives the probability of finding the oscillator with energy $n\hbar\omega$ if a measurement is made when the oscillator is in state $|\alpha\rangle$. If we substitute (2.5.3) in (2.5.2) and use (2.2.10), we have

$$a|\alpha\rangle = \sum_{n=1}^{\infty} c_n(\alpha)\sqrt{n}|n-1\rangle = \sum_{n=0}^{\infty} \alpha c_n(\alpha)|n\rangle. \tag{2.5.4}$$

The first sum runs from 1 to ∞ since the $n = 0$ term gives zero. We may therefore shift indices and let $n \to n + 1$ so this becomes

$$\sum_{n=0}^{\infty} c_{n+1}(\alpha)\sqrt{n+1}|n\rangle = \sum_{n=0}^{\infty} \alpha c_n(\alpha)|n\rangle. \tag{2.5.5}$$

If we multiply both sides from the left by $\langle m|$, since $\langle m|n\rangle = \delta_{nm}$, we obtain the simple difference equation

$$c_{n+1}(\alpha)\sqrt{n+1} = \alpha c_n(\alpha) \tag{2.5.6}$$

or

$$\begin{aligned} c_1 &= \frac{\alpha}{\sqrt{1}} c_0 \\ c_2 &= \frac{\alpha}{\sqrt{2}} c_1 = \frac{\alpha^2}{\sqrt{2!}} c_0 \\ c_3 &= \frac{\alpha^3}{\sqrt{3!}} c_0, \end{aligned} \tag{2.5.7}$$

so that

$$c_n(\alpha) = \frac{\alpha^n}{\sqrt{n!}} c_0. \tag{2.5.8}$$

Therefore, we see that

$$|\alpha\rangle = c_0 \sum_{n=0}^{\infty} \frac{\alpha^n}{\sqrt{n!}} |n\rangle. \tag{2.5.9}$$

We choose c_0 so that

$$\begin{aligned} \langle\alpha|\alpha\rangle = 1 &= |c_0|^2 \sum_{n=0}^{\infty} \sum_{m=0}^{\infty} \frac{\alpha^{*m}\alpha^n}{\sqrt{n!\,m!}} \langle m|n\rangle \\ &= |c_0|^2 \sum_{n=0}^{\infty} \frac{(|\alpha|^2)^n}{n!} \\ &= |c_0|^2 \exp|\alpha|^2. \end{aligned} \tag{2.5.10}$$

Thus

$$\langle n|\alpha\rangle = c_n(\alpha) = e^{-\frac{1}{2}|\alpha|^2} \frac{\alpha^n}{\sqrt{n!}}, \tag{2.5.11}$$

and the coherent state is given by

$$|\alpha\rangle = e^{-\frac{1}{2}|\alpha|^2} \sum_{n=0}^{\infty} \frac{\alpha^n}{\sqrt{n!}} |n\rangle. \tag{2.5.12a}$$

If we use (2.2.11), we have

$$\begin{aligned} |\alpha\rangle &= e^{-\frac{1}{2}|\alpha|^2} \sum_{n=0}^{\infty} \frac{(\alpha a^\dagger)^n}{n!} |0\rangle \\ &= e^{-\frac{1}{2}|\alpha|^2} e^{\alpha a^\dagger} |0\rangle, \end{aligned} \tag{2.5.12b}$$

where we could carry out the sum on n since the "vacuum" or ground state $|0\rangle$ is independent of n. We shall find both forms extremely useful.

We next note that

$$|\langle n|\alpha\rangle|^2 = \frac{e^{-\frac{1}{2}|\alpha|^2}(|\alpha|^2)^n}{n!} \tag{2.5.13}$$

represents a Poisson distribution over the photon number states.

From the normalization, we easily see from (2.5.12b) that

$$\langle 0|e^{\alpha^* a}e^{\alpha a^\dagger}|0\rangle = e^{|\alpha|^2}. \tag{2.5.14}$$

We have tacitly been assuming the eigenvalues α to be complex since a is nonhermitian.

Next we demonstrate that the coherent states are *not* orthogonal. Nevertheless, they are extremely useful as we shall see. We have by (2.5.12a) and its adjoint that

$$\begin{aligned}\langle\beta|\alpha\rangle &= e^{-\frac{1}{2}(|\alpha|^2+|\beta|^2)}\sum_{n=0}^{\infty}\sum_{m=0}^{\infty}\frac{\beta^{*n}}{\sqrt{n!}}\frac{\alpha^m}{\sqrt{m!}}\langle n|m\rangle \\ &= e^{-\frac{1}{2}(|\alpha|^2+|\beta|^2)}\sum_{n=0}^{\infty}\frac{(\alpha\beta^*)^n}{n!} \\ &= e^{-\frac{1}{2}(|\alpha|^2+|\beta|^2)+\alpha\beta^*}.\end{aligned} \tag{2.5.15}$$

If the states were orthogonal this would be zero for $\alpha \neq \beta$. From this it follows that

$$|\langle\beta|\alpha\rangle|^2 = e^{-|\alpha-\beta|^2}, \tag{2.5.16}$$

so that they become approximately orthogonal as $|\alpha - \beta|^2$ increases.

The coherent states do form a complete set of states. Otherwise they would not be very useful. In fact they form what mathematicians call an overcomplete set [8].

Since as we shall show α is complex, the completeness relation is written as

$$\int |\alpha\rangle\langle\alpha|\frac{d^2\alpha}{\pi} = 1, \tag{2.5.17}$$

where 1 is the identity operator. The integration is over the entire complex plane. If we let $\alpha = x + iy = re^{i\theta}$, then $d^2\alpha = dx\,dy = r\,dr\,d\theta$.

To verify (2.5.17) we use (2.5.12a) and its adjoint in the left side of (2.5.17). This gives

$$\int |\alpha\rangle\langle\alpha|\frac{d^2\alpha}{\pi} = \sum_{n=0}^{\infty}\sum_{m=0}^{\infty}\frac{|n\rangle\langle m|}{\pi\sqrt{n!\,m!}}\int e^{-|\alpha|^2}\alpha^{*m}\alpha^n d^2\alpha.$$

If we change to polar coordinates, this becomes

$$\int |\alpha\rangle\langle\alpha|\frac{d^2\alpha}{\pi} = \sum_{n,m=0}^{\infty}\frac{|n\rangle\langle m|}{\pi\sqrt{n!\,m!}}\int_0^{\infty} r\,dr\,e^{-r^2}r^{n+m}\int_0^{2\pi}d\theta\,e^{i(n-m)\theta}. \tag{2.5.18}$$

Since

$$\int_0^{2\pi}d\theta\,e^{i(n-m)\theta} = 2\pi\delta_{nm}, \tag{2.5.19}$$

we have

$$\int |\alpha\rangle\langle\alpha| \frac{d^2\alpha}{\pi} = \sum_{n=0}^{\infty} \frac{|n\rangle\langle n|}{n!} \int_0^{\infty} d\xi \, e^{-\xi} \, \xi^n,$$

where we let $\xi = r^2$. But the integral equals $n!$ so we have

$$\int |\alpha\rangle\langle\alpha| \frac{d^2\alpha}{\pi} = \sum_0^{\infty} |n\rangle\langle n| = 1 \qquad \text{Q.E.D.,} \tag{2.5.20}$$

which follows from the completeness relation for the number representation.

We next show that a coherent state is a minimum uncertainty state. From the relations between $a, a^\dagger$ and q, p we see that the expectation values of p, q, p^2, and q^2 in state $|\alpha\rangle$ are

$$\begin{aligned}
\langle q\rangle &= \sqrt{\frac{\hbar}{2\omega}} \langle\alpha|(a + a^\dagger)|\alpha\rangle = \sqrt{\frac{\hbar}{2\omega}} (\alpha + \alpha^*) \\
\langle p\rangle &= i\sqrt{\frac{\hbar\omega}{2}} \langle\alpha|(a^\dagger - a)|\alpha\rangle = i\sqrt{\frac{\hbar\omega}{2}} (\alpha^* - \alpha) \\
\langle q^2\rangle &= \frac{\hbar}{2\omega} \langle\alpha|(a^{\dagger 2} + a^2 + aa^\dagger + a^\dagger a)|\alpha\rangle \\
&= \frac{\hbar}{2\omega} (\alpha^{*2} + \alpha^2 + 2\alpha^*\alpha + 1) \\
\langle p^2\rangle &= -\frac{\hbar\omega}{2} \langle\alpha|(a^{\dagger 2} + a^2 - aa^\dagger - a^\dagger a)|\alpha\rangle \\
&= -\frac{\hbar\omega}{2} (\alpha^{*2} + \alpha^2 - 2\alpha^*\alpha - 1),
\end{aligned} \tag{2.5.21}$$

where we used $a|\alpha\rangle = \alpha|\alpha\rangle$ and its adjoint. The variances are therefore

$$\begin{aligned}
(\Delta q)^2 &= \langle q^2\rangle - \langle q\rangle^2 = \frac{\hbar}{2\omega} \\
(\Delta p)^2 &= \langle p^2\rangle - \langle p\rangle^2 = \frac{\hbar\omega}{2},
\end{aligned} \tag{2.5.22}$$

so that

$$\Delta p \, \Delta q = \frac{\hbar}{2}, \tag{2.5.23}$$

which is the minimum value allowed by the uncertainty principle.

From (2.5.21) we see that the eigenvalues α are given by

$$\alpha = \frac{1}{\sqrt{2\hbar\omega}} [\omega\langle q\rangle + i\langle p\rangle]. \tag{2.5.24}$$

Since $\langle q \rangle$ and $\langle p \rangle$ are real and may have any value from $-\infty$ to $+\infty$, α may lie anywhere in the complex plane.

We next obtain the transformation function from the coherent state to the coordinate representation, $\langle q'|\alpha\rangle$. We have from the definition of a in terms of q and p

$$\begin{aligned} a|\alpha\rangle &= \alpha|\alpha\rangle \\ &= \frac{1}{\sqrt{2\hbar\omega}}[\omega q + ip]|\alpha\rangle. \end{aligned} \tag{2.5.25}$$

If we multiply both sides from the left by $\langle q'|$, an eigenbra of q, we see by techniques which are now familiar that

$$\langle q'|(\omega q + ip)|\alpha\rangle = \sqrt{2\hbar\omega}\,\alpha\langle q'|\alpha\rangle$$

or

$$\left[\frac{\omega}{\hbar}q' + \frac{d}{dq'}\right]\langle q'|\alpha\rangle = \sqrt{\frac{2\omega}{\hbar}}\,\alpha\langle q'|\alpha\rangle. \tag{2.5.26}$$

We may rewrite this as

$$\frac{d\langle q'|\alpha\rangle}{\langle q'|\alpha\rangle} = \left[\sqrt{\frac{2\omega}{\hbar}}\,\alpha - \frac{\omega}{\hbar}q'\right] dq' \tag{2.5.27}$$

or on integrating, we obtain

$$\langle q'|\alpha\rangle = N \exp\left[-\frac{\omega}{2\hbar}q'^2 + \sqrt{\frac{2\omega}{\hbar}}\,\alpha q'\right], \tag{2.5.28}$$

where N is a constant of integration. We choose it so that

$$\int_{-\infty}^{\infty} |\langle q'|\alpha\rangle|^2\, dq' = 1. \tag{2.5.29}$$

By (2.5.28) we see that

$$\begin{aligned} |\langle q'|\alpha\rangle|^2 &= |N|^2 \exp\left[-\frac{\omega}{\hbar}q'^2 + \sqrt{\frac{2\omega}{\hbar}}(\alpha + \alpha^*)q'\right] \\ &= |N|^2 \exp\left\{-\left[\sqrt{\frac{\omega}{\hbar}}\,q' - \frac{(\alpha + \alpha^*)}{\sqrt{2}}\right]^2 + \frac{(\alpha + \alpha^*)^2}{2}\right\}, \end{aligned} \tag{2.5.30}$$

where we completed the square in the exponent. Since

$$\int_{-\infty}^{\infty} e^{-(x-\xi)^2}\, dx = \sqrt{\pi}, \tag{2.5.31}$$

we obtain on integrating (2.5.30)

$$\int_{-\infty}^{\infty} dq'|\langle q'|\alpha\rangle|^2 = |N|^2\sqrt{\frac{\pi\hbar}{\omega}}\, e^{+\frac{1}{2}(\alpha+\alpha^*)^2} = 1. \tag{2.5.32}$$

Therefore,

$$N = \left(\frac{\omega}{\pi\hbar}\right)^{\frac{1}{4}} e^{-\frac{1}{4}(\alpha+\alpha^*)^2+i\mu} \tag{2.5.33}$$

where μ is an arbitrary real phase. Thus

$$\langle q'|\alpha\rangle = \left(\frac{\omega}{\pi\hbar}\right)^{\frac{1}{4}} \exp\left[-\tfrac{1}{4}(\alpha+\alpha^*)^2 + \sqrt{\frac{2\omega}{\hbar}}\,\alpha q' + i\mu\right]. \tag{2.5.34}$$

If we use (2.5.24) and its conjugate to eliminate α and α^* we obtain after minor algebra

$$\langle q'|\alpha\rangle = \left(\frac{\omega}{\pi\hbar}\right)^{\frac{1}{4}} \exp\left[-\frac{\omega}{2\hbar}[q' - \langle q\rangle]^2 + i\frac{\langle p\rangle}{\hbar}q' + i\mu\right]. \tag{2.5.35}$$

If we compare this with (1.12.23), we see that it is indeed a minimum uncertainty wave packet state but with a fixed Δq and Δp given by (2.5.22). In (1.12.23) Δq could be arbitrary but satisfied (2.5.23). Thus (2.5.34) represents a single infinity of minimum uncertainty states while (1.12.23) was a double infinity of such states.

With no loss of generality let us choose the phase μ so that

$$N = e^{-\frac{1}{2}(|\alpha|^2+\alpha^2)}\left(\frac{\omega}{\pi\hbar}\right)^{\frac{1}{4}}. \tag{2.5.36}$$

This is satisfactory since then

$$|N|^2 = [\exp[-\tfrac{1}{2}(|\alpha|^2 + \alpha^2 + |\alpha|^2 + \alpha^{*2}]\left(\frac{\omega}{\pi\hbar}\right)^{\frac{1}{2}}$$

$$= e^{-\frac{1}{2}(\alpha+\alpha^*)^2}\sqrt{\frac{\omega}{\pi\hbar}}, \tag{2.5.37}$$

which satisfies (2.5.32). Then we have

$$\langle q'|\alpha\rangle = \left(\frac{\omega}{\pi\hbar}\right)^{\frac{1}{4}} \exp\left[-\frac{\omega}{2\hbar}q'^2 + \sqrt{\frac{2\omega}{\hbar}}\,\alpha q' - \tfrac{1}{2}|\alpha|^2 - \tfrac{1}{2}\alpha^2\right]. \tag{2.5.38}$$

This form will prove useful later as we shall see.

PART II. ORBITAL ANGULAR MOMENTUM; ELECTRON SPIN

2.6 EIGENVALUES AND EIGENVECTORS OF ANGULAR MOMENTUM

Orbital angular momentum plays an essential role in quantum mechanics just as it does in classical mechanics. Classically, the angular momentum about a point 0 is defined by

$$\mathbf{l} = \mathbf{r} \times \mathbf{p}, \tag{2.6.1}$$

where $\mathbf{r}$ is the radius vector from 0 to the particle and $\mathbf{p}$ is its linear momentum. Since $\mathbf{l}$ is an observable, we postulate that $\mathbf{l}$ is a hermitian operator defined by (2.6.1) where $\mathbf{r}$ and $\mathbf{p}$ are the coordinate and momentum operators. We let $[q_1, q_2, q_3]$ be the three coordinate operators corresponding to $\mathbf{r}$ and $[p_1, p_2, p_3] = \mathbf{p}$ be the corresponding momentum operators. We postulate as in (1.10.3) that these operators obey the commutation relations

$$[q_i, p_j] = i\hbar\delta_{ij}; \qquad [q_i, q_j] = 0 = [p_i, p_j], \tag{2.6.2}$$

where i and $j = 1, 2$, or 3. This says that q_1 and p_2 commute, for example. In other words, measurements of a coordinate in one direction does not interfere with the measurement of the momentum in an orthogonal direction as it does in the same direction.

From (2.6.1) and (2.6.2), we see that

$$\begin{aligned} l_1 &= q_2p_3 - q_3p_2 \\ l_2 &= q_3p_1 - q_1p_3 \\ l_3 &= q_1p_2 - q_2p_1. \end{aligned} \tag{2.6.3}$$

If we use (2.6.2), we may easily show that

$$[l_x, l_y] = i\hbar l_z; \qquad [l_y, l_z] = i\hbar l_x; \qquad [l_z, l_x] = i\hbar l_y, \tag{2.6.4}$$

so that no additional postulates are needed to quantize $\mathbf{l}$. Note also since, for example, q_2 and p_3 commute as do q_3 and p_2, we do not have to worry about ordering of the separate factors in l_1, l_2, and l_3.

The total angular momentum is

$$\mathbf{l}^2 = l_1^2 + l_2^2 + l_3^2. \tag{2.6.5}$$

We leave as an exercise to show that

$$[\mathbf{l}^2, l_i] = 0. \tag{2.6.6}$$

That is, each component of the angular momentum separately commutes with $\mathbf{l}^2$.

It is convenient to define two nonhermitian operators $l_\pm$ by

$$l_\pm = l_1 \pm i l_2, \tag{2.6.7}$$

or

$$\begin{aligned} l_1 &= \tfrac{1}{2}(l_- + l_+) \\ l_2 &= \tfrac{1}{2}i(l_- - l_+). \end{aligned} \tag{2.6.8}$$

Since l_1 and l_2 are hermitian it follows that

$$l_+ = l_-^\dagger. \tag{2.6.9}$$

In terms of $l_\pm$, we see that

$$\mathbf{l}^2 = l_3^2 + \tfrac{1}{2}(l_+l_- + l_-l_+). \tag{2.6.10}$$

It is left as an exercise to show that

$$[\mathbf{l}^2, l_\pm] = 0 \tag{2.6.11a}$$

$$[l_z, l_\pm] = \pm\hbar l_\pm \tag{2.6.11b}$$

$$[l_+, l_-] = 2\hbar l_3. \tag{2.6.11c}$$

If we alternatively add and subtract (2.6.11c) from (2.6.10) we obtain

$$l_+l_- = \mathbf{l}^2 - l_3^2 + \hbar l_3 \tag{2.6.12a}$$

$$l_-l_+ = \mathbf{l}^2 - l_3^2 - \hbar l_3. \tag{2.6.12b}$$

We have shown that $\mathbf{l}^2$ commutes with l_1, l_2, and l_3 but the components do not commute with each other. We have shown that any two operators that commute may be simultaneously diagonalized. We may therefore obtain a representation in which both are diagonal; that is, we may find eigenvectors which are simultaneous eigenvectors of two commuting operators. Consider the eigenvalue problems

$$l_z|\mu; \nu\rangle = \mu\hbar|\mu; \nu\rangle \tag{2.6.13a}$$

$$\mathbf{l}^2|\mu; \nu\rangle = \nu\hbar^2|\mu; \nu\rangle. \tag{2.6.13b}$$

Since $[l_z, \mathbf{l}^2] = 0$ we see that

$$\begin{aligned} \mathbf{l}^2 l_z|\mu; \nu\rangle &= \mu\hbar\mathbf{l}^2|\mu; \nu\rangle = \mu\nu\hbar^3|\mu, \nu\rangle \\ &= l_z\mathbf{l}^2|\mu, \nu\rangle. \end{aligned} \tag{2.6.14}$$

We wish to obtain the eigenvalues μ and ν by techniques similar to those used for the harmonic oscillator.

By (2.6.11a) we have

$$\mathbf{l}^2 l_\pm = l_\pm \mathbf{l}^2. \tag{2.6.15}$$

By (2.6.13) we see that

$$\mathbf{l}^2\{l_\pm|\mu;\nu\rangle\} = \nu\hbar^2\{l_\pm|\mu;\nu\rangle\}. \tag{2.6.16}$$

This says that if $|\mu;\nu\rangle$ is an eigenket of $\mathbf{l}^2$ with eigenvalue $\nu\hbar^2$, then $l_+|\mu;\nu\rangle$ and $l_-|\mu;\nu\rangle$ are also eigenkets with the *same* eigenvalue.

Consider next (2.6.11b):

$$l_z l_\pm = l_\pm l_z \pm \hbar l_\pm. \tag{2.6.17}$$

It therefore follows from (2.6.13) that

$$l_z\{l_\pm|\mu;\nu\rangle\} = (\mu \pm 1)\hbar\{l_\pm|\mu;\nu\rangle\}. \tag{2.6.18}$$

Thus if $|\mu;\nu\rangle$ is an eigenvector of l_z with eigenvalue $\mu\hbar$ then $l_+|\mu;\nu\rangle$ is an eigenvector of l_z with eigenvalue $(\mu+1)\hbar$ and $l_-|\mu;\nu\rangle$ is an eigenvector of l_z with eigenvalue $(\mu-1)\hbar$, both with the same ν by (2.6.16). We have thus generated two additional eigenvectors of l_z from the original whose eigenvalues differ by $\pm\hbar$. We may obviously continue this process and obtain the infinite sequence

$$\begin{array}{llll} |\mu;\nu\rangle & l_+|\mu;\nu\rangle & l_+^2|\mu;\nu\rangle & l_+^3|\mu;\nu\rangle \cdots \\ \mu\hbar & (\mu+1)\hbar & (\mu+2)\hbar & (\mu+3)\hbar \\ |\mu;\nu\rangle & l_-|\mu;\nu\rangle & l_-^2|\mu;\nu\rangle & l_-^3|\mu;\nu\rangle \\ \mu\hbar & (\mu-1)\hbar & (\mu-2)\hbar & (\mu-3)\hbar, \end{array} \tag{2.6.19}$$

where ν is unchanged.

Since the norm of a vector must be greater than or equal to zero, we assume that

$$\langle\mu;\nu|\mu;\nu\rangle > 0. \tag{2.6.20}$$

That is, the original vector exists. Then since $l_- = l_+^\dagger$, we have

$$\begin{aligned} \langle\mu;\nu|l_-l_+|\mu;\nu\rangle &= \langle\mu;\nu|(\mathbf{l}^2 - l_z{}^2 - \hbar l_z)|\mu;\nu\rangle \\ &= \hbar^2(\nu - \mu^2 - \mu)\langle\mu;\nu|\mu;\nu\rangle \geq 0, \end{aligned} \tag{2.6.21}$$

where we used (2.6.12b) and (2.6.13). By (2.6.20), it follows that

$$\nu - \mu^2 - \mu \geq 0, \tag{2.6.22}$$

where ν is fixed. This tells us that for any given ν if μ gets arbitrarily big either positively or negatively the vector $l_+|\mu;\nu\rangle$ would develop a negative norm which is forbidden. We must therefore anticipate an upper and lower bound on μ for each ν. The equality is satisfied above for

$$\mu = -\tfrac{1}{2} \pm \sqrt{\tfrac{1}{4} + \nu}, \tag{2.6.23}$$

which are the two bounds on μ for fixed ν. Let l be the largest value μ may have so that

$$l \equiv -\tfrac{1}{2} + \sqrt{\tfrac{1}{4} + \nu} \tag{2.6.24}$$

or

$$\nu = l(l + 1). \tag{2.6.25}$$

When $\mu = l$, we have

$$l_+|l; \nu\rangle = 0, \tag{2.6.26}$$

since otherwise we would generate an eigenvector with eigenvalue $\mu = l + 1$ which would violate (2.6.22).

If we start with state $|l; \nu\rangle$ and apply l_- k times then we generate state $|l - k; \nu\rangle$ by (2.6.19). The length of $l_-|l - k; \nu\rangle$ is

$$\begin{aligned}
&\langle l - k; \nu| l_+ l_- |l - k; \nu\rangle \\
&\quad = \langle l - k; \nu|(\mathbf{l}^2 - l_3{}^2 + \hbar l_3)|l - k; \nu\rangle \\
&\quad = \{\nu - (l - k)^2 + (l - k)\}\hbar^2\langle l - k; \nu|l - k; \nu\rangle \geq 0.
\end{aligned} \tag{2.6.27}$$

If $|l - k; \nu\rangle \neq 0$, then since $\nu = l(l + 1)$, we conclude

$$l(l + 1) - (l - k)^2 + (l - k) \geq 0. \tag{2.6.28}$$

This puts a limit on the size of k

$$(l - k) = \tfrac{1}{2} \pm \sqrt{\tfrac{1}{4} + l(l + 1)}, \tag{2.6.29}$$

or $k_{\max}$ is determined by

$$\begin{aligned}
l - k_{\max} &= \tfrac{1}{2} - \sqrt{\tfrac{1}{4} + l(l + 1)} \\
&= \tfrac{1}{2} - \sqrt{(l + \tfrac{1}{2})^2},
\end{aligned} \tag{2.6.30a}$$

so that

$$k_{\max} = 2l. \tag{2.6.30b}$$

But k is a positive integer so $2l$ must be a positive integer. Therefore, l may have only the values

$$l = 0, \tfrac{1}{2}, 1, \tfrac{3}{2}, 2, \tfrac{5}{2}, \ldots. \tag{2.6.31}$$

When $k_{\max} = 2l$, we conclude that

$$l_-|l - k_{\max}, \nu\rangle = l_-|-l; \nu\rangle = 0. \tag{2.6.32}$$

Therefore μ ranges between $+l$ and $-l$ in unit steps. It is conventional to let $\mu = m$ and to designate ν by l since $\nu = l(l + 1)$ and write

$$\begin{aligned}
l_3|m; l\rangle &= m\hbar|m; l\rangle \\
\mathbf{l}^2|m; l\rangle &= l(l + 1)\hbar^2|m; l\rangle,
\end{aligned} \tag{2.6.33}$$

where the eigenvalues are

$$\begin{aligned} l &= 0, \tfrac{1}{2}, 1, \tfrac{3}{2}, 2, \ldots \\ m &= -l, -l+1, -l+2, \ldots, l-2, l-1, l. \end{aligned} \tag{2.6.34}$$

The eigenvectors are orthogonal since $\mathbf{l}^2$ and l_3 are hermitian so that

$$\langle m'; l'|m; l\rangle = \delta_{ll'}\delta_{mm'}, \tag{2.6.35}$$

so that the matrix elements of l_3 and $\mathbf{l}^2$ are

$$\begin{aligned} \langle m'; l'|l_3|m; l\rangle &= m\hbar\delta_{ll'}\delta_{mm'} \\ \langle m'; l'|\mathbf{l}^2|m; l\rangle &= l(l+1)\hbar^2\delta_{ll'}\delta_{mm'}. \end{aligned} \tag{2.6.36}$$

Let us next obtain the matrix elements of $l_\pm$ in the representation in which l_3 and $\mathbf{l}^2$ are diagonal. We have shown that $l_+|m; l\rangle$ is an eigenvector of $\mathbf{l}^2$ with eigenvalue $l(l+1)\hbar^2$ and also an eigenvector of l_3 with eigenvalue $(m+1)\hbar$. Therefore, $l_+|m; l\rangle$ can differ from $|m+1; l\rangle$ by a complex constant. We may therefore write

$$l_+|m; l\rangle = \lambda_{l,m}\hbar|m+1; l\rangle. \tag{2.6.37}$$

So that

$$\langle m+1; l|l_+|m; l\rangle = \lambda_{l,m}\hbar. \tag{2.6.38}$$

If we take the complex conjugate of both sides, we have

$$\langle m; l|l_-|m+1; l\rangle = \lambda_{l,m}^*\hbar. \tag{2.6.39}$$

This relation is satisfied if

$$l_-|m+1; l\rangle = \lambda_{l,m}^*\hbar|m; l\rangle. \tag{2.6.40}$$

Consider next

$$\begin{aligned} l_-l_+|m; l\rangle &= l_-\hbar\lambda_{l,m}|m+1; l\rangle \\ &= \hbar^2|\lambda_{l,m}|^2|m; l\rangle \\ &= [\mathbf{l}^2 - l_3{}^2 - \hbar l_3]|m; l\rangle \\ &= [l(l+1) - m^2 - m]\hbar^2|m; l\rangle, \end{aligned} \tag{2.6.41}$$

where we used (2.6.37), (2.6.40), (2.6.12b), and (2.6.33). Therefore, we conclude that

$$\begin{aligned} \lambda_{m,l} &= \sqrt{l(l+1) - m(m+1)} \\ &= \sqrt{(l-m)(l+m+1)}, \end{aligned} \tag{2.6.42}$$

and

$$\begin{aligned} l_+|m; l\rangle &= \hbar\sqrt{(l-m)(l+m+1)}|m+1; l\rangle \\ l_-|m; l\rangle &= \hbar\sqrt{(l-m+1)(l+m)}|m-1; l\rangle, \end{aligned} \tag{2.6.43}$$

so the matrix elements are

$$\begin{aligned}\langle m', l'|l_+|m; l\rangle &= \hbar\sqrt{(l-m)(l+m+1)}\,\delta_{m',m+1}\delta_{ll'}\\ \langle m', l'|l_-|m; l\rangle &= \hbar\sqrt{(l-m+1)(l+m)}\,\delta_{m',m-1}\delta_{ll'}.\end{aligned} \tag{2.6.44}$$

If we use (2.6.8), we obtain the nonvanishing matrix elements

$$\begin{aligned}\langle m+1; l|l_1|m; l\rangle &= \frac{\hbar}{2}\sqrt{(l-m)(l+m+1)}\\ \langle m-1; l|l_1|m; l\rangle &= \frac{\hbar}{2}\sqrt{(l-m+1)(l+m)}\\ \langle m+1; l|l_2|m; l\rangle &= -\tfrac{1}{2}i\hbar\sqrt{(l-m)(l+m+1)}\\ \langle m-1; l|l_2|m; l\rangle &= +\tfrac{1}{2}i\hbar\sqrt{(l-m+1)(l+m)}.\end{aligned} \tag{2.6.45}$$

We shall write out a few of these explicitly. For $l = 0$, we have explicitly the null matrices

$$l_3 = 0 \qquad \mathbf{l}^2 = 0 \qquad l_1 = 0 = l_2.$$

Next for $l = \frac{1}{2}$, $m = \pm\frac{1}{2}$, the matrix elements are

$$\begin{aligned}l_3 &= \frac{\hbar}{2}\begin{pmatrix}1 & 0\\ 0 & -1\end{pmatrix} \qquad & \mathbf{l}^2 &= \hbar^2\frac{3}{4}\begin{pmatrix}1 & 0\\ 0 & 1\end{pmatrix}\\ l_1 &= \frac{\hbar}{2}\begin{pmatrix}0 & 1\\ 1 & 0\end{pmatrix} \qquad & l_2 &= \frac{\hbar}{2}\begin{pmatrix}0 & i\\ -i & 0\end{pmatrix},\end{aligned} \tag{2.6.46}$$

while the state vectors become

$$|+\tfrac{1}{2}; \tfrac{1}{2}\rangle = \begin{bmatrix}1\\ 0\end{bmatrix} \qquad |-\tfrac{1}{2}; \tfrac{1}{2}\rangle = \begin{bmatrix}0\\ 1\end{bmatrix}. \tag{2.6.47}$$

For $l = 1$, $m = -1, 0, +1$, and we have

$$\begin{aligned}l_3 &= \hbar\begin{bmatrix}1 & 0 & 0\\ 0 & 0 & 0\\ 0 & 0 & -1\end{bmatrix}; \qquad & \mathbf{l}^2 &= 2\hbar^2\begin{bmatrix}1 & 0 & 0\\ 0 & 1 & 0\\ 0 & 0 & 1\end{bmatrix}\\ l_1 &= \frac{\hbar}{\sqrt{2}}\begin{bmatrix}0 & 1 & 0\\ 1 & 0 & 1\\ 0 & 1 & 0\end{bmatrix}; \qquad & l_2 &= \frac{\hbar}{\sqrt{2}}\begin{bmatrix}0 & -i & 0\\ i & 0 & -i\\ 0 & i & 0\end{bmatrix},\end{aligned} \tag{2.6.48}$$

while

$$|+1;1\rangle = \begin{bmatrix} 1 \\ 0 \\ 0 \end{bmatrix}; \qquad |0;1\rangle = \begin{bmatrix} 0 \\ 1 \\ 0 \end{bmatrix}; \qquad |-1;1\rangle = \begin{bmatrix} 0 \\ 0 \\ 1 \end{bmatrix}. \tag{2.6.49}$$

The reader may easily proceed. For $l = \frac{3}{2}$, $m = \pm\frac{3}{2}$, and $\pm\frac{1}{2}$ so the matrices are 4×4.

We have chosen $\mathbf{l}^2$ and l_3 as the two commuting operators to diagonalize. We could have also chosen $\mathbf{l}^2$ and l_1 or $\mathbf{l}^2$ and l_2. We say that the 3 or z-axis is the axis of quantization when we diagonalize $\mathbf{l}^2$ and l_3. There is obviously nothing unique about the z-axis here. If we, for example, applied a uniform magnetic field in a certain direction, then it would usually be advantageous to choose this direction as the axis of quantization.

We have shown that l_z may have integral or half-integral multiples of $\hbar$ as its eigenvalues. Simply this result arises because of the commutation relations (2.6.4) and (2.6.6) and has nothing to do with the definitions (2.6.3) of $\mathbf{l}$ in terms of the coordinates and momentum. However, if $\mathbf{l}$ is to represent orbital angular momentum, then the eigenvectors of $\mathbf{l}^2$ and l_3 must have coordinate or momentum representatives. That is, we must be able to express the $\mathbf{l}$ matrices in terms of coordinate and momentum matrices. We show in the next section that this is only possible if we restrict the eigenvalues of l_z to be an *integer* times $\hbar$. The half-integer values do not have a classical analog in that $|m; l\rangle$ does not have a coordinate representative when m is a half-integer.

We have no reason for throwing out half-integral values if we say these correspond to intrinsically quantum mechanical effects which have no classical analog. We call such intrinsic angular momentum *spin* angular momentum. It turns out indeed that some particles have not only orbital angular momentum but in addition are born with spin angular momentum. Electrons are born with a spin $l = \frac{1}{2}$ so $m = \pm\frac{1}{2}$. Effects due to this can be measured experimentally and all attempts to explain these effects classically have failed.

2.7 PARTICLE IN A CENTRAL FORCE FIELD

The hamiltonian for a particle of mass μ in a central force field is given by

$$H = \frac{1}{2\mu}\mathbf{p}^2 + V(r), \tag{2.7.1}$$

where $V(r)$ is a spherically symmetric potential, and

$$r^2 = q_1^{\ 2} + q_2^{\ 2} + q_3^{\ 2}. \tag{2.7.2}$$

We would like to solve the energy eigenvalue problem

$$H|E\rangle = E|E\rangle. \tag{2.7.3}$$

To begin we prove that each component l_i of the orbital angular momentum as well as $\mathbf{l}^2$ commutes with H and are therefore constants of the motion for spherically symmetric potentials. To show this, we first use the commutation relations (2.6.2) to generalize (1.10.6) to three dimensions:

$$[q_i, F(\mathbf{q}, \mathbf{p})] = i\hbar \frac{\partial F}{\partial p_i} \tag{2.7.4a}$$

$$[p_i, F(\mathbf{q}, \mathbf{p})] = -i\hbar \frac{\partial F}{\partial q_i}. \tag{2.7.4b}$$

Next if we let i, j, and k form a cyclic permutation of 1, 2, 3, then (2.6.3) may be written as

$$l_i = q_j p_k - p_k q_j. \tag{2.7.5}$$

Next we shall need the identity

$$[AB, C] \equiv A[B, C] + [A, C]B, \tag{2.7.6}$$

which the reader may readily verify.

We next show that

$$[l_i, \mathbf{p}^2] = 0. \tag{2.7.7}$$

If we use (2.7.4)–(2.7.6), we have

$$\begin{aligned} [l_i, \mathbf{p}^2] &= [q_j p_k, \mathbf{p}^2] - [p_k q_j, \mathbf{p}^2] \\ &= [q_j, \mathbf{p}^2] p_k - p_k [q_j, \mathbf{p}^2] \\ &= i\hbar \left\{ \left(\frac{\partial \mathbf{p}^2}{\partial p_j}\right) p_k - p_k \left(\frac{\partial \mathbf{p}^2}{\partial p_j}\right) \right\} \\ &= 2i\hbar [p_j p_k - p_k p_j] \equiv 0. \qquad \text{Q.E.D.} \end{aligned} \tag{2.7.8}$$

Next we have

$$[l_i, V(r)] = 0, \tag{2.7.9}$$

since in a similar way we have

$$\begin{aligned} [l_i, V(r)] &= [(q_j p_k - p_k q_j), V] \\ &= q_j [p_k, V] - [p_k, V] q_j \\ &= -i\hbar \left\{ q_j \frac{\partial V}{\partial q_k} - \frac{\partial V}{\partial q_k} q_j \right\}. \end{aligned} \tag{2.7.10}$$

But by (2.7.2)

$$r \frac{\partial r}{\partial q_k} = q_k, \tag{2.7.11}$$

so

$$\frac{\partial V}{\partial q_k} = \frac{dV}{dr}\frac{\partial r}{\partial q_k} = \frac{q_k}{r}\frac{dV}{dr}, \tag{2.7.12}$$

and (2.7.10) becomes

$$[l_i, V(r)] = -i\hbar\frac{dV}{dr}(q_j q_k - q_k q_j) = 0. \tag{2.7.13}$$

If V is not spherically symmetric, this result does not follow. The reader may note that since i, j, k in (2.7.10) are a permutation of 1, 2, 3 it may be written in vector form as

$$[\mathbf{l}, V(\mathbf{q})] = -i\hbar\mathbf{q} \times \nabla V(\mathbf{q}). \tag{2.7.14}$$

It is now very simple to show that

$$\begin{aligned} [\mathbf{l}^2, \mathbf{p}^2] &= 0 \\ [\mathbf{l}^2, V(r)] &= 0. \end{aligned} \tag{2.7.15}$$

For if we use (2.7.6), we have

$$\begin{aligned} [\mathbf{l}^2, \mathbf{p}^2] &= \sum_i \{l_i[l_i, \mathbf{p}^2] + [l_i, \mathbf{p}^2]l_i\} = 0 \\ [\mathbf{l}^2, V(r)] &= \sum_i \{l_i[l_i, V(r)] + [l_i, V(r)]l_i\} = 0, \end{aligned} \tag{2.7.16}$$

where we used (2.7.7) and (2.7.9). If we use now (2.7.1) and the results above we see that

$$\begin{aligned} [l_i, H] &= 0 \\ [\mathbf{l}^2, H] &= 0, \end{aligned} \tag{2.7.17}$$

which shows that each l_i and $\mathbf{l}^2$ are constants of the motion. However, by (2.6.4) the l_i do not commute with each other although $[l_i, \mathbf{l}^2] = 0$. Therefore, we may simultaneously diagonalize H, l_3, and $\mathbf{l}^2$ as we shall now show [see Section 1.9]. We write the eigenvalue problems as

$$\begin{aligned} l_3|E, \mu, \nu\rangle &= \mu\hbar|E, \mu, \nu\rangle \\ \mathbf{l}^2|E, \mu, \nu\rangle &= \nu\hbar^2|E, \mu, \nu\rangle \\ H|E, \mu, \nu\rangle &= E|E, \mu, \nu\rangle. \end{aligned} \tag{2.7.18}$$

Since $[l_i, H] = [l_\pm, H] = 0$, all the analysis of the previous section may be taken over directly. For example,

$$\begin{aligned} l_+|E, m, l\rangle &= \hbar\sqrt{(l - m)(l + m + 1)}|E, m + 1, l\rangle \\ l_-|E, m, l\rangle &= \hbar\sqrt{(l - m + 1)(l + m)}|E, m - 1, l\rangle. \end{aligned} \tag{2.7.19}$$

That is, $l_+|E, m, l\rangle$ is also an eigenvector of l_3, $\mathbf{l}^2$, and H with the same E and

l but with m increased by 1 only. Therefore, we have

$$l_3|E, m, l\rangle = m\hbar|E, m, l\rangle \tag{2.7.20}$$

$$\mathbf{l}^2|E, m, l\rangle = l(l+1)\hbar^2|E, m, l\rangle \tag{2.7.21}$$

$$H|E, m, l\rangle = E|E, m, l\rangle. \tag{2.7.22}$$

We obviously cannot find the energy eigenvalues until we specify the explicit $V(r)$.

We next wish to show that we must exclude the half-integer eigenvalues of l_3 when $\mathbf{l}$ is true ordinary angular momentum. This can only be done by obtaining the transformation function from the energy representation $|E, m, l\rangle$ to the coordinate representation since true angular momentum must be expressible in terms of coordinates and momenta. If we multiply (2.7.20)–(2.7.22) from the left by $\langle\mathbf{q}|$ where

$$q_i|q_1', q_2', q_3'\rangle = q_i'|q_1', q_2', q_3'\rangle, \tag{2.7.23}$$

we obtain when we use the definitions of l_3, $\mathbf{l}^2$, and H in terms of $\mathbf{p}$ and $\mathbf{q}$

$$\frac{\hbar}{i}\left[q_1'\frac{\partial}{\partial q_2'} - q_2'\frac{\partial}{\partial q_1'}\right]\langle\mathbf{q}'|E, m, l\rangle = m\hbar\langle\mathbf{q}'|E, m, l\rangle \tag{2.7.24}$$

$$\begin{aligned}-\hbar^2\bigg\{(q_2'^2 + q_3'^2)\frac{\partial^2}{\partial q_1'^2} + (q_3'^2 + q_1'^2)\frac{\partial^2}{\partial q_2'^2} + (q_1'^2 + q_2'^2)\frac{\partial^2}{\partial q_3'^2}\\ - 2\left(q_1'\frac{\partial}{\partial q_1'} + q_2'\frac{\partial}{\partial q_2'} + q_3'\frac{\partial}{\partial q_3'}\right)\bigg\}\langle\mathbf{q}'|E, m, l\rangle\\ = l(l+1)\hbar^2\langle\mathbf{q}'|E, m, l\rangle\end{aligned} \tag{2.7.25}$$

$$\left\{-\frac{\hbar^2}{2\mu}\left[\frac{\partial^2}{\partial q_1'^2} + \frac{\partial^2}{\partial q_2'^2} + \frac{\partial^2}{\partial q_3'^2}\right] + V(r)\right\}\langle\mathbf{q}'|E, m, l\rangle = E\langle\mathbf{q}'|E, m, l\rangle. \tag{2.7.26}$$

In (2.7.25) we use the fact that

$$\begin{aligned}\mathbf{l}^2 = (q_2^2 + q_3^2)p_1^2 + (q_3^2 + q_1^2)p_2^2 + (q_1^2 + q_2^2)p_3^2\\ + 2i\hbar(q_1p_1 + q_2p_2 + q_3p_3),\end{aligned} \tag{2.7.27}$$

as the reader may verify. We have also generalized (1.10.46) and (1.10.47) to three dimensions:

$$\langle\mathbf{q}'|F(\mathbf{p}, \mathbf{q}) = F\left(\frac{\hbar}{i}\nabla', \mathbf{q}'\right)\langle\mathbf{q}'|, \tag{2.7.28}$$

where

$$\nabla' = \left[\frac{\partial}{\partial q_1'}, \frac{\partial}{\partial q_2'}, \frac{\partial}{\partial q_3'}\right]. \tag{2.7.29}$$

In applying (2.7.28), care must be exercised in maintaining the same ordering as appeared in F. For example,

$$\langle \mathbf{q}'|p_1q_1p_1 = \frac{\hbar}{i}\frac{\partial}{\partial q_1'}\, q_1' \frac{\hbar}{i}\frac{\partial}{\partial q_1'} \langle \mathbf{q}'|. \tag{2.7.30}$$

We have obtained three simultaneous partial differential equations that the energy eigenfunctions [transformation functions from the energy to coordinate representation] must satisfy. They are most easily solved by transforming to polar coordinates where

$$\begin{array}{l|l} q_1' = r\sin\theta\cos\varphi & r^2 = q_1'^2 + q_2'^2 + q_3'^2 \\ q_2' = r\sin\theta\sin\varphi & \cos\theta = \dfrac{q_3'}{\sqrt{q_1'^2 + q_2'^2 + q_3'^2}} \\ q_3' = r\cos\theta & \tan\varphi = \dfrac{q_2'}{q_1'}. \end{array} \tag{2.7.31}$$

We have by the usual rules

$$\frac{\partial}{\partial q_i'} = \frac{\partial r}{\partial q_i'}\frac{\partial}{\partial r} + \frac{\partial \theta}{\partial q_i'}\frac{\partial}{\partial \theta} + \frac{\partial \varphi}{\partial q_i'}\frac{\partial}{\partial \varphi}. \tag{2.7.32}$$

From (2.7.31) we obtain

$$\begin{aligned} &\frac{\partial r}{\partial q_1'} = \sin\theta\cos\varphi; && \frac{\partial \theta}{\partial q_1'} = \frac{1}{r}\cos\theta\cos\varphi; && \frac{\partial \varphi}{\partial q_1'} = \frac{\sin\varphi}{r\sin\theta} \\ &\frac{\partial r}{\partial q_2'} = \sin\theta\sin\varphi; && \frac{\partial \theta}{\partial q_2'} = \frac{1}{r}\cos\theta\sin\varphi; && \frac{\partial \varphi}{\partial q_2'} = \frac{\cos\varphi}{r\sin\theta} \\ &\frac{\partial r}{\partial q_3'} = \cos\theta; && \frac{\partial \theta}{\partial q_3'} = -\frac{\sin\theta}{r}; && \frac{\partial \varphi}{\partial q_3'} = 0. \end{aligned} \tag{2.7.33}$$

If we use (2.7.32) and (2.7.33) and preserve the order of factors, we see that

$$\begin{aligned} q_1'\frac{\partial}{\partial q_2'} - q_2'\frac{\partial}{\partial q_1'} &= r\sin\theta\cos\varphi\left(\sin\theta\sin\varphi\frac{\partial}{\partial r} + \frac{\cos\theta\sin\varphi}{r}\frac{\partial}{\partial\theta} + \frac{\cos\varphi}{r\sin\theta}\frac{\partial}{\partial\varphi}\right) \\ &\quad - r\sin\theta\sin\varphi\left(\sin\theta\sin\varphi\frac{\partial}{\partial r} + \frac{\cos\theta\cos\varphi}{r}\frac{\partial}{\partial\theta} - \frac{\sin\varphi}{r\sin\theta}\frac{\partial}{\partial\varphi}\right) = \frac{\partial}{\partial\varphi}. \end{aligned} \tag{2.7.34}$$

Therefore, in polar coordinates, (2.7.24) becomes

$$\frac{\hbar}{i}\frac{\partial}{\partial\varphi}\langle r, \varphi, \theta|E, m, l\rangle = m\hbar\langle r, \varphi, \theta|E, m, l\rangle. \tag{2.7.35}$$

A similar but more tedious calculation gives for (2.7.25) and (2.7.26)

$$-\hbar^2\left[\frac{1}{\sin\theta}\frac{\partial}{\partial\theta}\left(\sin\theta\frac{\partial}{\partial\theta}\right) + \frac{1}{\sin^2\theta}\frac{\partial^2}{\partial\varphi^2}\right]\langle r, \varphi, \theta|E, m, l\rangle$$
$$= l(l+1)\hbar^2\langle r, \varphi, \theta|E, m, l\rangle \tag{2.7.36}$$

$$\left\{-\frac{\hbar^2}{2\mu}\left[\frac{1}{r^2}\frac{\partial}{\partial r}\left(r^2\frac{\partial}{\partial r}\right) + \frac{1}{r^2\sin\theta}\frac{\partial}{\partial\theta}\left(\sin\theta\frac{\partial}{\partial\theta}\right) + \frac{1}{r^2\sin^2\theta}\frac{\partial^2}{\partial\varphi^2}\right] + V(r)\right\}$$
$$\times \langle r, \varphi, \theta|E, m, l\rangle = E\langle r, \varphi, \theta|E, m, l\rangle. \tag{2.7.37}$$

If we use (2.7.36) in (2.7.37), we easily obtain

$$\left\{-\frac{\hbar^2}{2\mu}\left[\frac{1}{r^2}\frac{\partial}{\partial r}\left(r^2\frac{\partial}{\partial r}\right)\right] - \frac{l(l+1)}{r^2} + V(r)\right\}\langle r, \varphi, \theta|E, m, l\rangle$$
$$= E\langle r, \varphi, \theta|E, m, l\rangle. \tag{2.7.38}$$

Next from (2.7.35), we have

$$\frac{\partial^2}{\partial\varphi^2}\langle r, \varphi, \theta|E, m, l\rangle = \frac{\partial}{\partial\varphi}im\langle r, \varphi, \theta|E, m, l\rangle$$
$$= -m^2\langle r, \varphi, \theta|E, m, l\rangle. \tag{2.7.39}$$

If we use this, (2.7.36) becomes

$$\left[\frac{1}{\sin\theta}\frac{\partial}{\partial\theta}\left(\sin\theta\frac{\partial}{\partial\theta}\right) - \frac{m^2}{\sin^2\theta}\right]\langle r, \theta, \varphi|E, m, l\rangle$$
$$= -l(l+1)\langle r, \theta, \varphi|E, m, l\rangle. \tag{2.7.40}$$

We then have succeeded in obtaining three equations, (2.7.35), (2.7.38), and (2.7.40), that the eigenfunctions must satisfy. The solution to (2.7.35) is

$$\langle r, \varphi, \theta|E, m, l\rangle = e^{im\varphi}f(r, \theta; E, m, l), \tag{2.7.41}$$

where f is as yet arbitrary but independent of φ. At this point the flag goes up. If we allow m to be a half-integer, the eigenfunction would not be single valued which is physically required for orbital angular momentum. The half-integer values of m cannot have a coordinate representation and therefore have no classical analog. They do exist but correspond to *spin* angular momentum.

When we restrict m to integers, l must also be an integer since m runs from $-l$ to $+l$ in integer steps. Then (2.7.40) is the equation for associated Legendre polynomials $P_l^m(\cos\theta)$ and its solution which is finite when

$\theta = 0$ and π is

$$\langle r, \theta, \varphi | E, m, l \rangle = e^{im\varphi} P_l^{\,m}(\cos\theta) R(r; E, m, l), \tag{2.7.42}$$

where R is an arbitrary function of r which satisfies (2.7.38), namely,

$$\left\{-\frac{\hbar^2}{2\mu}\left[\frac{1}{r^2}\frac{\partial}{\partial r}\left(r^2\frac{\partial}{\partial r}\right)\right] - \frac{l(l+1)}{r^2} + V(r)\right\} R(r) = ER(r). \tag{2.7.43}$$

This equation is independent of m so that R and E will depend on the quantum number l as well as another which we call n. Therefore, we have

$$\langle r, \theta, \varphi | n, m, l \rangle = N_{nml} e^{im\varphi} P_l^{\,m}(\cos\theta) R_{nl}(r), \tag{2.7.44}$$

where N is a normalization constant. Since in spherical coordinates the eigenfunctions have factored, we may write

$$\langle r, \theta, \varphi | n, m, l \rangle = \langle r | n \rangle \langle \theta, \varphi | m, l \rangle,$$

and let

$$Y_{lm}(\theta, \varphi) \equiv \langle \theta, \varphi | m, l \rangle = \sqrt{\frac{(2l+1)!\,(l-|m|)!}{4\pi(l+|m|)!}}\, P_{lm}(\cos\theta) e^{im\varphi}, \tag{2.7.45}$$

where we normalized so that

$$\int_0^{2\pi} d\varphi \int_0^{\pi} \sin\theta\, d\theta\, Y_{lm}^*(\theta, \varphi) Y_{l',m'}(\theta, \varphi) = \delta_{ll'}\delta_{mm'}; \tag{2.7.46}$$

$Y_{lm}(\theta, \varphi)$ are called spherical harmonics and are the transformation functions from the l_3, $\mathbf{l}^2$ representation to the angular ("coordinate") representation.

To proceed one must specify $V(r)$.

2.8 PAULI SPIN OPERATORS

We have seen in Section 2.6 that the commutation relations

$$[l_i, l_j] = i\hbar l_k, \tag{2.8.1}$$

where i, j, and k form a cyclic permutation of 1, 2, 3, and

$$[\mathbf{l}^2, l_i] = 0 \tag{2.8.2}$$

leads to integer and half-integer eigenvalues of l_3. In Section 2.7 we have shown that only integer values correspond to orbital angular momentum. If we consider an electron of charge $-|e|$, mass μ, and with angular momentum $\mathbf{l}$, it has a magnetic moment associated with it given by [9]

$$\mathbf{m} = -\frac{|e|}{2\mu}\mathbf{l} \equiv -\frac{\beta}{\hbar}\mathbf{l} \tag{2.8.3}$$

in mks units. Here β is called the Bohr magneton. The electron must be in some field of force to make it go in a curved orbit in order that $\mathbf{l} = \mathbf{r} \times \mathbf{p}$ does not vanish. In this case if we put the electron in a magnetic field $\mathbf{F}$, it will have in addition to its energy

$$H_0 = \frac{\mathbf{p}^2}{2\mu} + V(r), \tag{2.8.4}$$

for a central field of force an amount given by

$$H_1 = -\mathbf{m} \cdot \mathbf{F} = \frac{|e|}{2\mu} \mathbf{l} \cdot \mathbf{F}. \tag{2.8.5}$$

If the field is a uniform field along the z-axis, this reduces to

$$H_1 = \frac{|e|\, F_0}{2\mu}\, l_3. \tag{2.8.6}$$

Since l_3 and $\mathbf{l}^2$ commute with each other as well as H_0 and H_1, the eigenvectors of H_0 are also eigenvectors of H_1. Thus

$$H_1|n, m, l\rangle = \frac{|e|\, \hbar F_0}{2\mu}\, m|n, m, l\rangle. \tag{2.8.7}$$

So the expectation value for $H = H_0 + H_1$ is*

$$\langle n, m, l|H|n, m, l\rangle = E_{nl} + \frac{|e|\, \hbar}{2\mu} F_0 m, \tag{2.8.8}$$

where

$$H_0|n, m, l\rangle = E_{nl}|n, m, l\rangle, \tag{2.8.9}$$

and m is an integer which runs from $-l$ to $+l$ in unit integer steps. Therefore, if we measure the energy of the electron in the magnetic field, we should obtain the possible values given by (2.8.8). However, experimentally it is found to have an additional amount given by

$$H_{\text{spin}} = \frac{|e|\, \hbar}{2\mu} F_0 s = \beta F_0 s, \tag{2.8.10}$$

where $s = \pm\frac{1}{2}$ and it is independent of n, l, and m. To explain this, Uhlenbeck and Goudsmit [10] postulated that the electron has in addition to its orbital angular momentum an additional nonclassical spin angular momentum of $s\hbar$.

* This hamiltonian is not exactly correct, as we show in Section 2.10.

Let **s** correspond to the spin angular momentum of the electron so it will not be confused with **l** which we restrict to orbital angular momentum. For convenience we define an operator $\boldsymbol{\sigma}$ by means of

$$\mathbf{s} = \tfrac{1}{2}\hbar\boldsymbol{\sigma}. \tag{2.8.11}$$

Associated with this spin momentum, we assume the electron has a magnetic moment given by

$$\mathbf{m}_s = -\frac{|e|}{\mu}\mathbf{s} = -\frac{|e|\,\hbar}{2\mu}\boldsymbol{\sigma} \equiv -\beta\boldsymbol{\sigma}. \tag{2.8.12}$$

Note that this twice as large as $\mathbf{m}_l$ given by (2.8.3). The necessity for this definition is justified from experimental measurement of the energy. Since **s** has no classical analog, neither does $\mathbf{m}_s$ so we must define it to give results in agreement with experiment.

We now postulate that **s** or equivalently $\boldsymbol{\sigma}$ obey the same commutation relations as **l** since half-integer eigenvalues are included there. So we postulate that

$$[\sigma_i, \sigma_j] = 2i\sigma_k, \tag{2.8.13}$$

where i, j, k form an even permutation of 1, 2, 3. Also we require

$$[\sigma_i, \boldsymbol{\sigma}^2] = 0. \tag{2.8.14}$$

These postulates lead to eigenvalues of σ_3 that are integer or half-integer multiples of $\hbar$. To restrict the eigenvalues to $\pm\frac{1}{2}\hbar$ only, we postulate that they obey the *anticommutation* relations

$$\{\sigma_i, \sigma_j\}_+ = 2\delta_{ij}. \tag{2.8.15}$$

This is all that is needed as we shall show below to restrict the eigenvalues to $\pm\frac{1}{2}\hbar$.

From (2.8.15) we see that if $i = j$

$$\sigma_i^2 = 1, \tag{2.8.16}$$

where 1 is the identity operator. Note that (2.8.14) is therefore identically satisfied.

If we next add (2.8.13) and (2.8.15), we find that

$$\sigma_i\sigma_j = i\sigma_k, \tag{2.8.17}$$

where i, j, k are an even permutation of 1, 2, 3 so that $i \neq j$.

If we define the two nonhermitian operators $\sigma_\pm$ by

$$\begin{aligned} \sigma_\pm &= \tfrac{1}{2}(\sigma_1 \pm i\sigma_2) \\ \sigma_+ &= \sigma_-^\dagger \end{aligned} \tag{2.8.18}$$

we may use (2.8.13) and (2.8.15)–(2.8.17) to derive the following commutation and anticommutation relations:

$$[\sigma_{\pm}, \sigma_1] = \pm\sigma_3 \tag{2.8.19}$$

$$[\sigma_{\pm}, \sigma_2] = i\sigma_3 \tag{2.8.20}$$

$$[\sigma_{\pm}, \sigma_3] = \mp 2\sigma_{\pm} \tag{2.8.21}$$

$$[\sigma_+, \sigma_-] = \sigma_3 \tag{2.8.22}$$

$$\{\sigma_{\pm}, \sigma_1\}_+ = 1 \tag{2.8.23}$$

$$\{\sigma_{\pm}, \sigma_2\}_+ = \pm i \tag{2.8.24}$$

$$\{\sigma_{\pm}, \sigma_3\}_+ = 0 \tag{2.8.25}$$

$$\{\sigma_+, \sigma_-\}_+ = 1, \tag{2.8.26}$$

where 1 is the identity operator. Also

$$\begin{aligned} \sigma_+^2 &= 0 \\ \sigma_-^2 &= 0. \end{aligned} \tag{2.8.27}$$

We look for solutions of the eigenvalue problem

$$\sigma_3|\sigma_3'\rangle = \sigma_3'|\sigma_3'\rangle, \tag{2.8.28}$$

where

$$\sigma_3{}^2 = 1. \tag{2.8.29}$$

We have actually already solved this eigenvalue problem in Section 1.5. By virtue of the postulate (2.8.15) and therefore (2.8.16), $\sigma_3' = \pm 1$. This result has not used (2.8.13) or (2.8.14). Yet it is included as a special case of the results of Section 2.6 based on (2.8.13). That is, the anticommutation postulate has picked out only one of the eigenvalues of σ_3 allowed by (2.8.13) and (2.8.14). Therefore, we may take over (2.6.46) and (2.6.47) directly for this case since $\mathbf{s} = \frac{1}{2}\hbar\boldsymbol{\sigma}$

$$\begin{gathered} \sigma_1 = \begin{pmatrix} 0 & 1 \\ 1 & 0 \end{pmatrix}; \quad \sigma_2 = \begin{pmatrix} 0 & -i \\ +i & 0 \end{pmatrix}; \quad \sigma_3 = \begin{pmatrix} 1 & 0 \\ 0 & -1 \end{pmatrix} \\ |+\rangle = \begin{bmatrix} 1 \\ 0 \end{bmatrix} \quad |-\rangle = \begin{bmatrix} 0 \\ 1 \end{bmatrix}. \end{gathered} \tag{2.8.30}$$

Since $\boldsymbol{\sigma}^2 = \sigma_1{}^2 + \sigma_2{}^2 + \sigma_3{}^2 = 3$ always, there is no need to specify $\boldsymbol{\sigma}^2$. The σ_i are called the Pauli spin matrices in the representation in which σ_3 is diagonal. Also

$$\sigma_+ = \begin{pmatrix} 0 & 1 \\ 0 & 0 \end{pmatrix}; \quad \sigma_- = \begin{pmatrix} 0 & 0 \\ 1 & 0 \end{pmatrix}. \tag{2.8.31}$$

It is also easy to verify

$$\begin{aligned} &\sigma_1|+\rangle = |-\rangle; \qquad \sigma_2|+\rangle = i|-\rangle; \qquad \sigma_3|+\rangle = |+\rangle \\ &\sigma_1|-\rangle = |+\rangle; \qquad \sigma_2|-\rangle = -i|+\rangle; \qquad \sigma_3|-\rangle = -|+\rangle. \end{aligned} \tag{2.8.32}$$

Also

$$\begin{aligned} &\sigma_+|+\rangle = 0; \qquad \sigma_-|+\rangle = |-\rangle \\ &\sigma_+|-\rangle = |+\rangle; \qquad \sigma_-|-\rangle = 0. \end{aligned} \tag{2.8.33}$$

Let us now show how this explains the experimental results given by (2.8.8) and (2.8.10). The total angular momentum $\mathbf{j}$ is given by

$$\mathbf{j} = \mathbf{l} + \mathbf{s} = \mathbf{l} + \frac{\hbar}{2}\boldsymbol{\sigma} \tag{2.8.34}$$

while the total magnetic moment is

$$\begin{aligned} \mathbf{m} &= \mathbf{m}_l + \mathbf{m}_s \\ &= -\beta\left(\frac{\mathbf{l}}{\hbar} + \boldsymbol{\sigma}\right). \end{aligned} \tag{2.8.35}$$

Since $\mathbf{l}$ is orbital and $\boldsymbol{\sigma}$ spin momentum, $\mathbf{l}$ and $\boldsymbol{\sigma}$ commute

$$[l_i, \sigma_j] = 0. \tag{2.8.36}$$

Therefore, the total hamiltonian when we apply a magnetic field is

$$H = H_0 + \beta\left(\frac{\mathbf{l}}{\hbar} + \boldsymbol{\sigma}\right)\cdot\mathbf{F}. \tag{2.8.37}$$

The eigenvectors are $|n, m, l, s\rangle$ since

$$\begin{aligned} H_0|n, m, l, s\rangle &= E_{nl}|n, m, l, s\rangle \\ \sigma_3|n, m, l, s\rangle &= s|n, m, l, s\rangle \\ l_3|n, m, l, s\rangle &= m\hbar|n, m, l, s\rangle \\ \mathbf{l}^2|n, m, l, s\rangle &= l(l+1)\hbar^2|n, m, l, s\rangle. \end{aligned} \tag{2.8.38}$$

That is, since σ_3 commutes with all coordinate and momentum operators, we may simultaneously diagonalize H_0, σ_3, l_3, and $\mathbf{l}^2$. Therefore, if $\mathbf{F}$ is uniform along the z axis, we have

$$\langle n, m, l, s|H|n, m, l, s\rangle = E_{nl} + \beta F_0(m + s), \tag{2.8.39}$$

which agrees with experiment. The spectral lines are split by an amount $2\beta F_0$ due to the spin which is seen experimentally.

We have noted that the spin momentum eigenvectors do not possess a coordinate representation. We have then

$$\langle \mathbf{q}'|n, m, l, +1\rangle = \begin{pmatrix} \langle \mathbf{q}'|n, m, l\rangle \\ 0 \end{pmatrix} \tag{2.8.40}$$

$$\langle \mathbf{q}'|n, m, l, -1\rangle = \begin{pmatrix} 0 \\ \langle \mathbf{q}'|n, m, l\rangle \end{pmatrix}. \tag{2.8.41}$$

Also we have by Section 1.5

$$\begin{aligned} \langle s|s'\rangle &= \delta_{ss'} \\ |+1\rangle\langle +1| + |-1\rangle\langle -1| &= 1. \end{aligned} \tag{2.8.42}$$

As in the case of the harmonic oscillator, there is another formal interpretation of the energy of the electron spin in a d-c magnetic field. By (2.8.26) and (2.8.27)

$$\begin{aligned} \{\sigma_+, \sigma_-\}_+ &= I \\ \sigma_+^2 = \sigma_-^2 &= 0. \end{aligned} \tag{2.8.43}$$

But these are identical to the fermion operators b and $b^\dagger$ in (2.3.2). Since

$$\tfrac{1}{2}(\sigma_3 + 1) = \sigma_+\sigma_-, \tag{2.8.44}$$

then

$$H_{\text{spin}} = \beta F_0 \sigma_3 = \beta F_0(2\sigma_+\sigma_- - 1), \tag{2.8.45}$$

which aside from a trivial constant is the same as (2.3.1). Thus we may think of σ_+ and σ_- as creation and annihilation operators for a particle of spin $\frac{1}{2}$. Then since $\sigma_+\sigma_-$ is a number operator with eigenvalues 0 or 1, we have

$$\begin{aligned} \sigma_+\sigma_-|s\rangle &= \tfrac{1}{2}(\sigma_3 + 1)|s\rangle \\ &= \tfrac{1}{2}(s + 1)|s\rangle. \end{aligned} \tag{2.8.46}$$

The single "fermion" state is occupied or empty in agreement with the exclusion principle.

2.9 SPIN OPERATORS IN THE HEISENBERG PICTURE

We consider a free electron with spin in a d-c magnetic field along the z axis. The spin hamiltonian is

$$H - \beta F_0 \sigma_z. \tag{2.9.1}$$

The Heisenberg equations of motion (1.15.5) for σ_z, σ_+, and σ_- are

$$\begin{aligned}
\frac{d\sigma_z(t)}{dt} &= 0 \\
\frac{d\sigma_+(t)}{dt} &= \frac{1}{i\hbar}[\sigma_+(t), H] = i\frac{2\beta F_0}{\hbar}\sigma_+(t) \\
\frac{d\sigma_-(t)}{dt} &= \frac{1}{i\hbar}[\sigma_-(t), H] = -i2\frac{\beta F_0}{\hbar}\sigma_-(t),
\end{aligned} \tag{2.9.2}$$

where we used (2.8.21) and (2.9.1). The solutions are

$$\begin{aligned}
\sigma_z(t) &= e^{i\omega t\sigma_z/2}\sigma_z e^{-i\omega t\sigma_z/2} = \sigma_z \\
\sigma_+(t) &= e^{i\omega t\sigma_z/2}\sigma_+ e^{-i\omega t\sigma_z/2} = \sigma_+ e^{i\omega t} \\
\sigma_-(t) &= e^{i\omega t\sigma_z/2}\sigma_- e^{-i\omega t\sigma_z/2} = \sigma_- e^{-i\omega t},
\end{aligned} \tag{2.9.3}$$

where the separation between the two energy levels is

$$\hbar\omega = 2\beta F_0, \tag{2.9.4}$$

and we have used (1.15.3) with

$$U(t, 0) = e^{-i\omega t\sigma_z/2}. \tag{2.9.5}$$

The operators $\sigma_z(t)$, $\sigma_+(t)$, and $\sigma_-(t)$ in (2.9.3) are in the Heisenberg picture, and σ_z, σ_+, and σ_- are in the Schrödinger picture.

The state vector at time t is

$$|\psi(t)\rangle = e^{-i\omega t\sigma_z/2}|\psi(0)\rangle, \tag{2.9.6}$$

while an arbitrary initial state, by the completeness relation (2.8.42), may be written

$$\begin{aligned}
|\psi(0)\rangle &= |+1\rangle\langle+1|\psi(0)\rangle + |-1\rangle\langle-1|\psi(0)\rangle \\
&\equiv c_1|+1\rangle + c_2|-1\rangle,
\end{aligned} \tag{2.9.7}$$

where c_1 and c_2 are arbitrary constants. If we use (2.8.32), (2.9.6), and (2.9.7), the state vector at time t is

$$|\psi(t)\rangle = c_1 e^{-i\omega t/2}|+1\rangle + c_2 e^{i\omega t/2}|-1\rangle. \tag{2.9.8}$$

In order that $\langle\psi(t)|\psi(t)\rangle = \langle\psi(0)|\psi(0)\rangle = 1$, c_1 and c_2 must satisfy

$$|c_1|^2 + |c_2|^2 = 1. \tag{2.9.9}$$

It is left as an exercise to show that the expectation values of σ_z and $\sigma_\pm$ for the state $|\psi(t)\rangle$ are

$$\begin{aligned} \langle\sigma_3\rangle &= \langle\psi(t)|\sigma_z|\psi(t)\rangle = |c_1|^2 - |c_2|^2 \\ \langle\sigma_+\rangle &= c_1^*c_2e^{i\omega t} \qquad \langle\sigma_x\rangle = c_1^*c_2e^{i\omega t} + c_1c_2^*e^{-i\omega t} \\ \langle\sigma_-\rangle &= c_1c_2^*e^{-i\omega t} \qquad \langle\sigma_y\rangle = -ic_1^*c_2e^{i\omega t} + ic_1c_2^*e^{-i\omega t}, \end{aligned} \tag{2.9.10}$$

so that $\langle\sigma_x\rangle$ and $\langle\sigma_y\rangle$ will precess about the z axis at frequency ω while $\langle\sigma_z\rangle$ is left unchanged. This behavior is also typical of a classical magnet suspended on a gimble ring in a magnetic field.

PART III. ELECTRONS IN ELECTRIC AND MAGNETIC FIELDS

In Chapter 5, we give the quantum theory of the interaction of quantized radiation with matter. We include here a brief discussion of this interaction when the fields are unquantized.

2.10 HAMILTONIAN FOR ELECTRON IN ELECTROMAGNETIC FIELD

We consider an electron of charge e, mass μ, and spin $\mathbf{s}$ in a spherical potential as well as an electromagnetic field. If $\mathbf{A}$ is the vector potential defined by

$$\mathbf{B} = \text{curl}\,\mathbf{A} = \mu_0\mathbf{F}, \tag{2.10.1}$$

where μ_0 is the permeability of free space, the hamiltonian is (in mks units)†

$$\begin{aligned} H &= \frac{1}{2\mu}(\mathbf{p} - e\mathbf{A})^2 + V(r) + \beta\boldsymbol{\sigma}\cdot\mathbf{F} \\ &= \frac{1}{2\mu}\mathbf{p}^2 + V(r) + \frac{e^2}{2\mu}\mathbf{A}^2 - \frac{e}{2\mu}(\mathbf{p}\cdot\mathbf{A} + \mathbf{A}\cdot\mathbf{p}) + \beta\boldsymbol{\sigma}\cdot\mathbf{F}. \end{aligned} \tag{2.10.2}$$

From the commutation relations (2.7.4), we have

$$\sum_{i=1}^{3} [p_i, A_i] = \frac{\hbar}{i}\sum_i \frac{\partial A_i}{\partial q_i} = \frac{\hbar}{i}\,\text{div}\,\mathbf{A}. \tag{2.10.3}$$

Therefore, we may rewrite (2.10.2) as

$$H = \frac{1}{2\mu}\mathbf{p}^2 + V(r) + \frac{e^2}{2\mu}\mathbf{A}^2 - \frac{e}{\mu}\mathbf{A}\cdot\mathbf{p} + i\frac{e\hbar}{2\mu}\,\text{div}\,\mathbf{A} + \beta\boldsymbol{\sigma}\cdot\mathbf{F}. \tag{2.10.4}$$

† The generalized canonical momentum $\mathbf{p} = \mu\mathbf{v} + e\mathbf{A}$ where μ and $\mathbf{v}$ are the electron mass and velocity, respectively.

We are free to work in the Coulomb gauge for which

$$\operatorname{div} \mathbf{A} = 0. \tag{2.10.5}$$

Consider the special case in which

$$\mathbf{A} = \tfrac{1}{2}F_0[-y, x, 0]. \tag{2.10.6}$$

Then

$$\begin{aligned} \operatorname{div} \mathbf{A} &= 0 \\ \mathbf{F} &= [0, 0, F_0], \end{aligned} \tag{2.10.7}$$

and (2.10.4) becomes

$$H = \frac{1}{2\mu}\mathbf{p}^2 + V(r) + \frac{e^2F_0^2}{8\mu}(x^2 + y^2) - \frac{eF_0}{2\mu}[xp_y - yp_x] + \beta F_0\sigma_z. \tag{2.10.8}$$

Since $\beta = |e|\hbar/2\mu$ and $l_z = xp_y - yp_x$, we may rewrite this as $[e = -|e|]$

$$H = \frac{1}{2\mu}\mathbf{p}^2 + V(r) + \frac{e^2F_0^2}{8\mu}(x^2 + y^2) + \beta F_0\left(\frac{l_z}{\hbar} + \sigma_z\right). \tag{2.10.9}$$

If we compare this with (2.8.37), we see that we are in error unless the extra $\mathbf{A}^2$ term, namely, $e^2F_0^2(x^2 + y^2)/8\mu$, is negligible. However, the reader may show that $[l_3, x^2 + y^2] = 0$ but that $[\mathbf{l}^2, x^2 + y^2] \neq 0$, so that our results in Section 2.7 are not exact. In actual practice except for extremely strong fields the F_0 terms are small compared with the kinetic energy and the $F_0{}^2$ term is small compared with the F_0 term, so that the analysis in Section 2.7 is approximately valid for ordinary laboratory fields.

PROBLEMS

2.1 Write the matrix elements of p and q for a harmonic oscillator in the N-representation.

2.2 Show that the oscillator energy eigenfunction in the coordinate representation can be written as

$$u_n(q') = \sqrt{\frac{\delta^{2n+1}}{n!\sqrt{\pi}2^n}}\left(q' - \frac{\hbar}{\omega}\frac{d}{dq'}\right)^n e^{-\frac{1}{2}\delta^2 q'^2},$$

where $\delta = \sqrt{\omega/\hbar}$.

2.3 If n and m are integers and a and $a^\dagger$ are the boson annihilation and creation operators, respectively, show that

$$a^m|n\rangle = \sqrt{\frac{n!}{(n-m)!}}\,|n - m\rangle = \frac{\sqrt{n!}}{(n-m)!}(a^\dagger)^{n-m}|0\rangle$$

$$a^{\dagger m}|n\rangle = \frac{(a^\dagger)^{n+m}}{\sqrt{n!}}|0\rangle = \sqrt{\frac{(n+m)!}{n!}}\,|n + m\rangle,$$

where $a^\dagger a|n\rangle = n|n\rangle$.

2.4 If ξ is a parameter, show by a power-series expansion that

$$e^{\xi a}|0\rangle = |0\rangle$$

$$e^{\xi a^\dagger}|0\rangle = \sum_{n=0}^{\infty} \frac{\xi^n}{\sqrt{n!}} |n\rangle,$$

where $[a, a^\dagger] = 1$. If $\langle 0|0\rangle = 1$, normalize the vector $\exp(\xi a^\dagger)|0\rangle$.

2.5 Verify the commutation relations (2.6.4), (2.6.6), and (2.6.11).

2.6 Write the matrices for l_1, l_2, and l_3 when $l = \frac{3}{2}$.

2.7 Find the matrix elements of σ_+ and σ_- in a representation in which σ_x is diagonal.

2.8 Derive the relations given in (2.8.19)–(2.8.27).

2.9 Find the expectation values for σ_x and σ_y for the state given by (2.9.8).

2.10 Solve the Schrödinger equation for the hamiltonian (2.10.9) when $V(r) = 0$. Find a wave packet that localizes the particle on circular trajectories

REFERENCES

[1] P. A. M. Dirac, *The Principles of Quantum Mechanics*, 4th ed., Oxford: Clarendon, 1958, Chap. 11.

[2] L. I. Schiff, *Quantum Mechanics*, 3rd ed., New York: McGraw-Hill, 1968.

[3] See Ref. 1, Chaps. 9 and 10, and Ref. 2, Chaps. 14 and 15.

[4] S. S. Schweber, *An Introduction to Relativistic Quantum Field Theory*, New York: Harper & Row, 1961.

[5] W. H. Louisell, *Coupled Mode and Parametric Electronics*, New York: Wiley, 1961, Chap. 1; see *J. Appl. Phys.*, **33**, 2435 (1962).

[6] See Ref. 2, Chap. 14.

[7] See Ref. 2, Chap. 2.

[8] J. R. Klauder and E. C. G. Sudarshan, *Fundamentals of Quantum Optics*, New York: W. A. Benjamin, 1968; R. J. Glauber, *Phys. Rev.* **131,** 2766 (1963). The coherent state has been extensively used recently. See, for example, R. Horak, L. Mista, and J. Perina, *Phys. Lett. A* (*Netherlands*), **35A,** 400 (1971).

[9] H. Goldstein, *Classical Mechanics*, Reading, Mass.: Addison-Wesley, 1950, Chap. 5.

[10] G. E. Uhlenbeck and S. Goudsmit, *Naturwiss.*, **13,** 953 (1925); *Nature*, **117,** 264 (1926).

3

Operator Algebra

To solve problems in quantum mechanics, we have shown that operators are involved that obey a noncommutative algebra, which can sometimes be quite awkward. The first purpose of this chapter is to teach some of the techniques involved in dealing with operators [1].

The only difference between operators and *c*-numbers (commuting variables) is that the order in which operators are written down is significant while order is of no consequence with *c*-numbers. We therefore devote a major part of this chapter to the study of ordered operators which allows us to associate *c*-number functions with operators. In this way we may transform quantum problems to equivalent "classical" or *c*-number problems where the techniques of solution are usually more familiar. The equivalent "classical" problem contains all the quantum mechanical features and is therefore, strictly speaking, not a true classical problem. The major part of the chapter is devoted to the "classical" correspondence for boson operators, but we later generalize to arbitrary operators. To study this correspondence is the second major task of this chapter.

In Section 3.1 we begin by developing a few theorems which are valid for general noncommutative operators. Section 3.2 contains a discussion of normal and antinormal ordered boson operators. The algebraic properties of these operators are discussed in the following section. In Section 3.4 we introduce three characteristic functions for boson operators and derive the Wigner distribution function which is the Fourier transform of one of the characteristic functions. Since the Wigner distribution function as well as other distribution functions that we shall obtain may sometimes go negative and indeed fail to exist in certain instances, we see that our equivalent *c*-number representation is not a true classical probability distribution. We are not concerned here with these difficulties, however. In Sections 3.5 and 3.6, we discuss the Poisson and exponential distributions explicitly since they play such an important role in the statistical properties of radiation.

In Section 3.7 we derive the generalized Wick's theorem which allows us to write down practically by inspection the thermal average of the product of a large number of boson operators. In Section 3.8 we derive Wick's theorem for bosons which allows us to convert time-ordered operators to normal ordered operators. This makes the problem of taking matrix elements much simpler than would otherwise be the case and also allows us to make the transition to our associated c-number problem quite simply.

In Section 3.9 we generalize the ordering techniques to arbitrary operators which allows us to transform a quantum problem to an associated c-number ("classical") problem. We use this correspondence when we study the damping of atoms, and such, in later chapters.

In Section 3.10 we present an operator description of independent atoms which later proves quite useful.

In Part IV we give some elementary applications of the operator techniques developed. In Section 3.11 we solve the Schrödinger equation for a driven harmonic oscillator by the normal ordering technique. In Section 3.12 we discuss two weakly coupled oscillators while in the following two sections we obtain the distribution functions for a two level atom and for an oscillator. In Section 3.15 we obtain a generating function for oscillator eigenfunctions.

In Part V we discuss the principle of maximum entropy which is used to obtain the best estimate of the density operator for a quantum system subject only to partial knowledge about its state. We also obtain the density operator for spin-$\frac{1}{2}$ particles using this principle.

PART I. GENERAL OPERATORS

3.1 SOME GENERAL OPERATOR THEOREMS

In this section we derive several useful theorems involving two non-commuting operators A and B. We refer to functions of A or B, and we tacitly assume, without explicitly so stating in each theorem, that the functions may be expanded in a power series. Thus a function of B may be expanded as

$$F(B) = \sum_{n=0}^{\infty} c_n B^n, \tag{3.1.1}$$

where the c_n are constant expansion coefficients. The constants are called c-numbers. This is not the most general expansion of a function, but it suffices for our purposes. Some of the theorems are therefore true for more general functions, but we are not concerned with this.

Also, if functions such as (3.1.1) are to have meaning when applied to eigenvectors of B, for example, the series

$$F(b) = \sum_{n=0}^{\infty} c_n b^n, \tag{3.1.2}$$

where b is an eigenvalue of B and in general is a complex number, must converge, and $F(b)$ must be defined for complex b. If B is hermitian, then $F(b)$ would have to be defined only for real b. Again, we are not concerned about such matters.

We often use parameters that are c-numbers (not operators); we are not precise whether these parameters are real, pure imaginary, or complex, and tacitly assume that they can be complex, with other quantities defined adequately so that the meaning will be clear. For example, if ξ is complex and $F(A) = \exp(\xi A)$, we tacitly assume that $F(z)$ is a function defined for complex z.

THEOREM 1

If A and B are two noncommuting operators and ξ is a parameter, then, if n is an integer,

$$e^{\xi A} B^n e^{-\xi A} = (e^{\xi A} B e^{-\xi A})^n, \tag{3.1.3}$$

and

$$e^{\xi A} F(B) e^{-\xi A} = F(e^{\xi A} B e^{-\xi A}). \tag{3.1.4}$$

When $n = 1$, (3.1.3) is just an identity.

PROOF

To prove this theorem, we note first that

$$e^{\xi A} e^{-\xi A} = I. \tag{3.1.5}$$

We may write the right side of (3.1.3) as a product of n-factors

$$e^{\xi A} B e^{-\xi A} e^{\xi A} B e^{-\xi A} \cdots e^{+\xi A} B e^{-\xi A},$$

and by (3.1.5) all factors in the middle collapse to B^n and (3.1.3) is proved.

To prove (3.1.4), we use the expansion (3.1.1) to write the left side of (3.1.4) as

$$\begin{aligned} e^{\xi A} F(B) e^{-\xi A} &= \sum_n c_n e^{\xi A} B^n e^{-\xi A} \\ &= \sum_n c_n (e^{\xi A} B e^{-\xi A})^n, \end{aligned}$$

where we used (3.1.3) in the last step. From the expansion (3.1.1), with the argument replaced by $e^{\xi A} B e^{-\xi A}$, we see from the above that (3.1.4) follows.

As an application of this theorem, we let $A = ip/\hbar$, $B = F(q)$, where $[q, p] = i\hbar$. By (3.1.4),

$$\exp\left(\frac{i\xi p}{\hbar}\right) F(q) \exp\left(-\frac{i\xi p}{\hbar}\right) = F\left[\exp\left(\frac{i\xi p}{\hbar}\right) q \exp\left(-\frac{i\xi p}{\hbar}\right)\right] \tag{3.1.6}$$

If ξ is real, we have, by (1.10.8), (1.10.9), and (1.10.12),

$$\exp\left(\frac{i\xi p}{\hbar}\right) q \exp\left(-\frac{i\xi p}{\hbar}\right) = q + \xi, \tag{3.1.7}$$

so that (3.1.6) reduces to

$$f(\xi, p, q) = \exp\left(\frac{i\xi p}{\hbar}\right) F(q) \exp\left(-\frac{i\xi p}{\hbar}\right) = F(q + \xi). \tag{3.1.8}$$

There is an interesting alternative proof for (3.1.8). If we differentiate (3.1.8) partially with respect to ξ, we have

$$\frac{\partial f}{\partial \xi} = \frac{i}{\hbar}[p, f] = \frac{\partial f}{\partial q}, \tag{3.1.9}$$

where we used (1.10.6). This gives a partial differential equation that f must satisfy. By inspection, we see that any function of the form

$$f(\xi, p, q) = g(q + \xi) \tag{3.1.10}$$

is a solution of (3.1.9). To evaluate the form of the function g, we let $\xi = 0$ in (3.1.8) and (3.1.10) and see that

$$f(0, p, q) = F(q) = g(q),$$

so that (3.1.8) follows. We make repeated use of this technique in the remainder of the chapter.

THEOREM 2

If A and B are two noncommuting operators and if A^{-1} exists, we have

$$AB^nA^{-1} = (ABA^{-1})^n, \tag{3.1.11}$$

where n is an integer, and

$$AF(B)A^{-1} = F(ABA^{-1}). \tag{3.1.12}$$

The proof of this theorem follows the same pattern as that of Theorem 1 and is omitted. In Theorem 1 $\exp(\pm\xi A)$ always exists whereas here only operators A for which A^{-1} exists may be used.

An interesting case that often arises is $f(B) = \exp(B)$ so that (3.1.12) becomes

$$Ae^{B}A^{-1} = \exp(ABA^{-1}) \tag{3.1.13}$$

THEOREM 3

If A and B are two fixed noncommuting operators and ξ is a parameter, then

$$e^{\xi A}Be^{-\xi A} = B + \xi[A, B] + \frac{\xi^2}{2!}[A, [A, B]] + \frac{\xi^3}{3!}[A, [A, [A,B]]] + \cdots. \tag{3.1.14}$$

PROOF

We let

$$f(\xi) = e^{\xi A}Be^{-\xi A} \qquad f(0) = B,$$

and expand $f(\xi)$ in a Maclaurin series in powers of ξ. We then have

$$\frac{df}{d\xi} = [A, f(\xi)] \qquad \left.\frac{df}{d\xi}\right|_{\xi=0} = [A, B]$$

$$\frac{d^2f}{d\xi^2} = \left[A, \frac{df}{d\xi}\right] = [A, [A, f(\xi)]] \qquad \left.\frac{d^2f}{d\xi^2}\right|_{\xi=0} = [A, [,A\ B]].$$

If we continue in this way, (3.1.14) follows.

As one application of this theorem, we let $A = ip/\hbar$, $B = q$, and assume ξ is real. If $[p, q] = -i\hbar$, we have, from (3.1.14),

$$\exp\left(\frac{ip\xi}{\hbar}\right)q \exp\left(-\frac{ip\xi}{\hbar}\right) = q + \xi\left[\frac{ip}{\hbar}, q\right] = q + \xi, \tag{3.1.15}$$

since all the remaining commutators in (3.1.14) vanish. This is just (3.1.7), which we found by an alternative method.

As another illustration of the theorem, we let $A = q^2/2$ and $B = d/dq$, where q is an ordinary variable. It is easy to show that

$$[A, B] = \frac{1}{2}\left[q^2, \frac{d}{dq}\right] = -q \tag{3.1.16}$$

where we tacitly assume that both sides of (3.1.16) are to operate on some function $F(q)$. Also,

$$[A, [A, B]] = \tfrac{1}{2}[q^2, -q] = 0,$$

so that the series in (3.14) breaks off after the first two terms, and we have

$$e^{\xi q^2/2}\frac{d}{dq}e^{-\xi q^2/2} = \frac{d}{dq} - \xi q. \tag{3.1.17}$$

We again tacitly assume that we are to apply both sides of this operator equality to some function of q, say $F(q)$.

From (3.1.17) and Theorem 2 (Eq. 3.1.11) we find also that

$$e^{\xi q^2/2} \frac{d^n}{dq^n} e^{-\xi q^2/2} = \left(\frac{d}{dq} - \xi q\right)^n, \tag{3.1.18}$$

where n is an integer.

THEOREM 4

If A and B are two noncommuting operators that satisfy the conditions

$$[A, [A, B]] = [B, [A, B]] = 0, \tag{3.1.19}$$

then

$$e^{A+B} = e^A e^B e^{-\frac{1}{2}[A,B]} = e^B e^A e^{+\frac{1}{2}[A,B]}. \tag{3.1.20}$$

This is a special case of the Baker-Hausdorff theorem of group theory; the reader interested in pursuing the operator formalism to a greater degree of sophistication may find Refs. 2 to 5 of interest. The proof of this theorem given here is due to Glauber [6].

Any two operators whose commutator is a c-number, for example, $[q, p] = i\hbar$ and $[a, a^\dagger] = 1$, satisfy the conditions of the theorem; thus it is not surprising that this theorem has many applications.

PROOF

To prove the theorem, we consider the operator function

$$f(\xi) = e^{\xi A} e^{\xi B}, \tag{3.1.21}$$

where ξ is a (c-number) parameter. If we differentiate with respect to ξ, we have

$$\begin{aligned} \frac{df}{d\xi} &= A e^{\xi A} e^{\xi B} + e^{\xi A} e^{\xi B} B \\ &= (A + e^{\xi A} B e^{-\xi A}) f(\xi), \end{aligned} \tag{3.1.22}$$

since $\exp(\xi A)\exp(-\xi A) = I$. We may use Theorem 3 (Eq. 3.1.14) to expand the second term in the parentheses. By virtue of (3.1.19), all terms after the first two in (3.1.14) vanish, and so

$$e^{\xi A} B e^{-\xi A} = B + \xi[A, B]. \tag{3.1.23}$$

Equation 3.1.22 may be written

$$\frac{df}{d\xi} = \{(A + B) + \xi[A, B]\} f(\xi). \tag{3.1.24}$$

By (3.1.19) we see that the quantity $A + B$ commutes with $[A, B]$, and so we may treat these two quantities as ordinary commuting variables and integrate (3.1.24) subject to the initial condition [see (3.1.21)]

$$f(0) = 1. \tag{3.1.25}$$

The solution of (3.1.24) that satisfies (3.1.25) is

$$f(\xi) = e^{(A+B)\xi+(\xi^2/2)[A,B]} = e^{(A+B)\xi}e^{(\xi^2/2)[A,B]}, \tag{3.1.26}$$

where the last form follows since $A + B$ commutes with $[A, B]$. If we now equate (3.1.21) and (3.1.26), let $\xi = 1$, and multiply both sides from the right by $\exp(-\frac{1}{2}[A, B])$, then (3.1.20) follows. The proof of the second form of (3.1.20) is left as an exercise.

As an application of (3.1.20), we let $A = \lambda p$ and $B = \mu q$, where $[q, p] = i\hbar$ and λ and μ are parameters. Then by (3.1.20) we have

$$e^{\lambda p+\mu q} = e^{\lambda p}e^{\mu q}\exp\left(\frac{i\hbar\lambda\mu}{2}\right). \tag{3.1.27}$$

PART II. BOSON CREATION AND ANNIHILATION OPERATORS

In later chapters we are concerned with solutions of quantum problems involving creation and annihilation operators for bosons. This will include, for example, the calculation of expectation values of operators that are functions of a and $a^\dagger$, since these averages permit comparisons between theory and experiment. We also need to solve Schrödinger equations involving boson operators.

One example is a driven harmonic oscillator where the hamiltonian is

$$H = \hbar\omega(a^\dagger a + \tfrac{1}{2}) + \hbar f(t)(a + a^\dagger).$$

The function $f(t)$ is the driving term. Since this hamiltonian is time-dependent, the system is nonconservative. In such a simple problem as this, the only method of solution at our disposal at the moment is the cumbersome iterated solution given in Section 1.16. Our purpose here is to make the solution of such problems possible by much more powerful operator techniques. These techniques involve the concepts of ordered operators.

3.2 ORDERED BOSON OPERATORS

Any function $f(a, a^\dagger)$ of the boson operators a and $a^\dagger$ which satisfy the commutation relation $[a, a^\dagger] = 1$ is defined by its power series expansion in

a and $a^\dagger$. For example, $\sin a^\dagger a a^\dagger$ is defined by

$$\sin a^\dagger a a^\dagger = \sum_{l=0}^{\infty} (-1)^l \frac{(a^\dagger a a^\dagger)^{2l+1}}{(2l+1)!}.$$

More generally, $f(a, a^\dagger)$ will usually consist of sums of terms of the form

$$f(a, a^\dagger) = \sum_l \cdots \sum_n a^{\dagger l} a^m a^{\dagger r} \cdots a^n f(l, m, \ldots, n),$$

where $l, m, \ldots$ are positive integers or zero. We are always free to use the commutation relation $aa^\dagger - a^\dagger a = 1$ repeatedly to rearrange the a's and $a^\dagger$'s among themselves. This will yield different forms for $f(a, a^\dagger)$ but they will always be equal. For example, if $f = aa^\dagger$, then

$$f(a, a^\dagger) = aa^\dagger = a^\dagger a + 1.$$

The two functions have different forms, but they are equal. Suppose now we repeatedly use the commutation relation in all terms in f so that all a's in every term of the sum are to the right of all $a^\dagger$'s. The function then is said to be in *normal* order, and we may write it as

$$f(a, a^\dagger) = f^{(n)}(a, a^\dagger) = \sum_{r,s} f_{rs}^{(n)} a^{\dagger r} a^s. \tag{3.2.1}$$

We put a superscript n on the operator function to remind us that it is in normal order and since we have faithfully used the commutation relation to put f into normal order, $f = f^{(n)}$. The $f_{rs}^{(n)}$ are expansion coefficients which are independent of a and $a^\dagger$.

As another alternative we may commute all a's to the left so that the same function $f(a, a^\dagger)$ can also be written as

$$f(a, a^\dagger) = f^{(a)}(a, a^\dagger) = \sum_{r,s} f_{rs}^{(a)} a^r a^{\dagger s}. \tag{3.2.2}$$

In this case the function is said to be in *antinormal* order which we indicate by a superscript a. In general, $f_{rs}^{(a)} \neq f_{rs}^{(a)}$ although

$$f(a, a^\dagger) = f^{(n)}(a, a^\dagger) = f^{(a)}(a, a^\dagger). \tag{3.2.3}$$

These orderings are possible for all functions which may be expanded in a power series. However, in all but a few trivial cases, the ordering will be a very tedious procedure. In the next section, we develop some very useful techniques for obtaining the ordered operators by indirect means in a number of useful cases.

Since the normal and antinormal forms of a function which can be expanded in a power series are unique, we can establish a one-to-one correspondence between either $f^{(n)}(a, a^\dagger)$ or $f^{(a)}(a, a^\dagger)$ and ordinary functions $\bar{f}^{(n)}(\alpha, \alpha^*)$ or

$\bar{f}^{(a)}(\alpha, \alpha^*)$ of a complex variable α which will prove extremely useful. This correspondence may be accomplished as follows.

We define an operator $\mathcal{N}^{-1}$ which transforms the operator function $f^{(n)}(a, a^\dagger)$ to an ordinary function $\bar{f}^{(n)}(\alpha, \alpha^*)$ of the complex variable α by replacing a by α and $a^\dagger$ by α^*. That is,

$$\mathcal{N}^{-1}\{a^{\dagger l}a^m\} = \alpha^{*l}\alpha^m, \tag{3.2.4}$$

where l and m are integers. The operator is defined to be linear so that

$$\mathcal{N}^{-1}\{cf^{(n)}(a, a^\dagger)\} = c\bar{f}^{(n)}(\alpha, \alpha^*), \tag{3.2.5}$$

where c is any complex number, and

$$\mathcal{N}^{-1}\{f_1^{(n)}(a, a^\dagger) + f_2^{(n)}(a, a^\dagger)\} = \bar{f}_1(\alpha, \alpha^*) + \bar{f}_2(\alpha, \alpha^*) \tag{3.2.6}$$

$$\mathcal{N}^{-1}\{cI\} = c, \tag{3.2.7}$$

where I is the identity operator. From this definition, it follows from (3.2.1) that

$$\mathcal{N}^{-1}\{f^{(n)}(a, a^\dagger)\} = \sum_{r,s} f_{rs}^{(n)}\alpha^{*r}\alpha^s = \bar{f}^{(n)}(\alpha, \alpha^*). \tag{3.2.8}$$

We have used a bar over the function to indicate that we have an ordinary function, and the superscript n indicates the function is associated with the normally ordered form of the operator $f^{(n)}(a, a^\dagger)$. Since the normal form of the operator is unique, there is a one-to-one correspondence between $f^{(n)}(a, a^\dagger)$ and $\bar{f}^{(n)}(\alpha, \alpha^*)$.

To obtain $\bar{f}^{(n)}(\alpha, \alpha^*)$ we see that we first put $f(a, a^\dagger)$ into normal order and then replace a by α and $a^\dagger$ by α^*.

We may also define the normal ordering operator $\mathcal{N}$ by

$$\mathcal{N}\{\alpha^l\alpha^{*m}\} = a^{\dagger m}a^l \tag{3.2.9a}$$

$$\mathcal{N}\{c\bar{f}_1^{(n)}(\alpha, \alpha^*)\} = cf_1^{(n)}(a, a^\dagger) \tag{3.2.9b}$$

$$\mathcal{N}\{\bar{f}_1^{(n)}(\alpha, \alpha^*) + \bar{f}_2^{(n)}(\alpha, \alpha^*)\} = f_1^{(n)}(a, a^\dagger) + f_2^{(n)}(a, a^\dagger) \tag{3.2.9c}$$

$$\mathcal{N}\{c\} = cI \tag{3.2.9d}$$

$$\mathcal{N}\mathcal{N}^{-1} = \mathcal{N}^{-1}\mathcal{N} = 1. \tag{3.2.9e}$$

The normal ordering operator $\mathcal{N}$ replaces α^* by $a^\dagger$, α by a with all $a^\dagger$'s to the left of all a's.

In a similar way we may define the operator $\mathcal{A}^{-1}$ which is applied to antinormally ordered operators by means of

$$\mathcal{A}^{-1}\{a^l a^{\dagger m}\} = \alpha^l\alpha^{*m} \tag{3.2.10a}$$

$$\mathcal{A}^{-1}\{cf^{(a)}(a, a^\dagger)\} = c\bar{f}^{(a)}(\alpha, \alpha^*) \tag{3.2.10b}$$

$$\mathcal{A}^{-1}\{f_1^{(a)}(a, a^\dagger) + f_2^{(a)}(a, a^\dagger)\} = \bar{f}_1^{(a)}(\alpha, \alpha^*) + \bar{f}_2^{(a)}(\alpha, \alpha^\dagger) \tag{3.2.10c}$$

$$\mathcal{A}^{-1}(cI) = c. \tag{3.2.10d}$$

We define $\mathscr{A}$ as the inverse of $\mathscr{A}^{-1}$:

$$\mathscr{A}\mathscr{A}^{-1} = \mathscr{A}^{-1}\mathscr{A} = 1. \tag{3.2.11}$$

Therefore, when $\mathscr{A}$ is applied to a function of α and α^* we have

$$\mathscr{A}\{\alpha^l\alpha^{*m}\} = a^l a^{\dagger m} \tag{3.2.11a}$$

$$\mathscr{A}\{c\bar{f}^{(a)}(\alpha, \alpha^*)\} = cf^{(a)}(a, a^\dagger) \tag{3.2.11b}$$

$$\mathscr{A}\{\bar{f}_1^{(a)}(\alpha, \alpha^*) + \bar{f}_2^{(a)}(\alpha, \alpha^*)\} = f_1^{(a)}(a, a^\dagger) + f_2^{(a)}(a, a^\dagger) \tag{3.2.11c}$$

$$\mathscr{A}\{c\} = cI. \tag{3.2.11d}$$

Again $\mathscr{A}$ replaces α by a, α^* by $a^\dagger$ and puts all $a^\dagger$'s to the *right* of all a's.

To gain further insight into the ordinary associated functions $\bar{f}^{(n)}(\alpha, \alpha^*)$ and $\bar{f}^{(a)}(\alpha, \alpha^*)$, let us see how they are connected with the coherent state representation given in Section 2.5.

THEOREM 1*

The associated normal function $\bar{f}^{(n)}(\alpha, \alpha)$ is given by the diagonal matrix element of $f(a, a^\dagger)$ in the coherent state representation:

$$\bar{f}^{(n)}(\alpha, \alpha^*) = \langle\alpha|f(a, a^\dagger)|\alpha\rangle = \mathrm{Tr}\,[|\alpha\rangle\langle\alpha|f(a, a^\dagger)], \tag{3.2.12}$$

where $a|\alpha\rangle = \alpha|\alpha\rangle$.

The proof follows easily. From Sec. 2.5 we know that since $a|\alpha\rangle = \alpha|\alpha\rangle$ and $\langle\alpha|a^\dagger = \alpha^*\langle\alpha|$, then

$$\begin{aligned} F(a)|\alpha\rangle &= F(\alpha)|\alpha\rangle \\ \langle\alpha|\mathscr{G}(a^\dagger) &= \mathscr{G}(\alpha^*)\langle\alpha| \\ \langle\alpha|\alpha\rangle &= 1. \end{aligned} \tag{3.2.13}$$

If we therefore take the diagonal matrix element of both sides of (3.2.1) and use (3.2.13) we obtain

$$\langle\alpha|f(a, a^\dagger)|\alpha\rangle = \sum_{rs} f_{rs}^{(n)}\langle\alpha|a^{\dagger r}a^s|\alpha\rangle = \sum_{rs} f_{rs}^{(n)}\alpha^{*r}\alpha^s = \bar{f}^{(n)}(\alpha, \alpha^*),$$

where in the last step we used (3.2.8). This proves the first half of the theorem. Since $\mathrm{Tr}\,|u\rangle\langle v| = \langle v|u\rangle$, the last form follows if we let $|u\rangle = |\alpha\rangle$ and $\langle v| = \langle v|f(a, a^\dagger)$.

* Theorems referred to in this section are those in the section unless otherwise noted.

THEOREM 2

We may put a function $f(a, a^\dagger)$ into normal form by means of

$$f^{(n)}(a, a^\dagger) = \mathcal{N}\{\langle\alpha|f(a, a^\dagger)|\alpha\rangle\} = \mathcal{N}\left\{\bar{f}\left(\alpha + \frac{\partial}{\partial\alpha^*}, \alpha^*\right)\cdot 1\right\}, \tag{3.2.14}$$

where $\bar{f}(\alpha + (\partial/\partial\alpha^*, \alpha^*)$ is obtained by replacing a by $\alpha + (\partial/\partial\alpha^*)$ and $a^\dagger$ by α^* in the original power series expansion of the function $f(a, a^\dagger)$. Note that if f is already in normal form

$$\bar{f}^{(n)}\left(\alpha + \frac{\partial}{\partial\alpha^*}, \alpha^*\right)\cdot 1 = \bar{f}^{(n)}(\alpha, \alpha^*), \tag{3.2.15}$$

since all the terms $[\alpha + (\partial/\partial\alpha^*)]$ will appear on the right and $\partial/\partial\alpha^* \cdot 1 = 0$. Actually this theorem merely replaces commutation relations by differentiations and is just as tedious to apply in general. It will prove useful in certain special situations.

To prove the theorem we shall need the following lemma.

LEMMA

If $|\alpha\rangle$ is a coherent state vector, then

$$|\alpha\rangle\langle\alpha|a = \left(\alpha + \frac{\partial}{\partial\alpha^*}\right)|\alpha\rangle\langle\alpha|. \tag{3.2.16}$$

PROOF

By (2.5.12b) and its adjoint, we have

$$|\alpha\rangle\langle\alpha| = e^{-\alpha^*\alpha}e^{\alpha a^\dagger}|0\rangle\langle 0|e^{\alpha^* a}. \tag{3.2.17}$$

Then we see that $|\alpha\rangle\langle\alpha|a$ may be written as

$$|\alpha\rangle\langle\alpha|a = e^{-\alpha^*\alpha}\frac{\partial}{\partial\alpha^*}\{e^{\alpha a^\dagger}|0\rangle\langle 0|e^{\alpha^* a}\}, \tag{3.2.18}$$

since α and α^* may be visualized as independent variables and the differentiation with respect to α^* will bring down the desired operator a on the right side since a commutes with $\exp(\alpha^* a)$. We may obviously rewrite the right side of (3.2.18) as

$$\begin{aligned}|\alpha\rangle\langle\alpha|a &= \left(\frac{\partial}{\partial\alpha^*} + \alpha\right)e^{-\alpha^*\alpha}e^{\alpha a^\dagger}|0\rangle\langle 0|e^{\alpha^* a} \\ &= \left(\frac{\partial}{\partial\alpha^*} + \alpha\right)|\alpha\rangle\langle\alpha|. \qquad \text{Q.E.D.}\end{aligned}$$

By repeated application of the lemma, we see that

$$|\alpha\rangle\langle\alpha|a^l = \left(\alpha + \frac{\partial}{\partial\alpha^*}\right)^l |\alpha\rangle\langle\alpha|. \tag{3.2.19}$$

Similarly, since $\langle\alpha|a^\dagger = \alpha^*\langle\alpha|$, we see that

$$|\alpha\rangle\langle\alpha|a^{\dagger m} = \alpha^{*m}|\alpha\rangle\langle\alpha|. \tag{3.2.20}$$

We also note that

$$\mathrm{Tr}\,|\alpha\rangle\langle\alpha| = \langle\alpha|\alpha\rangle = 1. \tag{3.2.21}$$

Consider now a function expanded in a power series in arbitrary order:

$$f(a, a^\dagger) = \sum \cdots a^{\dagger l} a^m \cdots a^u a^{\dagger v} \cdots. \tag{3.2.22}$$

Then if we take diagonal matrix elements of both sides in the coherent state representation, we have by Theorem 1

$$\begin{aligned} f^{(n)}(\alpha, \alpha^*) &= \langle\alpha|f(a, a^\dagger)|\alpha\rangle = \mathrm{Tr}\,\{|\alpha\rangle\langle\alpha|f(a, a^\dagger)\} \\ &= \mathrm{Tr}\,\{|\alpha\rangle\langle\alpha| \textstyle\sum \cdots a^{\dagger l} a^m \cdots a^u a^{\dagger v} \cdots\}. \end{aligned}$$

If we repeatedly use (3.1.19) and (3.1.20), we may move the projection operator $|\alpha\rangle\langle\alpha|$ through each term as follows:

$$\begin{aligned} f^{(n)}(\alpha, \alpha^*) &= \mathrm{Tr}\,\{\textstyle\sum \cdots \alpha^{*l}|\alpha\rangle\langle\alpha|a^m \cdots a^u a^{\dagger v} \cdots\} \\ &= \mathrm{Tr}\left\{\sum \cdots \alpha^{*l}\left(\alpha + \frac{\partial}{\partial\alpha^*}\right)^m \cdots \left(\alpha + \frac{\partial}{\partial\alpha^*}\right)^u \alpha^{*v} \cdots |\alpha\rangle\langle a|\right\} \\ &= \sum \cdots \alpha^{*l}\left(\alpha + \frac{\partial}{\partial\alpha^*}\right)^m \cdots \left(\alpha + \frac{\partial}{\partial\alpha^*}\right)^u \alpha^{*v} \cdots \cdot 1, \end{aligned}$$

where in the last step we used (3.2.21). If we compare this with the original expansion of the function (3.2.22), we see that we obtain the normal associated function by replacing $a^\dagger$ everywhere by α^* and a by $\alpha + (\partial/\partial\alpha^*)$ and let it operate on unity. The theorem then follows if we apply the operator $\mathcal{N}$ to both sides. We have already noted that this method of putting a function into normal order is completely equivalent to using the commutation relations repeatedly.

THEOREM 3

Any function $f(a, a^\dagger)$ which is suitably behaved may be represented by the integral

$$\begin{aligned} f(a, a^\dagger) = f^{(a)}(a, a^\dagger) &= \int \frac{d^2\alpha}{\pi}\, |\alpha\rangle\langle\alpha|\, \bar{f}^{(a)}(\alpha, \alpha^*) \\ &= \mathscr{A}\{\bar{f}^{(a)}(\alpha, \alpha^*)\}, \end{aligned} \tag{3.2.23}$$

where $\bar{f}^{(a)}(\alpha, \alpha^*)$ is the antinormal associated function.

This theorem states that the antinormal ordering operator is equivalent to the integral operator $\int d^2\alpha|\alpha\rangle\langle\alpha|/\pi$. Glauber [7] has called $\bar{f}^{(a)}(\alpha, \alpha^*)/\pi$ the P-representation of the operator $f(a, a^\dagger)$.

The proof of this theorem follows from the completeness relation (2.5.17)

$$\int \frac{d^2\alpha}{\pi} |\alpha\rangle\langle\alpha| = 1. \tag{3.2.24}$$

From (3.2.2) we have

$$\begin{aligned} f(a, a^\dagger) &= \sum_{rs} f_{rs}^{(a)} a^r a^{\dagger s} = \int \frac{d^2\alpha}{\pi} \sum_{rs} f_{rs}^{(a)} a^r |\alpha\rangle\langle\alpha| a^{\dagger s} \\ &= \int \frac{d^2\alpha}{\pi} |\alpha\rangle\langle\alpha| \sum_{rs} f_{rs}^{(a)} \alpha^r \alpha^{*s} = \int \frac{d^2\alpha}{\pi} |\alpha\rangle\langle\alpha| \bar{f}^{(a)}(\alpha, \alpha^*). \end{aligned}$$

Here we have inserted the completeness relation between a^r and $a^{\dagger s}$, and used (3.2.13) and the associated antinormal function which proves the first part of the theorem. Since

$$f(a, a^\dagger) = f^{(a)}(a, a^\dagger) = \mathscr{A}\{\bar{f}^{(a)}(\alpha, \alpha^*)\},$$

and the second half of the theorem follows.

THEOREM 4

The antinormal form of $f(a, a^\dagger)$ may be obtained by

$$f^{(a)}(a, a^\dagger) = \mathscr{A}\left\{\bar{f}\left(\alpha, \alpha^* - \frac{\partial}{\partial\alpha}\right)\cdot 1\right\} = \int \frac{d^2\alpha}{\pi} |\alpha\rangle\langle\alpha| \bar{f}\left(\alpha, \alpha^* - \frac{\partial}{\partial\alpha}\right)\cdot 1, \tag{3.2.25}$$

where $\bar{f}(\alpha, \alpha^* - (\partial/\partial\alpha))$ is obtained from the original power series expansion of $f(a, a^\dagger)$ and replacing each a by α and each $a^\dagger$ by $\alpha^* - (\partial/\partial\alpha)$. Note that if f is in antinormal order, then

$$\begin{aligned} \bar{f}\left(\alpha, \alpha^* - \frac{\partial}{\partial\alpha}\right)\cdot 1 &= \sum_{rs} f_{rs}^{(a)} \alpha^r \left(\alpha^* - \frac{\partial}{\partial\alpha}\right)^s \cdot 1 \\ &= \sum_{rs} f_{rs}^{(a)} \alpha^r \alpha^{*s} = \bar{f}^{(a)}(\alpha, \alpha^*), \end{aligned} \tag{3.2.26}$$

since $(\partial/\partial\alpha)\cdot 1 = 0$. This is just the analog of Theorem 2 (3.2.14) for antinormal ordering. It again replaces commutation relations by differentiations of ordinary complex functions.

The proof of the theorem is rather tedious. If we use the completeness relation

(3.2.24), we may write

$$f(a, a^\dagger) = \int f(a, a^\dagger)|\alpha\rangle\langle\alpha| \frac{d^2\alpha}{\pi}$$

$$= \sum \int \cdots a^{\dagger l} a^m \cdots a^u a^{\dagger r} a^v |\alpha\rangle\langle\alpha| \frac{d^2\alpha}{\pi}. \tag{3.2.27}$$

We next need the following lemma to move the projection operator $|\alpha\rangle\langle\alpha|$ to the left.

LEMMA

$$a^\dagger|\alpha\rangle\langle\alpha| = \left(\alpha^* + \frac{\partial}{\partial\alpha}\right)|\alpha\rangle\langle\alpha| \tag{3.2.28a}$$

To prove this, we have

$$a^\dagger|\alpha\rangle\langle\alpha| = e^{-\alpha^*\alpha} \frac{\partial}{\partial\alpha} e^{\alpha a^\dagger}|0\rangle\langle 0|e^{\alpha^* a}$$

$$= \left(\frac{\partial}{\partial\alpha} + \alpha^*\right)|\alpha\rangle\langle\alpha|. \qquad \text{Q.E.D.}$$

By repeated application of the lemma we see that

$$a^{\dagger r}|\alpha\rangle\langle\alpha| = \left(\alpha^* + \frac{\partial}{\partial\alpha}\right)^r |\alpha\rangle\, \langle\alpha|. \tag{3.2.28b}$$

Since α^* and α are independent variables, we see that

$$\left[\alpha^*, \frac{\partial}{\partial\alpha}\right] = 0,$$

so that we may expand the right side of (3.2.28b) by the binomial theorem, which gives

$$a^{\dagger r}|\alpha\rangle\langle\alpha| = \sum_{s=0}^{r} \frac{r!}{s!\,(r-s)!} \alpha^{*r-s} \frac{\partial^s}{\partial\alpha^s} |\alpha\rangle\langle\alpha|. \tag{3.2.28c}$$

Consider now the last two terms in (3.2.27)

$$I = \int a^{\dagger r} a^v |a\rangle\langle\alpha| \frac{d^2\alpha}{\pi} = \int \alpha^v a^{\dagger r} |\alpha\rangle\langle\alpha| \frac{d^2\alpha}{\pi},$$

since $a^v|\alpha\rangle\langle\alpha| = \alpha^v|\alpha\rangle\langle\alpha|$. If we use (3.2.28c) this becomes

$$I = \sum_{s=0}^{r} \frac{r!}{s\,!(r-s)!} \int \alpha^{*r-s}\alpha^v \frac{\partial^s}{\partial\alpha^s} |\alpha\rangle\langle\alpha| \frac{d\alpha\, d\alpha^*}{\pi},$$

where we have written $d^2\alpha = d\alpha\, d\alpha^*$. (To be correct, we should change to real variables, but the reader may verify that our procedure gives the correct result without the extra labor.) We now integrate this expression for I by parts s-times with respect to α. Since the integrated parts vanish at infinity due to the presence of the factor $\exp(-\alpha^*\alpha)$ in $|\alpha\rangle\langle\alpha|$ which goes to zero faster than any finite power of α and α^* then, I becomes

$$I = \sum_{s=0} \frac{r!}{s!\,(r-s)!} \int |\alpha\rangle\langle\alpha| \frac{d^2\alpha}{\pi} \alpha^{*r-s}\left(-\frac{\partial}{\partial\alpha}\right)^s \alpha^v.$$

We may sum the series by the binomial theorem, and we have

$$I = \int |\alpha\rangle\langle\alpha| \frac{d^2\alpha}{\pi}\left(\alpha^* - \frac{\partial}{\partial\alpha}\right)^r \alpha^v \cdot 1.$$

If we proceed in this way to move $|\alpha\rangle\langle\alpha|$ through the other operators we see that

$$\int f(a, a^\dagger)|\alpha\rangle\langle\alpha| \frac{d^2\alpha}{\pi} = \int |\alpha\rangle\langle\alpha| \frac{d^2\alpha}{\pi} \sum \cdots \left(\alpha^* - \frac{\partial}{\partial\alpha}\right)^l \alpha^m \cdots \left(\alpha^* - \frac{\partial}{\partial\alpha}\right)^r \alpha^v \cdot 1,$$

and the theorem is proved.

THEOREM 5

The "R-representation" [7] is defined by

$$f(a, a^\dagger) = \iint d^2\alpha\, d^2\alpha' |\alpha\rangle\langle\alpha'| R(\alpha, \alpha'^*), \tag{3.2.29}$$

where

$$R(\alpha, \alpha'^*) = \frac{1}{\pi^2} \bar{f}^{(n)}(\alpha, \alpha'^*) \exp\left[-\tfrac{1}{2}|\alpha|^2 - \tfrac{1}{2}|\alpha'|^2 + \alpha'\alpha^*\right]. \tag{3.2.30}$$

To obtain this theorem we apply the completeness relation (3.2.24) twice:

$$f(a, a^\dagger) = \iint \frac{d^2\alpha}{\pi} |\alpha\rangle\langle\alpha| f^{(n)}(a, a^\dagger)|\alpha'\rangle\langle\alpha'| \frac{d^2\alpha'}{\pi}. \tag{3.2.31}$$

Since $f^{(n)}$ is in normal order, we have

$$\langle\alpha|f^{(n)}(a, a^\dagger)|\alpha'\rangle = \langle\alpha| \sum f^{(n)} a^{\dagger r} a^s |\alpha'\rangle = \sum_{rs} f_{rs}^{(n)} \alpha^{*r} \alpha'^s \langle\alpha|\alpha'\rangle \equiv \bar{f}^{(n)}(\alpha', \alpha^*),$$

where we used (3.2.13) and note that $\bar{f}^{(n)}(\alpha', \alpha^*)$ is obtained from $f^{(n)}(a, a^\dagger)$ by replacing a by α' and $a^\dagger$ by α^*. By (2.5.15)

$$\langle\alpha|\alpha'\rangle = \exp\left[-\tfrac{1}{2}|\alpha|^2 - \tfrac{1}{2}|\alpha'|^2 + \alpha'\alpha^*\right].$$

If we use these last two results in (3.1.31), the theorem follows. It turns out [8] that the R-function exists in cases where the P-function does not exist, but we shall not consider these finer mathematical points.

The associated functions turn out to be very useful since they allow us to transform quantum operators into ordinary functions whose manipulations are often more familiar.

A problem which often arises is the evaluation of traces of functions of a and $a^\dagger$. We show in the next two theorems how traces may be converted to ordinary integrals by means of the associated functions.

THEOREM 6

The trace of $f(a, a^\dagger)$ is given by

$$\operatorname{Tr} f(a, a^\dagger) = \int \bar{f}^{(a)}(\alpha, \alpha^*) \frac{d^2\alpha}{\pi} \tag{3.2.32}$$

$$\operatorname{Tr} f(a, a^\dagger) = \int \bar{f}^{(n)}(\alpha, \alpha^*) \frac{d^2\alpha}{\pi}. \tag{3.2.33}$$

That is, the trace may be found by integrating either the antinormal or normal associated functions over the complex α-plane. The traces exist if the integrals exist.

The proof of (3.2.32) follows directly if we take the trace of both sides of (3.2.23) and note that $\operatorname{Tr} |\alpha\rangle\langle\alpha| = \langle\alpha|\alpha\rangle = 1$. To prove (3.2.33) we introduce the completeness relation in the trace:

$$\begin{aligned}\operatorname{Tr} f(a, a^\dagger) &= \operatorname{Tr} \int f(a, a^\dagger)|\alpha\rangle\langle\alpha| \frac{d^2\alpha}{\pi} \\ &= \int \langle\alpha|f(a, a^\dagger)|\alpha\rangle \frac{d^2\alpha}{\pi},\end{aligned}$$

since $\operatorname{Tr} |u\rangle\langle v| = \langle v|u\rangle$, and we let $|u\rangle = f|\alpha\rangle$ and $\langle v| = \langle\alpha|$. The theorem then follows if we use (3.2.12).

THEOREM 7

If $f(a, a^\dagger)$ and $\rho(a, a^\dagger)$ are two functions of a and $a^\dagger$, then

$$\operatorname{Tr} \rho(a, a^\dagger) f(a, a^\dagger) = \int \bar{\rho}^{(a)}(\alpha, \alpha^*) \bar{f}^{(n)}(\alpha, \alpha^*) \frac{d^2\alpha}{\pi} \tag{3.2.34}$$

or

$$\operatorname{Tr} \rho(a, a^\dagger) f(a, a^\dagger) = \int \bar{\rho}^{(n)}(\alpha, \alpha^*) \bar{f}^{(a)}(\alpha, \alpha^*) \frac{d^2\alpha}{\pi}. \tag{3.2.35}$$

To prove (3.2.34) we expand $\rho = \rho^{(a)}$ and $f = f^{(n)}$ in power series and use the cyclic property of traces: Tr ABC = Tr CAB = Tr BCA. Thus

$$\begin{aligned}\text{Tr}\,\rho^{(a)}f^{(n)} &= \text{Tr}\sum_{rs}\rho_{rs}^{(a)}a^{r}a^{\dagger s}\sum_{lm}f_{lm}^{(n)}a^{\dagger l}a^{m}\\ &= \sum_{rs}\sum_{lm}\rho_{rs}^{(a)}f_{lm}^{(n)}\,\text{Tr}\,(a^{\dagger(s+l)}a^{m+r}).\end{aligned}$$

Since $a^{\dagger(s+l)}a^{(m+r)}$ is in normal order we use (3.2.33) and obtain

$$\begin{aligned}\text{Tr}\,\rho^{(a)}f^{(n)} &= \int\sum_{rs}\rho_{rs}^{(a)}\alpha^{*s}\alpha^{r}\sum_{lm}f_{lm}^{(n)}\alpha^{*l}\alpha^{m}\frac{d^2\alpha}{\pi}\\ &= \int\bar{\rho}^{(a)}(\alpha,\alpha^*)\bar{f}^{(n)}(\alpha,\alpha^*)\frac{d^2\alpha}{\pi}.\end{aligned}$$ Q.E.D.

A similar proof holds for (3.2.35).

The associated functions $\bar{f}^{(n)}(\alpha, \alpha^*)$ and $\bar{f}^{(a)}(\alpha, \alpha^*)$ of the complex variable α can be thought of as functions of two real variables x and y where $\alpha = x + iy$. As such we may define a two-dimensional Fourier transform of them. It will sometimes happen that a problem is simpler to solve in the transform variables than in the original variables x and y (or α and α^*). Accordingly, we define the Fourier transforms as

$$\bar{F}^{(n)}(\xi,\xi^*) = \int e^{-i(\xi\alpha+\xi^*\alpha^*)}\bar{f}^{(n)}(\alpha,\alpha^*)\frac{d^2\alpha}{\pi} \tag{3.2.36}$$

$$\bar{F}^{(a)}(\xi,\xi^*) = \int e^{-i(\xi\alpha+\xi^*\alpha^*)}\bar{f}^{(a)}(\alpha,\alpha^*)\frac{d^2\alpha}{\pi}. \tag{3.2.37}$$

(In real variables we have

$$\bar{F}^{(n)}(k_x,k_y) = \frac{1}{\pi}\iint_{-\infty}^{\infty} e^{-i(k_x x+k_y y}\bar{f}^{(n)}(x,y)\,dx\,dy$$

where

$$\begin{aligned}k_x &= \xi + \xi^*\\ k_y &= i(\xi - \xi^*).\end{aligned}$$

However, it is just as simple to work in the complex space directly with ξ and ξ^* independent variables as well as α and α^*.)

We can now prove the following theorem.

THEOREM 8

$$\text{Tr}\,f(a,a^\dagger) = \bar{F}^{(n)}(0,0) = \bar{F}^{(a)}(0,0) \tag{3.2.38}$$

That is, the trace of $f(a, a^\dagger)$ is found by evaluating the Fourier transform of either associated functions at $\xi = \xi^* = 0$.

The theorem may be proved by letting $\xi = \xi^* = 0$ in (3.2.36) and (3.2.37) and comparing with Theorem 6 above.

We can invert (3.2.36) and (3.2.37). We multiply both sides by $\exp i[\xi\alpha' + \xi^*\alpha'^*]$ and integrate over all ξ-space:

$$\int e^{i(\xi\alpha'+\xi^*\alpha'^*)}\bar{F}(\xi, \xi^*)\, d^2\xi = \int \bar{f}(\alpha, \alpha^*)\frac{d^2\alpha}{\pi}\int e^{i[\xi(\alpha'-\alpha)+\xi^*(\alpha'^*-\alpha^*)]}\, d^2\xi\,, \tag{3.2.39}$$

where we interchanged the order of integration. But*

$$\int e^{i(\xi\alpha+\xi^*\alpha^*)}\, d^2\xi = \pi^2\, \delta(\alpha)\, \delta(\alpha^*). \tag{3.2.40}$$

When we use this in (3.2.39) we have

$$\begin{aligned}\int e^{i(\xi\alpha'+\xi^*\alpha'^*)}\bar{F}(\xi, \xi^*)\, d^2\xi &= \pi\int \bar{f}(\alpha, \alpha^*)\, \delta(\alpha' - \alpha)\, \delta(\alpha'^* - \alpha^*)\, d^2\alpha \\ &= \pi\bar{f}(\alpha', \alpha'^*).\end{aligned}$$

We therefore have shown that

$$\begin{aligned}\bar{f}^{(n)}(\alpha, \alpha^*) &= \int e^{i(\xi\alpha+\xi^*\alpha^*)}\bar{F}^{(n)}(\xi, \xi^*)\frac{d^2\xi}{\pi} \\ \bar{f}^{(a)}(\alpha, \alpha^*) &= \int e^{i(\xi\alpha+\xi^*\alpha^*)}\bar{F}^{(a)}(\xi, \xi^*)\frac{d^2\xi}{\pi}\,.\end{aligned} \tag{3.2.41}$$

We may apply the normal and antinormal ordering operators to the first and second, respectively, of these equations and obtain the following interesting theorem.

THEOREM 9

$$\begin{aligned}f(a, a^\dagger) = f^{(n)}(a, a^\dagger) &= \int e^{i\xi^* a^\dagger}e^{i\xi a}\bar{F}^{(n)}(\xi, \xi^*)\frac{d^2\xi}{\pi} \\ f(a, a^\dagger) = f^{(a)}(a, a^\dagger) &= \int e^{i\xi a}e^{i\xi^* a^\dagger}\bar{F}^{(a)}(\xi, \xi^*)\frac{d^2\xi}{\pi}\,,\end{aligned} \tag{3.2.42}$$

which we shall call the Fourier representation of the operators. (Note that we used the fact that

$$\begin{aligned}\mathcal{N}\{e^{i(\xi\alpha+\xi^*\alpha^*)}\} &= e^{i\xi^* a^\dagger}e^{i\xi a} \\ \mathcal{A}\{e^{i(\xi\alpha+\xi^*\alpha^*)}\} &= e^{i\xi a}e^{i\xi^* a^\dagger}\end{aligned} \tag{3.2.43}$$

as may be seen from the power series expansions.)

* We may verify this by changing to real variables.

THEOREM 10

$$\bar{F}^{(n)}(\xi, \xi^*) = \mathrm{Tr}\,[e^{-i\xi a}e^{-i\xi^* a^\dagger}f(a, a^\dagger)] \tag{3.2.44}$$

$$\bar{F}^{(a)}(\xi, \xi^*) = \mathrm{Tr}\,[e^{-i\xi^* a^\dagger}e^{-i\xi a}f(a, a^\dagger)] \tag{3.2.45}$$

This gives the Fourier transforms in terms of the original function.

To prove this theorem which is closely related to the characteristic function to be discussed later, we note that in (3.2.44) the first two operators are in antinormal order. If f is in normal order, we may use Theorem 7 which leads directly to (3.2.36) and the theorem is proved. A similar argument may be used to obtain (3.2.45).

The following theorem gives a direct relation between the two associated functions.

THEOREM 11

$$\bar{f}^{(n)}(\alpha', \alpha'^*) = \int \bar{f}^{(a)}(\alpha, \alpha^*)e^{-|\alpha-\alpha'|^2}\frac{d^2\alpha}{\pi} \tag{3.2.46}$$

Therefore, if $\bar{f}^{(a)}$ is known, in principle we may obtain $\bar{f}^{(n)}$.

To prove the theorem, we take diagonal matrix elements of both sides of (3.2.23) in the coherent state representation. This gives

$$\langle\alpha'|f(a, a^\dagger)|\alpha'\rangle = \int \frac{d^2\alpha}{\pi}\, \bar{f}^{(a)}(\alpha, \alpha^*)|\langle\alpha'|\alpha\rangle|^2.$$

The left side by Theorem 1 is just $\bar{f}^{(n)}(\alpha', \alpha'^*)$. By (2.5.16)

$$|\langle\alpha'|\alpha\rangle|^2 = \exp -|\alpha - \alpha'|^2, \tag{3.2.47}$$

and the theorem follows.

3.3 ALGEBRAIC PROPERTIES OF BOSON OPERATORS

In this section, we prove a number of theorems which will aid us in putting functions of boson operators into normal and antinormal order.

THEOREM 1

If l is an integer, then

$$\begin{aligned} [a, a^{\dagger l}] &= l a^{\dagger l-1} = \frac{\partial a^{\dagger l}}{\partial a^\dagger} \\ [a^\dagger, a^l] &= -l a^{l-1} = -\frac{\partial a^l}{\partial a}. \end{aligned} \tag{3.3.1}$$

The theorem may be proved by induction since $[a, a^\dagger] = 1$. We leave this as an exercise.

THEOREM 2

Let x be a c-number (not an operator) and $f(a, a^\dagger)$ be a function which may be expanded in a power series in a and $a^\dagger$. Then

$$e^{xa} f(a, a^\dagger) e^{-xa} = f(a, a^\dagger + x) \tag{3.3.2a}$$

$$e^{-xa^\dagger} f(a, a^\dagger) e^{xa^\dagger} = f(a + x, a^\dagger). \tag{3.3.2b}$$

The theorem is true regardless of the order of f and is therefore true if f is in normal or antinormal order since $f = f^{(n)} = f^{(a)}$.

To prove the theorem, we have by (3.1.4)

$$e^{xa} f(a, a^\dagger) e^{-xa} = f[e^{xa} a e^{-xa}, e^{xa} a^\dagger e^{-xa}] = f[a, e^{xa} a^\dagger e^{xa}], \tag{3.3.3}$$

where we have assumed f may be expanded in a power series. The second step follows since a and $\exp xa$ commute.

Next, if we use (3.1.14) we see that

$$e^{xa} a^\dagger e^{-xa} = a^\dagger + x, \tag{3.3.4}$$

since

$$[a, a^\dagger] = 1,$$

and higher order commutators vanish. When we use (3.3.4) in (3.3.3), we have

$$e^{xa} f(a, a^\dagger) e^{-xa} = f(a, a^\dagger + x), \tag{3.3.5}$$

and (3.3.2a) follows. A similar proof holds for (3.3.2b).

Suppose that we have somehow managed to put f into normal order. However, $e^{xa} f^{(n)}$ and $f^{(n)} e^{xa^\dagger}$ will not be in normal order. By (3.3.2) we have that

$$\begin{aligned} e^{xa} f^{(n)}(a, a^\dagger) &= f^{(n)}(a, a^\dagger + x) e^{xa} \\ f^{(n)}(a, a^\dagger) e^{xa^\dagger} &= e^{xa^\dagger} f^{(n)}(a + x, a^\dagger). \end{aligned} \tag{3.3.6}$$

Since the right sides of both equations are in normal order, we have been able to put $e^{xa} f^{(n)}$ and $f^{(n)} e^{xa^\dagger}$ into normal order. If we take the diagonal elements of both sides of (3.3.6) in the coherent state representation, we have by (3.2.12) that

$$\begin{aligned} \langle \alpha | e^{xa} f(a, a^\dagger) | \alpha \rangle &= e^{x\alpha} \bar{f}^{(n)}(\alpha, \alpha^* + x) \\ \langle \alpha | f(a, a^\dagger) e^{xa^\dagger} | \alpha \rangle &= e^{x\alpha^*} \bar{f}^{(n)}(\alpha + x, \alpha^*). \end{aligned} \tag{3.3.7a}$$

From the definition of the normal ordering operator, we also have that

$$\begin{aligned} e^{xa}f(a, a^\dagger) &= \mathcal{N}\{e^{x\alpha}\bar{f}^{(n)}(\alpha, \alpha^* + x)\} \\ f(a, a^\dagger)e^{xa^\dagger} &= \mathcal{N}\{e^{x\alpha^*}\bar{f}^{(n)}(\alpha + x, \alpha^*)\}. \end{aligned} \tag{3.3.7b}$$

It is simple to show that the theorem also holds in the following case

$$\begin{aligned} e^{xa}f(a, a^\dagger - x)e^{-xa} &= f(a, a^\dagger) \\ e^{-xa^\dagger}f(a - x, a^\dagger)e^{xa^\dagger} &= f(a, a^\dagger). \end{aligned} \tag{3.3.8a}$$

If we assume f has somehow been put into antinormal order, we have by (3.2.23),

$$\begin{aligned} f^{(a)}(a, a^\dagger)e^{xa} &= e^{xa}f^{(a)}(a, a^\dagger - x) = \mathscr{A}\{e^{x\alpha}\bar{f}^{(a)}(\alpha, \alpha^* - x)\} \\ e^{xa^\dagger}f^{(a)}(a, a^\dagger) &= f^{(a)}(a - x, a^\dagger)e^{xa^\dagger} = \mathscr{A}\{e^{x\alpha^*}\bar{f}^{(a)}(\alpha - x, \alpha^*)\}, \end{aligned} \tag{3.3.8b}$$

since the middle forms are already in antinormal order. By (3.2.23) we may also write these as

$$\begin{aligned} f(a, a^\dagger)e^{xa} &= \int \frac{d^2\alpha}{\pi} |\alpha\rangle\langle\alpha|\{e^{x\alpha}\bar{f}^{(a)}(\alpha, \alpha^* - x)\} \\ e^{xa^\dagger}f(a, a^\dagger) &= \int \frac{d^2\alpha}{\pi} |\alpha\rangle\langle\alpha|\{e^{x\alpha^*}\bar{f}^{(a)}(\alpha - x, \alpha^*)\} \end{aligned} \tag{3.3.8c}$$

THEOREM 3

If $f(a, a^\dagger)$ is a function which may be expanded in a power series in a and $a^\dagger$, then

$$[a, f(a, a^\dagger)] = \frac{\partial f}{\partial a^\dagger} \tag{3.3.9a}$$

$$[a^\dagger, f(a, a^\dagger)] = -\frac{\partial f}{\partial a}. \tag{3.3.9b}$$

We prove (3.3.9b). Since $f = f^{(a)}$, we have on expanding $f^{(a)}$

$$[a^\dagger, f] = \sum_{r,s} f_{rs}^{(a)}[a^\dagger, a^r a^{\dagger s}].$$

If A, B, and C are any noncommuting operating operators, it is easy to verify by expanding the commutators that

$$[A, BC] = [A, B]C + B[A, C].$$

Therefore, we may write the commutator above as

$$[a^\dagger, f] = \sum_{r,s} f_{rs}^{(a)}\{[a^\dagger, a^r]a^{\dagger s} + a^r[a^\dagger, a^{\dagger s}]\}.$$

The second commutator here on the right obviously vanishes, and the first is found by (3.3.1) so that

$$[a^\dagger, f] = -\sum_{r,s} f_{rs}^{(a)} r a^{r-1} a^{\dagger s} = -\frac{\partial f^{(a)}}{\partial a}.$$

Since $f = f^{(a)}$, (3.3.9b) follows. The proof for (3.3.9a) is similar.

If f is in normal order, then $af^{(n)}$ and $f^{(n)}a^\dagger$ are not, but we can put them into normal order by the theorem. Therefore,

$$\begin{aligned} af^{(n)}(a, a^\dagger) &= f^{(n)}(a, a^\dagger)a + \frac{\partial f^{(n)}}{\partial a^\dagger} = \mathcal{N}\left\{\left(\alpha + \frac{\partial}{\partial \alpha^*}\right)\bar{f}^{(n)}(\alpha, \alpha^*)\right\} \\ f^{(n)}(a, a^\dagger)a^\dagger &= a^\dagger f^{(n)}(a, a^\dagger) + \frac{\partial f^{(n)}}{\partial a} = \mathcal{N}\left\{\left(\alpha^* + \frac{\partial}{\partial \alpha}\right)\bar{f}^{(n)}(\alpha, \alpha^*)\right\}. \end{aligned} \tag{3.3.10}$$

The right-hand expressions follow since the middle terms are in normal order. Clearly

$$\begin{aligned} \frac{\partial}{\partial a^\dagger} f^{(n)} &= \frac{\partial}{\partial a^\dagger}\left\{\sum_{rs} f_{rs}^{(n)} a^{\dagger r} a^s\right\} = \sum_{rs} f_{rs}^{(n)} r a^{\dagger r-1} a^s \\ &= \mathcal{N}\left\{\frac{\partial}{\partial \alpha^*} \bar{f}^{(n)}(\alpha, \alpha^*)\right\} = \mathcal{N}\left\{\sum_{rs} f_{rs}^{(n)} r \alpha^{*r-1}\alpha^s\right\}. \end{aligned}$$

Alternatively, we have by (3.2.12) that

$$\begin{aligned} \langle\alpha|af(a, a^\dagger)|\alpha\rangle &= \left(\alpha + \frac{\partial}{\partial \alpha^*}\right)\bar{f}^{(n)}(\alpha, \alpha^*) \\ \langle\alpha|f(a, a^\dagger)a^\dagger|\alpha\rangle &= \left(\alpha^* + \frac{\partial}{\partial \alpha}\right)\bar{f}^{(n)}(\alpha, \alpha^*). \end{aligned} \tag{3.3.11}$$

In case f is in antinormal order, we easily show that

$$\begin{aligned} f(a, a^\dagger)a &= \mathcal{A}\left\{\left(\alpha - \frac{\partial}{\partial \alpha^*}\right)\bar{f}^{(a)}(\alpha, \alpha^*)\right\} \\ &= \int |\alpha\rangle\langle\alpha| \frac{d^2\alpha}{\pi}\left\{\left(\alpha - \frac{\partial}{\partial \alpha^*}\right)\bar{f}^{(a)}(\alpha, \alpha^*)\right\} \end{aligned} \tag{3.3.12a}$$

$$\begin{aligned} a^\dagger f(a, a^\dagger) &= \mathcal{A}\left\{\left(\alpha^* - \frac{\partial}{\partial \alpha}\right)\bar{f}^{(a)}(\alpha, \alpha^*)\right\} \\ &= \int |\alpha\rangle\langle\alpha| \frac{d^2\alpha}{\pi}\left\{\left(\alpha^* - \frac{\partial}{\partial \alpha}\right)\bar{f}^{(a)}(\alpha, \alpha^*)\right\}. \end{aligned} \tag{3.3.12b}$$

THEOREM 4

If m is an integer and $f = f^{(n)} = f^{(a)}$,

$$a^m f(a, a^\dagger) = \mathcal{N}\left\{\left(\alpha + \frac{\partial}{\partial\alpha^*}\right)^m \bar{f}^{(n)}(\alpha, \alpha^*)\right\} = \mathcal{N}\{\langle\alpha|a^m f(a, a^\dagger)|\alpha\rangle\}$$

$$f(a, a^\dagger)a^{\dagger m} = \mathcal{N}\left\{\left(\alpha^* + \frac{\partial}{\partial\alpha}\right)^m \bar{f}^{(n)}(\alpha, \alpha^*)\right\} = \mathcal{N}\{\langle\alpha|f(a, a^\dagger)a^{\dagger m}|\alpha\rangle\}, \tag{3.3.13}$$

while

$$f(a, a^\dagger)a^m = \mathcal{A}\left\{\left(\alpha - \frac{\partial}{\partial\alpha^*}\right)^m \bar{f}^{(a)}(\alpha, \alpha^*)\right\}$$

$$a^{\dagger m} f(a, a^\dagger) = \mathcal{A}\left\{\left(\alpha^* - \frac{\partial}{\partial\alpha}\right)^m \bar{f}^{(a)}(\alpha, \alpha^*)\right\}. \tag{3.3.14}$$

For simplicity we omit the P-representation of (3.3.14). This theorem may be proved by induction and using Theorem 3 (3.3.9).

THEOREM 5

If $f(a, a^\dagger)$ may be expanded in a power series then

$$e^{xa^\dagger a} f(a, a^\dagger) e^{-xa^\dagger a} = f(ae^{-x}, a^\dagger e^x). \tag{3.3.15}$$

In particular

$$e^{xa^\dagger a} a e^{-xa^\dagger a} = ae^{-x} \tag{3.3.16a}$$

$$e^{xa^\dagger a} a^\dagger e^{-xa^\dagger a} = a^\dagger e^x. \tag{3.3.16b}$$

We first derive (3.3.16a) by expanding

$$g(x) = e^{xa^\dagger a} a e^{-xa^\dagger a}$$

in a power series in x. We have that $g(0) = a$ and

$$\frac{dg}{dx} = e^{xa^\dagger a}[a^\dagger a, a]e^{-xa^\dagger a}.$$

If we use (3.3.9a), this becomes

$$\frac{dg}{dx} = -e^{xa^\dagger a} a e^{-xa^\dagger a} = -g(x)$$

so that

$$\frac{d^l g}{dx^l} = (-1)^l g(x),$$

and

$$\left.\frac{d^l g}{dx^l}\right|_{x=0} = (-1)^l a.$$

Therefore,

$$g(x) = a\left[1 - x + \frac{x^2}{2!} - \frac{x^3}{3!} + \cdots\right] = ae^{-x}. \qquad \text{Q.E.D.}$$

A similar proof holds for (3.3.16b).

To prove the theorem, we expand f in a power series, insert $e^{xa^\dagger a}e^{-xa^\dagger a} = 1$ between each factor of a and $a^\dagger$ and resume the series. This gives

$$e^{xa^\dagger a}f(a, a^\dagger)e^{-xa^\dagger a} = f(e^{xa^\dagger a}ae^{-xa^\dagger a}, e^{xa^\dagger a}a^\dagger e^{-xa^\dagger a}).$$

If we use (3.3.16), the theorem follows. We also prove the following lemma.

LEMMA

If $|0\rangle$ is the vacuum state such that $a|0\rangle = 0$ and x and y are parameters, then

$$e^{xa^\dagger a}e^{ya^\dagger}|0\rangle = \exp(ye^x a^\dagger)|0\rangle. \tag{3.3.17}$$

By (3.3.15), we have

$$e^{xa^\dagger a}e^{ya^\dagger} = \exp(ya^\dagger e^x)e^{xa^\dagger a}.$$

Since

$$e^{xa^\dagger a}|0\rangle = |0\rangle,$$

the lemma follows if we apply both sides of the expression above to the vacuum state.

THEOREM 6

If f is a function of $a^\dagger a$, its normal form is

$$f^{(n)}(a^\dagger a) = \mathcal{N}\left\{\sum_{n=0}^{\infty} f(n)e^{-\alpha^*\alpha}\frac{(\alpha^*\alpha)^n}{n!}\right\} \equiv \mathcal{N}\{\bar{f}^{(n)}(\alpha, \alpha^*)\}. \tag{3.3.18}$$

PROOF

The completeness relation of the energy eigenstates is

$$\sum_{n=0}^{\infty} |n\rangle\langle n| = 1, \tag{3.3.19}$$

where $a^\dagger a|n\rangle = n|n\rangle$. Since

$$f(a^\dagger a)|n\rangle = f(n)|n\rangle, \tag{3.3.20}$$

it follows that

$$f(a^\dagger a) = \sum_0^\infty f(a^\dagger a)|n\rangle\langle n| = \sum_0^\infty f(n)|n\rangle\langle n|. \tag{3.3.21}$$

If we take the diagonal matrix element on both sides in the coherent state representation, we have by (3.2.12)

$$f^{(n)}(\alpha, \alpha^*) = \langle\alpha|f(a^\dagger a)|\alpha\rangle = \sum_0^\infty f(n)|\langle\alpha|n\rangle|^2. \tag{3.3.22}$$

By (2.5.13), we have

$$f^{(n)}(\alpha, \alpha^*) = \sum_{n=0}^\infty f(n)e^{-\alpha^*\alpha}\frac{(\alpha^*\alpha)^n}{n!}, \tag{3.3.23}$$

from which the theorem follows when we apply the normal ordering operator to both sides.

LEMMA 1

If x is a parameter, the normal form of exp $-xa^\dagger a$ is

$$\begin{aligned}[e^{-xa^\dagger a}]^{(n)} &= \sum_{l=0}^\infty \frac{(e^{-x} - 1)^l}{l!} a^{\dagger l}a^l \\ &= \mathcal{N}\{\exp[(e^{-x} - 1)\alpha^*\alpha]\}.\end{aligned} \tag{3.3.24}$$

PROOF

By the theorem we have

$$\begin{aligned}\langle\alpha|e^{-xa^\dagger a}|\alpha\rangle &= \sum_{n=0}^\infty e^{-\alpha^*\alpha}\frac{(\alpha^*\alpha e^{-x})^n}{n!} \\ &= \exp[(e^{-x} - 1)]\alpha^*\alpha].\end{aligned}$$

When we apply the normal ordering operator, the lemma follows.

An alternate proof of this lemma which is more general illustrates the use of the associated function to manipulate operators.

Let

$$f(a^\dagger a) = e^{-xa^\dagger a}. \tag{3.3.25}$$

If we differentiate both sides with respect to x, we have

$$\frac{\partial f}{\partial x} = -a^\dagger af. \tag{3.3.26}$$

Note that since f is a function of $a^\dagger a$ only, it follows that

$$[a^\dagger a, f] = 0.$$

Assume that f is in normal order so that $f = f^{(n)}$. If we use (3.3.10), we may rewrite (3.3.26) as

$$\frac{\partial f^{(n)}}{\partial x} = -a^\dagger\left[f^{(n)}a + \frac{\partial f^{(n)}}{\partial a^\dagger}\right]. \tag{3.3.27}$$

Then since $f^{(n)}$ is in normal order by assumption, if we take the coherent state diagonal matrix elements of both sides since $a|\alpha\rangle = \alpha|\alpha\rangle$, $\langle\alpha|a^\dagger = \alpha^*\langle\alpha|$, and

$$\bar{f}^{(n)}(\alpha, \alpha^*) = \langle\alpha|f(a, a^\dagger)|\alpha\rangle, \tag{3.3.28}$$

it follows from (3.3.27) that

$$\frac{\partial \bar{f}^{(n)}}{\partial x} = -\left[\alpha^*\alpha + \alpha^*\frac{\partial}{\partial\alpha^*}\right]\bar{f}^{(n)}, \tag{3.3.29}$$

which is the partial differential equation obeyed by the normal associated function. If we make the change of variable

$$\begin{aligned} x + \ln\alpha^* &= \xi \qquad & x &= \tfrac{1}{2}(\xi + \eta) \\ x - \ln\alpha^* &= \eta \qquad & \alpha^* &= \exp\tfrac{1}{2}(\xi - \eta), \end{aligned} \tag{3.3.30}$$

by the usual rules of differentiation we have

$$\begin{aligned} \frac{\partial}{\partial x} &= \frac{\partial}{\partial\xi} + \frac{\partial}{\partial\eta} \\ \alpha^*\frac{\partial}{\partial\alpha^*} &= \frac{\partial}{\partial\xi} - \frac{\partial}{\partial\eta}, \end{aligned} \tag{3.3.31}$$

with this change of variable, (3.3.29) becomes

$$2\frac{\partial \bar{f}^{(n)}}{\partial\xi} = -\alpha e^{\frac{1}{2}(\xi-\eta)}\bar{f}^{(n)}.$$

If we integrate, we have

$$2\ln\bar{f}^{(n)} = -2\alpha e^{\frac{1}{2}(\xi-\eta)} + g(\eta), \tag{3.3.32}$$

where $g(\eta)$ is an arbitrary function of integration which may be determined as follows. If $x = 0$ in (3.3.25), $f(0, a^\dagger a) = 1$ so that $\bar{f}^{(n)}(0, \alpha, \alpha^*) = 1$ and $\ln\bar{f}^{(n)}(0, \alpha, \alpha^*) = 0$. When $x = 0$, by (3.3.30) we have $\eta = -\xi$ and (3.3.32) reduces to

$$0 = -2\alpha e^{-\eta} + g(\eta), \tag{3.3.33}$$

which gives $g(\eta)$. If we put this in (3.3.32) and change variables back to x and α^*, we find that the normal associated function is

$$\bar{f}^{(n)}(x, \alpha, \alpha^*) = \exp\left[(e^{-x} - 1)\alpha^*\alpha\right],$$

as in the previous case.

LEMMA 2

The vacuum state projection operator $|0\rangle\langle 0|$ is given by

$$|0\rangle\langle 0| = \lim_{\epsilon\to 1} \mathcal{N}\{e^{-\epsilon\alpha^*\alpha}\} = \lim_{\epsilon\to 1} \sum_{l=0}^{\infty} \frac{(-\epsilon)^l a^{\dagger l} a^l}{l!} . \tag{3.3.34}$$

PROOF

By (3.2.12) and (2.5.13) we have, when we take the diagonal coherent state matrix element of $|0\rangle\langle 0|$,

$$\begin{aligned} \bar{f}^{(n)}(\alpha, \alpha^*) &= \langle\alpha|0\rangle\langle 0|\alpha\rangle = |\langle\alpha|0\rangle|^2 \\ &= e^{-\alpha^*\alpha}. \end{aligned} \tag{3.3.35}$$

We must write this as

$$\langle\alpha|0\rangle\langle 0|\alpha\rangle = \bar{f}^{(n)}(\alpha, \alpha^*) = \lim_{\epsilon\to 1} e^{-\epsilon\alpha^*\alpha} = \lim_{\epsilon\to 1} \sum_{l=0}^{\infty} \frac{(-\epsilon)^l}{l!} (\alpha^*\alpha)^l. \tag{3.3.36}$$

Let us apply the normal ordering operator to the above. This gives

$$|0\rangle\langle 0| = \lim_{\epsilon\to 1} \sum_{l=0}^{\infty} \frac{(-\epsilon)^l}{l!} a^{\dagger l} a^l. \tag{3.3.37}$$

When we take matrix element in the number representation, we obtain

$$\langle n|0\rangle\langle 0|m\rangle = \lim_{\epsilon\to 1} \sum_{l=0}^{\infty} \frac{(-\epsilon)^l}{l!} \langle n|a^{\dagger l} a^l|m\rangle. \tag{3.3.38}$$

By repeated application of (2.2.10) we have for $l \leq m$

$$a^l|m\rangle = \sqrt{\frac{m!}{(m-l)!}}\,|m-l\rangle, \tag{3.3.39a}$$

and for $l \leq n$

$$\langle n|a^{\dagger l} = \sqrt{\frac{n!}{(n-l)!}}\,\langle n-l|. \tag{3.3.39b}$$

Therefore, (3.2.38) becomes

$$\begin{aligned} \langle n|0\rangle\langle 0|m\rangle &= \lim_{\epsilon\to 1} \sum_{l=0}^{\infty} \frac{(-\epsilon)^l m!}{l!\,(m-l)!}\,\delta nm \\ &= \lim_{\epsilon\to 1} (1-\epsilon)^m \delta_{nm}, \end{aligned} \tag{3.3.40}$$

since

$$\langle n-l|m-l\rangle = \delta_{nm},$$

and we have summed the series. Since

$$\lim_{\epsilon \to 1} (1 - \epsilon)^m = \delta_{m0},$$

both sides of (3.3.40) are equal.

LEMMA 3

The normal form of the operator $|n\rangle\langle m|$ is

$$\begin{aligned} |n\rangle\langle m| &= \lim_{\varepsilon \to 1} \mathcal{N}\left\{e^{-\epsilon\alpha^*\alpha} \frac{\alpha^{*n}\alpha^m}{\sqrt{n!\, m!}}\right\} \\ &= \lim_{\epsilon \to 1} \sum_{l=0}^{\infty} \frac{(-\epsilon)^l}{l!\sqrt{n!\, m!}} a^{\dagger l+n} a^{l+m}. \end{aligned} \tag{3.3.41}$$

This is proved by the same procedure as Lemma 2.

LEMMA 4

The normal form of the projection operator $|\alpha\rangle\langle\alpha|$ is

$$|\alpha\rangle\langle\alpha| = \lim_{\epsilon \to 1} \mathcal{N}_\beta\{e^{-\epsilon|\alpha-\beta|^2}\} = e^{-\alpha^*\alpha} e^{\alpha a^\dagger} |0\rangle\langle 0| e^{\alpha^* a}, \tag{3.3.42}$$

where the normal ordering operator is understood to be applied to β and β^*.

PROOF

By (3.2.12), (3.2.9), and (2.5.16), we have

$$\begin{aligned} |\alpha\rangle\langle\alpha| &= \mathcal{N}_\beta\{\langle\beta|\alpha\rangle\langle\alpha|\beta\rangle\} = \lim_{\epsilon \to 1} \mathcal{N}_\beta\{e^{-\epsilon(\alpha^*-\beta^*)(\alpha-\beta)}\} \\ &= e^{-\alpha^*\alpha} e^{\alpha a^\dagger} \lim_{\epsilon \to 1} \mathcal{N}_\beta\{e^{-\epsilon\beta^*\beta}\} e^{\alpha^* a} \\ &= e^{-\alpha^*\alpha} e^{\alpha a^\dagger} |0\rangle\langle 0| e^{\alpha^* a}, \end{aligned}$$

where we used Lemma 2 above.

THEOREM 7

The antinormal form of $\exp(-xa^\dagger a)$ is given by

$$\begin{aligned} (e^{-xa^\dagger a})^{(a)} &= e^{+x} \mathcal{A}\{\exp(1 - e^{+x})\alpha^*\alpha\} \\ &= e^{+x} \sum_{l=0}^{\infty} \frac{(1 - e^{+x})^l}{l!} a^l a^{\dagger l}. \end{aligned} \tag{3.3.43}$$

This theorem is most easily proved by using the differential equation technique for the antinormal associated function as in Lemma 1 above. The proof is left as an exercise.

THEOREM 8

The normal form of $f(a^\dagger a)a^\dagger$ is

$$f(a^\dagger a)a^\dagger = a^\dagger f(a^\dagger a + 1) = \mathcal{N}\left\{\sum_{n=0}^{\infty} \frac{f(n+1)}{n!} (\alpha^*\alpha)^n e^{-\alpha^*\alpha}\alpha^*\right\}, \qquad (3.3.44a)$$

while

$$af(a^\dagger a) = f(a^\dagger a + 1)a = \mathcal{N}\left\{\sum_{n=0}^{\infty} \frac{f(n+1)}{n!} (\alpha^*\alpha)^n e^{-\alpha^*\alpha}\alpha\right\}. \qquad (3.3.44b)$$

PROOF

We prove (3.3.44a) and leave (3.3.44b) as an exercise. By (3.3.10) and (3.3.18) we have

$$\begin{aligned}
f(a^\dagger a)a^\dagger &= \mathcal{N}\left\{\left(\alpha^* + \frac{\partial}{\partial\alpha}\right) f^{(n)}(\alpha, \alpha^*)\right\} \\
&= \mathcal{N}\left\{\left(\alpha^* + \frac{\partial}{\partial\alpha}\right) \sum_{n=0}^{\infty} f(n) e^{-\alpha\alpha^*} \frac{(\alpha^*\alpha)^n}{n!}\right\} \\
&= \mathcal{N}\left\{\alpha^* \sum_{n=0}^{\infty} f(n) e^{-\alpha\alpha^*} \frac{(\alpha^*\alpha)^n}{n!} + \sum_{n=0}^{\infty} \frac{f(n)}{n!} e^{-\alpha^*\alpha}[n\alpha^{*n}\alpha^{n-1} - \alpha^{*n+1}\alpha^n]\right\} \\
&= \mathcal{N}\left\{\sum_{n=1}^{\infty} \frac{f(n)}{(n-1)!} e^{-\alpha^*\alpha}\alpha^{*n}\alpha^{n-1}\right\},
\end{aligned}$$

since the first and last sums cancel in the next to last form. If we let $n \to n+1$ we have

$$\begin{aligned}
f(a^\dagger a)a^\dagger &= \mathcal{N}\left\{\sum_{n=0}^{\infty} \frac{f(n+1)}{n!} (\alpha^*\alpha)^n e^{-\alpha^*\alpha}\alpha^*\right\} \\
&= a^\dagger \mathcal{N}\left\{\sum_{n=0}^{\infty} \frac{f(n+1)}{n!} (\alpha^*\alpha)^n e^{-\alpha^*\alpha}\right\}. \qquad (3.3.44c)
\end{aligned}$$

If we use (3.3.18), the theorem follows.

The evaluation of traces plays a very important role in radiation theory, since mean values of observables are to be compared with experimental measurements. For example, if L is an observable, the mean value of $f(L)$ when the system is in a state $|\psi\rangle$ is

$$\langle f(L)\rangle = \langle\psi|f(L)|\psi\rangle = \text{Tr}\,[|\psi\rangle\langle\psi|f(L)],$$

where the last step follows since $\mathrm{Tr}\,|u\rangle\langle v| = \langle v|u\rangle$. We have already given three ways of evaluating traces of functions of a and $a^\dagger$. The first involves taking the diagonal matrix elements of $f(a, a^\dagger)$ in the $|n\rangle$-representation and summing:

$$\mathrm{Tr}\,f(a, a^\dagger) = \sum_{n=0}^{\infty} \langle n|f(a, a^\dagger)|n\rangle.$$

A second method consists of obtaining either the normal or antinormal associated function for f and integrating [see (3.2.32) and (3.2.33)]:

$$\mathrm{Tr}\,f(a, a^\dagger) = \int \bar{f}^{(n)}(\alpha, \alpha^*)\frac{d^2\alpha}{\pi} = \int \bar{f}^{(a)}(\alpha, \alpha^*)\frac{d^2\alpha}{\pi}.$$

The third technique follows from the Fourier transforms of the associated functions. From (3.2.38) we have

$$\mathrm{Tr}\,f(a, a^\dagger) = \bar{F}^{(n)}(0, 0) = \bar{F}^{(a)}(0, 0).$$

In the following theorem we give another method which is often quite useful in evaluating the trace of $f(a, a^\dagger)\exp(-\lambda a^\dagger a)$. Physically this corresponds to an ensemble average for a system in thermal equilibrium (Boltzmann distribution) as we show later.

THEOREM 9

If $f(a, a^\dagger)$ is a function of a and $a^\dagger$, then

$$\begin{aligned}\langle f(a, a^\dagger)\rangle_0 &\equiv (1 - e^{-\lambda})\,\mathrm{Tr}\,f(a, a^\dagger)e^{-\lambda a^\dagger a} \\ &= \langle 0, 0|f[\sqrt{1+\bar{n}}\,a + \sqrt{\bar{n}}\,b^\dagger, \sqrt{1+\bar{n}}\,a^\dagger + \sqrt{\bar{n}}\,b]|0, 0\rangle, \qquad (3.3.45)\end{aligned}$$

where λ is a parameter, and

$$\bar{n} = \frac{1}{e^\lambda - 1}, \qquad (3.3.46)$$

where b and $b^\dagger$ are boson operators which commute with a and $a^\dagger$, and $|0, 0\rangle$ is the vacuum state for a and b:

$$\begin{aligned} a|0, 0\rangle &\equiv a|0\rangle_a|0\rangle_b = 0 \\ b|0, 0\rangle &\equiv b|0\rangle_b|0\rangle_a = 0. \end{aligned} \qquad (3.3.47)$$

This theorem allows us to convert the evaluation of such thermal averages to the evaluation of the double, vacuum expectation value of the function with a replaced by $\sqrt{1+\bar{n}}\,a + \sqrt{\bar{n}}\,b^\dagger$ and $a^\dagger$ by the adjoint of this linear combination [9].

PROOF

The trace in the number representation $|n\rangle$ where

$$a^\dagger a|n\rangle_a = n|n\rangle_a,$$

in (3.3.45) is given by

$$\langle f\rangle_0 \equiv (1 - e^{-\lambda}) \sum_{n=0}^{\infty} {}_a\langle n|f(a, a^\dagger)e^{-\lambda a^\dagger a}|n\rangle_a \qquad (3.3.48)$$

Since we are evaluating the trace in a representation in which $a^\dagger a$ is diagonal, we see that any term in $f(a, a^\dagger)$ which does not have equal powers of a and $a^\dagger$ will give no contribution to a diagonal matrix element (see Problem 3.13). Accordingly, in the trace above, we are free to replace a by ka and $a^\dagger$ by $a^\dagger/k$ in the function where k is a parameter with no change in the value of the diagonal matrix element. We may therefore write (3.3.48) as

$$\langle f\rangle_0 = (1 - e^{-\lambda}) \sum_{n=0}^{\infty} {}_a\langle n|f\left(ka, \frac{1}{k}a^\dagger\right)e^{-\lambda a^\dagger a}|n\rangle_a. \qquad (3.3.49)$$

We determine k later.

To proceed, we would like to carry out the sum over n in (3.3.49). This may be accomplished as follows. We rewrite (3.3.49) as

$$\frac{\langle f\rangle_0}{(1 - e^{-\lambda})} = \sum_{n=0}^{\infty}\sum_{m=0}^{\infty} {}_a\langle n|f\left(ka, \frac{1}{k}a^\dagger\right)e^{-\lambda a^\dagger a}|m\rangle_a\delta_{nm}. \qquad (3.3.50)$$

This is identical with the previous equation since the sum over m contributes only when $n = m$. We next introduce a representation for the Kronecker δ by means of a new set of boson operators b and $b^\dagger$ where $[b, b^\dagger] = 1$ together with their eigenstates

$$b^\dagger b|n\rangle_b = n|n\rangle_b. \qquad (3.3.51)$$

The b and $b^\dagger$ are independent of a and $a^\dagger$ and therefore commute with each other. Since the eigenkets $|n\rangle_b$ form a complete orthornormal set, we may write the Kronecker δ as

$$\delta_{nm} = {}_b\langle n|m\rangle_b = {}_b\langle 0|\frac{b^n}{\sqrt{n!}}\frac{b^{\dagger m}}{\sqrt{m!}}|0\rangle_b, \qquad (3.3.52)$$

where $|0\rangle_b$ is the vacuum state for the b-operator (see Eq. 2.2.11).

Since (3.3.52) is zero unless $n = m$, we see that if μ is a real parameter, we may rewrite the equation above as

$$\delta_{nm} = {}_b\langle 0|\frac{(\mu b)^n}{\sqrt{n!}}\left(\frac{b^\dagger}{\mu}\right)^n\frac{1}{\sqrt{m!}}|0\rangle_b. \qquad (3.3.53)$$

Also since by (2.2.11) of Chapter 1 we may write

$$
\begin{aligned}
{}_a\langle n| &= {}_a\langle 0| \frac{a^n}{\sqrt{n!}} \\
|m\rangle_a &= \frac{a^{\dagger m}}{\sqrt{m!}} |0\rangle_a,
\end{aligned} \tag{3.3.54}
$$

and since the a's and b's commute, we may use the last two results to rewrite (3.3.50) as

$$
\begin{aligned}
\frac{\langle f\rangle_0}{(1-e^{-\lambda})} &= \sum_{n=0}^{\infty}\sum_{m=0}^{\infty} \langle 0, 0| \frac{(\mu ab)^n}{n!} f\left(ka, \frac{1}{k}a^\dagger\right) e^{-\lambda a^\dagger a} \frac{1}{m!}\left(\frac{a^\dagger b^\dagger}{\mu}\right)^m |0, 0\rangle \\
&= \langle 0, 0| e^{\mu ab} f\left(ka, \frac{1}{k}a^\dagger\right) e^{-\lambda a^\dagger a} e^{a^\dagger b^\dagger/\mu}|0, 0\rangle.
\end{aligned} \tag{3.3.55}
$$

We have carried out the sums over n and m at the expense of introducing the dummy operators b and $b^\dagger$ and thereby succeeded in transforming a trace to a vacuum expectation value.

We should note that up to this point we could have omitted the terms $(1 - e^{-\lambda})$ and $\exp -\lambda a^\dagger a$ so that we have proved the lemma.

LEMMA

$$
\operatorname{Tr} f(a, a^\dagger) = \langle 0, 0| e^{\mu ab} f\left(ka, \frac{1}{k}a^\dagger\right) e^{a^\dagger b^\dagger/\mu} |0, 0\rangle, \tag{3.3.56}
$$

which is still another technique for evaluating the trace of a function of a and $a^\dagger$. The μ and k are arbitrary.

We may next use (3.3.2a) to commute $\exp \mu ab$ through $f(a, a^\dagger)$ since b commutes with both a and $a^\dagger$. This gives for (3.3.55) on inserting $\exp(-\mu ab)\exp(\mu ab)$ to the right of f

$$
\frac{\langle f\rangle_0}{(1-e^{-\lambda})} = \langle 0, 0| f\left[ka, \frac{1}{k}(a^\dagger + \mu b)\right] e^{\mu ab} e^{-\lambda a^\dagger a} e^{a^\dagger b^\dagger/\mu}|0, 0\rangle.
$$

By (3.3.17), we have ($x \to -\lambda$, $y \to b^\dagger/\mu$ since $b^\dagger$ commutes with a and $a^\dagger$)

$$
e^{-\lambda a^\dagger a} e^{a^\dagger b^\dagger/\mu}|0, 0\rangle = \exp \frac{1}{\mu} b^\dagger e^{-\lambda} a^\dagger |0, 0\rangle,
$$

so that

$$
\langle f\rangle_0 = (1 - e^{-\lambda})\langle 0, 0| f\left[ka, \frac{1}{k}(a^\dagger + \mu b) e^{\mu ab} \exp\left(\frac{e^{-\lambda} a^\dagger b^\dagger}{\mu}\right)\right] |0, 0\rangle. \tag{3.3.57}
$$

We show next that

$$
e^{\mu ab} e^{\xi a^\dagger b^\dagger}|0, 0\rangle = \frac{1}{1 - \xi\mu} e^{\xi a^\dagger b^\dagger/(1-\xi\mu)}|0, 0\rangle. \tag{3.3.58}
$$

To show this we look for a solution of the form

$$e^{\mu ab}e^{\xi a^\dagger b^\dagger}|0, 0\rangle = e^{\mathscr{G}(\mu, a^\dagger, b^\dagger)}|0, 0\rangle, \tag{3.3.59}$$

where $\mathscr{G}$ is only a function of $a^\dagger$ and $b^\dagger$. If we differentiate both sides with respect to μ, we have

$$abe^{\mu ab}e^{\xi a^\dagger b^\dagger}|0, 0\rangle = abe^{\mathscr{G}}|0, 0\rangle = e^{\mathscr{G}} \frac{\partial \mathscr{G}}{\partial \mu}|0, 0\rangle. \tag{3.3.60}$$

Note that $\partial \mathscr{G}/\partial \mu$ commutes with $\exp \mathscr{G}$ since $\mathscr{G}$ is a function of $a^\dagger$ and $b^\dagger$ only. If we multiply from the left by $\exp -\mathscr{G}$, we have

$$e^{-\mathscr{G}}\, ab\, e^{\mathscr{G}}|0, 0\rangle = \frac{\partial \mathscr{G}}{\partial \mu}|0, 0\rangle,$$

or on inserting $e^{\mathscr{G}}e^{-\mathscr{G}} = 1$ between a and b we have

$$e^{-\mathscr{G}}ae^{\mathscr{G}}e^{-\mathscr{G}}be^{\mathscr{G}}|0, 0\rangle = \frac{\partial \mathscr{G}}{\partial \mu}|0, 0\rangle. \tag{3.3.61}$$

If we use (3.3.9a) we have if we let $f = e^{\mathscr{G}}$

$$\begin{aligned} e^{-\mathscr{G}}ae^{\mathscr{G}} &= e^{-\mathscr{G}}\left\{e^{\mathscr{G}}a + \frac{\partial}{\partial a^\dagger} e^{\mathscr{G}}\right\} \\ &= a + \frac{\partial \mathscr{G}}{\partial a^\dagger} \end{aligned} \tag{3.3.62}$$

since $e^{\mathscr{G}}$ commutes with $\partial \mathscr{G}/\partial a^\dagger$. Similarly,

$$e^{-\mathscr{G}}be^{\mathscr{G}} = b + \frac{\partial \mathscr{G}}{\partial b^\dagger},$$

so that (3.3.61) becomes

$$\left(a + \frac{\partial \mathscr{G}}{\partial a^\dagger}\right)\left(b + \frac{\partial \mathscr{G}}{\partial b^\dagger}\right)|0, 0\rangle = \frac{\partial \mathscr{G}}{\partial \mu}|0, 0\rangle. \tag{3.3.63}$$

However,

$$b|0, 0\rangle = 0,$$

so that the above reduces to

$$\left(a\frac{\partial \mathscr{G}}{\partial b^\dagger} + \frac{\partial \mathscr{G}}{\partial a^\dagger}\frac{\partial \mathscr{G}}{\partial b^\dagger}\right)|0, 0\rangle = \frac{\partial \mathscr{G}}{\partial \mu}|0, 0\rangle. \tag{3.3.64}$$

If we again use (3.3.9a) and let $f = \partial \mathscr{G}/\partial b^\dagger$, we have

$$a\frac{\partial \mathscr{G}}{\partial b^\dagger} = \frac{\partial \mathscr{G}}{\partial b^\dagger}a + \frac{\partial^2 \mathscr{G}}{\partial a^\dagger\, \partial b^\dagger},$$

and since $a|0, 0\rangle = 0$, (3.3.64) becomes

$$\left[\frac{\partial^2 \mathscr{G}}{\partial a^\dagger \partial b^\dagger} + \frac{\partial \mathscr{G}}{\partial a^\dagger}\frac{\partial \mathscr{G}}{\partial b^\dagger}\right]|0, 0\rangle = \frac{\partial \mathscr{G}}{\partial \mu}|0, 0\rangle. \tag{3.3.65}$$

Since $\mathscr{G}$ contains only $a^\dagger$ and $b^\dagger$ which commute, we may look for a solution of (3.3.65) of the form

$$\mathscr{G} = A(\mu) + B(\mu)a^\dagger b^\dagger, \tag{3.3.66}$$

such that by (3.3.59) when $\mu = 0$

$$\mathscr{G}(0, a^\dagger, b^\dagger) = \xi a^\dagger b^\dagger \tag{3.3.67a}$$

or

$$A(0) = 0; \qquad B(0) = \xi. \tag{3.3.67b}$$

When we put (3.3.66) into (3.3.65) and equate equal powers of $a^\dagger b^\dagger$, we see that A and B satisfy the equations

$$\frac{dB}{d\mu} = B^2$$

$$\frac{dA}{d\mu} = B,$$

so that by (3.3.67b)

$$B(\mu) = \frac{\xi}{1 - \xi\mu}$$

$$A(\mu) = -\log(1 - \xi\mu),$$

so that

$$e^{\mathscr{G}}|0, 0\rangle = \frac{1}{1 - \xi\mu}\, e^{\xi a^\dagger b^\dagger/(1-\xi\mu)}|0, 0\rangle,$$

which is (3.3.58).

If we use this with $\mu\xi = e^{-\lambda}$ the thermal average (3.3.57) becomes

$$\langle f\rangle_0 = \langle 0, 0|f\left[ka, \frac{1}{k}\left(a^\dagger + \mu b\right) \exp\left(\frac{\bar{n}}{\mu}\, a^\dagger b^\dagger\right)\right]|0, 0\rangle, \tag{3.3.68}$$

where we have let

$$\bar{n} = \frac{1}{e^\lambda - 1}.$$

Since it is obvious that since $\langle 0, 0|a^\dagger = 0$ and $\langle 0, 0|b^\dagger = 0$,

$$\langle 0, 0|\exp\left(-\frac{\bar{n}}{\mu}\, a^\dagger b^\dagger\right) = \langle 0, 0|$$

and we may rewrite $\langle f\rangle_0$ as

$$\langle f\rangle_0 = \langle 0,0|e^{-\bar{n}a^\dagger b^\dagger/\mu} f\left[ka, \frac{1}{k}(a^\dagger + \mu b)\right] e^{\bar{n}a^\dagger b^\dagger/\mu}|0,0\rangle$$
$$\equiv \langle 0,0|f\left[e^{-\bar{n}a^\dagger b^\dagger/\mu} ka e^{\bar{n}a^\dagger b^\dagger/\mu}, e^{-\bar{n}a^\dagger b^\dagger/\mu}\frac{1}{k}(a^\dagger + \mu b)e^{\bar{n}a^\dagger b^\dagger/\mu}\right]|0,0\rangle. \tag{3.3.70}$$

The last step follows since we imagine f expanded in a power series and we insert $\exp(-\bar{n}a^\dagger b^\dagger/\mu)\exp(\bar{n}a^\dagger b^\dagger/\mu) = 1$ between each power of a and $(a^\dagger + \mu b)$. The series may be resummed to give (3.3.70).

If we use (3.3.2), we have

$$\langle f\rangle_0 = \langle 0,0|f\left\{k\left[a + \frac{\bar{n}}{\mu}b^\dagger\right], \frac{1}{k}[(1+\bar{n})a^\dagger + \mu b]\right\}|0,0\rangle. \tag{3.3.71}$$

Finally, we may choose k and μ in any convenient way. Let us choose them so that the two linear combinations are adjoints, that is, so that

$$c = ka + \frac{\bar{n}k}{\mu}b^\dagger$$
$$c^\dagger = \frac{(1+\bar{n})}{k}a^\dagger + \frac{\mu}{k}b.$$

This requires that

$$k = \sqrt{1+\bar{n}} \qquad \mu = \sqrt{\bar{n}}\,k.$$

Therefore (3.3.71) reduces to

$$\langle f\rangle_0 = \langle 0,0|f[\sqrt{1+\bar{n}}\,a + \sqrt{\bar{n}}\,b^\dagger, \sqrt{1+\bar{n}}\,a^\dagger + \sqrt{\bar{n}}\,b]|0,0\rangle,$$

which completes the proof.

THEOREM 10

If ξ is a complex parameter and η is real,

$$e^{i\eta(\xi a + \xi^* a^\dagger)} = e^{i\eta\xi^* a^\dagger}e^{i\eta\xi a}e^{-\frac{1}{2}\eta^2|\xi|^2}$$
$$= e^{i\eta\xi a}e^{i\eta\xi^* a^\dagger}e^{\frac{1}{2}\eta^2|\xi|^2}. \tag{3.3.72}$$

To prove this, let

$$X(\eta) = e^{i\eta(\xi a + \xi^* a^\dagger)} \tag{3.3.73}$$

$$X(0) = 1. \tag{3.3.74}$$

If we differentiate X with respect to $i\eta$, we have

$$\frac{\partial X}{\partial(i\eta)} = (\xi a + \xi^* a^\dagger)X. \tag{3.3.75}$$

Assume that by some means we have put $X(\mu, a, a^\dagger)$ into normal form where $X = X^{(n)}$. Then by (3.3.10), we have

$$\frac{\partial X^{(n)}}{\partial(i\eta)} = \xi\left(X^{(n)}a + \frac{\partial X^{(n)}}{\partial a^\dagger}\right) + \xi^* a^\dagger X^{(n)}. \tag{3.3.76}$$

Since both sides are now in normal order, we may take diagonal matrix elements of both sides in the coherent state representation. This transforms the operator equation (3.3.76) to the c-number partial differential equation

$$\frac{\partial \bar{X}}{\partial(i\eta)}(\eta, \alpha, \alpha^*) = (\xi\alpha + \xi^*\alpha^*)\bar{X}^{(n)} + \xi\frac{\partial \bar{X}^{(n)}}{\partial \alpha^*}\,. \tag{3.3.77}$$

We may easily solve this subject to the initial condition (3.3.74) by the substitution

$$\bar{X}^{(n)} = e^{\mathscr{G}}, \tag{3.3.78}$$

where

$$\mathscr{G} = A(\eta) + B(\eta)\alpha + C(\eta)\alpha^*, \tag{3.3.79}$$

such that

$$A(0) = B(0) = C(0) = 0. \tag{3.3.80}$$

If we substitute this into (3.3.77) and equate equal powers of α and α^*, we find that A, B, and C satisfy the simple equations

$$\frac{dA}{d(i\eta)} = \xi C$$

$$\frac{dB}{d(i\eta)} = \xi$$

$$\frac{dC}{d(i\eta)} = \xi^*.$$

The solutions which satisfy (3.3.80) are easily found so that

$$\bar{X}^{(n)}(\eta, \alpha) = \exp\left[-\tfrac{1}{2}\eta^2|\xi|^2 + i\eta\xi\alpha + i\eta\xi^*\alpha^*\right].$$

If we apply the normal ordering operator to both sides, we obtain the first form of (3.3.72).

We could repeat the procedure by putting both sides of (3.2.75) into antinormal order. An alternative simpler proof follows directly from (3.1.20).

3.4 CHARACTERISTIC FUNCTIONS [10]; THE WIGNER DISTRIBUTION FUNCTION

We have shown in Section 1.20 that the expectation value of an observable M may be evaluated in the SP or HP by [see (1.20.5)]

$$\langle M(t)\rangle = \text{Tr } M_S(t_0)\rho_S(t) = \text{Tr } M_H(t)\rho_H(t_0), \tag{3.4.1a}$$

where

$$\begin{aligned} \rho_S(t) &= |\psi_S(t)\rangle\langle\psi_S(t)| \\ \rho_H(t_0) &= |\psi_H(t_0)\rangle\langle\psi_H(t_0)|, \end{aligned} \tag{3.4.2a}$$

and $\rho_S(t)$ satisfies

$$i\hbar\frac{\partial\rho_S}{\partial t} = [H_S, \rho_S]. \tag{3.4.3}$$

The averages in (3.4.1) are quantum averages and assume we know that the system is in state $|\psi_H(t_0)\rangle$ at $t = t_0$. If we only know that the system is in state $|\psi_H(t_0)\rangle$ with probability p_ψ, we must take a further average over the probability distribution p_ψ. Thus we generalize (3.4.1) to

$$\begin{aligned} \langle\langle M(t)\rangle\rangle &= \sum_\psi p_\psi\langle\psi_S(t)|M_S|\psi_S(t)\rangle \\ &= \sum_\psi p_\psi\langle\psi_H(t_0)|M_H(t)|\psi(t_0)\rangle. \end{aligned} \tag{3.4.1b}$$

If we generalize ρ to be

$$\begin{aligned} \rho_S(t) &= \sum_\psi p_\psi|\psi_S(t)\rangle\langle\psi_S(t)| \\ \rho_H(t_0) &= \sum_\psi p_\psi|\psi_H(t_0)\rangle\langle\psi_H(t_0)|, \end{aligned} \tag{3.4.2b}$$

we may rewrite this as

$$\langle\langle M(t)\rangle\rangle = \text{Tr } \rho_S(t)M_S = \text{Tr } \rho_H(t_0)M_H(t), \tag{3.4.1c}$$

where $\rho_S(t)$ still satisfies (3.4.3) since p_ψ is the initial distribution and is time independent.

Often one is interested in the moments of some system operator A. Its lth moment in the Schrödinger picture is given by

$$\langle\langle A^l\rangle\rangle = \text{Tr } \rho(t)A^l. \tag{3.4.4}$$

It is sometimes easier to evaluate the average

$$C_A(\xi, t) \equiv \langle\langle e^{i\xi A(t)}\rangle\rangle = \text{Tr } \rho(t)e^{i\xi A} = \text{Tr } \rho(t_0)e^{i\xi A(t)}, \tag{3.4.5}$$

where ξ is a real parameter than to evaluate the moments (3.4.4) directly. Furthermore, all the moments of A can be found from $C_A(\xi, t)$ since it is clear that

$$\langle\langle A^l(t)\rangle\rangle = \frac{\partial^l}{\partial(i\xi)^l}C_A(\xi, t)\bigg|_{\xi=0}. \tag{3.4.6}$$

That is, the lth moment is obtained by differentiating C l times and then letting $\xi = 0$. $C_A(\xi, t)$ is called a moment generating function or the characteristic function for A. Furthermore, we may evaluate the characteristic function in either of the two pictures. The choice is a matter of convenience.

If we know the characteristic function, we may obtain the diagonal matrix elements of the density operator in the representation in which A is diagonal. (For the electromagnetic field, we show below that we may obtain the entire density operator from the characteristic function.)

To prove the statement above we write (3.4.5) as

$$C_A(\xi, t) = \text{Tr}\, \rho(t)e^{i\xi A} = \sum_{\psi} p_\psi \langle \psi(t)|e^{i\xi A}|\psi(t)\rangle. \tag{3.4.7}$$

Let us assume that A is a hermitian operator which satisfied the eigenvalue equation

$$A|A'\rangle = A'|A'\rangle. \tag{3.4.8}$$

To be definite assume that the A' are continuous from $-\infty$ to $+\infty$. The eigenvectors form a complete orthonormal set:

$$\int_{-\infty}^{\infty} dA'|A'\rangle\langle A'| = 1$$
$$\langle A'|A''\rangle = \delta(A' - A''). \tag{3.4.9}$$

If we insert the completeness relation twice in (3.4.7) on either side of the exponential, we have

$$C_A(\xi, t) = \sum_{\psi} p_\psi \iint_{-\infty}^{\infty} \langle \psi(t)|A'\rangle\langle A'|e^{i\xi A}|A''\rangle\langle A''|\psi(t)\rangle\, dA'\, dA''. \tag{3.4.10}$$

Since

$$\langle A'|e^{i\xi A}|A''\rangle = e^{i\xi A'}\langle A'|A''\rangle = e^{i\xi A'}\,\delta(A' - A''), \tag{3.4.11}$$

we may carry out the integration over A'' in (3.4.10) and obtain

$$C_A(\xi, t) = \sum_{\psi} p_\psi \int_{-\infty}^{\infty} dA'\, |\langle \psi(t')|A'\rangle|^2 e^{i\xi A'}, \tag{3.4.12}$$

since

$$\langle A'|\psi(t)\rangle = \langle \psi(t)|A'\rangle^*. \tag{3.4.13}$$

However, $|\langle \psi(t)|A'\rangle|^2\, dA'$ is the probability that a measurement of A when the system is in state $|\psi(t)\rangle$ will yield the value between A' and $A' + dA'$. The ensemble average when the system is in state $|\psi(t)\rangle$ with probability p_ψ is

$$P(A', t)\, dA' = \sum_{\psi} p_\psi\, |\langle \psi(t)|A'\rangle|^2\, dA' \equiv \langle A'|\rho(t)|A'\rangle\, dA', \tag{3.4.14}$$

where we have used (3.4.2b). This is the ensemble probability distribution function for the observable A and is the diagonal matrix element of the density operator $\rho(t)$ in the A-representation. When we substitute (3.4.14) into (3.4.12), we have

$$C_A(\xi, t) = \int_{-\infty}^{\infty} e^{i\xi A'} P(A', t)\, dA', \tag{3.4.15}$$

where ξ and A' are real. Thus the characteristic function for A is just the Fourier transform of the probability distribution function $P(A', t)$. Accordingly, we may invert (3.4.15) to obtain

$$\begin{aligned} P(A', t) &= \frac{1}{2\pi} \int_{-\infty}^{\infty} C_A(\xi, t) e^{-i\xi A'}\, d\xi \\ &= \langle A' | \rho(t) | A' \rangle. \end{aligned} \tag{3.4.16}$$

Therefore, the characteristic function for A determines the diagonal matrix elements for the density operator in the A-representation. This is all the information needed to determine the moments of A.

Let us next consider the special case of a harmonic oscillator described by the boson operators a and $a^\dagger$. The density operator will be a function of a and $a^\dagger$ with unit trace:

$$\mathrm{Tr}\, \rho(a, a^\dagger, t) = 1. \tag{3.4.17}$$

There are three useful ways to define characteristic functions depending on the moments of interest. These are

$$C^{(n)}(\xi, t) = \langle\langle e^{i\eta\xi^* a^\dagger} e^{i\eta\xi a} \rangle\rangle = \mathrm{Tr}\, \rho(t) e^{i\eta\xi^* a^\dagger} e^{i\eta\xi a} \tag{3.4.18a}$$

$$C^{(a)}(\xi, t) = \langle\langle e^{i\eta\xi a} e^{i\eta\xi^* a^\dagger} \rangle\rangle = \mathrm{Tr}\, \rho(t) e^{i\eta\xi a} e^{i\eta\xi^* a^\dagger} \tag{3.4.18b}$$

$$C^{(w)}(\xi, t) = \langle\langle e^{i\eta(\xi a + \xi^* a^\dagger)} \rangle\rangle = \mathrm{Tr}\, \rho(t) e^{i\eta(\xi a + \xi^* a^\dagger)}, \tag{3.4.18c}$$

where ξ is a complex parameter and η is real. We call these the antinormal, normal, and Wigner [11] characteristic functions, respectively. By (3.4.17) we see that

$$C^{(n)}(0, t) = C^{(a)}(0, t) = C^{(w)}(0, t) = 1. \tag{3.4.19}$$

The first of these is most useful if we want the normally ordered moments

$$\begin{aligned} \langle\langle a^{\dagger l}(t) a^m(t) \rangle\rangle &= \frac{\partial^{(l+m)}}{\partial(i\eta\xi^*)^l\, \partial(i\eta\xi)^m} C^{(n)}(\xi, t) \bigg|_{\eta=0} \\ &= \mathrm{Tr}\, \rho(t) a^{\dagger l} a^m. \end{aligned} \tag{3.4.20}$$

The second form yields directly the antinormally ordered moments

$$\langle\langle a^l(t)a^{\dagger m}(t)\rangle\rangle = \frac{\partial^{(l+m)}}{\partial(i\eta\xi)^l\,\partial(i\eta\xi^*)^m}\,C^{(a)}(\xi, t)\bigg|_{\eta=0}$$
$$= \operatorname{Tr}\rho(t)a^l a^{\dagger m}. \qquad (3.4.21)$$

while the third form yields the symmetric moments

$$\langle\langle(\xi a + \xi^* a^\dagger)^l\rangle\rangle = \frac{\partial^l}{\partial(i\eta)^l}\,C^{(w)}(\xi, t)\bigg|_{\eta=0}$$
$$= \operatorname{Tr}\rho(t)[\xi a + \xi^* a^\dagger]^l. \qquad (3.4.22)$$

In these, we have treated ξ and ξ^* as independent variables.

We may use (3.3.72)

$$e^{i\eta(\xi a+\xi^* a^\dagger)} = e^{i\xi^*\eta a^\dagger}e^{i\xi\eta a}e^{-\frac{1}{2}\eta^2|\xi|^2}$$
$$= e^{i\xi\eta a}e^{i\xi^*\eta a^\dagger}e^{\frac{1}{2}\eta^2|\xi|^2}, \qquad (3.4.23)$$

to show the connection among these characteristic functions. If we use this and the definitions (3.4.18), we see that

$$C^{(w)}(\xi, t) = e^{-\frac{1}{2}\eta^2|\xi|^2}C^{(n)}(\xi, t) = e^{\frac{1}{2}\eta^2|\xi|^2}C^{(a)}(\xi, t), \qquad (3.4.24)$$

so that we may easily obtain all three characteristic functions if one is known.

We next show that the density operator for a mode of the radiation field may be determined uniquely from the characteristic function [10] (3.4.18). By (3.2.34) since the exponential operators in (3.4.18a) are in normal order, it follows that

$$C^{(n)}(\xi, t) = \int \bar{\rho}^{(a)}(\alpha, t)e^{i\eta(\xi^*\alpha^*+\xi\alpha)}\,\frac{d^2\alpha}{\pi}, \qquad (3.4.25)$$

where $\bar{\rho}^{(a)}$ is the antinormal associated function for the density operator. (It is a function of α, α^*, and t.) Therefore, $C^{(n)}(\xi, t)$ is the double Fourier transform of $\bar{\rho}^{(a)}$. It follows from (3.2.37) and (3.2.41) that

$$\bar{\rho}^{(a)}(\alpha, t) = \eta^2\int e^{-i\eta(\xi a+\xi^*\alpha^*)}C^{(n)}(\xi, t)\,\frac{d^2\xi}{\pi}. \qquad (3.4.26)$$

Similarly, we have

$$\bar{\rho}^{(n)}(\alpha, t) = \eta^2\int e^{-i\eta(\xi\alpha+\xi^*\alpha^*)}C^{(a)}(\xi, t)\,\frac{d^2\xi}{\pi}. \qquad (3.4.27)$$

Therefore, the density operator is given by

$$\rho(a, a^\dagger, t) = \mathscr{A}\{\bar{\rho}^{(a)}(\alpha, t)\} = \mathscr{N}\{\bar{\rho}^{(n)}(\alpha, t)\}. \qquad (3.4.28)$$

We may use (3.2.23) and (3.4.26) to obtain the direct relation

$$\rho(a, a^\dagger, t) = \int \frac{d^2\alpha}{\pi} |\alpha\rangle\langle\alpha| \bar{\rho}^{(a)}(\alpha, t)$$

$$= \eta^2 \int \frac{d^2\xi}{\pi} C^{(n)}(\xi, t) \int e^{-i\eta(\xi\alpha+\xi^*\alpha^*)} |\alpha\rangle\langle\alpha| \frac{d^2\alpha}{\pi}. \tag{3.4.29}$$

We next use (3.4.18c) to obtain the Wigner distribution function [11]. This form is particularly appropriate if we are interested in obtaining moments of p and q directly where by (2.1.20)

$$\begin{aligned} q &= \sqrt{\frac{\hbar}{2\omega}}(a^\dagger + a) \qquad & a &= \frac{1}{\sqrt{2\hbar\omega}}(\omega q + ip) \\ p &= i\sqrt{\frac{\hbar\omega}{2}}(a^\dagger - a) \qquad & a^\dagger &= \frac{1}{\sqrt{2\hbar\omega}}(\omega q - ip). \end{aligned} \tag{3.4.30}$$

If we substitute for a and $a^\dagger$ in (3.4.18c), define two new real parameters λ and μ by the relations

$$\begin{aligned} \lambda &= \eta(\xi + \xi^*)\sqrt{\frac{\omega}{2\hbar}} \qquad & \xi\eta &= \sqrt{\frac{\hbar}{2\omega}}\lambda - i\sqrt{\frac{\hbar\omega}{2}}\mu \\ \mu &= i\eta(\xi - \xi^*)\frac{1}{\sqrt{2\hbar\omega}} \qquad & \xi^*\eta &= \sqrt{\frac{\hbar}{2\omega}}\lambda + i\sqrt{\frac{\hbar\omega}{2}}\mu, \end{aligned} \tag{3.4.31}$$

then (3.4.18c) becomes

$$C^{(w)}(\lambda, \mu, t) = \operatorname{Tr} \rho(p, q, t) e^{i(\lambda q + \mu p)}. \tag{3.4.32}$$

From (3.1.27), we may write

$$E \equiv e^{i(\lambda q + \mu p)} = e^{i\lambda q} e^{i\mu p} e^{\frac{1}{2} i\hbar\lambda\mu}, \tag{3.4.33}$$

since $[q, p] = i\hbar$. This may be rewritten as

$$E = e^{\frac{1}{2} i\mu p} e^{-\frac{1}{2} i\mu p} e^{i\lambda q} e^{\frac{1}{2} i\mu p} e^{\frac{1}{2} i\mu p} e^{\frac{1}{2} i\hbar\lambda\mu}. \tag{3.4.34}$$

By (3.1.4), we see that

$$e^{-\frac{1}{2} i\mu p} e^{i\lambda q} e^{\frac{1}{2} i\mu p} = \exp i\lambda e^{-\frac{1}{2} i\mu p}\, q e^{\frac{1}{2} i\mu p}$$

$$= \exp i\lambda\left(q - \frac{\mu\hbar}{2}\right), \tag{3.4.35}$$

where we used (3.1.7). If we use this in (3.4.34), we see that (3.4.32) becomes

$$C^{(w)}(\lambda, \mu, t) = \operatorname{Tr} \rho\,(p, q, t) e^{\frac{1}{2} i\mu p} e^{i\lambda q} e^{\frac{1}{2} i\mu p}. \tag{3.4.36}$$

Let us evaluate the trace in the q-representation where

$$
\begin{aligned}
q|q'\rangle &= q'|q'\rangle \\
\langle q'|q''\rangle &= \delta(q' - q'') \\
\int_{-\infty}^{\infty} dq'|q'\rangle\langle q'| &= 1.
\end{aligned}
\tag{3.4.37}
$$

Furthermore, it follows from (1.10.17) and its adjoint that

$$
\begin{aligned}
e^{\frac{1}{2}i\mu p}|q'\rangle &= \left|q' - \frac{\mu\hbar}{2}\right\rangle \\
\langle q'|e^{\frac{1}{2}i\mu p} &= \left\langle q' + \frac{\mu\hbar}{2}\right|.
\end{aligned}
\tag{3.4.38}
$$

Since Tr $AB =$ Tr BA, we may rewrite (3.4.36) as

$$
\begin{aligned}
C^{(w)}(\lambda, \mu, t) &= \text{Tr}\, e^{\frac{1}{2}i\mu p}\rho(p, q, t)e^{\frac{1}{2}i\mu p}e^{i\lambda q} \\
&= \int_{-\infty}^{\infty} dq''\langle q''|e^{\frac{1}{2}i\mu p}\rho(p, q, t)e^{\frac{1}{2}i\mu p}e^{i\lambda q}|q''\rangle \\
&= \int_{-\infty}^{\infty} e^{i\lambda q''}\left\langle q'' + \frac{\mu\hbar}{2}\right|\rho(p, q, t)\left|q'' - \frac{\mu\hbar}{2}\right\rangle dq'',
\end{aligned}
\tag{3.4.39}
$$

where we evaluated the trace in the q-representation and used (3.4.38).

As in the case of $C^{(a)}$ and $C^{(n)}$, we may obtain the Fourier transform of $C^{(w)}$, which is

$$
W(\alpha, \alpha^*, t) = \eta^2 \int e^{-i\eta(\xi\alpha+\xi^*\alpha^*)}\, C^{(w)}(\xi, t)\,\frac{d^2\xi}{\pi}. \tag{3.4.40}
$$

This is the Wigner distribution function and is the analog of the associated functions $\bar{\rho}^{(a)}$ and $\bar{\rho}^{(n)}$ for the density operator. Since

$$
\int e^{-i\eta(\xi\alpha+\xi^*\alpha^*)}\, d^2\alpha = \pi^2\,\delta(\xi\eta)\,\delta(\xi^*\eta), \tag{3.4.41}
$$

if we integrate both sides of (3.4.40), we see that

$$
\int W(\alpha, \alpha^*, t)\,\frac{d^2\alpha}{\pi} = C^{(w)}(0, t) = 1. \tag{3.4.42}
$$

We shall express the Wigner distribution function in the notation common in the literature and point out a few of its properties. For this purpose, we introduce a change of variables. Let

$$\alpha = \frac{1}{\sqrt{2\hbar\omega}}(\omega q' + ip')$$
$$\alpha^* = \frac{1}{\sqrt{2\hbar\omega}}(\omega q' + ip'). \tag{3.4.43}$$

We define $P(p', q', t)$ in terms of $W(\alpha, \alpha^*, t)$ so that

$$\int W(\alpha, \alpha^*, t)\frac{d^2\alpha}{\pi} = \iint_{-\infty}^{\infty} P(p', q', t)\, dp'\, dq' = 1. \tag{3.4.44}$$

Because

$$d^2\alpha = d\,\mathrm{Re}\,\alpha\, d\,\mathrm{Im}\,\alpha = \sqrt{\frac{\omega}{2\hbar}}\frac{1}{\sqrt{2\hbar\omega}}\, dq'\, dp' = \frac{1}{2\hbar}\, dp'\, dq', \tag{3.4.45}$$

we have on using (3.4.43), (3.4.31), and (3.4.40) that

$$P(p', q', t) = W(\alpha, \alpha', t)\frac{1}{2\pi\hbar}$$
$$= \frac{1}{4\pi^2}\int e^{-i(\lambda q' + \mu p')} C^{(w)}(\lambda, \mu, t)\, d\lambda\, d\mu, \tag{3.4.46}$$

since $\eta^2\, d^2\xi = \hbar\, d\lambda\, d\mu/2$. If we use (3.4.39), carry out the integral over λ first and then over q'', we find that (3.4.46) reduces to

$$P(p', q', t) = \frac{1}{2\pi}\int_{-\infty}^{\infty} e^{-i\mu p'}\left\langle q' + \frac{\mu\hbar}{2}\right|\rho\left|q' - \frac{\mu\hbar}{2}\right\rangle d\mu, \tag{3.4.47}$$

where we used the integral

$$\int_{-\infty}^{\infty} e^{i\lambda(q'' - q')}\, d\lambda = 2\pi\,\delta(q' - q''). \tag{3.4.48}$$

Finally, we let

$$y = \frac{\hbar}{2}\mu, \tag{3.4.49}$$

then P reduces to

$$P(p', q', t) = \frac{1}{\pi\hbar}\int_{-\infty}^{\infty} e^{-(2i/\hbar)p'y}\langle q' + y|\rho|q' - y\rangle\, dy. \tag{3.4.50}$$

The density operator is

$$\rho(t) = \sum_{\psi} p_{\psi}|\psi(t)\rangle\langle\psi(t)|, \tag{3.4.51}$$

while the scalar product,

$$\langle q'|\psi(t)\rangle \equiv \psi(q', t), \tag{3.4.52}$$

is the Schrödinger wave function in the q-representation. If we substitute (3.4.51) into (3.4.50) and use (3.4.52), the Wigner distribution function becomes

$$P(p', q', t) = \sum_{\psi} P_{\psi} \frac{1}{\pi\hbar} \int_{-\infty}^{\infty} e^{-(2i/\hbar)p'y} \psi^*(q' - y, t)\psi(q' + y, t)\, dy. \quad (3.4.53)$$

If we integrate $P(p', q', t)$ over all p', we have that

$$\int_{-\infty}^{\infty} P(p', q', t)\, dp' = \begin{cases} \sum_{\psi} P_{\psi} \, |\psi(q', t)|^2 & \text{(mixed state)} \\ |\psi(q', t)|^2 & \text{(pure state).*} \end{cases} \quad (3.4.54)$$

Here we used the result

$$\int_{-\infty}^{\infty} e^{-(2i/\hbar)p'y}\, dp' = \pi\hbar\, \delta(y). \quad (3.4.55)$$

Therefore, if we integrate the Wigner distribution over all p', we obtain the probability that a measurement of q will yield the value q' at time t.

If we next integrate (3.4.53) over all q' we find

$$\int_{-\infty}^{\infty} P(p', q', t)\, dq' = \begin{cases} \sum_{\psi} P_{\psi} \, |\Phi(p', t)|^2 & \text{(mixed state)} \\ |\Phi(p', t)|^2 & \text{(pure state)} \end{cases} \quad (3.4.56)$$

where

$$\Phi(p', t) = \frac{1}{\sqrt{2\pi\hbar}} \int_{-\infty}^{\infty} e^{-(i/\hbar)p'q'} \psi(q', t)\, dq', \quad (3.4.57)$$

is the Schrödinger wave function in the p-representation.

To prove this, we change variables in (3.4.53) and let

$$u = q' - y \qquad q' = \tfrac{1}{2}(v + u), \quad (3.4.58a)$$

$$v = q' + y \qquad p' = \tfrac{1}{2}(v - u), \quad (3.4.58b)$$

so that

$$dq'\, dy = \begin{vmatrix} \frac{1}{2} & \frac{1}{2} \\ \frac{1}{2} & -\frac{1}{2} \end{vmatrix} du\, dv = \tfrac{1}{2}\, du\, dv. \quad (3.4.59)$$

Therefore,

$$\int_{-\infty}^{\infty} P(p', q', t)\, dq' = \sum_{\psi} P_{\psi} \iint_{-\infty}^{\infty} \frac{dq'\, dy}{\pi\hbar} e^{-(2i/\hbar)p'y} \psi^*(q' - y)\psi(q' + y)$$

$$= \sum_{\psi} P_{\psi} \frac{1}{2\pi\hbar} \iint_{-\infty}^{\infty} du\, dv e^{-(ip'/\hbar)(v-u)} \psi^*(u)\, \psi(v)$$

$$= \sum_{\psi} P_{\psi} \left| \frac{1}{\sqrt{2\pi\hbar}} \int_{-\infty}^{\infty} e^{-(i/\hbar)p'v} \psi(v, t)\, dv \right|^2. \quad \text{Q.E.D.} \quad (3.4.60)$$

* In a pure state, we assume we know the initial $\psi(q', 0)$ (see Section 3.16).

Therefore, if we integrate the Wigner distribution function over all values of q', we obtain the probability that a measurement of p on the system in state $|\psi(t)\rangle$ will give a value between p' and $p' + dp'$.

3.5 THE POISSON DISTRIBUTION

If a harmonic oscillator is initially in a coherent state $|\alpha'\rangle$, the density operator is

$$\rho(a, a^\dagger, 0) = |\alpha'\rangle\langle\alpha'|. \tag{3.5.1}$$

Such a state represents a coherent signal in the cavity mode, since

$$\begin{aligned}\langle a(0)\rangle &= \operatorname{Tr} \rho(0)a = \langle\alpha'|a|\alpha'\rangle = \alpha' \equiv |\alpha'|\, e^{-i\varphi} \\ \langle a^\dagger(0)\rangle &= \operatorname{Tr} \rho(0)a^\dagger = \langle\alpha'|a^\dagger|\alpha'\rangle = \alpha'^* \equiv |\alpha'|\, e^{i\varphi},\end{aligned} \tag{3.5.2}$$

and the field has a definite amplitude and phase. We have normalized so that $\langle\alpha'|\alpha'\rangle = 1$. The mean number of photons is by definition

$$\bar{n} = \langle a^\dagger(0)a(0)\rangle = \operatorname{Tr} \rho(0)a^\dagger a = \langle\alpha'|a^\dagger a|\alpha'\rangle = |\alpha'|^2. \tag{3.5.3}$$

The state is a minimum uncertainty state as we have previously shown.

In the number representation, the diagonal matrix elements of (3.5.1) are

$$\langle n|\rho(a, a^\dagger, 0)|n\rangle = |\langle n|\alpha'\rangle|^2 = e^{-|\alpha'|^2}\frac{(|\alpha'|^2)^n}{n!}, \tag{3.5.4}$$

where we have used (2.5.13). This is just a Poisson distribution. If we measure $a^\dagger a$ when the oscillator is in state $|\alpha'\rangle$, this gives us the probability of obtaining n photons.

We next obtain the antinormal and normal associated functions for the coherent state $|\alpha'\rangle\langle\alpha'|$. We first show that

$$\begin{aligned}|\alpha'\rangle\langle\alpha'| &= \mathscr{A}\left\{\lim_{\epsilon\to\infty} \epsilon e^{-\epsilon(\alpha^*-\alpha'^*)(\alpha-\alpha')}\right\} \\ &\equiv \mathscr{A}\{\pi\, \delta(\alpha^* - \alpha'^*)\, \delta(\alpha - \alpha')\} \\ &= \int \pi\, \delta(\alpha^* - \alpha'^*)\, \delta(\alpha - \alpha')|\alpha\rangle\langle\alpha|\frac{d^2\alpha}{\pi},\end{aligned} \tag{3.5.5}$$

so that

$$\bar{\rho}^{(a)}(\alpha, \alpha^*, 0) = \lim_{\epsilon\to\infty} \epsilon e^{-\epsilon(\alpha^*-\alpha'^*)(\alpha-\alpha')} = \pi\, \delta(\alpha^* - \alpha'^*)\, \delta(\alpha - \alpha'). \tag{3.5.6}$$

To prove the result above, we must first show that

$$\lim_{\epsilon\to\infty} \epsilon e^{-\epsilon(\alpha^*-\alpha'^*)(\alpha-\alpha')} = \pi\, \delta(\alpha - \alpha'^*)\, \delta(\alpha - \alpha'). \tag{3.5.7}$$

We note first that if $\alpha^* \neq \alpha'^*$ and $\alpha \neq \alpha'$, the left side vanishes as $\epsilon \to \infty$ in agreement with the right side, while if $\alpha^* = \alpha'^*$ or $\alpha = \alpha'$, both sides become infinite as $\epsilon \to \infty$. Finally, we note that

$$\lim_{\epsilon\to\infty} \epsilon \int e^{-\epsilon(\alpha^*-\alpha'^*)(\alpha-\alpha')} \frac{d^2\alpha}{\pi} = \lim_{\epsilon\to\infty} \epsilon \frac{1}{\epsilon} \to 1, \tag{3.5.8}$$

while

$$\int \pi\, \delta(\alpha^* - \alpha'^*)\, \delta(\alpha - \alpha') \frac{d^2\alpha}{\pi} = 1, \tag{3.5.9}$$

which proves (3.5.7). We recall that

$$\begin{aligned}\rho(a, a^\dagger, 0) &= \int \bar{\rho}^{(a)}(\alpha, \alpha^*, 0)|\alpha\rangle\langle\alpha| \frac{d^2\alpha}{\pi} \\ &= \mathscr{A}\{\bar{\rho}^{(a)}(\alpha, \alpha^*)\}.\end{aligned} \tag{3.5.10}$$

If we use (3.5.6) here, (3.5.5) follows. As a special case, we note that for $\alpha' = \alpha'^* = 0$

$$|0\rangle\langle 0| = \mathscr{A}\left\{\lim_{\epsilon\to\infty} \epsilon e^{-\epsilon\alpha^*\alpha}\right\} = \mathscr{A}\{\pi\, \delta(\alpha)\, \delta(\alpha^*)\}. \tag{3.5.11}$$

To obtain $\bar{\rho}^{(n)}(\alpha, \alpha^*, 0)$ we have by (3.3.42)

$$|\alpha'\rangle\langle\alpha'| = \mathscr{N}\{\bar{\rho}^{(n)}(\alpha, \alpha^*, 0)\} = \lim_{\epsilon\to 1} \mathscr{N}\{e^{-\epsilon|\alpha-\alpha'|^2}\}, \tag{3.5.12}$$

so that

$$\bar{\rho}^{(n)}(\alpha, \alpha^*, 0) = \lim_{\epsilon\to 1} e^{-\epsilon|\alpha-\alpha'|^2} = \mathscr{N}^{-1}\{|\alpha'\rangle\langle\alpha'|\}. \tag{3.5.13}$$

So far we have considered a pure coherent state in which the amplitude and phase are known. Suppose, however, that the phase of the oscillators were random. The density operator then becomes

$$\rho(a, a^\dagger, 0) = \overline{|\alpha'\rangle\langle\alpha'|}, \tag{3.5.14}$$

where the bar indicates that we average over the phases. That is, if we let

$$\alpha' = |\alpha'|\, e^{i\varphi}, \tag{3.5.15}$$

then

$$\rho(a, a^\dagger, 0) = \overline{|\alpha'\rangle\langle\alpha'|} = \frac{\int_0^{2\pi} d\varphi |\alpha'\rangle\langle\alpha'|}{2\pi}. \tag{3.5.16}$$

Since by (2.5.12b)

$$\begin{aligned}|\alpha'\rangle\langle\alpha'| &= e^{-|\alpha'|^2} e^{\alpha' a^\dagger} |0\rangle\langle 0| e^{\alpha'^* a} \\ &= e^{-|\alpha'|^2} \sum_{n=0}^{\infty} \sum_{m=0}^{\infty} \frac{(\alpha' a^\dagger)^n}{n!} |0\rangle\langle 0| \frac{(\alpha'^* a)^m}{m!},\end{aligned} \tag{3.5.17}$$

the density operator becomes

$$\rho(a, a^\dagger, 0) = e^{-|\alpha'|^2} \sum_{n,m=0}^{\infty} \frac{(|\alpha'|\, a^\dagger)^n}{n!} |0\rangle\langle 0| \frac{(|\alpha'|\, a)^m}{m!} \int_0^{2\pi} e^{i(m-n)\varphi} \frac{d\varphi}{2\pi}. \quad (3.5.18)$$

Since

$$\int_0^{2\pi} e^{i(m-n)\varphi} \frac{d\varphi}{2\pi} = \delta_{nm}, \quad (3.5.19)$$

(3.5.18) reduces to

$$\begin{aligned} \rho(a, a^\dagger, 0) &= e^{-|\alpha'|^2} \sum_{n=0}^{\infty} \frac{|\alpha'|^{2n}}{(n!)^2} a^{\dagger n}|0\rangle\langle 0|a^n \\ &= e^{-|\alpha'|^2} \sum_{n=0}^{\infty} \frac{(|\alpha'|^2)^n}{n!} |n\rangle\langle n|. \end{aligned} \quad (3.5.20)$$

This is also seen to be a Poisson distribution since

$$\langle n|\rho(a, a^\dagger, 0)|n\rangle = e^{-|\alpha'|^2} \frac{(|\alpha'|^2)^n}{n!}. \quad (3.5.21)$$

However, all phase information has been lost. One easily sees that Tr $\rho^2(0) \neq 1$ so that we no longer have a pure state (see Section 3.16). This corresponds to giving up all information on the phase. It means that we make no measurements on the phase although the oscillators may still have a definite phase.

The normal associated function for (3.5.14) is given by (3.5.20), (3.2.12), and (2.5.13)

$$\begin{aligned} \bar{\rho}^{(n)}(\alpha, \alpha^*, 0) &= \langle\alpha|\rho(a, a^\dagger, 0)|\alpha\rangle = |\langle\alpha|\alpha'\rangle|^2 = e^{-|\alpha'-\alpha|^2} \\ &= e^{-\bar{n}} \sum_{n=0}^{\infty} \frac{(\bar{n})^n}{n!} |\langle\alpha|n\rangle|^2 \\ &= e^{-\bar{n}-|\alpha|^2} \sum_{n=0}^{\infty} \frac{(\bar{n}\,|\alpha|^2)^n}{(n!)^2}, \end{aligned} \quad (3.5.22)$$

where

$$\bar{n} = \text{Tr}\, a^\dagger a\, \rho(a, a^\dagger, 0) = |\alpha'|^2. \quad (3.5.23)$$

However, it is well-known that the modified zero order Bessel function $I_0(z)$ is given by

$$I_0(z) = \sum_{n=0}^{\infty} \frac{(z/2)^n}{n!^2}. \quad (3.5.24)$$

We therefore have that

$$\bar{\rho}^{(n)}(\alpha, \alpha^*, 0) = e^{-\bar{n}-|\alpha|^2} I_0(2\bar{n}\,|\alpha|^2), \quad (3.5.25)$$

which differs greatly from (3.5.13) for the pure coherent state. We call $\overline{|\alpha'\rangle\langle\alpha'|}$ the random phase coherent state.

We next obtain the characteristic function $C^{(n)}(\xi, t)$ for this state. It is by (3.5.20) that

$$\begin{aligned} C^{(n)}(\xi, t) &= \text{Tr}\ \rho(t)e^{i\xi^*\eta a^\dagger}e^{i\xi\eta a} \\ &= \text{Tr}\ e^{-\bar{n}}\left\{\sum_{n=0}^{\infty}\frac{\bar{n}^n}{n!}|n\rangle\langle n|e^{i\xi^*\eta a^\dagger}e^{i\xi\eta a}\right\} \\ &= e^{-\bar{n}}\sum_{n=0}^{\infty}\frac{\bar{n}^n}{n!}\langle n|e^{i\xi^*\eta a^\dagger}e^{i\xi\eta a}|n\rangle. \end{aligned} \tag{3.5.26}$$

However,

$$\begin{aligned} e^{i\xi\eta a}|n\rangle &= \sum_{l=0}^{\infty}\frac{(i\xi\eta)^l}{l!}a^l|n\rangle \\ &= \sum_{l=0}^{n}\frac{(i\xi\eta)^l}{l!}\sqrt{\frac{n!}{(n-l)!}}\,|n-l\rangle, \end{aligned} \tag{3.5.27a}$$

while

$$\langle n|e^{i\xi^*\eta a^\dagger} = \sum_{m=0}^{n}\frac{(i\xi^*\eta)^m}{m!}\sqrt{\frac{n!}{(n-m)!}}\,\langle n-m|. \tag{3.5.27b}$$

When we use these and the orthogonality relations, we obtain

$$C^{(n)}(\xi, t) = e^{-\bar{n}}\sum_{n=0}^{\infty}\frac{\bar{n}^n}{n!}\sum_{l=0}^{n}\frac{(-|\xi|^2\eta^2)^l n!}{(l!)^2(n-l)!}. \tag{3.5.28}$$

The Laguerre polynomials are given by the power series

$$L_n(z) = \sum_{l=0}^{n}\frac{n!^2(-z)^l}{(l!)^2(n-l)!}, \tag{3.5.29}$$

so that

$$C^{(n)}(\xi, t) = e^{-\bar{n}}\sum_{n=0}^{\infty}\frac{\bar{n}^n}{n!^2}L_n(|\xi|^2\eta^2). \tag{3.5.30}$$

Since the Laguerre polynomials have the well-known generating function,

$$J_0(2\sqrt{xt}) = e^{-t}\sum_{n=0}^{\infty}\frac{t^n}{n!^2}L_n(x), \tag{3.5.31}$$

where J_0 is the ordinary Bessel function of the first kind of order zero, $C^{(n)}$ becomes

$$C^{(n)}(\xi, t) = J_0(2\sqrt{\bar{n}}\,\eta\,|\xi|). \tag{3.5.32a}$$

The other characteristic functions then are

$$C^{(a)}(\xi, t) = e^{-\eta^2|\xi|^2}J_0(2\sqrt{\bar{n}}\,\eta\,|\xi|) \tag{3.5.32b}$$

$$C^{(w)}(\xi, t) = e^{-\frac{1}{2}\eta^2|\xi|^2}J_0(2\sqrt{\bar{n}}\,\eta\,|\xi|). \tag{3.5.32c}$$

One easily sees that $\langle a\rangle = \langle a^\dagger\rangle = 0$ for this state while $\langle a^\dagger a\rangle = \bar{n}$.

The coherent state represents a coherent signal with a definite phase while the state $\overline{|\alpha'\rangle\langle\alpha'|}$ represents a coherent signal with no knowledge of the phase. At optical frequencies phase measurements are difficult, and the latter situation represents the field of a laser above threshold to a good approximation as we shall see.

3.6 THE EXPONENTIAL DISTRIBUTION

Consider the density operator

$$\rho(a, a^\dagger) = (1 - e^{-\lambda})e^{-\lambda a^\dagger a}, \tag{3.6.1}$$

where $\lambda = \beta\hbar\omega$. In a later section we show that this density operator will maximize the entropy subject to the constraints that $\text{Tr}\,\rho = 1$ and that the average energy of the system is known.

If we use the completeness relation in the number representation, ρ may be written as

$$\begin{aligned}\rho(a, a^\dagger) &= (1 - e^{-\lambda}) \sum_{n=0}^{\infty} e^{-\lambda a^\dagger a}|n\rangle\langle n| \\ &= (1 - e^{-\lambda}) \sum_{0}^{\infty} e^{-\lambda n}|n\rangle\langle n|,\end{aligned} \tag{3.6.2}$$

from which it follows that

$$\text{Tr}\,\rho = (1 - e^{-\lambda}) \sum_{0}^{\infty} (e^{-\lambda})^n = 1, \tag{3.6.3}$$

since $\text{Tr}\,|n\rangle\langle n| = \langle n|n\rangle = 1$ and

$$\sum_{0}^{\infty} x^n = \frac{1}{1 - x}, \tag{3.6.4}$$

if $|x| < 1$.

The average energy of the oscillator is

$$\begin{aligned}\langle H\rangle &= \hbar\omega\langle a^\dagger a\rangle = \hbar\omega\,\text{Tr}\,a^\dagger a\rho \\ &= \hbar\omega(1 - e^{-\lambda}) \sum_{n=0}^{\infty} ne^{-\lambda n},\end{aligned} \tag{3.6.5}$$

where we used (3.6.2). If we differentiate both sides of (3.6.4) with respect to x, we obtain

$$\sum_{0}^{\infty} nx^{n-1} = \frac{1}{(1 - x)^2},$$

or on multiplying both sides by x,

$$\sum_{0}^{\infty} nx^n = \frac{x}{(1 - x)^2}. \tag{3.6.6}$$

Therefore, (3.6.5) becomes

$$\langle H \rangle = \hbar\omega \langle a^\dagger a \rangle = \frac{\hbar\omega}{e^{\beta\hbar\omega} - 1} = \hbar\omega\langle n \rangle. \tag{3.6.7}$$

As $\hbar \to 0$ in the classical limit this becomes

$$\langle H \rangle \to \frac{\omega}{\beta\omega} = \frac{1}{\beta} = kT. \tag{3.6.8}$$

That is, by the law of equipartition of energy, the average energy is $\frac{1}{2}kT$ per degree of freedom. Therefore,

$$\langle a^\dagger a \rangle = \frac{\langle H \rangle}{\hbar\omega} = \frac{1}{e^\lambda - 1} \equiv \bar{n}, \tag{3.6.9}$$

where

$$\lambda = \frac{\hbar\omega}{kT}. \tag{3.6.10}$$

This is the Planck distribution law for radiation in a cavity to be in thermal equilibrium at temperature T. It is therefore described by the density operator in (3.6.1).

In the number representation (3.6.2), we see that

$$p(n) \equiv (1 - e^{-\lambda})e^{-\lambda n} = \langle n | \rho(a, a^\dagger) | n \rangle \tag{3.6.11}$$

gives the probability of finding n photons when we measure the oscillator energy. The form suggests we call it an exponential distribution. By (3.6.5), we see that

$$\bar{n} = \sum_0^\infty n p(n) = \frac{1}{e^\lambda - 1}, \tag{3.6.12}$$

or

$$e^{-\lambda} \equiv \frac{\bar{n}}{1 + \bar{n}}, \tag{3.6.13}$$

so that we may also write $p(n)$ as

$$p(n) = \frac{\bar{n}^n}{(1 + \bar{n})^{n+1}}. \tag{3.6.14}$$

The reader may show for this distribution that

$$\begin{aligned} \langle a \rangle &= \operatorname{Tr} \rho a = 0 \\ \langle a^\dagger \rangle &= \operatorname{Tr} \rho a^\dagger = 0. \end{aligned} \tag{3.6.15}$$

3.7 GENERALIZED WICK'S THEOREM FOR BOSON OPERATORS

We have already given several methods for evaluating traces. The method given in Theorem 9, Section 3.3 is especially useful for evaluating thermal averages, and it may be generalized to include many independent oscillators (see Problem 3.17). In this section we give yet another method of evaluating thermal averages which is called the generalized Wick's theorem [12].

Let $\psi_1, \psi_2, \ldots$ be either boson creation or annihilation operators corresponding to various independent oscillators whose hamiltonian is

$$H_0 = \sum_j \hbar\omega_j a_j^\dagger a_j. \tag{3.7.1}$$

When the field is in thermal equilibrium, its density operator is the Boltzmann distribution

$$\rho = \frac{e^{-\beta H_0}}{\operatorname{Tr} e^{-\beta H_0}} = \prod_j (1 - e^{-\lambda_j})e^{-\lambda_j a_j^\dagger a_j}, \tag{3.7.2}$$

where

$$\beta = \frac{1}{kT} \qquad \lambda_j = \hbar\omega_j\beta. \tag{3.7.3}$$

We would like to evaluate thermal averages of the form

$$\langle \psi_1\,\psi_2 \cdots \psi_{2n}\rangle_0 \equiv \frac{\operatorname{Tr} \psi_1\,\psi_2 \cdots \psi_{2n}\, e^{-\beta H_0}}{\operatorname{Tr} e^{-\beta H_0}}, \tag{3.7.4}$$

when there are an even number of creation and annihilation operators.

When there are two operators, we first show that

$$\langle \psi_1\,\psi_2\rangle_0 = \frac{\operatorname{Tr} \psi_1\,\psi_2\, e^{-\beta H_0}}{\operatorname{Tr} e^{-\beta H_0}} = \frac{[\psi_1, \psi_2]}{1 - e^{\pm\lambda_j}}, \tag{3.7.5}$$

where we use $+\lambda_j$ if $\psi_1 = a_j^\dagger$ and $-\lambda_j$ if $\psi_1 = a_j$.

To prove this, we note that $\langle \psi_1\,\psi_2\rangle_0$ vanishes if (*a*) both ψ_1 and ψ_2 are creation operators (as does the commutator $[\psi_1, \psi_2]$ on the right), if (*b*) both are annihilation operators (as does $[\psi_1, \psi_2]$), and if (*c*) ψ_1 and ψ_2 are a creation and annihilation operator referring to two different modes of the field. The only nonvanishing cases then are when (*a*) $\psi_1 = a_j^\dagger$, $\psi_2 = a_j$, or (*b*) $\psi_1 = a_j$ and $\psi_2 = a_j^\dagger$ for a single mode *j*. But in case (*a*), we easily see that

$$\langle a_j^\dagger\, a_j\rangle_0 = \frac{1}{e^{\lambda_j} - 1} \equiv \frac{[a_j^\dagger, a_j]}{1 - e^{\lambda_j}}, \tag{3.7.6}$$

while in case b, since $a_j a_j^\dagger = a_j^\dagger a_j + 1$,

$$\langle a_j a_j^\dagger \rangle_0 = 1 + \langle a_j^\dagger a_j \rangle_0 = \frac{[a_j, a_j^\dagger]}{1 - e^{-\lambda_j}}, \tag{3.7.7}$$

which verifies (3.7.5). By introducing the commutator $[\psi_1, \psi_2]$ which is zero, plus 1, or minus 1, all possibilities are included.

We next show that

$$\langle \psi_1 \psi_2 \psi_3 \psi_4 \rangle_0 = \langle \psi_1 \psi_2 \rangle_0 \langle \psi_3 \psi_4 \rangle_0 + \langle \psi_1 \psi_3 \rangle_0 \langle \psi_2 \psi_4 \rangle_0 + \langle \psi_1 \psi_4 \rangle_0 \langle \psi_2 \psi_3 \rangle_0, \tag{3.7.8}$$

which says that the thermal average of any product of four creation and annihilation operators is the sum of the products of all averages taken in pairs.

To prove (3.7.8), we note that if u, v, and w are any noncommuting operators, then

$$[u, vw] = [u, v]w + v[u, w], \tag{3.7.9}$$

which may be verified by expanding all commutators. If we use this, we see that

$$[\psi_1, \psi_2 \psi_3 \psi_4] = [\psi_1, \psi_2]\psi_3 \psi_4 + \psi_2[\psi_1, \psi_3 \psi_4], \tag{3.7.10}$$

where we let $v = \psi_2$ and $w = \psi_3 \psi_4$. If we use (3.7.9) again on the last commutator, we have (this time $u = \psi_1$, $v = \psi_3$, and $w = \psi_4$)

$$[\psi_1, \psi_2 \psi_3 \psi_4] = [\psi_1, \psi_2]\psi_3 \psi_4 + \psi_2[\psi_1, \psi_3]\psi_4 + \psi_2 \psi_3[\psi_1, \psi_4]. \tag{3.7.11}$$

Since $[\psi_i, \psi_j]$ for two boson operators is a c-number, we may move operators through the commutators, and write this as

$$\begin{aligned} [\psi_1, \psi_2 \psi_3 \psi_4] &= [\psi_1, \psi_2]\psi_3 \psi_4 + [\psi_1, \psi_3]\psi_2 \psi_4 + [\psi_1, \psi_4]\psi_2 \psi_3 \\ &= \psi_1 \psi_2 \psi_3 \psi_4 - \psi_2 \psi_3 \psi_4 \psi_1, \end{aligned} \tag{3.7.12}$$

where the last form is just the original commutator written out.

Consider next the thermal average

$$\begin{aligned} \langle \psi_2 \psi_3 \psi_4 \psi_1 \rangle_0 &= \frac{\operatorname{Tr} \{\psi_2 \psi_3 \psi_4 \psi_1 e^{-\beta H_0}\}}{\operatorname{Tr} e^{-\beta H_0}} \\ &\equiv \frac{\operatorname{Tr} \{\psi_2 \psi_3 \psi_4 e^{-\beta H_0} e^{+\beta H_0} \psi_1 e^{-\beta H_0}\}}{\operatorname{Tr} e^{-\beta H_0}}, \end{aligned} \tag{3.7.13}$$

where we inserted $\exp -\beta H_0 \exp + \beta H_0 = 1$ between ψ_4 and ψ_1. However, from (3.3.16) we have

$$\begin{aligned} e^{+\lambda a^\dagger a} a^\dagger e^{-\lambda a^\dagger a} &= e^{\lambda} a^\dagger \\ e^{\lambda a^\dagger a} a e^{-\lambda a^\dagger a} &= e^{-\lambda} a, \end{aligned} \tag{3.7.14}$$

so that we may write

$$e^{\beta H_0}\,\psi_1\, e^{-\beta H_0} = \psi_1\, e^{\pm\lambda_j}, \tag{3.7.15}$$

where we use $+\lambda_j$ if $\psi_1 = a_j^\dagger$ and $-\lambda_j$ if $\psi_1 = a_j$. Therefore, (3.7.13) becomes

$$\begin{aligned}\langle\psi_2\,\psi_3\,\psi_4\,\psi_1\rangle_0 &= \frac{\mathrm{Tr}\,[\psi_2\,\psi_3\,\psi_4\, e^{-\beta H_0}\,\psi_1] e^{\pm\lambda_j}}{\mathrm{Tr}\, e^{-\beta H_0}} \\ &= \frac{\mathrm{Tr}\,[\psi_1\,\psi_2\,\psi_3\,\psi_4\, e^{-\beta H_0}]}{\mathrm{Tr}\, e^{-\beta H_0}}\, e^{\pm\lambda_j} \\ &= e^{\pm\lambda_j}\langle\psi_1\,\psi_2\,\psi_3\,\psi_4\rangle_0,\end{aligned} \tag{3.7.16}$$

where we have used the cyclic properties of traces in the second step.

Let us return to (3.7.12) and take the thermal average of both sides and use (3.7.16). We then find after minor algebra that

$$\langle\psi_1\,\psi_2\,\psi_3\,\psi_4\rangle_0 = \frac{1}{1 - e^{\pm\lambda_j}} \times \{[\psi_1, \psi_2]\langle\psi_3\,\psi_4\rangle_0 + [\psi_1, \psi_3]\langle\psi_2\,\psi_4\rangle_0 + [\psi_1, \psi_4]\langle\psi_2\,\psi_3\rangle_0\}. \tag{3.7.17}$$

Since the commutators are *not* operators they are not affected by the thermal average. If we use (3.7.5) here, we see that (3.7.8) follows.

The generalized Wick's theorem we wish to prove is

$$\langle\psi_1\,\psi_2 \cdots \psi_{2n}\rangle_0 = \langle\psi_1\,\psi_2\rangle_0\langle\psi_3\,\psi_4 \cdots \psi_{2n}\rangle_0 + \langle\psi_1\,\psi_3\rangle_0\langle\psi_2\,\psi_4 \cdots \psi_{2n}\rangle_0 + \cdots + \langle\psi_1\,\psi_{2n}\rangle_0\langle\psi_2\,\psi_3 \cdots \psi_{2n-1}\rangle_0. \tag{3.7.18}$$

By repeated application, this reduces to sums of products of all pairs of thermal averages.

We have proved (3.7.18) for $n = 2$ (four operators). Let us assume it is true for $2(n - 1)$ factors and show that it is then true for $2n$, and the proof will be complete, since it is true for $n = 2$.

We again evaluate by the trick used to show (3.7.8). To do this we take the thermal average of the commutator

$$\begin{aligned}\langle[\psi_1, \psi_2 \cdots \psi_{2n}]\rangle_0 &= \langle\psi_1\,\psi_2 \cdots \psi_{2n}\rangle_0(1 - e^{\pm\lambda_j}) \\ &= [\psi_1, \psi_2]\langle\psi_3 \cdots \psi_{2n}\rangle_0 + [\psi_1, \psi_3]\langle\psi_2\,\psi_4 \cdots \psi_{2n}\rangle_0 + \cdots \\ &\quad + [\psi_1, \psi_{2n}]\langle\psi_2\,\psi_3 \cdots \psi_{2n-1}\rangle_0.\end{aligned}$$

We have of course used the generalization of (3.7.13) and (3.7.9) repeatedly. Again, if $\psi_1 = a_j^\dagger$, we use $+\lambda_j$ and if $\psi_1 = a_j$ we use $-\lambda_j$. When we divide through by $(1 - e^{\pm\lambda_j})$ and use (3.7.5), we see that (3.7.18) follows. Thus if the theorem is true for $2(n - 1)$ factors (all thermal averages above now involve

$2n - 2$ factors), it is then true for $2n$ and the theorem follows, since it is true for 4 factors. This theorem is easily generalized to the case where the operators are fermions rather than bosons [12].

We leave as an exercise to show that if $\psi_i(t)$ is a creation or annihilation operator in the interaction picture, then the thermal average of the time ordered product of $2n$ operators is given by

$$\begin{aligned}\langle P\{\bar{\psi}_1(t_1)\cdots\bar{\psi}_{2n}(t_{2n})\}\rangle_0 &= \langle P\{\bar{\psi}_1\,\bar{\psi}_2\}\rangle_0\langle P\{\bar{\psi}_3\cdots\bar{\psi}_{2n}\}\rangle_0\\ &\quad+\langle P\{\bar{\psi}_1\,\bar{\psi}_3\}\rangle_0\langle P\{\bar{\psi}_2\,\bar{\psi}_4\cdots\bar{\psi}_{2n}\}\rangle_0+\cdots\\ &\quad+\langle P\{\bar{\psi}_1\,\bar{\psi}_{2n}\}\rangle_0\langle P\{\bar{\psi}_2\cdots\bar{\psi}_{2n-1}\}\rangle_0,\end{aligned} \tag{3.7.19}$$

where

$$\langle P\{\bar{\psi}_j(t_j)\,\bar{\psi}_k(t_k)\}\rangle_0 = \begin{cases}\langle\psi_j(t_j)\,\psi_k(t_k)\rangle_0 & (t_k < t_j)\\ \langle\psi_k(t_k)\,\psi_j(t_j)\rangle_0 & (t_j < t_k),\end{cases} \tag{3.7.20}$$

and

$$\langle\psi_l(t_l)\,\psi_m(t_m)\rangle_0 = \frac{[\psi_l(t_l),\,\psi_m(t_m)]}{1 - e^{\pm\lambda_l}}. \tag{3.7.21}$$

3.8 WICK'S THEOREM FOR BOSON OPERATORS

In Chapter 1 (1.16.53), we developed a perturbation theory solution for the Schrödinger equation in the interaction picture given by

$$\begin{aligned}U(t) &= P\left\{\exp -\frac{i}{\hbar}\int_0^t \bar{V}_I(t')\,dt'\right\}\\ &= 1 + \sum_{n=1}^{\infty}\left(\frac{1}{i\hbar}\right)^n\frac{1}{n!}\int_0^t\cdots\int_0^t dt_1\cdots dt_n\,P\{\bar{V}_I(t_1)\cdots\bar{V}_I(t_n)\},\end{aligned} \tag{3.8.1}$$

where

$$V_I(t) = e^{(i/\hbar)H_0t}V_Se^{-(i/\hbar)H_0t}, \tag{3.8.2}$$

and P is the Dyson time ordering operator. If again H_0 is the free field hamiltonian

$$H_0 = \sum_j \hbar\omega_j a_j^\dagger a_j, \tag{3.8.3}$$

then in the interaction picture

$$\begin{aligned}a_j(t) &= a_je^{-i\omega_jt}\\ a_j^\dagger(t) &= a_j^\dagger e^{i\omega_jt},\end{aligned} \tag{3.8.4}$$

so that boson operators obey the same commutation relations in the interaction picture as they do in the Schrödinger picture.

We now proceed to develop Wick's theorem [13] which allows us to convert the time-ordered operators appearing in (3.8.1) into normally ordered operators. We must first develop some notation.

Let $\psi_1, \psi_2, \ldots$ represent either creation or annihilation operators for various modes of the radiation field in the interaction picture and let $\bar{\psi}_1, \bar{\psi}_2, \ldots$ be the c-number associated functions in the interaction picture. For example, if $\psi_1 = a_j^\dagger(t_1)$ and $\psi_2 = a_k(t_2)$

$$a_j^\dagger(t_1)\, a_k(t_2) \equiv \mathcal{N}\{\alpha_j^*(t_1)\, \alpha_k(t_2)\}, \tag{3.8.5}$$

or in this case

$$\psi_1\, \psi_2 \equiv \mathcal{N}\{\bar{\psi}_1\, \bar{\psi}_2\}. \tag{3.8.6}$$

We next show that

$$P\{\bar{\psi}_1(t_1)\, \bar{\psi}_2(t_2)\} = \mathcal{N}\{\bar{\psi}_1(t)\, \bar{\psi}_2(t_2)\} + \langle 0, 0|P\{\bar{\psi}_1(t_1)\, \bar{\psi}_2(t_2)\}|0, 0\rangle, \tag{3.8.7}$$

where P is the Dyson time-ordering operator defined by

$$P\{\bar{\psi}_1(t_1)\, \bar{\psi}_2(t_2)\} = \begin{cases} \psi_1(t_1)\, \psi_2(t_2) & \text{if} \quad t_2 < t_1 \\ \psi_2(t_2)\, \psi_1(t_1) & \text{if} \quad t_1 < t_2, \end{cases} \tag{3.8.8}$$

and $\mathcal{N}$ is the normal ordering operator defined by

$$\mathcal{N}\{\bar{\psi}_1(t_1)\, \bar{\psi}_2(t_2)\} = \begin{cases} \psi_1(t_1)\, \psi_2(t_2) & \text{if} \quad \psi_2 = a_k(t_2). \\ \psi_2(t_2)\, \psi_1(t_1) & \text{if} \quad \psi_1 = a_k(t_1) \end{cases} \tag{3.8.9}$$

The last term in (3.8.7) is the vacuum matrix element for modes j and k.

PROOF.

Assume $t_2 < t_1$. Then

$$P\{\bar{\psi}_1(t_1)\, \bar{\psi}_2(t_2)\} = \psi_1(t_1)\, \psi_2(t_2).$$

If $\psi_1\, \psi_2$ is in normal order, then

$$\mathcal{N}\{\bar{\psi}_1(t_1)\, \bar{\psi}_2(t_2)\} = \psi_1(t_1)\, \psi_2(t_2),$$

and

$$\langle 0, 0|\psi_1(t_1)\, \psi_2(t_2)|0, 0\rangle = 0,$$

since the vacuum matrix element of operators in normal order vanish and (3.8.7) follows. If $\psi_1\, \psi_2$ is not in normal order, then we may write

$$\psi_1(t_1)\, \psi_2(t_2) = \psi_2(t_2)\, \psi_1(t_1) + [\psi_1(t_1), \psi_2(t_2)],$$

and since $\psi_2\, \psi_1$ is in normal order, we have

$$\psi_1(t_1)\, \psi_2(t_2) = \mathcal{N}\{\bar{\psi}_1(t_1)\, \bar{\psi}_2(t_2)\} + [\psi_1(t_1), \psi_2(t_2)]. \tag{3.8.10}$$

If we take the vacuum matrix element of both sides and note that such matrix elements vanish when the operator is in normal order, we have

$$\langle 0, 0|\psi_1(t_1)\, \psi_2(t_2)|0, 0\rangle = [\psi_1(t_1), \psi_2(t_2)],$$

so that (3.8.10) becomes

$$\psi_1(t_1)\, \psi_2(t_2) = \mathscr{N}\{\bar{\psi}_1(t_1)\, \bar{\psi}_2(t_2)\} + \langle 0, 0|\psi_1(t_1)\, \psi_2(t_2)|0, 0\rangle,$$

and (3.8.7) again follows when $t_2 < t_1$. A similar argument goes through when $t_1 < t_2$.

If we adopt the short-hand notation for the vacuum matrix element

$$\langle 0, 0|P\{\bar{\psi}_1(t_1)\, \bar{\psi}_2(t_2)\}|0, 0\rangle \equiv \underbrace{\psi_1(t_1)\, \psi_2(t_2)}, \tag{3.8.11}$$

which is called a contraction of ψ_1 and ψ_2, we may rewrite (3.8.7) as

$$P\{\bar{\psi}_1(t_1)\, \bar{\psi}_2(t_2)\} = \mathscr{N}\{\bar{\psi}_1(t_1)\, \bar{\psi}_2(t_2)\} + \underbrace{\psi_1(t_1)\, \psi_2(t_2)}, \tag{3.8.12}$$

which converts a time-ordered product of two operators into a normal product plus a contraction which is a c-number.

To develop the notation further, we define contractions inside the normal ordering operator by

$$\mathscr{N}\{\underbrace{\bar{\psi}_1\, \underbrace{\bar{\psi}_2\, \bar{\psi}_3}\, \bar{\psi}_4}\, \bar{\psi}_5 \cdots\} = \underbrace{\psi_1\, \psi_4}\, \underbrace{\psi_2\, \psi_3}\, \mathscr{N}\{\bar{\psi}_5 \cdots\} \tag{3.8.13}$$

$$\mathscr{N}\{\underbrace{\bar{\psi}_1\, \underbrace{\bar{\psi}_2\, \bar{\psi}_3}\, \underbrace{\bar{\psi}_4\, \bar{\psi}_5}\, \bar{\psi}_6}\} = \underbrace{\psi_1\, \psi_6}\, \underbrace{\psi_2\, \psi_3}\, \underbrace{\psi_4\, \psi_5}, \tag{3.8.14}$$

and the extension involving more contractions is straightforward.

We next prove the following lemma.

LEMMA

Let ψ_{n+1} be an operator at a time t_{n+1} which is *earlier* than the times of any of the operators $\psi_1, \psi_2, \ldots, \psi_n$. Then

$$\begin{aligned}\mathscr{N}\{\bar{\psi}_1\, \bar{\psi}_2 \cdots \bar{\psi}_n\}\, \psi_{n+1} &= \mathscr{N}\{\bar{\psi}_1 \cdots \bar{\psi}_n\, \bar{\psi}_{n+1}\} + \mathscr{N}\{\bar{\psi}_1\, \bar{\psi}_2 \cdots \bar{\psi}_{n-1}\, \underbrace{\bar{\psi}_n\, \bar{\psi}_{n+1}}\} \\ &\quad + \mathscr{N}\{\bar{\psi}_1 \cdots \underbrace{\bar{\psi}_{n-1}\bar{\psi}_n\, \bar{\psi}_{n+1}}\} + \cdots + \mathscr{N}\{\underbrace{\bar{\psi}_1\, \bar{\psi}_2 \cdots \bar{\psi}_n\, \bar{\psi}_{n+1}}\}.\end{aligned} \tag{3.8.15}$$

To prove this we note first that if ψ_{n+1} is an annihilation operator earlier than the times of all other operators, all contractions with it vanish, leaving only the first term on the right above which, by definition of the $\mathscr{N}$ operator, is identically equal to the left-hand side. We therefore must only prove that (3.8.15) is true when ψ_{n+1} is a creation operator.

Since the $\bar{\psi}_i$ are all c-numbers, we can therefore assume with no loss of generality that we have written them in normal order so that

$$\mathcal{N}\{\bar{\psi}_1 \bar{\psi}_2 \cdots \bar{\psi}_n\} = \psi_1 \psi_2 \cdots \psi_n, \tag{3.8.16}$$

but $\mathcal{N}\{\bar{\psi}_1 \cdots \bar{\psi}_n\} \psi_{n+1}$ will not be in normal order. We may put it into normal order as follows. We write the identity

$$\psi_1 \psi_2 \cdots \psi_n \psi_{n+1} = \psi_{n+1} \psi_1 \psi_2 \cdots \psi_n - [\psi_{n+1}, \psi_1 \psi_2 \cdots \psi_n], \tag{3.8.17}$$

since $[A, B] = -[B, A]$. Note that the first term on the right is in normal order. We may use (3.7.9) repeatedly and write the commutator as

$$\begin{aligned}
[\psi_{n+1}, \psi_1 \psi_2 \cdots \psi_n] &= [\psi_{n+1}, \psi_1] \psi_2 \cdots \psi_n + \psi_1 [\psi_{n+1}, \psi_2 \cdots \psi_n] \\
&= [\psi_{n+1}, \psi_1] \psi_2 \cdots \psi_n + \psi_1 [\psi_{n+1}, \psi_2] \psi_3 \cdots \psi_n + \psi_1 \psi_2 [\psi_{n+1}, \psi_3 \cdots \psi_n] \\
&= [\psi_{n+1}, \psi_1] \psi_2 \cdots \psi_n + [\psi_{n+1}, \psi_2] \psi_1 \psi_3 \cdots \psi_n \\
&\quad + \cdots + [\psi_{n+1}, \psi_n] \psi_1 \psi_2 \cdots \psi_{n-1}.
\end{aligned} \tag{3.8.18}$$

We used the fact that the commutator is a c-number and moved all operators through the commutators in the last step.

Accordingly, when we use (3.8.18) and change $[\psi_{n+1}, \psi_k]$ ($k = 1, 2, \ldots, n$) to $-[\psi_k, \psi_{n+1}]$, (3.8.17) becomes

$$\begin{aligned}
\psi_1 \psi_2 \cdots \psi_n \psi_{n+1} &= \psi_{n+1} \psi_1 \psi_2 \cdots \psi_n + [\psi_1, \psi_{n+1}] \psi_2 \cdots \psi_n \\
&\quad + [\psi_2, \psi_{n+1}] \psi_1 \psi_3 \cdots \psi_n + \cdots + [\psi_n, \psi_{n+1}] \psi_1 \psi_2 \cdots \psi_{n-1}.
\end{aligned} \tag{3.8.19a}$$

Since ψ_{n+1} is a creation operator at a time earlier than any other operator and $\psi_1 \cdots \psi_n$ is in normal order, all terms on the right are either in normal order or are commutators, which are c-numbers, and we may write this as

$$\begin{aligned}
\mathcal{N}\{\bar{\psi}_1 \bar{\psi}_2 \cdots \bar{\psi}_n\} \psi_{n+1} &= \mathcal{N}\{\bar{\psi}_1 \cdots \bar{\psi}_n \bar{\psi}_{n+1}\} + [\psi_1, \psi_{n+1}] \mathcal{N}\{\bar{\psi}_2 \cdots \bar{\psi}_n\} \\
&\quad + [\psi_2, \psi_{n+1}] \mathcal{N}\{\bar{\psi}_1 \bar{\psi}_3 \cdots \bar{\psi}_n\} + \cdots + [\psi_n, \psi_{n+1}] \mathcal{N}\{\bar{\psi}_1 \bar{\psi}_2 \cdots \bar{\psi}_{n-1}\},
\end{aligned} \tag{3.8.19b}$$

when we use the definition of the $\mathcal{N}$-operator. Now by (3.8.8)

$$P\{\bar{\psi}_k(t_k) \bar{\psi}_{n+1}(t_{n+1})\} = \psi_k(t_k) \psi_{n+1}(t_{n+1}),$$

since $t_{n+1} < t_k$ ($k = 1, 2, \ldots, n$). Since ψ_{n+1} is a creation operator, we may put this into normal order by

$$\psi_k \psi_{n+1} = \psi_{n+1} \psi_k + [\psi_k, \psi_{n+1}] = P\{\bar{\psi}_k(t_k) \bar{\psi}_{n+1}(t_{n+1})\}. \tag{3.8.20}$$

If we take vacuum matrix elements, we therefore see that

$$\begin{aligned}
\langle 0, 0| P\{\bar{\psi}_k \bar{\psi}_{n+1}\} |0, 0\rangle &= [\psi_k, \psi_{n+1}] \\
&= \underbrace{\psi_k \psi_{n+1}},
\end{aligned} \tag{3.8.21}$$

since $\langle 0, 0|\psi_{n+1}\,\psi_k|0, 0\rangle = 0$. If we use this result in (3.8.19b) and use the notation (3.8.13) and (3.8.14), we see that (3.8.15) follows when ψ_{n+1} is either a creation or annihilation operator at a time earlier than all other operators and the lemma follows.

We may generalize the lemma as follows. Since the contraction $\underbrace{A_1\,A_2}$ is a c-number, we may insert it inside each of the ordering operators in (3.8.15). Therefore, we have the generalization

$$\begin{aligned}\mathcal{N}\{\underbrace{A_1\,A_2}\,\bar{\psi}_1\cdots\bar{\psi}_n\}\,\psi_{n+1} &= \mathcal{N}\{\underbrace{A_1\,A_2}\,\bar{\psi}_1\cdots\bar{\psi}_{n+1}\} + \mathcal{N}\{\underbrace{A_1\,A_2}\,\bar{\psi}_1\cdots\underbrace{\bar{\psi}_n\,\bar{\psi}_{n+1}}\}\\ &\quad + \mathcal{N}\{\underbrace{A_1\,A_2}\,\bar{\psi}_1\cdots\underbrace{\bar{\psi}_{n-1}\,\bar{\psi}_n\,\bar{\psi}_{n+1}}\} + \cdots + \mathcal{N}\{\underbrace{A_1\,A_2}\,\underbrace{\bar{\psi}_1\cdots\bar{\psi}_n\,\bar{\psi}_{n+1}}\rangle.\end{aligned} \tag{3.8.22}$$

In fact, we may insert as many contractions as we like in each term above.

We next state Wick's theorem which we shall prove by induction. A Dyson time-ordered product of boson creation and annihilation operators in the interaction picture can be expressed as a sum of normal ordered operators by means of

$$\begin{aligned}P\{\bar{\psi}_1\,\bar{\psi}_2\cdots\bar{\psi}_n\} &= \mathcal{N}\{\bar{\psi}_1\,\bar{\psi}_2\cdots\bar{\psi}_n\} + \mathcal{N}\{\underbrace{\psi_1\,\psi_2}\,\bar{\psi}_3\cdots\bar{\psi}_n\} + \mathcal{N}\{\underbrace{\psi_1\,\bar{\psi}_2\,\psi_3}\cdots\bar{\psi}_n\}\\ &\quad + \cdots + \mathcal{N}\{\bar{\psi}_1\cdots\underbrace{\bar{\psi}_{n-1}\,\bar{\psi}_n}\} + \mathcal{N}\{\underbrace{\psi_1\,\psi_2}\,\underbrace{\psi_3\,\psi_4}\cdots\psi_n\}\\ &\quad + \cdots + \mathcal{N}\{\bar{\psi}_1\,\bar{\psi}_2\cdots\underbrace{\psi_{n-3}\,\psi_{n-2}}\,\underbrace{\psi_{n-1}\,\psi_n}\}\\ &\quad + \cdots + \mathcal{N}\{\underbrace{\psi_1\,\psi_2\,\psi_3\,\psi_4\cdots\psi_{n-2}}\,\underbrace{\psi_{n-1}\,\psi_n}\}.\end{aligned} \tag{3.8.23}$$

We should understand clearly the notation used in the theorem since it is a great labor saver in nth order perturbation theory. The time-ordered operator equals (a) a term in which all operators are placed directly in normal order (b) plus a sum of all normal products times all possible contractions between every pair of operators; (c) plus a sum of the normal products containing all possible contractions of two pairs of operators . . . (d) and so on until we have a term in which all operator pairs are contracted.

To prove (3.8.23) by induction, we assume it is true as it stands when n factors are involved. We then shall show it is also true for $n + 1$. Since we have shown it true for two factors in (3.8.12), it will be true for three and so on.

Since (3.8.23) is true, we multiply each term in the equation from the right by the operator $\psi_{n+1}(t_{n+1})$ where t_{n+1} is the earliest time of any operator. Then, by definition of the P-operator, we have that

$$P\{\bar{\psi}_1\,\bar{\psi}_2\cdots\bar{\psi}_n\}\,\psi_{n+1} = P\{\bar{\psi}_1\,\bar{\psi}_2\cdots\bar{\psi}_n\,\bar{\psi}_{n+1}\}, \tag{3.8.24}$$

and (3.8.23) becomes

$$\begin{aligned}P\{\bar{\psi}_1\,\bar{\psi}_2\cdots\bar{\psi}_n\,\bar{\psi}_{n+1}\} &= \mathcal{N}\{\bar{\psi}_1\,\bar{\psi}_2\cdots\bar{\psi}_n\}\psi_{n+1} + \mathcal{N}\{\underbrace{\psi_1\,\psi_2}\,\bar{\psi}_3\cdots\bar{\psi}_n\}\,\psi_{n+1}\\ &\quad + \mathcal{N}\{\underbrace{\psi_1\,\bar{\psi}_2\,\psi_3}\cdots\bar{\psi}_n\}\,\psi_{n+1} + \cdots.\end{aligned} \tag{3.8.25}$$

We now apply the lemma (3.8.15) and its generalization (3.8.22) and so on to each term on the right of (3.8.25). We then have

$$
\begin{aligned}
P\{\bar{\psi}_1\,\bar{\psi}_2\cdots\bar{\psi}_n\,\bar{\psi}_{n+1}\} = {} & \mathcal{N}\{\bar{\psi}_1\cdots\bar{\psi}_n\,\bar{\psi}_{n+1}\} + \mathcal{N}\{\bar{\psi}_1\,\bar{\psi}_2\cdots\underbracket{\bar{\psi}_n\,\bar{\psi}_{n+1}}\} \\
& + \cdots + \mathcal{N}\{\underbracket{\psi_1\,\bar{\psi}_2\cdots\bar{\psi}_n\,\psi_{n+1}}\} + \mathcal{N}\{\underbracket{\psi_1\,\psi_2}\,\bar{\psi}_3\cdots\bar{\psi}_n\,\bar{\psi}_{n+1}\} \\
& + \mathcal{N}\{\underbracket{\psi_1\,\psi_2}\,\bar{\psi}_3\cdots\underbracket{\psi_n\,\psi_{n+1}} \cdots + \mathcal{N}\{\underbracket{\psi_1\,\psi_2}\,\underbracket{\psi_3\cdots\bar{\psi}_n\,\psi_{n+1}}\} \\
& + \cdots,
\end{aligned}
$$

and we see that the theorem is true for $n + 1$ factors. Q.E.D.

There is a minor difficulty that remains to be discussed. There may be cases when two or more operators are evaluated at the same time so that the P-operator is undefined. Consider, for example, a term like

$$P\{\bar{\psi}_1\,\bar{\psi}_2\,\mathcal{N}\{\bar{\psi}_3(t)\,\bar{\psi}_4(t)\,\bar{\psi}_5(t)\}\,\bar{\psi}_6\},$$

where ψ_3, ψ_4, and ψ_5 are operators at the same time. In this case we maintain the normal ordering of these operators. We may define the P-operator in these cases if we agree to evaluate the creation operators at a slightly later time $t + \Delta t$ so that the P-operator will be equivalent to the $\mathcal{N}$-operator. After we apply the theorem, we then let $\Delta t \to 0$.

Let us assume that ψ_3 and ψ_4 are creation operators and that ψ_5 is an annihilation operator. We then see that

$$P\{\bar{\psi}_1\,\bar{\psi}_2\,\mathcal{N}\{\bar{\psi}_3\,\bar{\psi}_4\,\bar{\psi}_5\}\,\psi_6\} = \lim_{\Delta t \to 0} P\{\bar{\psi}_1\,\bar{\psi}_2\,\bar{\psi}_3(t + \Delta t)\,\bar{\psi}_4(t + \Delta t)\,\bar{\psi}_5(t)\,\bar{\psi}_6\}.$$

We then apply the theorem and we have contractions such as $\underbracket{\psi_3(t+\Delta t)\,\psi_5(t)}$ among the almost equal time operators. However, these will all vanish since

$$\lim_{\Delta t \to 0} \underbracket{\psi_3(t + \Delta t)\,\psi_5(t)} = \lim_{\Delta t \to 0} \langle 0, 0|P\{\bar{\psi}_3(t + \Delta t)\,\bar{\psi}_5(t)\}|0, 0\rangle,$$

because the operators will all be in normal order. Therefore, we may as well omit all equal time contractions when using the theorem from the start.

PART III. ARBITRARY OPERATORS

3.9 GENERALIZATION OF ORDERING TECHNIQUES TO ARBITRARY QUANTUM OPERATORS [14]

In the case of bosons we have developed a one-to-one correspondence between operators in various orders and c-number associated functions. In this section we generalize this correspondence to include any set of quantum operators.

Let us consider a complete set of noncommuting operators $a_1, a_2, \ldots, a_f$ in the Schrödinger picture which obey some set of commutation or anticommutation relations. The anticommutator of two operators is defined by $[a_i, a_j]_+ \equiv a_i a_j + a_j a_i$. Suppose we have some function Q of these operators which may be expanded in a power series. We may use the commutation (or anticommutation) relations to reorder the terms in the function into some predetermined or chosen order. Let this chosen order be $a_1, a_2, a_3, \ldots, a_f$. Therefore, Q in chosen order becomes

$$Q = Q^c(a_1, \ldots, a_f) = \sum_{r_1} \cdots \sum_{r_f} Q^c_{r_1,\ldots,r_f} a_1^{r_1} a_2^{r_2} \cdots a_f^{r_f}. \tag{3.9.1}$$

We put a superscript c to indicate we have put the function in chosen order which of course is equal to the function in the original order. Once we have put Q in chosen order, we may define an associated c-number function by means of

$$\bar{Q}^c(\alpha_1, \alpha_2, \ldots, \alpha_f) = \sum_{r_1} \cdots \sum_{r_f} Q^c_{r_1,r_2,\ldots,r_f} \alpha_1^{r_1} \alpha_2^{r_2} \cdots \alpha_f^{r_f}, \tag{3.9.2}$$

where we replace the operator a_i by the c-number α_i which is real or complex depending on whether a_i is hermitian or not. We put a bar to remind us we are now dealing with a quasi-classical or c-number function.

We next define a linear chosen ordering operator $\mathscr{C}$ by means of

$$Q^c(a_1, a_2, \ldots, a_f) = \mathscr{C}\{\bar{Q}^c(\alpha_1, \ldots, \alpha_f)\}, \tag{3.9.3}$$

where $\mathscr{C}$ tells us to replace each α_i by the corresponding operator and write all terms in chosen order. We may also define the inverse operator $\mathscr{C}^{-1}$ (compare with Section 3.2).

We may give a formal representation to the $\mathscr{C}$ operator by means of

$$\mathscr{C}\{\bar{Q}^c(\alpha_1, \ldots, \alpha_f)\} = \int \cdots \int \bar{Q}^c(\alpha_1, \ldots, \alpha_f) \prod_{i=1}^{f} \delta(\alpha_i - a_i)\, d\alpha_i, \tag{3.9.4}$$

where the δ-functions are operators *in the chosen order*. In the integration, we formally replace each α_i by the corresponding a_i and put all terms in the chosen order. But since this is exactly what $\mathscr{C}$ tells us to do, the two expressions in (3.9.4) are formally equivalent.

If the operator a_i is hermitian and the α_i real, we may represent the δ-function by means of

$$\delta(\alpha - a) = \frac{1}{2\pi} \int_{-\infty}^{\infty} e^{-i\xi(\alpha - a)}\, d\xi, \tag{3.9.5}$$

while if it is nonhermitian, the operator and its adjoint will both be present and we then let

$$\delta(\alpha - a)\,\delta(\alpha^* - a^\dagger) = \frac{1}{\pi^2}\iint e^{-i\xi(\alpha-a)}e^{-i\xi^*(\alpha^*-a^\dagger)}\,d^2\xi, \qquad (3.9.6)$$

if the chosen order is a, $a^\dagger$. Here we integrate over the entire complex ξ-plane and $d^2\xi = d(\mathrm{Re}\,\xi)\,d(\mathrm{Im}\,\xi)$. In this formal presentation we use (3.9.5) for simplicity.

If we use (3.9.5), (3.9.4) may be rewritten as

$$Q^c(a_1, \ldots, a_f) = \left(\frac{1}{2\pi}\right)^f \int\cdots\int d\alpha_1 \cdots d\alpha_f \bar{Q}^c(\alpha_1, \ldots, \alpha_f)$$
$$\times \int\cdots\int d\xi_1 \cdots d\xi_f e^{-i\xi_1(\alpha_1-a_1)} \cdots e^{-i\xi_f(\alpha_f-a_f)}. \qquad (3.9.7)$$

If we interchange the order of integration of the ξ's and α's, we may write this as

$$Q^c(a_1, \ldots, a_f) = \int\cdots\int d\xi_1 \cdots d\xi_f e^{i\xi_1 a_1} \cdots e^{i\xi_f a_f} F(\xi_1, \ldots, \xi_f), \qquad (3.9.8a)$$

where

$$F(\xi_1, \ldots, \xi_f) = \left(\frac{1}{2\pi}\right)^f \int\cdots\int d\alpha_1 \cdots d\alpha_f e^{-i(\xi_1\alpha_1+\cdots+\xi_f\alpha_f)}\bar{Q}^c(\alpha_1, \ldots, \alpha_f). \qquad (3.9.8b)$$

Therefore, $F(\xi_1, \ldots, \xi_f)$ is just the Fourier transform of the associated c-number function $\bar{Q}^c(\alpha_1, \ldots, \alpha_f)$. If we invert (3.9.8b) by taking the inverse Fourier transform, we obtain

$$\bar{Q}^c(\alpha_1, \ldots, \alpha_f) = \int\cdots\int e^{i(\xi_1\alpha_1+\cdots+\xi_f\alpha_f)} F(\xi_1, \ldots, \xi_f)\, d\xi_1 \cdots d\xi_f, \qquad (3.9.8c)$$

which gives us the associated function in terms of F. If we apply the $\mathscr{C}$-operator to both sides of (3.9.8c), we obtain (3.9.8a) exactly.

We have seen this same result already in the case of bosons (see Eqs. 3.2.36 and 3.2.42). For if we let $a_1 = a^\dagger$ and $a_2 = a$ where a and $a^\dagger$ are boson operators, the chosen order is normal order and we have (on using 3.9.6)

$$Q^{(n)}(a, a^\dagger) = \mathscr{C}\{\bar{Q}^{(n)}(\alpha, \alpha^*)\} = \iint d^2\alpha\, \bar{Q}^{(n)}(\alpha, \alpha^*)\,\delta(\alpha^* - a^\dagger)\,\delta(\alpha - a)$$
$$= \frac{1}{\pi^2}\iint d^2\alpha\, \bar{Q}^{(n)}(\alpha, \alpha^*)\iint d^2\xi e^{-i\xi^*(\alpha^*-a^\dagger)}e^{-i\xi(\alpha-a)}$$
$$= \frac{1}{\pi}\iint d^2\xi e^{i\xi^* a^\dagger} e^{i\xi a} F^{(n)}(\xi, \xi^*), \qquad (3.9.9a)$$

where

$$F^{(n)}(\xi, \xi^*) = \frac{1}{\pi} \iint d^2\alpha\, \bar{Q}^{(n)}(\alpha, \alpha^*) e^{-i(\xi\alpha+\xi^*\alpha^*)}, \tag{3.9.9b}$$

as we have found earlier.

As an example of (3.9.8) let Q be the density operator $\rho(a_1, \ldots, a_f, t)$ in the Schrödinger picture. Then by (3.9.8)

$$\rho^c(a_1, \ldots, a_f, t) = \int \cdots \int e^{i\xi_1 a_1} \cdots e^{i\xi_f a_f} \mathscr{G}(\xi_1, \ldots, \xi_f, t)\, d\xi_1 \cdots d\xi_f, \tag{3.9.10a}$$

where

$$\mathscr{G}(\xi_1, \ldots, \xi_f, t) = \left(\frac{1}{2\pi}\right)^f \int \cdots \int d\alpha_1 \cdots d\alpha_f e^{-i(\xi_1\alpha_1+\cdots+\xi_f\alpha_f)} \bar{\rho}^c(\alpha_1, \ldots, \alpha_f, t), \tag{3.9.10b}$$

or

$$\bar{\rho}^c(\alpha_1, \ldots, \alpha_f, t) = \int \cdots \int d\xi_1 \cdots d\xi_f e^{i(\alpha_1\xi_1+\cdots+\alpha_f\xi_f)} \mathscr{G}(\xi_1, \ldots, \xi_f, t), \tag{3.9.10c}$$

so that

$$\rho^c(a_1, \ldots, a_f, t) = \mathscr{C}\{\bar{\rho}^{(c)}(\alpha_1, \ldots, \alpha_f, t)\}. \tag{3.9.10d}$$

We are often interested in the density operator in the interaction picture $\chi(t)$ which is related to the Schrödinger picture by

$$\rho(t) = e^{-(i/\hbar)H_0 t} \chi(t) e^{(i/\hbar)H_0 t}, \tag{3.9.11}$$

while operators in the two pictures are related by

$$a_i^I(t) = e^{(i/\hbar)H_0 t} a_i^S e^{-(i/\hbar)H_0 t}. \tag{3.9.12}$$

If we use (3.9.11), we have on solving for χ

$$i\hbar \frac{\partial \chi}{\partial t} = e^{(i/\hbar)H_0 t} \left\{ [\rho, H_0] + i\hbar \frac{\partial \rho}{\partial t} \right\} e^{-(i/\hbar)H_0 t}. \tag{3.9.13}$$

If we use (3.9.10a), we have for the equation of motion for the density operator in the interaction picture

$$i\hbar \frac{\partial \chi}{\partial t} = e^{(i/\hbar)H_0 t} \int \cdots \int d\xi_1 \cdots d\xi_f \Big\{ [e^{i\xi_1 a_1} \cdots e^{i\xi_f a_f}, H_0] \mathscr{G}(\xi_1, \ldots, \xi_f, t)$$
$$+ e^{i\xi_1 a_1} \cdots e^{i\xi_f a_f} i\hbar \frac{\partial \mathscr{G}}{\partial t} (\xi_1, \ldots, \xi_f, t) \Big\} e^{-(i/\hbar)H_0 t}. \tag{3.9.14}$$

If we use the result (3.1.13)

$$A e^B A^{-1} = \exp ABA^{-1}, \tag{3.9.15}$$

and insert

$$e^{(i/\hbar)H_0t}e^{-(i/\hbar)H_0t} = 1 \tag{3.9.16}$$

between all exponential factors in (3.9.14), it follows from (3.9.12) that

$$i\hbar\frac{\partial\chi}{\partial t} = \int\cdots\int d\xi_1\cdots d\xi_f\Big\{[e^{i\xi_1 a_1{}^I(t)}\cdots e^{i\xi_f a_f{}^I(t)}, H_0]\mathscr{G}(\xi_1,\ldots,\xi_f,t)$$
$$+ e^{i\xi_1 a_1{}^I(t)}\cdots e^{i\xi_f a_f{}^I(t)} i\hbar\frac{\partial\mathscr{G}}{\partial t}(\xi_1,\ldots,\xi_f,t)\Big\}, \tag{3.9.17}$$

where all operators are in the interaction picture. We should note that although ρ in (3.9.10a) was in chosen order, $(\partial\chi/\partial t)$ as it stands above is not in chosen order because of the commutator term in the integrand. In practice, we must put this into chosen order.

If we use (3.9.10a), we see by (3.9.11) that χ becomes

$$\chi^c(a_1,\ldots,a_f,t) = \int\cdots\int e^{i\xi_1 a_1{}^I(t)}\cdots e^{i\xi_f a_f{}^I(t)}\mathscr{G}(\xi_1,\ldots,\xi_f,t)\,d\xi_1\cdots d\xi_f, \tag{3.9.18}$$

which gives the density operator in the interaction picture directly in chosen order if $\mathscr{G}(\underset{\sim}{\xi}, t)$ is known.

Consider next the expectation value of Q

$$\langle Q^c(a_1,\ldots,a_f,t)\rangle = \mathrm{Tr}\,[\rho(t)Q^c(a_1,\ldots,a_f)]. \tag{3.9.19}$$

If we use (3.9.8), we have

$$\langle Q^c(a_1,\ldots,a_f,t)\rangle = \int\cdots\int d\xi_1\cdots d\xi_f C(\xi_1,\ldots,\xi_f,t)F(\xi_1,\ldots,\xi_f), \tag{3.9.20}$$

where

$$C(\xi_1,\ldots,\xi_f,t) = \langle e^{i\xi_1 a_1}\cdots e^{i\xi_f a_f}\rangle$$
$$= \mathrm{Tr}\,[\rho(t)e^{i\xi_1 a_1}\cdots e^{i\xi_f a_f}], \tag{3.9.21}$$

is a characteristic function and $F(\xi_1,\ldots,\xi_f)$ is given by (3.9.8b) and is the Fourier transform of the associated function $\bar{Q}^c(\alpha_1,\ldots,\alpha_f)$. In (3.9.20) we have converted the quantum expectation value to an integral over c-number functions. We may express this in another way as follows. By (3.9.4) we have an alternate expression equivalent to (3.9.20)

$$\langle Q^c(a_1,\ldots,a_f,t)\rangle = \int\cdots\int d\alpha_1\cdots d\alpha_f\bar{Q}^c(\alpha_1,\ldots,\alpha_f)P(\alpha_1,\ldots,\alpha_f,t) \tag{3.9.22}$$

where we have let

$$P(\alpha_1, \ldots, \alpha_f, t) = \langle \delta(\alpha_1 - a_1) \cdots \delta(\alpha_f - a_f) \rangle$$
$$= \text{Tr}\, \rho(t)\, \delta(\alpha_1 - a_1) \cdots \delta(\alpha_f - a_f)$$
$$= \left(\frac{1}{2\pi}\right)^f \int \cdots \int d\xi_1 \cdots d\xi_f e^{-i(\xi_1 \alpha_1 + \cdots + \xi_f \alpha_f)} C(\xi_1, \ldots, \xi_f, t), \tag{3.9.23}$$

and where we have used (3.9.5) and (3.9.21). We call P a distribution function which is the Fourier transform of the characteristic function. If we apply the inverse Fourier transform to (3.9.23), we have

$$C(\xi_1, \ldots, \xi_f, t) = \int \cdots \int d\alpha_1 \cdots d\alpha_f e^{i(\xi_1 \alpha_1 + \cdots + \xi_f \alpha_f)} P(\alpha_1, \ldots, \alpha_f, t)$$
$$= \text{Tr}\, e^{i\xi_1 a_1} \cdots e^{i\xi_f a_f} \rho(a_1, \ldots, a_f, t). \tag{3.9.24}$$

The advantage of using the distribution function stems from the fact that expectation values of operators in chosen order are evaluated just like classical averages if P is considered as a classical probability. We may see this as follows. We have by (3.9.24)

$$\left. \frac{\partial^{r_1 + r_2 + \cdots r_f} C}{\partial(i\xi_1)^{r_1}\, \partial(i\xi_2)^{r_2} \cdots \partial(i\xi_f)^{r_f}} \right|_{\xi_i = 0}$$
$$= \langle a_1^{r_1} a_2^{r_2} \cdots a_f^{r_f} \rangle$$
$$= \int \cdots \int \alpha_1^{r_1} \alpha_2^{r_2} \cdots \alpha_f^{r_f} P(\alpha_1, \ldots, \alpha_f, t)\, d\alpha_1 \cdots d\alpha_f. \tag{3.9.25}$$

There will be corrections to this quantum-classical correspondence for operator means which are not in chosen order as we show later. For example,

$$\langle a_2 a_1 \rangle \neq \iint \alpha_1 \alpha_2 P(\alpha_1, \alpha_2, t)\, d\alpha_1\, d\alpha_2 = \langle a_1 a_2 \rangle, \tag{3.9.26}$$

if a_1, a_2 is the chosen order.

The advantage of using the $\bar{\rho}^c(\alpha_1, \ldots, \alpha_f, t)$ associated function of (3.9.10) rather than the P-function is that we may very simply recover the density operator itself directly at any stage of the calculation by (3.9.10d). This is somewhat more difficult to do with the P-function. We may obtain a direct connection between the P-function and $\bar{\rho}^c$ quite easily. By (3.9.23) we have

$$P(\alpha_1, \ldots, \alpha_f, t) = \text{Tr}\, \rho(t)\, \delta(\alpha_1 - a_1) \cdots \delta(\alpha_f - a_f), \tag{3.9.27}$$

while by (3.9.4), we have

$$\rho^c(a_1, \ldots, a_f, t)$$
$$= \int \cdots \int d\beta_1 \cdots d\beta_f \bar{\rho}^c(\beta_1, \ldots, \beta_f, t)\, \delta(\beta_1 - a_1) \cdots \delta(\beta_f - a_f)$$
$$= \mathscr{C}\{\bar{\rho}^c(\beta_1, \ldots, \beta_f, t)\}. \tag{3.9.28}$$

If we substitute (3.9.28) into (3.9.27) we have

$$P(\underset{\sim}{\alpha}, t) = \int \cdots \int d\underset{\sim}{\beta}\, \bar{\rho}^c(\underset{\sim}{\beta}, t) K(\underset{\sim}{\beta}, \underset{\sim}{\alpha}), \tag{3.9.29a}$$

where the kernel is

$$K(\underset{\sim}{\beta}, \underset{\sim}{\alpha}) = \text{Tr}\,\{\delta(\beta_1 - a_1) \cdots \delta(\beta_f - a_f)\, \delta(\alpha_1 - a_1) \cdots \delta(\alpha_f - a_f)\}, \tag{3.9.29b}$$

and where $\underset{\sim}{\alpha} \equiv \alpha_1, \alpha_2, \ldots, \alpha_f$, $d\underset{\sim}{\beta} = d\beta_1 \cdots d\beta_f$, and so on. From the kernel and $\bar{\rho}^c$ we may obtain P in principle. The kernel may be evaluated by using a representation for the δ-functions such as (3.9.5).

From (3.9.10) and (3.9.22), we see that

$$\langle Q \rangle = \langle Q^c \rangle = \text{Tr}\, \rho^c\, Q = \text{Tr}\, \rho\, Q^c$$
$$= \int \cdots \int d\xi_1 \cdots d\xi_f\, \mathscr{G}(\xi_1, \ldots, \xi_f, t)\, \text{Tr}\,(Q e^{i\xi_1 a_1} \cdots e^{i\xi_f a_f})$$
$$= \int \cdots \int d\alpha_1 \cdots d\alpha_f \bar{Q}^c(\alpha_1 \cdots \alpha_f) P(\alpha_1, \ldots, \alpha_f, t), \tag{3.9.30}$$

which offers two independent ways of calculating the mean value of Q. In general, for practical reasons, the last form is generally preferred.

3.10 OPERATOR DESCRIPTION OF INDEPENDENT ATOMS

In this section we present an operator description of a single atom and then generalize it to cover the case of N atoms which do not interact with each other directly. Such a description will be adequate to discuss the interaction of an atom with a radiation field, an atom with a reservoir of oscillators (elastic vibrations in a solid) and so on.

For simplicity let us consider the hamiltonian for a single atom given by

$$H_A = \frac{\mathbf{p}^2}{2m} + U(\mathbf{r}). \tag{3.10.1}$$

This hamiltonian satisfies the energy eigenvalue problem

$$H_A|l\rangle = \epsilon_l|l\rangle, \tag{3.10.2}$$

where ϵ_l are the energy eigenvalues and $|l\rangle$ are the eigenvectors. The reader should not confuse the vectors $|l\rangle$ with those of the harmonic oscillator which are just a special case when $H = \hbar\omega a^\dagger a$.

The vectors $|l\rangle$ form a complete orthonormal set so that

$$\sum_l |l\rangle\langle l| = 1 \tag{3.10.3a}$$

$$\langle l|m\rangle = \delta_{lm}. \tag{3.10.3b}$$

We may use the completeness relation to express an arbitrary atomic operator in terms of the operators $|k\rangle\langle l|$. To illustrate, let us apply H_A to both sides of (3.10.3a) and use (3.10.2). This gives

$$H_A = \sum_l H_A|l\rangle\langle l| = \sum_l \epsilon_l|l\rangle\langle l|. \tag{3.10.4}$$

We have therefore expressed the hamiltonian in terms of the projection operators $|l\rangle\langle l|$. The expansion coefficients ϵ_l are just the diagonal matrix elements of H_A, namely,

$$\epsilon_l = \langle l|H_A|l\rangle. \tag{3.10.5}$$

Consider next any function $Q(\mathbf{p}, \mathbf{r})$. If we use the completeness relation twice, we have

$$\begin{aligned} Q &= \sum_l |l\rangle\langle l|Q \sum_m |m\rangle\langle m| \\ &\equiv \sum_{l,m} Q_{lm}|l\rangle\langle m|, \end{aligned} \tag{3.10.6}$$

where the expansion coefficients are just the matrix elements of Q in the energy representation

$$Q_{lm} = \langle l|Q|m\rangle. \tag{3.10.7}$$

If Q involves operators for another system such as the a and $a^\dagger$ of the radiation field, then Q_{kl} will still contain these operators obviously. In particular, if Q is the interaction energy between the atom and some other system, we may expand it as in (3.10.6).

The state vector $|\psi\rangle$ may also be expanded as

$$|\psi\rangle \equiv \sum_l |l\rangle\langle l|\psi\rangle = \sum_l c_l|l\rangle, \tag{3.10.8a}$$

where

$$c_l = \langle l|\psi\rangle, \tag{3.10.8b}$$

and

$$|c_l|^2 = \langle\psi|l\rangle\langle l|\psi\rangle \tag{3.10.8c}$$

is the probability the atom is in state $|l\rangle$. It is just the expectation value of the projection operator, $|l\rangle\langle l|$.

Due to the orthogonality of the energy eigenvectors, we always reduce products to a bilinear operator, since

$$|k\rangle\langle l|m\rangle\langle n| = \delta_{lm}|k\rangle\langle n|. \tag{3.10.9}$$

This property makes the algebra of these operators relatively simple.

Consider next the Heisenberg equation of motion for an arbitrary operator Q which is a function of atomic operators. If we use a superscript H or S to represent Heisenberg or Schrödinger pictures, we have

$$i\hbar\frac{dQ^H}{dt} = [Q^H, H_A{}^H]. \tag{3.10.10}$$

Let us express both $H_A{}^H$ and Q^H by (3.10.4) and (3.10.6), respectively. Then, (3.10.10) becomes

$$i\hbar\frac{d}{dt}\sum_{l,m} Q_{lm}(|l\rangle\langle m|)^H = \sum_{k,l,m} Q_{lm}\epsilon_k\{(|l\rangle\langle m|)^H(|k\rangle\langle k|)^H - (|k\rangle\langle k|)^H(|l\rangle\langle m|)^H\}. \tag{3.10.11}$$

Since the orthogonality relations must be the same in both pictures, the equation above reduces to

$$\begin{aligned} i\hbar\frac{d}{dt}\sum_{l,m} Q_{lm}(|l\rangle\langle m|)^H &= \sum_{k,l,m} Q_{lm}\epsilon_k\{(|l\rangle\langle k|)^H\delta_{mk} - \delta_{kl}(|k\rangle\langle m|)^H\} \\ &= \sum_{l,m} Q_{lm}(\epsilon_m - \epsilon_l)(|l\rangle\langle m|)^H, \end{aligned} \tag{3.10.12}$$

where we carried out the sum over k. Since the function Q was arbitrary, let us choose it so that it has only nonvanishing matrix elements are Q_{rs}. The sums then reduce to the single term

$$i\hbar\frac{d}{dt}(|r\rangle\langle s|)^H = (\epsilon_r - \epsilon_s)(|r\rangle\langle s|)^H. \tag{3.10.13}$$

The solution is clearly

$$(|r\rangle\langle s|)^H = e^{i\omega_{rs}t}(|r\rangle\langle s|)^S, \tag{3.10.14}$$

where at $t = 0$ the two pictures coincide and we have defined

$$\omega_{rs} = \frac{\epsilon_r - \epsilon_s}{\hbar}. \tag{3.10.15}$$

The reader may quickly verify that (3.10.13) may be written directly as

$$i\hbar\frac{d}{dt}(|r\rangle\langle s|)^H = [(|r\rangle\langle s|)^H, H_A{}^H], \tag{3.10.16}$$

so that the operators $|r\rangle\langle s|$ obey the familiar Heisenberg equation of motion.

When the atom has only two levels or is interacting with a system that causes it to make transitions between only two levels, say $|1\rangle$ and $|2\rangle$, where $\epsilon_2 > \epsilon_1$ to be definite, the operators $|1\rangle\langle 1|$, $|2\rangle\langle 2|$, $|1\rangle\langle 2|$, and $|2\rangle\langle 1|$ can be put in a one-to-one correspondence with the Pauli spin-$\frac{1}{2}$ operators σ_x, σ_y, σ_z, and the identity operator.

To carry out this correspondence let us assume the following correspondence, namely,

$$\sigma_+ = \tfrac{1}{2}(\sigma_x + i\sigma_y) = |2\rangle\langle 1| \tag{3.10.17a}$$

$$\sigma_- = \tfrac{1}{2}(\sigma_x - i\sigma_y) = |1\rangle\langle 2| = \sigma_+^\dagger \tag{3.10.17b}$$

$$\sigma_z = |2\rangle\langle 2| - |1\rangle\langle 1| \tag{3.10.17c}$$

$$1 = |2\rangle\langle 2| + |1\rangle\langle 1|. \tag{3.10.17d}$$

To prove the equivalence, we must merely show that the operators satisfy the same algebra as the spin operators, namely,

$$[\sigma_i, \sigma_j] = 2i\sigma_k, \tag{3.10.18}$$

where i, j, k form an even permutation of x, y, and z, and

$$[\sigma_i, \sigma_j]_+ \equiv \sigma_i\sigma_j + \sigma_j\sigma_i = 2\delta_{ij} \qquad i, j = x, y, \text{ or } z. \tag{3.10.19}$$

If we use the completeness and orthogonality relations (3.10.3), we see that the operators in (3.10.17) satisfy (3.10.18) and (3.10.19).

Thus we may visualize the operator $\sigma_+ = |2\rangle\langle 1|$ as raising an atom from state $|1\rangle$ to state $|2\rangle$ and σ_- as the inverse lowering operator. It effectively annihilates a particle in the upper level $|2\rangle$ and simultaneously creates the particle in the lower level $|1\rangle$.

Let us derive a few theorems for the operators of a two level atom which will prove useful in our later work.

THEOREM 1

If n is a positive integer, then

$$(\sigma_+\sigma_-)^n = \sigma_+\sigma_- = \tfrac{1}{2}(\sigma_z + 1) \tag{3.10.20}$$

PROOF.

By (3.10.17) and the orthogonality relations

$$\sigma_+\sigma_- = |2\rangle\langle 1|1\rangle\langle 2| = |2\rangle\langle 2|,$$

so that

$$(\sigma_+\sigma_-)^n = (|2\rangle\langle 2|)^n = |2\rangle\langle 2| = \sigma_+\sigma_-. \qquad \text{Q.E.D.}$$

THEOREM 2

If $f(\sigma_+\sigma_-)$ is any function of $\sigma_+\sigma_-$ that can be expanded in a power series, then

$$f(\sigma_+\sigma_-) = f(0) + [f(1) - f(0)]\sigma_+\sigma_-. \tag{3.10.21}$$

PROOF.

We expand f as

$$f(z) = \sum_{n=0}^{\infty} f_n z^n = f(0) + \sum_1^{\infty} f_n z^n.$$

But if $z = \sigma_+\sigma_-$, by Theorem 1 above $z^n = z$. So

$$f(z) = f(0) + \sum_1^{\infty} f_n z,$$

and

$$f(1) = f(0) + \sum_1^{\infty} f_n,$$

and the theorem follows directly, provided the sum exists.

As an example, we see that

$$e^{-\xi\sigma_+\sigma_-} = 1 + [e^{-\xi} - 1]\sigma_+\sigma_-. \tag{3.10.22}$$

THEOREM 3

If ξ is a parameter, then

$$e^{\frac{1}{2}\xi\sigma_z}\sigma_+ e^{-\frac{1}{2}\xi\sigma_z} = \sigma_+ e^{\xi} \tag{3.10.23a}$$

$$e^{\frac{1}{2}\xi\sigma_z}\sigma_- e^{-\frac{1}{2}\xi\sigma_z} = \sigma_- e^{-\xi}. \tag{3.10.23b}$$

From these and (3.10.17), it immediately follows that

$$\begin{aligned} e^{\frac{1}{2}\xi\sigma_z}\sigma_x e^{-\frac{1}{2}\xi\sigma_z} &= \sigma_+ e^{\frac{1}{2}\xi} + \sigma_- e^{-\frac{1}{2}\xi} = \sigma_x \cosh \tfrac{1}{2}\,\xi + i\sigma_y \sinh \tfrac{1}{2}\xi \\ e^{\frac{1}{2}\xi\sigma_z}\sigma_y e^{-\frac{1}{2}\xi\sigma_z} &= -i\sigma_+ e^{\frac{1}{2}\xi} + i\sigma_- e^{-\frac{1}{2}\xi} = -i\sigma_x \sinh \tfrac{1}{2}\,\xi + \sigma_y \cosh \tfrac{1}{2}\,\xi. \end{aligned} \tag{3.10.24}$$

PROOF.

Let

$$f(\xi) = e^{\frac{1}{2}\xi\sigma_z}\sigma_+ e^{-\frac{1}{2}\xi\sigma_z}, \tag{3.10.25}$$

so that

$$f(0) = \sigma_+. \tag{3.10.26}$$

Then

$$\frac{\partial f}{\partial \xi} = \tfrac{1}{2}e^{\frac{1}{2}\xi\sigma_z}[\sigma_z, \sigma_+]e^{-\frac{1}{2}\xi\sigma_z}$$

$$= e^{\xi\sigma_z}\sigma_+ e^{-\xi\sigma_z} = f(\xi), \tag{3.10.27}$$

where we used (2.8.21). When we integrate, we obtain

$$f(\xi) = f(0)e^{\xi} = \sigma_+ e^{\xi}. \tag{3.10.28}$$

A similar proof holds for (3.10.23b).

THEOREM 4

If ξ and η are parameters, then

$$e^{i(\xi\sigma_z + \eta\sigma_x)} = e^{i\xi(2\sigma_+\sigma_- - 1) + i\eta(\sigma_+ + \sigma_-)}$$

$$= \cos\sqrt{\xi^2 + \eta^2} + \frac{i \sin\sqrt{\xi^2 + \eta^2}}{\sqrt{\xi^2 + \eta^2}}(\xi\sigma_z + \eta\sigma_x). \tag{3.10.29}$$

We leave the proof of this and the following theorems as exercises.

THEOREM 5

If ξ and η are parameters, then

$$e^{i(\xi\sigma_+ + \eta\sigma_-)} = \cos\sqrt{\xi\eta} + i\frac{\sin\sqrt{\xi\eta}}{\sqrt{\xi\eta}}(\xi\sigma_+ + \eta\sigma_-). \tag{3.10.30}$$

THEOREM 6

$$e^{i\xi\sigma_\pm} = 1 + i\xi\sigma_\pm \tag{3.10.31}$$

$$e^{i\xi\sigma_i} = \cos\xi + i\sigma_i \sin\xi \qquad i = 1, 2, 3. \tag{3.10.32}$$

THEOREM 7

If $i, j = 1, 2,$ or 3, then

$$\begin{aligned} \operatorname{Tr} \sigma_i \sigma_j &= 2\delta_{ij} \\ \operatorname{Tr} \sigma_i &= 0. \end{aligned} \tag{3.10.33}$$

The density operator describing a single atom in the Schrödinger picture may obviously also be expanded in the energy representation as

$$\rho(t) = \sum_{k,l} \rho_{kl}(t)|k\rangle\langle l|, \tag{3.10.34a}$$

where

$$\rho_{kl}(t) = \langle k|\rho(t)|l\rangle. \tag{3.10.34b}$$

The expectation value of Q is given by

$$\begin{aligned}\langle Q\rangle &= \operatorname{Tr} \rho(t)Q = \operatorname{Tr} \sum_{k,l} \rho_{kl}(t)|k\rangle\langle l| \sum_{mn} |m\rangle\langle n|Q_{mn} \\ &= \sum_{\substack{k,l\\ m,m}} Q_{mn}\rho_{kl}(t) \operatorname{Tr} |k\rangle\langle n|\delta_{lm}.\end{aligned} \tag{3.10.35}$$

Since

$$\operatorname{Tr} |k\rangle\langle n| = \langle n|k\rangle = \delta_{nk}, \tag{3.10.36}$$

we have

$$\langle Q(t)\rangle = \sum_{m,n} Q_{mn}\rho_{nm}(t) \equiv \operatorname{Tr} Q\, \rho(t), \tag{3.10.37}$$

which is simply the trace of the product of the two matrices in the Schrödinger picture.

All of the algebra above may be repeated if we identify

$$\begin{aligned}|k\rangle &\rightarrow a_k^\dagger \\ \langle l| &\rightarrow a_l,\end{aligned} \tag{3.10.38}$$

and require that these operators obey the relations

$$a_k a_l^\dagger = \delta_{lk} \tag{3.10.39a}$$

$$a_k^\dagger a_l a_m^\dagger a_n = \delta_{lm} a_k^\dagger a_n \tag{3.10.39b}$$

$$\operatorname{Tr} a_k^\dagger a_l = \delta_{kl}. \tag{3.10.39c}$$

We may call a_l an annihilation operator for a particle in state $|l\rangle$ and $a_k^\dagger$ a creation operator for a particle in state $|k\rangle$. However, to ask whether these operators obey commutation or anticommutation relations is a meaningless question. All that is implied is simply expressed by (3.10.38). It is usually simply more convenient to write $a_k^\dagger$ than it is to write $|k\rangle$ so that we use the notation above. The Heisenberg equation of motion (3.10.16)

$$\frac{d}{dt} a_r^\dagger(t)a_s(t) = \frac{1}{i\hbar} [a_r^\dagger(t)a_s(t), H_A{}^H] \tag{3.10.40}$$

looks more familiar in this notation where we write

$$H_A = \sum_l \epsilon_l a_l^\dagger a_l. \tag{3.10.41}$$

If we have N independent atoms, we take our hamiltonian to be

$$H_A = \sum_{m=1}^{N} \sum_l \epsilon_l^{(m)} (a_l^\dagger a_l)_m, \tag{3.10.42}$$

and if all atoms are identical, we have

$$\epsilon_l^{(m)} = \epsilon_l. \tag{3.10.43}$$

That is, the energy levels of all atoms are the same. Therefore, the generalization to N independent atoms is straightforward.

PART IV. ELEMENTARY APPLICATIONS

3.11 SOLUTION OF THE SCHRÖDINGER EQUATION BY NORMAL ORDERING; DRIVEN HARMONIC OSCILLATOR [15]

The Schrödinger equation

$$i\hbar \frac{\partial|\psi(t)\rangle}{\partial t} = H|\psi(t)\rangle \tag{3.11.1}$$

has a solution of the form

$$|\psi(t)\rangle = U(t, t_0)|\psi(t_0)\rangle, \tag{3.11.2}$$

where U satisfies

$$i\hbar \frac{\partial U}{\partial t} = HU \tag{3.11.3}$$

subject to the initial condition

$$U(t_0, t_0) = 1. \tag{3.11.4}$$

Let us assume the hamiltonian is of the form

$$H(a, a^\dagger, t) = \sum_{l,m} h_{lm}(t) a^{\dagger l} a^m, \tag{3.11.5}$$

where we have put it into normal order. The $h_{lm}(t)$ are c-number expansion coefficients. Then (3.11.3) becomes

$$i\hbar \frac{\partial U}{\partial t} = \sum_{l,m} h_{lm}(t) a^{\dagger l} a^m U. \tag{3.11.6}$$

If we use (3.3.13), we may rewrite this as

$$i\hbar \frac{\partial U}{\partial t} = \sum_{l,m} h_{lm}(t) a^{\dagger l} \mathcal{N}\left\{\left(\alpha + \frac{\partial}{\partial \alpha^*}\right)^m \bar{U}^{(n)}(\alpha, \alpha^*, t)\right\}. \tag{3.11.7}$$

where

$$\bar{U}^{(n)}(\alpha, \alpha^*, t) = \langle\alpha|U(\alpha, \alpha^*, t)|\alpha\rangle. \tag{3.11.8}$$

If we take diagonal coherent state matrix elements of both sides of (3.11.7), we obtain the c-number equation

$$i\hbar \frac{\partial \bar{U}^{(n)}}{\partial t} = \sum_{l,m} h_{lm}(t)\alpha^{*l}\left(\alpha + \frac{\partial}{\partial \alpha^*}\right)^m \bar{U}^{(n)}, \tag{3.11.9}$$

since the right side is in normal order. When we solve (3.11.9), we obtain $|\psi(t)\rangle$ by

$$|\psi(t)\rangle = \mathcal{N}\{U^{(n)}(\alpha, \alpha^*, t)\}|\psi(t_0)\rangle. \tag{3.11.10}$$

We may also first transform (3.11.1) to the interaction picture and proceed as an alternative.

As a simple example let us consider a driven harmonic oscillator described by

$$H = \hbar\omega a^\dagger a + \hbar[f(t)a + f^*(t)a^\dagger]. \tag{3.11.11}$$

In this case, we see that (3.11.9) becomes

$$i\hbar \frac{\partial \bar{U}^{(n)}}{\partial t} = \left\{\hbar\omega\alpha^*\left(\alpha + \frac{\partial}{\partial \alpha^*}\right) + \hbar\left[f(t)\left(\alpha + \frac{\partial}{\partial \alpha^*}\right) + f^*(t)\alpha^*\right]\right\}\bar{U}^{(n)}. \tag{3.11.12}$$

If we let

$$\bar{U}^{(n)} = e^{G(\alpha,\alpha^*,t)}, \tag{3.11.13}$$

where

$$G = A(t) + B(t)\alpha + C(t)\alpha^* + D(t)\alpha^*\alpha, \tag{3.11.14}$$

then (3.11.12) becomes

$$i\left[\frac{dA}{dt} + \frac{dB}{dt}\alpha + \frac{dC}{dt}\alpha^* + \frac{dD}{dt}\alpha^*\alpha\right]$$
$$= \omega\alpha^*\alpha + \omega\alpha^*(C + D\alpha) + f(t)\alpha + f^*(t)\alpha^* + f(t)(C + D\alpha). \tag{3.11.15}$$

Equating coefficients of $\alpha^*\alpha$, α, and α^* we have

$$\begin{aligned} i\frac{dD}{dt} &= \omega(D + 1) \\ i\frac{dB}{dt} &= f(t)(D + 1) \\ i\frac{dC}{dt} &= \omega C + f^*(t) \\ i\frac{dA}{dt} &= f(t)C. \end{aligned} \tag{3.11.16}$$

Since $U(t=0)=1$, $A(0)=B(0)=C(0)=D(0)=0$. The solutions of (3.11.16) are easily seen to be

$$
\begin{aligned}
D(t) &= e^{-i\omega t}-1 \\
B(t) &= -i\int_0^t e^{-i\omega t'} f(t')\,dt' \\
C(t) &= -i\int_0^t e^{i\omega(t'-t)} f^*(t')\,dt' = -e^{-i\omega t}B^*(t) \\
A(t) &= -\int_0^t dt''\, f(t'')\int_0^{t''} e^{i\omega(t'-t'')} f(t')\,dt'.
\end{aligned}
\tag{3.11.17}
$$

Therefore,

$$
\begin{aligned}
|\psi(t)\rangle &= U(t)|\psi(t_0)\rangle = \mathcal{N}\{e^{A+B\alpha+C\alpha^*+D\alpha^*\alpha}\}|\psi(t_0)\rangle \\
&= e^{A(t)}e^{C(t)a^\dagger}\mathcal{N}\{e^{D(t)\alpha^*\alpha}\}e^{B(t)a}|\psi(t_0)\rangle.
\end{aligned}
\tag{3.11.18}
$$

If we desire the solution to be in normal order, there is no advantage in going to the IP for this simple example.

If at $t=t_0$, $|\psi(t_0)\rangle=|\alpha\rangle$, the coherent state, then it follows from (3.11.18) and $f(a)|\alpha\rangle=f(\alpha)|\alpha\rangle$ that

$$
|\psi(t)\rangle = e^{A(t)+B(t)\alpha}e^{C(t)a^\dagger}e^{D(t)\alpha a^\dagger}|\alpha\rangle. \tag{3.11.19}
$$

If we express $|\alpha\rangle$ as

$$
|\alpha\rangle = e^{-\frac{1}{2}|\alpha|^2}e^{\alpha a^\dagger}|0\rangle, \tag{3.11.20}
$$

then we see that (3.11.19) may be written as

$$
|\psi(t)\rangle = e^{A(t)+B(t)\alpha-\frac{1}{2}|\alpha|^2}e^{\{[1+D(t)]\alpha+C(t)\}a^\dagger}|0\rangle, \tag{3.11.21}
$$

so that a coherent state will always remain a coherent state.

If $|\psi(t_0)\rangle=|0\rangle$, then we see that (let $\alpha=0$ in 3.11.21)

$$
|\psi(t)\rangle = e^{A(t)+C(t)a^\dagger}|0\rangle, \tag{3.11.22}
$$

so that a driven oscillator starting from the vacuum state develops into a coherent state. This, of course, assumes the oscillator is at absolute zero.

It should be noted that antinormal ordering may also be used in the same way to solve the Schrödinger equation.

3.12 TWO WEAKLY COUPLED OSCILLATORS

We may apply the techniques of the last section to solve the Schrödinger equation for two weakly coupled oscillators described by the hamiltonian

$$
H = \hbar\omega_1 a^\dagger a + \hbar\omega_2 b^\dagger b + \hbar\kappa(a^\dagger b + b^\dagger a). \tag{3.12.1}
$$

We have on assuming U is in normal order

$$i\hbar \frac{\partial U^{(n)}}{\partial t} = \hbar\omega_1 a^\dagger \left(U^{(n)} a + \frac{\partial U^{(n)}}{\partial a^\dagger} \right) + \hbar\omega_2 b^\dagger \left(U^{(n)} b + \frac{\partial U^{(n)}}{\partial b^\dagger} \right)$$
$$+ \hbar\kappa a^\dagger \left(U^{(n)} b + \frac{\partial U^{n}}{\partial b^\dagger} \right) + \hbar\kappa b^\dagger \left(U^{(n)} a + \frac{\partial U^{(n)}}{\partial a^\dagger} \right). \quad (3.12.2)$$

Since both sides are in normal order and since

$$\bar{U}^{(n)}(\alpha, \alpha^*, \beta, \beta^*, t) = \langle \alpha, \beta | U^{(n)} | \alpha, \beta \rangle, \quad (3.12.3)$$

we have when we let

$$\bar{U}^n = e^{G(\alpha, \alpha^*, \beta, \beta^*, t)} \quad (3.12.4)$$

$$i \frac{\partial G}{\partial t} = \omega_1 \alpha^* \alpha + \omega_2 \beta^* \beta + \omega_1 \alpha^* \frac{\partial G}{\partial \alpha^*} + \omega_2 \beta^* \frac{\partial G}{\partial \beta^*}$$
$$+ \kappa \alpha^* \beta + \kappa \alpha \beta^* + \kappa \alpha^* \frac{\partial G}{\partial \beta^*} + \kappa \beta^* \frac{\partial G}{\partial \alpha^*}. \quad (3.12.5)$$

If we assume G of the form

$$G = A(t)\alpha^*\alpha + B(t)\beta^*\beta + C(t)\alpha^*\beta + D(t)\alpha\beta^*, \quad (3.12.6)$$

we see the coefficients satisfy the equations

$$i \frac{dA}{dt} = \omega_1(A + 1) + \kappa D \quad (3.12.7)$$

$$i \frac{dB}{dt} = \omega_2(B + 1) + \kappa C \quad (3.12.8)$$

$$i \frac{dC}{dt} = \omega_1 C + (B + 1)\kappa \quad (3.12.9)$$

$$i \frac{dD}{dt} = \omega_2 D + (A + 1)\kappa, \quad (3.12.10)$$

where $A(0) = B(0) = C(0) = D(0) = 0$. From (3.12.7) and (3.12.10) we obtain

$$\frac{d}{dt}(A + 1) + i\omega_1(A + 1) = -i\kappa D \quad (3.12.11)$$

$$\frac{dD}{dt} + i\omega_2 D = -i\kappa(A + 1), \quad (3.12.12)$$

or on substituting D from (3.12.11) into (3.12.12) we have

$$\left(\frac{d}{dt} + i\omega_2\right)\left(\frac{d}{dt} + i\omega_1\right)(A + 1) = -\kappa^2(A + 1),$$

so that

$$\begin{aligned} A(t) + 1 &= e^{-i(\omega_1+\omega_2)t/2} \cos \Gamma t \\ D(t) &= e^{-i(\omega_1+\omega_2)t/2} \sin \Gamma t, \end{aligned} \tag{3.12.13}$$

where

$$\Gamma = \sqrt{\left(\frac{\omega_1 - \omega_2}{2}\right)^2 + \kappa^2}. \tag{3.12.14}$$

Similarly, one finds that

$$\begin{aligned} A(t) &= B(t) \\ C(t) &= D(t). \end{aligned} \tag{3.12.15}$$

Thus

$$U(t) = \mathcal{N}\{e^{A\alpha^*\alpha + B\beta^*\beta + C\alpha^*\beta + D\alpha\beta^*}\}. \tag{3.12.16}$$

If at $t = 0$, $|\psi(0)\rangle = |\alpha, \beta\rangle$, then

$$\begin{aligned} |\psi(t)\rangle &= e^{A\alpha a^\dagger + B\beta b^\dagger + C\beta a^\dagger + D\alpha b^\dagger}|\alpha, \beta\rangle \\ &= e^{-\frac{1}{2}|\alpha|^2 - \frac{1}{2}|\beta|^2} e^{[(A+1)\alpha + C\beta]a^\dagger + [(B+1)\beta + D\alpha]b^\dagger}|0, 0\rangle, \end{aligned} \tag{3.12.17}$$

so again we find that a coherent state remains a coherent state. If $\alpha = \beta = 0$, the vacuum state remains a vacuum state since these oscillators are not driven.

3.13 DISTRIBUTION FUNCTION FOR TWO-LEVEL ATOM

To illustrate the techniques developed in Sections 3.9 and 3.10 let us obtain the distribution function for a two-level atom. Let

$$N_l = |l\rangle\langle l| \qquad l = 1, 2 \tag{3.13.1}$$

be the projection operator for the atom in levels 1 and 2, respectively, and let

$$M = |1\rangle\langle 2| \qquad M^\dagger = |2\rangle\langle 1| \tag{3.13.2}$$

be the dipole moment operators. The hamiltonian by (3.10.4) is

$$H_0 = \epsilon_1|1\rangle\langle 1| + \epsilon_2|2\rangle\langle 2|, \tag{3.13.3}$$

where we assume $\epsilon_2 > \epsilon_1$.

Since

$$\langle l|m\rangle = \delta_{lm}, \tag{3.13.4}$$

it follows directly from (3.13.1) and (3.13.2) that

$$N_l^2 = N_l \qquad M^2 = M^{\dagger 2} = 0. \tag{3.13.5}$$

The reader may therefore verify the following commutation relations (see problem 3.20) which we shall need

$$\begin{aligned}
[e^{i\xi M}, N_1] &= -i\xi M \equiv -i\xi M e^{i\xi M} \\
[e^{i\xi^* M^\dagger}, N_1] &= i\xi^* M^\dagger \equiv i\xi^* M^\dagger e^{i\xi^* M^\dagger} \\
[e^{i\xi M}, N_2] &= i\xi M \equiv i\xi M e^{i\xi M} \\
[e^{i\xi^* M^\dagger}, N_2] &= -i\xi^* M^\dagger \equiv -i\xi^* M^\dagger e^{i\xi^* M^\dagger} \\
[N_1, N_2] &= 0.
\end{aligned} \tag{3.13.6}$$

We must first decide on a chosen ordering for these operators. We pick the ordering $M^\dagger$, N_1, N_2, M. According to (3.9.23), the distribution function is defined by

$$P(\alpha, t) = \mathrm{Tr}\ \rho(t) \prod_{i=1}^{4} \delta(\alpha_i - a_i), \tag{3.13.7}$$

where we let

$$\begin{aligned}
a_1 &= M^\dagger = a_4^\dagger \\
a_2 &= N_1 \\
a_3 &= N_2 \\
a_4 &= M = a_1^\dagger.
\end{aligned} \tag{3.13.8}$$

The density operator satisfies the equation

$$i\hbar \frac{\partial \rho}{\partial t} = [H_0, \rho], \tag{3.13.9}$$

so that when we use this in (3.13.7) we obtain

$$i\hbar \frac{\partial P}{\partial t}(\underset{\sim}{\alpha}, t) = \mathrm{Tr} \left\{ [H_0, \rho] \prod_{i=1}^{4} \delta(\alpha_i - a_i) \right\}. \tag{3.13.10}$$

If we use the cyclic property of traces, namely, Tr AB = Tr BA, this may be rewritten as

$$i\hbar \frac{\partial P}{\partial t}(\underset{\sim}{\alpha}, t) = \mathrm{Tr}\ \rho(t) \left[\prod_{i=1}^{4} \delta(\alpha_i - a_i), H_0 \right], \tag{3.13.11}$$

we represent the δ functions (see Eqs. 3.9.5 and 3.9.6) as

$$\prod_{i=1}^{4} \delta(\alpha_i - a_i) = \frac{1}{(2\pi)^2} \frac{1}{\pi^2} \int \cdots \int d^2\xi\, d\xi_1\, d\xi_2$$
$$\times\, e^{-i\xi^*(\mathscr{M}^* - M^\dagger)} e^{-i\xi_1(\mathscr{N}_1 - N_1)} e^{-i\xi_2(\mathscr{N}_2 - N_2)} e^{-i\xi(\mathscr{M} - M)}, \tag{3.13.12}$$

where we let

$$\begin{aligned} \alpha_1 &\equiv \mathscr{M}^* \qquad \alpha_3 = \mathscr{N}_2 \\ \alpha_2 &= \mathscr{N}_1 \qquad \alpha_4 = \mathscr{M}. \end{aligned} \tag{3.13.13}$$

Thus when we use (3.13.12) and (3.13.13), (3.13.11) becomes

$$i\hbar \frac{\partial P}{\partial t}(\underset{\sim}{\alpha}, t) = \frac{1}{(2\pi)^2\pi^2} \int \cdots \int d^2\xi \, d\xi_1 \, d\xi_2 e^{-i[\xi^* \mathscr{M}^* + \xi_1 \mathscr{N}_1 + \xi_2 \mathscr{N}_2 + \xi \mathscr{M}]} I, \tag{3.13.14}$$

where

$$I \equiv \text{Tr}\,\{\rho(t)[e^{i\xi^* M^\dagger} e^{i\xi_1 N_1} e^{i\xi_2 N_2} e^{i\xi M}, (\epsilon_1 N_1 + \epsilon_2 N_2)]\}. \tag{3.13.15}$$

We note that the δ-function is in chosen order but the commutator terms are not. Our first task is to use the commutation relations (3.13.6) to put all terms in chosen order. Since

$$[AB, C] \equiv A[B, C] + [A, C]B, \tag{3.13.16}$$

we have on repeated application of (3.13.16),

$$\begin{aligned} [e^{i\xi^* M^\dagger} e^{i\xi_1 N_1} e^{i\xi_2 N_2} e^{i\xi M}, N_1] &= e^{i\xi^* M^\dagger}[e^{i\xi_1 N_1} e^{i\xi_2 N_2} e^{i\xi M}, N_1] \\ &\quad + [e^{i\xi^* M^\dagger}, N_1] e^{i\xi_1 N_1} e^{i\xi_2 N_2} e^{i\xi M} \\ &= -e^{i\xi^* M^\dagger} e^{i\xi_1 N_1} e^{i\xi_2 N_2} e^{i\xi M} i\xi M \\ &\quad + i\xi^* M^\dagger e^{i\xi^* M^\dagger} e^{i\xi_1 N_1} e^{i\xi_2 N_2} e^{i\xi M}, \end{aligned} \tag{3.13.17}$$

which is now in chosen order. Similarly, we see that

$$[e^{i\xi^* M^\dagger} e^{i\xi_1 N_1} e^{i\xi_2 N_2} e^{i\xi M}, N_2] = -[e^{i\xi^* M^\dagger} e^{i\xi_1 N_1} e^{i\xi_2 N_2} e^{i\xi M}, N_1]. \tag{3.13.18}$$

We have used the fact that N_1 and N_2 commute. Therefore, (3.13.15) becomes

$$\begin{aligned} I = -(\epsilon_2 - \epsilon_1)\,\text{Tr}\,\rho(t)\{ & i\xi^* M^\dagger e^{i\xi^* M^\dagger} e^{i\xi_1 N_1} e^{i\xi_2 N_2} e^{i\xi M} \\ & - e^{i\xi^* M^\dagger} e^{i\xi_1 N_1} e^{i\xi_2 N_2} e^{i\xi M} i\xi M\}. \end{aligned} \tag{3.13.19}$$

Since both operators in the curly brackets are in chosen order, we may use (3.9.4) and write

$$\begin{aligned} I = -(\epsilon_2 - \epsilon_1) & \int \cdots \int d\underset{\sim}{\alpha}' \,\text{Tr}\, \rho(t) \\ & \times \left\{ (i\xi^* \mathscr{M}^{*\prime} - i\xi \mathscr{M}') e^{i(\xi^* \mathscr{M}^{*\prime} + \xi_1 \mathscr{N}_1' + \xi_2 \mathscr{N}_2' + \xi \mathscr{M}')} \prod_{i=1}^{4} \delta[\alpha_i' - a_i] \right\}, \end{aligned} \tag{3.13.20}$$

where

$$d\alpha' \equiv d\alpha_1'\, d\alpha_2'\, d\alpha_3'\, d\alpha_4', \tag{3.13.21}$$

where

$$\begin{aligned} \alpha_1' &\equiv \mathscr{M}^{*\prime} \qquad \alpha_3' \equiv \mathscr{N}_2' \\ \alpha_2' &\equiv \mathscr{N}_1' \qquad \alpha_4' \equiv \mathscr{M}', \end{aligned} \tag{3.13.22}$$

and the a_i are given by (3.13.8). From the definition of the distribution function (3.13.7), we see that (3.13.20) may be written as

$$I = -(\epsilon_2 - \epsilon_1)\int\cdots\int d\underset{\sim}{\alpha}'\, P(\underset{\sim}{\alpha}', t)(i\xi^*\mathscr{M}'^* - i\xi\mathscr{M}')e^{i(\xi^*\mathscr{M}'^* + \xi_1\mathscr{N}_1' + \xi_2\mathscr{N}_2' + \xi\mathscr{M}')}$$

$$\equiv -(\epsilon_2 - \epsilon_1)\int\cdots\int d\underset{\sim}{\alpha}'\, P(\underset{\sim}{\alpha}', t) \times \left[\mathscr{M}'^*\frac{\partial}{\partial\mathscr{M}'^*} - \mathscr{M}'\frac{\partial}{\partial\mathscr{M}'}\right]e^{i(\xi^*\mathscr{M}'^* + \xi_1\mathscr{N}_1' + \xi_2\mathscr{N}_2' + \xi\mathscr{M}')}. \quad (3.13.23)$$

We next integrate the first term by parts with respect to $\mathscr{M}'^*$ and the second with respect to $\mathscr{M}'$. This gives

$$I = +(\epsilon_2 - \epsilon_1)\int\cdots\int d\underset{\sim}{\alpha}' \times e^{i(\xi^*\mathscr{M}'^* + \xi_1\mathscr{N}_1' + \xi_2\mathscr{N}_2' + \xi\mathscr{M}')}\left\{\frac{\partial}{\partial\mathscr{M}'^*}\mathscr{M}'^* - \frac{\partial}{\partial\mathscr{M}'}\mathscr{M}'\right\}P(\underset{\sim}{\alpha}', t), \quad (3.13.24)$$

where we assume the distribution function vanishes at the limits of integration.

We now substitute (3.13.24) into (3.13.14) and carry out the ξ integration. This gives

$$\frac{\partial P}{\partial t}(\underset{\sim}{\alpha}, t) = -i\frac{(\epsilon_2 - \epsilon_1)}{\hbar}\int\cdots\int d\underset{\sim}{\alpha}' \times \delta(\mathscr{M}^* - \mathscr{M}'^*)\,\delta(\mathscr{N}_1 - \mathscr{N}_1')\,\delta(\mathscr{N}_2 - \mathscr{N}_2')\,\delta(\mathscr{M} - \mathscr{M}') \times \left\{\frac{\partial}{\partial\mathscr{M}'^*}[\mathscr{M}'^*P(\underset{\sim}{\alpha}', t)] - \frac{\partial}{\partial\mathscr{M}'}[\mathscr{M}'P(\underset{\sim}{\alpha}', t)]\right\}$$

$$= -i\frac{(\epsilon_2 - \epsilon_1)}{\hbar}\left\{\left[\frac{\partial}{\partial\mathscr{M}^*}[\mathscr{M}^*P(\underset{\sim}{\alpha}, t)] - \frac{\partial}{\partial\mathscr{M}}[\mathscr{M}P(\underset{\sim}{\alpha}, t)]\right]\right\}, \quad (3.13.25)$$

where $\underset{\sim}{\alpha} \equiv (\mathscr{M}^*, \mathscr{N}_1, \mathscr{N}_2, \mathscr{M})$. We have therefore obtained the c-number equation obeyed by the distribution function which is equivalent to the density operator equation.

We may obtain solutions of (3.13.25) very easily by the method of characteristics (See Appendix A). We may rewrite (3.13.25) as

$$\frac{\partial P}{\partial t} + i\omega_{21}\mathscr{M}^*\frac{\partial P}{\partial\mathscr{M}^*} - i\omega_{21}\mathscr{M}\frac{\partial P}{\partial\mathscr{M}} + 0\frac{\partial P}{\partial\mathscr{N}_1} + 0\frac{\partial P}{\partial\mathscr{N}_2} = 0. \quad (3.13.26)$$

The characteristic equations are therefore

$$\frac{dt}{1} = \frac{d\mathscr{M}^*}{i\omega_{21}\mathscr{M}^*} = \frac{d\mathscr{M}}{-i\omega_{21}\mathscr{M}} = \frac{d\mathscr{N}_1}{0} = \frac{d\mathscr{N}_2}{0}. \quad (3.13.27)$$

We therefore have that

$$d\mathcal{N}_1 = d\mathcal{N}_2 = 0, \tag{3.13.28}$$

or

$$\mathcal{N}_1 = \mathcal{N}_{10}; \qquad \mathcal{N}_2 = \mathcal{N}_{20} \tag{3.13.29}$$

$$\begin{aligned} \mathcal{M}(t) &= \mathcal{M}(0)e^{-i\omega_{21}t} \\ \mathcal{M}^*(t) &= \mathcal{M}(0)e^{+i\omega_{21}t}. \end{aligned} \tag{3.13.30}$$

Therefore, any arbitrary function of $\mathcal{M}(0)$, $\mathcal{M}^*(0)$, $\mathcal{N}_{10}$, and $\mathcal{N}_{20}$ such as

$$P = P[\mathcal{M}(t)e^{i\omega_{21}t}, \mathcal{M}^*(t)e^{-i\omega_{21}t}, \mathcal{N}_{10}, \mathcal{N}_{20}] \tag{3.13.31}$$

satisfies (3.13.26). The functional form of P must be determined by the initial conditions. If we know that $\mathcal{N}_1 = \mathcal{N}_{10}$, $\mathcal{N}_2 = \mathcal{N}_{20}$, $\mathcal{M} = \mathcal{M}(0)$, and $\mathcal{M}^* = \mathcal{M}^*(0)$ at $t = 0$, then

$$\begin{aligned} P = \delta[\mathcal{M}(0) - \mathcal{M}e^{i\omega_{21}t}]\, \delta[\mathcal{M}^*(0) - \mathcal{M}^*e^{-i\omega_{21}t}] \\ \times \delta(\mathcal{N}_{10} - \mathcal{N}_1)\, \delta(\mathcal{N}_{20} - \mathcal{N}_2). \end{aligned} \tag{3.13.32}$$

From (3.9.22), it follows that the averages of M, $M^\dagger$, N_1, and N_2 are given by

$$\left.\begin{aligned} \langle M(t)\rangle &= \int \cdots \int d^2\mathcal{M}\, d\mathcal{N}_1\, d\mathcal{N}_2\, \mathcal{M}\, P \\ &= \mathcal{M}(0)e^{-i\omega_{21}t} \\ \langle M^\dagger(t)\rangle &= \mathcal{M}^*(0)e^{i\omega_{21}t} \\ \langle N_1(t)\rangle &= \mathcal{N}_{10} \\ \langle N_2(t)\rangle &= \mathcal{N}_{20}, \end{aligned}\right\} \tag{3.13.33}$$

when measurements were made at $t = 0$. Therefore, we see that the average number of atoms in levels 1 and 2 does not change with time while the dipole moment oscillates at frequency ω_{21}.

3.14 DISTRIBUTION FUNCTION FOR HARMONIC OSCILLATOR

The hamiltonian for an oscillator is

$$H_0 = \hbar\omega a^\dagger a.$$

Let us consider the case in which the operators are in normal order and obtain the equation of motion for the distribution function

$$P(\alpha, \alpha^*, t) = \mathrm{Tr}\, \rho(t)\, \delta(\alpha^* - a^\dagger)\, \delta(\alpha - a). \tag{3.14.1}$$

Then

$$i\hbar \frac{\partial P}{\partial t} = \text{Tr}\,[H_0, \rho]\,\delta(\alpha^* - a^\dagger)\,\delta(\alpha - a)$$
$$= \text{Tr}\,\{\rho(t)[\delta(\alpha^* - a^\dagger)\,\delta(\alpha - a), \hbar\omega a^\dagger a]\}$$
$$= \frac{\hbar\omega}{\pi^2}\iint d^2\xi\, e^{-i(\xi\alpha+\xi^*\alpha^*)} I, \tag{3.14.2}$$

where

$$I = \text{Tr}\,\rho(t)[e^{i\xi^* a^\dagger} e^{i\xi a}, a^\dagger a]. \tag{3.14.3}$$

We have used (3.13.9), the cyclic properties of traces and (3.9.6). By (3.13.16) we may rewrite I as

$$I = \text{Tr}\,\rho(t)\{e^{i\xi a}[e^{i\xi^* a^\dagger}, a^\dagger a] + [e^{i\xi a}, a^\dagger a]e^{i\xi^* a^\dagger}\}. \tag{3.14.4}$$

Also since

$$[A, BC] \equiv [A, B]C + B[A, C], \tag{3.14.5}$$

we have

$$I = \text{Tr}\,\rho(t)\{e^{i\xi a}([e^{i\xi^* a^\dagger}, a^\dagger]a + a^\dagger[e^{i\xi^* a^\dagger}, a])$$
$$+ ([e^{i\xi a}, a^\dagger]a + a^\dagger[e^{i\xi a}, a])e^{i\xi^* a^\dagger}\}. \tag{3.14.6}$$

But

$$\begin{aligned}[a, e^{i\xi^* a^\dagger}] &= \frac{\partial}{\partial a^\dagger} e^{i\xi^* a^\dagger} = e^{i\xi^* a^\dagger} i\xi^* \\ [a^\dagger, e^{i\xi a}] &= -\frac{\partial}{\partial a} e^{i\xi a} = -i\xi e^{i\xi a}.\end{aligned} \tag{3.14.7}$$

Therefore,

$$I = \text{Tr}\,\rho(t)\{-e^{i\xi a} a^\dagger i\xi^* e^{i\xi^* a^\dagger} + e^{i\xi a} i\xi a e^{i\xi^* a^\dagger}\}.$$

Since these are in chosen (normal) order, we have by (3.9.4)

$$I = \iint d^2\alpha'\, \text{Tr}\,\rho(t)\{(i\xi\alpha' - i\xi^*\alpha'^*)e^{i(\xi\alpha'+\xi^*\alpha'^*)}\,\delta(\alpha^{*\prime} - a^\dagger)\,\delta(\alpha' - a)\}$$
$$= \iint d^2\alpha'\, P(\underset{\sim}{\alpha}', t)(i\xi\alpha' - i\xi^*\alpha'^*)e^{i(\xi\alpha'+\xi^*\alpha'^*)}, \tag{3.14.8}$$

where we used (3.14.1). This may be rewritten as

$$I = \iint d^2\alpha'\, P(\alpha', \alpha'^*, t)\left[\alpha' \frac{\partial}{\partial\alpha'} - \alpha'^* \frac{\partial}{\partial\alpha'^*}\right] e^{i(\xi\alpha'+\xi^*\alpha'^*)}$$
$$= +\iint d^2\alpha'\, e^{i(\xi\alpha'+\xi^*\alpha'^*)}\left[\frac{\partial}{\partial\alpha'^*}\alpha'^* - \frac{\partial}{\partial\alpha'}\alpha'\right] P(\alpha', \alpha'^*, t), \tag{3.14.9}$$

where we integrated by parts. When we put this into (3.14.2) and integrate over $d^2\xi$, we obtain

$$\frac{\partial P}{\partial t}(\alpha, \alpha^*, t) = -i\omega \iint d^2\alpha' \times \left\{\left[\frac{\partial}{\partial \alpha'^*}\alpha'^* - \frac{\partial}{\partial \alpha'}\alpha'\right] P(\alpha', \alpha'^*, t)\right\} \delta(\alpha' - \alpha)\, \delta(\alpha'^* - \alpha^*),$$

or the distribution function obeys the equation

$$\begin{aligned}\frac{\partial P}{\partial t}(\alpha, \alpha^*, t) &= i\omega\left[\frac{\partial}{\partial \alpha}(\alpha P) - \frac{\partial}{\partial \alpha^*}(\alpha^* P)\right] \\ &= i\omega\left[\alpha\frac{\partial P}{\partial \alpha} - \alpha^*\frac{\partial P}{\partial \alpha^*}\right].\end{aligned} \tag{3.14.10}$$

By Appendix A, the associated characteristic equations are

$$\frac{dt}{1} = \frac{d\alpha}{-i\omega\alpha} = \frac{d\alpha^*}{+i\omega\alpha^*} \tag{3.14.11}$$

so that

$$\begin{aligned}\alpha &= \alpha_0 e^{-i\omega t} \\ \alpha^* &= \alpha_0^* e^{i\omega t},\end{aligned} \tag{3.14.12}$$

and the solution of (3.14.10) is

$$P(\alpha, \alpha^*, t) = g[\alpha e^{i\omega t}, \alpha^* e^{-i\omega t}], \tag{3.14.13}$$

where g is an arbitrary function.

3.15 GENERATING FUNCTION FOR OSCILLATOR EIGENFUNCTIONS

The coherent state is given by (2.5.12) as

$$\begin{aligned}|\alpha\rangle &= e^{-\frac{1}{2}|\alpha|^2} e^{\alpha a^\dagger}|0\rangle \\ &= e^{-\frac{1}{2}|\alpha|^2}\sum_{n=0}^{\infty}\frac{\alpha^n}{\sqrt{n!}}|n\rangle.\end{aligned} \tag{3.15.1}$$

If $|q'\rangle$ is a coordinate eigenvector, then

$$\langle q'|\alpha\rangle = e^{-\frac{1}{2}|\alpha|^2}\sum_{0}^{\infty}\frac{\alpha^n}{\sqrt{n!}}u_n(q'), \tag{3.15.2}$$

where

$$u_n(q') = \langle q'|n\rangle \tag{3.15.3}$$

are the oscillator energy eigenfunctions in the coordinate representation. If we use (2.5.38), we have

$$\left(\frac{\omega}{\pi\hbar}\right)^{1/4} e^{-(\omega/2\hbar)q'^2} e^{-\frac{1}{2}\alpha^2+\sqrt{(2\omega/\hbar)}\alpha q'} = \sum_0^\infty \frac{\alpha^n}{\sqrt{n!}} u_n(q'). \tag{3.15.4}$$

Therefore, by expanding the left side in powers of α, the coefficient of α^n is $u_n(q')/\sqrt{n!}$. However, a generating function for hermite polynomials is given by

$$e^{-z^2+2xz} = \sum_0^\infty H_n(x)\frac{z^n}{n!} \tag{3.15.5}$$

This may be shown as follows. We first differentiate both sides of (3.15.5) with respect to x twice and with respect to z once. This gives

$$e^{-z^2+2xz}2z = \sum_0^\infty \frac{dH_n}{dx}\frac{z^n}{n!} \tag{3.15.6}$$

$$e^{-z^2+2xz}4z^2 = \sum_0^\infty \frac{d^2H_n}{dx^2}\frac{z^n}{n!} \tag{3.15.7}$$

$$e^{-z^2+2xz}(2x-2z) = \sum_0^\infty H_n(x)\frac{nz^{n-1}}{n!}. \tag{3.15.8}$$

If we multiply both sides of (3.15.6) by $-2x$, (3.15.7) by 1, and (3.15.8) by $2z$, and add, we have

$$0 = \sum_{n=0}^\infty \left\{\frac{d^2H_n}{dx^2} - 2x\frac{dH_n}{dx} + 2nH_n\right\}\frac{z^n}{n!}. \tag{3.15.9}$$

But the hermite polynomials satisfy the equation

$$\frac{d^2H_n}{dx^2} - 2x\frac{dH_n}{dx} + 2nH_n = 0. \qquad \text{Q.E.D.} \tag{3.15.10}$$

If we make the identification

$$z = \frac{\alpha}{\sqrt{2}}$$
$$x = \sqrt{\frac{\omega}{\hbar}}\,q', \tag{3.15.11}$$

and use (3.15.5), (3.15.4) becomes

$$\left(\frac{\omega}{\pi\hbar}\right)^{1/4} e^{-(\omega/2\hbar)q'^2} \sum_0^\infty \frac{H_n(\sqrt{(\omega/\hbar)}\,q')}{(\sqrt{2})^n n!}\alpha^n = \sum_0^\infty \frac{\alpha^n}{\sqrt{n!}} u_n(q'). \tag{3.15.12}$$

Therefore, the normalized oscillator eigenfunctions are

$$u_n(q') = \left(\frac{\omega}{\pi\hbar}\right)^{1/4} \frac{e^{-(\omega/2\hbar)q'^2}}{\sqrt{2^{n/2}n!}} H_n\left(\sqrt{\frac{\omega}{\hbar}}\, q'\right). \tag{3.15.13}$$

Note from (3.15.2) that the minimum uncertainty wave packet state $\langle q'|\alpha\rangle$ is a generating function for the $\{u_n(q')\}$.

PART V. PRINCIPLE OF MAXIMUM ENTROPY

3.16 DEFINITION OF ENTROPY

The entropy of an ensemble of systems is defined in classical statistical mechanics by the relation

$$S = -k \sum_l p_l \ln p_l, \tag{3.16.1}$$

where k is Boltzmann's constant and p_l is the probability of finding the system in state l. The p_l's satisfy the conditions

$$\sum_l p_l = 1 \qquad 0 \leq p_l \leq 1 \qquad p_l = p_l^*. \tag{3.16.2}$$

Entropy may be visualized physically as a measure of the lack of knowledge of the system. If we know the system is in a definite state i, then $p_l = \delta_{li}$, and we see by (3.16.1) that the entropy is zero. In this case we have complete knowledge about the system; it is in a definite state.

On the other hand, if we know nothing about the system, it is equally likely to find the system in any of its possible states l, subject only to the constraint that $\sum p_l = 1$. We show that the entropy is a maximum under these conditions. We therefore maximize S, subject to the constraint $\sum p_l = 1$, by the method of Lagrange multipliers. If we vary the p_l's, the variation in S is

$$\delta S = -k \sum_l (1 + \ln p_l)\, \delta p_l = 0, \tag{3.16.3}$$

where we set $\delta S = 0$ to find its maximum. At the same time, the variation in the constraint is

$$\sum_l \delta p_l = 0. \tag{3.16.4}$$

To apply the method of Lagrange multipliers, we multiply (3.16.4) by an undetermined parameter λ and add to (3.16.3). Then we have

$$\sum (1 + \ln p_l + \lambda)\, \delta p_l = 0.$$

Each δp_l is now independent, and this equation will be satisfied if and only if each term is zero:

$$\ln p_l = -(1 + \lambda).$$

From this we conclude that each p_l must be a constant independent of l, the state of the system; that is, the probability of finding the system in any of its states is equally likely. We then have no information about the state of the system. Entropy is therefore a measure of the lack of information about the states of the elements of the ensemble, as we stated. This is the starting point of Shannon's theory of communication [16]. Jaynes [17, 18] has proposed that entropy be used as a fundamental postulate of statistical mechanics.

In quantum statistical mechanics [19] entropy is defined in terms of the density matrix as

$$S = -k \operatorname{Tr} \rho \ln \rho, \tag{3.16.5}$$

subject to the constraint

$$\operatorname{Tr} \rho = 1. \tag{3.16.6}$$

It should be noted that both ρ's appearing in (3.16.5) are evaluated at the same time.

For purposes of evaluating the entropy, we must evaluate the trace in some representation. Let the set $\{|n\rangle\}$ be some complete orthonormal set. Then (3.16.5) means

$$S = -k \sum_{n,m} \langle n|\rho|m\rangle \langle m| \ln \rho |n\rangle. \tag{3.16.7}$$

If we transform from the representation $\{|n\rangle\}$ to a representation in which ρ is diagonal by means of a similarity transformation, then (3.16.7) reduces to

$$S = -k \sum_{\alpha} p_\alpha \ln p_\alpha, \tag{3.16.8}$$

where $p_\alpha = \langle \alpha|\rho|\alpha\rangle$, and $\{|\alpha\rangle\}$ is the representation in which ρ is diagonal and p_α are the diagonal matrix elements.

The question now arises as to how to find the density matrix already discussed in detail. We know that the entropy (3.16.5) is a measure of lack of knowledge about the states of the elements of the ensemble, just as in the classical case. For if we maximize S subject to (3.16.6), just as in the classical case we find

$$\Sigma(1 + \ln \rho + \lambda)\, \delta\rho = 0, \tag{3.16.9}$$

or $\rho =$ const. This tells us that the entropy is a maximum when the probability of finding the system in any of its possible states is the same. On the other hand, if we know the system is in a pure state $|\psi\rangle$, then $S = 0$ since $p_{\psi'} = \delta_{\psi\psi'}$.

We now suppose that we know something about the system, for example, its average energy. The average energy is

$$\langle E \rangle = \text{Tr}\ \rho H, \tag{3.16.10}$$

where H is the hamiltonian. This knowledge about the system must be reflected in the choice of a density operator to describe the ensemble. We may regard it as an added constraint and choose ρ so that it maximizes the entropy subject to the constraints (3.16.6) and (3.16.10). Then, when we vary ρ,

$$\begin{aligned} \text{Tr}\ (1 + \ln \rho)\ \delta\rho &= 0 \\ \text{Tr}\ \delta\rho &= 0 \\ \text{Tr}\ H\ \delta\rho &= 0. \end{aligned}$$

If we multiply the second of these by the undetermined multiplier λ and the third by β and add to the first, we have

$$\text{Tr}\ (1 + \lambda + \ln \rho + \beta H)\ \delta\rho = 0.$$

Since $\delta\rho$ is arbitrary and all variations are now independent, this will be satisfied if and only if

$$\ln \rho = -1 - \lambda - \beta H$$

or

$$\rho = e^{-(1+\lambda)} e^{-\beta H}. \tag{3.16.11}$$

We may determine λ as follows: we take the trace of both sides of (3.16.11), use (3.16.6), and find that

$$e^{1+\lambda} = \text{Tr}\ e^{-\beta H} \equiv Z,$$

so that (3.16.11) becomes

$$\rho = \frac{e^{-\beta H}}{\text{Tr}\ e^{-\beta H}} \equiv \frac{e^{-\beta H}}{Z}, \tag{3.16.12}$$

where Z is called the partition function. To determine β, we use the constraint (3.16.10). We have

$$\langle E \rangle = \frac{\text{Tr}\ H e^{-\beta H}}{\text{Tr}\ e^{-\beta H}} \equiv -\frac{\partial}{\partial \beta} \ln Z. \tag{3.16.13}$$

From this, we can, in principle, solve for β in terms of the average energy of the system.

As a particular example, we consider a cavity filled with electromagnetic radiation* in thermal equilibrium with the walls at temperature T. The

* In the following chapter we show that the radiation field in a cavity is equivalent to a set of fictitious harmonic oscillators. Each normal mode of frequency ω_l is associated with a harmonic oscillator with hamiltonian $\hbar\omega_l a_l^\dagger a_l$.

average energy contained in one mode of the cavity at frequency ω may be written as $\bar{n}\hbar\omega$, where $\bar{n}$ is the average number of quanta. Since the hamiltonian for this mode is

$$H = \hbar\omega a^\dagger a$$

we have, by (3.16.13),

$$\bar{n}\hbar\omega = \langle E\rangle = \frac{\hbar\omega \sum_{n=0}^{\infty} \langle n|a^\dagger a \exp[-xa^\dagger a]|n\rangle}{\sum_{m=0}^{\infty} \langle m|\exp[-xa^\dagger a]|m\rangle} \tag{3.16.14}$$

where

$$x = \exp(\beta\hbar\omega) \tag{3.16.15}$$

and $a^\dagger a|n\rangle = n|n\rangle$. We have written the traces in (3.16.13) in the $\{|n\rangle\}$ representation.

The sums in (3.16.14) may be carried out easily. We obtain for the average number of quanta in the mode

$$\bar{n} = \frac{\sum_0^\infty nx^{-n}}{\sum_0^\infty x^{-m}} = \frac{1}{x-1} \equiv \frac{1}{\exp(\hbar\omega\beta) - 1} = \frac{\langle E\rangle}{\hbar\omega}. \tag{3.16.16}$$

From this we may solve for β in terms of $\bar{n}$. However, from the correspondence principle, as $\hbar \to 0$ the energy E must become the average classical energy contained in a cavity mode; that is, $E \to kT$ as $\hbar \to 0$ since from the classical equipartition-of-energy theorem, we get $\frac{1}{2}kT$ per degree of freedom. The electric and magnetic fields each correspond to one degree of freedom. Therefore, in the limit, as $\hbar \to 0$, we have by (3.16.16)

$$\langle E\rangle \to kT \underset{\hbar\to 0}{\to} \frac{\hbar\omega}{\hbar\omega\beta} = \frac{1}{\beta}, \tag{3.16.17}$$

so that $\beta = 1/kT$.

The density operator (3.16.12) that maximizes the entropy subject to the constraints that $\operatorname{Tr}\rho = 1$ and average energy $\langle E\rangle$ is therefore

$$\rho = \frac{\exp(-H/kT)}{\operatorname{Tr}\exp(-H/kT)}. \tag{3.16.18}$$

This density operator can be considered as describing an ensemble of harmonic oscillators in thermodynamic equilibrium with a heat bath at temperature T.

In the energy representation, where $H|E_n\rangle = E_n|E_n\rangle$, the matrix elements of ρ, from (3.16.18), are

$$\langle E_n|\rho|E_m\rangle = \frac{e^{-\beta E_m}\langle E_n|E_m\rangle}{\sum\limits_{E_m} e^{-\beta E_m}}. \tag{3.16.19}$$

The probability of finding the system in state E_m (or of finding one element of the ensemble in E_m) is therefore

$$p_m = \frac{e^{-\beta E_m}}{\sum\limits_{E_m} e^{-\beta E_m}}. \tag{3.16.20}$$

(This is the diagonal matrix element of ρ in the energy representation.) This corresponds to a Maxwell-Boltzmann probability distribution. In this representation, we may therefore write the density operator as

$$\rho = \sum_{E_m} |E_m\rangle p_m \langle E_m|, \tag{3.16.21}$$

where p_m is given by (3.16.20).

We also observe that

$$\text{Tr}\,\rho^2 = \sum_{E_m} p_m^2 \neq 1, \tag{3.16.22}$$

so that the ensemble represents a mixed state; that is, knowledge of the average energy of the system is not sufficient to determine the state of the system completely.

The maximum entropy for this ensemble is

$$S_{\max} = -k\,\text{Tr}\,\rho \ln \rho = -k\,\text{Tr}\left(\frac{e^{-\beta H}}{Z}\right)\ln\frac{e^{-\beta H}}{Z} = \frac{\langle E\rangle}{T} + k \ln Z, \tag{3.16.23}$$

where

$$Z = \text{Tr}\, e^{-\beta H} = \sum e^{-\beta E_m} \tag{3.16.24}$$

is the partition function and

$$\langle E\rangle = \text{Tr}\,\rho H = \sum \frac{E_m e^{-\beta E_m}}{Z} = -\frac{\partial}{\partial\beta}\ln Z. \tag{3.16.25}$$

The partition function determines the thermodynamic properties of the system.

We repeat, for emphasis, that when the average energy is known the ensemble may be chosen to be made up of a large number of elements, each of which is in a state, say $|E_m\rangle$, and weighted with probability p_m given by (3.16.20).

If additional measurements are made on other variables in the problem, we add their averages as additional constraints and again maximize the entropy. In this way we obtain the best estimate of the density operator possible, subject to our knowledge of the system.

3.17 DENSITY OPERATOR FOR SPIN-½ PARTICLES [20]

We use the principle of entropy maximization presented in the previous section to obtain a density operator for a beam (ensemble) of spin-½ particles. Let

$$\mathbf{s} = \text{Tr}\ \rho\boldsymbol{\sigma}, \tag{3.17.1}$$

where σ_x, σ_y, and σ_z be the Pauli spin matrices,

$$\sigma_x = \begin{pmatrix} 0 & 1 \\ 1 & 0 \end{pmatrix}; \quad \sigma_y = \begin{pmatrix} 0 & -i \\ i & 0 \end{pmatrix}; \quad \sigma_z = \begin{pmatrix} 1 & 0 \\ 0 & -1 \end{pmatrix}, \tag{3.17.2}$$

in the σ_z representation. That is, s represents the ensemble average of $\boldsymbol{\sigma}$. We would like to obtain ρ by maximizing the entropy subject to the constraints Tr $\rho = 1$ and the knowledge of s. This gives

$$\text{Tr}\ \delta\rho[1 + \ln \rho + \lambda + \boldsymbol{\alpha} \cdot \boldsymbol{\sigma}] = 0, \tag{3.17.3}$$

where α_x, α_y, α_z are Lagrange multipliers. Therefore, we have that

$$\rho = \frac{\exp -(\boldsymbol{\alpha} \cdot \boldsymbol{\sigma})}{\text{Tr} \exp(-\boldsymbol{\alpha} \cdot \boldsymbol{\sigma})}; \tag{3.17.4}$$

we must determine the α_x, α_y, and α_z in terms of the measured ensemble averages s_x, s_y, and s_z.

We leave it as an exercise for the reader to show that

$$e^{-\boldsymbol{\alpha}\cdot\boldsymbol{\sigma}} = I \cosh |\boldsymbol{\alpha}| - \frac{\sinh |\boldsymbol{\alpha}|}{|\boldsymbol{\alpha}|} \boldsymbol{\alpha} \cdot \boldsymbol{\sigma}, \tag{3.17.5}$$

since $(\boldsymbol{\alpha} \cdot \boldsymbol{\sigma})^2 = \boldsymbol{\alpha}^2 I$ where I is the identity matrix

$$I = \begin{pmatrix} 1 & 0 \\ 0 & 1 \end{pmatrix} \tag{3.17.6}$$

and

$$|\boldsymbol{\alpha}| = \sqrt{\alpha_x^{\ 2} + \alpha_y^{\ 2} + \alpha_z^{\ 2}}. \tag{3.17.7}$$

Since Tr $\boldsymbol{\sigma} = 0$ and Tr $I = 2$, we see that

$$\text{Tr}\ e^{-\boldsymbol{\alpha}\cdot\boldsymbol{\sigma}} = 2 \cosh |\boldsymbol{\alpha}|, \tag{3.17.8}$$

so that we may write our optimum density operator as

$$\rho = \frac{1}{2}\left[I - \frac{\tanh|\boldsymbol{\alpha}|}{|\boldsymbol{\alpha}|}\,\boldsymbol{\alpha}\cdot\boldsymbol{\sigma}\right]. \tag{3.17.9}$$

We next have when we use this in (3.17.1) that

$$\mathbf{s} = \mathrm{Tr}\,\frac{1}{2}\left[\boldsymbol{\sigma} - \frac{\tanh|\boldsymbol{\alpha}|}{|\boldsymbol{\alpha}|}(\boldsymbol{\alpha}\cdot\boldsymbol{\sigma})\boldsymbol{\sigma}\right]$$

$$= -\frac{1}{2}\frac{\tanh|\boldsymbol{\alpha}|}{|\boldsymbol{\alpha}|}\,\mathrm{Tr}\,\boldsymbol{\sigma}(\boldsymbol{\alpha}\cdot\boldsymbol{\sigma}). \tag{3.17.10}$$

Since $\mathrm{Tr}\,\sigma_i\sigma_j = 0$ if $i \neq j$ and $\mathrm{Tr}\,\sigma_i^2 = 2$, (3.17.10) reduces to

$$\mathbf{s} = -\frac{\tanh|\boldsymbol{\alpha}|}{|\boldsymbol{\alpha}|}\,\boldsymbol{\alpha}. \tag{3.17.11}$$

Then (3.17.9) becomes

$$\rho = \tfrac{1}{2}(I + \mathbf{s}\cdot\boldsymbol{\sigma}). \tag{3.17.12}$$

This density operator maximizes the entropy subject to $\mathrm{Tr}\,\rho = 1$ and measurements of $\boldsymbol{\sigma}$ on the ensemble. In the σ_z-representation it becomes

$$\rho = \frac{1}{2}\begin{bmatrix} 1 + s_z & s_x - is_y \\ s_x + is_y & 1 - s_z \end{bmatrix} = \rho^\dagger. \tag{3.17.13}$$

In the representation in which ρ is diagonal, we have

$$\rho' = \frac{1}{2}\begin{bmatrix} 1 + |\mathbf{s}| & 0 \\ 0 & 1 - |\mathbf{s}| \end{bmatrix}, \tag{3.17.14}$$

where $|\mathbf{s}| = \sqrt{s_x^2 + s_y^2 + s_z^2}$. (*Prove this.*) Then we easily see that

$$\mathrm{Tr}\,\rho^2 = \mathrm{Tr}\,\rho'^2 = \tfrac{1}{2}(1 + \mathbf{s}^2) \leq 1, \tag{3.17.15}$$

where the inequality follows from the general theory.

If a beam is unpolarized, then $s_x = s_y = s_z = 0$ since it is equally likely that the spins are pointing in any direction. Thus

$$\rho = \tfrac{1}{2}I, \tag{3.17.16}$$

for an unpolarized beam. In case the beam is completely polarized, then it is in a pure state so that by (3.17.15) the equality holds and

$$|\mathbf{s}| = \pm 1, \tag{3.17.17}$$

so that

$$\rho' = \begin{bmatrix} 1 & 0 \\ 0 & 0 \end{bmatrix} \quad \text{or} \quad \begin{bmatrix} 0 & 0 \\ 0 & 1 \end{bmatrix}. \tag{3.17.18}$$

When the beam is in a mixed state (partially polarized) $|\mathbf{s}| < 1$ and we may define the *degree* of polarization, P, in the direction $\mathbf{s}$ by

$$P = |\mathbf{s}|; \tag{3.17.19}$$

$P = 0$ is unpolarized and $P = 1$ is completely polarized.

Let us assume at $t = 0$, we have measured $\mathbf{s}$ so that $\rho(0)$ is given by (3.17.13). If we pass our beam into a region in which there is a magnetic field, $\mathbf{H}(t)$, the hamiltonian is

$$H = \frac{\gamma\hbar}{2}\,\boldsymbol{\sigma}\cdot\mathbf{H}(t). \tag{3.17.20}$$

The density operator in the SP satisfies

$$i\hbar\frac{\partial\rho}{\partial t} = \frac{\gamma\hbar}{2}\,[\boldsymbol{\sigma}\cdot\mathbf{H}(t), \rho]. \tag{3.17.21}$$

We would like to find $\rho(t)$.

Since any 2×2 matrix can be expanded in terms of $\boldsymbol{\sigma}$ and I, we write

$$\rho(t) = \tfrac{1}{2}[s_0(t)I + \mathbf{s}(t)\cdot\boldsymbol{\sigma}], \tag{3.17.22}$$

where by (3.17.12) $\mathbf{s}(0) = \mathbf{s}$ and $s_0(0) = 1$. If we use this in (3.17.21), we obtain

$$i\hbar\left[\frac{ds_0(t)}{dt}I + \frac{d\mathbf{s}(t)}{dt}\cdot\boldsymbol{\sigma}\right] = \frac{\gamma\hbar}{2}\,[\boldsymbol{\sigma}\cdot\mathbf{H}(t), \boldsymbol{\sigma}\cdot\mathbf{s}(t)], \tag{3.17.23}$$

since I commutes with $\boldsymbol{\sigma}\cdot\mathbf{H}$. We may rewrite the commutator as

$$\begin{aligned}[\boldsymbol{\sigma}\cdot\mathbf{H}, \boldsymbol{\sigma}\cdot\mathbf{s}] &= (\boldsymbol{\sigma}\cdot\mathbf{H})(\boldsymbol{\sigma}\cdot\mathbf{s}) - (\boldsymbol{\sigma}\cdot\mathbf{s})(\boldsymbol{\sigma}\cdot\mathbf{H}) \\ &= \sum_{i=1}^{3}\sum_{j=1}^{3} H_i s_j(\sigma_i\sigma_j - \sigma_j\sigma_i).\end{aligned} \tag{3.17.24}$$

However,

$$[\sigma_i, \sigma_j] = 2i\sigma_k, \tag{3.17.25}$$

where i, j, k form an even permutation of 1, 2, 3. Thus

$$[\boldsymbol{\sigma}\cdot\mathbf{H}, \boldsymbol{\sigma}\cdot\mathbf{s}] = 2i\sum_{[i,j,k]} H_i s_j \sigma_k = 2i\mathbf{H}\times\mathbf{s}\cdot\boldsymbol{\sigma}, \tag{3.17.26}$$

since in (3.17.24) the $i = j$ terms vanish. Therefore, (3.17.23) becomes

$$\frac{ds_0}{dt}I + \frac{d\mathbf{s}}{dt}\cdot\boldsymbol{\sigma} = \gamma\mathbf{H}\times\mathbf{s}\cdot\boldsymbol{\sigma}. \tag{3.17.27}$$

If we take the trace of both sides since $\mathrm{Tr}\, I = 2$, $\mathrm{Tr}\, \sigma_i = 0$ we see that

$$\frac{ds_0(t)}{dt} = 0, \tag{3.17.28}$$

so from the initial conditions we see that

$$s_0(t) = 1 \tag{3.17.29}$$

for all time. If we next multiply both sides of (3.17.27) from the left by σ_i, trace and note that $\mathrm{Tr}\, \sigma_i\sigma_j = 2\delta_{ij}$, we obtain the three equations for $\mathbf{s}(t)$

$$\frac{d\mathbf{s}(t)}{dt} = +\gamma\mathbf{H}(t) \times \mathbf{s}(t). \tag{3.17.30}$$

To proceed further we must specify $\mathbf{H}(t)$. An interesting case to consider is

$$\mathbf{H}(t) = [h_1 \cos \omega t,\ h_1 \sin \omega t,\ H_0]. \tag{3.17.31}$$

In this case we have

$$\frac{ds_x}{dt} = -\omega_0 s_y + \gamma h_1 \sin \omega t\ s_z \tag{3.17.32}$$

$$\frac{ds_y}{dt} = +\omega_0\, s_x - \gamma h_1 \cos \omega t\ s_z \tag{3.17.33}$$

$$\frac{ds_z}{dt} = \gamma h_1(\cos \omega t\ s_y - \sin \omega t\ s_x), \tag{3.17.34}$$

where $\omega_0 \equiv \gamma H_0$.

If we let

$$s_\pm = s_x \pm i s_y \tag{3.17.35}$$

we may rewrite these as

$$\frac{ds_+}{dt} = i\omega_0 s_+ - i\gamma h_1 e^{i\omega t} s_z \tag{3.17.36}$$

$$\frac{ds_-}{dt} = -i\omega_0 s_- + i\gamma h_1 e^{-i\omega t} s_z \tag{3.17.37}$$

$$\frac{ds_z}{dt} = \tfrac{1}{2} i\gamma h_1[s_- e^{i\omega t} - s_+ e^{-i\omega t}]. \tag{3.17.38}$$

If we let

$$s_\pm(t) = e^{\pm i\omega t} S_\pm(t), \tag{3.17.39}$$

these reduce to equations with constant coefficients:

$$\frac{dS_+}{dt} + i\,\Delta\omega S_+ = -i\gamma h_1 s_z \tag{3.17.40}$$

$$\frac{dS_-}{dt} - i\,\Delta\omega S_- = i\gamma h_1 s_z \tag{3.17.41}$$

$$\frac{ds_z}{dt} = \tfrac{1}{2} i\gamma h_1 [S_- - S_+]. \tag{3.17.42}$$

where $\Delta\omega = \omega - \omega_0$. If we add the first two of these we have that

$$\frac{d}{dt}(S_+ + S_-) = i\,\Delta\omega(S_- - S_+)$$

$$= \frac{2\,\Delta\omega}{\gamma h_1}\frac{ds_z}{dt}, \tag{3.17.43}$$

where we used (3.17.42). From this it follows that one integral of the equations of motion is

$$S_+(t) + S_-(t) - \frac{2\,\Delta\omega}{\gamma h_1} s_z(t) = 2\left[s_x(0) - \frac{\Delta\omega}{\gamma h_1} s_z(0)\right] \equiv c_1. \tag{3.17.44}$$

We may use this to eliminate s_z in (3.17.40) and (3.17.41):

$$\left(\frac{d}{dt} + i\,\Delta\omega\right) S_+ = -\frac{i(\gamma h_1)^2}{2\,\Delta\omega}(S_+ + S_- - c_1) \tag{3.17.45}$$

$$\left(\frac{d}{dt} - i\,\Delta\omega\right) S_- = \frac{i(\gamma h_1)^2}{2\,\Delta\omega}(S_+ + S_- - c_1). \tag{3.17.46}$$

From these we easily see that $S_\pm$ satisfy

$$\left[\frac{d^2}{dt^2} + \Omega^2\right] S_\pm = \tfrac{1}{2}(\gamma h_1)^2 c_1, \tag{3.17.47}$$

where

$$\Omega^2 = (\Delta\omega)^2 + (\gamma h_1)^2, \tag{3.17.48}$$

so that the solutions may be written as

$$S_+(t) = \left[s_+(0) - \frac{1}{2}\left(\frac{\gamma h_1}{\Omega}\right)^2 c_1\right]\cos\Omega t + A\sin\Omega t + \tfrac{1}{2}c_1\left(\frac{\gamma h_1}{\Omega}\right)^2 \tag{3.17.49}$$

$$S_-(t) = \left[s_-(0) - \frac{1}{2}\left(\frac{\gamma h_1}{\Omega}\right)^2 c_1\right]\cos\Omega t + A^*\sin\Omega t + \tfrac{1}{2}c_1\left(\frac{\gamma h_1}{\Omega}\right)^2, \tag{3.17.50}$$

and A is a constant of integration yet to be determined. The amplitudes of the $\cos \Omega t$ term were chosen so that $S_\pm(0) = s_\pm(0)$. If we use (3.17.44) and the above we see that

$$s_z(t) = \frac{\gamma h_1}{2\,\Delta\omega}\left\{\left[2s_x(0) - c_1\left(\frac{\gamma h_1}{\Omega}\right)^2\right] \cos \Omega t + (A + A^*) \sin \Omega t - c_1\left(\frac{\Delta\omega}{\Omega}\right)^2\right\}. \tag{3.17.51}$$

From (3.17.34) it follows that

$$\frac{ds_z(0)}{dt} = \gamma h_1 s_y(0), \tag{3.17.52}$$

so if we use (3.17.51) we see that

$$A + A^* = \frac{2\,\Delta\omega}{\Omega}\, s_y(0). \tag{3.17.53}$$

When we use this and (3.17.44) we have that

$$s_z(t) = \frac{\gamma h_1}{\Omega^2}\left\{[\Delta\omega s_x(0) + \gamma h_1 s_z(0)] \cos \Omega t + \Omega\, s_y(0) \sin \Omega t + \frac{(\Delta\omega)^2}{\gamma h_1}\, s_z(0) - \Delta\omega\, s_x(0)\right\}. \tag{3.17.54}$$

From (3.17.40) we see that

$$\frac{dS_+(0)}{dt} = -i[\gamma h_1 s_z(0) + \Delta\omega\, s_+(0)]. \tag{3.17.55}$$

If we use this and (3.17.49), we see that

$$A = -i\Omega^{-1}[\gamma h_1 s_z(0) + \Delta\omega s_+(0)]. \tag{3.17.56}$$

After minor algebra we therefore obtain

$$\begin{aligned} s_x(t) = {} & \Omega^{-2} \cos \omega t\{\Delta\omega[\gamma h_1 s_z(0) + \Delta\omega s_x(0)] \cos \Omega t \\ & + \Delta\omega\Omega\, s_y(0) \sin \Omega t - \Delta\omega\gamma h_1 s_z(0) + (\gamma h_1)^2 s_x(0)\} \\ & + \frac{\sin \omega t}{\Omega}\{[\gamma h_1 s_z(0) + \Delta\omega s_x(0)] \sin \Omega t - \Omega\, s_y(0) \cos \Omega t\} \end{aligned} \tag{3.17.57}$$

$$\begin{aligned} s_y(t) = {} & -\frac{\cos \omega t}{\Omega}\{[\gamma h_1 s_z(0) + \Delta\omega s_x(0)] \sin \Omega t - \Omega\, s_y(0) \cos \Omega t\} \\ & + \frac{\sin \omega t}{\Omega^2}\{\Delta\omega[\gamma h_1 s_z(0) + \Delta\omega\, s_x(0)] \cos \Omega t \\ & + \Delta\omega\Omega\, s_y(0) \sin \Omega t - \gamma h_1\, \Delta\omega\, s_z(0) + (\gamma h_1)^2 s_x(0)\}. \end{aligned} \tag{3.17.58}$$

The magnetic moment is

$$\boldsymbol{\mu} = -\tfrac{1}{2}\gamma\hbar\boldsymbol{\sigma}. \tag{3.17.59}$$

Its expectation value in the SP is

$$\langle\boldsymbol{\mu}\rangle = -\tfrac{1}{2}\gamma\hbar \operatorname{Tr} \rho(t)\boldsymbol{\sigma} = -\tfrac{1}{4}\gamma\hbar \operatorname{Tr}[I + \mathbf{s}(t) \cdot \boldsymbol{\sigma}]\boldsymbol{\sigma}. \tag{3.17.60}$$

Since $\operatorname{Tr} \sigma_i = 0$ and $\operatorname{Tr} \sigma_i\sigma_j = 2\delta_{ij}$, we see that

$$\langle\boldsymbol{\mu}(t)\rangle = -\tfrac{1}{2}\gamma\hbar\mathbf{s}(t). \tag{3.17.61}$$

Therefore, we may give a direct physical interpretation to the "components" of the density operator, $\mathbf{s}(t)$. If we restrict ourselves to resonance, $\Delta\omega = 0$ and by (3.17.54), (3.17.57), and (3.17.58), we see that

$$\langle\mu_x(t)\rangle = -\tfrac{1}{2}\gamma\hbar\{s_x(0) \cos \omega_0 t + \sin \omega_0 t[s_z(0) \sin \gamma h_1 t - s_y(0) \cos \gamma h_1 t]\} \tag{3.17.62}$$

$$\langle\mu_y(t)\rangle = -\tfrac{1}{2}\gamma\hbar\{-\cos \omega_0 t[s_z(0) \sin \gamma h_1 t - s_y(0) \cos \gamma h_1 t] + \sin \omega_0 t s_x(0)\} \tag{3.17.63}$$

$$\langle\mu_z(t)\rangle = -\tfrac{1}{2}\gamma\hbar\{s_z(0) \cos \gamma h_1 t + s_y(0) \sin \gamma h_1 t\}. \tag{3.17.64}$$

If the beam is polarized along the z axis, $s_x(0) = s_y(0) = 0$ and $s_z(0) = 1$. In this case

$$\begin{aligned}\langle\mu_x(t)\rangle &= -\tfrac{1}{2}\gamma\hbar \sin \gamma h_1 t \sin \omega_0 t\\ \langle\mu_y(t)\rangle &= +\tfrac{1}{2}\gamma\hbar \sin \gamma h_1 t \cos \omega_0 t\\ \langle\mu_z(t)\rangle &= -\tfrac{1}{2}\gamma\hbar \cos \gamma h_1 t.\end{aligned} \tag{3.17.65}$$

This represents a precession of μ_x and μ_y about the z axis and a "nutation" of $\langle\mu_z\rangle$ between "up" and "down."

By virtue of (3.17.61) and (3.17.30), we see that $\langle\boldsymbol{\mu}(t)\rangle$ obeys the equation of motion

$$\frac{d}{dt}\langle\boldsymbol{\mu}\rangle = -\gamma\langle\boldsymbol{\mu}\rangle \times \mathbf{H}(t), \tag{3.17.66}$$

which is the equation of motion of a classical dipole in a magnetic field and represents an example of Ehrenfest's theorem.

We may also obtain the Heisenberg equation of motion for $\boldsymbol{\sigma}$:

$$i\hbar\frac{d\boldsymbol{\sigma}}{dt} = [\boldsymbol{\sigma}, H] = \tfrac{1}{2}\gamma\hbar[\boldsymbol{\sigma}, \boldsymbol{\sigma} \cdot \mathbf{H}(t)]. \tag{3.17.67}$$

We may easily show that this reduces to

$$\frac{d\boldsymbol{\sigma}}{dt} = -\gamma\boldsymbol{\sigma} \times \mathbf{H}(t). \tag{3.17.68}$$

PROBLEMS

3.1 If A and B are two noncommuting operators, show that

$$e^{A}e^{B}e^{-A} = \exp\,[e^{A}Be^{-A}].$$

3.2 If $A = xy + \partial^2/\partial x\,\partial y$ and ξ is a real variable that commutes with x and y, show that

$$e^{i\xi A}xe^{-i\xi A} = x\cosh\xi + i\sinh\xi\,\frac{\partial}{\partial y}$$

$$e^{i\xi A}ye^{-i\xi A} = y\cosh\xi + i\sinh\xi\,\frac{\partial}{\partial x}\,.$$

3.3 If ξ and η are parameters independent of y, show that

$$\exp\left[\xi\left(\frac{\partial}{\partial y} - \eta y\right)\right] = e^{\frac{1}{2}\eta y^2}e^{\xi(\partial/\partial y)}e^{-\frac{1}{2}\eta y^2} = e^{-\frac{1}{2}\xi^2\eta}e^{-\xi\eta y}e^{\xi(\partial/\partial y)}.$$

3.4 Show that

$$[a, e^{-xa^\dagger a}] = (e^{x} - 1)e^{-xa^\dagger a}a = \frac{\partial}{\partial a^\dagger}\,e^{-xa^\dagger a}$$

$$[a^\dagger, e^{-xa^\dagger a}] = (e^{x} - 1)e^{-a^\dagger a}a^\dagger = -\frac{\partial}{\partial a}\,e^{-xa^\dagger a}.$$

3.5 Show that if m is an integer

$$[a^\dagger a, a^{\dagger m}] = ma^{\dagger m}$$

$$[a^\dagger a, a^{m}] = -ma^{m}.$$

3.6 If $|0\rangle$ is the boson vacuum state, show that

$$e^{xa}f(a^\dagger)|0\rangle = f(a^\dagger + x)|0\rangle,$$

where x is a parameter and $f(a^\dagger)$ is any function of $a^\dagger$ that may be expanded in a Taylor series.

3.7 Show that

$$e^{-xa^\dagger a}f(a^\dagger)|0\rangle = f(a^\dagger e^{-x})|0\rangle.$$

3.8 If x is a parameter and m is an integer, verify that

$$e^{xa^\dagger}a^{m} = (a - x)^{m}e^{xa^\dagger} = \mathscr{A}\{(\alpha - x)^{m}e^{x\alpha^*}\} = \mathscr{N}\{e^{x\alpha^*}\alpha^{m}\}$$

$$e^{xa}a^{\dagger m} = (a^\dagger + x)^{m}e^{xa} = \mathscr{A}\{e^{x\alpha}\alpha^{*m}\} = \mathscr{N}\{(\alpha^* + x)^{m}e^{x\alpha}\}.$$

3.9 If ξ and η are parameters and $(a, a^\dagger)$ and $(b, b^\dagger)$ are two independent sets of boson operators, find the normal form of

$$f = e^{\xi ab}e^{\eta a^\dagger b^\dagger} = \mathscr{A}\{e^{\xi\alpha\beta+\eta\alpha^*\beta^*}\}.$$

3.10 Express $e^{-xa^\dagger a}$ in (*a*) its diagonal representation, (*b*) in the coherent state representation, and (*c*) in the *R*-representation.

3.11 By means of Theorem 6, Section 3.2, show that

$$\mathrm{Tr}\, e^{i\xi^* a^\dagger} e^{i\xi a} = \pi\, \delta[\mathrm{Re}\, \xi]\, \delta[\mathrm{Im}\, \xi],$$

where Re and Im mean real and imaginary parts, respectively.

3.12 Find the Fourier transform of the antinormal associated function for $e^{-\lambda a^\dagger a}$ where λ is real. Use the result to evaluate $\mathrm{Tr}\, e^{-\lambda a^\dagger a}$.

3.13 Evaluate the thermal average

$$\langle a^{\dagger l} a^m \rangle_0 \equiv (1 - e^{-\lambda})\, \mathrm{Tr}\, a^{\dagger l} a^m e^{-\lambda a^\dagger a}$$

(*a*) in the number representation, (*b*) by Theorem 7, Section 3.2, and (*c*) by Theorem 9, Section 3.3.

3.14 If an ensemble of harmonic oscillators is in a coherent state $|\alpha'\rangle\langle\alpha'|$, obtain the three characteristic functions $C^{(a)}$, $C^{(n)}$, and $C^{(w)}$ as well as the three associated functions $\bar{\rho}^{(a)}$, $\bar{\rho}^{(n)}$, and P of Section 3.4.

3.15 If the oscillators of Problem 3.14 are described by $\rho = \overline{|\alpha'\rangle\langle\alpha'|}$ where the bar indicates we are averaging over random phases, find $\bar{\rho}^{(a)}$ and P.

3.16 If $\rho = (1 - e^{-\lambda})e^{-\lambda a^\dagger a}$, find $C^{(a)}$, $C^{(n)}$, $C^{(w)}$ as well as P.

3.17 Generalize Theorem 9 of Section 3.3 for the case

$$\langle f[a_1, a_1^\dagger; a_2, a_2^\dagger; \ldots] \rangle_0 = \mathrm{Tr} \left\{ f[a_1, a_1^\dagger; a_2, a_2^\dagger; \ldots] \prod_j (1 - e^{-\lambda_j}) e^{-\lambda_j a_j^\dagger a_j} \right\}$$

where the $(a_i, a_i^\dagger)$ are independent boson operators such that

$$[a_i, a_j] = [a_i^\dagger, a_j^\dagger] = 0$$
$$[a_i, a_j^\dagger] = \delta_{ij}$$

3.18 Show that the generalized Wick theorem of Section 3.7 for bosons is unmodified in form if the operators ψ_i are in the interaction picture

$$\psi_i(t_i) = e^{iH_0 t_i/\hbar} \psi_i e^{-iH_0 t_i/\hbar} = \psi_i e^{\pm i\omega_i t_i}$$

provided we let

$$\langle \psi_j(t_j)\, \psi_k(t_k) \rangle_0 = \frac{[\psi_j(t_j), \psi_k(t_k)]}{1 - e^{\pm\lambda_j}}.$$

3.19 Show that the generalized Wick theorem for time-ordered boson operators becomes

$$\begin{aligned}\langle P\{\bar{\psi}_1(t_1) \cdots \bar{\psi}_{2n}(t_{2n})\}\rangle_0 &= \langle P\{\bar{\psi}_1(t_1)\bar{\psi}_2(t_2)\}\rangle_0 \langle P\{\bar{\psi}_3(t_3) \cdots \bar{\psi}_{2n}(t_{2n})\}\rangle_0 \\ &+ \langle P\{\bar{\psi}_1(t_1)\, \bar{\psi}_3(t_3)\}\rangle_0 \langle P\{\bar{\psi}_2(t_2)\bar{\psi}_4(t_4) \cdots \bar{\psi}_{2n}(t_{2n})\}\rangle_0 \\ &+ \cdots + \langle P\{\bar{\psi}_1(t_1)\, \bar{\psi}_{2n}(t_{2n})\}\rangle_0 \langle P\{\bar{\psi}_2(t_2)\, \bar{\psi}_3(t_3) \cdots \bar{\psi}_{2n-1}(t_{2n-1})\}\rangle_0.\end{aligned}$$

3.20 For a two level atom, show that the operators $N_i = |i\rangle\langle i|$ $(i = 1, 2)$, $M^\dagger = |2\rangle\langle 1|$ and $M = |1\rangle\langle 2|$ satisfy the following commutation relations

$$[e^{i\xi M}, N_1] = -i\xi M e^{i\xi M}$$

$$[e^{i\xi^* M^\dagger}, N_1] = i\xi^* M^\dagger e^{i\xi^* M^\dagger}$$

$$[e^{i\xi^* M^\dagger}, N_2] = -i\xi^* M^\dagger e^{i\xi^* M^\dagger}$$

$$[e^{i\xi M}, N_2] = i\xi M e^{i\xi M}$$

$$[N_1, N_2] = 0.$$

3.21 Show that $U(t)$ in (3.11.13) is unitary.

3.22 Prove Theorems 4–7. Section 3.10.

REFERENCES

[1] The author is deeply indebted to Dr. L. R. Walker of Bell Telephone Laboratories for teaching him many of the operator techniques given in this chapter. See R. Kubo, *J. Phys. Soc. Japan*, **7**, 1100 (1962); F. Coester and H. Kümmel, *Nucl. Phys.*, **17**, 477 (1960).

[2] G. Weiss and A. Maradudin, *J. Math. Phys.*, **3**, 771 (1962).

[3] W. Magnus, *Commun. Pure Appl. Math.*, **7**, 649 (1954).

[4] E. Wichman, *J. Math. Phys.*, **2**, 876 (1961).

[5] D. Finkelstein, *Commun. Pure Appl. Math.*, **8**, 245 (1955).

[6] A. Messiah, *Quantum Mechanics*, Vol. 1, New York: Interscience, 1961, p. 442.

[7] R. J. Glauber, *Phys. Rev.*, **131**, 2766 (1963). See also N. Chandra and H. Prakash, *Indian J. Pure Appl. Phys.*, **9**, 409 (1971), **9**, 677 (1971), and **9**, 688 (1971); H. Haken and H. D. Vollmer, *Z. Phys.*, **242**, 416 (1971); I. D. Dryugin and V. N. Kurashov, *Opt. Spectrosc.*, **29**, 183 (1970) and **29**, 345 (1970).

[8] J. R. Klauder, J. McKenna, and D. G. Currie, *J. Math. Phys.*, **6**, 743 (1965).

[9] L. R. Walker, private communication.

[10] A. E. Glassgold and D. Halliday, *Phys. Rev.*, **139**, A1717 (1965).

[11] E. Wigner, *Phys. Rev.*, **40**, 749 (1932); M. T. Raiford, *Phys. Rev.*, **2**, A1541 (1970); I. A. Deryugin, V. N. Kurashov, and A. I. Mashchenko, *Opt. Spectrosc.*, **30**, 507 (1971).

[12] M. Gaudin, *Nucl. Phys.*, **15**, 89 (1960).

[13] G. C. Wick, *Phys. Rev.*, **80**, 268 (1950); F. Dyson, *Phys. Rev.* **75**, 486 and 1736 (1949).

[14] M. Lax and H. Yuen, *Phys. Rev.*, **172**, 362 (1968).

[15] H. Heffner and W. H. Louisell, *J. Math. Phys.*, **6**, 474 (1965).

[16] C. E. Shannon and W. Weaver, *The Mathematical Theory of Communication*, Urbana, Ill.: University of Illinois Press, 1949.

[17] E. T. Jaynes, *Phys. Rev.*, **106**, 620 (1957).

[18] E. T. Jaynes, *Phys. Rev.*, **108**, 171 (1957).

[19] R. C. Tolman, *The Principles of Statistical Mechanics*, Oxford: Clarendon, 1938.

[20] H. A. Tolhoek and S. R. de Groot, *Physica*, **15**, 833 (1951).

4

Quantization of the Electromagnetic Field

Light has wave-like properties in interference and diffraction experiments and particle-like properties when it is absorbed or emitted by atoms. A satisfactory theory of radiation must explain in a unified way these two apparently paradoxical properties. By quantizing the electromagnetic field, Dirac [1] was able to bring about the first successful synthesis of these two aspects of radiation.

It will be sufficient for our purposes to give a noncovariant (nonrelativistic) formulation of Dirac's theory of radiation along the lines of Fermi's [2] classic paper and Heitler's book [3]. We shall not quantize charges and currents which are the sources of the radiation field.

Since we are interested in a phenomenological use of quantized radiation theory, we begin with the study of a classical lossless *LC* circuit with a voltage generator and show, by analogy with a harmonic oscillator, how to treat such a circuit quantum-mechanically. In the next section we show how to quantize a classical lossless transmission line. After these introductory examples, we begin a more general systematic study of the problem of quantizing an electromagnetic field in a cavity. In Section 4.3 we show that a classical radiation field in vacuum is equivalent to an infinite set of uncoupled harmonic oscillators. This equivalence suggests that we quantize the radiation field in the same way we quantize a harmonic oscillator; the quantization is carried out in Section 4.4. The density of modes in a cavity is obtained in Section 4.5.

Section 4.6 gives the commutation relations for fields in a vacuum since these relations are closely connected with the theory of measurement and the uncertainty principle. The zero-point field fluctuations are discussed in Section 4.7 and are the source of the natural line width of atoms, the Lamb shift, and quantum noise.

In the next two sections we present a simplified treatment of a radiation field interacting with charges and currents. This should give the reader some idea of how to treat quantum-mechanically empirical models in quantum electronics. A very general and thorough treatment of phenomenological quantum electrodynamics has been given by Jauch and Watson [4].

References 5 and 6 give a more thorough treatment of some of the topics of this chapter.

4.1 QUANTIZATION OF AN *LC* CIRCUIT WITH A SOURCE

We utilize the classical analogy between a lossless LC circuit in series with a voltage generator and a driven harmonic oscillator as a simple example of the method to be presented in Section 4.3 for quantizing the radiation field in a multimode lossless cavity. We do not quantize the generator since the reaction of the circuit back on the generator may be neglected in both the classical and quantum theories.

The classical "equation of motion" of a lossless LC circuit in series with a voltage generator, $e(t)$, is

$$\frac{d^2q}{dt^2} + \omega_0^{\ 2} q = \frac{1}{L} e(t), \tag{4.1.1}$$

where $q(t)$ is the charge and $\omega_0^{\ 2} = (LC)^{-1}$ is the circuit resonant frequency. The conjugate variable in the circuit, $p(t)$, is given by

$$L\frac{dq}{dt} = p(t). \tag{4.1.2a}$$

We may use this to write (4.1.1) as

$$\frac{dp}{dt} = -C^{-1}q + e(t). \tag{4.1.2b}$$

Equations (4.1.1) or (4.1.2a) and (4.1.2b) are equivalent ways of writing Kirchhoff's law for an LC circuit with a generator. If we let the charge $q(t)$ be the analog of the coordinate and the current $p(t)$ be the analog of the momentum, (4.1.1) or (4.1.2) describe equally well a driven harmonic oscillator. We may utilize this analogy and visualize (4.1.2) as the hamiltonian form of the equations of motion for an LC circuit. If we let the hamiltonian be

$$H(t) = \frac{1}{2L} p^2 + \frac{1}{2C} q^2 - e(t)q, \tag{4.1.3}$$

and use the classical equations of motion (1.15.12), we find

$$\begin{aligned} \frac{dq}{dt} &= \frac{\partial H}{\partial p} = \frac{1}{L} p \\ \frac{dp}{dt} &= -\frac{\partial H}{\partial q} = -\frac{1}{C} q + e(t) \end{aligned} \tag{4.1.4}$$

which agree with (4.1.2). This result justifies the choice of (4.1.3) for the hamiltonian $H(t)$ since (4.1.4) and (4.1.2) are the same. The hamiltonian and the energy are not always the same.

We may pursue the analogy between the circuit and material oscillator further to quantize the circuit. We associate hermitian operators with the charge q and the current p and require that they satisfy the commutation relation

$$[q, p] = i\hbar, \tag{4.1.5a}$$

and we have quantized the LC circuit. We may also introduce the non-hermitian operators a and $a^\dagger$ in terms of the charge and current by means of

$$\begin{aligned} a &= (2\hbar\omega_0 L)^{-1/2}[\omega_0 L q + ip] \\ a^\dagger &= (2\hbar\omega_0 L)^{-1/2}[\omega_0 L q - ip], \end{aligned} \tag{4.1.6a}$$

or

$$\begin{aligned} p &= i\sqrt{\frac{\hbar\omega_0 L}{2}}\,[a^\dagger - a] \\ q &= \sqrt{\frac{\hbar\omega_0 C}{2}}\,[a^\dagger + a], \end{aligned} \tag{4.1.6b}$$

where $\omega_0 L = (\omega_0 C)^{-1}$. From (4.1.5a) it follows that

$$[a, a^\dagger] = 1. \tag{4.1.5b}$$

In terms of a and $a^\dagger$, the hermitian operator associated with the hamiltonian (4.1.3) for this nonconservative system is, by (4.1.6),

$$H(t) = \hbar\omega_0(a^\dagger a + \tfrac{1}{2}) - \frac{\hbar e(t)}{\sqrt{2\hbar\omega_0 L}}(a^\dagger + a) \tag{4.1.7}$$

If we let

$$f(t) = -\frac{e(t)}{\sqrt{2\hbar\omega_0 L}} = f^*(t) \tag{4.1.8}$$

Equation (4.1.7) is the hamiltonian given in (3.11.11) for a driven harmonic oscillator. As we have already solved the Schrödinger equation in that case, we may take over completely the solution in Chapter 3 [Eq. 3.11.18]. The formalism is the same but the physical meaning of the operators is different. The number operator $a^\dagger a$ for quanta of the oscillator becomes the number operator for photons in the LC circuit, and a and $a^\dagger$ are circuit photon annihilation and creation operators, respectively. Now p and q are current and charge rather than momentum and position. Therefore, from (3.11.18), the state of the circuit at time t is related to the state at time $t_0 = 0$ by

$$|\psi(t)\rangle = e^{A(t)}e^{C(t)a^\dagger}\mathcal{N}\{e^{D(t)a^*a}\}e^{B(t)a}|\psi(0)\rangle, \tag{4.1.9}$$

where A, B, C, and D are given by (3.11.17).

In Section 3.11 we showed that if the circuit is initially in a vacuum state, then at time t it will be in a coherent (minimum uncertainty wave packet) state given by (3.11.22):

$$|\psi(t)\rangle = e^{A(t)}e^{C(t)a^\dagger}|0\rangle. \tag{4.1.10}$$

It is left to the reader to show that

$$A(t) + A^*(t) = -|C(t)|^2, \tag{4.1.11}$$

so that

$$\langle\psi(t)|\psi(t)\rangle = 1. \tag{4.1.12}$$

This minimum uncertainty state may be considered in another way. From the general theory of Chapter 1, the probability that a measurement of $a^\dagger a$ will yield m photons at time t when it is known with certainty the system is in the vacuum state at $t = 0$ is

$$P_{m,0}(t) = |\langle m|\psi(t)\rangle|^2 = e^{-|C(t)|^2}|\langle m|e^{Ca^\dagger}|0\rangle|^2, \tag{4.1.13}$$

where we used (4.1.10) and (4.1.11). Since from Section 2.5 we know that

$$\begin{aligned} e^{Ca^\dagger}|0\rangle &= \sum_{l=0}^{\infty}\frac{C^l a^{\dagger l}}{l!}|0\rangle \\ &= \sum_{l=0}^{\infty}\frac{C^l}{\sqrt{l!}}|l\rangle, \end{aligned} \tag{4.1.14}$$

it follows from the orthogonality relation $\langle m|l\rangle = \delta_{me}$ that (4.1.13) reduces to

$$P_{m,0}(t) = e^{-|C(t)|^2}\frac{[|C(t)|^2]^m}{m!}. \tag{4.1.15}$$

This probability distribution function is called a Poisson distribution over the photon energy eigenstates. It is normalized to unity for all time, since

$$\sum_{m=0}^{\infty}P_{m,0}(t) = e^{-|C(t)|^2}\sum_{m=0}^{\infty}\frac{[|C(t)|^2]^m}{m!} = 1. \tag{4.1.16a}$$

A Poisson distribution is generated by the voltage generator.

The average number of photons in the circuit at time t is

$$\bar{m}(t) = \sum_{m=0}^{\infty} mP_{m,0}(t) = |C(t)|^2. \tag{4.1.16b}$$

Therefore $\bar{m}(t) = \langle a^\dagger a \rangle = |C(t)|^2$; this shows that the average number of photons calculated in two ways is the same. The mean-square number of photons is

$$\overline{m^2(t)} = \sum_{m=0}^{\infty} m^2 P_{m,0}(t) = \bar{m}(t)[\bar{m}(t) + 1], \tag{4.1.16c}$$

which is characteristic of a Poisson distribution. The variance of the minimum uncertainty state is

$$(\Delta m)^2 = \overline{m^2(t)} - \overline{[m(t)]}^2 = \bar{m}(t). \tag{4.1.16d}$$

The state $|\psi(t)\rangle$ will be referred to either as a minimum uncertainty state or as a Poisson state; we see that it may be produced by the application of a generator to a circuit initially in a vacuum state. It should be emphasized that the wave packet is in terms of charge and current—not physical position and momentum.

We have discussed the state of the circuit at time t in great detail; it is worthwhile to study the properties of the assumed initial vacuum state with more care. We discuss the zero-point energy and zero-point fluctuations of the charge and current at $t = 0$.

Before the generator is turned on, the hamiltonian is given by (4.1.3) with $e(t) = 0$. If the state $|\psi(0)\rangle = |0\rangle$, the average values of the charge, current, and energy are easily found to be

$$\begin{aligned}
&\langle q \rangle = \langle 0|q|0\rangle = 0 && \langle p \rangle = \langle 0|p|0\rangle = 0 \\
&\langle q^2 \rangle = \frac{\hbar\omega_0 C}{2} && \langle p^2 \rangle = \frac{\hbar\omega_0 L}{2} \\
&\langle a^\dagger a \rangle = 0 && \langle H \rangle = \frac{\hbar\omega_0}{2} \\
&(\Delta q)^2 = \frac{\hbar\omega_0 C}{2} && (\Delta p)^2 = \frac{\hbar\omega_0 L}{2}.
\end{aligned} \tag{4.1.17}$$

We see that, in the vacuum state, the average value of the charge and current is zero but the average of their squares is not zero. Therefore the charge has a zero-point fluctuation given by $(\Delta q)^2 = \hbar\omega_0 C/2$ and the current has a zero-point fluctuation of $(\Delta p)^2 = \hbar\omega_0 L/2$ while the zero-point energy is $\hbar\omega_0/2$. If $\langle q^2 \rangle$ and $\langle p^2 \rangle$ were zero, as they are classically when the current is initially unexcited, then the uncertainty principle would clearly be violated. We know that $\Delta q\, \Delta p = \hbar/2$ is a direct consequence of the commutation relation $[q, p] = i\hbar$ so that the zero-point fluctuations of q and p result because the circuit has been quantized.

We consider next the zero-point energy $\hbar\omega_0/2$. This also arises because q and p do not commute. However, since energy is not absolute, it may be measured from the zero-point level. This means that the hamiltonian (4.1.3) can be redefined by [with $e = 0$]

$$H = \frac{1}{2}\left[\frac{1}{L}p^2 + \frac{1}{C}q^2\right] - \frac{\hbar\omega_0}{2} = \hbar\omega_0 a^\dagger a. \tag{4.1.18}$$

In this case we still find that for state $|0\rangle$

$$\begin{aligned} \langle p\rangle &= \langle 0|p|0\rangle = 0 & \langle q\rangle &= \langle 0|q|0\rangle = 0 \\ \langle p^2\rangle &= \frac{\hbar\omega_0 L}{2} & \langle q^2\rangle &= \frac{\hbar}{2\omega_0 L} = \frac{\hbar\omega_0 C}{2}. \end{aligned} \tag{4.1.19}$$

From this, the average energy is

$$\hbar\omega_0\langle a^\dagger a\rangle = 0.$$

By the redefinition, the zero-point energy has disappeared, but the zero-point charge and current fluctuations, $\langle q^2\rangle$ and $\langle p^2\rangle$, have not disappeared; they are always present to ensure that the uncertainty principle is not violated. Since the zero-point energy may be subtracted out in this trivial way, we neglect it.

4.2 QUANTIZATION OF A LOSSLESS TRANSMISSION LINE

Another simple circuit example easily treated quantum-mechanically is a lossless transmission line whose classical "equations of motion" are

$$\begin{aligned} \frac{\partial V}{\partial z} &= -L\frac{\partial I}{\partial t} \\ \frac{\partial I}{\partial z} &= -C\frac{\partial V}{\partial t}, \end{aligned} \tag{4.2.1}$$

where $V(z, t)$ is the voltage, $I(z, t)$ is the current, and L and C are the inductance and capacitance per unit length, respectively. These equations may be decoupled in the usual way, and both V and I satisfy the wave equation

$$\frac{\partial^2 V}{\partial z^2} = \frac{1}{c^2}\frac{\partial^2 V}{\partial t^2} \qquad \frac{\partial^2 I}{\partial z^2} = \frac{1}{c^2}\frac{\partial^2 I}{\partial t^2}, \tag{4.2.2}$$

where c is the velocity of propagation given by

$$c^2 = \frac{1}{LC}. \tag{4.2.3}$$

It is well known that the wave equation (4.2.2) has forward- and backward-wave solutions. By proper choice of boundary conditions, we may have standing waves or traveling waves. Since the next section deals with the wave equation in great detail, for simplicity we consider here only one forward-plane-wave solution of (4.2.2). We write a forward wave solution in the form

$$V(z, t) = \sqrt{\frac{\hbar\omega}{2Cz_0}}\,(ae^{-i\omega t+ikz} + a^*e^{i\omega t-ikz}) = \sqrt{\frac{L}{C}}\,I(z, t), \qquad (4.2.4)$$

where z_0 is the length of the transmission line under study and a and its complex conjugate a^* are arbitrary constants. The factor $\sqrt{\hbar\omega/2Cz_0}$ is put first for normalization purposes. Furthermore, the propagation constant k is

$$k = \frac{\omega}{c} = \omega\sqrt{LC} = \frac{2\pi}{\lambda}, \qquad (4.2.5)$$

where $\sqrt{L/C} = Z_0$ is the characteristic impedance of the line and λ is the wavelength of the wave on the line.

We assume that ω is given and choose the length of line z_0 to be a fixed integral number of wavelengths

$$z_0 = m\lambda = \frac{2\pi}{k}\,m, \qquad (4.2.6)$$

where m is a fixed integer. In this way, we restrict ourselves to one forward mode only.

The energy contained in this length of line is

$$\begin{aligned} H &= \frac{1}{2}\int_0^{z_0} [CV^2(z, t) + LI^2(z, t)]\,dz \\ &= \int_0^{2\pi m/k} CV^2(z, t)\,dz, \end{aligned} \qquad (4.2.7)$$

where we used (4.2.4) and (4.2.6). If we use (4.2.4), the energy is given by

$$H = \hbar\omega a^* a. \qquad (4.2.8)$$

The choice of the normalization factor of V and I is responsible for the $\hbar\omega$ in this classical result. We are just expressing the energy in units of $\hbar\omega$.

Let us consider the significance of the arbitrary constants a and a^*. If we specify them, by (4.2.4) we specify the mode completely. We may therefore regard a and a^* as the quantities that describe the state of the system. In fact, a and a^* may define two new real quantities q and p by means of

$$\begin{aligned} a &= \frac{1}{\sqrt{2\hbar\omega}}(\omega q + ip) \\ a^* &= \frac{1}{\sqrt{2\hbar\omega}}(\omega q - ip); \end{aligned} \qquad (4.2.9)$$

so that the energy contained in a length of line of m wavelengths is, by (4.2.8) and (4.2.9),

$$H = \tfrac{1}{2}(p^2 + \omega^2 q^2). \tag{4.2.10}$$

We have therefore shown classically that a single forward mode on a transmission line m wavelengths long is completely equivalent to a single harmonic oscillator. The mode is completely specified if a and a^* (or p and q) are given; they indicate the state of excitation of the mode.

We now see how to quantize the transmission line. We let p and q be hermitian operators related to the nonhermitian operators a and $a^\dagger$ by means of (4.2.9). The voltage and current are therefore operators. We impose the quantum conditions

$$[q, p] = i\hbar \qquad [a, a^\dagger] = 1, \tag{4.2.11}$$

and the rest is straightforward. The operators $a(t)$ and $a^\dagger(t)$ in the Heisenberg picture are

$$\begin{aligned} a(t) &= e^{i\omega t a^\dagger a} a e^{-i\omega t a^\dagger a} = a e^{-i\omega t} \\ a^\dagger(t) &= e^{i\omega t a^\dagger a} a^\dagger e^{-i\omega t a^\dagger a} = a^\dagger e^{i\omega t}, \end{aligned} \tag{4.2.12}$$

where the hamiltonian (without the zero-point energy) is

$$H = \hbar\omega a^\dagger a. \tag{4.2.13}$$

The voltage and current operators may therefore be written in the Heisenberg picture by (4.2.4) and (4.2.12) as

$$V_H(z, t) = \sqrt{\frac{\hbar\omega}{2Cz_0}}\,[a(t)e^{ikz} + a^\dagger(t)e^{-ikz}] = \sqrt{\frac{L}{C}}\,I_H(z, t). \tag{4.2.14}$$

It is clear that the voltage and current for this single mode commute.

Let us derive the Heisenberg-operator equations of motion for $V_H(z, t)$ and $I_H(z, t)$. We have

$$\begin{aligned} i\hbar\frac{\partial V_H}{\partial t} &= [V_H, H_H] \\ i\hbar\frac{\partial IH}{\partial t} &= [I_H, H_H], \end{aligned} \tag{4.2.15}$$

where the hamiltonian in the Heisenberg picture is

$$H_H = \hbar\omega a_H^\dagger a_H \equiv \hbar\omega a^\dagger a. \tag{4.2.16}$$

Since

$$[a, a^\dagger a] = a \qquad [a^\dagger, a^\dagger a] = -a^\dagger \tag{4.2.17}$$

in both the Schrödinger and Heisenberg pictures, from (4.2.14), (4.2.16), and (4.2.17), Eqs. (4.2.15) reduce to

$$i\hbar \frac{\partial V_H}{\partial t} = \hbar\omega\sqrt{\frac{\hbar\omega}{2Cz_0}}\,[a(t)e^{ikz} - a^\dagger(t)e^{-ikz}] = -\frac{i\hbar}{C}\frac{\partial I_H}{\partial z} \qquad (4.2.18a)$$

$$i\hbar \frac{\partial I_H}{\partial t} = -\frac{i\hbar}{L}\frac{\partial V_H}{\partial z}\,. \qquad (4.2.18b)$$

These are the classical equations (4.2.1). Again we have shown the equivalence in form between the classical "equations of motion" and the Heisenberg-operator equations of motion.

The state of the system (one mode of the transmission line) at time t is given as usual by

$$|\psi(t)\rangle = e^{-i\omega t a^\dagger a}|\psi(0)\rangle, \qquad (4.2.19)$$

where $|\psi(0)\rangle$ is its initial state of excitation and $a^\dagger a$ is the photon-number operator for the number of photons in the length of line.

Since the hermitian operators V_H and I_H commute for this mode, they can be measured simultaneously. However, it is easy to show that $V_H(z, t)$ and $\partial I_H/\partial z$ do not commute.

4.3 EQUIVALENCE OF CLASSICAL RADIATION FIELD IN CAVITY TO INFINITE SET OF OSCILLATORS

In this section we review briefly the classical theory of radiation in a source-free cavity. We cast the theory in canonical form and show the equivalence of the field to an infinite set of harmonic oscillators. In the following section, we quantize the field in the cavity by quantizing the oscillators as done in the previous two sections.

We use mks units throughout; symbols that are universally used will not be defined since their meaning is assumed to be well known to the reader.

Maxwell Field Equations

If no sources are present, the electromagnetic field is determined by the equations

$$\operatorname{div} \mathbf{B} = 0 \qquad (4.3.1)$$

$$\operatorname{curl} \mathbf{E} = -\frac{\partial \mathbf{B}}{\partial t} \qquad (4.3.2)$$

$$\operatorname{div} \mathbf{D} = 0 \qquad (4.3.3)$$

$$\operatorname{curl} \mathbf{H} = \frac{\partial \mathbf{D}}{\partial t}, \qquad (4.3.4)$$

where

$$\mathbf{B} = \mu_0 \mathbf{H} \qquad \mathbf{D} = \epsilon_0 \mathbf{E} \tag{4.3.5}$$

and $\mu_0 \epsilon_0 = c^{-2}$, where μ_0 and ϵ_0 are for free space.

We may satisfy (4.3.1) identically if we let

$$\mathbf{B} = \operatorname{curl} \mathbf{A}, \tag{4.3.6}$$

and (4.3.2) will be satisfied identically if we let

$$\mathbf{E} = -\frac{\partial \mathbf{A}}{\partial t} - \operatorname{grad} V, \tag{4.3.7}$$

where $\mathbf{A}$ and V are the scalar and vector potentials.

Since Maxwell's equations are gauge invariant, it is easy to show that when no sources are present we may work in the *Coulomb gauge* in which

$$\begin{aligned} \operatorname{div} \mathbf{A} &= 0 \\ V &= 0, \end{aligned} \tag{4.3.8}$$

so that both $\mathbf{B}$ and $\mathbf{E}$ are determined by $\mathbf{A}$ alone. In this gauge, the fields are given by

$$\begin{aligned} \mathbf{B} &= \operatorname{curl} \mathbf{A} = \mu_0 \mathbf{H} \qquad & \operatorname{div} \mathbf{A} &= 0 \\ \mathbf{E} &= -\frac{\partial \mathbf{A}}{\partial t} & V &= 0. \end{aligned} \tag{4.3.9}$$

If we substitute these in (4.3.4) and use (4.3.5), we find that $\mathbf{A}(\mathbf{r}, t)$ satisfies the wave equation

$$\nabla^2 \mathbf{A} = \frac{1}{c^2} \frac{\partial^2 \mathbf{A}}{\partial t^2}. \tag{4.3.10}$$

The fields in vacuum in the Coulomb gauge are determined by this wave equation.

Energy and Momentum of the Field

The energy H contained in the field inside a cavity is

$$\begin{aligned} H &= \frac{1}{2} \int (\epsilon_0 \mathbf{E}^2 + \mu_0 \mathbf{H}^2)\, d\tau \\ &= \frac{1}{2} \int \left[\epsilon_0 \left(\frac{\partial \mathbf{A}}{\partial t} \right)^2 + \frac{1}{\mu_0} (\operatorname{curl} \mathbf{A})^2 \right] d\tau, \end{aligned} \tag{4.3.11}$$

where $d\tau = dx\, dy\, dz$ is a volume element and the integration is carried out over the volume of the cavity. We may also associate a momentum $\mathbf{G}$ with the classical field in the cavity by means of Poynting's theorem given by

$$\mathbf{G} = \frac{1}{c^2}\int(\mathbf{E}\times\mathbf{H})\,d\tau = -\epsilon_0\int\left(\frac{\partial\mathbf{A}}{\partial t}\times\text{curl}\,\mathbf{A}\right)d\tau, \tag{4.3.12}$$

where we have used (4.3.9).

Expansion of A(r, *t*) in Cavity Normal Modes

The electric and magnetic fields in the Coulomb gauge in a vacuum are determined by giving the values of A_x, A_y, A_z at each point (x, y, z) at time t. If we think of A_x, A_y, and A_z as the variables used to describe the field, we see that there will be an uncountably infinite number. By the following procedure we may describe the field by another infinite set of variables that are enumerable or countable.

We assume that the radiation field is contained in a cavity with perfectly conducting walls. For simplicity, we let the cavity be a cube of total volume $\tau = L^3$. The procedure is independent of the shape of the cavity. If we are interested in a radiation field in free space, we may let $\tau \to \infty$ after the calculations are complete. If we impose boundary conditions on the fields, the solution of the wave equation will have an infinite discrete set of normal-mode solutions orthogonal to one another and complete in the sense that any arbitrary field in the cavity can be expressed as a sum of these normal modes with suitable amplitudes. The amplitudes of each mode can then be used instead of A_x, A_y, and A_z to describe the field in the cavity. Since the modes are discrete and infinite, the amplitudes (new field "variables") are then countably infinite. The boundary conditions force a discrete set of modes that allows description of the field by a countable set of variables.

By the familiar procedure of separation of variables we may assume a solution of (4.3.10) of the form

$$\mathbf{A}(\mathbf{r}, t) = \frac{1}{\sqrt{\epsilon_0}}\sum_l q_l(t)\mathbf{u}_l(\mathbf{r}), \tag{4.3.13}$$

where the coefficient $\epsilon_0^{-1/2}$ is used for normalization purposes. If we substitute this into the wave equation, then for each l we have

$$\nabla^2\mathbf{u}_l(\mathbf{r}) + \frac{\omega_l^2}{c^2}\mathbf{u}_l(\mathbf{r}) = 0 \tag{4.3.14}$$

$$\frac{d^2q_l}{dt^2} + \omega_l^2 q_l = 0, \tag{4.3.15}$$

where ω_l^2 is the separation constant for each l.

We may obtain standing-wave solutions if we require that at the walls the tangential component of **E** and the normal component of **B** vanish. From

(4.3.9) and (4.3.13), we have

$$\mathbf{u}_l\big|_{\tan} = 0 \qquad \operatorname{curl} \mathbf{u}_l\big|_{\text{norm}} = 0 \qquad \text{(on walls).} \tag{4.3.16}$$

From (4.3.8),

$$\operatorname{div} \mathbf{u}_l(\mathbf{r}) = 0 \tag{4.3.17}$$

everywhere in the cavity.

The solutions of (4.3.14) that satisfy (4.3.16) give a discrete set of normal modes that are orthogonal and normalized to unity:

$$\int_{\text{cavity}} \mathbf{u}_l(\mathbf{r}) \cdot \mathbf{u}_m(\mathbf{r})\, d\tau = \delta_{lm}. \tag{4.3.18}$$

The $\mathbf{u}_l(\mathbf{r})$ and $q_l(t)$ are real in order that $\mathbf{A}$ be real.

The $\mathbf{u}_l(\mathbf{r})$ are $\sin \mathbf{k}_l \cdot \mathbf{r}$ or $\cos \mathbf{k}_l \cdot \mathbf{r}$ for the cubical cavity. For other geometries the $\mathbf{u}_l$'s will be other complete sets of functions. The boundary conditions and geometry determine the different modes which are distinguished by the index l. In general, three or four numbers are needed to specify a mode; l is an abbreviation for this set of numbers, as we shall see.

The normalization condition (4.3.18) removes all freedom from choosing the amplitude of $\mathbf{u}_l$. The $\mathbf{u}_l$'s are completely specified known functions. The amplitude of each normal mode in (4.3.13) needed to specify a particular field configuration is $q_l(t)$. If each $q_l(t)$ is given, the field is just as completely determined as if the value of A_x, A_y, and A_z at each point in space at time t were given. The q_l are taken as a new set of variables to describe the field.

The amplitudes $q_l(t)$ satisfy (4.3.15), the equation of motion of a harmonic oscillator. Therefore with each mode of the field there may be associated a radiation oscillator of frequency ω_l. We next show that the energy contained in the total field is just the energy of the infinite set of uncoupled radiation oscillators.

If we substitute (4.3.13) into (4.3.11), we have for the energy in the cavity

$$H = \tfrac{1}{2} \sum_{l,m} \dot{q}_l \dot{q}_m \int_{\text{cavity}} \mathbf{u}_l \cdot \mathbf{u}_m\, d\tau + \frac{c^2}{2} \sum_{l,m} q_l q_m \int_{\text{cavity}} \operatorname{curl} \mathbf{u}_l \cdot \operatorname{curl} \mathbf{u}_m\, d\tau. \tag{4.3.19}$$

If we use the orthonormality condition (4.3.18), the first double sum in (4.3.19) reduces to the single sum $\frac{1}{2} \sum_l \dot{q}_l^2$. By the vector identity,

$$\operatorname{curl} \mathbf{u}_l \cdot \operatorname{curl} \mathbf{u}_m = \mathbf{u}_m \cdot \operatorname{curl} \operatorname{curl} \mathbf{u}_l + \operatorname{div} (\mathbf{u}_m \times \operatorname{curl} \mathbf{u}_l), \tag{4.3.20}$$

the last integral in (4.3.19) reduces by means of Gauss's theorem to

$$\int_{\text{cavity}} \mathbf{u}_m \cdot \operatorname{curl} \operatorname{curl} \mathbf{u}_l\, d\tau + \int_{\text{walls}} (\mathbf{u}_m \times \operatorname{curl} \mathbf{u}_l) \cdot d\mathbf{S}, \tag{4.3.21}$$

where $d\mathbf{S}$ is an element of area on the cavity wall. The surface integral vanishes because of (4.3.16). The first integral in (4.3.21) is, by a well-known identity,

$$\int \mathbf{u}_m \cdot (\text{grad div } \mathbf{u}_l - \nabla^2 \mathbf{u}_l)\, d\tau = \frac{\omega_l^2}{c^2} \int \mathbf{u}_l \cdot \mathbf{u}_m \, d\tau = \frac{\omega_l^2}{c^2}\, \delta_{lm}. \tag{4.3.22}$$

We have used (4.3.14), (4.3.17) and (4.3.18). When we put this in (4.3.19), the hamiltonian for the field is

$$H = \tfrac{1}{2} \sum_l (\dot{q}_l^2 + \omega_l^2 q_l^2) \equiv \sum_l H_l. \tag{4.3.23}$$

But H_l is the energy of a harmonic oscillator of frequency ω_l, and so the field energy is equivalent to an infinite set of uncoupled radiation oscillators.

The hamiltonian equations of motion for the lth oscillator are

$$\frac{\partial H_l}{\partial q_l} = -\dot{p}_l = \omega_l^2 q_l \qquad \frac{\partial H_l}{\partial p_l} = \dot{q}_l = p_l. \tag{4.3.24}$$

Here p_l and q_l are called canonically conjugate variables.

We are, as usual, able to define two complex variables a_l and $a_l^\dagger$ for each mode in the now familiar way by

$$\begin{aligned} q_l &= \sqrt{\frac{\hbar}{2\omega_l}}\,(a_l^\dagger + a_l) & a_l &= \frac{1}{\sqrt{2\hbar\omega_l}}\,(\omega_l q_l + i p_l) \\ p_l &= i\sqrt{\frac{\hbar\omega_l}{2}}\,(a_l^\dagger - a_l) & a_l^\dagger &= \frac{1}{\sqrt{2\hbar\omega_l}}\,(\omega_l q_l - i p_l). \end{aligned} \tag{4.3.25}$$

The equations for a_l and $a_l^\dagger$ are

$$\frac{da_l(t)}{dt} = -i\omega_l a_l(t) \qquad \frac{da_l^\dagger(t)}{dt} = i\omega_l a_l^\dagger(t), \tag{4.3.26}$$

and have solutions

$$a_l(t) = a_l e^{-i\omega_l t} \qquad a_l^\dagger(t) = a_l^\dagger e^{i\omega_l t}. \tag{4.3.27}$$

In terms of a_l and $a_l^\dagger$, (4.3.23) is given by

$$H = \tfrac{1}{2} \sum \hbar\omega_l (a_l^\dagger a_l + a_l a_l^\dagger). \tag{4.3.28}$$

In this purely classical theory, a_l and $a_l^\dagger$ commute. We have preserved the order for the quantum treatment in the following section.

To summarize, we may write the fields in standing waves as

$$\mathbf{A} = \frac{1}{\sqrt{\epsilon_0}} \sum_l q_l(t)\mathbf{u}_l(\mathbf{r})$$

$$\mathbf{E} = -\frac{1}{\sqrt{\epsilon_0}} \sum_l p_l(t)\mathbf{u}_l(\mathbf{r}) \tag{4.3.29}$$

$$\mathbf{H} = \frac{1}{\mu_0\sqrt{\epsilon_0}} \sum q_l(t) \operatorname{curl} \mathbf{u}_l(\mathbf{r}).$$

Plane-Wave Representation of the Fields

It is often convenient to represent the fields in terms of plane traveling waves rather than standing waves.

We write the vector potential as a linear superposition of plane waves in the form

$$\mathbf{A}(\mathbf{r}, t) = \sum_l \sum_{\sigma=1}^{2} \sqrt{\frac{\hbar}{2\omega_l \epsilon_0 \tau}}\, \hat{\mathbf{e}}_{l\sigma}\{a_{l\sigma} \exp[i(\mathbf{k}_l \cdot \mathbf{r} - \omega_l t)] + a_{l\sigma}^{\dagger} \exp[-i(\mathbf{k}_l \cdot \mathbf{r} - \omega_l t)]\}. \tag{4.3.30}$$

We now explain the meaning of the symbols. The vector $\hat{\mathbf{e}}_{l\sigma}$ and the numbers $a_{l\sigma}$ and $a_{l\sigma}^{\dagger}$ are constants.

The vector $\mathbf{k}_l$ is the propagation constant and if

$$\mathbf{k}_l^2 = \frac{\omega_l^2}{c^2}, \tag{4.3.31}$$

then each term in the series (4.3.30) satisfies the wave equation, which we assume is familiar to the reader.

From the Coulomb-gauge condition div $\mathbf{A} = 0$, we see that

$$\hat{\mathbf{e}}_{l\sigma} \cdot \mathbf{k}_l = 0. \tag{4.3.32}$$

This is called the transversality condition. Here $\mathbf{k}_l$ is the direction of propagation of the plane wave. Since $\mathbf{E} = -\partial\mathbf{A}/\partial t$, we see from (4.3.30) and (4.3.32) that $\mathbf{E}$ and $\mathbf{A}$ are transverse to the direction of propagation in the absence of sources. It is an auxiliary condition imposed by working in the Coulomb gauge.

The vectors $\hat{\mathbf{e}}_{l1}$ and $\hat{\mathbf{e}}_{l2}$ are unit vectors used to specify the polarization of the plane wave. Since each polarization is independent, the total field must be summed over both polarizations in (4.3.30). For convenience, we shall choose $\hat{\mathbf{e}}_{l1}$ perpendicular to $\hat{\mathbf{e}}_{l2}$ so that

$$\hat{\mathbf{e}}_{l1} \cdot \hat{\mathbf{e}}_{l2} = 0. \tag{4.3.33}$$

It is often convenient to define a unit vector in the direction $\mathbf{k}_l$ as

$$\hat{\mathbf{k}}_l = \frac{\mathbf{k}_l}{|\mathbf{k}_l|}. \tag{4.3.34}$$

Equations (4.3.32) and (4.3.33) may then be written

$$\hat{\mathbf{e}}_{l\sigma} \cdot \hat{\mathbf{e}}_{l\sigma'} = \delta_{\sigma\sigma'} \qquad \sigma, \sigma' = 1, 2 \qquad \hat{\mathbf{k}}_l \cdot \hat{\mathbf{e}}_{l\sigma} = 0. \tag{4.3.35}$$

It is convenient to require that the vector potential satisfy periodic boundary conditions on opposite faces of the cavity in order to make the modes discrete. If $\hat{\mathbf{i}}$, $\hat{\mathbf{j}}$, and $\hat{\mathbf{k}}$ are three unit vectors along the cube edges, the position vector is $\mathbf{r} = x\hat{\mathbf{i}} + y\hat{\mathbf{j}} + z\hat{\mathbf{k}}$ and the propagation vector is $\mathbf{k}_l = k_{lx}\hat{\mathbf{i}} + k_{ly}\hat{\mathbf{j}} + k_{lz}\hat{\mathbf{k}}$. The periodic boundary conditions require that

$$\mathbf{A}(\mathbf{r} + L\hat{\mathbf{i}}, t) = \mathbf{A}(\mathbf{r} + L\hat{\mathbf{j}}, t) = \mathbf{A}(\mathbf{r} + L\hat{\mathbf{k}}, t) = \mathbf{A}(\mathbf{r}, t),$$

and are satisfied if

$$\mathbf{k}_l = \frac{2\pi}{L}(l_1\hat{\mathbf{i}} + l_2\hat{\mathbf{j}} + l_3\hat{\mathbf{k}}) \tag{4.3.36}$$

where l_1, l_2, and l_3 are integers from $-\infty$ to $+\infty$. That is, the propagation constants are restricted to a discrete set of values by virtue of the boundary conditions.

For each triple of integers (l_1, l_2, l_3) there are two traveling modes (one for each polarization σ), according to (4.3.30). If we let (l_1, l_2, l_3) go to $(-l_1, -l_2, -l_3)$ (which is designated simply by $l \to -l$), from (4.3.36),

$$\mathbf{k}_{-l} = -\mathbf{k}_l. \tag{4.3.37}$$

From (4.3.31),

$$\omega_l \equiv \omega_{-l}. \tag{4.3.38}$$

Therefore, from (4.3.30), (4.3.37), and (4.3.38), we see that, if we change l to $-l$, the two corresponding plane waves travel in opposite directions. We therefore have included in (4.3.30) forward and backward modes.

The sum $\sum_l$ is now understood as a shorthand notation for

$$\sum_l \equiv \sum_{l_1=-\infty}^{\infty} \sum_{l_2=-\infty}^{\infty} \sum_{l_3=-\infty}^{\infty}. \tag{4.3.39}$$

The set (l_1, l_2, l_3, σ) gives a mode of a given polarization.

It is easy to verify that the vector potential is real.

We next show that the vector potential $\mathbf{A}(\mathbf{r}, t)$ and the electric field $\mathbf{E} = -\partial\mathbf{A}/\partial t$ may be expressed in terms of canonically conjugate variables. For this purpose, we let

$$a_{l\sigma}(t) = a_{l\sigma}e^{-i\omega_l t} \qquad a_{l\sigma}^{\dagger}(t) = a_{l\sigma}^{\dagger}e^{i\omega_l t}, \tag{4.3.40}$$

and

$$\mathbf{u}_{l\sigma}(\mathbf{r}) = \frac{\hat{\mathbf{e}}_{l\sigma} \exp(i\mathbf{k}_l \cdot \mathbf{r})}{\sqrt{\tau}} \qquad \mathbf{u}_{l\sigma}^* = \frac{\hat{\mathbf{e}}_{l\sigma} \exp(-i\mathbf{k}_l \cdot \mathbf{r})}{\sqrt{\tau}}, \tag{4.3.41}$$

which satisfy the orthonormality relations

$$\int_{\text{cavity}} \mathbf{u}_{l\sigma}^*(\mathbf{r}) \cdot \mathbf{u}_{l'\sigma'}(\mathbf{r})\, d\tau = \delta_{ll'}\delta_{\sigma\sigma'}. \tag{4.3.42}$$

The variables $a_{l\sigma}$ and $a_{l\sigma}^\dagger$ may be used to describe the field, and we may again introduce real variables $p_{l\sigma}$ and $q_{l\sigma}$ by

$$\begin{aligned} a_{l\sigma} &= \frac{1}{\sqrt{2\hbar\omega_l}}(\omega_l q_{l\sigma} + ip_{l\sigma}) \\ a_{l\sigma}^\dagger &= \frac{1}{\sqrt{2\hbar\omega_l}}(\omega_l q_{l\sigma} - ip_{l\sigma}). \end{aligned} \tag{4.3.43}$$

The electric field $\mathbf{E} = -\partial\mathbf{A}/\partial t$ so that, by (4.3.30) and (4.3.40),

$$\mathbf{E}(\mathbf{r}, t) = i \sum_{l,\sigma} \sqrt{\frac{\hbar\omega_l}{2\epsilon_0\tau}}\, \hat{\mathbf{e}}_{l\sigma}[a_{l\sigma}(t) \exp(i\mathbf{k}_l \cdot \mathbf{r}) - a_{l\sigma}^\dagger(t) \exp(-i\mathbf{k}_l \cdot \mathbf{r})], \tag{4.3.44a}$$

while the magnetic field $\mathbf{H} = \text{curl}\, \mathbf{A}/\mu_0$ is

$$\mathbf{H}(\mathbf{r}, t) = -\frac{i}{c\mu_0} \sum_{l,\sigma} \sqrt{\frac{\hbar\omega_l}{2\epsilon_0\tau}}\, (\hat{\mathbf{e}}_{l\sigma} \times \hat{\mathbf{k}}_l)[a_{l\sigma}(t) \exp(i\mathbf{k}_l \cdot \mathbf{r}) - a_{l\sigma}^\dagger(t) \exp(-i\mathbf{k}_l \cdot \mathbf{r})], \tag{4.3.44b}$$

where we used the fact that

$$\begin{aligned} \text{curl}\, [\hat{\mathbf{e}}_{l\sigma} \exp(\pm i\mathbf{k}_l \cdot \mathbf{r})] &= \mp i(\hat{\mathbf{e}}_{l\sigma} \times \mathbf{k}_l) \exp(\pm i\mathbf{k}_l \cdot \mathbf{r}) \\ &\equiv \mp i\frac{\omega_l}{c}(\hat{\mathbf{e}}_{l\sigma} \times \hat{\mathbf{k}}_l) \exp(\pm i\mathbf{k}_l \cdot \mathbf{r}). \end{aligned}$$

Here $|\mathbf{k}_l| = \omega_l/c$ and $\hat{\mathbf{k}}_l|\mathbf{k}_l| = \mathbf{k}_l$.

The hamiltonian for the field in the cavity is

$$\begin{aligned} H &= \tfrac{1}{2}\int_{\text{cavity}} (\epsilon_0\mathbf{E}^2 + \mu_0\mathbf{H}^2)\, d\tau \\ &= \tfrac{1}{2}\sum_{l,\sigma} \hbar\omega_l(a_{l\sigma}a_{l\sigma}^\dagger + a_{l\sigma}^\dagger a_{l\sigma}) \\ &= \tfrac{1}{2}\sum_{l,\sigma} ({p_{l\sigma}}^2 + \omega_l^2 {q_{l\sigma}}^2). \end{aligned} \tag{4.3.45}$$

The derivation of this result is given in Appendix B. We have kept the order of factors of $aa^\dagger$ and $a^\dagger a$ so that the analysis will also be valid when we treat these as noncommuting operators, although they commute classically.

Again, as in the treatment for standing waves, from (4.3.28) a radiation oscillator may be associated with each mode of the cavity and $p_{l\sigma}$ and $q_{l\sigma}$ are canonically conjugate variables. The total field energy is just the sum of the energies of each independent normal mode of the field.

The canonical hamiltonian equations of motion for the field, from (4.3.45), are

$$\frac{\partial H}{\partial p_{l\sigma}} = \dot{q}_{l\sigma} = p_{l\sigma} \qquad \frac{\partial H}{\partial q_{l\sigma}} = -\dot{p}_{l\sigma} = \omega_l^{\,2} q_{l\sigma}, \tag{4.3.46}$$

which, as expected, are the oscillator equations of motion.

The multiplying constant in (4.3.30) $(\hbar/2\omega_l\epsilon_0\tau)^{+1/2}$ was chosen so that H would be measured in units of $\hbar\omega_l$.

Momentum of the Field

We evaluate the momentum of the field

$$\mathbf{G} = \frac{1}{c^2}\int_{\text{cavity}} (\mathbf{E} \times \mathbf{H})\, d\tau, \tag{4.3.47}$$

in Appendix C. It is

$$\mathbf{G} = \tfrac{1}{2}\sum_{l,\sigma} \hbar\mathbf{k}_l(a_{l\sigma}a_{l\sigma}^\dagger + a_{l\sigma}^\dagger a_{l\sigma}). \tag{4.3.48}$$

Again the units have been chosen to measure the momentum in units of $\hbar\mathbf{k}_l$ although the present analysis is still classical.

4.4 QUANTIZATION OF THE RADIATION FIELD IN VACUUM

The method of quantizing the radiation field is now straightforward. We associate hermitian operators with the classical field variables $p_{l\sigma}$ and $q_{l\sigma}$. It is experimentally known that photons are bosons so that we postulate that $q_{l\sigma}$ and $p_{l\sigma}$ satisfy the boson commutation relations. Quantization is necessary to show the particle nature of light. In terms of the nonhermitian operators $a_{l\sigma}$ and $a_{l\sigma}^\dagger$, these relations are

$$\begin{aligned} [a_{l\sigma}, a_{l'\sigma'}^\dagger] &= \delta_{ll'}\delta_{\sigma\sigma'} \\ [a_{l\sigma}, a_{l'\sigma'}] &= 0 = [a_{l\sigma}^\dagger, a_{l'\sigma'}^\dagger]. \end{aligned} \tag{4.4.1}$$

That is, since the radiation oscillators are all independent, $a_{l\sigma}$ and $a_{l'\sigma'}^\dagger$ commute for different oscillators.

The hamiltonian for the field (without the zero-point energy) is

$$H = \sum_{l,\sigma} \hbar\omega_l a_{l\sigma}^\dagger a_{l\sigma}. \tag{4.4.2}$$

We have already shown that it is possible to change the level from which the energy is measured.

The momentum is, by (4.3.48) and (4.4.1),

$$\mathbf{G} = \sum_{l,\sigma} \hbar \mathbf{k}_l (a_{l\sigma}^\dagger a_{l\sigma} + \tfrac{1}{2}).$$

But since $\mathbf{k}_{-l} = -\mathbf{k}_l$, the sum $\sum_l \hbar \mathbf{k}_l = 0$. Therefore,

$$\mathbf{G} = \sum_{l,\sigma} \hbar \mathbf{k}_l a_{l\sigma}^\dagger a_{l\sigma}. \tag{4.4.3}$$

We may now take over all the previously derived results for bosons. There is the slight generalization in that there are now an infinite number of independent field oscillators instead of one oscillator. We interpret $a_{l\sigma}$ and $a_{l\sigma}^\dagger$ as an annihilation and a creation operator, respectively, for a photon in direction $\mathbf{k}_l$, polarization σ, and frequency ω_l. Here $N_{l\sigma} = a_{l\sigma}^\dagger a_{l\sigma}$ is the photon-number operator for the number of photons in the l, σ mode. The eigenvalues of $N_{l\sigma}$ are $n_{l\sigma} = 0, 1, 2, \ldots, \infty$. The energy of the photon in state l, σ is $\hbar\omega_l$, by (4.4.2), and the photon has a momentum $\hbar\mathbf{k}_l$, by (4.4.3). The photon is therefore seen to exhibit particle-like properties, and travel with velocity c.

Since each cavity mode is independent, a complete set of state vectors may be written as a simple product of the state vectors for each mode; that is, a state vector for the radiation field may be written as

$$|n_1\rangle|n_2\rangle \cdots |n_\infty\rangle = |n_1, n_2, \ldots, n_\infty\rangle \tag{4.4.4}$$

where each subscript $1, 2, \ldots$ stands for the quartet of integers (l_1, l_2, l_3, σ).

The state vector for an assembly of noninteracting bosons must be symmetric under the interchange of any two of the bosons. It can be shown [1, 5] that, by specifying the number of bosons in each state, we obtain a wave function with the correct symmetry so that (4.4.4) correctly describes an assembly of noninteracting bosons. We do not go into this question although such considerations justify the interpretation of $a_{l\sigma}$ and $a_{l\sigma}^\dagger$ as boson annihilation and creation operators [1].

The effect of $a_{l\sigma}$ and $a_{l\sigma}^\dagger$ on the state vectors (4.4.4) is given by

$$\begin{aligned} a_{l\sigma}^\dagger|\ldots, n_{l\sigma}, \ldots\rangle &= \sqrt{n_{l\sigma}+1}\,|\ldots, n_{l\sigma}+1, \ldots\rangle \\ a_{l\sigma}|\ldots, n_{l\sigma}, \ldots\rangle &= \sqrt{n_{l\sigma}}\,|\ldots, n_{l\sigma}-1, \ldots\rangle \\ a_{l\sigma}|\ldots, 0, \ldots\rangle &= 0 \\ N_{l\sigma}|\ldots, n_{l\sigma}, \ldots\rangle &= n_{l\sigma}|\ldots, n_{l\sigma}, \ldots\rangle. \end{aligned} \tag{4.4.5}$$

With this choice, these state vectors are normalized to unity.

As in the case of the single oscillator, these operators are in the Schrödinger picture. They may also easily be generalized to the Heisenberg or interaction picture. For example, the Heisenberg equations of motion for $a_{l\sigma}(t)$

are

$$i\hbar \frac{da_{l\sigma}(t)}{dt} = [a_{l\sigma}(t), H_H] = -i\omega_l a_{l\sigma}(t). \tag{4.4.6}$$

The Schrödinger equation is

$$i\hbar \frac{\partial|\psi_S(t)\rangle}{\partial t} = H_S|\psi_S(t)\rangle. \tag{4.4.7}$$

If H_S is given by (4.4.2) for an infinite number of noninteracting oscillators, then

$$|\psi_S(t)\rangle = \exp\left(-i\sum_{l,\sigma} a_{l\sigma}^\dagger a_{l\sigma}\omega_l t\right)|\psi(0)\rangle. \tag{4.4.8}$$

We may express an arbitrary initial state of the system by

$$|\psi(0)\rangle = \sum_{n_1,n_2,\ldots,n_\infty} c(n_1, n_2, \ldots, n_\infty, 0)|n_1, n_2, \ldots, n_\infty), \tag{4.4.9}$$

where we have abbreviated l, σ by l. Therefore, (4.4.8) is

$$|\psi(t)\rangle = \sum_{n_1,n_2,\ldots,n_\infty} c(n_1, n_2, \ldots, n_\infty, 0)\exp\left(-i\sum_{l,\sigma} n_l\omega_l t\right)|n_1, n_2, \ldots, n_\infty\rangle. \tag{4.4.10}$$

We also have the completeness and orthogonality relations

$$\begin{gathered}\sum_{n_1,n_2,\ldots,n_\infty} |n_1, n_2, \ldots, n_\infty\rangle\langle n_1, n_2, \ldots, n_\infty| = I \\ \langle n_1, n_2, \ldots | n_1', n_2', \ldots\rangle = \delta_{n_1n_1'}\delta_{n_2n_2'}\cdots.\end{gathered} \tag{4.4.11}$$

Since each mode of the radiation field is independent, we may introduce a coherent state as in Section 2.5 for each mode. If we again let l represent l and σ as above, we have

$$a_l|\alpha_l\rangle = \alpha_l|\alpha_l\rangle \tag{4.4.12}$$

$$\begin{aligned}|\alpha_l\rangle &= e^{-\frac{1}{2}|\alpha_l|^2}e^{\alpha_l a_l^\dagger}|0\rangle \\ &= e^{-\frac{1}{2}|\alpha_l|^2}\sum_{n_l=0}^{\infty}\frac{(\alpha_l)^{n_l}}{\sqrt{n_l!}}|n_l\rangle.\end{aligned} \tag{4.4.13}$$

All our previous results on the coherent state of an oscillator may be taken over directly. A classical source driving the cavity will excite a mode in a coherent state. We may write a state vector as

$$|\underset{\sim}{\alpha}\rangle \equiv |\alpha_1\rangle|\alpha_2\rangle\cdots|\alpha_\infty\rangle \equiv |\alpha_1, \alpha_2, \ldots, \alpha_\infty\rangle. \tag{4.4.14}$$

If we take the expectation value of the free electric field (4.3.44a) when the cavity is in a state described by (4.4.14), we have

$$\langle \underset{\sim}{\alpha}|\mathbf{E}(\mathbf{r}, t)|\alpha\rangle \equiv \mathscr{E}^{(+)}(\mathbf{r}, t) + \mathscr{E}^{(-)}(\mathbf{r}, t) \tag{4.4.15a}$$

where

$$\begin{aligned}\mathscr{E}^{+}(\mathbf{r}, t) &= i \sum_{l} \sqrt{\hbar\omega_l/2\epsilon_0\tau}\, \alpha_l e^{-i\omega_l t} e^{i\mathbf{k}_l\cdot\mathbf{r}} \hat{e}_l \\ &= [\mathscr{E}^{(-)}(\mathbf{r}, t)]^*\end{aligned} \tag{4.4.15b}$$

Therefore, when the field is in a coherent state, its average value looks like a classical field. In contrast, if the field is in the state $|\underset{\sim}{n}\rangle \equiv |n_1, n_2, \ldots, n_\infty\rangle$, we have that

$$\langle \underset{\sim}{n}|\mathbf{E}(\mathbf{r}, t)|\underset{\sim}{n}\rangle = 0 \tag{4.4.16}$$

no matter how highly excited the modes are; that is, no matter how large the n_l's are. By the correspondence principle when a system is highly excited its energy is large compared to $\hbar\omega$ and the system should behave classically. Therefore, the eigenstate $|\underset{\sim}{n}\rangle$ is very poor to show the classical nature of the field whereas the coherent state is ideally suited for this purpose.

We may expand an arbitrary initial state by means of the completeness relations:

$$\begin{aligned}|\psi(0)\rangle &= \int |\underset{\sim}{\alpha}\rangle \left(\frac{d^2\underset{\sim}{\alpha}}{\pi}\right) \langle \underset{\sim}{\alpha}|\psi(0)\rangle \\ &= \int \left(\frac{d^2\underset{\sim}{\alpha}}{\pi}\right) \psi(\underset{\sim}{\alpha}, 0)|\underset{\sim}{\alpha}\rangle,\end{aligned} \tag{4.4.17}$$

where

$$\frac{d^2\underset{\sim}{\alpha}}{\pi} \equiv \prod_l \frac{d^2\alpha_l}{\pi}, \tag{4.4.18}$$

and

$$\psi(\underset{\sim}{\alpha}, 0) = \psi(\alpha_1, \alpha_2 \cdots; 0) = \langle \underset{\sim}{\alpha}|\psi(0)\rangle. \tag{4.4.19}$$

At time t, we have from (4.4.8) and the above that

$$\begin{aligned}|\psi(t)\rangle &= \int \frac{d^2\underset{\sim}{\alpha}}{\pi} \psi(\underset{\sim}{\alpha}, 0)\left[\prod_l e^{-i\omega_l t a_l^\dagger a_l}|\alpha_l\rangle\right] \\ &= \prod_l \int \frac{d^2\alpha_l}{\pi} \langle \alpha_l|\psi(0)\rangle e^{-i\omega_l t a_l^\dagger a_l}|\alpha_l\rangle.\end{aligned} \tag{4.4.20}$$

If we write the exponential in normal order, we have

$$e^{-i\omega_l t a_l^\dagger a_l} = \mathscr{N}\{\exp[(e^{-i\omega_l t} - 1)\alpha_l^* \alpha_l]\}, \tag{4.4.21}$$

so that

$$|\psi(t)\rangle = \prod_l \int \frac{d^2\alpha_l}{\pi} \exp\,[(e^{-i\omega_l t} - 1)\alpha_l a_l^\dagger]|\alpha_l\rangle\langle\alpha_l|\psi(0)\rangle. \qquad (4.4.22)$$

If at $t = 0$ the free field is in a coherent state, this shows that it will remain in a coherent state. [Prove this.]

4.5 DENSITY OF MODES

In the ensuing work it will be necessary to know the number of normal modes in a given frequency range contained in a cavity of volume τ. This information is found in (4.3.36):

$$\mathbf{k}_l = \frac{2\pi}{L}(l_1\hat{\mathbf{i}} + l_2\hat{\mathbf{j}} + l_3\hat{\mathbf{k}}). \qquad (4.5.1)$$

Each set of integers (l_1, l_2, l_3) corresponds to two traveling-wave modes, since there are two polarizations. We may represent each mode of a given polarization by a dot in a three-dimensional space, as shown in Figure 4.1. In a small element of volume $dl_1\, dl_2\, dl_3$, the number of normal modes is

$$dN = 2dl_1\, dl_2\, dl_3. \qquad (4.5.2)$$

If we use (4.5.1), this becomes

$$dN = 2\left(\frac{L}{2\pi}\right)^3 dk_x\, dk_y\, dk_z. \qquad (4.5.3)$$

In summations of discrete values of l (l_1, l_2, l_3), when we let $L \to \infty$ (free-space limit), l_1/L, l_2/L, and l_3/L become practically continuous variables. We may replace sums by integrals so that

$$\frac{1}{L^3}\sum_l (\ \ ) \xrightarrow[L\to\infty]{} \frac{1}{(2\pi)^3}\iiint_{-\infty}^{\infty} dk_x\, dk_y\, dk_z\, (\ \ ). \qquad (4.5.4)$$

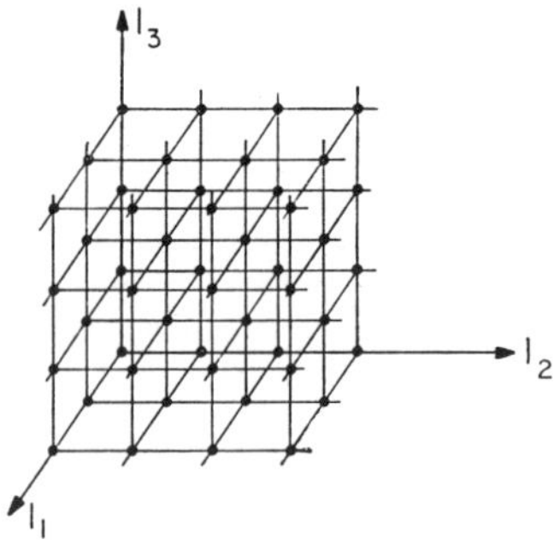

Figure 4.1 Diagram of normal modes in a cavity. Each triple of integers (l_1, l_2, l_3) corresponds to a mode of fixed polarization.

We may transform from rectangular coordinates (k_x, k_y, k_z) to polar coordinates by means of

$$\mathbf{k} = k(\sin\theta\cos\varphi, \sin\theta\sin\varphi, \cos\theta), \tag{4.5.5}$$

so that the element of volume in **k** space is

$$dk_x\,dk_y\,dk_z = k^2\,dk\,\sin\theta\,d\theta\,d\varphi \equiv k^2\,dk\,d\Omega, \tag{4.5.6}$$

where $d\Omega$ is an element of solid angle about the direction of propagation **k**. We may then write (4.5.3) as

$$dN = 2\left(\frac{L}{2\pi}\right)^3 k^2\,dk\,d\Omega. \tag{4.5.7}$$

Obviously, the total number of modes is

$$N = \int dN = 2\left(\frac{L}{2\pi}\right)^3 \int_0^\infty k^2\,dk \int_0^{4\pi} d\Omega \to \infty. \tag{4.5.8}$$

In (4.5.7) dN gives the number of modes in volume L^3 in a solid angle $d\Omega$ about the direction of propagation **k** in the range between k and $k + dk$. Since $\omega^2 = c^2\mathbf{k}^2$, we may also write the number of modes in L^3 in the frequency range ω to $\omega + d\omega$. We have

$$k^2\,dk = \frac{\omega^2}{c^3}\,d\omega, \tag{4.5.9}$$

and so

$$dN = 2\left(\frac{L}{2\pi c}\right)^3 \omega^2\,d\omega\,d\Omega = 2\left(\frac{L}{c}\right)^3 \nu^2\,d\nu\,d\Omega, \tag{4.5.10}$$

where $\omega = 2\pi\nu$. Therefore, the number of oscillators per unit volume with angular frequency between ω and $\omega + d\omega$ in solid angle $d\Omega$ is

$$g(\omega)\,d\omega = \frac{2\omega^2}{(2\pi c)^3}\,d\omega, \tag{4.5.11}$$

where $g(\omega)$ is called the mode density.

4.6 COMMUTATION RELATIONS FOR FIELDS IN VACUUM AT EQUAL TIMES

We have seen in Section 1.12 that the commutation relations for observables are closely associated with the problem of measurement of these observables. The physical observables of electromagnetic fields are $\mathbf{D} = \epsilon_0\mathbf{E}$ and $\mathbf{B} = \mu_0\mathbf{H}$. From (4.3.44), these fields are represented by hermitian operators, since $a_{l\sigma}^\dagger$ is the adjoint of $a_{l\sigma}$. It is of interest to use the commutation relations for these operators to find the commutation relations for the physical observables **E** and **H** to see the quantum restrictions on their measurement imposed by the uncertainty relations. Such questions are of great importance in quantum electronics.

By (4.3.30) the vector potential is also a hermitian operator although classically it is not a physical observable. However, it obviously plays a very useful role in the theory, and so we begin this section by evaluating the commutator of **D** and **A**.

Commutation Relations for D and A at Equal Times

The vector potential in the Schrödinger picture is, by (4.3.30),

$$\mathbf{A}_S(\mathbf{r}, 0) = \sum_{l,\sigma} \sqrt{\frac{\hbar}{2\omega_l \epsilon_0 \tau}}\, \hat{\mathbf{e}}_{l\sigma}[a_{l\sigma} \exp(i\mathbf{k}_l \cdot \mathbf{r}) + a_{l\sigma}^\dagger \exp(-i\mathbf{k}_l \cdot \mathbf{r})], \quad (4.6.1a)$$

where we use the convention that $a_{l\sigma}(t)$ is in the Heisenberg picture while $a_{l\sigma} \equiv a_{l\sigma}(0)$ is in the Schrödinger picture at $t = 0$. The fixed time for the Schrödinger picture could be taken as t_0 but nothing is gained by this choice.

The electric displacement operator $\mathbf{D} = \epsilon_0 \mathbf{E}$ in the Schrödinger picture is, by (4.3.44a),

$$\mathbf{D}_S(\mathbf{r}, 0) = i \sum_{l,\sigma} \sqrt{\frac{\epsilon_0 \hbar \omega_l}{2\tau}}\, \hat{\mathbf{e}}_{l\sigma}[a_{l\sigma} \exp(i\mathbf{k}_l \cdot \mathbf{r}) - a_{l\sigma}^\dagger \exp(-i\mathbf{k}_l \cdot \mathbf{r})]. \quad (4.6.1b)$$

The commutation relations for $a_{l\sigma}$ and $a_{l\sigma}^\dagger$ are given by (4.4.1) and must be the same in both the Heisenberg and Schrödinger pictures.

Let $A_i(\mathbf{r})$ be the ith component of $\mathbf{A}_S(\mathbf{r}, 0)$ in (4.6.1a) and $D_j(\mathbf{r})$ be the jth component of $\mathbf{D}_S(\mathbf{r}, 0)$ in (4.6.1b). If we use the commutation relations (4.4.1), then we see that

$$[A_i(\mathbf{r}), D_j(\mathbf{r}')] = -\frac{i\hbar}{2\tau} \sum_{l,\sigma} (\hat{\mathbf{e}}_{l\sigma})_i (\hat{\mathbf{e}}_{l\sigma})_j [\exp(i\mathbf{k}_l \cdot \boldsymbol{\rho}) + \exp(-i\mathbf{k}_l \cdot \boldsymbol{\rho})], \quad (4.6.2a)$$

where

$$\boldsymbol{\rho} = \mathbf{r} - \mathbf{r}'. \quad (4.6.2b)$$

We may next carry out the sum over the polarization index σ. We know that $\hat{\mathbf{k}}_l$, $\hat{\mathbf{e}}_{l1}$, and $\hat{\mathbf{e}}_{l2}$ are three mutually perpendicular unit vectors. The number $(\hat{\mathbf{e}}_{l1})_i$ is the component of the vector $\hat{\mathbf{e}}_{l1}$ on the cartesian axis x_i and is just the direction cosine of the angle between $\hat{\mathbf{e}}_{l1}$ and the x_i axis. Similarly, $(\hat{\mathbf{e}}_{l2})_j$ is the direction cosine between $\hat{\mathbf{e}}_{l2}$ and the x_j axis, and $(\hat{\mathbf{k}}_l)_i$ is the direction cosine between $\hat{\mathbf{k}}_l$ and x_i axis. By the well-known properties of direction cosines, we then have

$$(\hat{\mathbf{e}}_{l1})_i (\hat{\mathbf{e}}_{l1})_j + (\hat{\mathbf{e}}_{l2})_i (\hat{\mathbf{e}}_{l2})_j + (\hat{\mathbf{k}}_l)_i (\hat{\mathbf{k}}_l)_j = \delta_{ij},$$

or

$$\sum_{\sigma=1}^{2}(\hat{\mathbf{e}}_{l\sigma})_i(\hat{\mathbf{e}}_{l\sigma})_j = \delta_{ij} - (\hat{\mathbf{k}}_l)_i(\hat{\mathbf{k}}_l)_j \equiv \delta_{ij} - \frac{(\mathbf{k}_l)_i(\mathbf{k}_l)_j}{k_l^{\,2}}. \tag{4.6.3}$$

If we substitute this in (4.6.2), we have for the commutation relations in the *cavity*

$$[A_i(\mathbf{r}), D_j(\mathbf{r}')] = -\frac{i\hbar}{\tau}\sum_l [\delta_{ij} - (\hat{\mathbf{k}}_l)_i(\hat{\mathbf{k}}_l)_j] \exp(i\hat{\mathbf{k}}_l \cdot \boldsymbol{\rho}). \tag{4.6.4}$$

We have been able to combine $\exp(-i\mathbf{k}_l \cdot \boldsymbol{\rho})$ and $\exp(+i\mathbf{k}_l \cdot \boldsymbol{\rho})$ since $\hat{\mathbf{k}}_{-l} = -\hat{\mathbf{k}}_{+l}$ and the sum over l is for all positive and negative integers.

The commutator $A_i(\mathbf{r})$ and $D_j(\mathbf{r}')$ in a cavity is therefore given by the unusual-looking result in (4.6.4). For free space, we may let the cavity become infinitely large ($L^3 = \tau \to \infty$). When we do this, we may use (4.5.4) in the right side of (4.6.4) to replace the sums by integrals, and we have in *free space*

$$[A_i(\mathbf{r}), \mathbf{D}_j(\mathbf{r}')] = -i\hbar\delta_{ij}{}^T(\mathbf{r} - \mathbf{r}'), \tag{4.6.5}$$

where we define

$$\delta_{ij}{}^T(\boldsymbol{\rho}) = \frac{1}{(2\pi)^3}\iiint_{-\infty}^{\infty} d\mathbf{k} \exp(i\mathbf{k}\cdot\boldsymbol{\rho})(\delta_{ij} - \hat{k}_i\hat{k}_j), \tag{4.6.6}$$

and $d\mathbf{k} = dk_x\, dk_y\, dk_z$ is a volume element in k-space. Here $\delta_{ij}{}^T$ is called the transverse δ function and is *not* the ordinary Dirac δ function, which is

$$\delta(\boldsymbol{\rho}) = \frac{1}{(2\pi)^3}\iiint_{-\infty}^{\infty} d\mathbf{k} \exp(i\mathbf{k}\cdot\boldsymbol{\rho}) \equiv \delta(\rho_x)\,\delta(\rho_y)\,\delta(\rho_z). \tag{4.6.7}$$

We derive some very useful properties of the transverse δ function in Appendix D. For example, we show there that [Eq. D.6)]

$$\sum_{i=1}^{3}\frac{\partial\delta_{ij}{}^T(\boldsymbol{\rho})}{\partial x_i} = 0, \tag{4.6.8}$$

where $\boldsymbol{\rho} = \mathbf{r} - \mathbf{r}'$ and x_1, x_2, x_3 are the components of $\mathbf{r}$. From this, if we take the divergence of both sides of (4.6.5), we have

$$\left[\sum_{i=1}^{3}\frac{\partial A_i(\mathbf{r})}{\partial x_i}, D_j(\mathbf{r}')\right] = -i\hbar\sum_{i=1}^{3}\frac{\partial\delta_{ij}{}^T}{\partial x_i} = 0. \tag{4.6.9}$$

Since $D_j(\mathbf{r}') \neq 0$, then

$$\operatorname{div}\mathbf{A} \equiv \sum_i \frac{\partial A}{\partial x_i} = 0.$$

This, however, is the requirement for using the Coulomb gauge, so that the commutation relation (4.6.5) is valid for this gauge.

We may use the fact that commutation relations are identical in both the Heisenberg and Schrödinger pictures to write from (4.6.5)

$$[A_i^H(\mathbf{r}, t), D_j^H(\mathbf{r}', t)] = -i\hbar\delta_{ij}{}^T(\mathbf{r} - \mathbf{r}'), \tag{4.6.10}$$

where t is the *same* time in A_i and D_j.

We may next show that

$$[A_i(\mathbf{r}, t), A_j(\mathbf{r}', t)] = 0. \tag{4.6.11}$$

To do this, we work in the Schrödinger picture. By (4.6.1a) and the commutation relations (4.4.1), we have

$$[A_i(\mathbf{r}), A_j(\mathbf{r}')] = \sum_{l,\sigma} \frac{\hbar}{2\omega_l\epsilon_0\tau} (\hat{\mathbf{e}}_{l\sigma})_i(\hat{\mathbf{e}}_{l\sigma})_j[\exp(i\mathbf{k}_l \cdot \boldsymbol{\rho}) - \exp(-i\mathbf{k}_l \cdot \boldsymbol{\rho})],$$

where $\boldsymbol{\rho} = \mathbf{r} - \mathbf{r}'$. We may use (4.6.3) to carry out the sum over σ so that the above commutator becomes

$$[A_i(\mathbf{r}), A_j(\mathbf{r}')] = i\hbar \sum_l \frac{1}{\omega_l\epsilon_0\tau} [\delta_{ij} - (\hat{\mathbf{k}}_l)_i(\hat{\mathbf{k}}_l)_j] \sin \mathbf{k}_l \cdot \boldsymbol{\rho}. \tag{4.6.12}$$

Since $\omega_l = c|\mathbf{k}_l|$ and $\mathbf{k}_{-l} = -\mathbf{k}_l$, we see that $\omega_l^{-1}[\delta_{ij} - (\hat{\mathbf{k}}_l)_i(\hat{\mathbf{k}}_l)_j]$ is even under an interchange of $-l$ for l while $\sin \mathbf{k}_l \cdot \boldsymbol{\rho}$ is odd, and so the sum vanishes. Therefore, (4.6.11) is true in the Schrödinger picture and hence in the Heisenberg picture.

Commutation Relations for D and B at Equal Times

The physical observables are **D** and **B** (or **E** and **H**). It is therefore very important to know their commutation relations at equal times.

By the same procedure used to prove (4.6.11), it is easily shown that

$$[D_i(\mathbf{r}, t), D_j(\mathbf{r}', t)] = 0 \tag{4.6.13a}$$

$$[B_i(\mathbf{r}, t), B_j(\mathbf{r}', t)] = 0, \tag{4.6.13b}$$

where B_i in the Schrödinger picture is given by

$$\mathbf{B}_S(\mathbf{r}, 0) = -i \sum_{l,\sigma} \sqrt{\frac{\hbar\omega_l\mu_0}{2\tau}} (\hat{\mathbf{e}}_{l\sigma} \times \hat{\mathbf{k}}_l)[a_{l\sigma} \exp(i\mathbf{k}_l \cdot \mathbf{r}) - a_{l\sigma}^\dagger \exp(-i\mathbf{k}_l \cdot \mathbf{r})]. \tag{4.6.14}$$

In the proof of (4.6.13a), we need the result

$$\sum_\sigma (\hat{\mathbf{e}}_{l\sigma} \times \hat{\mathbf{k}}_l)_i(\hat{\mathbf{e}}_{l\sigma} \times \hat{\mathbf{k}}_l)_j = \sum_\sigma (\hat{\mathbf{e}}_{l\sigma})_i(\hat{\mathbf{e}}_{l\sigma})_j. \tag{4.6.15}$$

The proof of this follows easily if we let $\hat{\mathbf{e}}_{l1}$, $\hat{\mathbf{e}}_{l2}$, and $\hat{\mathbf{k}}_l$ form a right-hand system with

$$\begin{aligned} \hat{\mathbf{e}}_{l\sigma} \cdot \hat{\mathbf{e}}_{l\sigma'} &= \delta_{\sigma\sigma'} \qquad \sigma, \sigma' = 1, 2 \\ \hat{\mathbf{e}}_{l\sigma} \cdot \hat{\mathbf{k}}_l &= 0 \\ \hat{\mathbf{e}}_{l1} \times \hat{\mathbf{e}}_{l2} &= \hat{\mathbf{k}}_l \end{aligned} \tag{4.6.16}$$

Then, $\hat{\mathbf{e}}_{l1} \times \hat{\mathbf{k}}_l = -\hat{\mathbf{e}}_{l2}$ and $\hat{\mathbf{e}}_{l2} \times \hat{\mathbf{k}}_l = \hat{\mathbf{e}}_{l1}$, and the proof is evident.

From the theory of Section 1.12, it may be concluded from (4.6.13) that any two compounds of **D** (or **E**) may be simultaneously measured without any mutual interference. The same remark applies for any two components of **B** (or **H**) since all components commute.

In Appendix E, we derive the commutation relations

$$[D_i(\mathbf{r}, t), B_i(\mathbf{r}', t)] = 0 \qquad i = 1, 2, 3, \tag{4.6.17}$$

and

$$[D_i(\mathbf{r}, t), B_j(\mathbf{r}', t)] = \begin{cases} -i\hbar \dfrac{\partial}{\partial x_k} \delta(\boldsymbol{\rho}) & \text{(4.6.18a)} \\ +i\hbar \dfrac{\partial}{\partial x_k} \delta(\boldsymbol{\rho}), & \text{(4.6.18b)} \end{cases}$$

where we use (4.6.18a) if i, j, and k form a cyclic permutation of 1, 2, and 3 and we use (4.6.18b) if i, j, and k form a cyclic permutation of 1, 3, 2.

From (4.6.17) we conclude that parallel components of **D** and **B** may be measured simultaneously without mutual interference, and from (4.6.18) we see that perpendicular components cannot be measured simultaneously without interference.

Heisenberg Equations of Motion for D and B

Since **D**, **B**, **E**, and **H** are operators, they satisfy the Heisenberg equations of motion. The hamiltonian for the field may be written as

$$H = \frac{1}{2}\int \left[\frac{1}{\epsilon_0} \mathbf{D}^2(\mathbf{r}', t) + \frac{1}{\mu_0} \mathbf{B}^2(\mathbf{r}', t)\right] d\tau'. \tag{4.6.19}$$

It is a hermitian operator given here in the Heisenberg picture. In Appendix F, we show that

$$i\hbar \frac{d\mathbf{D}_H}{dt} = [\mathbf{D}_H, H] = i\hbar \operatorname{curl} \mathbf{H}_H \tag{4.6.20a}$$

$$i\hbar \frac{d\mathbf{B}_H}{dt} = [\mathbf{B}_H, H] = -i\hbar \operatorname{curl} \mathbf{E}_H. \tag{4.6.20b}$$

These are the Heisenberg equations of motion for the *operators* in the Heisenberg picture, but they are identical in form with the classical Maxwell "equations of motion"; thus we have developed a self-consistent quantum theory in that it will become classical theory when $\hbar \to 0$. From (4.6.18), as $\hbar \to 0$, the operators commute and behave like classical variables.

The remaining two Maxwell equations are also satisfied. Since $\hat{\mathbf{e}}_{l\sigma} \cdot \hat{\mathbf{k}}_l = 0$, it is obvious from (4.6.1b) that $\operatorname{div} \mathbf{D} = 0$. Also, since $(\hat{\mathbf{e}}_{l\sigma} \times \hat{\mathbf{k}}_l) \cdot \hat{\mathbf{k}}_l = 0$, (4.6.14) shows that $\operatorname{div} \mathbf{B} = 0$.

4.7 ZERO-POINT FIELD FLUCTUATIONS

In Section 4.1 we discussed the zero-point energy and the zero-point fluctuations of an LC circuit. In this section we discuss the zero-point fluctuations of the fields.

We may write the jth component of the electric displacement vector $D_j(\mathbf{r}, t)$ in the Heisenberg picture from (4.6.1b) as

$$D_j(\mathbf{r}, t) = D_j^{(+)}(\mathbf{r}, t) + D_j^{(-)}(\mathbf{r}, t), \tag{4.7.1}$$

where

$$\begin{aligned} D_j^{(+)}(\mathbf{r}, t) &= i \sum_{l,\sigma} \sqrt{\frac{\epsilon_0 \hbar \omega_l}{2\tau}} (\hat{\mathbf{e}}_{l\sigma})_j a_{l\sigma}(t) \exp(i\mathbf{k}_l \cdot \mathbf{r}) \\ D_j^{(-)}(\mathbf{r}, t) &= [D_j^{(+)}(\mathbf{r}, t)]^\dagger. \end{aligned} \tag{4.7.2}$$

Here $D_j^{(+)}$ contains only annihilation operators, and its adjoint $D_j^{(-)}$ contains only creation operators.

The average value of $\mathbf{D}$ is

$$\langle D_j \rangle = \langle \psi(0) | D_j(\mathbf{r}, t) | \psi(0) \rangle, \tag{4.7.3}$$

where $|\psi(0)\rangle$ is the state of the field at $t = 0$. In general, $|\psi(0)\rangle$ is given by (4.4.9) as an expansion in the complete set of eigenkets of the energy (or photon-number operator). For simplicity, we consider the case in which the field has been prepared in a pure energy eigenstate at $t = 0$ and there are $n_{l\sigma}$ quanta in the $l\sigma$ mode, that is,

$$|\psi(0)\rangle = |n_1, n_2, \ldots, n_l, \ldots, n_\infty\rangle, \tag{4.7.4}$$

where we have abbreviated $l\sigma$ by l; that is, the modes are labeled $1, 2, \ldots, \infty$. If we use (4.4.5) and the orthogonality relations (4.4.11), the average of D_j for the energy eigenstate (4.7.4) is

$$\langle D_j \rangle = 0, \tag{4.7.5}$$

that is, the average value of the electric displacement vector is zero when the field is in an energy eigenstate. It is also easily shown that the average of $\mathbf{B}$ is

$$\langle B_j \rangle = 0 \tag{4.7.6}$$

for (4.7.4). This is the analog of the classical result that the time average of $\mathbf{D}(\mathbf{r}, t)$ and $\mathbf{B}(\mathbf{r}, t)$ is zero for each harmonic component.

Next we consider the expectation value of D_j^2 for the state $|\psi(0)\rangle$ in (4.7.4). From (4.7.1), we have

$$D_j^2 = D_j^{(+)2} + D_j^{(-)2} + D_j^{(-)}D_j^{(+)} + D_j^{(+)}D_j^{(-)}. \tag{4.7.7}$$

If we calculate $\langle D_j^2\rangle$, we have by (4.4.5), (4.4.11), (4.4.1), and (4.7.2), after minor algebra,

$$\langle D_j^2(\mathbf{r}, t)\rangle = \sum_{l,\sigma} \frac{\epsilon_0 \hbar\omega_l}{\tau} (\hat{\mathbf{e}}_{l\sigma})_j^2 (n_{l\sigma} + \tfrac{1}{2}). \tag{4.7.8}$$

Since

$$\mathbf{D}^2 = \sum_{j=1}^{3} D_j^2 \tag{4.7.9}$$

we have

$$\langle \mathbf{D}^2\rangle = \frac{1}{\tau} \sum_{l,\sigma} \epsilon_0 \hbar\omega_l (n_{l\sigma} + \tfrac{1}{2}) \sum_{j=1}^{3} (\hat{\mathbf{e}}_{l\sigma})_j^2. \tag{4.7.10}$$

But $(\hat{\mathbf{e}}_{l\sigma})_j$ is just the direction cosine between $\hat{\mathbf{e}}_{l\sigma}$ and the x_j axis, and so $\sum (\hat{\mathbf{e}}_{l\sigma})_j^2 = 1$. Therefore, the average value of the total field is

$$\langle \mathbf{D}^2\rangle = \frac{\epsilon_0}{\tau} \sum_{l,\sigma} \hbar\omega_l (n_{l\sigma} + \tfrac{1}{2}). \tag{4.7.11}$$

If the $n_{l\sigma} = 0$ (the vacuum state), then

$$\langle \mathbf{D}^2\rangle = \frac{\epsilon_0}{\tau} \sum_{l,\sigma} \frac{\hbar\omega_l}{2}. \tag{4.7.12}$$

This is called the zero-point field fluctuation. Each mode of the field contributes an amount $\hbar\omega_l/2$; since there are an infinite number of modes in the cavity, the zero-point field fluctuation is infinite.

This infinity is not fundamental since we have not specified the means by which we shall measure the field. In electrical engineering terminology, the bandwidth of the detecting instrument would be finite and so it would not detect an infinite value for these zero-point field fluctuations. We might envision an electric charge interacting with the field as a means of detection. In this case the charge would occupy some small volume of space, ΔV, and it would take some small amount of time, Δt, for a measurable effect to occur. This would imply that, to obtain physically meaningful results, we would have to compute a space-time average:

$$\bar{D}_j(\Delta V, \Delta t) \equiv \frac{1}{\Delta V\, \Delta t} \int_{\Delta V} d\tau \int_{\Delta t} dt\, D_j^H(\mathbf{r}, t). \tag{4.7.13}$$

The expectation value of D_j^2 in the vacuum state is then

$$\langle 0|[\bar{D}_j(\Delta V, \Delta t)]^2|0\rangle = \frac{1}{(\Delta V)^2(\Delta t)^2}\int_{\Delta V,\Delta t} d\tau\, d\tau'\, dt\, dt'\, \langle 0|\, D_j(\mathbf{r}, t)\, D_j(\mathbf{r}', t')|0\rangle. \tag{4.7.14}$$

From (4.7.2) we see that

$$\langle 0|\, D_j(\mathbf{r}, t)\, D_j(\mathbf{r}', t')|0\rangle = \sum_{l,\sigma} \frac{\epsilon_0 \hbar\omega_l}{2\tau} (\hat{\mathbf{e}}_{l\sigma})_j^{\,2} \exp\{i[\mathbf{k}_l \cdot (\mathbf{r} - \mathbf{r}') - \omega_l(t - t')]\}. \tag{4.7.15}$$

If we sum over j to get the total field, since $\sum_{j=1}^{3} (\hat{\mathbf{e}}_{l\sigma})_j^2 = 1$, we have for (4.7.14)

$$\langle 0|\sum_j [\bar{D}_j(\Delta V, \Delta t)]^2|0\rangle = \frac{\epsilon_0\hbar}{2(\Delta V)^2(\Delta t)^2}\int d\tau\, d\tau'\, dt\, dt' \times \sum_l \frac{\omega_l}{\tau} \exp\{i[\mathbf{k}_l \cdot \boldsymbol{\rho} - \omega_l(t - t')]\}, \tag{4.7.16}$$

where $\boldsymbol{\rho} = \mathbf{r} - \mathbf{r}'$. We may carry out the integral over dt and dt' and change the sum over l to an integral over $\mathbf{k}$ by means of (4.5.4), and we have

$$\langle 0|\sum_j [\bar{D}_j(\Delta V, \Delta t)]^2|0\rangle = \frac{\hbar\epsilon_0}{16\pi^3(\Delta V)^2(\Delta t)^2 c} \times \iiint \frac{\mathbf{dk}\, d\tau\, d\tau'}{k} \exp(i\mathbf{k}\cdot\boldsymbol{\rho}) 4\sin^2 \tfrac{1}{2}ck\,\Delta t, \tag{4.7.17}$$

where $\omega_l \to c|\mathbf{k}| \equiv ck$. As long as Δt and ΔV are finite, the average value of the field in the region measured will be finite, from the above integral, whereas if ΔV and $\Delta t \to 0$, it becomes infinite. Restricting knowledge of the field to a limited region of space time has effectively placed a bandwidth limitation on the detecting instrument.

The zero-point field fluctuations give rise to measurable effects in quantum mechanics. In particular, they account for the Lamb shift of the $2P_{1/2}$-$2S_{1/2}$ energy levels of atomic hydrogen, where the zero-point fluctuations interact with the electron. One might say picturesquely that they "induce" spontaneous emission of the electron in the $2P_{1/2}$ state. The zero-point fluctuations also give rise to spontaneous emission in lasers, parametric amplifiers, attenuators, and such, and are the source of quantum noise. They may also be said to be the "source" of the natural line width of atoms. We discuss these questions in more detail later, after we discuss the interaction of electromagnetic fields with matter.

4.8 CLASSICAL RADIATION FIELD WITH SOURCES [7]

We next turn our attention to the problem of an electromagnetic field interacting with a given charge and current distribution. This is the field generalization of an *LC* circuit with a voltage generator. This and the next section should give some insight into the way to proceed in such problems.

The Maxwell equations with sources in mks units are

$$\text{div } \mathbf{B} = 0 \tag{4.8.1}$$

$$\text{curl } \mathbf{E} = -\frac{\partial \mathbf{B}}{\partial t} \tag{4.8.2}$$

$$\text{div } \mathbf{D} = \rho \tag{4.8.3}$$

$$\text{curl } \mathbf{H} = \frac{\partial \mathbf{D}}{\partial t} + \mathbf{J}, \tag{4.8.4}$$

where

$$\mathbf{B} = \mu_0 \mathbf{H} \qquad \mathbf{D} = \epsilon_0 \mathbf{E}. \tag{4.8.5}$$

The charge and current must satisfy the continuity equation

$$\text{div } \mathbf{J} + \frac{\partial \rho}{\partial t} = 0, \tag{4.8.6}$$

where $\mathbf{J}$ and ρ are functions of $\mathbf{r}$ and t. This follows directly if we take the divergence of both sides of (4.8.4) and use (4.8.3).

As in the source-free case, (4.8.1) and (4.8.2) are identically satisfied if we let

$$\mathbf{B} = \text{curl } \mathbf{A} \tag{4.8.7a}$$

$$\mathbf{E} = -\frac{\partial \mathbf{A}}{\partial t} - \nabla V. \tag{4.8.7b}$$

Since it is well known that the Maxwell equations are gauge invariant, we shall work in the Coulomb gauge as we did in the source-free field case, in which

$$\text{div } \mathbf{A} = 0. \tag{4.8.8}$$

However, when ρ and $\mathbf{J}$ are not zero, we may no longer let $V = 0$.

If we now substitute (4.8.7a) and (4.8.7b) into (4.8.4) and use div $\mathbf{A} = 0$, and (4.8.5), we obtain, instead of the wave equation (4.3.10),

$$\nabla^2 \mathbf{A} - \frac{1}{c^2}\frac{\partial^2 \mathbf{A}}{\partial t^2} = -\mu_0 \mathbf{J} + \nabla\left(\frac{1}{c^2}\frac{\partial V}{\partial t}\right), \tag{4.8.9}$$

while if we put (4.8.7b) into (4.8.3) and use (4.8.5) and (4.8.8), then V satisfies the Poisson equation

$$\nabla^2 V = -\frac{1}{\epsilon_0}\rho. \tag{4.8.10}$$

We are assuming that ρ and $\mathbf{J}$ are given functions. Therefore, the potential V is determined by the well-known result that

$$V(\mathbf{r}, t) = \frac{1}{4\pi\epsilon_0}\int \frac{\rho(\mathbf{r}', t)}{|\mathbf{r} - \mathbf{r}'|}\, d\tau', \tag{4.8.11}$$

where $d\tau' = dx'\, dy'\, dz'$.

It is known from vector analysis that any vector $\mathbf{A}$ may always be written as the sum of its transverse components $\mathbf{A}^T$ and its longitudinal components $\mathbf{A}^L$.

$$\mathbf{A} = \mathbf{A}^T + A^L, \tag{4.8.12}$$

where, by definition,

$$\operatorname{div} \mathbf{A}^T = 0 \qquad \operatorname{curl} \mathbf{A}^L = 0. \tag{4.8.13}$$

In the Coulomb gauge, since $\operatorname{div} \mathbf{A} = 0$, we have, by (4.8.12) and (4.8.13),

$$\operatorname{div} \mathbf{A}^L = 0 \tag{4.8.14}$$

If both $\operatorname{div} \mathbf{C} = \operatorname{curl} \mathbf{C} = 0$, then we may let $\mathbf{C} = 0$. Therefore in the *Coulomb gauge*

$$\mathbf{A}^L = 0, \tag{4.8.15}$$

and the vector potential is purely transverse.

We may also decompose the current density as

$$\mathbf{J} = \mathbf{J}^T + \mathbf{J}^L \tag{4.8.16}$$

where

$$\operatorname{div} \mathbf{J}^T = 0 \tag{4.8.17a}$$

$$\operatorname{curl} \mathbf{J}^L = 0. \tag{4.8.17b}$$

Then, from the continuity equation (4.8.6),

$$\operatorname{div} \mathbf{J}^L = -\frac{\partial \rho}{\partial t}, \tag{4.8.18}$$

that is, $\partial\rho/\partial t$ gives rise only to a longitudinal current.

We may satisfy (4.8.17b) identically if we let

$$\mathbf{J}^L = \nabla\psi, \tag{4.8.19}$$

where ψ is any scalar function since $\operatorname{curl} \operatorname{grad} \psi \equiv 0$. If we substitute this into (4.8.18), we see that

$$\nabla^2 \psi = -\frac{\partial \rho}{\partial t}. \tag{4.8.20}$$

At each instant this is Poisson's equation whose solution is

$$\psi = \frac{1}{4\pi}\frac{\partial}{\partial t}\int \frac{\rho(\mathbf{r}', t)}{|\mathbf{r} - \mathbf{r}'|}\, d\tau'. \tag{4.8.21}$$

By (4.8.19) and (4.8.11),

$$\mathbf{J}^L = \epsilon_0 \nabla \frac{\partial V}{\partial t}. \tag{4.8.22}$$

If we substitute (4.8.12) and (4.8.16) into (4.8.9) and use (4.8.22), we have

$$\nabla^2 \mathbf{A}^T - \frac{1}{c^2}\frac{\partial^2 \mathbf{A}^T}{\partial t^2} = -\mu_0 \mathbf{J}^T, \tag{4.8.23}$$

which is the equation that $\mathbf{A}^T$ must satisfy while the potential is given by (4.8.11). Classical fields with sources must satisfy (4.8.10) and (4.8.23) in the Coulomb gauge.

4.9 QUANTIZATION OF FIELD WITH CLASSICAL SOURCES

To quantize the field, we must find a hamiltonian such that the Heisenberg equations of motion for the field quantities, considered as operators reduce to the Maxwell equations. The hamiltonian is not necessarily the same thing as the energy in general. In treating empirical models, trial-and-error methods must be used to find a hamiltonian such that the correct Heisenberg equations of motion have the same form as the classical equations of motion for systems having a classical analog.

Since the potential V is determined by the given charge distribution (4.8.11), it cannot be treated as an independent field variable. Therefore V is not a field operator in the quantum theory. The vector potential as well as **B**, **H**, **D**, and **E** will be treated as operators in the quantum case.

We begin by taking as the hamiltonian

$$H = \frac{1}{2}\int\left[\epsilon_0\left(\frac{\partial \mathbf{A}}{\partial t}\right)^2 + \frac{1}{\mu_0}(\text{curl } \mathbf{A})^2\right] d\tau$$
$$- \int \mathbf{J}\cdot\mathbf{A}\, d\tau + \frac{1}{8\pi\epsilon_0}\iint \frac{\rho(\mathbf{r}, t)\rho(\mathbf{r}', t)}{|\mathbf{r} - \mathbf{r}'|}\, d\tau\, d\tau'. \tag{4.9.1}$$

As noted above, the justification for this hamiltonian follows when we obtain from it the correct equations of motion. We may, however, identify the first integral as the energy contained in the field in the absence of sources. The second integral is the interaction energy between the field and current, and the last term gives the Coulomb energy between the charges present.

We now postulate the following commutation relations based on the results in the source-free case (Section 4.6). In the Schrödinger picture we have

$$\begin{aligned}[A_k(\mathbf{r}), D_l(\mathbf{r}')] &= -i\hbar\delta_{kl}{}^T(\mathbf{r}-\mathbf{r}') \\ [A_k(\mathbf{r}), A_l(\mathbf{r}')] &= [D_k(\mathbf{r}), D_l(\mathbf{r}')] = 0,\end{aligned} \tag{4.9.2}$$

where δ^T is the transverse δ function. We must show that these commutation relations together with the hamiltonian (4.9.1) give the correct equations of motion.

We know that $\mathbf{D}$ and $\mathbf{A}$ are related by

$$\epsilon_0{}^{-1}\mathbf{D} = -\frac{\partial \mathbf{A}}{\partial t} - \text{grad } V. \tag{4.9.3}$$

Since V is a c-number, grad V is also, and so

$$\begin{aligned}[A_k(\mathbf{r}), D_l(\mathbf{r}')] &= -\epsilon_0\left[A_k(\mathbf{r}), \dot{A}_l(\mathbf{r}') + \frac{\partial V}{\partial x_l'}(\mathbf{r}', t)\right] \\ &= -\epsilon_0[A_k(\mathbf{r}), \dot{A}_l(\mathbf{r}')] = -i\hbar\delta_{kl}{}^T(\mathbf{r}-\mathbf{r}'). \end{aligned} \tag{4.9.4}$$

Since grad V is a c-number, the commutation relations for the source-free field and the field with sources are identical. We may therefore use the source-free expansions for $\mathbf{A}$ and $\dot{\mathbf{A}}$ in the Schrödinger picture when sources are present. Therefore, from (4.6.1a) we have

$$\mathbf{A}_S(\mathbf{r}, 0) = \sum_{l,\sigma}\sqrt{\frac{\hbar}{2\omega_l\epsilon_0\tau}}\,\hat{\mathbf{e}}_{l\sigma}[a_{l\sigma}\exp(i\mathbf{k}_l\cdot\mathbf{r}) + a_{l\sigma}^\dagger\exp(-i\mathbf{k}_l\cdot\mathbf{r})], \tag{4.9.5}$$

and we take for $\dot{\mathbf{A}}_S(\mathbf{r}, 0)$ the value $-D_S(\mathbf{r}, 0)/\epsilon_0$ in (4.6.1b):

$$\dot{\mathbf{A}}_S(\mathbf{r}, 0) = -i\sum_{l,\sigma}\sqrt{\frac{\hbar\omega_l}{2\epsilon_0\tau}}\,\hat{\mathbf{e}}_{l\sigma}[a_{l\sigma}\exp(i\mathbf{k}_l\cdot\mathbf{r}) - a_{l\sigma}^\dagger\exp(-i\mathbf{k}_l\cdot\mathbf{r})]. \tag{4.9.6}$$

When sources are present, $\mathbf{D}$ is given by (4.9.3) and $\dot{\mathbf{A}}_S$ is still given by (4.9.6).

Since the commutation relations (4.9.2) must be the same in the Heisenberg and Schrödinger pictures, we may now derive the Heisenberg equations of motion for $\mathbf{A}(\mathbf{r}, t)$. We have

$$i\hbar\frac{dA_k}{dt} = [A_k, H]. \tag{4.9.7}$$

In (4.9.1), since ρ is a c-number, all components of $\mathbf{A}$ commute with the Coulomb term in $\mathbf{H}$. Also, $\mathbf{J}$ is a c-number, and by (4.9.2) all components of $\mathbf{A}$ commute with the interaction term in (4.9.1). Furthermore, A_k commutes with all spatial derivatives of $\mathbf{A}$ and therefore commutes with the term $(\text{curl }\mathbf{A})^2$ in (4.9.1). The only term with which A_k fails to commute is the term involving $(\dot{\mathbf{A}})^2$ in (4.9.1). In Appendix G we show that

$$i\hbar\dot{A}_k = \left[A_k, \int \frac{\epsilon_0}{2}(\dot{\mathbf{A}})^2\, d\tau\right] = i\hbar\dot{A}_k^T, \tag{4.9.8}$$

from which we conclude that the longitudinal component of $\mathbf{A}^L = 0$ while the transverse part gives an identity. This is in agreement with the classical result (4.8.15).

The next Heisenberg equation of motion to be considered is

$$i\hbar\frac{d\dot{A}_k}{dt} = [\dot{A}_k, H]. \tag{4.9.9}$$

We see immediately that $\dot{A}_k$ commutes with all terms in the hamiltonian (4.9.1) except the curl $\mathbf{A}$ term and the $\mathbf{J}\cdot\mathbf{A}$ term. We leave as an exercise to show that

$$\left[\dot{A}_k(\mathbf{r}), \int \frac{d\tau'}{2\mu_0}[\text{curl}'\ \mathbf{A}(\mathbf{r}')]^2\right] = i\hbar c^2\nabla^2 A_k{}^T(\mathbf{r}) \tag{4.9.10}$$

and

$$\left[\dot{A}_k(\mathbf{r}), \int \mathbf{J}(\mathbf{r}')\cdot\mathbf{A}(\mathbf{r}')\, d\tau'\right] = -\frac{i\hbar}{\epsilon_0}J_k{}^T(\mathbf{r}). \tag{4.9.11}$$

All quantities are tacitly assumed to apply at the same time.

When we put (4.9.10) and (4.9.11) into (4.9.9), we obtain (4.8.23), the correct field equation. We have therefore justified the use of the hamiltonian (4.9.1).

If we substitute the field expansions (4.9.5) and (4.9.6) into the hamiltonian (4.9.1), we find that H becomes

$$\begin{aligned} H = \sum_{l,\sigma}\hbar\omega_l a_{l\sigma}^\dagger a_{l\sigma} &+ \frac{1}{8\pi\epsilon_0}\iint\frac{\rho(\mathbf{r})\rho(\mathbf{r}')}{|\mathbf{r}-\mathbf{r}'|}\, d\tau\, d\tau' \\ &- \frac{1}{L^{3/2}}\sum_{l,\sigma}\sqrt{\frac{\hbar}{2\omega_l\epsilon_0}}\left[a_{l\sigma}\int \hat{\mathbf{e}}_{l\sigma}\cdot\mathbf{J}(\mathbf{r})\exp(i\mathbf{k}_l\cdot\mathbf{r})\, d\tau\right. \\ &\left. + a_{l\sigma}^\dagger\int\hat{\mathbf{e}}_{l\sigma}\cdot\mathbf{J}(\mathbf{r})\exp(-i\mathbf{k}_l\cdot\mathbf{r})\right] d\tau. \end{aligned} \tag{4.9.12}$$

The first term is the familiar source-free field energy, the next is the Coulomb energy, and the last is the interaction energy. The time dependence in H in

(4.9.12) is contained in the explicit time dependence of ρ and J whereas the operators $a_{l\sigma}$ and $a_{l\sigma}^{\dagger}$ are time-independent in the Schrödinger picture.

4.10 DENSITY OPERATOR FOR RADIATION FIELD

We use the principle of entropy maximization introduced in Chapter 3 to determine the density operator which describes a mode of the radiation field when we have made various measurements. Let us assume we measure the average energy and the average electric and magnetic field of a mode. We therefore have

$$S = -k \operatorname{Tr} \rho \ln \rho \tag{4.10.1}$$

$$\operatorname{Tr} \rho = 1 \tag{4.10.2}$$

$$\langle H \rangle = \operatorname{Tr} \rho H = \hbar\omega \operatorname{Tr} \rho a^{\dagger} a \tag{4.10.3}$$

$$\langle p \rangle = \operatorname{Tr} \rho p = \operatorname{Tr} \rho i \sqrt{\frac{\hbar\omega}{2}} (a^{\dagger} - a) \tag{4.10.4}$$

$$\langle q \rangle = \operatorname{Tr} \rho q = \operatorname{Tr} \rho \sqrt{\frac{\hbar}{2\omega}} (a^{\dagger} + a). \tag{4.10.5}$$

If we maximize the entropy subject to the measured values, we have on using Lagrange multipliers

$$\operatorname{Tr} (1 + \ln \rho + \lambda_1 + \beta H - \lambda_2 p - \lambda_3 q)\, \delta\rho = 0. \tag{4.10.6}$$

This will be satisfied by

$$\rho = e^{-(1+\lambda_1)} e^{-\beta H + \lambda_2 p + \lambda_3 q}. \tag{4.10.7}$$

When we use (4.10.2), we determine the normalizing constant $\exp -(1 + \lambda_1)$ so that

$$\rho = \frac{e^{-\beta H + \lambda_2 p + \lambda_3 q}}{\operatorname{Tr} e^{-\beta H + \lambda_2 p + \lambda_3 q}}. \tag{4.10.8}$$

Next we express H, p, and q in terms of a and $a^{\dagger}$ and let

$$\begin{aligned} \lambda &= \beta\hbar\omega \\ \lambda w &= \sqrt{\frac{\hbar}{2\omega}} (\lambda_3 + i\omega\lambda_2) \\ \lambda w^* &= \sqrt{\frac{\hbar}{2\omega}} (\lambda_3 - i\omega\lambda_2). \end{aligned} \tag{4.10.9}$$

Then we may rewrite ρ as

$$\rho = \frac{e^{-\lambda(a^{\dagger} - w^*)(a - w)}}{\operatorname{Tr} e^{-\lambda(a^{\dagger} - w^*)(a - w)}}. \tag{4.10.10}$$

We must next determine the Lagrange multipliers λ, w, and w^* in terms of the measured ensemble averages $\langle E\rangle$, $\langle p\rangle$, and $\langle q\rangle$. For this purpose, let us introduce the new operators

$$\begin{aligned} c &= a - w \\ c^\dagger &= a^\dagger - w^*, \end{aligned} \tag{4.10.11}$$

where w and w^* are c-numbers. We see that $[c, c^\dagger] = 1$ so we may establish a set of basis vectors

$$\begin{aligned} c^\dagger c|n\rangle_c &= n|n\rangle_c \\ c|n\rangle_c &= \sqrt{n}\,|n-1\rangle_c \\ c^\dagger|n\rangle_c &= \sqrt{n+1}\,|n+1\rangle_c, \end{aligned} \tag{4.10.12}$$

and use these to evaluate the trace in (4.10.10). Thus

$$\rho = \frac{e^{-\lambda c^\dagger c}}{\operatorname{Tr} e^{-\lambda c^\dagger c}} = (1 - e^{-\lambda})e^{-\lambda a^\dagger a} \tag{4.10.13}$$

(see Section 3.6). Also, we have

$$a^\dagger a = (c^\dagger + w^*)(c + w), \tag{4.10.14}$$

so that

$$\begin{aligned} \langle H\rangle &= \hbar\omega\langle a^\dagger a\rangle = \hbar\omega \operatorname{Tr}\{[c^\dagger c + cw^* + c^*w + |w|^2]\} \\ &= \hbar\omega\left[\frac{1}{e^\lambda - 1} + |w|^2\right], \end{aligned} \tag{4.10.15}$$

while

$$\begin{aligned} \langle p\rangle &= i\sqrt{\frac{\hbar\omega}{2}} \operatorname{Tr} \rho(c^\dagger - c + w^* - w) = i\sqrt{\frac{\hbar\omega}{2}}(w^* - w) \\ \langle q\rangle &= \sqrt{\frac{\hbar}{2\omega}} \operatorname{Tr} \rho(c^\dagger + c + w^* + w) = \sqrt{\frac{\hbar}{2\omega}}(w^* + w), \end{aligned} \tag{4.10.16}$$

which follow since

$$\operatorname{Tr} e^{-\lambda c^\dagger c}c = 0 = \operatorname{Tr} e^{-\lambda c^\dagger c}c^\dagger. \tag{4.10.17}$$

Thus w and w^* are directly expressible in terms of the (ensemble) average of the electric and magnetic field and λ is expressible in terms of the average energy.

Let us consider first the case in which the cavity is filled with thermal radiation only in equilibrium with the walls at temperature T. Then the average electric and magnetic fields are zero:

$$\langle p\rangle = \langle q\rangle = 0 \rightarrow w = w^* = 0, \tag{4.10.18a}$$

and

$$\langle E\rangle = \frac{\hbar\omega}{e^{\hbar\omega/kT} - 1} = \frac{\hbar\omega}{e^\lambda - 1}, \tag{4.10.18b}$$

so that $\lambda = \hbar\omega/kT$. Then

$$\rho = (1 - e^{-\lambda})e^{-\lambda a^\dagger a} = (1 - e^{-\lambda})\mathfrak{N}\{-(1 - e^{-\lambda})\alpha^*\alpha\} \qquad (4.10.19)$$

corresponds to pure noise. We used (3.3.24). As $T \to 0$, $(1 - e^{-\lambda}) \to 1$ and by (3.3.34) we see that

$$\rho \xrightarrow[T\to 0]{} |0\rangle\langle 0|. \qquad (4.10.20)$$

That is, when the cavity walls go to zero temperature, ρ approaches the pure vacuum state with zero entropy.

If we again use (3.3.24) for $c^\dagger c$, we see that

$$\rho = (1 - e^{-\lambda})\mathfrak{N}\{\exp - (1 - e^{-\lambda})(\alpha^* - w^*)(\alpha - w)\}, \qquad (4.10.21)$$

for the case in which $\langle p\rangle$ and $\langle q\rangle$ are not zero. If we have signal plus noise:

$$\langle H\rangle = \frac{\hbar\omega}{e^{\hbar\omega/kT} - 1} + E_{\text{signal}} = \frac{\hbar\omega}{e^\lambda - 1} + \hbar\omega\,|w|^2. \qquad (4.10.22)$$

and (4.10.21) describes signal plus noise. If we again let $T \to 0$ and use (3.3.42), we see that (4.10.21) becomes

$$\rho \xrightarrow[\lambda\to\infty]{} e^{-w^2} e^{wa^\dagger}|0\rangle\langle 0|e^{w^*a} \equiv |w\rangle\langle w|, \qquad (4.10.23)$$

which is a pure noiseless coherent state with zero entropy.

For the general signal plus noise density operator the entropy is in the $c^\dagger c$-representation

$$S = -k\,\mathrm{Tr}\,\rho\ln\rho = -k(1 - e^{-\lambda})\sum_0^\infty e^{-\lambda n}[-\lambda n + \ln(1 - e^{-\lambda})],$$

or

$$S = k\left\{\frac{\lambda}{e^\lambda - 1} - \ln(1 - e^{-\lambda})\right\}, \qquad (4.10.24)$$

where

$$\frac{\hbar\omega}{e^\lambda - 1} = \langle E\rangle - \hbar\omega\,|w|^2$$

$$= \tfrac{1}{2}[\langle p^2\rangle - \langle p\rangle^2] + \tfrac{1}{2}\omega^2[\langle q^2\rangle - \langle q\rangle^2]. \qquad (4.10.25)$$

Let us next consider a case in which the signal $w = |w|e^{i\varphi}$ has a random phase. We may use (3.3.24) to write the density operator as

$$\rho = \epsilon e^{-\lambda(a^\dagger - w^*)(a - w)} = \epsilon\mathfrak{N}\{e^{-\epsilon(\alpha^* - w^*)(\alpha - w)}\}$$

$$= \epsilon e^{-\epsilon|w|^2} e^{\epsilon w a^\dagger}\mathfrak{N}\{e^{-\epsilon\alpha^*\alpha}\}e^{\epsilon w^* a}$$

$$= \epsilon e^{-\epsilon|w|^2} e^{\epsilon w a^\dagger} e^{-\lambda a^\dagger a} e^{\epsilon w^* a}, \qquad (4.10.26)$$

where $\epsilon = 1 - e^{-\lambda}$. Then, we have

$$\rho = \epsilon e^{-\epsilon|w|^2}\mathfrak{N}\{e^{-\epsilon\alpha^*\alpha+\epsilon w\alpha^*+\epsilon w^*\alpha}\}$$
$$= \epsilon e^{-\epsilon|w|^2}\mathfrak{N}\left\{e^{-\epsilon\alpha^*\alpha}\sum_{l=0}^{\infty}\frac{(\epsilon w\alpha^*)^l}{l!}\sum_{m=0}^{\infty}\frac{(\epsilon w^*\alpha)^m}{m!}\right\}. \tag{4.10.27}$$

If we average over phases, since

$$\frac{1}{2\pi}\int_0^{2\pi} e^{i(l-m)\varphi}\,d\varphi = \delta_{lm},$$

then

$$\bar{\rho} = \epsilon e^{-\epsilon|w|^2}\mathfrak{N}\left\{e^{-\epsilon\alpha^*\alpha}\sum_{l=0}^{\infty}\frac{(\epsilon^2\,|w|^2)^l}{l!}\frac{\alpha^{*l}}{\sqrt{l!}}\frac{\alpha^l}{\sqrt{l!}}\right\}$$
$$= \epsilon e^{-\epsilon|w|^2}\sum_{l=0}^{\infty}\frac{(\epsilon^2\,|w|^2)^l}{l!}\frac{a^{\dagger l}}{\sqrt{l!}}\mathfrak{N}\{e^{-\epsilon\alpha^*\alpha}\}\frac{a^l}{\sqrt{l!}}\,. \tag{4.10.28}$$

If we again use (3.3.24) this becomes

$$\bar{\rho} = \epsilon e^{-\epsilon|w|^2}\sum_{l=0}^{\infty}\frac{(\epsilon^2\,|w|^2)^l}{l!}\frac{a^{\dagger l}}{\sqrt{l!}}e^{-\lambda a^\dagger a}\frac{a^l}{\sqrt{l!}}, \tag{4.10.29}$$

which cannot be summed. In case $\lambda \to \infty$ and $\epsilon \to 1$, this becomes

$$\bar{\rho}\xrightarrow[T\to 0]{} e^{-|w|^2}\sum_{l=0}^{\infty}\frac{(|w|^2)^l}{l!}\frac{a^{\dagger l}}{\sqrt{l!}}|0\rangle\langle 0|\frac{a^l}{\sqrt{l!}}$$
$$= e^{-|w|^2}\sum_{l=0}^{\infty}\frac{(|w|^2)^l}{l!}|l\rangle\langle l|, \tag{4.10.30}$$

which describes a system of signals with random phase and zero thermal noise. In this case the entropy is *not* zero since it does not correspond to a pure state. This follows, since

$$\mathrm{Tr}\,\bar{\rho}^2 \neq 1.$$

Note that $\bar{\rho}$ represents a Poisson distribution over the eigenstates $|n\rangle$ where $a^\dagger a|n\rangle = n|n\rangle$.

PROBLEMS

4.1 Show that, when no sources are present, in the Coulomb gauge we may determine the fields entirely from the vector potential.

4.2 Using (4.3.14) and the boundary conditions (4.3.16), show that the normal modes in a cavity are orthogonal. Discuss the case in which modes are degenerate.

4.3 Verify (4.6.20b).

4.4 Find the commutator $[D_x(\mathbf{r}, t), B_y(\mathbf{r}', t')]$, where $t \neq t'$.

4.5 Calculate $[V_H(z, t), \partial I_H(z', t)/\partial z']$, where V_H and I_H are given by (4.2.14).

4.6 Using the notation of (4.7.1) and (4.7.2), find $\mathfrak{N}\{\overline{\mathbf{D}^2}\}$, where $\mathfrak{N}$ is the normal-ordering operator. Evaluate $\langle 0|\mathfrak{N}\{\overline{\mathbf{D}^2}\}|0\rangle$, where $|0\rangle$ is the field vacuum state. Compare this result with (4.7.12).

4.7 Carry out the commutation relations indicated to verify (4.9.10) and (4.9.11).

REFERENCES

[1] P. A. M. Dirac, *Proc. Roy. Soc. (London), Ser. A*, **114,** 243 (1927).

[2] E. Fermi, *Rev. Mod. Phys.*, **4,** 87 (1932).

[3] W. Heitler, *The Quantum Theory of Radiation*, 2nd ed., Fair Lawn, N.J.: Oxford University Press, 1944, Chaps. 1 and 2.

[4] J. M. Jauch and K. M. Watson, *Phys. Rev.*, **74,** 950 (1948); **74,** 1485 (1948); and **75,** 1249 (1949).

[5] S. S. Schweber, *An Introduction to Relativistic Quantum Field Theory*, New York: Harper & Row, 1961.

[6] A. I. Akhiezer and V. B. Berestetsky, "Quantum Electrodynamics," *U.S. At. Energy Comm. Transl.* 2876.

[7] The treatment given here was motivated by lecture notes of D. Walecka of Stanford University.

5

Interaction of Radiation with Matter

In Dirac's theory of radiation, he considers an atom and the radiation field with which it interacts as a single system whose energy is represented by (*a*) the energy of the atom alone, (*b*) the energy of the radiation field alone, and (*c*) a small term equal to the coupling energy between the atom and the field.

The interaction term is obviously necessary if the atom and field are to affect one another. A very simple "model" due to Fermi [1] will illustrate the interaction. We consider a pendulum of resonant frequency ω_0, which corresponds to the atom, and a vibrating string of resonant frequency ω_1, which corresponds to the radiation field. When they are uncoupled, they vibrate independently, and the energy is the sum of the energy of the pendulum and the energy of the string. If we connect the two by a small massless elastic thread, we may transfer energy between the two systems. If, at $t = 0$, the string is vibrating and the pendulum is at rest, after a time t some of the energy will be transferred to the pendulum. If the two frequencies are the same, complete transfer of energy can take place. This corresponds to absorption of radiation by an atom.

In the reverse process, in which the pendulum is initially excited and the string at rest, if the frequencies of the string and the pendulum are the same, the transfer of energy to the string corresponds to emission of radiation by an atom.

In Section 5.1 we present the nonrelativistic hamiltonian for a one-electron atom in the presence of a radiation field, in line with the mechanical model discussed above. The solution of the Schrödinger equation in this case is impossible and we must resort to time-dependent perturbation theory to obtain approximate solutions. In Section 5.2 we use perturbation theory to explain absorption and emission of radiation by an atom. In Section 5.3 we present the Wigner-Weisskopff theory of natural line width and obtain the Lamb shift. In the following four sections we obtain the Kramers-Heisenberg scattering cross section and study the special cases of Rayleigh, Thomson, and Raman scattering. In Section 5.8 we discuss resonance fluorescence.

The Doppler effect, which is the change in frequency of light emitted from a moving source, is very simply explained by the wave theory of light. In Section 5.9 we show that in the quantum theory the Doppler effect is explained from the conservation of energy and momentum of the emitting atom and emitted photon. In Section 5.10 we give the quantum explanation of a typical wave-like phenomenon such as the propagation of light in a vacuum.

The semiclassical theory of a two-level spin-resonance experiment presented in Section 5.11, followed by a short account of the effect of collision broadening on the line width of the resonance in Section 5.12. The spin-resonance experiment is again discussed in the final section where we quantize the radiation field.

5.1 HAMILTONIAN OF AN ATOM IN A RADIATION FIELD

According to Dirac, the energy of an atom interacting with a radiation field is considered a single system. For simplicity, the atom is assumed to have a single electron of charge e and mass m in a potential $V(\mathbf{r})$, where $\mathbf{r}$ is the position of the electron. The electron momentum is $\mathbf{p}$. The electron spin is neglected, and in this book we are not concerned with energies sufficiently large that relativistic effects need be considered.

The radiation field may be described by the vector potential $\mathbf{A}(\mathbf{r}, t)$ in the Coulomb gauge, div $\mathbf{A} = 0$. For simplicity, the source of the radiation field (charges and currents) is not considered.

The nonrelativistic hamiltonian for the atom and radiation field in mks units is [compare with (2.10.4)]

$$H = \frac{1}{2m}(\mathbf{p} - e\mathbf{A})^2 + eV(\mathbf{r}) + H_r \tag{5.1.1}$$

where H_r is the energy of the radiation field in the absence of the atom. Since div $\mathbf{A} = 0$, we may write (5.1.1) as

$$H = \frac{\mathbf{p}^2}{2m} + eV(\mathbf{r}) + H_r - \frac{e}{m}\mathbf{A}\cdot\mathbf{p} + \frac{e^2}{2m}\mathbf{A}^2. \tag{5.1.2}$$

The first two terms give the energy of the free atom, namely,

$$H_a = \frac{\mathbf{p}^2}{2m} + eV(\mathbf{r}). \tag{5.1.3}$$

The term H_r is the energy of the quantized source-free radiation field in the absence of the atom,

$$H = \sum_{l\sigma} \hbar\omega_l(a_{l\sigma}^\dagger a_{l\sigma} + \tfrac{1}{2}), \tag{5.1.4}$$

which is familiar from Chapter 4. We combine them into H_0, defined by

$$H_0 = H_a + H_r, \tag{5.1.5}$$

as the unperturbed hamiltonian.

The next term in (5.1.2), which is of first order in the coupling constant, e, we call

$$H_1 = -\frac{e}{m}\mathbf{A}\cdot\mathbf{p}. \tag{5.1.6}$$

This term is small compared with H_a and H_r and represents the interaction between the electron momentum $\mathbf{p}$ and radiation field $\mathbf{A}$. It is large compared with the last term of order e^2 in (5.1.2)

$$H_2 = \frac{e^2}{2m}\mathbf{A}^2, \tag{5.1.7}$$

which represents the energy of mutual interaction between different radiation oscillators of the radiation field through the coupling of the electron to the field.

The hamiltonian is given in the Schrödinger picture so that we must use the expansion for the vector potential in H_1 and H_2 in the Schrödinger picture given by (4.6.1a), namely,

$$\mathbf{A}(\mathbf{r}) = \sum_{l,\sigma}\sqrt{\frac{\hbar}{2\omega_l\epsilon_0\tau}}\,\hat{\mathbf{e}}_{l\sigma}[a_{l\sigma}\exp(i\mathbf{k}_l\cdot\mathbf{r}) + a_{l\sigma}^{\dagger}\exp(-i\mathbf{k}_l\cdot\mathbf{r})], \tag{5.1.8}$$

where $a_{l\sigma}$ and $a_{l\sigma}^{\dagger}$ obey the boson commutator relations and we assume that the atom and field are contained in a cubic cavity of volume τ.

The hamiltonian (5.1.2), together with the Schrödinger equation of motion

$$i\hbar\frac{\partial|\psi(t)\rangle}{\partial t} = H|\psi(t)\rangle, \tag{5.1.9}$$

and the vector potential (5.1.8) constitute the entire formulation of the Dirac theory of radiation for the interaction of a one-electron atom with an electromagnetic field that has been quantized in the nonrelativistic limit. There is left only the problem of solving (5.1.9); this cannot be done exactly, and so perturbation calculations must be used.

5.2 ABSORPTION AND EMISSION OF RADIATION BY AN ATOM

We use the perturbation theory results of Section 1.21 to show how the theory explains the absorption and emission of a quantum of energy from the radiation field.

We identify system A in Section 1.21 with the atom and system B with the radiation field. The unperturbed hamiltonian is given by (5.1.5). The free atom satisfies the eigenvalue problem

$$H_a|s\rangle = \epsilon_s|s\rangle, \tag{5.2.1}$$

where the atomic state vectors are orthogonal and complete for the description of the atom. The number operator of a radiation oscillator of polarization $\hat{e}_{l\sigma}$, momentum $\hbar\mathbf{k}_l$ and energy $\hbar\omega_l$ satisfies the eigenvalue equation

$$a_{l\sigma}^\dagger a_{l\sigma}|n_{l\sigma}\rangle = n_{l\sigma}|n_{l\sigma}\rangle. \tag{5.2.2}$$

These are orthogonal and complete for describing the field. An energy eigenstate for the free atom and field is

$$|s; n_{l_1\sigma_1}, n_{l_2\sigma_2}, \ldots\rangle = |s\rangle \prod_{l_i\sigma_i} |n_{l_i\sigma_i}\rangle \equiv |s; \{n_{l\sigma}\}\rangle, \tag{5.2.3}$$

with energy $\epsilon_s + \sum\limits_{l,\sigma} \hbar\omega_l n_{l\sigma}$.

The interaction energy between the atom and field in the SP by (5.1.6) and (5.1.8) is

$$V_{ar}{}^S = -\frac{e}{m}\sum_{l,\sigma}\sqrt{\frac{\hbar}{2\epsilon_0\omega_l L^3}}\,[a_{l\sigma}e^{i\mathbf{k}_l\cdot\mathbf{r}} + a_{l\sigma}^\dagger e^{-i\mathbf{k}_l\cdot\mathbf{r}}]\hat{e}_{l\sigma}\cdot\mathbf{p}. \tag{5.2.4}$$

If we take matrix elements between atomic states $|s\rangle$ and $|s'\rangle$ of both sides, we shall need

$$\langle s|e^{\pm i\mathbf{k}_l\cdot\mathbf{r}}\mathbf{p}|s'\rangle = \int \psi_s^*(\mathbf{r})e^{\pm i\mathbf{k}_l\cdot\mathbf{r}}\frac{\hbar}{i}\nabla\psi_{s'}(\mathbf{r})\,d\mathbf{r}, \tag{5.2.5}$$

where we have expressed this in the coordinate representation and

$$\psi_{s'}(\mathbf{r}) = \langle\mathbf{r}|s'\rangle. \tag{5.2.6}$$

These atomic wave functions vanish when r is greater than about 10^{-8} cm, the approximate atomic diameter. The wave vector $k_l = 2\pi/\lambda \sim 10^5\ \text{cm}^{-1}$ at optical wavelengths. Therefore, in the range of integration over which the wave functions are nonzero $\exp \pm ik_l\mathbf{r} \sim \exp \pm i \times 10^{-3} \approx 1$. Therefore, in the dipole approximation, we let

$$\langle s|e^{\pm i\mathbf{k}_l\cdot\mathbf{r}}\mathbf{p}|s'\rangle \cong \langle s|\mathbf{p}|s'\rangle, \tag{5.2.7}$$

so that from (5.2.4) we obtain

$$\langle s|V_{ar}{}^S|s'\rangle \cong -\sum_{l,\sigma}\sqrt{\frac{\hbar}{2\epsilon_0\omega_l L^3}}\,(a_{l\sigma} + a_{l\sigma})\hat{e}_{l\sigma}\cdot\langle s|\frac{\mathbf{p}}{m}|s'\rangle e. \tag{5.2.8}$$

We next show that

$$\frac{e}{m}\langle s|\mathbf{p}|s'\rangle = i\omega_{ss'}e\mathbf{x}_{ss'}, \tag{5.2.9a}$$

where

$$\boldsymbol{\mu}_{ss'} \equiv e\mathbf{x}_{ss'} \equiv e\langle s|\mathbf{x}|s'\rangle = e\langle s'|\mathbf{x}|s\rangle^* \tag{5.2.9b}$$

is the atomic dipole matrix element between states $|s\rangle$ and $|s'\rangle$ and

$$\hbar\omega_{ss'} = \epsilon_s - \epsilon_{s'} = -\hbar\omega_{s's}. \tag{5.2.9c}$$

To show this, we write the Heisenberg equation of motion for $x_i (i = 1, 2, 3)$ for the free atom. It is

$$\frac{dx_i}{dt} = \frac{1}{i\hbar}[x_i, H_a] = \frac{1}{i\hbar}\left[x_i, \frac{\mathbf{p}^2}{2m} + eV(\mathbf{x})\right]. \tag{5.2.10}$$

Since x_i commutes with $V(\mathbf{x})$ and

$$[x_i, p_j^2] = 2i\hbar p_j \delta_{ij}, \tag{5.2.11}$$

we have

$$\frac{dx_i}{dt} = \frac{p_i}{m} = \frac{1}{i\hbar}[x_i, H_a]. \tag{5.2.12}$$

If we take the s, s' matrix element of both sides of this equation and use (5.2.1) and its adjoint, (5.2.9) follows.

Therefore, if we use (5.2.9), (5.2.8) becomes

$$\langle s|V_{ar}{}^S|s'\rangle = -i\omega_{ss'} \sum_{l,\sigma} \sqrt{\frac{\hbar}{2\epsilon_0\omega_l L^3}}(a_{l\sigma} + a_{l\sigma}^\dagger)\hat{e}_{l\sigma} \cdot \boldsymbol{\mu}_{ss'}. \tag{5.2.13}$$

If we let

$$\begin{aligned} U_0^{(r)}(t) &= \exp -\frac{i}{\hbar}\sum_{l,\sigma} \hbar\omega_l a_{l\sigma}^\dagger a_{l\sigma} t \\ &= \prod_{l,\sigma} \exp\,[-i\omega_l t a_{l\sigma}^\dagger a_{l\sigma}], \end{aligned} \tag{5.2.14}$$

then it easily follows that

$$U_0^{(r)\dagger}(t)\langle s|V|s'\rangle U_0^{(r)}(t) = -i\omega_{ss'} \sum_{l,\sigma} \sqrt{\frac{\hbar}{2\omega_l\epsilon_0 L^3}}(a_{l\sigma}e^{-i\omega_l t} + a_{l\sigma}^\dagger e^{i\omega_l t})\hat{e}_{l\sigma} \cdot \boldsymbol{\mu}_{ss'}. \tag{5.2.15}$$

The probability of finding the atom in state s at time t given that it was in state s' at time 0 by (1.21.27) and the above is

$$\begin{aligned} &|c_a^{(1)}(s, t|s', 0)|^2 \\ &= \frac{\omega_{ss'}^2}{2\hbar\epsilon_0 L^3} \sum_{\substack{l',\sigma' \\ l,\sigma}} \frac{(\hat{e}_{l\sigma} \cdot \boldsymbol{\mu}_{ss'})(\hat{e}_{l'\sigma'} \cdot \boldsymbol{\mu}_{ss'}^*)}{\sqrt{\omega_l\omega_{l'}}} \int_0^t dt_1 \int_0^t dt_2\, e^{i\omega_{ss'}(t_1 - t_2)}\, \mathrm{Tr}\, \rho_r(0) \\ &\times \{[\delta_{ll'}\,\delta_{\sigma\sigma'} + a_{l'\sigma'}^\dagger a_{l\sigma}]e^{i(\omega_{l'}t_1 - \omega_l t_2)} + a_{l\sigma}^\dagger a_{l'\sigma'} e^{i(\omega_l t_2 - \omega_{l'} t_1)} \\ &+ a_{l\sigma}a_{l'\sigma'}e^{-i(\omega_l t_2 + \omega_{l'} t_1)} + a_{l\sigma}^\dagger a_{l'\sigma'}^\dagger e^{i(\omega_l t_2 + \omega_{l'} t_1)}\}, \end{aligned} \tag{5.2.16}$$

where we have used the commutation relation $[a_{l\sigma}, a_{l'\sigma'}^\dagger] = \delta_{ll'}\delta_{\sigma\sigma'}$.

As yet we have not specified the state of the field at $t = 0$ when the interaction was turned on. We consider several cases.

CASE 1 MONOCHROMATIC PLANE POLARIZED FIELD

We assume that we have measured the energy in each mode of the radiation field and found that only one mode was excited so that all others are in the vacuum state. We recall that a mode is specified completely by giving its wave vectors $\mathbf{k}_l$ and its polarization σ. Its frequency $\omega_l = c|\mathbf{k}_l|$ is then determined. The density operator is then symbolically written as

$$\rho_r(0) = |n_{l_0\sigma_0}\rangle\langle n_{l_0\sigma_0}| \times \{|0\rangle\}\{\langle 0|\}. \tag{5.2.17}$$

In this case the traces are easily evaluated since

$$\{\langle 0|\} < n_{l_0\sigma_0}|a^\dagger_{l'\sigma'}a_{l\sigma}|n_{l_0\sigma_0}\rangle\{|0\rangle\} = n_{l_0\sigma_0}\,\delta_{l'l}\,\delta_{\sigma'\sigma}\,\delta_{ll_0}\,\delta_{\sigma\sigma_0}, \tag{5.2.18}$$

and the a^2 and $a^{\dagger 2}$ terms all vanish. Then (5.2.16) reduces to

$$|c_a^{(1)}(s, t|s', 0)|^2 = \frac{\omega_{ss'}^2}{2\hbar\epsilon_0 L^3} \sum_{l,\sigma} \frac{|\hat{e}_{l\sigma} \cdot \boldsymbol{\mu}_{ss'}|^2}{\omega_l}$$
$$\times \left\{(1 + n_{l_0\sigma_0}\,\delta_{ll_0}\,\delta_{\sigma\sigma_0}) \frac{4 \sin^2 \frac{1}{2}(\omega_{ss'} + \omega_l)t}{(\omega_{ss'} + \omega_l)^2} + n_{l_0\sigma_0}\,\delta_{ll_0}\,\delta_{\sigma\sigma_0} \frac{4 \sin^2 \frac{1}{2}(\omega_{ss'} - \omega_l)t}{(\omega_{ss'} - \omega_l)^2}\right\}, \tag{5.2.19}$$

where the sum over l, σ, l', σ' has reduced to the sum over l, σ due to the Kronecker δ's.

Absorption

Assume at $t = 0$ the atom is in its ground state $|g\rangle$ with energy ϵ_g and we ask for the probability that at time t it is found in an excited state $|e\rangle$ with energy $\epsilon_e > \epsilon_g$. Then

$$\hbar\omega_{ss'} \equiv \hbar\omega_{eg} = \epsilon_e - \epsilon_g > 0 \tag{5.2.20}$$

Since all $\omega_l > 0$, then if

$$\omega_{eg} \cong \omega_{l_0}, \tag{5.2.21}$$

the last (sine)^2 term in (5.2.19) will be much larger than the first and we have

$$|c_a^{(1)}(e, t|g, 0)|^2 \cong \frac{\omega_{eg}}{2\hbar\epsilon_0 L^3} n_{l_0\sigma_0} |\hat{e}_{l_0\sigma_0} \cdot \boldsymbol{\mu}_{eg}|^2 \frac{4 \sin^2 \frac{1}{2}(\omega_{l_0} - \omega_{eg})t}{(\omega_{l_0} - \omega_{eg})^2}. \tag{5.2.22}$$

The transition probability per second is

$$w_{\text{abs}} = \frac{d}{dt} |c_a^{(1)}(e, t|g, 0)|^2 = \frac{\omega_{eg}}{\hbar\epsilon_0 L^3} n_{l_0\sigma_0} |\hat{e}_{l_0\sigma_0} \cdot \boldsymbol{\mu}_{eg}|^2 \frac{\sin (\omega_{l_0} - \omega_{eg})t}{(\omega_{l_0} - \omega_{eg})}. \tag{5.2.23}$$

We note that

$$\lim_{t\to\infty} \frac{\sin xt}{\pi x} = \delta(x), \tag{5.2.24}$$

since (*a*) if $x = 0$:

$$\lim_{t\to\infty} \lim_{x\to 0} \frac{\sin xt}{\pi x} = \lim_{t\to\infty} \lim_{x\to 0} \frac{t \cos xt}{\pi} = \infty \tag{5.2.25}$$

and (*b*) if $x \neq 0$:

$$\lim_{t\to\infty} \frac{\sin xt}{\pi x} \underset{|x|\to\infty}{\longrightarrow} 0, \tag{5.2.26}$$

because of the oscillations of the sine, and (*c*)

$$\lim_{t\to\infty} \int_{-\infty}^{\infty} \frac{\sin xt}{\pi x}\, dx = 1 = \int_{-\infty}^{\infty} \delta(x)\, dx. \tag{5.2.27}$$

Therefore, as $t \to \infty$, (5.2.23) becomes

$$w_{\text{abs}} \to \frac{\pi}{\hbar} \frac{\omega_{eg}}{\epsilon_0 L^3} \left|\hat{e}_{l_0\sigma_0} \cdot \boldsymbol{\mu}_{eg}\right|^2 n_{l_0\sigma_0}\, \delta(\omega_{l_0} - \omega_{eg}). \tag{5.2.28}$$

This gives the probability per second for the absorption of a photon by the atom when the field is perfectly monochromatic and plane polarized. We have assumed that the atomic energy levels have no linewidth (no damping) so that only if $\omega_{l_0} = \omega_{eg}$ (energy conservation) will the atom be able to absorb a photon and be excited. The process is shown schematically in Figure 5.1. If $n_{l_0\sigma_0} = 0$, the atom will not be excited. The probability is linearly proportional to $\omega_{eg} = \omega_{l_0}$. Obviously this is an unrealistic case since it is impossible to have a completely monochromatic wave. In addition, the reader should note that for the case we have considered in which we measured the energy precisely of the mode, the average electric fields will be zero.

Emission

Consider next the case in which at $t = 0$, the atom is in the excited state $|e\rangle$ and let us determine the probability of finding it in its ground state $|g\rangle$ at

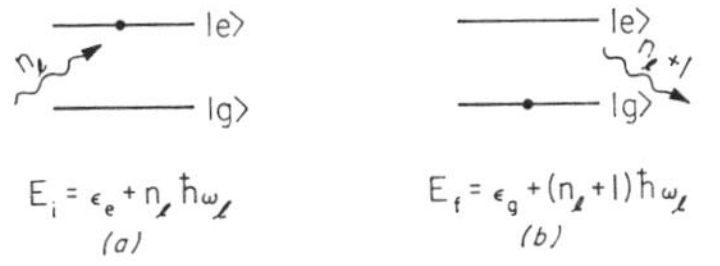

Figure 5.1 Schematic representation of a one-photon absorption by an atom. (*a*) Initial state; (*b*) final state.

time t. Then

$$\hbar\omega_{ss'} \equiv -\hbar\omega_{eg}, \tag{5.2.29}$$

and the possibility now exists that the first (sine)^2 term in (5.2.19) can be large. There are two different terms. Suppose that $n_{l_0\sigma_0} = 0$, that is, that the atom is excited and *no* radiation field is present. Then (5.2.19) becomes

$$|c_a^{(1)}(g, t|e, 0)|^2 = \frac{\omega_{eg}^{\ 2}}{2\hbar\epsilon_0 L^3}\sum_{l,\sigma}\frac{|\hat{e}_{l\sigma}\cdot\boldsymbol{\mu}_{eg}|^2}{\omega_l}\frac{4\sin^2\frac{1}{2}(\omega_l - \omega_{eg})^2 t}{(\omega_l - \omega_{eg})^2}. \tag{5.2.30}$$

This gives the probability of finding the atom in the ground state at time t given that it was in its excited state at $t = 0$ even when no radiation was initially present. That is, the atom may decay *spontaneously* with the emission of radiation. This probability is very small unless the emitted radiation satisfies energy conservation

$$\omega_l \simeq \omega_{eg}. \tag{5.2.31}$$

It may be emitted in all directions and in all polarizations, however, so that we must sum over all modes and both polarizations. By (4.5.4) and (4.5.10), we replace the sum over l for a *single* polarization by

$$\sum_l\{\ \ \} = \left(\frac{L}{2\pi c}\right)^3\int_0^\infty \omega_l^{\ 2}\, d\omega_l \int d\Omega\{\ \ \}, \tag{5.2.32}$$

so that

$$|c_a^{(1)}(g, t|e, 0)|^2 = \frac{\omega_{eg}^{\ 2}}{2\hbar\epsilon_0(2\pi c)^3}\sum_{\sigma=1}^{2}\int_0^\infty \omega_l\, d\omega_l \int d\Omega\, |\hat{e}_{l\sigma}\cdot\boldsymbol{\mu}_{eg}|^2 \frac{4\sin^2\frac{1}{2}(\omega_l - \omega_{eg})^2 t}{(\omega_l - \omega_{eg})^2}. \tag{5.2.33}$$

Now if we carry out the sum on polarizations, we have

$$\sum_{\sigma=1}^{2}|\hat{e}_{l\sigma}\cdot\boldsymbol{\mu}_{eg}|^2 = |\boldsymbol{\mu}_{eg}|^2(\cos^2\alpha + \cos^2\beta), \tag{5.2.34}$$

where α is the angle between $\boldsymbol{\mu}_{eg}$ and $\hat{e}_{l_1}$ and β the angle between $\boldsymbol{\mu}_{eg}$ and $\hat{e}_{l_2}$. If θ is the angle between $\boldsymbol{\mu}_{eg}$ and $\hat{k}_l$, then by the law of cosines,

$$\sum_{\sigma=1}^{2}|\hat{e}_{l\sigma}\cdot\boldsymbol{\mu}_{eg}|^2 = |\boldsymbol{\mu}_{eg}|^2(1 - \cos^2\theta) \tag{5.2.35}$$

(see Figure 5.2). If we take our z-axis for the angular integration in (5.2.33) to be along $\boldsymbol{\mu}_{eg}$, then

$$d\Omega = \sin\theta\, d\theta\, d\varphi, \tag{5.2.36}$$

and

$$\int_0^{2\pi} d\varphi \int_0^{\pi}\sin\theta\, d\theta(1 - \cos^2\theta) = \frac{8\pi}{3}, \tag{5.2.37}$$

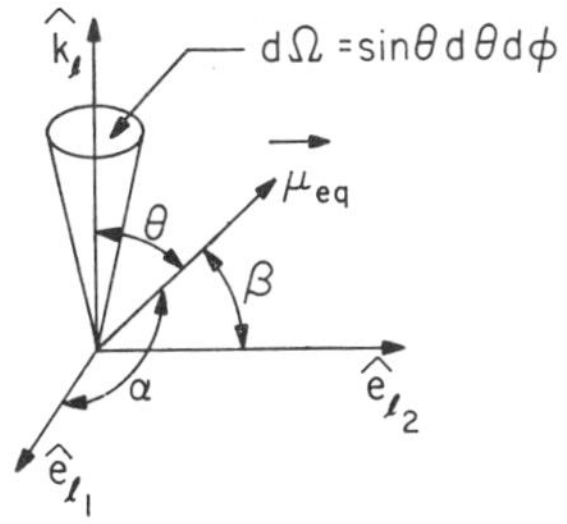

Figure 5.2 Angles used in (5.2.37).

and (5.2.33) becomes

$$|c_a^{(1)}(g, t|e, 0)|^2 = \frac{\omega_{eg}{}^2}{2\hbar\epsilon_0(2\pi c)^3}\,\frac{8\pi}{3}\,|\boldsymbol{\mu}_{eg}|^2 \int_0^\infty \frac{\omega_l\, d\omega_l\, 4\sin^2 \frac{1}{2}(\omega_l - \omega_{eg})^2 t}{(\omega_l - \omega_{eg})^2}. \tag{5.2.38}$$

Since the $\sin^2 \frac{1}{2}xt/x^2$ term is so highly peaked at $x = \omega_l - \omega_{eg} \cong 0$, we may extend the lower limit to $-\infty$, and since

$$\int_{-\infty}^{\infty} \frac{\sin^2 \frac{1}{2}xt}{x^2}\, dx = \frac{\pi t}{2}, \tag{5.2.39}$$

the probability per second for spontaneous emission is

$$w_{\text{emiss}}^{\text{spon}} = \frac{d}{dt}\,|c_a^{(1)}(g, t|e, 0)|^2 = \frac{\omega_{eg}{}^3\, |\boldsymbol{\mu}_{eg}|^2}{3\pi\hbar\epsilon_0 c^3}. \tag{5.2.40}$$

Note that the cavity volume, L^3, has canceled and that the probability is proportional to $\omega_{eg}{}^3$.

The remaining term proportional to $n_{l_0\sigma_0}$ in (5.2.19) corresponds to *induced* emission. It vanishes unless radiation is initially present. In this case we have

$$|c_a^{(1)}(g, t|e, 0)|^2 = \frac{\omega_{eg}{}^2}{2\epsilon_0\hbar L^3}\,\frac{|\hat{e}_{l_0\sigma_0}\cdot\boldsymbol{\mu}_{eg}|^2}{\omega_{l_0}}\, n_{l_0\sigma_0}\, \frac{4\sin^2 \frac{1}{2}(\omega_{l_0} - \omega_{eg})t}{(\omega_{l_0} - \omega_{eg})^2}, \tag{5.2.41}$$

which is identical to the absorption term (5.2.22) so the probability per second for the atom to decay with the induced emission of a photon by (5.2.28) is

$$w_{\text{emiss}}^{\text{ind}} = w_{\text{abs}} = \frac{\pi}{\hbar}\,\frac{\omega_{eg}}{\epsilon_0 L^3}\,|\hat{e}_{l_0\sigma_0}\cdot\boldsymbol{\mu}_{eg}|^2\, n_{l_0\sigma_0}\, \delta(\omega_{l_0} - \omega_{eg}). \tag{5.2.42}$$

It is induced to emit into the identical mode from which it was able to absorb and with the *same* polarization. This case is shown schematically in Figure 5.3.

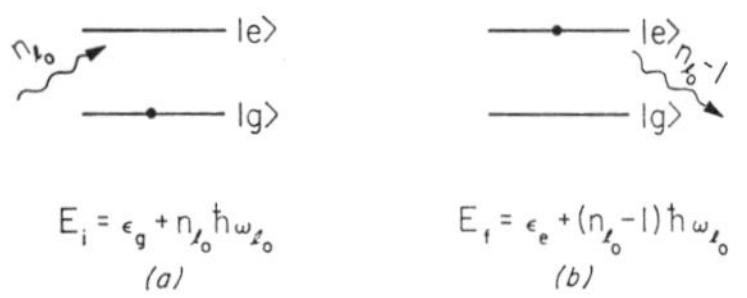

Figure 5.3 Schematic representation of a one-photon emission by an atom in an excited state. (*a*) Initial state; (*b*) final state.

As we have noted it is impossible to obtain a perfectly monochromatic plane polarized wave. Let us consider the following slightly more realistic case.

CASE 2 PLANE POLARIZED WAVES WITH FREQUENCY SPREAD

Let us assume that the electromagnetic field has a range of frequencies so that many modes are excited at $t = 0$. If we measure the energy of each mode, the density operator is given by

$$\rho_r(0) = \prod_{l,\sigma} |n_{l\sigma}\rangle\langle n_{l\sigma}|. \tag{5.2.43}$$

The traces in (5.2.16) may be easily evaluated and we may do the time integrals. This yields

$$|c_a^{(1)}(s, t|s', 0)|^2 = \frac{\omega_{ss'}^2}{2\hbar\epsilon_0 L^3} \sum_{l,\sigma} \frac{|\hat{e}_{l\sigma} \cdot \boldsymbol{\mu}_{ss'}|^2}{\omega_l}$$

$$\times \left\{(1 + n_{l\sigma}) \frac{4\sin^2 \frac{1}{2}(\omega_l + \omega_{ss'})t}{(\omega_l + \omega_{ss'})^2} + n_{l\sigma} \frac{4\sin^2 \frac{1}{2}(\omega_l - \omega_{ss'})t}{(\omega_l - \omega_{ss'})^2}\right\}. \tag{5.2.44}$$

Again we consider absorption and emission separately.

Absorption

At $t = 0$ the atom is in the ground state $|s'\rangle \equiv |g\rangle$ and at time t we ask for the probability it is in an excited state $|s\rangle \equiv |e\rangle$. Since $\omega_{ss'} = \omega_{eg}$, only the last term in (5.2.44) may be large when

$$\omega_l \cong \omega_{eg}, \tag{5.2.45}$$

and we have

$$|c_a^{(1)}(e, t|g, 0)|^2 \cong \frac{\omega_{eg}{}^2}{2\hbar\epsilon_0 L^3} \sum_{l,\sigma} \frac{|\hat{e}_{l\sigma} \cdot \boldsymbol{\mu}_{eg}|^2}{\omega_l} n_{l\sigma} \frac{4\sin^2 \frac{1}{2}(\omega_l - \omega_{eg})t}{(\omega_l - \omega_{eg})^2}. \tag{5.2.46}$$

In contrast to the single mode case in which

$$n_{l\sigma} = n_{l_0\sigma_0}\delta_{ll_0}\,\delta_{\sigma\sigma_0}, \tag{5.2.47}$$

many modes are excited. We assume that they are closely spaced in frequency so that we may again use (4.5.4) and (4.5.10) to convert the sum over l to an integral. For a single polarization

$$\sum_l \{\ \} \to \left(\frac{L}{2\pi c}\right)^3 \int_0^\infty \omega_l^2\, d\omega_l \int_{4\pi} d\Omega \{\ \}, \tag{5.2.48}$$

so that (5.2.46) becomes

$$|c_a^{(1)}(e, t|g, 0)|^2 \cong \frac{\omega_{eg}^2}{2\hbar\epsilon_0(2\pi c)^3} \sum_{\sigma=1}^{2} \int_0^\infty \omega_l\, d\omega_l \int d\Omega |\hat{e}_{l\sigma} \cdot \boldsymbol{\mu}_{eg}|^2\, n_\sigma(\omega_l, \Omega) \frac{4 \sin^2 \frac{1}{2}(\omega_l - \omega_{eg})t}{(\omega_l - \omega_{eg})^2}, \tag{5.2.49}$$

where we have let $n_{l\sigma} \to n_\sigma(\omega_l, \Omega)$. That is, a given mode is completely specified by giving three integers that fix $\mathbf{k}_l$ and a polarization σ. The frequency $\omega_l = c\,|\mathbf{k}_l|$ is then determined for these plane wave modes. In polar coordinates then, $n_{l\sigma}$ is a function of ω_l and the two polar angles are (θ, φ) and σ. Our original measurements on the field must determine this dependence completely.

We are assuming here that the atomic energy levels are infinitely sharp so that as t increases, the sine^2 term is very highly peaked at $\omega_l = \omega_{eg}$. We may then let the lower limit on the ω_l integral go to $-\infty$ and remove the slowly varying factors. This gives the absorption probability per second as

$$w_{eg} = \frac{d}{dt} |c^{(1)}(e, t|g, 0)|^2 = \frac{\omega_{eg}^3 2\pi}{2\hbar\epsilon_0(2\pi c)^3} \sum_{\sigma=1}^{2} \int d\Omega\, |\hat{e}_{l\sigma} \cdot \boldsymbol{\mu}_{eg}|^2\, n_l(\omega_{eg}, \Omega). \tag{5.2.50}$$

If the radiation field has a Lorentzian spectrum with line width γ centered at ω_0, we have

$$n_\sigma(\omega_{eg}, \Omega) = \frac{n_0}{\pi} \frac{(\gamma/2)}{(\omega_{eg} - \omega_0)^2 + (\gamma/2)^2} f_\sigma(\Omega) \equiv n(\omega_{eg}) f_\sigma(\Omega). \tag{5.2.51}$$

This will be largest when $\omega_{eg} \cong \omega_0$. If the radiation is polarized and we accept only a certain small solid angle $\Delta\Omega$, then (5.2.50) reduces to

$$\begin{aligned} w_{eg}^{\text{pol}} &= \left(\frac{\omega_{eg}}{c}\right)^3 \frac{1}{8\pi^2\hbar\epsilon_0} |\hat{e}_{l1} \cdot \boldsymbol{\mu}_{eg}|^2\, n(\omega_{eg})\, \Delta\Omega \\ &= \left(\frac{\omega_{eg}}{c}\right)^3 \frac{n(\omega_{eg})}{8\pi^2\hbar\epsilon_0} |\boldsymbol{\mu}_{eg}|^2 \cos^2\alpha\, \Delta\Omega, \end{aligned} \tag{5.2.52}$$

where α is the angle between $\boldsymbol{\mu}_{eg}$ and the polarization direction, $\hat{e}_{l1}$. If in addition our atoms are in a gaseous state, $\boldsymbol{\mu}_{eg}$ will be randomly oriented and we should average over all directions of $\boldsymbol{\mu}_{eg}$ with respect to $\hat{e}_{l1}$. This gives

$$\overline{\cos^2 \alpha} = \frac{\int_0^{2\pi} d\varphi \int_0^{\pi} \cos^2 \alpha \sin \alpha \, d\alpha}{4\pi} = \frac{1}{3}, \tag{5.2.53}$$

so that

$$\bar{w}_{eg}^{\text{pol}} = \left(\frac{\omega_{eg}}{c}\right)^3 \frac{n(\omega_{eg}) \, |\boldsymbol{\mu}_{eg}|^2}{24\pi^2 \hbar \epsilon_0} \Delta\Omega. \tag{5.2.54}$$

When the light is unpolarized, $n_\sigma(\omega_l, \Omega)$ is independent of σ; that is, we have equal intensities for both polarizations. If we again only accept a small solid angle $\Delta\Omega$, we find

$$w_{eg}^{\text{unpol}} = \frac{\omega_{eg}{}^3 n(\omega_{eg})}{\hbar \epsilon_0 c^3 8\pi^2} |\boldsymbol{\mu}_{eg}|^2 (\cos^2 \alpha + \cos^2 \beta) \, \Delta\Omega, \tag{5.2.55}$$

where α and β are the angles between $\boldsymbol{\mu}_{eg}$ and $\hat{e}_{l1}$ and $\hat{e}_{l2}$, respectively. For a gas, we again average over all directions for $\boldsymbol{\mu}_{eg}$. Since

$$\overline{\cos^2 \alpha + \cos^2 \beta} = \frac{1}{4\pi} \int_0^{2\pi} d\varphi \int_0^{\pi} \sin^3 \theta \, d\theta = \frac{2}{3}, \tag{5.2.56}$$

we have

$$\bar{w}_{eg}^{\text{unpol}} = \left(\frac{\omega_{eg}}{c}\right)^3 \frac{n(\omega_{eg}) \, |\boldsymbol{\mu}_{eg}|^2 \, \Delta\Omega}{12\pi^2 \hbar \epsilon_0}, \tag{5.2.57}$$

which is twice as large as the unpolarized case as expected.

Emission

In case the atom is in state $|e\rangle$ at $t = 0$ and state g at time t, the spontaneous emission term in (5.2.44) is exactly the same as it was in Case 1. The induced emission term becomes identical to the absorption and we have

$$w_{eg} = w_{ge} \tag{5.2.58}$$

That is, the probability per second for induced emission and absorption is the same. Comparison of (5.2.57) with (5.2.40) shows that

$$\bar{w}_{eg}^{\text{unpol}} = n(\omega_{eg}) \left(\frac{\Delta\Omega}{4\pi}\right) w_{ge}^{\text{spon}} = \bar{w}_{ge}^{\text{unpol}}. \tag{5.2.59}$$

CASE 3 BLACKBODY (THERMAL) RADIATION

Let us next consider the case in which the radiation in the cavity is in thermal equilibrium with the walls at temperature T. The density operator for the field at

$t = 0$ is then given by the Boltzmann distribution

$$\rho_r(0) = \prod_{l,\sigma} (1 - e^{-\lambda_l}) e^{-\lambda_l a^\dagger_{l\sigma} a_{l\sigma}}, \tag{5.2.60}$$

where

$$\lambda_l = \frac{\hbar\omega_l}{kT}. \tag{5.2.61}$$

We may use the results of Section 3.6 to evaluate the traces in (5.2.16). From (3.6.9) and (3.6.15), we easily see that

$$\begin{aligned} \langle a^\dagger_{l'\sigma'} a_{l\sigma} \rangle_0 &\equiv \text{Tr}\ \rho_r(0) a^\dagger_{l'\sigma'} a_{l\sigma} \\ &= \bar{n}_l\, \delta_{ll'}\, \delta_{\sigma\sigma'} = \langle a^\dagger_{l\sigma} a_{\,l'\sigma'} \rangle_0, \end{aligned} \tag{5.2.62}$$

where

$$\bar{n}_l = [\exp(\lambda_l) - 1]^{-1}. \tag{5.2.63}$$

The others all vanish. This frequency dependence should be contrasted with the Lorentzian shape discussed earlier. The reader might also note that the average electric field is again zero in this case as it was in the two previous cases we have considered. In contrast to the two prior cases, we have *not* determined the energy of each mode of the radiation field. Rather the energy in each mode is distributed over the energy eigenstates with a Boltzmann distribution as seen by (3.6.11).

When we use these traces and carry out the time integrals, (5.2.16) becomes

$$\begin{aligned} |c_a^{(1)}(s, t|s', 0)|^2 &= \frac{\omega_{ss'}^2}{2\hbar\epsilon_0 L^3} \sum_l \frac{1}{\omega_l} \left\{ \sum_\sigma |\hat{e}_{l\sigma} \cdot \boldsymbol{\mu}_{eg}|^2 \right\} \\ &\times \left\{ (1 + \bar{n}_l) \frac{4 \sin^2 \frac{1}{2}(\omega_{ss'} + \omega_l)t}{(\omega_{ss'} + \omega_l)^2} + \bar{n}_l \frac{4 \sin^2 \frac{1}{2}(\omega_{ss'} - \omega_l)t}{(\omega_{ss'} - \omega_l)^2} \right\}. \end{aligned} \tag{5.2.64}$$

This differs from (5.2.44) in that $n_{l\sigma}$ is replaced by $\bar{n}_l$, independent of σ. The black-body radiation filling the cavity is unpolarized and isotropic.

Absorption

In this case (5.2.64) becomes as in the former cases

$$\begin{aligned} &|c_a^{(1)}(e, t|g, 0)|^2 \\ &= \frac{\omega_{eg}{}^2 |\boldsymbol{\mu}_{eg}|^2}{2\hbar\epsilon_0 (2\pi c)^3} \int_0^\infty d\omega_l \omega_l \int_0^\pi \sin\theta\, d\theta \int_0^{2\pi} \sin^2\theta \bar{n}(\omega_l) \frac{4 \sin^2 \frac{1}{2}(\omega_l - \omega_{eg})t}{(\omega_l - \omega_{eg})^2}, \end{aligned} \tag{5.2.65}$$

where we must integrate over all solid angles since the radiation is isotropic. Again we see the (sine)^2 term is highly peaked so that on integrating we obtain the absorption probability per second

$$w_{eg} = \frac{\omega_{eg}{}^2 |\boldsymbol{\mu}_{eg}|^2 \bar{n}(\omega_{eg})}{3\pi\hbar\epsilon_0 c^3}. \tag{5.2.66}$$

We next calculate the energy density of the radiation in the cavity. By (4.5.10) the number of modes per unit volume of both polarizations with frequency lying between ω_l and $\omega_l + d\omega_l$ in all directions is

$$\frac{\Delta N}{L^3} = 2\left(\frac{1}{2\pi c}\right)^3 \omega_l^2\, d\omega_l \int_{4\pi} d\Omega = \frac{\omega_l^2\, d\omega_l}{c^3\pi^2} = \frac{8\pi}{c^3}\nu_l^2\, d\nu_l. \tag{5.2.67}$$

The average energy per mode is

$$\langle \hbar\omega_l a_{l\sigma}^\dagger a_{l\sigma}\rangle_\sigma = \hbar\omega_l \bar{n}(\omega_l) = h\nu_l \frac{1}{(e^{h\nu_l/kT} - 1)}. \tag{5.2.68}$$

If we let $u_{\nu_l}\, d\nu_l$ be the average energy per unit volume lying between ν_l and $\nu_l + d\nu_l$, then

$$u_{\nu_l}\, d\nu_l = \frac{\Delta N}{L^3}\hbar\omega_l \bar{n}(\omega_l) = \frac{8\pi}{c^3} h\nu_l^3 \bar{n}(\nu_l)\, d\nu_l, \tag{5.2.69}$$

which is the Planck radiation formula. Since $2\pi\nu_l \cong \omega_{eg}$, we may rewrite (5.2.66) as

$$w_{eg} = \frac{|\boldsymbol{\mu}_{eg}|^2}{6\hbar^2\epsilon_0} u_{\nu_{eg}} \equiv B_{eg} u_{\nu_{eg}}, \tag{5.2.70}$$

where B_{eg} is called the Einstein coefficient for absorption. Einstein postulated that the transition probability per second for absorption was proportional to u_ν where B_{eg} was independent of frequency and T. We now have an explicit expression which depends only on the dipole moment matrix elements for the atom.

Emission

The spontaneous emission term is identical with the prior cases so that

$$w_{ge}^{\text{spon}} = \frac{\omega_{eg}^3 |\boldsymbol{\mu}_{eg}|^2}{3\pi\hbar\epsilon_0 c^3} \equiv A_{ge}, \tag{5.2.71}$$

which is Einstein's coefficient for spontaneous emission.

A similar analysis to that above shows that for induced emission

$$w_{ge}^{\text{ind}} = w_{eg}^{\text{abs}} = B_{eg} u_\nu = B_{ge} u_\nu, \tag{5.2.72}$$

so that

$$B_{eg} = B_{ge}. \tag{5.2.73}$$

From (5.2.70) and (5.2.71) we easily see that

$$\frac{A_{ge}}{B_{ge}} = \frac{2\hbar\omega_{eg}^3}{\pi c^3} = \frac{8\pi\hbar\nu_{eg}^3}{c^3}. \tag{5.2.74}$$

Einstein was able to derive the Planck radiation formula without the use of quantum mechanics. It may be of interest to give his argument.

Suppose we have many identical atoms in a cavity filled with electromagnetic radiation in thermal equilibrium with the walls at temperature T. For simplicity, let us consider only the two lowest energy levels $\epsilon_e > \epsilon_g$. If an atom is in level e, it will spontaneously decay to level g and emit a quantum of energy with a probability A_{ge} independent of whether there is radiation in the cavity. In addition, if the atom is in state e there is a probability $B_{ge}u_\nu$ that the atom will be induced to decay and emit a quantum due to the presence of the electromagnetic field. This probability is obviously proportional to the density of radiation u_ν present in the cavity. Also, if an atom is in state g, the lower state, it will have a probability $B_{eg}u_\nu$ to absorb a quantum of energy and be excited to state e.

The atoms are in thermal equilbrium at temperature T and the probability that an atom is in state g is proportional to the Boltzmann factor $\exp -(\epsilon_g/kT)$ while the probability it is in state e is proportional to $\exp -(\epsilon_e/kT)$. In thermal equilibrium, we must have

$$(A_{ge} + B_{eg}u_\nu)e^{-\epsilon_e/kT} = B_{ge}u_\nu e^{-\epsilon_g/kT}. \tag{5.2.75}$$

That is, the spontaneous plus induced emission probability per second times the probability the atom is in the upper state e equals the probability of absorption per second times the probability the atom is in the lower state g. The A and B coefficients are determined by the properties of the atom and are independent of the frequency and temperature. As the temperature of the cavity walls gets very large, the radiation density will get large so that

$$B_{eg}u_\nu \gg A_{eg}, \tag{5.2.76}$$

and it then follows from (5.2.75) as $T \to \infty$ that

$$B_{eg} = B_{ge}. \tag{5.2.77}$$

Since these probabilities are independent of the radiation frequency and the temperature, we see that the probability for induced emission of radiation per second equals the probability per second for absorption. If we use this result and solve (5.2.75) for u_ν we obtain

$$u_\nu = \frac{A/B}{\exp(h\nu/kT) - 1} \tag{5.2.78}$$

where we have let

$$h\nu = \epsilon_e - \epsilon_g > 0,$$

and have omitted the subscripts on the Einstein A and B coefficients. Equation (5.2.78) is the Planck radiation formula. We obtain A and B from a correct quantum treatment as we have shown. Note that if there were no spontaneous emission, $A \to 0$ and Planck's formula would not follow in violation of experimental results. Therefore, spontaneous emission is necessary to explain blackbody radiation and this necessitates a quantum treatment of the radiation field.

CASE 4 COHERENT STATE

We consider next the case in which a single mode of the radiation field is excited in a coherent state. This will, of course, be another case of a purely monochromatic wave with zero line width, but a laser output in single mode operation can approximate this situation due to its high degree of monochromaticity (small linewidth) and high degree of collimation. It may also be plane polarized. Thus

$$\rho_r(0) = |\alpha_{l_0\sigma_0}\rangle\langle\alpha_{l_0\sigma_0}| \times \{|0\rangle\langle 0|\}, \tag{5.2.79}$$

where

$$a_{l_0\sigma_0}|\alpha_{l_0\sigma_0}\rangle = \alpha_{l_0\sigma_0}|\alpha_{l_0\sigma_0}\rangle. \tag{5.2.80}$$

In this case the traces of all terms exist in (5.2.16). When we evaluate them and carry out the time integrals we obtain

$$\begin{aligned} |c_a^{(1)}(s, t|s', 0)|^2 = \frac{\omega_{ss'}^2}{2\epsilon_0\hbar L^3}\sum_{l,\sigma}\frac{|\hat{e}_{l\sigma}\cdot\boldsymbol{\mu}_{ss'}|^2}{\omega_l}&\left\{\frac{4\sin^2\frac{1}{2}(\omega_l+\omega_{ss'})t}{(\omega_l+\omega_{ss'})^2}(1+|\alpha_{l\sigma}|^2\,\delta_{ll_0}\,\delta_{\sigma\sigma_0})\right.\\ &+|\alpha_{l\sigma}|^2\,\delta_{ll_0}\,\delta_{\sigma\sigma_0}\frac{4\sin^2\frac{1}{2}(\omega_l-\omega_{ss'})t}{(\omega_l-\omega_{ss'})^2}\\ &\left.+2\delta_{ll_0}\,\delta_{\sigma\sigma_0}\,\mathrm{Re}\,\frac{\alpha_{l\sigma}^2(1+e^{-2i\omega_l t}-e^{-i\omega_l t}\,2\cos\omega_{ss'}t)}{\omega_{ss'}^2-\omega_l^2}\right\}. \end{aligned} \tag{5.2.81}$$

If we let

$$\alpha_{l\sigma} = |\alpha_{l\sigma}|e^{i\xi} \tag{5.2.82}$$

and differentiate the above with respect to t, we obtain

$$\begin{aligned} w = \frac{\omega_{ss'}^2}{\epsilon_0\hbar L^3}\sum_{l,\sigma}\frac{|\hat{e}_{l\sigma}\cdot\boldsymbol{\mu}_{ss'}|^2}{\omega_l}&\left\{\frac{\sin(\omega_l+\omega_{ss'})t}{(\omega_l+\omega_{ss'})}(1+|\alpha_{l\sigma}|^2\,\delta_{ll_0}\,\delta_{\sigma\sigma_0})\right.\\ &+\delta_{ll_0}\,\delta_{\sigma\sigma_0}\frac{\sin(\omega_l-\omega_{ss'})t}{(\omega_l-\omega_{ss'})}|\alpha_{l\sigma}|^2\\ &+|\alpha_{l\sigma}|^2\,\delta_{ll_0}\,\delta_{\sigma\sigma_0}\left[\frac{\sin[(\omega_{ss'}-\omega_l)t+2\xi]}{\omega_{ss'}+\omega_l}+\frac{\sin[(\omega_{ss'}+\omega_l)t-2\xi]}{\omega_{ss'}-\omega_l}\right.\\ &\left.\left.-\frac{2\omega_l\sin 2(\omega_l t-\xi)}{(\omega_{ss'}+\omega_l)(\omega_{ss'}-\omega_l)}\right]\right\}. \end{aligned} \tag{5.2.83}$$

The reader will recognize the spontaneous emission term which is just as it has appeared in all prior cases. When the atom is in state $|g\rangle$ at $t = 0$ and $|e\rangle$ at t, we have the following result for absorption.

Absorption

$$w_{eg}^{\text{abs}} = \frac{\omega_{eg}{}^{2}}{\epsilon_0 \hbar L^3} \frac{|\hat{e}_{l\sigma} \cdot \boldsymbol{\mu}_{eg}|^2}{\omega_l} |\alpha_{l\sigma}|^2$$
$$\times \left\{ \frac{\sin(\omega_l + \omega_{eg})t}{\omega_l + \omega_{eg}} + \frac{\sin(\omega_l - \omega_{eg})t}{\omega_l - \omega_{eg}} + \frac{\sin[(\omega_{eg} - \omega_l)t + 2\xi]}{\omega_{eg} + \omega_l} \right.$$
$$\left. + \frac{\sin[(\omega_{eg} + \omega_l)t - 2\xi]}{\omega_{eg} - \omega_l} - \frac{2\omega_l \sin 2(\omega_l t - \xi)}{(\omega_{eg} + \omega_l)(\omega_{eg} - \omega_l)} \right\} \tag{5.2.84}$$

As $t \to \infty$, the first term becomes $\pi\, \delta(\omega_l + \omega_{eg})$. But $\omega_{eg} > 0$ and $\omega_l > 0$ so $\omega_l + \omega_{eg} \neq 0$ and we neglect it. The second becomes $\pi\, \delta(\omega_l - \omega_{eg})$ which is large when $\omega_l = \omega_{eg}$. In this case the third term reduces to $\sin 2\xi/2\omega_{eg}$, independent of t and is negligible compared with $\pi\, \delta(\omega_l - \omega_{eg})$. The fourth and fifth terms exactly cancel when $\omega_l \to \omega_{eg}$, for all t. Thus

$$w_{eg}^{\text{abs}} = \frac{\pi \omega_{eg}}{\epsilon_0 \hbar L^3} |\hat{e}_{l\sigma} \cdot \boldsymbol{\mu}_{eg}|^2 \delta(\omega_l - \omega_{eg}) |\alpha_{l\sigma}|^2, \tag{5.2.85}$$

which is identical with (5.2.28) except the "exact" number of photons in the excited mode $n_{l\sigma}$ has been replaced by the mean number $\langle \alpha_{l\sigma} | a_{l\sigma}^\dagger a_{l\sigma} | \alpha_{l\sigma} \rangle = |\alpha_{l\sigma}|^2$. In the former case there were no interference terms like the last three in (5.2.84). Why?

Emission

A similar argument shows that for induced emission,

$$w_{ge}^{\text{emiss}} \equiv w_{eg}^{\text{abs}}. \tag{5.2.86}$$

5.3 WIGNER–WEISSKOPF THEORY OF NATURAL LINEWIDTH [2]; LAMB SHIFT

In the theory of emission and absorption of radiation by an atom presented in the previous section, we assumed that the atomic energy levels were infinitely sharp whereas we know from experiment that the observed emission and absorption lines have a finite width. There are many interactions which may broaden an atomic line, but the most fundamental one is the reaction of the radiation field on the atom. That is, when an atom decays spontaneously from an excited state radiatively, it emits a quantum of energy into the radiation field. This radiation may be reabsorbed by the atom. The reaction of the field on the atom gives the atom a linewidth and causes the original level to be shifted as we show. This is the source of the natural linewidth and the Lamb shift. Spontaneous emission is also the source of quantum noise due to the random times at which emission occurs.

If we believe the uncertainty relation

$$\Delta E\,\Delta t \sim \hbar, \tag{5.3.1}$$

it states that to measure the energy of the atom to an accuracy ΔE, a time of order Δt is needed. An atom in an excited state decays spontaneously to a lower state with the emission of radiation. To make the measurement, we must wait sufficiently long to be sure the atom has decayed, namely, a time of the order of the lifetime τ of the excited state. Therefore, the energy is known only to an accuracy

$$\Delta E \sim \frac{\hbar}{\tau}, \tag{5.3.2}$$

which gives a linewidth. We shall show that τ^{-1} is equal to the transition probability per second for spontaneous emission.

The time-dependent perturbation theory which led to Fermi's golden rule is valid only for times sufficiently short that the atom does not change its state significantly. It is clearly not adequate to show the decay of the initial state. Accordingly, we must obtain more accurate solutions of the Schrödinger equation for an atom coupled to a radiation field.

We begin by assuming an atom is in some excited state $|E\rangle$ at $t = 0$ and that there is no radiation field present. The initial state is $|E; 0\rangle$ of energy ϵ_E. We let $c(E, 0; t)$ be the probability amplitude for finding the atom in state E with no photons at time t. The interaction energy is in the SP

$$\begin{aligned} V &= -\frac{e}{m}\mathbf{A}\cdot\mathbf{p} \\ &= -\frac{e}{m}\sum_{l',\sigma'}\sqrt{\frac{\hbar}{2\omega_{l'}\epsilon_0 L^3}}\,[a_{l'\sigma'}e^{i\mathbf{k}_{l'}\cdot\mathbf{r}} + a^\dagger_{l'\sigma'}e^{-i\mathbf{k}_{l'}\cdot\mathbf{r}}]\hat{e}_{l'\sigma'}\cdot\mathbf{p}. \end{aligned} \tag{5.3.3}$$

If

$$c(E, 0; 0) = 1, \tag{5.3.4a}$$

the atom may decay to some state $|I\rangle$ with the emission of a photon of momentum $\hbar\mathbf{k}_l$ and polarization $\hat{e}_{l\sigma}$. At $t = 0$

$$c(I, 1_{l\sigma}; 0) = 0. \tag{5.3.4b}$$

The nonzero matrix elements which connect the initial states to any other are given by

$$\begin{aligned} \langle I, 1_{l\sigma}|V|E, 0\rangle &= -\frac{e}{m}\sqrt{\frac{\hbar}{2\omega_l\epsilon_0 L^3}}\,\langle I|e^{-i\mathbf{k}_l\cdot\mathbf{r}}\mathbf{p}|E\rangle\cdot\hat{e}_{l\sigma} \\ &\equiv \langle E, 0|V|I, 1_{l\sigma}\rangle^*. \end{aligned} \tag{5.3.5}$$

If we use the exact equations (1.16.59), we see that the probability amplitudes satisfy

$$i\hbar \frac{dc}{dt}(E, 0; t) = \sum_{I} \sum_{l,\sigma} V_{E,0;I,1_{l\sigma}} e^{-i(\omega_{EI}-\omega_l)t} c(I, 1_{l\sigma}; t) \tag{5.3.6}$$

$$i\hbar \frac{dc}{dt}(I, 1_{l\sigma}, t) = V_{I,1_{l\sigma};E,0} e^{-i(\omega_{EI}-\omega_l)t} c(E, 0; t), \tag{5.3.7}$$

where we have let

$$\omega_{EI} \equiv \frac{\epsilon_E - \epsilon_I}{\hbar}, \tag{5.3.8}$$

and we must sum over all states $|I\rangle$ for which the matrix elements are nonzero and over photons of all momenta and polarizations. It should be noted that the equations above are still exact. In first order perturbation theory we replace $c(I, 1_{l\sigma}, t)$ and $c(E, 0, t)$ on the right side by their initial values, but we must obtain better solutions now subject to the initial conditions. Unfortunately, they cannot be solved exactly. If we integrate both sides of (5.3.7) and use (5.3.4), we obtain

$$c(I, 1_{l\sigma}; t) = \frac{1}{i\hbar} \langle I, 1_{l\sigma}|V|E, 0\rangle \int_0^t e^{-i(\omega_{EI}-\omega_l)t'} c(E, 0; t')\, dt'. \tag{5.3.9}$$

We may substitute this into (5.3.6) to obtain the integro-differential equation

$$\frac{dc}{dt}(E, 0; t) = -\frac{1}{\hbar^2} \sum_{I} \sum_{l,\sigma} |V_{E,0;I,1_{l\sigma}}|^2 \int_0^t e^{i(\omega_{EI}-\omega_l)(t-t')} c(E, 0; t')\, dt'. \tag{5.3.10}$$

From our original expansion for the state vector (1.16.29b), we have

$$|\psi(E, 0, t)\rangle \simeq c(E, 0; t) e^{-(i/\hbar)\epsilon_E t} |E; 0\rangle. \tag{5.3.11}$$

We would expect an *approximate* solution of the exact equation (5.3.10) to behave as

$$c(E, 0; t) \simeq e^{-(i/\hbar)\Delta\epsilon_E t}, \tag{5.3.12a}$$

where

$$\Delta\epsilon_E = \hbar(\Delta\omega_E - i\tfrac{1}{2}\Gamma_E), \tag{5.3.12b}$$

which would show that the atom decays from its excited state with a lifetime Γ_E^{-1} and has a shift in its energy level due to its interaction with the many degrees of the radiation field. To obtain an approximate solution of this form, let us take the Laplace transform [3] of both sides. If we let

$$\bar{c}(s) = \int_0^\infty e^{-st} c(E, 0; t)\, dt, \tag{5.3.13}$$

then with the initial condition (5.3.4a), it follows on integration by parts that

$$\int_0^\infty e^{-st} \frac{dc}{dt}(E, 0; t)\, dt = s\bar{c}(s) - 1. \tag{5.3.14}$$

Also we note that

$$\begin{aligned}\int_0^\infty e^{-st}\, dt \int_0^t e^{i\Omega(t-t')} c(E, 0; t')\, dt' \\ &= \int_0^\infty c(E, 0; t') e^{-i\Omega t'}\, dt' \int_{t'}^\infty e^{-(s-i\Omega)t}\, dt \\ &= \int_0^\infty \frac{c(E, 0, t') e^{-st'}\, dt'}{s - i\Omega} = \frac{\bar{c}(s)}{s - i\Omega},\end{aligned} \tag{5.3.15}$$

where we interchanged the order of integration over t and t'. Thus if we multiply both sides of (5.3.10) by $e^{-st}\, dt$ and integrate from 0 to ∞ and use the results above, we obtain after minor algebra

$$\bar{c}(s) = \left(s + \frac{i}{\hbar^2} \sum_I \sum_{l,\sigma} \frac{|V_{E,0;I,1l\sigma}|^2}{\omega_{EI} - \omega_l + is}\right)^{-1} \equiv \frac{1}{\Delta(s)}. \tag{5.3.16}$$

The exact formal solution is given by [3]

$$c(E, 0; t) = \frac{1}{2\pi i} \int_{\epsilon - i\infty}^{\epsilon + i\infty} e^{st} \bar{c}(s)\, ds, \tag{5.3.17}$$

where the path of integration is a contour parallel to the imaginary s axis. To evaluate, we must know the poles in the half plane Re $(s) < 0$ of $\bar{c}(s)$. If $\bar{c}(s)$ has a simple pole at

$$s = -\frac{i}{\hbar} \Delta\epsilon_E = -\tfrac{1}{2}\Gamma_E - i\,\Delta\omega_E, \tag{5.3.18}$$

then (5.3.17) becomes

$$c(E, 0; t) = e^{-(i/\hbar)\Delta\epsilon_E t} = e^{-(\frac{1}{2}\Gamma_E + i\Delta\omega_E)t}, \tag{5.3.19}$$

which is the desired form.

We therefore make the Wigner-Weisskopf approximation to solve for the zeros of $\Delta(s)$ in (5.3.16). We note that if the atom-field interaction is small, as a zeroth approximation $\Delta(s) = 0$ if $s = 0$. As a next approximation we let $s \to 0$ in the denominator of the sum in $\Delta(s)$. In other words, we let the pole be shifted approximately to

$$\Delta(0) - s \simeq +\frac{i}{\hbar} \Delta\epsilon_E = \lim_{s \to 0^+} \frac{i}{\hbar^2} \sum_I \sum_{l,\sigma} \frac{|V_{E,0;I,1l\sigma}|^2}{\omega_{EI} - \omega_l + is} \equiv \tfrac{1}{2}\Gamma_E + i\,\Delta\omega_E. \tag{5.3.20}$$

Thus the W-W approximation consists in calculating the first order shift in the simple pole of $\bar{c}(s)$ due to the atom-field interaction. Now [4]

$$\lim_{s\to 0^+} \frac{1}{x+is} = \lim_{s\to 0^+}\left[\frac{x}{x^2+s^2} - \frac{is}{x^2+s^2}\right]$$

$$= \frac{1}{x} - i\pi\,\delta(x), \tag{5.3.21}$$

since

$$\lim_{s\to 0^+} \frac{s}{\pi(x^2+s^2)} = \begin{cases} 0 & x \neq 0 \\ \infty & x = 0 \end{cases} \tag{5.3.22}$$

$$\lim_{s\to 0^+} \int_{-\infty}^{\infty} \frac{s}{\pi(x^2+s^2)}\,dx = 1, \tag{5.3.23}$$

which are the required properties of a δ-function. If we use (5.3.21), (5.3.20) becomes

$$\mathrm{Re}\,(\Delta\epsilon_E) = \hbar\,\Delta\omega_E = \frac{1}{\hbar}\sum_I \sum_{l,\sigma} \frac{|V_{E,0;I,1_{l\sigma}}|^2}{\omega_{EI} - \omega_l} \tag{5.3.24}$$

$$\mathrm{Im}\,(\Delta\epsilon_E) = -\frac{\hbar}{2}\Gamma_E = -\frac{\pi}{\hbar}\sum_I \sum_{l,\sigma} |V_{E,0\;I,1_{l\sigma}}|^2\,\delta(\omega_{EI} - \omega_l). \tag{5.3.25}$$

We see that the atom is continually emitting and reabsorbing quanta of radiation. The energy level shift does not require energy to be conserved while the damping requires energy conservation. Thus damping is brought about by the emission and absorption of real photons while the photons emitted and absorbed which contribute to the energy shift are called virtual photons.

We may calculate Γ_E explicitly. For an atom emitting or absorbing radiation at optical frequencies, the atomic wave functions decay in a distance of the order of the atomic size, namely, $\sim 10^{-8}$ cm. So we are justified in making the dipole approximation in (5.3.5) so that

$$\Gamma_E = \frac{2\pi}{\hbar}\sum_{I\neq E}\sum_{l,\sigma} \frac{e^2}{m^2}\frac{\hbar}{2\omega_l\epsilon_0 L^3}\,|\mathbf{p}_{IE}\cdot\hat{e}_{l\sigma}|^2\,\delta(\omega_{EI} - \omega_l). \tag{5.3.26}$$

If we use the results of Section 4.5 to convert the sum over l to an integral, we have

$$\Gamma_E = \frac{\pi e^2}{\hbar m^2\epsilon_0 L^3}\sum_\sigma\sum_I \int_0^\infty \left(\frac{L}{2\pi c}\right)^3 \omega_l\,d\omega_l \int_{4\pi} d\Omega\,|\mathbf{p}_{EI}\cdot\hat{e}_{l\sigma}|^2\,\delta(\omega_{EI} - \omega_l). \tag{5.3.27}$$

By (5.2.35)–(5.2.37), we have

$$\sum_\sigma \int d\Omega\,|\mathbf{p}_{EI}\cdot\hat{e}_{l\sigma}|^2 = \frac{8\pi}{3}\,|\mathbf{p}_{EI}|^2 \tag{5.3.28}$$

so that

$$\Gamma_E = \sum_I \frac{e^2}{4\pi\epsilon_0\hbar c} \frac{4}{3} \frac{\omega_{EI}}{(mc^2)} |\mathbf{p}_{EI}|^2. \tag{5.3.29}$$

But by (5.2.9) this may be written as

$$\Gamma_E = \sum_{I \neq E} \frac{\omega_{EI}^3 |\boldsymbol{\mu}_{EI}|^2}{3\pi\hbar\epsilon_0 c^3} \equiv \sum_{I \neq E} w_{IE}^{\text{spon}}. \tag{5.3.30}$$

If we compare this with (5.2.40) we see that Γ_E which is the inverse lifetime of state $|E\rangle$ is just the sum of the probability per second for the atom to decay spontaneously to all lower states I which have nonzero dipole moment matrix elements. This is a very important result.

Before discussing the energy shift, let us obtain $c(I, 1_{l\sigma}; t)$. By (5.3.9) and (5.3.19) we have

$$c(I, 1_{l\sigma}; t) = \langle I, 1_{l\sigma}|V|E, 0\rangle \frac{\{e^{-[\frac{1}{2}\Gamma_E + i(\omega_{EI} - \omega_l + \Delta\omega)]t} - 1\}}{\hbar[\omega_{EI} - \omega_l + \Delta\omega - i\frac{1}{2}\Gamma_E]}, \tag{5.3.31}$$

while the probability of finding the atom in state I with one photon is

$$|c(I, 1_{l\sigma}; t)|^2 = \frac{|V_{I,1_{l\sigma};E,0}|^2 \{1 + e^{-\Gamma_E t} - 2e^{-\frac{1}{2}\Gamma_E t} \cos(\omega_{EI} - \omega_l + \Delta\omega)t\}}{\hbar^2\left[(\omega_{EI} - \omega_l + \Delta\omega)^2 + \left(\frac{\Gamma_E}{2}\right)^2\right]}, \tag{5.3.32}$$

which has a Lorentzian shape of half-width Γ_E. If we sum this over all states of the field, use (5.3.5), and convert the sum to an integral, we have

$$|c(I, t)|^2 \equiv \sum_{l,\sigma} |c(I, 1_{l\sigma}; t)|^2$$

$$= \frac{e^2 |p_{EI}|^2}{6\hbar m^2 \epsilon_0 c^3 \pi^2} \int_{-\omega_{EI}-\Delta\omega}^{\infty} x\, dx \frac{\{1 + e^{-\Gamma_E t} - 2e^{-\frac{1}{2}\Gamma_E t} \cos xt\}}{x^2 + (\Gamma_E/2)^2}, \tag{5.3.33}$$

where we have let

$$x = \omega_l - \omega_{EI} - \Delta\omega, \tag{5.3.34}$$

and have used (5.3.28). Since the integral is strongly peaked at $x = 0$, we may replace the lower limit by $-\infty$. Then since

$$\int_{-\infty}^{\infty} \frac{dx}{x^2 + a^2} = \frac{\pi}{a}$$

$$\int_{-\infty}^{\infty} \frac{x\, dx}{x^2 + a^2} = 0 = \int_{-\infty}^{\infty} \frac{x \cos xt}{x^2 + a^2}\, dx \tag{5.3.35}$$

$$\int_{-\infty}^{\infty} \frac{\cos xt\, dx}{x^2 + a^2} = \frac{\pi}{a} e^{-a|t|},$$

(5.3.33) becomes

$$|c(I, t)|^2 = \frac{\omega_{EI}^2 |\mathbf{p}_{EI}|^2}{6\hbar\epsilon_0 c^3 \pi^2} (\omega_{EI} + \Delta\omega) \frac{2\pi}{\Gamma_E} [1 - e^{-\Gamma_E t}]. \tag{5.3.36}$$

Finally, if we sum over all states I and neglect $\Delta\omega$, we have on using (5.3.30)

$$\sum_I \sum_{l,\sigma} |c(I, 1_{l\sigma}; t)|^2 \to 1 - e^{-\Gamma_E t}. \tag{5.3.37}$$

That is, as $t \to \infty$, the atom has decayed from its original state to one of the possible lower states.

We are now ready to return to the troublesome question of calculating the energy level shift given in (5.3.24). When we use (5.3.5), convert the sum to an integral and use (5.3.28) we obtain

$$\hbar\,\Delta\omega_E = \sum_I \frac{e^2 |\mathbf{p}_{EI}|^2}{6c^3\epsilon_0\pi^2 m^2} \int_0^\infty \frac{\omega_l\, d\omega_l}{\omega_{EI} - \omega_l}, \tag{5.3.38}$$

which diverges linearly and we have assumed the pole was only shifted a small amount. This is a standard problem in quantum electrodynamics and the removal of this divergence is based on the concept of *mass renormalization*. To remove the divergence completely, we must use the Dirac relativistic theory of the electron. We do not do this here and only are able to make the divergence logarithmic instead of linear which is an improvement.

We begin by arguing that contributions to the energy shift due to extremely high energy photons cannot be important since the nonrelativistic approximation for the electron must not be valid for photons of energy $\hbar\omega_l \sim mc^2$. Therefore, we only integrate to some $\omega_l^{\max}$ so that

$$\hbar\,\Delta\omega_E \cong -\sum_I \frac{e^2 |\mathbf{p}_{EI}|^2}{6c^3\epsilon_0\pi^2 m^2} \int_0^{\omega_l^{\max}} \frac{\omega_l\, d\omega_l}{\omega_l - \omega_{EI}} \sim -\sum_I \frac{e^2}{4\pi\epsilon_0\hbar c} \frac{|\mathbf{p}_{EI}|^2}{2m} \frac{4\hbar\omega_l^{\max}}{3\pi mc^2} \tag{5.3.39}$$

But this is very sensitive to the choice of $\omega_l^{\max}$. In an effort to improve on this, let us try to understand the physical origin of the divergence. Let us consider a free electron with momentum $\mathbf{p}$ interacting with the radiation field. Initially we assume no radiation present. The interaction energy is

$$\begin{aligned} V &= -\frac{e}{m} \mathbf{A} \cdot \mathbf{p} \\ &= -\frac{e}{m} \sum_{l,\sigma} \sqrt{\frac{\hbar}{2\epsilon_0\omega_l L^3}} [a_{l\sigma} e^{i\mathbf{k}_l\cdot\mathbf{r}} + a_{l\sigma}^\dagger e^{-i\mathbf{k}_l\cdot\mathbf{r}}] \hat{e}_{l\sigma} \cdot \mathbf{p}. \end{aligned} \tag{5.3.40}$$

Then nonzero matrix elements are

$$\langle \mathbf{p}', 1_{l\sigma}|V|\mathbf{p}, 0\rangle = -\frac{e}{m}\sqrt{\frac{\hbar}{2\epsilon_0\omega_l L^3}}\,\langle \mathbf{p}'|e^{-i\mathbf{k}_l\cdot\mathbf{r}}\mathbf{p}|\mathbf{p}\rangle \cdot \hat{e}_{l\sigma}$$

$$= -\frac{e}{m}\sqrt{\frac{\hbar}{2\epsilon_0\omega_l L^3}}\,(\mathbf{p}\cdot\hat{e}_{l\sigma})\cdot\langle \mathbf{p}'|e^{-i\mathbf{k}_l\cdot\mathbf{r}}|\mathbf{p}\rangle, \qquad (5.3.41)$$

which connect the initial state of momentum $\mathbf{p}$ and no photons to all other states. We may not make the dipole approximation in the present case because the electron eigenstates are plane waves and do not cut off like bound atomic states. Therefore, we have

$$\langle \mathbf{p}'|e^{-i\mathbf{k}_l\cdot\mathbf{r}}|\mathbf{p}\rangle = \frac{1}{L^3}\int e^{-(i/\hbar)\mathbf{p}'\cdot\mathbf{r}}e^{-i\mathbf{k}_l\cdot\mathbf{r}}e^{(i/\hbar)\mathbf{p}\cdot\mathbf{r}}\,d\mathbf{r} = \delta[\mathbf{p}';\mathbf{p}-\hbar\mathbf{k}_l], \quad (5.3.42)$$

since

$$\langle \mathbf{r}|\mathbf{p}\rangle = \frac{1}{L^{3/2}}\,e^{(i/\hbar)\mathbf{p}\cdot\mathbf{r}}. \qquad (5.3.43)$$

But (5.3.42) is just a statement of momentum conservation for the photon and electron:

$$\mathbf{p}' + \hbar\mathbf{k}_l = \mathbf{p} \qquad (5.3.44)$$

Thus the nonzero matrix elements are

$$\langle \mathbf{p}', 1_{l\sigma}|V|\mathbf{p}, 0\rangle = -\frac{e}{m}\sqrt{\frac{\hbar}{2\epsilon_0\omega_l L^3}}\,(\mathbf{p}\cdot\hat{e}_{l\sigma})\,\delta[\mathbf{p}';\mathbf{p}-\hbar\mathbf{k}_l]. \qquad (5.3.45)$$

The equations for the probability amplitudes corresponding to (5.3.6) and (5.3.7) become

$$i\hbar\frac{dc}{dt}(\mathbf{p}, 0; t) = \sum_{\mathbf{p}'}\sum_{l,\sigma} V_{\mathbf{p},0:\mathbf{p}',1_{l\sigma}}e^{-i\omega_{fi}t}c(\mathbf{p}', 1_{l\sigma}; t) \qquad (5.3.46)$$

$$i\hbar\frac{dc}{dt}(\mathbf{p}', 1_{l\sigma}; t) = V_{\mathbf{p}',1_{l\sigma};\mathbf{p},0}e^{+i\omega_{fi}t}c(\mathbf{p}, 0; t) \qquad (5.3.47)$$

where we have let

$$\hbar\omega_{fi} = \frac{\mathbf{p}'^2}{2m} + \hbar\omega_l - \frac{\mathbf{p}^2}{2m}. \qquad (5.3.48)$$

We proceed to solve in the same way. The electron emits and reabsorbs photons, and the energy shift is

$$\begin{aligned}\hbar\,\Delta\omega &= -\frac{1}{\hbar}\sum_{\mathbf{p}'}\sum_{l,\sigma}\frac{e^2}{m^2}\frac{\hbar}{2\epsilon_0\omega_l L^3}\frac{|\mathbf{p}\cdot\hat{e}_{l\sigma}|^2\,\delta[\mathbf{p}';\mathbf{p}-\hbar\mathbf{k}_l]}{\omega_{fi}} \\ &= -\frac{e^2\hbar}{m^2 2\epsilon_0 L^3}\sum_{l,\sigma}\frac{|\mathbf{p}\cdot\hat{e}_{l\sigma}|^2\,\omega_l^{-1}}{\frac{1}{2}(\mathbf{p}-\hbar\mathbf{k}_l)^2-(1/2m)\mathbf{p}^2+\hbar\omega_l}. \end{aligned} \tag{5.3.49}$$

In the nonrelativistic approximation

$$\frac{1}{2m}(\mathbf{p}-\hbar k_l)^2-\frac{1}{2m}\mathbf{p}^2+\hbar\omega_l\approx\hbar\omega_l, \tag{5.3.50}$$

so that

$$\begin{aligned}\hbar\,\Delta\omega &\cong -\frac{e^2}{m^2 2\epsilon_0 L^3}\sum_\sigma\int_0^\infty\left(\frac{L}{2\pi c}\right)^3 d\omega_l\int d\Omega\,|\mathbf{p}\cdot\hat{e}_{l\sigma}|^2 \\ &= -\frac{e^2}{4\pi\epsilon_0\hbar c}\frac{2}{3\pi}\frac{1}{(mc)^2}\int_0^\infty\hbar\,d\omega_l\,\mathbf{p}^2 = -\left(\frac{e^2}{4\pi\epsilon_0\hbar c}\right)\frac{2\hbar\omega_l^{\max}}{3\pi(mc)^2}\mathbf{p}^2 \\ &= -K\mathbf{p}^2, \end{aligned} \tag{5.3.51}$$

where K still diverges linearly. The problem is that we can never have a "bare" electron in real life. It always carries along with it an electromagnetic field which contributes to its kinetic energy since it is proportional to $\mathbf{p}^2$. The field shows up as a continual emission and reabsorption of virtual photons. This is just the self energy of the electron. When we measure the kinetic energy of a free electron, we can never separate the field contribution so we actually observe the kinetic energy

$$T_{\text{obs}} = \frac{\mathbf{p}^2}{2m_{\text{bare}}} - K\mathbf{p}^2, \tag{5.3.52}$$

where m_{bare} is the mass of a fictitious bare electron without its associated electromagnetic field which we have so far been using in the theory. We therefore define a new observable mass by the relation

$$\begin{aligned}T_{\text{obs}} &= \frac{1}{2m_{\text{obs}}}\mathbf{p}^2 = \frac{1}{2m_{\text{bare}}}(1-2m_{\text{bare}}K)\mathbf{p}^2 \\ &= \frac{1}{2m_{\text{bare}}}\left[1-\frac{e^2}{4\pi\epsilon_0\hbar c}\frac{4}{3\pi}\frac{\hbar\omega_l^{\max}}{m_{\text{bare}}c^2}\right]\mathbf{p}^2, \end{aligned} \tag{5.3.53}$$

where we used (5.3.51). For $\hbar\omega_l^{\max}\sim mc^2$, since $e^2/4\pi\epsilon_0\hbar c\sim 1/137$, we see that this is a small correction.

We now argue that we have solved the energy eigenvalue problem for an atom using the bare mass when in reality we should use m_{obs}. The hamiltonian for the free atom should be

$$H_a^{\text{new}} = \frac{1}{2m_{\text{obs}}}\,\mathbf{p}^2 + V(\mathbf{r})$$
$$= \frac{\mathbf{p}^2}{2m_{\text{bare}}} + V(\mathbf{r}) - K\mathbf{p}^2 = H_a^{\text{old}} - K\mathbf{p}^2. \tag{5.3.54}$$

Then we have for our new "correct" unperturbed energy levels

$$\epsilon_E^{\text{new}} = \langle E|H_a^{\text{old}} - K\mathbf{p}^2|E\rangle \cong \epsilon_E^{\text{old}} - K\langle E|\mathbf{p}^2|E\rangle. \tag{5.3.55}$$

When we recalculate the energy level shift, we must add back the term $+K\langle E|\mathbf{p}^2|E\rangle$ since it has already been accounted for. Thus by (5.3.39) and (5.3.51), the observed energy shift is

$$\hbar\,\Delta\omega_E^{\text{obs}} \cong \frac{e^2}{4\pi\epsilon_0\hbar c}\,\frac{2\hbar}{3\pi(m_b c)^2}\int_0^{\omega_l^{\max}} d\omega_l\left\{\omega_l\sum_I \frac{|\mathbf{p}_{EI}|^2}{\omega_{EI}-\omega_l} + \langle E|\mathbf{p}^2|E\rangle\right\}, \tag{5.3.56}$$

where to this degree of approximation the matrix elements are to be computed using the unperturbed or "bare" energy eigenvectors and we use the "bare" mass. This follows because although K actually diverges if we take an $\hbar\omega_l^{\max} \sim m_b c^2$, $K \ll 1$ and corrections to $|\mathbf{p}_{EI}|^2$ and the ω_{EI} in the denominator above would be of order K, and we are not keeping K^2 terms.

We simplify the integrand above as follows:

$$\sum_I |\mathbf{p}_{EI}|^2 \frac{\omega_l}{\omega_{EI}-\omega_l} = \sum_I |\mathbf{p}_{EI}|^2\left\{-1 + \frac{\omega_{EI}}{\omega_{EI}-\omega_l}\right\}. \tag{5.3.57}$$

If we use the completeness relation, we have

$$\sum_I |\mathbf{p}_{EI}|^2 = \sum_I \langle E|\mathbf{p}|I\rangle\cdot\langle I|\mathbf{p}|E\rangle = \langle E|\mathbf{p}^2|E\rangle. \tag{5.3.58}$$

Thus, by these two, (5.3.56) becomes

$$\hbar\,\Delta\omega_E^{\text{obs}} \cong \frac{e^2}{4\pi\epsilon_0\hbar c}\,\frac{2\hbar}{3\pi(m_b c)^2}\int_0^{\omega_l^{\max}} \sum_I \omega_{EI}\,|\mathbf{p}_{EI}|^2\,\frac{d\omega_l}{\omega_{EI}-\omega_l}$$
$$= -\frac{e^2}{4\pi\epsilon_0\hbar c}\,\frac{2\hbar}{3\pi(m_b c)^2}\sum_I \omega_{EI}\,|\mathbf{p}_{EI}|^2 \log\frac{\omega_l^{\max}-\omega_{EI}}{\omega_{IE}}. \tag{5.3.59}$$

But

$$\hbar\omega_l^{\max} \sim m_b c^2 \gg \omega_{EI}, \tag{5.3.60}$$

so that

$$\hbar\,\Delta\omega_E^{\text{obs}} \cong +\frac{e^2}{4\pi\epsilon_0\hbar c}\,\frac{2\hbar}{3\pi(m_b c)^2}\sum_I \omega_{IE}\,|\mathbf{p}_{EI}|^2 \log\left(\frac{\omega_l^{\max}}{\omega_{IE}}\right). \tag{5.3.61}$$

If we replace ω_{IE} in the log term by some average value, we may remove it from the sum and obtain

$$\hbar\,\Delta\omega_E^{\text{obs}} \cong \frac{e^2}{4\pi\epsilon_0\hbar c}\,\frac{2\hbar}{3\pi(m_b c)^2}\log\frac{\omega_l^{\max}}{\langle\omega_{IE}\rangle}\sum_I \omega_{IE}\,|\mathbf{p}_{EI}|^2. \tag{5.3.62}$$

We may now evaluate the sum as follows. The atomic hamiltonian is

$$H_a = \frac{\mathbf{p}^2}{2m_b} + V(\mathbf{r}), \tag{5.3.63}$$

so that

$$[\mathbf{p}, H_a] = [\mathbf{p}, V(\mathbf{r})] = \frac{\hbar}{i}\nabla V. \tag{5.3.64}$$

Also

$$[p_i, [p_i, H_a]] = -\hbar^2\,\nabla^2 V, \tag{5.3.65}$$

where we sum on i from 1 to 3. If we take the diagonal matrix elements of both sides for state E, we have on expanding the double commutator

$$\langle E|[p_i(p_iH_a - H_ap_i) - (p_iH_a - H_ap_i)p_i]|E\rangle = -\hbar^2\langle E|\nabla^2 V|E\rangle, \tag{5.3.66}$$

or

$$2\epsilon_E\langle E|p_ip_i|E\rangle - 2\langle E|p_iH_ap_i|E\rangle = -\hbar^2\langle E|\nabla^2 V|E\rangle. \tag{5.3.67}$$

If we insert the completeness relation after the first p_i in both terms, we have

$$\sum_I(\epsilon_E - \epsilon_I)\langle E|p_i|I\rangle\langle I|p_i|E\rangle = -\frac{\hbar^2}{2}\langle E|\nabla^2 V|E\rangle, \tag{5.3.68}$$

or

$$\sum_I \omega_{IE}\,|p_{EI}|^2 = +\frac{\hbar}{2}\langle E|\nabla^2 V|E\rangle. \tag{5.3.69}$$

As a specific example, let us consider the hydrogen atom. In this case,

$$\nabla^2 V = \frac{e^2\,\delta(\mathbf{r})}{\epsilon_0}, \tag{5.3.70}$$

where $\delta(\mathbf{r}) = \delta(x)\,\delta(y)\,\delta(z)$. Thus

$$\langle E|\nabla^2 V|E\rangle = \frac{e^2}{\epsilon_0}\int d\mathbf{r}\,|\psi_E(\mathbf{r})|^2\,\delta(\mathbf{r}) = \frac{e^2}{\epsilon_0}\,|\psi_E(0)|^2. \tag{5.3.71}$$

If one looks at the hydrogen wave functions [5], they all vanish at the origin unless the quantum number $l = 0$, that is, for s-states. Any state with a higher orbital angular momentum vanishes, so there is a Lamb shift only for s-states in this nonrelativistic theory. In this case

$$\frac{\hbar}{2}\langle E|\nabla^2 V|E\rangle = \frac{e^2\hbar}{\epsilon_0 2\pi n^3 a_0^3} \qquad \text{for } s\text{-states}, \tag{5.3.72}$$

where

$$a_0 = \frac{4\pi\epsilon_0\hbar^2}{me^2} = \frac{\hbar/mc}{e^2/4\pi\epsilon_0\hbar c}. \tag{5.3.73}$$

Thus

$$\hbar\,\Delta\omega_E^{\text{obs}} \sim \left(\frac{e^2}{4\pi\epsilon_0\hbar c}\right)\frac{2\hbar}{3\pi(mc)^2}\log\frac{\omega_l^{\max}}{\langle\omega_{EI}\rangle}\cdot\frac{e^2\hbar}{2\pi\epsilon_0 n^3}\left(\frac{mc}{\hbar}\right)^2\frac{1}{a_0}$$

$$= \frac{8}{3\pi}\left(\frac{e^2}{4\pi\epsilon_0\hbar c}\right)^3\left(\frac{e^2}{8\pi\epsilon_0 a_0}\right)\frac{1}{n^3}\log\frac{\omega_l^{\max}}{\langle\omega_{EI}\rangle}. \tag{5.3.74}$$

Now $e^2/8\pi\epsilon_0 a_0$ is just the ionization energy for an electron in the 1*s*-state. If we take $\hbar\omega_l^{\max} = mc^2$ and estimate $\hbar\langle\omega_{EI}\rangle = 17.8(e^2/8\pi\epsilon_0 a_0)$ for the 2*s*-state, $\Delta\omega_E$ turns out to be 1040 Mc/sec in very good agreement with the measured value of 1057 Mc/sec of Lamb and Retherford. Relativistic calculations of Bethe give exact agreement when the Dirac relativistic wave functions are used.

5.4 KRAMERS–HEISENBERG SCATTERING CROSS-SECTION

We next consider the scattering of light by one-electron atoms. Initially we assume the atom is in some bound state $|a\rangle$ and the radiation field is in state $|n_i, n_f\rangle$ with n_i photons of momentum $\hbar\mathbf{k}_i$ and polarization $\hat{e}_i$ and n_f photons of momentum $\hbar\mathbf{k}_f$ and polarization $\hat{e}_f$. The initial atom-field eigenvector is $|a; n_i, n_f\rangle$ of energy $\epsilon_a + n_i\hbar\omega_i + n_f\hbar\omega_f$. After the scattering is over, the atom is left in state $|b\rangle$ and the field is in state $|n_i - 1; n_f + 1\rangle$ so that the final energy is $\epsilon_b + \hbar\omega_i(n_i - 1) + \hbar\omega_f(n_f + 1)$. We would like to calculate the probability for such a scattering process to occur and obtain the differential scattering cross-section.

The atom and field are described by the hamiltonian (5.1.2)

$$H = H_a + H_f + H_1 + H_2, \tag{5.4.1}$$

where

$$H_1 = -\frac{e}{m}\mathbf{A}\cdot\mathbf{p} \tag{5.4.2}$$

$$H_2 = \frac{e^2}{2m}\mathbf{A}^2. \tag{5.4.3}$$

If we make the dipole approximation which is valid at optical frequencies and let $\exp \pm i\mathbf{k}_l \cdot \mathbf{r} \simeq 1$, and use (5.1.8), we have that

$$H_1 \simeq -\frac{e}{m}\sum_{l,\sigma}\sqrt{\frac{\hbar}{2\omega_l\epsilon_0 L^3}}(a_{l\sigma} + a_{l\sigma}^\dagger)(\hat{e}_{l\sigma}\cdot\mathbf{p}) \tag{5.4.4}$$

$$H_2 = \frac{e^2\hbar}{2m2\epsilon_0 L^3}\sum_{l,\sigma}\sum_{l',\sigma'}\frac{\hat{e}_{l\sigma}\cdot\hat{e}_{l'\sigma'}}{\sqrt{\omega_l\omega_l'}}[a_{l\sigma}a_{l'\sigma'}^\dagger + a_{l\sigma}^\dagger a_{l'\sigma'} + a_{l\sigma}a_{l'\sigma'} + a_{l\sigma}^\dagger a_{l'\sigma'}^\dagger]. \tag{5.4.5}$$

We wish to see if H_1 and H_2 can cause the atom-field system to make transitions between the initial state $|a; n_i, n_f\rangle$ and the final state $|b; n_i - 1, n_f + 1\rangle$. In first-order perturbation theory, by Problem 1.15, such transitions will *not* be induced to occur, since

$$\langle f|H_1|i\rangle = -\frac{e}{m}\sum_{l,\sigma}\sqrt{\frac{\hbar}{2\omega_l\epsilon_0 L^3}}\langle n_i - 1, n_f + 1|(a_{l\sigma} + a_{l\sigma}^\dagger)|n_i, n_f\rangle\langle b|\mathbf{p}|a\rangle\cdot\hat{e}_{l\sigma} \equiv 0 \tag{5.4.6}$$

That is, the first-order matrix elements vanish unless the *net number* of photons changes by unity. Here we lose one photon and gain another with no net change. [Note that we are interested only in photons $(\mathbf{k}_i, \hat{e}_i)$ and $(\mathbf{k}_f, \hat{e}_f)$.] We must therefore see if such a scattering process can occur in second order. According to Problem 1.15, we therefore need the nonzero matrix elements

$$\langle f|H_2|i\rangle = \frac{e^2\hbar}{4m\epsilon_0 L^3}\sum_{l,\sigma}\sum_{l',\sigma'}\frac{\hat{e}_{l\sigma}\cdot\hat{e}_{l'\sigma'}}{\sqrt{\omega_l\omega_{l'}}}\langle n_i - 1, n_f + 1|(a_{l\sigma}a_{l'\sigma'}^\dagger + a_{l\sigma}^\dagger a_{l'\sigma'} + a_{l\sigma}a_{l'\sigma'} + a_{l\sigma}^\dagger a_{l'\sigma'}^\dagger)|n_i, n_f\rangle\langle a|b\rangle. \tag{5.4.7}$$

The last two terms are clearly zero since they either annihilate or create two photons, a process we are not considering. However, for the first term we have

$$\langle n_i - 1, n_f + 1|a_{l\sigma}a_{l'\sigma'}^\dagger|n_i, n_f\rangle = \begin{cases}\sqrt{n_i(n_f+1)} & \text{if} \quad l'\sigma' \equiv f;\ l\sigma = i\\ 0 & \text{(otherwise).}\end{cases} \tag{5.4.8a}$$

That is, we may first create the final photon and then annihilate the initial one. For the second term, we have

$$\langle n_i - 1, n_f + 1|a_{l\sigma}a_{l'\sigma'}^\dagger|n_i, n_f\rangle = \begin{cases}\sqrt{n_i(n_f+1)} & l'\sigma' = i;\ l\sigma = f\\ 0 & \text{(otherwise).}\end{cases} \tag{5.4.8b}$$

If we use these, (5.4.7) becomes

$$\langle f|H_2|i\rangle = \frac{e^2\hbar}{2m\epsilon_0 L^3}\frac{\hat{e}_i\cdot\hat{e}_f}{\sqrt{\omega_i\omega_f}}\sqrt{n_i(n_f+1)}\,\delta_{a,b}, \tag{5.4.9}$$

which vanishes if the atom does not return to its original state after the scattering. Also if no photons of frequency ω_i are present, then $n_i = 0$, and this term vanishes. By Problem 1.15, the contribution to the probability amplitude of H_2 is

$$c_2^{(2)}(f, t/i, 0) = -\frac{e^2\,\delta_{ab}}{2m\epsilon_0 L^3}\,\hat{e}_i\cdot\hat{e}_f\sqrt{\frac{n_i(n_f+1)}{\omega_i\omega_f}}\left(\frac{e^{i\omega_{fi}t}-1}{\omega_{fi}}\right), \tag{5.4.10}$$

where

$$\hbar\omega_{fi} = \epsilon_b - \epsilon_a + \hbar(\omega_f - \omega_i). \tag{5.4.11}$$

We next consider the contribution of H_1 to the transition probability amplitude in second order. According to Problem 1.15 we need the nonzero matrix elements $\langle f|H_1|k\rangle\langle k|H_1|i\rangle$ where $|k\rangle$ is any intermediate atom-field state $|I, n_i', n_f'\rangle$ for which the individual matrix elements do not vanish. Under the dipole approximation, we have

$$\langle f|H_1|k\rangle\langle k|H_1|i\rangle = \frac{e^2\hbar}{2m^2\epsilon_0 L^3}\sum_{l,\sigma}\sum_{l'\sigma'}\frac{X\langle b|\mathbf{p}\cdot\hat{e}_{l\sigma}|I\rangle\langle I|\mathbf{p}\cdot\hat{e}_{l'\sigma'}|a\rangle}{\sqrt{\omega_l\omega_{l'}}}, \tag{5.4.12a}$$

where

$$X = \langle n_i - 1, n_f + 1|(a_{l\sigma} + a_{l\sigma}^\dagger)|n_i', n_f'\rangle\langle n_i', n_f'|(a_{l'\sigma'} + a_{l'\sigma'}^\dagger)|n_i, n_f\rangle. \tag{5.4.12b}$$

These are zero except for two cases $|n_i - 1, n_f\rangle$ and $|n_i, n_f + 1\rangle$. Then for the first case, the only nonzero term is

$$\begin{aligned} X &= \langle n_i - 1, n_f + 1|a_{l\sigma}^\dagger|n_i - 1, n_f\rangle\langle n_i - 1, n_f|a_{l'\sigma'}|n_i, n_f\rangle \\ &= \sqrt{n_i(n_f+1)} \qquad \text{if} \quad l, \sigma \equiv f \quad \text{and} \quad l', \sigma' \equiv i. \end{aligned} \tag{5.4.13a}$$

In the second case, the only nonzero term is

$$\begin{aligned} X &= \langle n_i - 1, n_f + 1|a_{l\sigma}|n_i, n_f + 1\rangle\langle n_i, n_f + 1|a_{l'\sigma'}^\dagger|n_i, n_f\rangle \\ &= \sqrt{n_i(n_f+1)} \qquad \text{if} \quad l, \sigma \equiv i \quad \text{and} \quad l', \sigma' \equiv f. \end{aligned} \tag{5.4.13b}$$

Therefore, we have

$$\begin{aligned} &\langle f|H_1|k\rangle\langle k|H_1|i\rangle \\ &\quad = \frac{e^2\hbar}{2m^2\epsilon_0 L^3}\sqrt{\frac{n_i(n_f+1)}{\omega_i\omega_f}}\begin{cases}(\mathbf{p}_{bI}\cdot\hat{e}_f)(\mathbf{p}_{Ia}\cdot\hat{e}_i) & \text{if} \quad |k\rangle = |n_i - 1, n_f\rangle \\ (\mathbf{p}_{bI}\cdot\hat{e}_i)(\mathbf{p}_{Ia}\cdot\hat{e}_f) & \text{if} \quad |k\rangle = |n_i, n_f + 1\rangle.\end{cases} \end{aligned} \tag{5.4.14}$$

We have used the notation

$$\mathbf{p}_{bI} \equiv \langle b|\mathbf{p}|I\rangle. \tag{5.4.15}$$

By Problem 1.15, the contribution of H_1 to the transition probability amplitude is

$$c_1^{(2)}(f, t/i, 0) = -\frac{e^2}{2m^2\epsilon_0 L^3 \hbar}\sqrt{\frac{n_i(n_f+1)}{\omega_i\omega_f}}$$
$$\times \sum_I \left\{(\mathbf{p}_{bI}\cdot\hat{e}_f)(\mathbf{p}_{Ia}\cdot\hat{e}_i)\int_0^t dt_1\, e^{i[\epsilon_b-\epsilon_a+\hbar\omega_f]t_1/\hbar}\int_0^{t_1} dt_2\, e^{i[\epsilon_I-\epsilon_a-\hbar\omega_i]t_2/\hbar}\right.$$
$$\left. + (\mathbf{p}_{bI}\cdot\hat{e}_i)(\mathbf{p}_{Ia}\cdot\hat{e}_f)\int_0^t dt_1\, e^{i[\epsilon_b-\epsilon_I-\hbar\omega_i]t_1/\hbar}\int_0^{t_1} dt_2\, e^{i[\epsilon_I-\epsilon_a+\hbar\omega_f]t_2/\hbar}\right\}. \tag{5.4.16}$$

When we summed over the two intermediate photon states when $|k\rangle = |I, n_i - 1, n_f\rangle$, then $\hbar\omega_{fk} = \epsilon_b - \epsilon_I + \hbar\omega_f$ and $\hbar\omega_{ki} = \epsilon_I - \epsilon_a - \hbar\omega_i$ and when $|k\rangle = |I, n_i, n_f + 1\rangle$, then $\hbar\omega_{fk} = \epsilon_b - \epsilon_I - \hbar\omega_i$ and $\hbar\omega_{ki} = \epsilon_I - \epsilon_a + \hbar\omega_f$.

We next carry out the integration over t_2. This gives

$$c_1^{(2)}(f, t/i, 0) = \frac{e^2 i}{2m^2\epsilon_0 L^3}\sqrt{\frac{n_i(n_f+1)}{\omega_i\omega_f}}$$
$$\times \sum_I \left\{\frac{(\hat{e}_i\cdot\mathbf{p}_{bI})(\hat{e}_f\cdot\mathbf{p}_{Ia})}{\epsilon_I-\epsilon_a+\hbar\omega_f}\int_0^t dt_1[e^{i\omega_{fi}t_1} - e^{(i/\hbar)(\epsilon_b-\epsilon_I-\hbar\omega_i)t_1}]\right.$$
$$\left. + \frac{(\hat{e}_f\cdot\mathbf{p}_{bI})(\hat{e}_i\cdot\mathbf{p}_{Ia})}{\epsilon_I-\epsilon_a-\hbar\omega_i}\int_0^t dt_1[e^{i\omega_{fi}t_1} - e^{i(\epsilon_b-\epsilon_I+\hbar\omega_f)t_1/\hbar}]\right\}, \tag{5.4.17}$$

where ω_{fi} is given by (5.4.11). The second term in each integral arises from $t_2 = 0$ and is a transient effect due to turning on the interaction suddenly at $t = 0$. It will give a small contribution to the transition probability, since, for example,

$$\left|\int_0^t e^{i\omega t_1}\, dt_1\right|^2 = \frac{4\sin^2\frac{1}{2}\omega t}{\omega^2}$$

is only large as $t \to \infty$ if $\omega = 0$. Since energy is conserved $\omega_{fi} = \epsilon_b - \epsilon_a + \hbar(\omega_f - \omega_i) \simeq 0$; thus $\epsilon_b - \epsilon_I + \hbar\omega_i$ or $\epsilon_b - \epsilon_I + \hbar\omega_f$ will never be small. We therefore omit the transient contributions and (5.4.17) becomes

$$c_1^{(2)}(f, t/i, 0) = \frac{e^2}{2m^2\epsilon_0 L^3}\sqrt{\frac{n_i(n_f+1)}{\omega_i\omega_f}}$$
$$\times \sum_I \left\{\frac{(\hat{e}_i\cdot\mathbf{p}_{bI})(\hat{e}_f\cdot\mathbf{p}_{Ia})}{\epsilon_I-\epsilon_a+\hbar\omega_f} + \frac{(\hat{e}_f\cdot\mathbf{p}_{bI})(\hat{e}_i\cdot\mathbf{p}_{Ia})}{\epsilon_I-\epsilon_a-\hbar\omega_i}\right\}\frac{(e^{i\omega_{fi}t}-1)}{\omega_{fi}}. \tag{5.4.18}$$

If we add this to the H_2 contribution (5.4.10), we obtain

$$|c^{(2)}(f, t/i, 0)|^2 = \left(\frac{e^2}{2m\epsilon_0 L^3}\right)^2 \frac{n_i(n_f+1)}{\omega_f \omega_i} \frac{4\sin^2 \frac{1}{2}\omega_{fi}t}{\omega_{fi}^2} \times \left| (\hat{e}_f \cdot \hat{e}_i)\,\delta_{ab} - \frac{1}{m}\sum_I \left[\frac{(\hat{e}_i \cdot \mathbf{p}_{bI})(\hat{e}_f \cdot \mathbf{p}_{Ia})}{\epsilon_I - \epsilon_a + \hbar\omega_f} + \frac{(\hat{e}_f \cdot \mathbf{p}_{bI})(\hat{e}_i \cdot \mathbf{p}_{Ia})}{\epsilon_I - \epsilon_a - \hbar\omega_i}\right]\right|^2. \tag{5.4.19}$$

Next we note that photons of any frequency may occur so we must sum over all final frequencies. Again by Section 4.5 we have in a solid angle $d\Omega$ about $\mathbf{k}_f$

$$\sum_f |c^{(2)}(f, t/i, 0)|^2 = \int_0^\infty \frac{L^3}{(2\pi c)^3}\,\omega_f^2\, d\omega_f\, d\Omega\, |c^{(2)}(f, t/i, 0)|^2 = \left(\frac{e^2}{4\pi\epsilon_0 mc^2}\right)^2 \frac{cn_i(n_f+1)}{2\pi L^3 \omega_i} \int_0^\infty \omega_f\, d\omega_f\, |M|^2 \frac{4\sin^2 \frac{1}{2}\omega_{fi}t}{\omega_{fi}^2}\, d\Omega, \tag{5.4.20}$$

where

$$|M|^2 = \left| (\hat{e}_f \cdot \hat{e}_i)\,\delta_{ab} - \frac{1}{m}\sum_I \left[\frac{(\hat{e}_i \cdot \mathbf{p}_{bI})(\hat{e}_f \cdot \mathbf{p}_{Ia})}{\epsilon_I - \epsilon_a + \hbar\omega_f} + \frac{(\hat{e}_f \cdot \mathbf{p}_{bI})(\hat{e}_i \cdot \mathbf{p}_{Ia})}{\epsilon_I - \epsilon_a - \hbar\omega_i}\right]\right|^2. \tag{5.4.21}$$

This gives the probability of finding the atom and field in state $|b, n_i - 1, n_f + 1\rangle$ at time t when initially it was in state $|a, n_i, n_f\rangle$. If we differentiate (5.4.20) with respect to time the probability per second is [see (5.2.24)]

$$w_{d\Omega} = \left(\frac{e^2}{4\pi\epsilon_0 mc^2}\right)^2 \frac{cn_i(n_f+1)\, d\Omega}{2\pi L^3 \omega_i} \int_0^\infty \omega_f\, d\omega_f\, |M|^2 \frac{2\sin \omega_{fi}t}{\omega_{fi}} \xrightarrow[t\to\infty]{} \left(\frac{e^2}{4\pi\epsilon_0 mc^2}\right)^2 \frac{cn_i(n_f+1)\, d\Omega}{L^3 \omega_i} \int_0^\infty d\omega_f\, \omega_f\, |M|^2\, \delta(\omega_{fi}) = \frac{c}{L^3}\left(\frac{e^2}{4\pi\epsilon_0 mc^2}\right)^2 \left[\frac{\omega_f}{\omega_i}|M|^2\right]_{\omega_f = \omega_i - [(\epsilon_b - \epsilon_a)/\hbar]} n_i(n_f+1)\, d\Omega. \tag{5.4.22}$$

The differential scattering cross-section is defined as the transition probability per second per incident number of photons per second per unit area (flux). But the flux for n_i photon in L^3 is $n_i c/L^3$. Thus

$$d\sigma = \frac{w_{d\Omega}}{(n_i c/L^3)} \tag{5.4.23}$$

or

$$\frac{d\sigma}{d\Omega} = \left(\frac{e^2}{4\pi\epsilon_0 mc^2}\right)^2 (n_f + 1) \frac{\omega_f}{\omega_i} \Bigg| (\hat{e}_i \cdot \hat{e}_f)\,\delta_{ab} - \frac{1}{m} \sum_I \left[\frac{(\hat{e}_i \cdot \mathbf{p}_{bI})(\hat{e}_f \cdot \mathbf{p}_{Ia})}{\epsilon_I - \epsilon_a + \hbar\omega_f} + \frac{(\hat{e}_f \cdot \mathbf{p}_{bI})(\hat{e}_i \cdot \mathbf{p}_{Ia})}{\epsilon_I - \epsilon_a - \hbar\omega_i}\right]\Bigg|^2_{\omega_f = \omega_i - [(\epsilon_b - \epsilon_a)/\hbar]}. \tag{5.4.24}$$

This is the Kramers-Heisenberg scattering cross-section for light scattered by atomic electrons. The quantity

$$r_0 = \frac{e^2}{4\pi\epsilon_0 mc^2} = \frac{e^2}{4\pi\epsilon_0 \hbar c} \cdot \frac{\hbar}{mc} = \frac{1}{137} \frac{\hbar}{mc} \approx 2.8 \times 10^{-15}\ \text{m} \tag{5.4.25}$$

so that $r_0^2 \sim 9 \times 10^{-30}\ \text{m}^2$. We shall apply this formula to several particular cases. The n_f term is stimulated and the 1 is spontaneous scattering.

5.5 RAYLEIGH SCATTERING

The first case we consider is elastic scattering in which the atom returns to its original state. Then $\omega_f = \omega_i \equiv \omega$ and the cross-section becomes

$$\frac{d\sigma}{d\Omega} = r_0^2 (n_f + 1) \left| \hat{e}_f \cdot \hat{e}_i - \frac{1}{m} \sum_I \left[\frac{(\hat{e}_i \cdot \mathbf{p}_{bI})(\hat{e}_f \cdot \mathbf{p}_{Ia})}{\epsilon_I - \epsilon_a + \hbar\omega} + \frac{(\hat{e}_f \cdot \mathbf{p}_{bI})(\hat{e}_i \cdot \mathbf{p}_{Ia})}{\epsilon_I - \epsilon_a - \hbar\omega}\right] \right|^2. \tag{5.5.1}$$

We would like to simplify this as follows. Consider the diagonal matrix element of the commutator $[\mathbf{x} \cdot \hat{e}_i, \mathbf{p} \cdot e_f]$ for state $|a\rangle$. We have

$$\sum_{k,l} [x_k e_k{}^i, p_l e_l{}^f] = {}_{k,l}\ e_k{}^i e_l{}^f [x_k, p_l]$$

$$= i\hbar \sum_{k,l} \delta_{kl} e_k{}^i e_l{}^f = i\hbar(\hat{e}_i \cdot \hat{e}_f). \tag{5.5.2}$$

Therefore,

$$\hat{e}_i \cdot \hat{e}_f = \frac{1}{i\hbar} [\langle a|(\mathbf{x} \cdot \hat{e}_i)(\mathbf{p} \cdot \hat{e}_f)|a\rangle - \langle a|(\mathbf{p} \cdot \hat{e}_f)(\mathbf{x} \cdot \hat{e}_i)|a\rangle].$$

If we insert a completeness relation, we have

$$\hat{e}_i \cdot \hat{e}_f = \frac{1}{i\hbar} \sum_I [(\hat{e}_i \cdot \mathbf{x}_{aI})(\hat{e}_f \cdot \mathbf{p}_{Ia}) - (\hat{e}_f \cdot \mathbf{p}_{aI})(\hat{e}_i \cdot \mathbf{x}_{Ia})]$$

$$= \frac{1}{m} \sum_I \frac{(e_i \cdot \mathbf{p}_{aI})(\hat{e}_f \cdot \mathbf{p}_{Ia})}{\epsilon_I - \epsilon_a} + \frac{1}{m} \sum_I \frac{(\hat{e}_f \cdot \mathbf{p}_{aI})(\hat{e}_i \cdot \mathbf{p}_{Ia})}{\epsilon_I - \epsilon_a}, \tag{5.5.3}$$

where we used (5.2.9). If we use this in (5.5.1), we obtain

$$\frac{d\sigma}{d\Omega}=\left(\frac{r_0\hbar\omega}{m}\right)^2$$
$$\times\left|\sum_I\left[\frac{(\hat{e}_i\cdot\mathbf{p}_{aI})(\hat{e}_f\cdot\mathbf{p}_{Ia})}{(\epsilon_I-\epsilon_a)(\epsilon_I-\epsilon_a+\hbar\omega)}-\frac{(\hat{e}_f\cdot\mathbf{p}_{aI})(\hat{e}_i\cdot\mathbf{p}_{Ia})}{(\epsilon_I-\epsilon_a)(\epsilon_I-\epsilon_a-\hbar\omega)}\right]\right|^2(n_f+1). \tag{5.5.4}$$

We next consider scattering for which $\hbar\omega=\hbar 2\pi c/\lambda \ll \epsilon_I-\epsilon_a$ and expand the denominators to first order:

$$(\epsilon_I-\epsilon_a\pm\hbar\omega)^{-1}\cong(\epsilon_I-\epsilon_a)^{-1}\mp\hbar\omega(\epsilon_I-\epsilon_a)^{-2}$$

so that

$$\frac{d\sigma}{d\Omega}=\left(\frac{r_0\hbar\omega}{m}\right)^2\left|\sum_I\frac{1}{(\epsilon_I-\epsilon_a)^2}\left\{[(\hat{e}_i\cdot\mathbf{p}_{aI})(\hat{e}_f\cdot\mathbf{p}_{Ia})-(\hat{e}_f\cdot\mathbf{p}_{aI})(\hat{e}_i\cdot\mathbf{p}_{Ia})]\right.\right.$$
$$\left.\left.-\frac{\hbar\omega}{(\epsilon_I-\epsilon_a)}[(\hat{e}_i\cdot\mathbf{p}_{aI})(\hat{e}_f\cdot\mathbf{p}_{Ia})-(\hat{e}_f\cdot\mathbf{p}_{aI})(\hat{e}_i\cdot\mathbf{p}_{Ia})]\right\}\right|^2(n_f+1). \tag{5.5.5}$$

We first show that the first square bracket term vanishes as follows.

$$\sum_I\frac{1}{(\epsilon_I-\epsilon_a)^2}[(\hat{e}_i\cdot\mathbf{p}_{aI})(\hat{e}_f\cdot\mathbf{p}_{Ia})-(\hat{e}_f\cdot\mathbf{p}_{aI})(\hat{e}_i\cdot\mathbf{p}_{Ia})]$$
$$=\sum_I\frac{m^2}{(\epsilon_I-\epsilon_a)^2}\left[\frac{(\epsilon_a-\epsilon_I)^2}{\hbar^2}(\hat{e}_i\cdot\mathbf{x}_{aI})(\hat{e}_f\cdot\mathbf{x}_{Ia})-\frac{(\epsilon_a-\epsilon_I)^2}{\hbar^2}(\hat{e}_f\cdot\mathbf{x}_{aI})(\hat{e}_i\cdot\mathbf{x}_{Ia})\right]$$
$$=\sum_I\frac{m^2}{\hbar^2}[\hat{e}_i\cdot\langle a|\mathbf{x}|I\rangle\langle I|\mathbf{x}|a\rangle\cdot\hat{e}_f-\hat{e}_f\cdot\langle a|\mathbf{x}|I\rangle\langle I|\mathbf{x}|a\rangle\cdot\hat{e}_i]$$
$$=\frac{m^2}{\hbar^2}\langle a|(\hat{e}_i\cdot\mathbf{x})(\hat{e}_f\cdot\mathbf{x})-(\hat{e}_f\cdot\mathbf{x})(\hat{e}_i\cdot\mathbf{x})|a\rangle\equiv 0. \tag{5.5.6}$$

We have used (5.2.9) and the completeness relation. Therefore, (5.5.5) becomes

$$\frac{d\sigma}{d\Omega}\cong(n_f+1)\left(\frac{r_0\hbar^2\omega^2}{m}\right)^2\left|\sum_I(\epsilon_I-\epsilon_a)^{-3}[(\hat{e}_i\cdot\mathbf{p}_{aI})(\hat{e}_f\cdot\mathbf{p}_{Ia})-(\hat{e}_f\cdot\mathbf{p}_{aI})(\hat{e}_i\cdot\mathbf{p}_{Ia})]\right|^2$$
$$=(r_0m)^2\omega^4\left|\sum_I(\epsilon_I-\epsilon_a)^{-1}[(\hat{e}_i\cdot\mathbf{x}_{aI})(\hat{e}_f\cdot\mathbf{x}_{Ia})-(\hat{e}_f\cdot\mathbf{x}_{aI})(\hat{e}_i\cdot\mathbf{x}_{Ia})]\right|^2, \tag{5.5.7}$$

where we used (5.2.9). Therefore, for long wavelengths for which $\hbar 2\pi c/\lambda \ll \epsilon_I-\epsilon_a$, the elastic scattering cross-section varies as λ^{-4} which is called Rayleigh's law.

We may visualize the scattering process as an absorption of a photon by the atom in which the atom goes to some virtual state and then decays and emits another photon of the same energy but with a different polarization and in a different direction. The intermediate atomic state is called virtual since the incident photon does not have enough energy to excite the electron to another real state which would conserve energy ($\hbar\omega \ll |\epsilon_I - \epsilon_a|$).

5.6 THOMSON SCATTERING

Consider again the case of elastic scattering in which $\omega_i = \omega_f \gg \hbar^{-1}|\epsilon_I - \epsilon_a|$. In order for the dipole approximation to be valid, $\omega_i \ll c/a_0$ where $a_0 = 4\pi\epsilon_0\hbar^2/me^2 \simeq 5.3 \times 10^{-11}$ m. In this case, we may neglect the two sums in (5.4.24) and we obtain

$$\frac{d\sigma}{d\Omega} = r_0^2 |\hat{e}_f \cdot \hat{e}_i|^2 (n_f + 1), \tag{5.6.1}$$

which is the Thomson scattering cross-section. The electron in the atom acts as if we were scattering from a free electron. The cross-section is independent of the frequency provided $|\epsilon_I - \epsilon_a| \ll \hbar\omega_i \ll c\hbar/a_0$.

Assume the incident photon is incident along the z axis and polarized along the x axis. Then (see Figure 5.4)

$$\frac{\mathbf{k}_i}{|\mathbf{k}_i|} = (0, 0, 1) \tag{5.6.2}$$

$$\hat{e}_i = (1, 0, 0). \tag{5.6.3}$$

The scattered photon has momentum $\hbar\mathbf{k}_f$ where $\omega_i = ck_i = \omega_f = ck_f$ and

$$\mathbf{k}_f = k_i[\cos\varphi \sin\theta, \sin\varphi \sin\theta, \cos\theta]. \tag{5.6.4}$$

Since $\hat{e}_f \cdot \mathbf{k}_f = 0$, we may take for $\hat{e}_f$

$$\hat{e}_f^{(1)} = \frac{\mathbf{k}_f \times \mathbf{k}_i}{|\mathbf{k}_f \times \mathbf{k}_i|} = [\sin\varphi, -\cos\varphi, 0]$$

$$\hat{e}_f^{(2)} = \frac{\mathbf{k}_f \times \hat{e}_f^{(1)}}{|\mathbf{k}_f \times \hat{e}_f^{(1)}|} = [\cos\varphi \cos\theta, \sin\varphi \cos\theta, -\sin\theta]. \tag{5.6.5}$$

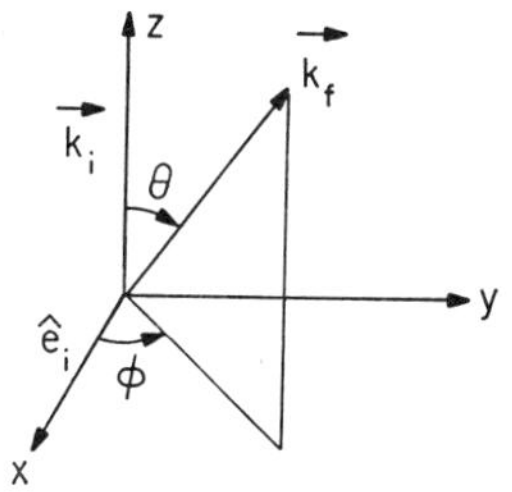

Figure 5.4 Polarizations and wave vectors for incident and scattered waves for Thomson scattering.

These satisfy the requirements $\hat{e}_f^{(1)} \cdot \hat{e}_f^{(2)} = 0$, $\hat{e}_f^{(1)} \cdot \hat{k}_f = 0$, and $\hat{e}_f^{(2)} \cdot \hat{k}_f = 0$ Then

$$\frac{d\sigma}{d\Omega} = \begin{cases} r_0^2 \sin^2 \varphi(n_f + 1) & \text{for} \quad \hat{e}_f^{(1)} \\ r_0^2 \cos^2 \varphi \cos^2 \theta(n_f + 1) & \text{for} \quad \hat{e}_f^{(2)}, \end{cases} \tag{5.6.6}$$

where $d\Omega = \sin\theta\, d\theta\, d\varphi$. If we do not measure the polarization of the scattered photon our detector picks up both polarizations and in this case we measure

$$\frac{d\sigma}{d\Omega} = r_0^2[\sin^2 \varphi + \cos^2 \varphi \cos^2 \theta](n_f + 1), \tag{5.6.7}$$

when the incident radiation is polarized. If we look in the direction $\theta = \pi/2$, all emitted photons will be polarized along $\hat{e}_f^{(1)}$. If we put in a polarizer which admits only photons polarized along $\hat{e}_f^{(1)}$, we measure only the first case in (5.6.6) while if our polarizer admits light to our detector polarized along $\hat{e}_f^{(2)}$, we measure the second case in (5.6.6).

Consider next the case in which the incident radiation is *unpolarized.* We may treat this by averaging (5.6.6) over all angles φ:

$$\left.\frac{d\sigma}{d\Omega}\right|_{\text{unpolarized } \omega_i} = \frac{1}{2\pi} \int_0^{2\pi} \frac{d\sigma}{d\Omega}\, d\varphi = (n_f + 1)\frac{r_0}{2} \begin{cases} 1 & \text{for} \quad \hat{e}_f^{(1)}, \\ \cos^2 \theta & \text{for} \quad \hat{e}_f^{(2)} \end{cases} \tag{5.6.8}$$

Then, if we admit both polarizations to our detector we measure

$$\left.\frac{d\sigma}{d\Omega}\right|_{\text{unpolarized } \omega_i} = \tfrac{1}{2} r_0^2(n_f + 1)(1 + \cos^2 \theta), \tag{5.6.9}$$

and

$$\sigma_{\text{Total, unpol.}} = (n_f + 1)\tfrac{1}{2} r_0^2 \int_0^{2\pi} d\varphi \int_0^{\pi} \sin\theta\, d\theta (1 + \cos^2 \theta)$$

$$= \frac{8\pi}{3} r_0^2(n_f + 1), \tag{5.6.10}$$

which is the total cross-section when the incident light is unpolarized and we accept both polarizations in the scattered light.

Note again that at $\theta = \pi/2$, the scattered light is completely polarized again along $\hat{e}_f^{(1)}$ even when the incident light is unpolarized as may be seen from (5.6.8).

5.7 RAMAN SCATTERING

So far we have considered elastic scattering in the two limits $\hbar\omega_i \ll |\epsilon_I - \epsilon_a|$ and $\hbar\omega_i \gg |\epsilon_I - \epsilon_a|$. Let us next consider inelastic Raman scattering.

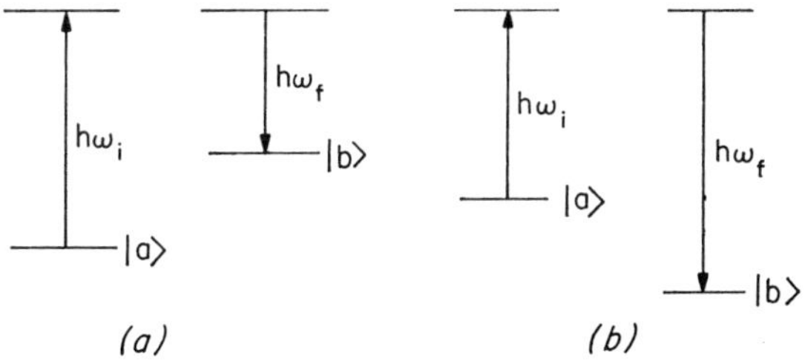

Figure 5.5 Raman scattering.

In this case, the first term in (5.4.24) makes no contribution. We consider first the case in which $\epsilon_b > \epsilon_a$. If the incident frequency is ω_L the scattered frequency $\omega_S = \omega_L - (\epsilon_b - \epsilon_a)/\hbar < \omega_L$ and this is called the Stokes line (see Figure 5.5). Then, the Stokes cross-section is

$$\left(\frac{d\sigma}{d\Omega}\right)_{\text{stokes}} = r_0^2\left(\frac{\omega_S}{\omega_L}\right)(n_S + 1)\,|\hat{e}_S \cdot \underset{\sim}{R} \cdot \hat{e}_L|^2, \tag{5.7.1}$$

where we define the Raman tensor as

$$\underset{\sim}{R} = \frac{1}{m}\sum_I \left[\frac{\mathbf{p}_{bI}\mathbf{p}_{Ia}}{\epsilon_I - \epsilon_a - \hbar\omega_L} + \frac{\mathbf{p}_{aI}^*\mathbf{p}_{Ib}^*}{\epsilon_I - \epsilon_a + \hbar\omega_S}\right], \tag{5.7.2}$$

and

$$\hbar\omega_S = \hbar\omega_L - (\epsilon_b - \epsilon_a) < \hbar\omega_L \tag{5.7.3}$$

In the second case, $\epsilon_a > \epsilon_b$ and the scattered frequency ω_A is higher than the incident frequency. This is called the anti-Stokes line. The cross-section is

$$\left(\frac{d\sigma}{d\Omega}\right)_{\text{antistokes}} = r_0^2\left(\frac{\omega_A}{\omega_L}\right)(n_A + 1)\,|\hat{e}_A \cdot \underset{\sim}{R} \cdot \hat{e}_L|^2, \tag{5.7.4}$$

where

$$\hbar\omega_A = \hbar\omega_L + \epsilon_a - \epsilon_b > \hbar\omega_L. \tag{5.7.5}$$

It is of interest to study the angular dependence of the Raman cross-section. In an arbitrary coordinate system (x', y', z'), the tensor R'_{ij} will have nine components. We may associate an ellipsoid with this tensor defined by

$$\varphi = x'_i R'_{ij} x'_j = \text{constant}, \tag{5.7.6}$$

where we use the summation convention and sum over repeated indices from 1 to 3. If α_{ij} is a rotation matrix then we may rotate to a new set of axes x_i given by [6]

$$x'_i = \alpha_{ij} x_j, \tag{5.7.7}$$

where the α_{ij} satisfy

$$\begin{aligned}\alpha_{il}\alpha_{jl} &= \delta_{ij} \\ \alpha_{li}\alpha_{lj} &= \delta_{ij}.\end{aligned} \tag{5.7.8}$$

Then, (5.7.6) becomes

$$\varphi = \alpha_{ik}x_k R'_{ij}\alpha_{jl}x_l = \text{constant}. \tag{5.7.9}$$

We choose the α_{ij} elements so that in the x_i coordinate system the tensor is diagonal. That is, we require

$$R_{kl} = \alpha_{ik}R'_{ij}\alpha_{jl} = R_{(k)}\delta_{kl}. \tag{5.7.10}$$

The parenthesis means that we are not considering the k on the right side a repeated index. The $R_{(k)}$ are the eigenvalues of the original tensor. If we use (5.7.8) we have

$$\alpha_{mk}\alpha_{ik}R'_{ij}\alpha_{jl} = R'_{mj}\alpha_{jl} = \alpha_{mk}R_{(k)}\delta_{kl}, \tag{5.7.11}$$

or

$$R'_{mj}\alpha_{jl} = R_{(l)}\alpha_{ml} \equiv R_{(l)}\,\delta_{mj}\alpha_{jl}. \tag{5.7.12}$$

This gives

$$[R'_{mj} - R_{(l)}\,\delta_{mj}]\alpha_{jl} = 0 \tag{5.7.13}$$

which is a set of homogeneous equations to determine the rotation matrix α_{jl} which will diagonalize R'_{ij}. It is necessary and sufficient for these equations to have a nontrivial solution that the determinant of the coefficients vanish:

$$\|R'_{mj} - R_{(l)}\,\delta_{mj}\| = 0. \tag{5.7.14}$$

This gives the three eigenvalues R_1, R_2, and R_3. We then have

$$\begin{aligned}(R'_{11} - R_l)\alpha_{1l} + R'_{12}\alpha_{2l} &= -R'_{13}\alpha_{3l} \\ R'_{21}\alpha_{1l} + (R'_{22} - R_l)\alpha_{2l} &= -R'_{23}\alpha_{3l},\end{aligned} \tag{5.7.15}$$

which allows us to solve for α_{1l} and α_{2l} in terms of α_{3l} for $l = 1, 2, 3$.

Let us assume the incident radiation has a wave vector along the z' axis so that

$$\mathbf{k}_L = k_L(0, 0, 1) \tag{5.7.16}$$

polarized along the x' axis so that

$$\hat{e}_L = (1, 0, 0). \tag{5.7.17}$$

If $\mathbf{k}_R$ has components along the (x', y', z') axes given by

$$\mathbf{k}_R = k_R[\cos\Phi \sin\Theta, \sin\Phi \sin\Theta, \cos\Theta], \tag{5.7.18}$$

we may take the two orthogonal polarization vectors for the Raman scattered light as

$$\hat{e}_R{}^1 = \frac{\mathbf{k}_R \times \mathbf{k}_L}{|\mathbf{k}_R \times \mathbf{k}_L|} = [\sin \Phi, -\cos \Phi, 0] \tag{5.7.19}$$

$$\hat{e}_R{}^2 = \frac{\mathbf{k}_R \times \hat{e}_R{}^1}{|\mathbf{k}_R \times \hat{e}_R{}^1|} = [\cos \Theta \cos \Phi, \cos \Theta \sin \Phi, -\sin \Theta]. \tag{5.7.20}$$

We then have in the primed coordinate system for $\hat{e}_{\hat{R}}{}^1$

$$\hat{e}_R{}^1 \cdot \underset{\sim}{R}' \cdot \hat{e}_L \equiv (\hat{e}_R{}^1)_i R'_{ij} (\hat{e}_L)_j = \sin \Phi \, R'_{11} - \cos \Phi \, R'_{21}. \tag{5.7.21}$$

If we use (5.7.8), we may invert (5.7.10):

$$\alpha_{mk}\alpha_{ik}R'_{ij}\alpha_{jl}\alpha_{nl} = R_{(k)}\,\delta_{kl}\alpha_{mk}\alpha_{nl}, \tag{5.7.22}$$

or

$$R'_{mn} = R_{(k)}\alpha_{mk}\alpha_{nk} \tag{5.7.23}$$

Therefore, (5.7.21) becomes

$$\begin{aligned} \hat{e}_R{}^1 \cdot \underset{\sim}{R}' \cdot \hat{e}_L = {} & \sin \Phi[\alpha_{11}{}^2 R_1 + \alpha_{12}{}^2 R_2 + \alpha_{13}{}^2 R_3] \\ & - \cos \Phi[\alpha_{21}\alpha_{11}R_1 + \alpha_{22}\alpha_{12}R_2 + \alpha_{23}\alpha_{13}R_3]. \end{aligned} \tag{5.7.24}$$

We may express the α_{ij} in terms of Euler angles (θ, φ, ψ) (see Figure 5.6). The elements are [6]

$$\begin{aligned} \alpha_{11} &= \cos \psi \cos \varphi - \cos \theta \sin \varphi \sin \psi \\ \alpha_{12} &= \cos \psi \sin \varphi + \cos \theta \cos \varphi \sin \psi \\ \alpha_{13} &= \sin \psi \sin \theta \\ \alpha_{21} &= -\sin \psi \cos \varphi - \cos \theta \sin \varphi \cos \psi \\ \alpha_{22} &= -\sin \psi \sin \varphi + \cos \theta \cos \varphi \cos \psi \\ \alpha_{23} &= \cos \psi \sin \theta \\ \alpha_{31} &= \sin \theta \sin \varphi \\ \alpha_{32} &= -\sin \theta \cos \varphi \\ \alpha_{33} &= \cos \theta. \end{aligned} \tag{5.7.25}$$

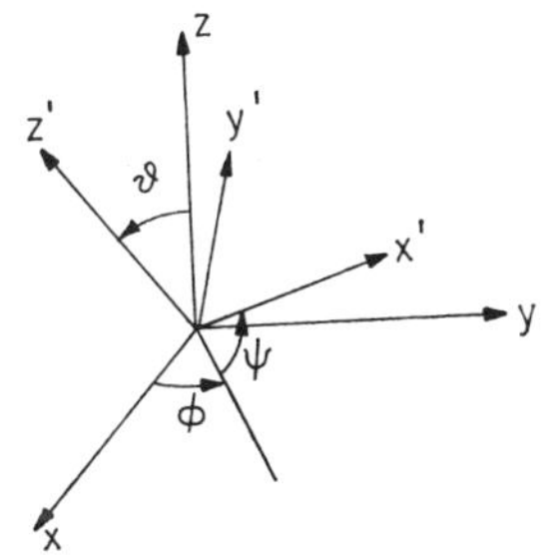

Figure 5.6 Euler angles.

As a simple example let us consider a molecule for which $R_1 = R_2$. Then we have

$$\begin{aligned}|\hat{e}_R{}^1 \cdot \underset{\approx}{R} \cdot \hat{e}_L|^2 &= \{\sin\Phi[R_1(\alpha_{11}{}^2 + \alpha_{12}{}^2) + R_3\alpha_{13}{}^2] \\ &\quad - \cos\Phi[R_1(\alpha_{11}\alpha_{21} + \alpha_{22}\alpha_{12}) + \alpha_{23}\alpha_{13}R_3]\}^2 \\ &= \{\sin\Phi[R_1(\cos^2\psi + \cos^2\theta\sin^2\psi) + R_3\sin^2\psi\sin^2\theta] \\ &\quad + \cos\Phi(R_1 - R_3)\sin^2\theta\sin\psi\cos\psi\}^2.\end{aligned} \tag{5.7.26}$$

In this case, we note that the angle φ has disappeared. When $R_1 = R_2$ the cross-section and the Raman tensor in this case are rotationally invariant about the z axis of the molecule (see Figure 5.6). The z axis is an axis of symmetry.

If we scatter light from a gas, then molecules are oriented at random so that we must average the cross-section over all angles (θ, φ, ψ). Thus for a gas we have

$$\begin{aligned}&\left(\frac{d\sigma}{d\Omega}\right)_{\text{gas pol.}\hat{e}_R{}^1} \\ &= \frac{1}{8\pi^2}\int_0^\pi \sin\theta\, d\theta \int_0^{2\pi} d\psi \int_0^{2\pi} d\varphi \left(\frac{d\sigma}{d\Omega}\right) \\ &= \frac{r_0{}^2}{4\pi}\left(\frac{\omega_R}{\omega_L}\right)(n_R + 1)\int_0^\pi \sin\theta\, d\theta \int_0^{2\pi} d\psi\{\cos\Phi(R_1 - R_3)\sin^2\theta\sin\psi\cos\psi \\ &\quad + \sin\Phi[R_1(\cos^2\psi + \cos^2\theta\sin^2\psi) + R_3\sin^2\psi\sin^2\theta]\} \\ &= \frac{r_0{}^2}{15}\left(\frac{\omega_R}{\omega_L}\right)(n_R + 1)\{(8R_1{}^2 + 3R_1R_3 + 3R_3{}^2)\sin^2\Phi + (R_1 - R_3)^2\cos^2\Phi\}\end{aligned} \tag{5.7.27}$$

A similar calculation for $\hat{e}_R{}^2$ gives

$$\begin{aligned}\left(\frac{d\sigma}{d\Omega}\right)_{\text{gas pol, }\hat{e}_R{}^2} &= \frac{r_0{}^2}{15}\left(\frac{\omega_R}{\omega_L}\right)(n_R + 1)\{(8R_1{}^2 + 3R_1R_3 + 3R_3{}^2)\cos^2\Theta\cos^2\Phi \\ &\quad + (R_1 - R_3)^2(\sin^2\Theta + \cos^2\Theta\sin^2\Phi)\}.\end{aligned} \tag{5.7.28}$$

If we do not measure the polarization, our detector collects the sum of the expressions above. Note that in contrast to Thomson scattering, there is no scattered direction in which the scattered radiation will be totally polarized.

5.8 RESONANCE FLUORESCENCE

We have considered elastic scattering in the limiting cases in which the incident frequency was small compared with $|\epsilon_I - \epsilon_a|/\hbar$ (Rayleigh scattering) and in the case when it was large compared with these atomic energy differences (Thomson scattering). In the event the incident frequency $\omega_i = (\epsilon_I - \epsilon_a)/\hbar$ for any two atomic levels the cross-section (5.4.24) becomes infinite. This arises because in the derivation we neglected the small but finite linewidth of the various atomic levels as well as their frequency shift. In this case the resonance term will still be extremely large, and we have

$$\frac{d\sigma}{d\Omega} \cong r_0^2 \frac{\omega_f}{\omega_i} \frac{(n_f + 1)}{m^2} \frac{|(\hat{e}_f \cdot \mathbf{p}_{bI})(\mathbf{p}_{Ia} \cdot \hat{e}_i)|^2}{(\epsilon_I - \epsilon_a - \hbar\omega_i)^2 + \frac{1}{4}\Gamma_I^2}, \tag{5.8.1}$$

where $\hbar\omega_i \approx \epsilon_I - \epsilon_a$. This, of course, assumes the sum of the nonresonant terms are negligible.

5.9 THE DOPPLER EFFECT [1]

It is well known that a moving source that emits radiation will have a change in frequency. This fact is explained by the wave theory of light. It may also be explained quite simply in the quantum theory of light by the conservation of energy and momentum.

When an atom initially in an excited state $|a\rangle$ is at rest and decays to state $|b\rangle$, it emits a quantum of light of frequency

$$\omega = \frac{E_a - E_b}{\hbar}, \tag{5.9.1}$$

with momentum $\hbar\mathbf{k} = \hbar\omega\hat{\mathbf{k}}/c$.

We now suppose that the atom is set in motion in excited state $|a\rangle$ with a nonrelativistic velocity $\mathbf{v}$. The initial energy is

$$E_a + \tfrac{1}{2}M\mathbf{v}^2. \tag{5.9.2}$$

At some instant the atom decays to state $|b\rangle$ and emits a quantum of frequency ω'. The atom recoils when the quantum is emitted; this changes the velocity of the atom to $\mathbf{v}'$ so that the energy of the atom is then $E_b + \frac{1}{2}M\mathbf{v}'^2$. By the law of conservation of energy,

$$\hbar\omega' = (E_a + \tfrac{1}{2}M\mathbf{v}^2) - (E_b + \tfrac{1}{2}M\mathbf{v}'^2) = \hbar\omega + \tfrac{1}{2}M(\mathbf{v}^2 - \mathbf{v}'^2). \tag{5.9.3}$$

The emitted quantum will have momentum $\hbar\mathbf{k}' \equiv \hbar\omega'\hat{\mathbf{k}}'/c$. If the momentum of atom and emitted quantum are conserved, we have

$$M\mathbf{v}' = M\mathbf{v} - \frac{\hbar\omega'}{c}\hat{\mathbf{k}}'. \tag{5.9.4}$$

If we square this and neglect the term of order $1/c^2$, then

$$\frac{M\mathbf{v}'^2}{2} \cong \frac{M\mathbf{v}^2}{2} - \frac{v}{c}\hbar\omega' \cos\vartheta, \tag{5.9.5}$$

where ϑ is the angle between $\mathbf{v}$ and the direction of the emitted quantum. If we substitute this into (5.9.3) and solve for ω', we find

$$\omega' = \omega\left(1 + \frac{v}{c}\cos\vartheta\right). \tag{5.9.6}$$

This is the same as the nonrelativistic result obtained for the Doppler frequency shift, using the wave theory of light. That is, when the electromagnetic field is considered to be composed of quanta of light of energy $\hbar\omega$ and momentum $\hbar\omega/c$, the conservation of energy and momentum for the atom and emitted photon gives the same frequency shift obtained when the light is considered as a wave phenomenon.

It was shown in Chapter 4 that light quanta in free space act like "particles" with the characteristics above. However, when the radiation field interacts with the atom, is there conservation of both energy and momentum? To answer this, we consider a hydrogen atom consisting of a proton of charge $+e$, mass m_p, and coordinates $\mathbf{x}_p$ and an electron of charge $-e$, mass m_e, and coordinates $\mathbf{x}_e$. The nonrelativistic hamiltonian for the atom and radiation field corresponding to (5.1.2) is

$$H = \frac{\mathbf{p}_1^{\,2}}{2m_p} + \frac{\mathbf{p}_2^{\,2}}{2m_e} - eV(|\mathbf{x}_p - \mathbf{x}_e|) + \sum_{l,\sigma} \hbar\omega_l a_{l\sigma}^\dagger a_{l\sigma}$$
$$- \frac{e}{m_p}\mathbf{A}(\mathbf{x}_p)\cdot\mathbf{p}_1 + \frac{e}{m_e}\mathbf{A}(\mathbf{x}_e)\cdot\mathbf{p}_2, \tag{5.9.7}$$

where we evaluate the vector potentials (5.1.8) at the position of the proton and electron, respectively. We must consider the atom nucleus, the proton in this case, since the recoil is taken up mainly by the nucleus; this recoil allows us to explain the Doppler effect.

We introduce new coordinates. Let

$$\boldsymbol{\xi} = \frac{m_p\mathbf{x}_p + m_e\mathbf{x}_e}{m_p + m_e} \tag{5.9.8}$$

be the coordinates of the center of gravity of the atom and

$$\boldsymbol{\rho} = \mathbf{x}_p - \mathbf{x}_e \tag{5.9.9}$$

be the relative coordinates for the two particles.

From the above we see that

$$\begin{aligned} m_p\dot{\mathbf{x}}_p + m_e\dot{\mathbf{x}}_e &= M\dot{\boldsymbol{\xi}} \\ \dot{\mathbf{x}}_p - \dot{\mathbf{x}}_e &= \dot{\boldsymbol{\rho}}, \end{aligned} \tag{5.9.10}$$

where $M = m_e + m_p$. If we solve for $\dot{\mathbf{x}}_p$ and $\dot{\mathbf{x}}_e$, we obtain

$$\begin{aligned} M\dot{\mathbf{x}}_p &= M\dot{\boldsymbol{\xi}} + m_e\dot{\boldsymbol{\rho}} \\ M\dot{\mathbf{x}}_e &= M\dot{\boldsymbol{\xi}} - m_p\dot{\boldsymbol{\rho}} \end{aligned} \tag{5.9.11}$$

Then the kinetic energy is

$$T = \tfrac{1}{2}m_p\dot{\mathbf{x}}_p^{\,2} + \tfrac{1}{2}m_e\dot{\mathbf{x}}_e^{\,2} = \frac{M}{2}\,\dot{\boldsymbol{\xi}}^2 + \frac{\mu}{2}\,\dot{\boldsymbol{\rho}}^2, \tag{5.9.12}$$

where $\mu = m_e m_p/M$ is the reduced mass. The momenta conjugate to $\boldsymbol{\xi}$ and $\boldsymbol{\rho}$ are given by

$$(p_\xi)_i = \frac{\partial T}{\partial \dot{\xi}_i} = M\dot{\xi}_i \tag{5.9.13}$$

$$(p_\rho)_i = \frac{\partial T}{\partial \dot{\rho}_i} = \mu\dot{\rho}_i. \tag{5.9.14}$$

If we make the familiar assumption that the wavelength of the radiation is long compared with the dimensions of the atom, we may replace $\mathbf{x}_p$ and $\mathbf{x}_e$ in the interaction terms by $\boldsymbol{\xi}$, the value of the field at the center of gravity of the atom. With these changes of variables, the hamiltonian (5.9.7) may be written

$$H = H_0 + H_1, \tag{5.9.15}$$

where

$$H_0 = \frac{\mathbf{p}_\xi^{\,2}}{2M} + \frac{\mathbf{p}_\rho^{\,2}}{2\mu} + eV(\rho) + \sum_{l,\sigma} \hbar\omega_l a_{l\sigma}^\dagger a_{l\sigma}, \tag{5.9.16}$$

and

$$H_1 = -\frac{e}{\mu}\,\mathbf{A}(\boldsymbol{\xi})\cdot\mathbf{p}_\rho. \tag{5.9.17}$$

The first term in H_0 represents a free particle of mass M and momentum $\mathbf{p}_\xi$. It gives the motion of the center of gravity. If we use box normalization, [5] its eigenfunctions are

$$L^{-3/2}\exp\left(\frac{i}{\hbar}\,\mathbf{p}'_\xi\cdot\boldsymbol{\xi}\right), \tag{5.9.18}$$

where $\mathbf{p}'_\xi$ is the momentum eigenvalue. These eigenvalues are discrete since this wave function must satisfy periodic boundary conditions on opposite walls of the box. The energy eigenvalue is $\mathbf{p}'^2_\xi/2M$ for this state.

The next two terms in (5.9.16) represent the internal motion of the electron relative to the protons. The energy eigenfunctions for these terms are the hydrogen wave functions [5], designated by $|n\rangle$, with associated energy E_n. The last term in (5.9.16) gives the energy of the free radiation field. A state vector will be designated by $|n_1, n_2, \ldots, n_l, \ldots\rangle$, with the associated energy eigenvalue $n_1\hbar\omega_1 + \cdots + \hbar\omega_l n_l + \cdots$. Therefore, an eigenket of H_0 will be designated by

$$|\mathbf{p}'_\xi, n, n_1, n_2, \ldots\rangle, \tag{5.9.19}$$

with energy eigenvalue

$$\frac{\mathbf{p}'^2_\xi}{2M} + E_n + n_1\hbar\omega_1 + \cdots + n_l\hbar\omega_l + \cdots. \tag{5.9.20}$$

The equations of motion for the probability amplitudes corresponding to (1.16.59) are

$$\begin{aligned} i\hbar \frac{dc}{dt}\langle(\mathbf{p}'_\xi, n, n_1, \ldots, n_l, \ldots) \\ = \sum_{p\xi'', m, m_1, \ldots, m_l, \ldots} \langle \mathbf{p}'_\xi, n, n_1, \ldots, n_l, \ldots | H_1 | \mathbf{p}''_\xi, m, m_1, \ldots, m_l, \ldots\rangle \\ \times c(\mathbf{p}''_\xi, m, m_1, \ldots, m_l, \ldots) e^{i\omega_{ab}t}, \end{aligned} \tag{5.9.21}$$

where

$$\begin{aligned} \hbar\omega_{ab} = \left(\frac{\mathbf{p}'^2_\xi}{2M} + E_n + \hbar\omega_1 n_1 + \cdots + \hbar\omega_l n_l + \cdots\right) \\ - \left(\frac{\mathbf{p}''^2_\xi}{2M} + E_m + \hbar\omega_1 m_1 + \cdots + \hbar\omega_l m_l + \cdots\right), \end{aligned} \tag{5.9.22}$$

and the interaction is

$$H_1 = -\frac{e}{\mu}\sum_{l,\sigma}\sqrt{\frac{\hbar}{2\omega_l\epsilon_0\tau}}\,[a_{l\sigma}\exp(i\mathbf{k}_l\cdot\boldsymbol{\xi}) + a^\dagger_{l\sigma}\exp(-i\mathbf{k}_l\cdot\boldsymbol{\xi})](\hat{\mathbf{e}}_{l\sigma}\cdot\mathbf{p}_\rho). \tag{5.9.23}$$

We may show the conservation of momentum directly by evaluating the matrix element

$$\langle\mathbf{p}'_\xi|H_1|\mathbf{p}''_\xi\rangle \sim \frac{1}{L^3}\iiint_0^L d\boldsymbol{\xi}\,\exp\left[\left(-\frac{i}{\hbar}\mathbf{p}'_\xi \pm i\mathbf{k}_l + \frac{i}{\hbar}\mathbf{p}''_\xi\right)\cdot\boldsymbol{\xi}\right].$$

This is obviously zero unless

$$\mathbf{p}''_\xi - \mathbf{p}'_\xi = \pm\hbar\mathbf{k}_l, \tag{5.9.24}$$

which is the law of conservation of momentum for the center of mass of the atom and the photon. If the matrix elements of H_1 are zero between any two states, no transitions are possible so that momentum must be conserved for allowed transitions. Therefore, momentum conservation is contained in the quantized radiation theory.

We may therefore write the nonzero matrix elements of H_1 as

$$\langle \mathbf{p}'_\xi, n, n_1, \ldots, n_l, \ldots | H_1 | \mathbf{p}'_\xi - \hbar\mathbf{k}_l, m, n_1, \ldots, n_l + 1, \ldots \rangle = -\frac{e}{\mu}\sqrt{\frac{\hbar(n_{l\sigma}+1)}{2\omega_l\epsilon_0\tau}}(\hat{\mathbf{e}}_{l\sigma}\cdot\mathbf{p}_{\rho nm}) \quad (5.9.25a)$$

$$\langle \mathbf{p}'_\xi, n, \ldots, n_l, \ldots | H_1 | \mathbf{p}'_\xi + \hbar\mathbf{k}_l, m, \ldots, n_l - 1, \ldots \rangle = -\frac{e}{\mu}\sqrt{\frac{\hbar n_{l\sigma}}{2\omega_l\epsilon_0\tau}}(\hat{\mathbf{e}}_{l\sigma}\cdot\mathbf{p}_{\rho nm}), \quad (5.9.25b)$$

where by (5.2.9), with a slight change in notation,

$$\mathbf{p}_{\rho nm} \equiv \langle n|\mathbf{p}_\rho|m\rangle = \frac{i\mu}{\hbar}(E_n - E_m)\langle n|\boldsymbol{\rho}|m\rangle. \quad (5.9.26)$$

The conservation of energy follows directly either from a perturbation-theory approximation or from the Wigner-Weisskopf approximation used in the theory of the natural line width. Therefore, the explanation of the Doppler effect is contained in the quantized radiation theory.

The Doppler shift in frequency causes the spectral line emitted by a moving atom to be broadened, in addition to the natural broadening discussed in the previous section. This additional broadening may be seen as follows: the center of mass of the molecules has a Maxwell probability distribution of momenta. If the gas is at temperature T, then

$$\exp\left(-\frac{\mathbf{p}'^2_\xi}{2MkT}\right) d\mathbf{p}'_\xi \quad (5.9.27)$$

will be the probability of finding the molecule with momentum between $\mathbf{p}'_\xi$ and $\mathbf{p}'_\xi + d\mathbf{p}'_\xi$. When this molecule emits a photon of frequency ω'_l, energy and momentum must be conserved; this puts a constraint on the range of allowed values of $\mathbf{p}'_\xi$. From (5.9.3), we have

$$\frac{1}{2M}(\mathbf{p}'^2_\xi - \mathbf{p}''^2_\xi) = \hbar(\omega' - \omega) \quad (5.9.28)$$

with an obvious change of notation, while from (5.9.4) we obtain, on squaring and neglecting the $1/c^2$ term,

$$\frac{\mathbf{p}''^2_\xi - \mathbf{p}'^2_\xi}{2M} \cong \frac{\hbar\omega}{cM}\mathbf{p}'_{\hat{\mathbf{k}}}, \quad (5.9.29)$$

where $\mathbf{p}'_{\hat{\mathbf{k}}}$ is the component of $\mathbf{p}'_\xi$ in the direction of emission. When we combine (5.9.28) and (5.9.29), we see that the component of $\mathbf{p}'_\xi$ in the direction of emission must remain constant if we are to obtain the intensity of the emitted light at a given frequency ω'. The components of $\mathbf{p}'_\xi$ normal to $\hat{\mathbf{k}}$, the direction of emission, may have any value from $-\infty$ to $+\infty$. The intensity of the emitted light at frequency ω' will therefore be proportional to

$$d\omega'_l \exp\left(-\frac{\mathbf{p}_{\hat{\mathbf{k}}}'^2}{2MkT}\right) = \exp\left(-\frac{Mc^2}{2kT}\frac{(\omega'_l - \omega_0)^2}{{\omega_0}^2}\right) d\omega'_l, \qquad (5.9.30)$$

where $\hbar\omega_0 = E_n - E_m$ and ω'_l is the frequency of the emitted quantum. The Doppler line shape is therefore gaussian in character.

Another factor of proportionality is the transition probability per second from an initial state, in which the atom has momentum $\mathbf{p}'_\xi$, is in excited state $|n\rangle$, and with no quanta present, to a final state $\mathbf{p}''_\xi$, ground state $|m\rangle$, and one quantum present. Since the matrix elements (5.9.25) do not involve $\mathbf{p}_\xi$, the integration of (5.9.27) over the components of $\mathbf{p}_\xi$ transverse to the direction of emission is not affected.

The breadth of the line (5.9.30) at half maximum is

$$\delta = 2\omega_0\sqrt{\frac{2kT}{Mc^2}\log_e 2}. \qquad (5.9.31)$$

Although $kT \ll Mc^2$, the effect of Doppler broadening is much greater than the broadening of the natural line width. However, the Lorentzian natural line shape has a value for ω' far from ω_0 whereas the Doppler effect is small. Far from the line center, the natural broadening is therefore more pronounced than the Doppler broadening.

5.10 PROPAGATION OF LIGHT IN VACUUM [1]

In this section we show how the quantum theory of radiation may be used to calculate the intensity of light propagated in free space and how the phase relations between the components of the radiation field give a finite velocity of propagation. This problem is related to that of a transmitter and receiver in communications and serves as a background for studying quantum-noise effects in such problems.

The transmitter will be represented by a single atom, which we shall call A, located at the origin of a coordinate system. At $t = 0$, atom A is assumed to be in an excited state; after a certain time, A will decay to its ground state with the emission of a photon. The receiver is another atom, B, located a distance r away on the z axis of the coordinate system; B is initially in its ground state (Figure 5.7). If the photon emitted by A interacts with B, the photon will be absorbed, leaving B in an excited state. Since the emitted photon travels with velocity c, the absorption by B can take place only after

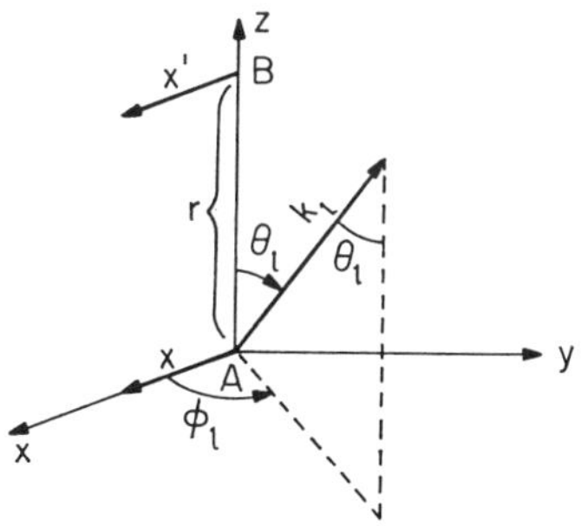

Figure 5.7 Atom A is excited and decays while atom B absorbs the emitted photon after time $t > r/c$.

a time r/c from emission by A. We show how these results are obtained from the radiation theory.

To simplify the analysis, we assume that the mean life of atom A is very short so that we may say that A emits at some precise time. If $1/\gamma_A$ is the mean life of atom A, γ_A is large, and from the theory of natural linewidth given in Section 5.3, we see that atom A has a very broad, practically continuous spectrum. On the other hand, we assume that the lifetime of atom B is very long (γ_B very small) so that atom B absorbs a very sharp spectral line. For our purposes, it essentially stays excited once it has absorbed a photon.

We also assume that the wavelength of the radiation emitted by A is long compared with the dimensions of both A and B. This allows us to use the dipole approximation for both atoms.

For simplicity, we assume that the dipole moments of both A and B between the ground state and excited state have a component only in the x direction.

The unperturbed hamiltonian for A, B, and the radiation field is given by

$$H_0 = H_A + H_B + H_r. \tag{5.10.1}$$

An eigenstate of H_0 is written

$$|E_0\rangle = |n; n'; n_{l\sigma}\rangle, \tag{5.10.2}$$

in which atom A is in state $|n\rangle$, B is in state $|n'\rangle$, and there are $n_{l\sigma}$ quanta in the field with momentum $\hbar\mathbf{k}_l$, energy $\hbar\omega_l$, and polarization σ. Primes distinguish B from A, which is unprimed. The energy associated with state $|E_0\rangle$ is

$$E_0 = E_n + E_{n'} + \sum_{l,\sigma} \hbar\omega_l n_{l\sigma}. \tag{5.10.3}$$

If $|n\rangle$ and $|s\rangle$ are two states of A and $|n'\rangle$ and $|s'\rangle$ are two states of B, then

$$\hbar\omega_{ns} = E_n - E_s \qquad \hbar\omega'_{n's'} = E_{n'} - E_{s'}. \tag{5.10.4}$$

Also, if m is the electron mass, (5.2.9) applied to A and B in our new notation becomes

$$\langle n|\mathbf{p}_A|s\rangle = im\omega_{ns}\mathbf{x}_{ns} \qquad \langle n'|\mathbf{p}_B|s'\rangle = im\omega_{n's'}\mathbf{x}_{n's'}, \tag{5.10.5}$$

where $\mathbf{p}_A$ and $\mathbf{p}_B$ are the momenta of A and B, respectively.

The interaction hamiltonian for the interaction of A and B with the radiation field is

$$H_1 = -\frac{e}{m}\mathbf{A}(0)\cdot\mathbf{p}_A - \frac{e}{m}\mathbf{A}(r\hat{\mathbf{k}})\cdot\mathbf{p}_B, \tag{5.10.6}$$

since the field $\mathbf{A}$ is to be evaluated at the origin for atom A and at a distance r along the z axis where atom B is located. Then $\mathbf{A}(\mathbf{r})$ is given by (5.1.8). Evaluation of the nonzero matrix elements of (5.10.6) between two eigenstates of H_0, which are needed for (1.16.59), is straightforward.

We may write the coupled equations (1.16.59), using (5.10.6), as follows: To begin, the nonzero matrix elements of H_1 are

$$\langle n; n'; n_{l\sigma}|H_1|s; s'; n_{l\sigma}+1\rangle = -\frac{e}{m}\sqrt{\frac{\hbar}{2\epsilon_0 L^3}}$$
$$\times\sqrt{\frac{n_{l\sigma}+1}{\omega_l}}\,\hat{\mathbf{e}}_{l\sigma}\cdot(\langle n|\mathbf{p}_A|s\rangle\langle n'|s'\rangle + \langle n'|e^{ik_l r\cos\theta_l}\mathbf{p}_B|s'\rangle\langle n|s\rangle), \tag{5.10.7}$$

and

$$\langle n; n'; n_{l\sigma}|H_1|s; s'; n_{l\sigma}-1\rangle = -\frac{e}{m}\sqrt{\frac{\hbar}{2\epsilon_0 L^3}}$$
$$\times\sqrt{\frac{n_{l\sigma}}{\omega_l}}\,\hat{\mathbf{e}}_{l\sigma}\cdot(\langle n|\mathbf{p}_A|s\rangle\langle n'|s'\rangle + \langle n'|e^{-ik_l r\cos\theta_l}\mathbf{p}_B|s'\rangle\langle n|s\rangle). \tag{5.10.8}$$

Here θ_l is the angle between $\mathbf{k}_l$ and the z axis where B is located. If we assume that the atoms are small compared with the wavelength, we may remove the $\exp(\pm ik_l r\cos\theta_l)$ from the integrals above and use (5.10.5) to write these matrix elements as

$$\langle n; n'; n_{l\sigma}|H_1|s; s'; n_{l\sigma}\pm 1\rangle$$
$$= -ie\sqrt{\frac{\hbar}{2\epsilon_0 L^3}}\begin{Bmatrix}\sqrt{\dfrac{n_{l\sigma}+1}{\omega_l}}\\ \sqrt{\dfrac{n_{l\sigma}}{\omega_l}}\end{Bmatrix}\hat{\mathbf{e}}_{l\sigma}\cdot(\omega_{ns}\mathbf{x}_{ns}\,\delta_{n's'} + e^{\pm ik_l r\cos\theta_l}\omega'_{n's'}\mathbf{x}'_{n's'}\,\delta_{ns}). \tag{5.10.9}$$

The first term in (5.10.9) corresponds to the emission or absorption of a photon by atom A while atom B does not change its state; the second term corresponds to the absorption or emission of a photon by B while A stays in the same state.

We may now use (5.10.9) to write the coupled equations (1.16.59). We have

$$
\begin{aligned}
i\hbar \frac{dc}{dt}(n; n'; n_{l\sigma}, t) = & -i \sum_{s,l,\sigma} e \sqrt{\frac{\hbar}{2\epsilon_0 L^3 \omega_l}} (\hat{e}_{l\sigma} \cdot \mathbf{x}_{ns}) \omega_{ns} \\
& \times [\sqrt{n_{l\sigma} + 1}\, e^{i(\omega_{ns} - \omega_l)t} c(s, n', n_{l\sigma} + 1, t) \\
& + \sqrt{n_{l\sigma}}\, e^{i(\omega_{ns} + \omega_l)t} c(s, n', n_{l\sigma} - 1, t)] \\
& -i \sum_{s' l \sigma} e \sqrt{\frac{\hbar}{2\epsilon_0 L^3 \omega_l}} (\hat{\mathbf{e}}_{l\sigma} \cdot \mathbf{x}'_{n's'}) \omega'_{n's'} \\
& \times [\sqrt{n_{l\sigma} + 1}\, e^{i(k_l r \cos\theta_l + \omega'_{n's'} - \omega_l)t} c(n, s', n_{l\sigma} + 1, t) \\
& + \sqrt{n_l}\, e^{-i(k_l r \cos\theta_l - \omega'_{n's'} - \omega_l)t} c(n, s', n_{l\sigma} - 1, t)].
\end{aligned}
\tag{5.10.10}
$$

For simplicity, the ground states of A and B are designated by 1, and the only excited state of interest is designated by 2. We make the following changes in notation:

$$
\begin{aligned}
-\omega_{12} = \omega_{21} &\equiv -\omega \qquad & -\omega'_{12} = +\omega'_{21} &= -\omega' \\
\mathbf{x}_{21} = \mathbf{x}_{12} &= \mathbf{x} \qquad & \mathbf{x}'_{12} = \mathbf{x}'_{21} &= \mathbf{x}'
\end{aligned}
, \tag{5.10.11}
$$

and, as noted earlier, we make the simplifying assumption that $\mathbf{x}$ and $\mathbf{x}'$ have a component in the x direction only.

At $t = 0$, the initial state of atoms A and B is given by $|2; 1; 0; t = 0\rangle$ so that

$$
c(2; 1; 0; t = 0) = 1, \tag{5.10.12}
$$

and all other c's are zero. The problem is to find the probability that, at time t, the final state in which atom A has decayed and the photon has been absorbed by atom B is given by $|1; 2; 0; t\rangle$. This probability is

$$
|c(1; 2; 0; t)|^2. \tag{5.10.13}
$$

Atom B is far enough away from A so that its effect on the spontaneous decay of A is negligible. Accordingly, we may use the analysis of Section 5.3 for natural linewidth. We assume that the initial state of the system (5.10.12) decays as

$$
c(2; 1; 0; t) = e^{-\gamma_A t/2}, \tag{5.10.14}
$$

where $1/\gamma_A$ is the half life of atom A. If we neglect the presence of atom B, then, as in Section 5.3, the solution of (5.10.10) for $c(1; 1; 1_{l\sigma}; t)$ in which atom A decays and emits a photon is

$$
c(1; 1; 1_{l\sigma}; t) = \frac{e}{\sqrt{2\hbar\epsilon_0 L^3}} \frac{\omega}{\sqrt{\omega_l}} (\hat{\mathbf{e}}_{l\sigma} \cdot \mathbf{x}) \frac{1}{-i(\omega_l - \omega) + \gamma_A/2}, \tag{5.10.15}
$$

after a time sufficiently long that atom A has certainly decayed to its ground state by (5.3.31).

In the transition from the state $|1; 1; 1_{l\sigma}\rangle$ to the final state $|1; 2; 0\rangle$, the matrix elements $\mathbf{x}_{11}$ are zero, and so that (5.10.10) reduces to

$$\frac{dc(1; 2; 0; t)}{dt} = -\sum_{l,\sigma} \frac{e}{\sqrt{2\hbar\epsilon_0 L^3}} \frac{\omega'}{\sqrt{\omega_l}} (\hat{\mathbf{e}}_{l\sigma} \cdot \mathbf{x}') e^{i[k_l r \cos\theta_l + (\omega' - \omega_l)t]} c(1; 1; 1_{l\sigma}; t) \tag{5.10.16}$$

since all other terms vanish. (There are only two atomic levels involved.) We may use the approximate result of (5.10.15) in (5.10.16) and integrate with respect to time to obtain

$$c(1; 2; 0; t) = -\frac{e^2}{2\hbar\epsilon_0 L^3}\omega\omega' \sum_{l,\sigma} \frac{(\hat{\mathbf{e}}_{l\sigma} \cdot \mathbf{x})(\hat{\mathbf{e}}_{l\sigma} \cdot \mathbf{x}') e^{ik_l r \cos\theta_l}[1 - e^{-i(\omega_l - \omega')t}]}{\omega_l[i(\omega - \omega_l) + \gamma_A/2]i(\omega_l - \omega')}, \tag{5.10.17}$$

where, at $t = 0$, $c(1; 2; 0; 0) = 0$. We must evaluate the sum in (5.10.17) by transforming it to an integral. This is slightly tedious; it is done in Appendix H [1], where it is shown that

$$c(1; 2; 0; t) = \begin{cases} 0 & t < \dfrac{r}{c} \\ -\dfrac{1}{r}\dfrac{e^2\,|\mathbf{x}|\,|\mathbf{x}'|\,\omega\omega'\mu_0 e^{i\omega' r/c}}{4\pi\hbar i[i(\omega - \omega') + \gamma_A/2]} & t > \dfrac{r}{c}. \end{cases} \tag{5.10.18}$$

Therefore, the probability of finding atom B excited is zero if $t < r/c$, the time needed for the emitted photon to reach atom B. For $t > r/c$, we have

$$|c(1; 2; 0; t)|^2 = \frac{1}{r^2}\left(\frac{e^2\,|\mathbf{x}|\,|\mathbf{x}'|\,\omega\omega'\mu_0}{4\pi\hbar}\right)^2 \frac{1}{(\omega - \omega')^2 + (\gamma_A/2)^2}. \tag{5.10.19}$$

This probability is inversely proportional to the square of the distance r. We conclude that the theory predicts correctly the velocity of propagation of light and also gives the correct decrease in intensity of the light from the source. Again we see the characteristic Lorentzian line shape.

5.11 SEMICLASSICAL THEORY OF ELECTRON-SPIN RESONANCE

The theory of electron- and nuclear-spin resonance and the theory of the two-level laser have in common the interaction of radiation with a two-level quantum system. Because of its fundamental nature, we present a rather complete analysis of the interaction of a magnetic dipole moment (spin-$\frac{1}{2}$) in a d-c magnetic field with an rf magnetic field. The case of an electric dipole moment (e.g., the ammonia molecular-beam maser) interacting with an rf electric field is formally identical with the ensuing analysis.

The rf magnetic field is treated classically in this section; we do not take into account the reaction of the spin back on the field [7]. In Section 5.13 we quantize the radiation field and thereby consider the reaction of the spin on the field. Fortunately, this simple problem may be solved completely.

The hamiltonian for a magnetic dipole $\boldsymbol{\mu}$ in a magnetic field $\mathbf{H}$ is

$$H = -\boldsymbol{\mu} \cdot \mathbf{H}. \tag{5.11.1}$$

For an electron of spin angular momentum $\frac{1}{2}\hbar$, the magnetic moment is related to the spin by

$$\boldsymbol{\mu} = -\frac{\gamma\hbar}{2}\boldsymbol{\sigma}, \tag{5.11.2}$$

where γ is the gyromagnetic ratio and σ_x, σ_y, σ_z are the Pauli spin operators introduced in Section 2.8. If we combine (5.11.2) and (5.11.1) and assume an applied magnetic field

$$\mathbf{H} = (H_1 \cos \omega t,\ H_1 \sin \omega t,\ H_0), \tag{5.11.3}$$

where H_0 is a d-c field in the z direction and the rf field is circularly polarized, the hamiltonian becomes

$$H = \frac{\gamma\hbar}{2}[H_0\sigma_z + H_1(\sigma_x \cos \omega t + \sigma_y \sin \omega t)]. \tag{5.11.4}$$

If we write σ_x and σ_y in terms of σ_+ and σ_- [see (2.8.18)], H may be written as

$$H = \frac{\gamma\hbar}{2}[H_0\sigma_z + H_1(\sigma_+ e^{-i\omega t} + \sigma_- e^{i\omega t})]. \tag{5.11.5}$$

The Schrödinger equation

$$H|\psi(t)\rangle = i\hbar \frac{\partial|\psi(t)\rangle}{\partial t} \tag{5.11.6}$$

may be solved exactly for H given by (5.11.5). When $H_1 = 0$, from Section 2.8 we know that the eigenvalues of σ_z are ± 1 corresponding to the stationary states $|\pm 1\rangle$. The unperturbed energy eigenvalues are therefore

$$E_\pm = \pm \frac{\hbar\omega_0}{2}, \tag{5.11.7}$$

where

$$\omega_0 = \gamma H_0. \tag{5.11.8}$$

The general solution of (5.11.6) when $H_1 = 0$ is (Section 2.9)

$$|\psi(t)\rangle = c_1 e^{-i\omega_0 t/2}|+1\rangle + c_2 e^{i\omega_0 t/2}|-1\rangle, \tag{5.11.9}$$

where $|c_1|^2$ is the probability that σ_z is in state $|+1\rangle$ and $|c_2|^2$ is the probability that it is in state $|-1\rangle$. Normalization to unity requires that

$$|c_1|^2 + |c_2|^2 = 1. \tag{5.11.10}$$

It is easily shown, by means of (2.8.32) and (2.8.33), and the orthogonality relations, that for $|\psi(t)\rangle$ given by (5.11.9)

$$\begin{aligned}\langle\sigma_x\rangle &= c_1^* c_2 e^{i\omega_0 t} + c_1 c_2^* e^{-i\omega_0 t}\\ \langle\sigma_y\rangle &= -i(c_1^* c_2 e^{i\omega_0 t} - c_1 c_2^* e^{-i\omega_0 t})\\ \langle\sigma_z\rangle &= |c_1|^2 - |c_2|^2\\ \langle\sigma_+\rangle &= c_1^* c_2 e^{i\omega_0 t}\\ \langle\sigma_-\rangle &= c_1 c_2^* e^{-i\omega_0 t}.\end{aligned} \tag{5.11.11}$$

Therefore, the expectation value of σ_z is a constant while σ_x and σ_y precess about the z axis with $\langle\sigma_x\rangle^2 + \langle\sigma_y\rangle^2 = 4\,|c_1|^2\,|c_2|^2$.

When $H_1 \neq 0$, we may think of H_1 as causing transitions between the two states $|+1\rangle$ and $|-1\rangle$. By a simple transformation, we may obtain a solution of (5.11.6) directly. We let

$$|\psi(t)\rangle = e^{-i\omega t\sigma_z/2}|\chi(t)\rangle. \tag{5.11.12}$$

This is *not* a transformation to the interaction picture. (Why?) When we substitute this into (5.11.6), where H is given by (5.11.5), we see that $|\chi(t)\rangle$ must satisfy

$$i\,\frac{\partial|\chi\rangle}{\partial t} = \tfrac{1}{2}\{(\omega_0 - \omega)\sigma_z + \gamma H_1[e^{-i\omega t}\sigma_+(t) + e^{i\omega t}\sigma_-(t)]\}\chi(t)\rangle, \tag{5.11.13}$$

where

$$\sigma_\pm(t) \equiv e^{i\omega t\sigma_z/2}\sigma_\pm e^{-i\omega t\sigma_z/2}. \tag{5.11.14}$$

This may be simplified by differentiating with respect to t. This gives

$$\begin{aligned}\frac{d\sigma_\pm}{dt} &= \tfrac{1}{2}i\omega e^{i\omega t\sigma_z/2}[\sigma_z, \sigma_\pm]e^{i\omega t\sigma_z/2}\\ &= \pm i\omega\sigma_\pm(t)\end{aligned} \tag{5.11.15}$$

where we used (2.8.21). The solution is

$$\sigma_\pm(t) = \sigma_\pm e^{\pm i\omega t}. \tag{5.11.16}$$

If we substitute this into (5.11.13), the transformation (5.11.12) leads to an equation with no explicit time dependence, namely,

$$i\frac{\partial|\chi\rangle}{\partial t} = \tfrac{1}{2}[(\omega_0 - \omega)\sigma_z + \gamma H_1(\sigma_+ + \sigma_-)]|\chi\rangle. \tag{5.11.17}$$

In this equation $\sigma_\pm$ and σ_z are in the Schrödinger picture. If we had used the interaction picture, the hamiltonian would still have had an explicit time dependence. We may therefore write the solution of (5.11.17) directly. If we let

$$\begin{aligned}\Omega\cos\theta &= \omega_0 - \omega\\ \Omega\sin\theta &= \gamma H_1,\end{aligned} \tag{5.11.18}$$

where $\Omega^2 = (\omega - \omega_0)^2 + (\gamma H_1)^2$ and use (2.8.18), the formal solution of (5.11.17) is

$$|\chi(t)\rangle = e^{-(i\Omega t/2)(\cos\theta\sigma_z + \sin\theta\sigma_x)}|\psi(0)\rangle. \tag{5.11.19}$$

It is easy to show that

$$\exp[-i(\xi\sigma_z + \eta\sigma_x)] = \cos\sqrt{\xi^2+\eta^2} - i\frac{\sin\sqrt{\xi^2+\eta^2}}{\sqrt{\xi^2+\eta^2}}(\xi\sigma_z + \eta\sigma_x),$$

which we leave as an exercise. If we let $\xi = (\Omega t/2)\cos\theta$ and $\eta = (\Omega t/2)\sin\theta$, then we have that

$$\begin{aligned}|\psi(t)\rangle &= e^{-i\omega t\sigma_z/2}[\cos\tfrac{1}{2}\Omega t - i\sin\tfrac{1}{2}\Omega t(\cos\theta\sigma_z + \sin\theta\sigma_x)]|\psi(0)\rangle\\ &\equiv U(t,0)|\psi(0)\rangle.\end{aligned} \tag{5.11.20}$$

This is the complete solution for the Schrödinger wave function for the hamiltonian (5.11.5).

Since the states $|+1\rangle$ and $|-1\rangle$ represent a complete orthonormal set, we may expand $|\psi(t)\rangle$ as in (5.11.9), where c_1 and c_2 are now functions of t. It follows that

$$|c_1(t)|^2 = |\langle +1|\psi(t)\rangle|^2 \qquad |c_2(t)|^2 = |\langle -1|\psi(t)\rangle|^2 \tag{5.11.21}$$

give the probabilities of finding the spin in the state $|+1\rangle$ and $|-1\rangle$, respectively, at time t.

As an example, we suppose that the system initially is in state $|\psi(0)\rangle = |+1\rangle$. If we use (2.8.32), (2.8.33), and (5.11.20), it follows directly that

$$|c_2(t)|^2 = \sin^2\theta\sin^2\tfrac{1}{2}\Omega t. \tag{5.11.22}$$

This is the probability that the spin in state $|+1\rangle$ at $t = 0$ will be in state $|-1\rangle$ at time t. If we use (5.11.18) to eliminate θ and Ω, this probability is

$$|c_2(t)|^2 = \frac{(\gamma H_1)^2}{(\omega-\omega_0)^2 + (\gamma H_1)^2}\sin^2\tfrac{1}{2}t\sqrt{(\omega-\omega_0)^2 + (\gamma H_1)^2}, \tag{5.11.23}$$

where, in general,

$$|c_2(t)|^2 \equiv 1 - |c_1(t)|^2, \tag{5.11.24}$$

from the normalization condition.

If the system is initially in state $|-1\rangle$, the probability of finding it in state $|+1\rangle$ at time t is also given by (5.11.23). Therefore, the probabilities of emission and of absorption are equal.

The transition probability (5.11.23) reaches a value of unity only at resonance, that is, when the applied frequency $\omega = \omega_0 \equiv \gamma H_0$. Far from resonance the probability of an rf field inducing a transition is negligible.

Before ending this section, we should consider the temporal behavior of $\sigma_\pm(t)$ and $\sigma_z(t)$. We may write the Heisenberg equations for these operators, or we may note that

$$\begin{aligned} \sigma_\pm{}^H(t) &= U^{-1}(t, 0)\sigma_\pm{}^S U(t, 0) \\ \sigma_z{}^H(t) &= U^{-1}(t, 0)\sigma_z{}^S U(t, 0), \end{aligned} \tag{5.11.25}$$

where $U(t, 0)$ is unitary and given by (5.11.20). By using the results of Section 2.9, we may obtain these operators in the Heisenberg picture. In particular, it is found that

$$\sigma_z(t) = (\cos^2\theta + \sin^2\theta\cos\Omega t)\sigma_z + \sin\theta\sin\Omega t\sigma_y + \sin 2\theta\sin^2\tfrac{1}{2}\Omega t\sigma_x. \tag{5.11.26}$$

This is left as an exercise.

We may show that if, at $t = 0$, $|\psi(0)\rangle = |+1\rangle$, then by (5.11.20), (2.8.32), and (2.8.33)

$$|\psi(t)\rangle = e^{-i\omega t/2}(\cos\tfrac{1}{2}\Omega t - i\sin\tfrac{1}{2}\Omega t\cos\theta)|+1\rangle - ie^{i\omega t/2}\sin\tfrac{1}{2}\Omega t\sin\theta|-1\rangle, \tag{5.11.27}$$

so that the expectation values of $\sigma_z(t)$, $\sigma_\pm(t)$ for this state are

$$\begin{aligned} \langle\psi(t)|\sigma_\pm|\psi(t)\rangle &= \mp ie^{\pm i\omega t}\sin\tfrac{1}{2}\Omega t\sin\theta(\cos\tfrac{1}{2}\Omega t \pm i\sin\tfrac{1}{2}\Omega t\cos\theta) \\ \langle\psi(t)|\sigma_z|\psi(t)\rangle &= \cos^2\tfrac{1}{2}\Omega t + \sin^2\tfrac{1}{2}\Omega t\cos 2\theta. \end{aligned} \tag{5.11.28}$$

At resonance, $\omega = \omega_0$, $\theta = \pi/2$, and $\Omega = \gamma H_1$, and it follows from (5.11.28) and (2.8.18) that

$$\begin{aligned} \langle\sigma_x\rangle &= \sin\omega_0 t\sin\gamma H_1 t \\ \langle\sigma_y\rangle &= -\cos\omega_0 t\sin\gamma H_1 t \\ \langle\sigma_z\rangle &= \cos\gamma H_1 t \end{aligned} \tag{5.11.29}$$

Therefore, $\langle\sigma_x\rangle$ and $\langle\sigma_y\rangle$ precess about H_0 with frequency ω_0 while $\langle\sigma_z\rangle$ nutates at frequency γH_1; this is the classical behavior of a magnetic moment in a d-c circularly polarized rf field, in agreement with Ehrenfest's principle. If

$H_1 \ll H_0$, the spin precesses through many cycles before $\langle \sigma_z \rangle$ changes very much.

5.12 COLLISION BROADENING OF TWO-LEVEL SPIN SYSTEM

The analysis in the previous section assumed that the interaction between the spin and the field continues indefinitely. In a gas at finite temperature there are random collisions of the gas molecules that interrupt the interaction between the spin and field. This effect is important in a gaseous laser and results in a broadening of the spectral lines. We have already studied two mechanisms that broaden a spectral line, namely, the natural line width and the Doppler effect. Let us consider the effect of collisions.

If no collisions occur, the instantaneous power flow between the spins and the radiation field is

$$P(t) = \hbar\omega_0 \frac{d}{dt} |c_2(t)|^2, \tag{5.12.1}$$

where $|c_2(t)|^2$ is given by (5.11.23). If there are random collisions, the probability that no collision has occurred in a time t is $\exp(-t/T_2)$, where T_2 is the average time between collisions, while the probability that a spin had its last collision between t and $t + dt$ is

$$\frac{dt}{T_2} e^{-t/T_2}$$

The average power transferred to or from the field is therefore

$$\begin{aligned} P_{\text{av}} &= \hbar\omega_0 \int_0^\infty \frac{d}{dt} |c_2(t)|^2 e^{-t/T_2} \frac{dt}{T_2} \\ &= \frac{\hbar\omega_0(\gamma H_1)^2}{T_2} \frac{1}{(\omega - \omega_0)^2 + (\gamma H_1)^2 + (1/T_2)^2}, \end{aligned} \tag{5.12.2}$$

where we used (5.11.23). Therefore collision broadening gives a Lorentz line shape to the spin-resonance line.

5.13 EFFECT OF FIELD QUANTIZATION ON SPIN RESONANCE [7]

In Section 5.11 we considered a spin-$\frac{1}{2}$ particle interacting with a classical radiation field, and we neglected the reaction of the spin on the field. We consider the same problem again but quantize the radiation field and take into account the reaction of the spin on the field.

We consider a small sample of material containing spin-$\frac{1}{2}$ particles located at a position in a cavity in which the magnetic field has a component only in the x direction for a particular mode. As shown in Chapter 4, the energy of this mode of the field is

$$H_{\text{field}} = \hbar\omega a^\dagger a, \tag{5.13.1}$$

where ω is the frequency of the mode under consideration.

By (5.11.1) and (5.11.2), the interaction energy is

$$\begin{aligned} H_{in} &= \int_{\text{sample}} \frac{\gamma\hbar}{2}\, \boldsymbol{\sigma}\cdot\mathbf{H}\, d\tau \\ &= \gamma H_0 \frac{\hbar}{2}\sigma_z + \frac{\gamma\hbar}{2}\int_{\text{sample}} \sigma_x H_x\, d\tau, \end{aligned} \tag{5.13.2}$$

where H_0 is the value of the d-c field in the z direction and H_x is the rf field at the sample. If we use (4.3.9) and (4.3.30) and assume that σ_x occupies a small volume of the cavity, we may write (5.13.2) as

$$H_{in} = \frac{\omega_0\hbar}{2}\sigma_z + \hbar\kappa(a + a^\dagger)\sigma_x, \tag{5.13.3}$$

where all quantities multiplying σ_x in (5.13.2) and not appearing explicitly in (5.13.3) are grouped into a coupling constant κ, which will be small. Since, by (2.8.18), $\sigma_x = \sigma_+ + \sigma_-$, we may add (5.13.1) to (5.13.3) and write the total hamiltonian as

$$H = \hbar\omega a^\dagger a + \frac{\hbar\omega_0}{2}\sigma_z + \hbar\kappa(a + a^\dagger)(\sigma_+ + \sigma_-). \tag{5.13.4}$$

This hamiltonian may be simplified by the following considerations. When $\kappa = 0$, that is, when there is no rf field, the Heisenberg operators have a time dependence given by

$$\begin{aligned} a(t) &= a(0)e^{-i\omega t} \qquad & \sigma_+(t) &= \sigma_+(0)e^{i\omega_0 t} \\ a^\dagger(t) &= a^\dagger(0)e^{i\omega t} \qquad & \sigma_-(t) &= \sigma_-(0)e^{-i\omega_0 t}, \end{aligned} \tag{5.13.5}$$

so that, near resonance ($\omega \approx \omega_0$), the interaction terms $a\sigma_+$ and $a^\dagger\sigma_-$ in (5.13.4) are practically d-c terms whereas the terms $a\sigma_-$ and $a^\dagger\sigma_+$ vary rapidly at frequencies $\pm(\omega + \omega_0)$. To a good degree of approximation for times of interest, the high-frequency terms average to zero, and we may write (5.13.4) approximately as

$$H = \hbar\omega a^\dagger a + \frac{\hbar\omega_0}{2}\sigma_z + \hbar\kappa(a^\dagger\sigma_- + a\sigma_+). \tag{5.13.6}$$

This approximation is equivalent to decomposing the linearly polarized rf cavity field into two opposite circularly polarized waves and keeping only the one rotating in the same sense as the spin precession. For small rf fields, little error is involved.*

The hamiltonian (5.13.6) is still hermitian and takes into account the effect of the spin on the field as well as the usual effect of the field on the spin.

The problem at this stage is to solve the Schrödinger equation for the hamiltonian given by (5.13.6). We present a solution due to Jaynes and Cummings [7].

For convenience, we define two new operators

$$S_+ = \sigma_+ a \qquad S_- = \sigma_- a^\dagger. \tag{5.13.7}$$

From the commutation relations (2.8.19)–(2.8.26) and $[a, a^\dagger] = 1$, it is left as an exercise to verify the following commutation relations:

$$[a^\dagger a, S_\pm] = \mp S_\pm \qquad [\sigma_z, S_\pm] = \pm 2S_\pm \qquad [S_+, S_-] = \frac{1+\sigma_z}{2} + a^\dagger a \sigma_z. \tag{5.13.8}$$

It will be found that the Heisenberg equations of motion for these operators are nonlinear. Rather than try to solve these equations, we resort to "trickery" in finding two constants of the motion, either by inspection or from first integrals of the nonlinear equations of motion. We may verify directly that the operators

$$C_1 = \omega(a^\dagger a + \tfrac{1}{2}\sigma_z) \qquad C_2 = \kappa(S_+ + S_-) - \frac{\Delta\omega}{2}\sigma_z, \tag{5.13.9}$$

where $\Delta\omega = \omega - \omega_0$, commute with the hamiltonian (5.13.6) and are therefore constants of the motion. It follows directly from (5.13.9) and (5.13.6) that

$$H = \hbar(C_1 + C_2). \tag{5.13.10}$$

Therefore, C_1 and C_2 are not independent of H, and we may show that

$$[C_1, C_2] = 0. \tag{5.13.11}$$

by using (5.13.8) so that C_1 and C_2 may be treated as c-numbers with respect to one another.

When there is no coupling between the spins and radiation field, we have a complete set of basis vectors, namely, the states $|n\rangle$ for the radiation field where $a^\dagger a|n\rangle = n|n\rangle$ and the states $|\pm 1\rangle$ for the spins where $\sigma_z|\pm\rangle = \pm|\pm 1\rangle$. We may use this complete set of states to expand the state vectors of the coupled system. We see that

$$C_1|n, \pm 1\rangle \equiv \omega(a^\dagger a + \tfrac{1}{2}\sigma_z)|n, \pm 1\rangle = \omega(n \pm \tfrac{1}{2})|n, \pm 1\rangle \tag{5.13.12}$$

* This is called the rotating wave approximation and is valid for weak coupling.

so that C_1 has eigenvalues $\omega(n \pm \frac{1}{2})$ with the corresponding eigenkets $|n, \pm 1\rangle$ and C_1 is therefore diagonal in this representation. However, C_2 is not diagonal in this representation, but since C_1 and C_2 commute, we may find a representation in which they are both diagonal (see Section 1.9) by taking linear combinations of the eigenkets of C_1. In this representation, by (5.13.10), H is also diagonal so that it may appropriately be called the energy representation. Once H has been diagonalized, the Schrödinger equation has been solved.

Let us choose two simple linear combinations of eigenvectors of C_1 and verify that they are eigenkets of C_2 (as well as H). We consider the state vectors

$$\begin{aligned} |\varphi(n, 1)\rangle &= \cos\theta_n|n+1, -1\rangle + \sin\theta_n|n, +1\rangle \\ |\varphi(n, 2)\rangle &= -\sin\theta_n|n+1, -1\rangle + \cos\theta_n|n, +1\rangle, \end{aligned} \tag{5.13.13}$$

where $|n+1, -1\rangle$ corresponds to $n+1$ quanta in the field with spin "down" and $|n, +1\rangle$ corresponds to n quanta with spin "up." There is another state that is not included, namely, no quanta in the field and spin down, $|0, -1\rangle$; this state must be considered separately and is called the ground state. The angle θ_n is a parameter at our disposal to make $|\varphi(n, 1)\rangle$ and $|\varphi(n, 2)\rangle$ eigenkets of C_2, and n may have any value from 0 to ∞.

We begin by observing that

$$\begin{aligned} \langle\varphi(n, 1)|\varphi(n, 2)\rangle &= 0 \\ \langle\varphi(n, 1)|\varphi(n, 1)\rangle &= \langle\varphi(n, 2)|\varphi(n, 2)\rangle = 1. \end{aligned} \tag{5.13.14}$$

Both these states are orthogonal to the ground state, $|0, -1\rangle$. We may also note that, by (5.13.12),

$$\begin{aligned} C_1|\varphi(n, 1)\rangle &= \omega(n + \tfrac{1}{2})|\varphi(n, 1)\rangle \\ C_1|\varphi(n, 2)\rangle &= \omega(n + \tfrac{1}{2})|\varphi(n, 2)\rangle \\ C_1|0, -1\rangle &= -\tfrac{1}{2}\omega|0, -1\rangle, \end{aligned} \tag{5.13.15}$$

so that the states $|\varphi(n, 1)\rangle$ and $|\varphi(n, 2)\rangle$ are degenerate eigenstates of C_1, no matter how we choose θ_n [8].

Let us now apply C_2 to $|\varphi(n, 1)\rangle$ and $|\varphi(n, 2)\rangle$. After minor simplification, we obtain

$$\begin{aligned} C_2|\varphi(n, 1)\rangle = &\left(\kappa\sqrt{n+1}\sin\theta_n + \frac{\Delta\omega}{2}\cos\theta_n\right)|n+1, -1\rangle \\ &+ \left(\kappa\sqrt{n+1}\cos\theta_n - \frac{\Delta\omega}{2}\sin\theta_n\right)|n, +1\rangle \end{aligned} \tag{5.13.16}$$

$$\begin{aligned} C_2|\varphi(n, 2)\rangle = &\left(\kappa\sqrt{n+1}\cos\theta_n - \frac{\Delta\omega}{2}\sin\theta_n\right)|n+1, -1\rangle \\ &- \left(\kappa\sqrt{n+1}\sin\theta_n + \frac{\Delta\omega}{2}\cos\theta_n\right)|n, +1\rangle, \end{aligned} \tag{5.13.17}$$

where we used the relations

$$\begin{aligned} S_+|n+1, -1\rangle &= \sqrt{n+1}|n, +1\rangle \\ S_+|n, +1\rangle &= 0 \\ S_-|n+1, -1\rangle &= 0 \\ S_-|n, +1\rangle &= \sqrt{n+1}|n+1, -1\rangle. \end{aligned} \tag{5.13.18}$$

In order for $|\varphi(n, 1)\rangle$ and $|\varphi(n, 2)\rangle$ to be eigenvectors of C_2, we must be able to choose θ_n so that

$$\begin{aligned} C_2|\varphi(n, 1)\rangle &= \lambda_n|\varphi(n, 1)\rangle \\ C_2|\varphi(n, 2)\rangle &= \lambda_n'|\varphi(n, 2)\rangle. \end{aligned} \tag{5.13.19}$$

From (5.13.16), (5.13.17), and (5.13.19), if we let

$$\tan\theta_n = \frac{\kappa\sqrt{n+1}}{\frac{1}{2}\Delta\omega + \lambda_n},$$

and

$$\lambda_n = +\sqrt{\left(\frac{\Delta\omega}{2}\right)^2 + \kappa^2(n+1)} \tag{5.13.20}$$

then $|\varphi(n, 1)\rangle$ and $|\varphi(n, 2)\rangle$ are eigenkets of C_2 with eigenvalues $\pm\lambda_n$. Thus $\lambda_n' = -\lambda_n$. By simple trigonometry, we may show that

$$\tan 2\theta_n = \frac{2\tan\theta_n}{1 - \tan^2\theta_n} = \frac{2\kappa\sqrt{n+1}}{\Delta\omega}. \tag{5.13.21}$$

For the ground state,

$$C_2|0, -1\rangle = \frac{\Delta\omega}{2}|0, -1\rangle, \tag{5.13.22}$$

which is therefore an eigenket of C_2 also.

We have succeeded in finding the eigenvalues of H, for by (5.13.10), (5.13.15), and (5.13.19) we have

$$\begin{aligned} H|\varphi(n, 1)\rangle &= \hbar[\omega(n + \tfrac{1}{2}) + \lambda_n]|\varphi(n, 1)\rangle \\ H|\varphi(n, 2)\rangle &= \hbar[\omega(n + \tfrac{1}{2}) - \lambda_n]|\varphi(n, 2)\rangle, \end{aligned} \tag{5.13.23a}$$

and for the ground state

$$H|0, -1\rangle = -\frac{\hbar\omega_0}{2}|0, -1\rangle. \tag{5.13.23b}$$

Note that eigenstates of H are mixtures of eigenstates of H_0, the unperturbed hamiltonian. The states $|\varphi(n, 1)\rangle$, $|\varphi(n, 2)\rangle$ and $|0, -1\rangle$ are complete so that

$$\sum_{n=0}^{\infty}\{|\varphi(n, 1)\rangle\langle\varphi(n, 1)| + |\varphi(n, 2)\rangle\langle\varphi(n, 2)|\} + |0, -1\rangle\langle 0, -1| = 1. \tag{5.13.24}$$

It is useful to have expressions for $|n, \pm 1\rangle$ in terms of $|\varphi(n, 1)\rangle$ and $|\varphi(n, 2)\rangle$.

From (5.13.13) and the orthogonality relations (5.13.14), we have ($n = 0, 1, 2, \ldots, \infty$)

$$\begin{aligned} |n+1, -1\rangle &= \cos\theta_n |\varphi(n, 1)\rangle - \sin\theta_n |\varphi(n, 2)\rangle \\ |n, +1\rangle &= \sin\theta_n |\varphi(n, 1)\rangle + \cos\theta_n |\varphi(n, 2)\rangle, \end{aligned} \tag{5.13.25}$$

and

$$\begin{aligned} \langle n+1, -1|\varphi(n, 1)\rangle &= \cos\theta_n = \langle n, +1|\varphi(n, 2)\rangle \\ \langle n, +1|\varphi(n, 1)\rangle &= \sin\theta_n = -\langle n+1, -1|\varphi(n, 2)\rangle. \end{aligned} \tag{5.13.26}$$

We now calculate the transition probability between an initial state with n quanta and spin "up" ($|n, +1\rangle$) and a final state with $n + 1$ quanta and spin "down" ($|n + 1, -1\rangle$) to compare the results of the quantized-field case with the unquantized case of Section 5.11.

Since the hamiltonian (5.13.6) does not contain the time explicitly, we may write the solution of the Schrödinger equation as

$$|\psi(t)\rangle = \exp\left(-\frac{iHt}{\hbar}\right)|\psi(0)\rangle = e^{-iC_1t}e^{-iC_2t}|\psi(0)\rangle, \tag{5.13.27}$$

where $|\psi(0)\rangle = |n, +1\rangle$ for the case of interest. The probability of finding the system with $n + 1$ quanta and spin "down" at time t is given by $|\langle n + 1, -1|\psi(t)\rangle|^2$, which, on using (5.13.25), (5.13.15), (5.13.14), and (5.13.19), becomes

$$\begin{aligned} &\left|\langle n+1, -1| \exp\left(-\frac{iHt}{\hbar}\right)|n, +1\rangle\right|^2 \\ &\quad = \sin^2 2\theta_n \sin^2 \lambda_n t \\ &\quad = \frac{4\kappa^2(n+1)}{(\Delta\omega)^2 + 4\kappa^2(n+1)} \sin^2 \tfrac{1}{2}t\sqrt{(\Delta\omega)^2 + 4\kappa^2(n+1)}. \end{aligned} \tag{5.13.28}$$

This is also equal to the transition probability from state $|n + 1, -1\rangle$ to state $|n, +1\rangle$.

If we compare this result with (5.11.23) for the unquantized-field case, we see that the two probabilities are remarkably similar. The square of the rf amplitude, H_1^2, in (5.11.23) is proportional to the number of quanta, n, in the field. The difference is the appearance of the spontaneous-emission term, 1, in (5.13.28). If $n = 0$, the spin will still flip and emit a quantum that may be reabsorbed. Jaynes and Cummings [7] have shown that, if in the semiclassical treatment the effect of the spin on the radiation field had been included, spontaneous emission could take place in the semiclassical theory.

PROBLEMS

5.1 The hamiltonian for a hydrogen atom in a Coulomb field is

$$H = \frac{1}{2m}\mathbf{p}^2 - \frac{e^2}{r},$$

where the symbols have the usual meaning. Write the Heisenberg equations of motion for $\mathbf{p}_H(t)$ and $\mathbf{x}_H(t)$.

5.2 Evaluate the induced dipole moment (5.2.9) for atomic hydrogen between (*a*) and 1*s*- and 2*s*-state and (*b*) the 1*s*- and 2*p*-state. The wave functions may be found in any quantum mechanics book.

5.3 If a set of two-level atoms with energies E_a and E_b are in thermal equilibrium at temperature T, the number of atoms in state $|a\rangle$ is proportional to $\exp(-E_a/kT)$ and for state $|b\rangle$ the number is proportional to $\exp(-E_b/kT)$, where k is Boltzmann's constant. Show that, if the atoms are in thermal equilibrium with a radiation field, the average number of quanta is given by the Planck distribution law

$$\bar{n} = \left[\exp\left(\frac{\hbar\omega}{kT}\right) - 1\right]^{-1},$$

where $\hbar\omega = E_a - E_b$.

5.4 Calculate the natural line width of the $2p - 1s$ transition in atomic hydrogen and compare it with the observed line width.

5.5 Calculate the Doppler line width for atomic hydrogen for the $2p - 1s$ transition at room temperature.

5.6 Solve the Heisenberg equations of motion for the spin operators $\sigma_\pm(t)$ and $\sigma_z(t)$, using the hamiltonian (5.11.5). Express the result in terms of the operators $\sigma_\pm(0)$ and $\sigma_z(0)$. *Hint:* Let $\sigma_\pm(t) = s_\pm(t)\exp(\pm i\omega t)$.

5.7 Verify (5.13.11).

5.8 Find the Heisenberg equations of motion for the operators $a^\dagger a$, σ_z, and $S_\pm$, using the hamiltonian (5.13.6). From these equations, find the two integrals of motion (5.13.9). Finally, from these equations eliminate $a^\dagger a$, $S_\pm$ and obtain the equation for σ_z alone. One integral of this equation is immediately apparent. From this integral, show that σ_z must involve elliptic functions.

5.9 For the hamiltonian (5.13.6), obtain the transition probability per sec for the system to make a transition from the initial state $|n, +1\rangle$ to the state $|n + 1, -1\rangle$ by using perturbation theory. Compare this with the exact result obtained from (5.13.28).

5.10 Evaluate the Rayleigh differential scattering cross-section if we assume the "atom" is a three-dimensional isotropic harmonic oscillator described by the hamiltonian

$$H_a = \frac{1}{2m}\mathbf{p}^2 + \frac{m}{2}\omega_0^2(x^2 + y^2 + z^2).$$

5.11 Calculate the cross-section for the scattering of a photon from a free electron.

REFERENCES

[1] E. Fermi, *Rev. Mod. Phys.*, **4,** 87 (1932).
[2] V. G. Weisskopf and E. Wigner, *Z. Phys.*, **63,** 54 (1930).
[3] R. V. Churchill, *Modern Operational Mathematics in Engineering*, New York: McGraw-Hill, 1944, pp. 157ff.

[4] W. Heitler, *Quantum Theory of Radiation*, 3rd ed., Fair Lawn, N.J.: Oxford University Press, 1954, pp. 66ff.

[5] L. I. Schiff, *Quantum Mechanics*, 3rd ed., New York: McGraw-Hill, 1968, p. 85.

[6] H. Goldstein, *Classical Mechanics*, Cambridge, Mass.: Addison-Wesley, 1950, pp. 97ff, 107ff.

[7] E. T. Jaynes and F. W. Cummings, *Proc. I.R.E.* **51,** 89 (1963).

[8] That is, the states $|\varphi(n, 1)\rangle$ and $|\varphi(n, 2)\rangle$ are different but the eigenvalues of C_1 are the same.

6

Quantum Theory of Damping—Density Operator Methods

Damping plays such an important role in so many physical problems that we devote two chapters to its study. In this chapter we are primarily concerned with the density operator formulation of the problem while we devote the following chapter to the Langevin formulation. We exploit the quantum-classical correspondence that we developed in Chapter 3 to separate the quantum and stochastic natures of the problem. The methods are illustrated by simple examples such as a damped harmonic oscillator and homogeneously broadened atoms. We also discuss a rotating wave van der Pol oscillator which plays an important role in the study of lasers and optical parametric oscillators.

We begin by presenting a simple explicit model for a damping mechanism which allows us to give some of the properties of a reservoir. In Section 6.2 we consider an arbitrary system which is damped. We discuss the reduced density operator introduced in Chapter 1 which describes the statistical properties of a system coupled to a reservoir. We derive its equation of motion under the Markoff approximation. We obtain the reduced density operator equation for several simple damped systems for illustrative purposes.

In Section 6.3 we develop the mean equations of motion for system operators under the Markoff approximation. We again illustrate these equations for a damped driven oscillator and an atom with linewidth. This section is useful for the Langevin approach of the following chapter.

The equations of motion of associated distribution functions are obtained in Section 6.4. This allows us to study the quantum statistical properties of systems in a c-number domain. We work out the classical analog of a large number of homogeneously broadened atoms which is useful for the theory of a laser. In this section we also obtain the operator equations of motion in a c-number domain.

In Section 6.5 we solve the Fokker–Planck equation for a damped driven mode of the radiation field by two different methods. We also obtain an explicit expression for a characteristic function.

In Section 6.6 we give a method of converting two time averages to one-time averages for Markoffian systems and obtain the spectrum of a damped oscillator. In the final section we discuss the rotating wave van der Pol oscillator.

6.1 MODEL FOR LOSS MECHANISM

The hamiltonian for a single mode of a quantized radiation field of frequency ω_c in a cavity is given by $H = \hbar\omega_c a^\dagger a$. We have shown in Chapter 2 that in the Heisenberg picture, the annihilation and creation operators are given by $a(t) = a \exp(-i\omega_c t)$ and $a^\dagger(t) = a^\dagger \exp i\omega_c t$, respectively. Suppose that we attempt to describe damping in this mode by introducing a phenomenological loss term in the cavity by analogy with a circuit resistance. Since the classical and Heisenberg equations have the same form, the operators would then have the damped solutions

$$a(t) = ae^{-[i\omega_c + (\gamma/2)]t}$$

$$a^\dagger(t) = a^\dagger e^{[i\omega_c - (\gamma/2)]t},$$

and the fields would decay in a time of order γ^{-1}. This result unfortunately violates a fundamental principle of quantum mechanics since the commutator

$$[a(t), a^\dagger(t)] = e^{-\gamma t}$$

approaches zero. This implies that the uncertainty principle is violated (see Section 1.12). For times short compared with the relaxation time ($t \ll \gamma^{-1}$), the violation is not serious and the model would be satisfactory.

The difficulty with the model lies in the fact that there are thermal fluctuations in the resistor which feed noise into the oscillator, and these fluctuations have not yet been taken into account. In an equivalent circuit model, we should therefore put in not only a resistance but also a noise generator. One might be tempted to argue that this would not be unnecessary if the resistor were at absolute zero. Classically, this argument would be correct since then the oscillator would be highly excited (to be in the classical regime at all), and we would not expect to see any quantum effects. The violation of the uncertainty principle would not concern us. However, even at absolute zero there will always be zero point motion in the resistor due to purely quantum mechanical effects and these fluctuations will couple into the circuit. At low excitation of the field in the cavity mode, these fluctuations are responsible for preventing the violation of the uncertainty principle. What we therefore need in an equivalent circuit model is a noise generator which has sufficient output even at abolute zero to preserve the commutation relation $[a(t),$

$a^\dagger(t)] = 1$: a quantum noise generator. We pursue this equivalent circuit approach in the next chapter but we work from a different point of view in this chapter.

A similar problem was encountered in the theory of emission and absorption of radiation by an atom. If the lifetime Γ^{-1} of an atom in an excited state is very long, we may adequately describe the emission and absorption by first-order perturbation theory and neglect the atomic linewidth (see Section 5.3). However, if the atomic lifetime is short, we must take into account the reaction of the field on the atom which gives the atom a linewidth. That is, the fluctuations of the radiation field must be taken into account (by analogy with the noise generator) to prevent the violation of the uncertainty principle.

To obtain some insight into the atomic linewidth problem, we may visualize the process as follows. We think of the atom as a single system coupled to the radiation field in a cavity. The coupling is through the charge on the atomic electrons and the electric field in the cavity. Quantum mechanically there will always be zero point fluctuations in the cavity modes even in the vacuum state (no modes excited). As we have seen, the field may be visualized as a large number of fictitious harmonic oscillators, one for each mode of the cavity. If the cavity is very large, the modes will be very closely spaced in frequency so that they form practically a continuum. There will therefore be a cavity mode of frequency ω_c which approximately matches the energy separation of two atomic levels, that is,

$$\hbar\omega_c \cong E_2 - E_1.$$

The weak coupling of the atom and the field causes a splitting of the levels into a band and this gives a linewidth to the originally sharp atomic energy levels which were obtained by neglecting, in first approximation, the interaction of the atom and field.

The same phenomenon also occurs classically. If we weakly couple two oscillators of the same resonant frequency, the coupled system will have two normal modes whose frequencies differ slightly [1] (see Figure 6.1). If one oscillator is excited initially, its energy will all go to the other oscillator at a

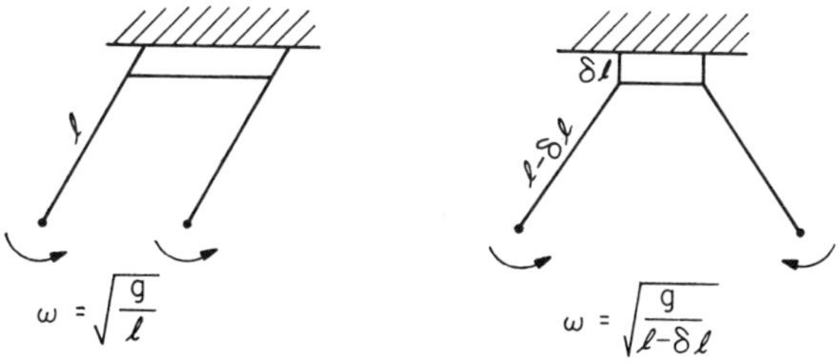

Figure 6.1 Normal modes of two weakly coupled oscillators.

later time because of the interference of the oscillations at the two new normal mode frequencies. We may visualize one oscillator as the atom and the other as a mode of the radiation field. The atom in an excited state corresponds to one oscillator excited. It decays giving its energy to the field mode. In turn the energy is reabsorbed by the atom from the field.

Similarly, if a large number of oscillators of the same resonant frequency are coupled, the normal mode frequencies of the coupled system will spread into a narrow band of frequencies near the original frequency. Energy initially in one oscillator will distribute itself among the other oscillators due to the interference of the various normal mode frequencies. Eventually all the energy will return to the original oscillator and the process will repeat. However, for sufficiently short times, the other oscillators absorb the energy of the original oscillator and act like a damping mechanism in addition to giving a linewidth. This is the analog of an atom coupled to modes of the radiation field. The field acts like a reservoir for the atomic energy. In addition the fluctuations present in the modes of the radiation field (reservoir) will couple to the atom.

With the foregoing discussion as a guide, let us consider some system described by a hamiltonian H. Next we assume there is a reservoir which may be taken as any large collection of systems with many degrees of freedom and which may be described by a hamiltonian R. If the reservoir consists of the modes of the radiation field, then we have

$$R = \sum_j \hbar\omega_j b_j^\dagger b_j, \tag{6.1.1}$$

where the b_j and $b_j^\dagger$ satisfy the boson commutation relations $[b_j, b_k^\dagger] = \delta_{jk}$. These might also be the quantized modes of elastic vibrations in a solid. The quanta of energy of elastic vibrations are called phonons and the creation and annihilation operators in this case also satisfy the boson commutation relations.

Next we assume there is a coupling between the system and reservoir which is described by the interaction energy V. The total hamiltonian for the system and reservoir is given by

$$H_T = H + R + V \equiv H_0 + V. \tag{6.1.2}$$

We may assume that the interaction is "turned on" at some time t_0. In a scattering experiment, for example, the system and reservoir are so far apart before t_0 that we may neglect the interaction V so that it is not unreasonable to imagine some experimental arrangement whereby $V = 0$ for $t < t_0$.

Other damping mechanisms may require different forms of R. However, the results are not very sensitive to the particular mechanism; thus we consider this simple model.

The statistical properties of a damped system are described by a density operator $\rho(t)$ which by (1.20.12) satisfies the equation of motion.

$$i\hbar\frac{\partial\rho}{\partial t} = [H_T, \rho] = [H + R + V, \rho] \equiv [H_0 + V, \rho], \qquad (6.1.3a)$$

where

$$\mathrm{Tr}_{R,S}\, \rho(t) = 1, \qquad (6.1.3b)$$

and where all operators are in the SP and we trace over both the system and reservoir in (6.1.3b). At $t = t_0$, the system and reservoir are uncoupled and the density operator factors as

$$\rho(t_0) = f_0(R)S(t_0), \qquad (6.1.4)$$

where $S(t_0)$ describes the initial state of the system and $f_0(R)$ describes the initial state of the reservoir. We must therefore solve (6.1.3) subject to the initial condition (6.1.4) and show that we have adequately described a damped system consistent with quantum mechanical principles. Actually, we must give some special properties to the reservoir in order to accomplish this goal. We do this in the following section.

Usually, the reservoir will be in thermal equilibrium at a temperature T initially. In this case the reservoir has a Boltzmann distribution described by the density operator

$$f_0(R) = \frac{e^{-\beta R}}{\mathrm{Tr}_R\, e^{-\beta R}}, \qquad (6.1.5a)$$

where

$$\beta = \frac{1}{kT}. \qquad (6.1.5b)$$

If the reservoir consists of independent oscillators, then we may factor the density operator as

$$f_0(R) = \prod_j \rho_j(t_0) \qquad (6.1.6a)$$

where

$$\rho_j(t_0) = (1 - e^{-\lambda_j})e^{-\lambda_j b_j{}^\dagger b_j}$$

$$\lambda_j = \frac{\hbar\omega_j}{kT}. \qquad (6.1.6b)$$

The reader should verify that

$$\mathrm{Tr}_R\, f_0(R) = 1. \qquad (6.1.7)$$

We leave the initial system density operator $S(t_0)$ arbitrary for the moment.

Let us consider in more detail a possible damping mechanism for a mode in a cavity. The mode is characterized by a frequency ω_c, polarization $\hat{e}$, and wave vector **k**. Light in this mode could be scattered by a gas atom in the cavity or from the walls, either elastically or inelastically, into other cavity modes and be lost from the desired mode. Conversely, thermal and zero point fluctuations in other modes could be scattered into the particular mode of interest. We may visualize the net effect of the scattering process from the desired mode as the annihilation of a photon at frequency ω_c and the simultaneous creation of a photon at frequency ω_j. Scattering into the mode consists of the annihilation of a photon at ω_j and creation of a photon at ω_c. We may therefore write an equivalent or effective interaction energy as

$$V = \hbar \sum_j (\kappa_j b_j a^\dagger + \kappa_j^* b_j^\dagger a). \tag{6.1.8}$$

This is an effective interaction since we have not taken into account the details of the scattering mechanism but it represents the net effect. We have summed over all modes that may scatter into or out of the desired mode. Actually, only those modes which conserve energy ($\omega_j \approx \omega_c$) and momentum ($\mathbf{k}_j \approx \mathbf{k}_c$) interact very strongly. The coupling coefficients κ_j denote the strength of the coupling and depends on the actual interaction mechanism. We have neglected the processes $b_j a + b_j^\dagger a^\dagger$ in which two quanta are annihilated and created since they are unimportant when the coupling is weak [1]. This neglect is called the rotating wave approximation (see Section 5.13). We may argue that they are unimportant as follows. In the absence of coupling the terms in (6.1.8) vary as $\exp \pm i(\omega_j - \omega_c)t$ which is approximately unity when $\omega_j \approx \omega_c$ while the omitted terms vary as $\exp \pm i(\omega_j + \omega_c)t$ which is rapidly varying. For times large compared with ω_c^{-1}, they will average approximately to zero and we neglect them.

The interaction energy (6.1.8) also describes loss of energy from an elastic mode of vibration in a solid into other modes.

To summarize, our reservoir may be thought of as any collection of a large number of quantum systems (many degrees of freedom) initially in thermal equilibrium. The system is weakly coupled to the reservoir and loses energy to the reservoir. The fluctuations in the reservoir also couples back into the system. Further properties of the reservoir needed to prevent the excitation energy initially in the system from returning completely from the reservoir back into the system are described in the next section.

6.2 THE MARKOFF APPROXIMATION IN THE SCHRÖDINGER PICTURE [2–5]

Reduced Density Operator for System with Loss; Master Equation

The density operator for a system coupled to a reservoir in the SP satisfies (6.1.3) where the total hamiltonian is given by (6.1.2). In general we are interested in the statistical properties of the system only. That is, if M is some function of the system operators only in the SP, we are often interested in its expectation value given by

$$\langle M(t)\rangle = \mathrm{Tr}_{R,S}\, M\rho(t) = \mathrm{Tr}_S\, M \mathrm{Tr}_R \rho(t), \tag{6.2.1}$$

where we must trace over both the system and reservoir. Here both $\rho(t)$ and M are in the SP. Since the system and reservoir variables are independent and since M is a function of system operators only, we may first trace $\rho(t)$ over the reservoir and then carry out the remaining trace over the system. We may therefore let

$$S(t) = \mathrm{Tr}_R\, \rho(t), \tag{6.2.2}$$

which will depend only on system operators since we have traced over all reservoir operators. The $S(t)$ is called the reduced density operator for the system in the SP (see Section 1.21). The one-time average (6.2.1) may then be written as

$$\langle M(t)\rangle = \mathrm{Tr}_S\, MS(t). \tag{6.2.3}$$

Therefore, we may obtain one-time averages of system operators from $S(t)$ without the necessity of knowing the full density operator $\rho(t)$. Of course, we could not obtain any reservoir averages from $S(t)$ but they are usually of no interest anyway. We would therefore like to remove the unnecessary information from (6.1.3) and obtain an equation of motion for $S(t)$ directly. For this purpose we first transform (6.1.3) to the IP to remove the rapidly varying unperturbed system motion from $\rho(t)$. We therefore let

$$\rho(t) = e^{-(i/\hbar)H_0(t-t_0)}\chi(t)e^{(i/\hbar)H_0(t-t_0)}, \tag{6.2.4a}$$

where $H_0 = H + R$ and $\rho(t)$ are in the SP and $\chi(t)$ is the density operator in the IP. Note that

$$\rho(t_0) = \chi(t_0) = S(t_0)f_0(R) \tag{6.2.4b}$$

so that the two pictures coincide at $t = t_0$. We have used (6.1.4) also. Since the system and reservoir are independent before coupling, their hamiltonians commute

$$[H, R] = 0.$$

In fact, since all system and reservoir operators commute at $t = t_0$ in the SP, they must commute whenever they are in the same picture. Accordingly, if we trace both sides of (6.2.4a) over the reservoir, we have

$$S(t) = e^{-(i/\hbar)H(t-t_0)}s(t)e^{(i/\hbar)H(t-t_0)}, \tag{6.2.5a}$$

where we used (6.2.2) and have let

$$s(t) = \mathrm{Tr}_R\, \chi(t) \tag{6.2.5b}$$

be the reduced density operator in the IP. We may see this as follows. In the R-representation, we have

$$R|R'\rangle = R'|R'\rangle$$

$$\langle R'|R''\rangle = \delta_{R',R''} \quad \text{or} \quad \delta(R' - R'') \tag{6.2.6}$$

$$\sum_{R'} |R'\rangle\langle R'| = 1 \quad \text{or} \quad \int |R'\rangle\, dR'\, \langle R'| = 1.$$

Then we have by (6.2.4a)

$$\begin{aligned}\mathrm{Tr}_R\, \rho(t) &= \sum_{R'} \langle R'|\rho(t)|R'\rangle = \sum_{R'} \langle R'|e^{-(i/\hbar)(H+R)(t-t_0)}\chi(t)e^{(i/\hbar)(H+R)(t-t_0)}|R'\rangle \\ &= \sum_{R'} e^{-(i/\hbar)R'(t-t_0)}\langle R'|e^{-(i/\hbar)H(t-t_0)}\chi(t)e^{(i/\hbar)H(t-t_0)}|R'\rangle e^{(i/\hbar)R'(t-t_0)} \\ &= e^{-(i/\hbar)H(t-t_0)} \sum_{R'} \langle R'|\chi(t)|R'\rangle e^{(i/\hbar)H(t-t_0)}\end{aligned} \tag{6.2.7}$$

The inner term in the last line is just $\mathrm{Tr}_R\, \chi(t)$.

We see by differentiating (6.2.5a) that

$$\frac{\partial S}{\partial t} = e^{-(i/\hbar)H(t-t_0)}\left\{\frac{1}{i\hbar}[H, s(t)] + \frac{\partial s}{\partial t}\right\}e^{(i/\hbar)H(t-t_0)}, \tag{6.2.8}$$

which relates the equation of motion for the reduced density operator in the two pictures.

We next obtain an equation of motion for $\chi(t)$. If we differentiate both sides of (6.2.4a) we obtain

$$i\hbar\frac{\partial \rho}{\partial t} = [H_0, \rho] + e^{-(i/\hbar)H_0(t-t_0)} i\hbar \frac{\partial \chi}{\partial t} e^{(i/\hbar)H_0(t-t_0)}. \tag{6.2.9}$$

When we compare this with (6.1.3) we see that

$$i\hbar e^{-(i/\hbar)H_0(t-t_0)} \frac{\partial \chi}{\partial t} e^{(i/\hbar)H_0(t-t_0)} = [V, \rho(t)] = V\rho - \rho V. \tag{6.2.10}$$

If we insert

$$e^{-(i/\hbar)H_0(t-t_0)}e^{+(i/\hbar)H_0(t-t_0)} = 1$$

between the V and ρ in both terms on the right above and again use (6.2.4a), we see that

$$\frac{\partial \chi}{\partial t} = \frac{1}{i\hbar}[V(t - t_0), \chi], \tag{6.2.11}$$

where we have let

$$V(t - t_0) = e^{(i/\hbar)H_0(t-t_0)} V e^{-(i/\hbar)H_0(t-t_0)} \tag{6.2.12}$$

be the interaction energy in the IP. The H_0 and V on the right are both in the SP. Equation (6.2.11) is therefore the equation of motion for the full density operator in the IP. Since in general we cannot solve it exactly, we assume that the interaction is weak and resort to perturbation theory. We iterate (6.2.11) up to second order in V as in Chapter 1. This gives

$$\chi(t) = \chi(t_0) + \frac{1}{i\hbar}\int_{t_0}^{t}[V(t'-t_0), \chi(t_0)]\,dt' + \left(\frac{1}{i\hbar}\right)^2\int_{t_0}^{t}dt'\int_{t_0}^{t'}dt''[V(t'-t_0), [V(t''-t_0), \chi(t_0)]] + \cdots. \tag{6.2.13}$$

Next we trace both sides over the reservoir. If we use (6.2.5b), (6.2.4b), and (6.1.7) this gives the reduced density operator in the IP to second order in the interaction:

$$s(t) - s(t_0) = \frac{1}{i\hbar}\int_{t_0}^{t}\mathrm{Tr}_R\,[V(t'-t_0), s(t_0)f_0(R)]\,dt' + \left(\frac{1}{i\hbar}\right)^2\int_{t_0}^{t}dt'\int_{t_0}^{t'}dt''\,\mathrm{Tr}_R\,[V(t'-t_0), [V(t''-t_0), s(t_0)f_0(R)]]. \tag{6.2.14}$$

To proceed, we assume that the interaction energy may be written as a sum of products of the form

$$V = \hbar\sum_i Q_iF_i, \tag{6.2.15}$$

where Q_i is a function of system operators only and F_i is a function of reservoir operators only and both are in the SP. For the example of the damped mode of the radiation field (6.1.8) we would make the identification

$$\begin{aligned} Q_1 &= a^\dagger & Q_2 &= a \\ F_1 &= \sum_j \kappa_j b_j & F_2 &= \sum_j \kappa_j^* b_j^\dagger. \end{aligned} \tag{6.2.16}$$

In the IP since system and reservoir operators commute, we have by (6.2.12) that

$$V(t-t_0) = \hbar\sum_i Q_i(t-t_0)F_i(t-t_0) \tag{6.2.17a}$$

$$\begin{aligned} Q_i(t-t_0) &= e^{(i/\hbar)H(t-t_0)}Q_ie^{-(i/\hbar)H(t-t_0)} \\ F_i(t-t_0) &= e^{(i/\hbar)R(t-t_0)}F_ie^{-(i/\hbar)R(t-t_0)}. \end{aligned} \tag{6.2.17b}$$

Note that

$$[Q_i(t'-t_0), F_j(t''-t_0)] = 0, \tag{6.2.18}$$

for all t' and t''. (Why?)

We next substitute (6.2.17) into (6.2.14). When we expand the double commutator and use the property (6.2.18), we obtain

$$\begin{aligned} s(t) - s(t_0) = -i\sum_i\int_{t_0}^{t}[Q_i', s(t_0)]\langle F_i'\rangle_R\,dt' \\ -\sum_{i,j}\int_{t_0}^{t}dt'\int_{t_0}^{t'}dt''\{[Q_i'Q_j''s(t_0) - Q_j''s(t_0)Q_i']\langle F_i'F_j''\rangle_R \\ - [Q_i's(t_0)Q_j'' - s(t_0)Q_j''Q_i']\langle F_j''F_i'\rangle_R\}, \end{aligned} \tag{6.2.19}$$

where for simplicity we have used primes to indicate $t' - t_0$ as a time argument and double primes to indicate $t'' - t_0$ as an argument. We also used the cyclic property of traces and (6.2.17b) to show that

$$\mathrm{Tr}_R f_0(R)F_i(t' - t_0) = \mathrm{Tr}_R f_0(R)F_i \tag{6.2.20a}$$

$$\begin{aligned}\mathrm{Tr}_R\, F_i'F_j''f_0(R) &= \mathrm{Tr}_R\, F_j''f_0(R)F_i' \\ &\equiv \langle F_i(t' - t_0)F_j(t'' - t_0)\rangle_R \\ &= \langle F_i(t' - t'')F_j\rangle_R \end{aligned}\tag{6.2.20b}$$

$$\begin{aligned}\mathrm{Tr}_R\, F_i'f_0(R)F_j'' &= \mathrm{Tr}_R f_0(R)F_j''F_i' \\ &\equiv \langle F_j(t'' - t_0)F_i(t' - t_0)\rangle_R \\ &= \langle F_jF_i(t' - t'')\rangle_R. \end{aligned}\tag{6.2.20c}$$

In Figure 6.2*a* we show the shaded area of integration in the double integral in (6.2.19). If we let

$$\tau = t' - t'' \qquad t' - t_0 = \tau + \xi \tag{6.2.21a}$$

the area transforms into that shown in Figure 6.2b. Since

$$dt'\,dt'' = d\tau\,d\xi, \tag{6.2.21b}$$

we have one using (6.2.20) and (6.2.21) in (6.2.19)

$$\begin{aligned} s(t) &- s(t_0) \\ &= -i\sum_i \langle F_i\rangle_R \int_{t_0}^{t} [Q_i(t' - t_0), s(t_0)]\,dt' - \sum_{i,j}\int_0^{t-t_0} d\xi \int_0^{t-t_0-\xi} d\tau \\ &\quad \times \{[Q_i(\tau + \xi)Q_j(\xi)s(t_0) - Q_j(\xi)s(t_0)Q_i(\tau + \xi)]\langle F_i(\tau)F_j\rangle_R \\ &\quad - [Q_i(\tau + \xi)s(t_0)Q_j(\xi) - s(t_0)Q_j(\xi)Q_i(\xi + \tau)]\langle F_jF_i(\tau)\rangle_R\}. \end{aligned}\tag{6.2.22}$$

We next assume that the Q_i are single system operators which are given in the IP by

$$Q_i(\lambda) = e^{(i/\hbar)H\lambda}Q_i^{\,S}e^{-(i/\hbar)H\lambda} = e^{i\omega_i\lambda}Q_i^{\,S}. \tag{6.2.23}$$

It is easy to generalize when this is not true, but it will hold for the examples we consider [see, (6.2.16)]. In this case (6.2.22) reduces to

$$\begin{aligned} s(t) - s(t_0) &= -i\sum_i \langle F_i\rangle_R[Q_i, s(t_0)]\int_0^{t-t_0} e^{i\omega_i\xi}\,d\xi \\ &\quad - \sum_{i,j}\int_0^{t-t_0} d\xi \Big\{[Q_iQ_js(t_0) - Q_js(t_0)Q_i]\int_0^{t-t_0-\xi} e^{i\omega_i\tau}\langle F_i(\tau)F_j\rangle_R\,d\tau \\ &\quad - [Q_is(t_0)Q_j - s(t_0)Q_jQ_i]\int_0^{t-t_0-\xi} e^{i\omega_i\tau}\langle F_jF_i(\tau)\rangle\,d\tau\Big\}e^{i(\omega_i+\omega_j)\xi}, \end{aligned}\tag{6.2.24}$$

where all system operators are in the SP. We also let $\xi = t' - t_0$ in the first integral above.

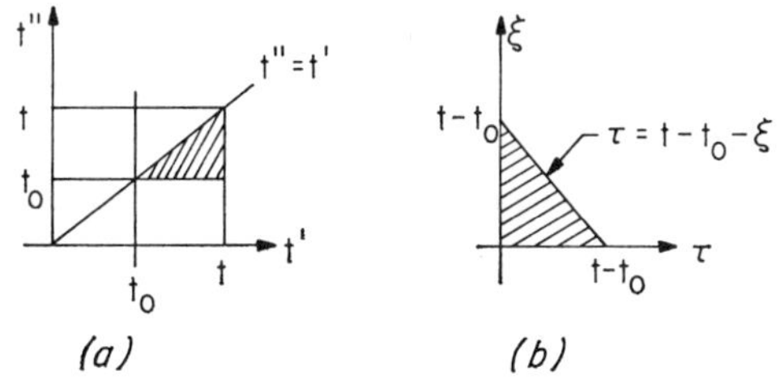

Figure 6.2 Region of integration in (6.2.22).

So far we have evaluated the reduced density operator in the IP to second order in the interaction. We would now like to make the Markoff approximation on $s(t)$ in the IP where we have removed the rapidly varying free motion of the system; $s(t) - s(t_0)$ represents the change in the reduced density operator during the time interval $t - t_0$. By definition, a system is Markoffian if its future is determined by the present and not its past [6]. In other words it loses all memory of its past. This is seen to be a sufficient condition to ensure that energy which goes into the reservoir will not return to the system. For otherwise, the system would develop memory.

Now in (6.2.24) we have two reservoir correlation functions $\langle F_i(\tau)F_j\rangle_R$ and $\langle F_jF_i(\tau)\rangle_R$. In Figure 6.3 we have sketched such a correlation function. Typically, it is nonzero over some time interval τ_c which is called the reservoir correlation time. As long as we require that $t - t_0 \gg \tau_c$, we may therefore extend the upper limits on the τ integrals to infinity with very little error. However, we require that $t - t_0$ be short compared with the system damping time, γ^{-1}. That is,

$$\tau_c \ll t - t_0 \ll \gamma^{-1}. \tag{6.2.25}$$

Since the bandwidth of a system is proportional to the reciprocal of the lifetime, the bandwidth of the reservoir must be large compared with the system bandwidth (see Appendix I). Physically we are smoothing out the fluctuations of the system on a time scale during which the reservoir is correlated but not on a scale during which the system is damped and so the system will lose its past memory on this time scale. Therefore, when (6.2.25) is satisfied (6.2.24)

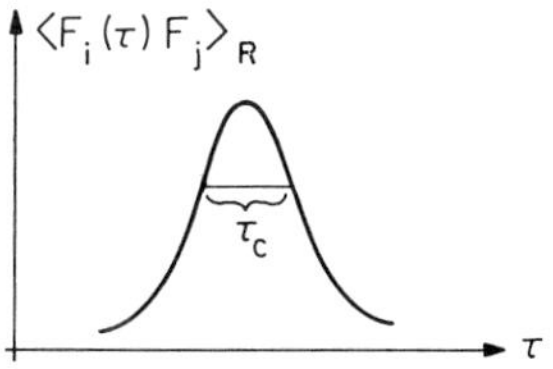

Figure 6.3 Typical reservoir correlation function behavior.

reduces to

$$s(t) - s(t_0) = -i \sum_i \langle F_i \rangle_R [Q_i, s(t_0)] I(\omega_i) - \sum_{i,j} \{[Q_i Q_j s(t_0) - Q_j s(t_0) Q_i] w_{ij}^+ - [Q_i s(t_0) Q_j - s(t_0) Q_j Q_i] w_{ji}^- \} I(\omega_i + \omega_j), \quad (6.2.26)$$

where we have let

$$w_{ij}^+ = \int_0^\infty e^{i\omega_i\tau} \langle F_i(\tau) F_j \rangle_R \, d\tau$$
$$w_{ji}^- = \int_0^\infty e^{i\omega_i\tau} \langle F_j F_i(\tau) \rangle_R \, d\tau \quad (6.2.27)$$

$$I(\omega_i) = \int_0^{t-t_0} e^{i\omega_i\xi} \, d\xi. \quad (6.2.28)$$

The $w^\pm$ are reservoir spectral densities (see Appendix I).

If we also require that

$$t - t_0 \gg \omega_i^{-1} \quad \text{or} \quad (\omega_i + \omega_j)^{-1}, \quad (6.2.29)$$

we are integrating over a time long compared with a period of the free motion of the system and we see that the integrand of (6.2.28) goes through many cycles and averages to zero unless $\omega_i = 0$. That is,

$$I(\omega_i) = (t - t_0)\delta(\omega_i, 0), \quad (6.2.30)$$

and

$$I(\omega_i + \omega_j) = (t - t_0)\delta(\omega_i, -\omega_j) \equiv \begin{cases} t - t_0 & \text{if} \quad \omega_i + \omega_j = 0 \\ 0 & \text{if} \quad \omega_i + \omega_j \neq 0. \end{cases} \quad (6.2.31)$$

Since none of our operators have zero frequency, we have that

$$I(\omega_i) = 0,$$

while (6.2.31) retains only the secular terms in (6.2.26). We therefore have that (6.2.26) reduces to

$$\frac{\Delta s}{\Delta t} = -\sum_{i,j} \delta(\omega_i, -\omega_j)\{[Q_i Q_j s(t_0) - Q_j s(t_0) Q_i] w_{ij}^+ - [Q_i s(t_0) Q_j - s(t_0) Q_j Q_i] w_{ji}^- \}, \quad (6.2.32)$$

where we have let

$$\Delta s = s(t) - s(t_0)$$
$$\tau_c \ll \Delta t = t - t_0 \ll \gamma^{-1}. \quad (6.2.33)$$

In the IP, the interval $\Delta t \to 0$ as far as system operators are concerned under the Markoff approximation. We therefore replace the exact (6.2.8) by the approximate

$$\frac{\partial S}{\partial t} \cong e^{-(i/\hbar)H(t-t_0)}\left[\lim_{t_0\to t}\left\{\frac{1}{i\hbar}[H, s(t)] + \frac{\Delta s}{\Delta t}\right\}\right]e^{+(i/\hbar)H(t-t_0)}$$

$$= e^{-(i/\hbar)H(t-t_0)}\left\{\frac{1}{i\hbar}[H, s(t)] - \sum_{i,j}\delta(\omega_i, -\omega_j)[(Q_iQ_js(t) - Q_js(t)Q_i)w_{ij}^{+}\right.$$

$$\left. - (Q_is(t)Q_j - s(t)Q_jQ_i)w_{ji}^{-}]\right\}e^{(i/\hbar)H(t-t_0)}, \quad (6.2.34a)$$

where we have used (6.2.32). If we insert between all factors

$$\exp + (i/h)H(t - t_0) \cdot \exp - (i/h)H(t - t_0) = 1,$$

use (6.2.23), and (6.2.5a), and take account of the fact that $\omega_i + \omega_j = 0$, we see that (6.2.34a) becomes

$$\frac{\partial S}{\partial t} \cong \frac{1}{i\hbar}[H, S(t)] - \sum_{i,j}\delta(\omega_i, -\omega_j)$$

$$\times \{(Q_iQ_jS(t) - Q_jS(t)Q_i)w_{ij}^{+} - [Q_iS(t)Q_j - S(t)Q_jQ_i]w_{ji}^{-}\}, \quad (6.2.34b)$$

which is the desired equation of motion for the reduced density operator under the Markoff approximation and we may think of all operators (only system operators appear) as being in the SP under the Markoff approximation. We could call this the Schrödinger-Markoff picture (SMP). The system operators obey the same commutation relations under this approximation as they did originally.

We note that the right side of (6.2.34b) no longer contains time integrals over $S(t')$ for times earlier than the present [as they did in Δs in (6.2.19)] so that the future is now indeed determined by the present. We have assumed that the reservoir correlation time is zero on a time scale in which the system loses an appreciable amount of its energy. Alternatively, the reservoir spectrum is infinite compared with the bandwidth of the system. We have smoothed the system fluctuations on a time scale in which the reservoir is correlated. One sometimes refers to the Markoff approximation as a coarse-grained averaging.

The reduced density operator equation (6.2.34b) is sometimes called the master equation. In the event the system hamiltonian has other terms such as W which do not depend on the reservoir, the master equation may be written as

$$\frac{\partial S}{\partial t} = \frac{1}{i\hbar}[H + W, S] - \sum_{i,j}\delta(\omega_i, -\omega_j)$$

$$\times \{[Q_iQ_jS - Q_jSQ_i]w_{ij}^{+} - [Q_iSQ_j - SQ_jQ_i]w_{ji}^{-}\}. \quad (6.2.35)$$

It should again be noted that all system operators in this form of the master equation are to be thought of as being in the SP. We therefore imagine that at $t = t_0$, the system is in the SP and propagates to time t according to (6.2.35). After the Markoff approximation, we can no longer derive (6.2.35) from a true hamiltonian according to an equation like (6.1.3).

We are free to make a transformation on (6.2.34) like

$$S(t) = e^{-(i/\hbar)H(t-t_0)}s(t)e^{(i/\hbar)H(t-t_0)}, \tag{6.2.36}$$

which is the IP on a time scale in which the reservoir correlation time is zero. In this case we see that (6.2.35) becomes

$$\frac{\partial s}{\partial t} = \frac{1}{i\hbar}[W^I, s(t)] - \sum_{i,j}\delta(\omega_i, -\omega_j)$$
$$\times \{[Q_i{}^I Q_j{}^I s - Q_j{}^I s Q_i{}^I]w_{ij}^+ - [Q_i{}^I s Q_j{}^I - s Q_j{}^I Q_i{}^I]w_{ji}^-\}, \tag{6.2.37}$$

where

$$\begin{aligned} Q_i^I &= e^{+(i/\hbar)H(t-t_0)}Q_i e^{-(i/\hbar)H(t-t_0)} \\ W^I &= e^{(i/\hbar)H(t-t_0)}W e^{-(i/\hbar)H(t-t_0)}, \end{aligned} \tag{6.2.38}$$

which corresponds to going to the interaction picture. We are not free to let $t - t_0$ approach zero here because on the present time scale, $t - t_0$ can be comparable to γ^{-1}.

Driven Damped Oscillator

Let us apply the analysis above to obtain the master equation for the reduced density operator in the SMP for a driven damped mode of the radiation field in a cavity. The unperturbed hamiltonian is

$$H_0 = \hbar\omega_c a^\dagger a + \sum_j \hbar\omega_j b_j^\dagger b_j. \tag{6.2.39}$$

We take the interaction energy to be (6.1.8)

$$V = \hbar\sum_l (\kappa_l b_l a^\dagger + \kappa_l^* b_l^\dagger a). \tag{6.2.40}$$

In the notation of (6.2.16), we have

$$\begin{aligned} Q_1 &= a^\dagger & Q_2 &= a = Q_1^\dagger \\ F_1 &= \sum \kappa_l b_l & F_2 &= \sum \kappa_l^* b_l^\dagger = F_1^\dagger. \end{aligned} \tag{6.2.41}$$

We take the driving term to be

$$W = \hbar[v(t)a^\dagger + v^*(t)a], \tag{6.2.42}$$

where $v(t)$ is a classical arbitrary function which excites the ω_c-mode by an external generator. In the IMP we easily see by (6.2.38) that

$$Q_1{}^I = a^\dagger e^{i\omega_c(t-t_0)} = Q_2{}^{\dagger I}$$
$$F_1{}^I = \sum_l \kappa_l b_l e^{-i\omega_l(t-t_0)} = F_2{}^{\dagger I} \tag{6.2.43}$$

$$W^I = \hbar[a^\dagger v(t)e^{i\omega_c(t-t_0)} + av^*(t)e^{-i\omega_c(t-t_0)}]. \tag{6.2.44}$$

Our first task is to compute the reservoir spectral densities w_{ij}^+ and w_{ji}^-. We calculate one in detail and leave the others as an exercise. We have by (6.2.27) and (6.2.43) that

$$w_{21}^+ = \int_0^\infty e^{i\omega_2\tau}\langle F_2(\tau)F_1\rangle_R\, d\tau. \tag{6.2.45}$$

In the notation of (6.2.43), we see that

$$\omega_2 = -\omega_c. \tag{6.2.46}$$

Also by (6.2.43)

$$\langle F_2(\tau)F_1\rangle_R = \left\langle \sum_l \kappa_l^* e^{i\omega_l\tau}\, b_l^\dagger \sum_m \kappa_m b_m \right\rangle_R$$
$$= \sum_{l,m} \kappa_l^* \kappa_m e^{i\omega_l\tau}\langle b_l^\dagger b_m\rangle_R. \tag{6.2.47}$$

But

$$\langle b_l^\dagger b_m\rangle_R = \frac{\mathrm{Tr}_R\, e^{-\beta R} b_l^\dagger b_m}{\mathrm{Tr}_R\, e^{-\beta R}}, \tag{6.2.48}$$

where $R = \sum_j \hbar\omega_j b_j^\dagger b_j$. The reader may show easily that

$$\langle b_l^\dagger b_m\rangle = \delta_{lm}\bar{n}_l, \tag{6.2.49a}$$

where

$$\bar{n}_l = \frac{1}{e^{\lambda_l} - 1} \tag{6.2.49b}$$

and

$$\lambda_l = \frac{\hbar\omega_l}{kT}. \tag{6.2.49c}$$

Therefore, (6.2.45) becomes on using these results

$$w_{21}^+ = \sum_l |\kappa_l|^2\, \bar{n}_l \int_0^\infty e^{i(\omega_l-\omega_c)\tau}\, d\tau, \tag{6.2.50}$$

where we interchanged the order of summation and integration. Since we assume the reservoir modes are closely spaced with $g(\omega_l)\, d\omega_l$ the number of modes between ω_l and $\omega_l + d\omega_l$, we may change the sum to an integral

$$\sum_l \{\quad\} \to \int_0^\infty d\omega_l\, g(\omega_l)\{\quad\}, \tag{6.2.51}$$

so that (6.2.50) becomes

$$w_{21}^{+} = \int_0^\infty d\omega_l\, g(\omega_l)\, |\kappa(\omega_l)|^2\, \bar{n}(\omega_l) I, \tag{6.2.52a}$$

where we let

$$I = \int_0^\infty e^{i(\omega_l - \omega_c)\tau}\, d\tau. \tag{6.2.52b}$$

Now

$$\int_0^\infty e^{\pm i\Omega\tau}\, d\tau = \pi\delta(\Omega) \pm i\mathscr{P}\frac{1}{\Omega}, \tag{6.2.53}$$

where $\mathscr{P}$ is the Cauchy principle part defined by

$$\int_{-a}^{+b} \mathscr{P}\frac{f(\Omega)}{\Omega}\, d\Omega = \lim_{\epsilon\to 0}\left\{\int_{-a}^{-\epsilon} \frac{f(\Omega)}{\Omega}\, d\Omega + \int_{\epsilon}^{b} \frac{f(\Omega)\, d\Omega}{\Omega}\right\}. \tag{6.2.54}$$

Therefore, (6.2.52a) becomes

$$\begin{aligned} w_{21}^{+} &= \int_0^\infty d\omega_l\, g(\omega_l)\, |\kappa(\omega_l)|^2\, \bar{n}(\omega_l)\left\{\pi\delta(\omega_l - \omega_c) + i\mathscr{P}\frac{1}{\omega_l - \omega_c}\right\} \\ &\equiv \left(\frac{\gamma}{2} - i\,\Delta\omega\right)\bar{n}, \end{aligned} \tag{6.2.55}$$

where we have let

$$\begin{aligned} \gamma &= 2\pi g(\omega_c)\, |\kappa(\omega_c)|^2 \\ \Delta\omega &= -\mathscr{P}\int_0^\infty \frac{d\omega_l\, g(\omega_l)\, |\kappa(\omega_l)|^2}{\omega_l - \omega_c}, \end{aligned} \tag{6.2.56}$$

and

$$\bar{n} = \frac{1}{e^{\hbar\omega_c/kT} - 1}. \tag{6.2.57}$$

The imaginary part of w_{21}^{+} represents the Lamb shift in the cavity frequency due to the coupling to the reservoir and may be generally neglected (see Section 5.3). The remaining w's are found by the same type arguments and are given by

$$\begin{aligned} w_{21}^{+} &= \left(\frac{\gamma}{2} - i\,\Delta\omega\right)\bar{n} = w_{21}^{-*} \\ w_{12}^{+} &= \left(\frac{\gamma}{2} + i\,\Delta\omega\right)(1 + \bar{n}) = w_{12}^{-*} \\ w_{11}^{\pm} &= w_{22}^{\pm} = 0. \end{aligned} \tag{6.2.58}$$

If we use these and (6.2.41) and (6.2.42) in (6.2.35), we find after minor algebra that

$$\frac{\partial S}{\partial t} = -i(\omega_c + \Delta\omega)[a^\dagger a, S] - iv(t)[a^\dagger, S] - iv^*(t)[a, S]$$
$$+ \frac{\gamma}{2}[2aSa^\dagger - a^\dagger aS - Sa^\dagger a] + \gamma\bar{n}[a^\dagger Sa + aSa^\dagger - a^\dagger aS - Saa^\dagger] \tag{6.2.59}$$

We therefore see that the only effect of $\Delta\omega$ is to change slightly the cavity resonant frequency ω_c. The $\Delta\omega$ is proportional to $|\kappa|^2$ by (6.2.56), which is the Lamb shift. We may redefine $\omega_c + \Delta\omega$ as ω_c' or neglect it. All operators in (6.2.59) are in the SMP.

The first term in (6.2.59) represents the free motion of the cavity mode. The next two terms are due to an external classical generator which drives or pumps the mode. The γ terms represent the loss of energy from the system to the reservoir, while the $\gamma\bar{n}$ terms represent the diffusion of fluctuations in the reservoir into the system mode. Note by (6.2.57) that as the reservoir approaches absolute zero, $\bar{n} \to 0$.

We may transform (6.2.59) to the IMP. If we let

$$H' = \hbar(\omega_c + \Delta\omega)a^\dagger a \equiv h\omega_c' a^\dagger a, \tag{6.2.60}$$

then (6.2.59) becomes by (6.2.36) in the IMP.

$$\frac{\partial s}{\partial t} = -iv(t)[a^I, s] - iv^*(t)[a^{\dagger I}, s]$$
$$+ \frac{\gamma}{2}[2asa^\dagger - a^\dagger as - sa^\dagger a] + \gamma\bar{n}[a^\dagger sa + asa^\dagger - a^\dagger as - saa^\dagger], \tag{6.2.61}$$

where all operators are now in the IMP. We omitted the superscript I in the γ and $\gamma\bar{n}$ terms for simplicity since all these terms contain both an a and an $a^\dagger$, and we note that $a^I a^{\dagger I} = aa^\dagger$ and no confusion is likely to arise.

The master equation (6.2.61) may be solved by various techniques [2]. We convert it to an associated c-number equation and solve it in a later section.

Single Atom with Linewidth [7] Pauli Equations

As a second example of a damped system, we derive the master equation for a single atom coupled to a reservoir and show how the interaction with the damping mechanism results in an atomic linewidth. If the reservoir consists of the modes of a radiation field in a cavity, we obtain a linewidth in agreement with the Wigner-Weisskopff approximation for the natural linewidth of the atom (see Section 5.3).

We may use the completeness relation for the energy eigenvectors for the free atom to represent the hamiltonian by

$$H = \sum \epsilon_l |l\rangle\langle l|, \tag{6.2.62}$$

where the kets $|l\rangle$ are the atomic energy eigenkets

$$\begin{aligned} H|l\rangle &= \epsilon_l |l\rangle \\ \langle l|m\rangle &= \delta_{lm} \\ \sum_l |l\rangle\langle l| &= 1. \end{aligned} \tag{6.2.63}$$

The reservoir is described by the hamiltonian R and V is the interaction energy between the atom and reservoir. We may use the atomic completeness relation twice to write V as

$$V = \hbar \sum_{k,l} f_{kl} |k\rangle\langle l|, \tag{6.2.64}$$

where

$$\hbar f_{kl} = \langle k|V|l\rangle \tag{6.2.65}$$

are the matrix elements of V in the atom energy representation. They will obviously contain reservoir operators.

As an example, let the reservoir be the radiation field in a cavity in which the atom is located. In Appendix J it is shown that in the dipole approximation the atom field interaction is given by

$$\begin{aligned} V &= -e\mathbf{r} \cdot \mathbf{E} \equiv -\boldsymbol{\mu} \cdot \mathbf{E} \\ &= -i \sum_{l,\sigma} \sqrt{\frac{\hbar\omega_l}{2\epsilon_0 L^3}} (b_{l\sigma} - b_{l\sigma}^{\dagger})(\hat{e}_{l\sigma} \cdot \boldsymbol{\mu}), \end{aligned} \tag{6.2.66}$$

where we have used (4.3.44a). In the SP, the field operators are evaluated at $t = 0$ and under the dipole approximation $\exp \pm i\mathbf{k}_l \cdot \mathbf{r} \simeq 1$. Thus

$$\hbar f_{kl} = -i \sum_j \sqrt{\frac{\hbar\omega_j}{2\epsilon_0 L^3}} (b_j - b_j^{\dagger})(\hat{e}_j \cdot \boldsymbol{\mu}_{kl}) \equiv \langle k|V|l\rangle, \tag{6.2.67}$$

where we have let $j \equiv (l_1, l_2, l_3, \sigma)$ and $\boldsymbol{\mu}_{kl}$ are the atomic dipole moment matrix elements. We see that f_{kl} contains the reservoir operators b_j and $b_j^{\dagger}$ explicitly as we noted. In this case we have that

$$R = \sum_j \hbar\omega_j b_j^{\dagger} b_j. \tag{6.2.68}$$

In order that V be hermitian, we see by (6.2.64) that

$$V^\dagger = \hbar \sum_{kl} f_{kl}^\dagger |l\rangle\langle k| \equiv \sum_{kl} f_{lk}^\dagger |k\rangle\langle l|$$
$$= V = \sum_{k,l} f_{kl} |k\rangle\langle l|.$$

It therefore follows that

$$f_{kl} = f_{lk}^\dagger. \tag{6.2.69}$$

In the IP we have that

$$V(t) = e^{(i/\hbar)(H+R)t} V e^{-(i/\hbar)(H+R)t}$$
$$= \hbar \sum_{k,l} f_{kl}(t) |k\rangle\langle l| e^{i\omega_{kl} t}, \tag{6.2.70}$$

where we used (3.10.14) and have let

$$f_{kl}(t) = e^{(i/\hbar)Rt} f_{kl} e^{-(i/\hbar)Rt}, \tag{6.2.71}$$

and

$$\hbar\omega_{kl} = \epsilon_k - \epsilon_l, \tag{6.2.72}$$

and have let $t_0 = 0$ for simplicity.

To write the master equation (6.2.37) (here $W = 0$), we make the following identifications according to the notation of (6.2.15):

$$\begin{aligned} &Q_i \to |k\rangle\langle l| \qquad && Q_j \to |m\rangle\langle n| \\ &F_i \to f_{kl} && F_j \to f_{mn}. \end{aligned} \tag{6.2.73}$$

In the IP the F's are given by (6.2.71) and

$$\begin{aligned} &Q_i(t) \to |k\rangle\langle l|^{i\omega_{kl} t} \\ &Q_j(t) \to |m\rangle\langle n| e^{i\omega_{mn} t}. \end{aligned} \tag{6.2.74}$$

The coefficients $w^\pm$ of (6.2.27) become

$$\begin{aligned} w_{klmn}^+ &= \int_0^\infty e^{i\omega_{kl}\tau} \langle f_{kl}(\tau) f_{mn} \rangle_R \, d\tau \\ w_{mnkl}^- &= \int_0^\infty e^{i\omega_{kl}\tau} \langle f_{mn} f_{kl}(\tau) \rangle_R \, d\tau. \end{aligned} \tag{6.2.75a}$$

We leave as an exercise to show that

$$w_{mnkl}^- = (w_{lknm}^+)^*. \tag{6.2.75b}$$

With these associations, the master equation (6.2.37) becomes

$$\frac{\partial s}{\partial t} = -\sum_{klmn} \delta(\omega_{kl}, -\omega_{mn}) \{[|k\rangle\langle n| \delta_{lm} s - |m\rangle\langle n| s |k\rangle\langle l|] w_{klmn}^+$$
$$- [|k\rangle\langle l| s |m\rangle\langle n| - s|m\rangle\langle l| \delta_{nk}] w_{mnkl}^- \}, \tag{6.2.76}$$

where we used the orthogonality relations (6.2.63). The secular terms are those for which

$$\hbar(\omega_{kl} + \omega_{mn}) = \epsilon_k - \epsilon_l + \epsilon_m - \epsilon_n = 0, \tag{6.2.77}$$

so that the atomic operators are the same in the IP and SP. If we assume the atomic energy levels are nondegenerate and unevenly spaced, the secular terms arise in the following three cases:

Case 1:	$k = n;$	$l = m$	$k \neq l$
Case 2:	$k = l;$	$m = n$	$k \neq m$
Case 3:	$k = l = m = n.$		

In Case 1, the secular terms in (6.2.76) are

$$\frac{\partial s^{(1)}}{\partial t} = \sum_{k,l}{}' \{[|l\rangle\langle k|s^{(1)}|k\rangle\langle l| - |k\rangle\langle k|s^{(1)}]w_{kllk}^{+} + [|k\rangle\langle l|s^{(1)}|l\rangle\langle k| - s^{(1)}|l\rangle\langle l|[w_{lkkl}^{-}\}. \tag{6.2.78}$$

The prime on the sum indicates the $k = l$ term is omitted. We may interchange the dummy indices k and l in the last two terms. This gives

$$\frac{\partial s^{(1)}}{\partial t} = \sum_{k,l} \{|l\rangle\langle k|s^{(1)}|k\rangle\langle l|(w_{kllk}^{+} + w_{kllk}^{-}) - |k\rangle\langle k|s^{(1)}w_{kllk}^{+} - s|k\rangle\langle k|w_{kllk}^{-}\}. \tag{6.2.79}$$

In Case 2 the secular terms are

$$\begin{aligned}\frac{\partial s^{(2)}}{\partial t} &= \sum_{k,m}{}' \{|m\rangle\langle m|s^{(2)}|k\rangle\langle k|w_{kkmm}^{+} + |k\rangle\langle k|s^{(2)}|m\rangle\langle m|w_{mmkk}^{-}\} \\ &= \sum_{k,m}{}' \{|m\rangle\langle m|s^{(2)}|k\rangle\langle k|(w_{kkmm}^{+} + w_{kkmm}^{-})\} \\ &\equiv \sum_{k,l}{}' \{|k\rangle\langle k|s^{(2)}|l\rangle\langle l|(w_{llkk}^{+} + w_{llkk}^{-}),\end{aligned} \tag{6.2.80}$$

where again we interchanged dummy indices in the second as well as the third sums.

In Case 3, we have

$$\frac{\partial s^{(3)}}{\partial t} = \sum_{m} \{[|m\rangle\langle m|s^{(3)}|m\rangle\langle m| - |m\rangle\langle m|s^{(3)}]w_{mmmm}^{+} + [|m\rangle\langle m|s^{(3)}|m\rangle\langle m| - s^{(3)}|m\rangle\langle m|]w_{mmmm}^{-}\}. \tag{6.2.81}$$

We see that the $k = l$ terms in (6.2.79) are identical to (6.2.81). If we therefore remove the prime on the sums in (6.2.79), (6.2.81) is automatically included, and if we added this to (6.2.80), we obtain

$$\frac{\partial s}{\partial t} = \sum_{k,l} \{|l\rangle\langle k|s|k\rangle\langle l|w_{lk} - |k\rangle\langle k|sw_{kllk}^{+} - s|k\rangle\langle k|w_{kllk}^{-}\} + \sum_{k,l}{}' |k\rangle\langle k|s|l\rangle\langle l|(w_{llkk}^{+} + w_{llkk}^{-}), \tag{6.2.82}$$

where we have let

$$w_{lk} = w^{+}_{kllk} + w^{-}_{kllk}. \tag{6.2.83}$$

From (6.2.75) it follows that

$$w_{lk} = w^{+}_{kllk} + (w^{+}_{kllk})^{*}, \tag{6.2.84}$$

so we see that the w_{lk} are real.

By inspection we see that the $k = l$ term in the first double sum is just the missing $k = l$ term in the last double sum by means of (6.2.83). Accordingly, we may write (6.2.82) in the alternative equivalent form

$$\frac{\partial s}{\partial t} = \sum_{k,l}{}' w_{lk}|l\rangle\langle k|s|k\rangle\langle l| + \sum_{k,l}\{|k\rangle\langle k|s|l\rangle\langle l|(w^{+}_{llkk} + w^{-}_{llkk}) - w^{+}_{kllk}|k\rangle\langle k|s - w^{-}_{kllk}s|k\rangle\langle k|\}. \tag{6.2.85}$$

If we take the j, i matrix element of both sides of this equation, use the orthogonality relations (6.2.63), and let

$$s_{ji} \equiv \langle j|s|i\rangle, \tag{6.2.86}$$

and

$$\Gamma_{ij}{}^{c} = \sum_{l}(w^{+}_{jllj} + w^{-}_{illi}) - w^{+}_{iijj} - w^{-}_{iijj}, \tag{6.2.87a}$$

we obtain

$$\frac{\partial s_{ji}}{\partial t} = \delta_{ij}\sum_{k}{}' w_{ik}s_{kk} - \Gamma_{ij}{}^{c}s_{ji}, \tag{6.2.88}$$

which are called the Pauli master equations. Note that

$$\Gamma_{ij}{}^{c} = \Gamma_{ji}{}^{c*}, \tag{6.2.87b}$$

which follows from (6.2.75b). In the SP, the Pauli master equation becomes

$$\frac{\partial S_{ji}}{\partial t} = \delta_{ij}\sum_{k}{}' w_{ik}S_{kk} - (\Gamma_{ij}{}^{c} - i\omega_{ij})S_{ji}, \tag{6.2.89}$$

as the reader may readily show.

Before discussing the Pauli equations, let us consider the constants w_{ik} and $\Gamma_{ij}{}^{c}$ to obtain some physical insight as to their significance.

From (6.2.84), (6.2.75), and (6.2.69), the reader may show that

$$w_{ik} = \int_{-\infty}^{\infty} e^{i\omega_{ki}\tau}\langle f_{ki}(\tau)f_{ik}\rangle_{R}\, d\tau. \tag{6.2.90}$$

We shall evaluate the trace in the R-representation where

$$b_j^{\dagger}b_j|n_j\rangle = n_j|n_j\rangle, \tag{6.2.91}$$

and j stands for l_1, l_2, l_3 and the polarization σ. We have

$$w_{ik} = \int_{-\infty}^{\infty} d\tau\, e^{i\omega_{ki}\tau}\sum_{\{n\}}\sum_{\{n'\}}\langle\{n\}|e^{iR\tau/\hbar}f_{ki}e^{-iR\tau/\hbar}|\{n'\}\rangle\langle\{n'\}|f_{ik}e^{-\beta R}|\{n\}\rangle[\mathrm{Tr}_{R}\, e^{-\beta R}]^{-1}, \tag{6.2.92}$$

where $|\{n\}\rangle \equiv |n_1, n_2, \ldots, n_\infty\rangle$,

$$\sum_{\{n\}} \equiv \sum_{n_1=0}^{\infty} \sum_{n_2=0}^{\infty} \cdots \sum_{n_\infty=0}^{\infty} \tag{6.2.93}$$

and we have inserted a completeness relation before f_{ik}. We also used (6.2.71). Since

$$g(R)|\{n\}\rangle = g\left[\sum_j \hbar\omega_j n_j\right]|\{n\}\rangle, \tag{6.2.94}$$

we have that

$$w_{ik} = \sum_{\{n\}} \sum_{\{n'\}} \left\{ \langle\{n\}|f_{ki}|\{n'\}\rangle\langle\{n'\}|f_{ik}|\{n\}\rangle \right. \\ \left. \times \left[\prod_l (1 - e^{-\lambda_l})e^{-\lambda_l n_l}\right] \int_{-\infty}^{\infty} \exp i\left[\omega_{ki} + \sum_m \omega_m(n_m - n'_m)\right]\tau \, d\tau, \right. \tag{6.2.95}$$

where

$$\lambda_l = \hbar\omega_l/kT. \tag{6.2.96}$$

If we note that

$$\int_{-\infty}^{\infty} e^{ix\tau} \, d\tau = 2\pi\delta(x), \tag{6.2.97}$$

and use (6.2.69), we obtain

$$w_{ik} = 2\pi \sum_{\{n\}} \sum_{\{n'\}} \left\{ |\langle\{n'\}|f_{ik}|\{n\}\rangle|^2 \left[\prod_l (1 - e^{-\lambda_l})e^{-\lambda_l n_l}\right] \right. \\ \left. \times \delta\left[\omega_{ki} + \sum_m \omega_m(n_m - n'_m)\right]\right\}. \tag{6.2.98}$$

The δ-function is just a statement of energy conservation between the atom and field:

$$\epsilon_i + \sum_m \hbar\omega_m n'_m = \epsilon_k + \sum_m \hbar\omega_m n_m. \tag{6.2.99}$$

If we next insert the interaction (6.2.67), we obtain

$$\begin{aligned} w_{ik} &= \frac{2\pi\hbar}{\epsilon_0 L^3} \sum_{\{n\}} \sum_{\{n'\}} \left\{ \left| \sum_j \sqrt{\omega_j}\,(\hat{e}_j \cdot \boldsymbol{\mu}_{ik})\langle\{n'\}|(b_j - b_j^\dagger)|\{n\}\rangle \right.\right. \\ &\qquad \left.\left. \times \left[\prod_l (1 - e^{-\lambda_l})e^{-\lambda_l n_l}\right] \delta\left[\omega_{ki} + \sum_m \omega_m(n_m - n'_m)\right]\right\} \right. \\ &= \frac{2\pi\hbar}{\epsilon_0 L^3} \sum_{\{n\}} \sum_{\{n'\}} \left\{ \left| \sum_j \sqrt{\omega_j}\,(\hat{e}_j \cdot \boldsymbol{\mu}_{ik})[\sqrt{n_j}\langle n'_1 \cdots n'_j \cdots | n_1 \cdots n_j - 1 \cdots\rangle \right.\right. \\ &\qquad \left. - \sqrt{n_j + 1}\,\langle n'_1 \cdots n'_j \cdots | n_j \cdots n_j + 1 \cdots\rangle] \right|^2 \\ &\qquad \left. \times \left[\prod_l (1 - e^{-\lambda_l})e^{-\lambda_l n_l}\right] \delta\left[\omega_{ki} + \sum_m \omega_m(n_m - n'_m)\right]\right\}. \end{aligned} \tag{6.2.100}$$

If we use the orthogonality relations, the reader may show that this reduces to

$$w_{ik} = \frac{2\pi\hbar}{\epsilon_0 L^3} \sum_{\{n\}} \sum_{\{n'\}} \Big\{ \sum_j \omega_j \, |\hat{e}_j \cdot \boldsymbol{\mu}_{ik}|^2 \, [n_j \langle n_1' \cdots n_j' \cdots | n_1 \cdots n_j - 1 \cdots \rangle$$
$$+ (n_j + 1) \langle n_1' \cdots n_j' \cdots | n_1 \cdots n_j + 1 \cdots \rangle]$$
$$\times \prod_l (1 - e^{-\lambda_l}) e^{-\lambda_l n_l} \delta \Big[\omega_{ki} + \sum_m \omega_m (n_m - n_m') \Big] \Big\}. \quad (6.2.101)$$

Next we may carry out the sums on all the $\{n'\}$. This gives because of the orthogonality relations

$$w_{ik} = \frac{2\pi\hbar}{\epsilon_0 L^3} \prod_l \sum_{\{n\}} \Big\{ \Big[\sum_j \omega_j \, |\hat{e}_j \cdot \boldsymbol{\mu}_{ik}|^2$$
$$\times \{ n_j \delta(\omega_{ki} + \omega_j) + (n_j + 1) \delta(\omega_{ki} - \omega_j) \Big] (1 - e^{-\lambda_l}) e^{-\lambda_l n_l} \Big\}. \quad (6.2.102)$$

We may next carry out the sum on all $\{n\}$. When $n_l \neq n_j$, we have

$$(1 - e^{-\lambda_l}) \sum_{n_l=0}^{\infty} e^{-\lambda_l n_l} = 1, \quad (6.2.103)$$

while for the jth term, we have

$$(1 - e^{-\lambda_j}) \sum_{n_j=0}^{\infty} n_j e^{-\lambda_j n_j} \equiv \bar{n}_j = [e^{\lambda_j} - 1]^{-1}. \quad (6.2.104)$$

Therefore,

$$w_{ik} = \frac{2\pi\hbar}{\epsilon_0 L^3} \sum_{l,\sigma} \omega_l \, |\hat{e}_{l\sigma} \cdot \boldsymbol{\mu}_{ik}|^2 \, \{ \bar{n}_l \delta(\omega_{ki} + \omega_l) + (\bar{n}_l + 1) \delta(\omega_{ki} - \omega_l) \}. \quad (6.2.105)$$

We have restored the sum over modes and polarizations.

Since $\omega_l > 0$, we have two cases: $\omega_{ki} < 0$ which corresponds to the absorption of radiation by the atom and the atom goes from state k to state i of higher energy and $\omega_{ki} > 0$ which corresponds to the atom going from state k to a state of lower energy i with the emission of radiation. Thus for $\epsilon_i > \epsilon_k$

$$w_{ik}^{\text{abs}} = \frac{2\tau\hbar}{\epsilon_0 L^3} \sum_{l,\sigma} \omega_l \, |\hat{e}_{l\sigma} \cdot \boldsymbol{\mu}_{ik}|^2 \, \bar{n}(\omega_l) \delta(\omega_{ik} + \omega_l) \quad (6.2.106a)$$

is the transition probability per second for the absorption of radiation while for $\epsilon_i < \epsilon_k$

$$w_{ik}^{\text{emiss}} = \frac{2\pi\hbar}{\epsilon_0 L^3} \sum_{l,\sigma} \omega_l \, |\hat{e}_{l\sigma} \cdot \boldsymbol{\mu}_{ik}|^2 \, [1 + \bar{n}(\omega_l)] \delta(\omega_{ki} - \omega_l) \quad (6.2.106b)$$

gives the transition probability per second for spontaneous and induced emission. These should be compared with the results of Section 5.2.

Consider next the constants $\Gamma_{ij}{}^c$. These will in general be complex and we may write them as

$$\Gamma_{ij}{}^c \equiv \Gamma_{ij} - i\,\Delta\omega_{ij} \tag{6.2.107}$$

where we separate the real and imaginary parts. We see by (6.2.89) that the imaginary part will cause a shift in the atomic frequencies. Since $\Delta\omega_{ij}$ is small of order V^2 in the perturbation, we either neglect it or define new atomic frequencies including the shift as we did in the case of the damped oscillator. This is just the Lamb shift.

If we use (6.2.87a) and (6.2.75a) we have

$$\Gamma_{ij}{}^c = \int_0^\infty d\tau\,\{-\langle f_{ii}(\tau)f_{jj}\rangle_R - \langle f_{ii}f_{jj}(\tau)\rangle_R + \sum_l [e^{i\omega_{jl}\tau}\langle f_{jl}(\tau)f_{lj}\rangle_R + e^{i\omega_{li}\tau}\langle f_{il}f_{li}(\tau)\rangle_R]\}. \tag{6.2.108}$$

If we again evaluate the traces in the R-representation and insert a completeness relation in each term, we have

$$\begin{aligned}\Gamma_{ij}{}^c = \int_0^\infty d\tau \sum_{\{n\}}\sum_{\{n'\}}\Bigg\{&-\langle\{n\}|f_{ii}|\{n'\}\rangle\langle\{n'\}|f_{jj}|\{n\}\rangle\left[\exp i\sum_i \omega_i(n_i - n_i')\tau + \text{cc}\right]\\ &+\sum_l [|\langle\{n\}|f_{jl}|\{n'\}\rangle|^2]\exp i\left[\omega_{jl} + \sum_i \omega_i(n_i - n_i')\right]\tau\\ &+|\langle\{n\}|f_{il}|\{n'\}\rangle|^2 \exp i\left[\omega_{li} - \sum_i \omega_i(n_i - n_i')\tau\right]\Bigg\}\\ &\times\left[\prod_m (1 - e^{-\lambda_m})e^{-\lambda_m n_m}\right]\end{aligned} \tag{6.2.109}$$

where cc means complex conjugate. Since

$$\begin{aligned}\int_0^\infty d\tau\,(e^{ix\tau} + e^{-ix\tau}) &= 2\pi\delta(x)\\ \int_0^\infty d\tau\, e^{\pm ix\tau} &= \pi\delta(x) \pm i\mathscr{P}\frac{1}{x},\end{aligned} \tag{6.2.110}$$

we have for $\text{Re}\,(\Gamma_{ij}{}^c) = \Gamma_{ij}$

$$\begin{aligned}\Gamma_{ij} = 2\pi\sum_{\{n\}}\sum_{\{n'\}}\Bigg\{&-\langle\{n\}|f_{ii}|\{n'\}\rangle\langle\{n'\}|f_{jj}|\{n\}\rangle\delta\left[\sum_m \omega_m(n_m - n_m')\right]\\ &+\frac{1}{2}\sum_l [|\langle\{n\}|f_{jl}|\{n'\}\rangle|^2]\delta\left[\omega_{jl} + \sum_m \omega_m(n_m - n_m')\right]\\ &+|\langle\{n\}|f_{il}|\{n'\}\rangle|^2\,\delta\left[\omega_{li} - \sum_m \omega_m(n_m - n_m')\right]\Bigg\}\\ &\times\left[\prod_r (1 - e^{-\lambda_r})e^{-\lambda_r n_r}\right],\end{aligned} \tag{6.2.111}$$

while the Lamb shifts are

$$\Delta\omega_{ij} = -\mathscr{P}\sum_{\{n\}}\sum_{\{n'\}}\sum_{l}\left\{\frac{|\langle\{n\}|f_{jl}|\{n'\}\rangle|^2}{\omega_{jl} + \sum_m \omega_m(n_m - n'_m)} + \frac{|\langle\{n\}|f_{il}|\{n'\}\rangle|^2}{\omega_{li} - \sum_m \omega_m(n_m - n'_m)}\right\}\left[\prod_r (1 - e^{-\lambda_r})e^{-\lambda_r n_r}\right]. \quad (6.2.112)$$

If we combine the $l = j$ term from the first l-sum in Γ_{ij} above and the $l = i$ term from the second l-sum with the first term and note that $\omega_{ii} = \omega_{jj} = 0$, we may rewrite Γ_{ij} as

$$\begin{aligned}\Gamma_{ij} = \pi\sum_{\{n\}}\sum_{\{n'\}}\Bigg\{&|\langle\{n\}|(f_{ii} - f_{jj})|\{n'\}\rangle|^2\,\delta\left[\sum_m \omega_m(n_m - n'_m)\right]\\ &+\sum_{l\neq j}|\langle\{n\}|f_{jl}|\{n'\}\rangle|^2\,\delta\left[\omega_{jl} + \sum_m \omega_m(n_m - n'_m)\right]\\ &+\sum_{l\neq i}|\langle\{n\}|f_{il}|\{n'\}\rangle|^2\,\delta\left[\omega_{il} + \sum_m \omega_m(n_m - n'_m)\right]\Bigg\}\\ &\times\prod_r (1 - e^{-\lambda_r})e^{-\lambda_r n_r}.\end{aligned} \quad (6.2.113)$$

When $i = j$, we have since $f_{il}^\dagger = f_{li}$

$$\Gamma_{ii} \equiv \Gamma_i = 2\pi\sum_{l\neq i}\sum_{\{n\}}\sum_{\{n'\}}|\langle\{n'\}|f_{li}|\{n\}\rangle|^2 \times \delta\left[\omega_{il} + \sum_m \omega_m(n_m - n'_m)\right]\prod_r (1 - e^{-\lambda_r})e^{-\lambda_r n_r}. \quad (6.2.114)$$

If we compare this with (6.2.98), we see that

$$\Gamma_i \equiv \Gamma_{ii} = \sum_{l\neq i} w_{li}. \quad (6.2.115)$$

That is, Γ_i is the sum of the probability per second for the atom going from state $|i\rangle$ to all other states. We may use this to rewrite (6.2.113) as

$$\Gamma_{ij} = \tfrac{1}{2}(\Gamma_i + \Gamma_j) + \Gamma_{ij}^{ph}, \quad (6.2.116)$$

where

$$\Gamma_{ij}^{ph} = \pi\sum_{\{n\}}\sum_{\{n'\}}\left\{|\langle\{n'\}|(f_{ii} - f_{jj})|\{n\}\rangle|^2 \times \delta\left[\sum_m \omega_m(n_m - n'_m)\right]\prod_r (1 - e^{-\lambda_r})e^{-\lambda_r n_r}\right\}. \quad (6.2.117)$$

For the particular interaction we have chosen $\Gamma_{ij}^{ph} = 0$. If we went to higher order in the interactions and took into account elastic scattering (see Chapter 5) we would obtain a $\Gamma_{ij}^{ph} \neq 0$. A phenomenological term of the form

$$|l\rangle\langle l|b_m^\dagger b_n \quad (6.2.118)$$

would correspond to the absorption of a photon of frequency ω_n and the emission of one at frequency ω_m with the atom remaining in state $|l\rangle$. This would correspond to the elastic scattering of a photon by an atom through a virtual (intermediate state). If we let

$$V = \hbar \sum_{l,m,n} \kappa_{lmn}|l\rangle\langle l|b_m^\dagger b_n, \tag{6.2.119}$$

then

$$\hbar f_{ii} = \hbar \sum_{m,n} \kappa_{imn} b_m^\dagger b_n, \tag{6.2.120}$$

and

$$\begin{aligned}\Gamma_{ij}{}^{ph} &= \pi \sum_{\{n\}} \sum_{\{n'\}} \left\{ |\langle\{n'\}| \sum_{l,m} (\kappa_{ilm} - \kappa_{jlm}) b_l^\dagger b_m |\{n\}\rangle|^2 \right. \\ &\quad \left. \times \delta\left[\sum_s \omega_s(n_s - n_s')\right] \prod_r (1 - e^{-\lambda_r}) e^{-\lambda_r n_r} \right\} \\ &= \pi \sum_{\{n\}} \sum_{\{n'\}} \left\{ \left| \sum_{l,m} (\kappa_{ilm} - \kappa_{jlm}) \sqrt{n_m(n_l + 1)} \langle\{n'\}| \cdots n_l + 1, \ldots, n_m - 1, \ldots\rangle \right| \right. \\ &\quad \left. \times \delta\left[\sum_s \omega_s(n_s - n_s')\right] \prod_r (1 - e^{-\lambda_r}) e^{-\lambda_r n_r} \right\},\end{aligned} \tag{6.2.121}$$

which is not zero in general. Such inelastic scattering interrupts the phase coherence in that the atom goes through some virtual state which interrupts its phase. Atomic collisions could also contribute to $\Gamma_{ij}{}^{ph}$.

By inspection of (6.2.116) and (6.2.117) we see that

$$\Gamma_{ij} = \Gamma_{ji} > 0. \tag{6.2.122}$$

Let us return to the master equation (6.2.89). When $i \neq j$, we have

$$\frac{\partial S_{ji}}{\partial t} = -(\Gamma_{ij} - i\omega_{ij}')S_{ji}, \tag{6.2.123}$$

where

$$\omega_{ij}' = \omega_{ij} + \Delta\omega_{ij}. \tag{6.2.124}$$

The solution of (6.2.123) is

$$S_{ji}(t) = e^{-(\Gamma_{ij} - i\omega_{ij}')t} S_{ji}(0), \tag{6.2.125}$$

so that the off-diagonal elements of the density matrix decay with a relaxation time Γ_{ij}^{-1} by atoms leaving both state j and i as well as by elastic processes which interrupt the phase of the atomic motion.

The diagonal elements $i = j$ of (6.2.89) satisfy the Pauli or rate equations

$$\begin{aligned}\frac{\partial S_{ii}}{\partial t} &= -\Gamma_i S_{ii} + \sum_k{}' w_{ik} S_{kk} \\ &= -\sum_k{}' w_{ki} S_{ii} + \sum_k{}' w_{ik} S_{kk},\end{aligned} \tag{6.2.126}$$

where we used (6.2.115). This states that the time rate of change of the probability of finding the atom in state i equals the rate of loss of atoms in state i going to all other allowed states plus the rate at which atoms in all other states are entering state i. Unfortunately, these represent an infinite set of coupled equations which we cannot solve in general.

In the steady state (superscript ss), the rates in and rates out of each level are equal. Then

$$\frac{\partial S_{ii}{}^{\text{ss}}}{\partial t} = 0, \tag{6.2.127}$$

so that

$$\Gamma_i S_{ii}{}^{\text{ss}} = \sum_k{}' w_{ik} S_{kk}{}^{\text{ss}} = \left(\sum_k w'_{ki}\right) S_{ii}{}^{\text{ss}}. \tag{6.2.128}$$

This will have a nontrivial solution if the determinant

$$|\Gamma_i\, \delta_{ik} - w_{ik}\,(1 - \delta_{ik})| = 0. \tag{6.2.129a}$$

That is,

$$\begin{vmatrix} \Gamma_1 & -w_{12} & -w_{13} & \cdots \\ -w_{21} & \Gamma_2 & -w_{23} & \cdots \\ -w_{31} & -w_{32} & \Gamma_3 & \cdots \end{vmatrix} = 0. \tag{6.2.129b}$$

However, we may easily show that this determinant is automatically zero and therefore does not represent any new relation between the Γ's and w's. To see this we note that since $\Gamma_i = \sum_{l \neq i} w_{li}$, if we add every row to the first row and leave all the other rows the same, every element in the first row equals zero.

In the steady state we expect the atom to come into thermal equilibrium with the reservoir at temperature T. We then have that

$$S_{ii}{}^{\text{ss}} = \left[\sum_i e^{-\beta\epsilon_i}\right]^{-1} e^{-\beta\epsilon_i}, \tag{6.2.130}$$

where $\beta = (kT)^{-1}$. That is, the atom will be distributed over its possible states with a Boltzmann distribution. Since Tr $S = 1$, we see that $\sum S_{ii}{}^{\text{ss}} = 1$. If we use (6.2.128), we see that

$$\Gamma_i = \sum_k{}' w_{ki} = \sum_k{}' w_{ik} e^{-\beta(\epsilon_k - \epsilon_i)}. \tag{6.2.131}$$

Driven Two Level Atom. The Bloch Equations

To gain further physical insight into the meaning of our master equation for a damped atom, let us specialize to the case of a damped two level atom in the presence of an externally applied electric field. The energy of interaction between the atom and field by (6.2.66) is given by

$$W = -e\mathbf{r} \cdot \mathbf{E}(\mathbf{r}, t), \tag{6.2.132}$$

where $e\mathbf{r}$ is the atomic dipole moment operator and $\mathbf{E}(\mathbf{r}, t)$ is the applied electric field:

$$\mathbf{E}(\mathbf{r}, t) = i\sqrt{\frac{\hbar\omega}{2\epsilon_0 L^3}}\,\hat{e}[be^{i(\mathbf{k}\cdot\mathbf{k}-\omega t)} - \text{cc}]. \tag{6.2.133}$$

We assume the field is so highly excited in the cavity that we may treat it classically. Also we assume that the wavelength is long compared with the atomic dimensions so that we may replace $\exp \pm ik \cdot \mathbf{r} \cong 1$. If we use the completeness relation twice for the two levels,

$$|1\rangle\langle 1| + |2\rangle\langle 2| = 1, \tag{6.2.134}$$

and assume that the atom has no permanent dipole moment

$$\langle 1|\mathbf{r}|1\rangle = \langle 2|\mathbf{r}|2\rangle = 0,$$

then (6.2.132) may be written as

$$\begin{aligned} W &= \sum_{k,l=1}^{2} |k\rangle\langle k|W|l\rangle\langle l| \\ &= -\boldsymbol{\mu}_{12}\cdot\mathbf{E}(t)|1\rangle\langle 2| - \boldsymbol{\mu}_{21}\cdot\mathbf{E}(t)|2\rangle\langle 1|, \end{aligned} \tag{6.2.135}$$

where

$$\boldsymbol{\mu}_{12} = e\langle 1|\mathbf{r}|2\rangle = e\langle 2|\mathbf{r}|1\rangle^*, \tag{6.2.136}$$

and

$$\mathbf{E}(t) = i\sqrt{\frac{\hbar\omega}{2\epsilon_0 L^3}}\,\hat{e}[be^{-i\omega t} - b^* e^{i\omega t}]. \tag{6.2.137}$$

The b and b^* are classical amplitudes.

In the IP, we have by (6.2.74) that

$$|1\rangle\langle 2| \rightarrow e^{-i\omega_{21}t}$$
$$|2\rangle\langle 1| \rightarrow e^{+i\omega_{21}t},$$

where

$$-\hbar\omega_{12} = \hbar\omega_{21} \equiv \epsilon_2 - \epsilon_1 > 0. \tag{6.2.138}$$

If we therefore retain only the terms in (6.2.135) of the form $\pm(\omega - \omega_{21})$ and neglect those of the form $\pm(\omega + \omega_{21})$ (rotating wave approximation), W reduces to

$$W \cong -\hbar[v^*(t)|1\rangle\langle 2| + v(t)|2\rangle\langle 1|], \tag{6.2.139}$$

where we let

$$v(t) = \hat{e}\cdot\boldsymbol{\mu}_{21}\sqrt{\frac{\omega}{2\hbar\epsilon_0 L^3}}\,be^{-i\omega t} \equiv v_0 e^{-i\omega t}. \tag{6.2.140}$$

If we use (6.2.35) and (6.2.85), the density operator equation becomes

$$\begin{aligned} \frac{\partial S}{\partial t} &= -\frac{i}{\hbar}\sum_l \epsilon_l[|l\rangle\langle l|, S] + iv^*(t)[|1\rangle\langle 2|, S] \\ &\quad + iv(t)[|2\rangle\langle 1|, S] + \sum_{k,l}{}' w_{lk}|l\rangle\langle k|S|k\rangle\langle l| \\ &\quad + \sum_{k,l}\{|k\rangle\langle k|S|l\rangle\langle l|(w^+_{llkk} + w^-_{llkk}) - w^+_{kllk}|k\rangle\langle k|S - w^-_{kllk}S|k\rangle\langle k|\}. \end{aligned} \tag{6.2.141}$$

Note by virtue of (6.2.74) and the retention of secular terms that the form of (6.2.85) is unchanged in the SP except for replacing s by S. If only two levels are involved the sums go from 1 to 2 only.

If we take the j, i matrix elements of (6.2.141) we obtain the equation S_{11} and S_{21}

$$\begin{aligned}\frac{\partial S_{11}}{\partial t} &= iv^*(t)S_{21} - iv(t)S_{12} - \Gamma_1 S_{11} + \Gamma_2 S_{22}\\ \frac{\partial S_{21}}{\partial t} &= -[i\omega_{21} + \Gamma_{21}]S_{21} - iv(t)(S_{22} - S_{11}).\end{aligned} \tag{6.2.142}$$

Here we have used (6.2.115) which for only two levels reduces to

$$\begin{aligned}\Gamma_1 &= w_{21}\\ \Gamma_2 &= w_{12}.\end{aligned} \tag{6.2.143}$$

In addition we have that

$$\begin{aligned}S_{21} &= S_{12}^*\\ 1 &= S_{11} + S_{22}.\end{aligned} \tag{6.2.144}$$

There are three operators of interest: the population difference;

$$\sigma_z = |2\rangle\langle 2| - |1\rangle\langle 1|, \tag{6.2.145}$$

and the dipole moment operators;

$$\begin{aligned}\sigma_+ &= |2\rangle\langle 1|\\ \sigma_- &= |1\rangle\langle 2|.\end{aligned} \tag{6.2.146}$$

We showed in (3.10.17) that these operators could be put in one-to-one correspondence with the Pauli spin-$\frac{1}{2}$ operators.

The mean value of these operators is given by

$$\begin{aligned}\langle\sigma_+\rangle &= \mathrm{Tr}\, S(t)|2\rangle\langle 1| = \langle 1|S(t)|2\rangle = S_{12}\\ \langle\sigma_-\rangle &= \mathrm{Tr}\, S(t)|1\rangle\langle 2| = S_{21}(t)\\ \langle\sigma_z\rangle &= \mathrm{Tr}\, S(t)[|2\rangle\langle 2| - |1\rangle\langle 1|] = S_{22}(t) - S_{11}(t)\\ 1 &= \mathrm{Tr}\, S(t)[|2\rangle\langle 2| + |1\rangle\langle 1|] = S_{22}(t) + S_{11}(t).\end{aligned} \tag{6.2.147}$$

If we use these and (6.2.142) and (6.2.144) above, we obtain the mean equations of motion

$$\begin{aligned}\frac{d\langle\sigma_+\rangle}{dt} &= +(i\omega_{21} - \Gamma_{21})\langle\sigma_+\rangle + iv^*(t)\langle\sigma_z\rangle\\ \frac{d\langle\sigma_-\rangle}{dt} &= -(i\omega_{21} + \Gamma_{21})\langle\sigma_-\rangle - iv(t)\langle\sigma_z\rangle\\ \frac{d\langle\sigma_z\rangle}{dt} &= 2i[v(t)\langle\sigma_+\rangle - v^*(t)\langle\sigma_-\rangle] - (\Gamma_1 + \Gamma_2)\left[\langle\sigma_z\rangle - \left(\frac{\Gamma_1 - \Gamma_2}{\Gamma_1 + \Gamma_2}\right)\right].\end{aligned} \tag{6.2.148}$$

These should be recognized as the Bloch equations. If we let

$$\sigma_{\pm} = \tfrac{1}{2}(\sigma_x \pm i\sigma_y) \qquad \text{or} \qquad \begin{cases} \sigma_+ + \sigma_- = \sigma_x \\ \sigma_+ - \sigma_- = i\sigma_y, \end{cases} \tag{6.2.149}$$

we may write them in the more familiar form

$$\begin{aligned} \frac{d\langle \boldsymbol{\sigma}\rangle_{x,y}}{dt} &= -\gamma[\langle \boldsymbol{\sigma}\rangle \times \mathbf{H}]_{x,y} - \frac{\langle \boldsymbol{\sigma}\rangle_{x,y}}{T_2} \\ \frac{d\langle \sigma_z\rangle}{dt} &= -\gamma[\langle \boldsymbol{\sigma}\rangle \times \mathbf{H}]_z - \frac{(\langle \sigma_z\rangle - \sigma_e)}{T_1}, \end{aligned} \tag{6.2.150}$$

if we make the identifications

$$\begin{aligned} \gamma\mathbf{H} &= [-(v^*(t) + v(t)), i(v^*(t) - v(t)), \omega_{21}] \\ \frac{1}{T_2} &= \Gamma_{21} = \frac{\Gamma_1 + \Gamma_2}{2} + \Gamma_{12}^{ph} \\ \frac{1}{T_1} &= \Gamma_1 + \Gamma_2 \\ \sigma_e &= \frac{\Gamma_1 - \Gamma_2}{\Gamma_1 + \Gamma_2} = \frac{w_{21} - w_{12}}{w_{21} + w_{21}}. \end{aligned} \tag{6.2.151}$$

[It is of interest to compare these with (3.17.30).] We therefore see that a two level atom in the presence of a classical driving field coupled to a reservoir is equivalent to a spin-$\frac{1}{2}$ particle in the presence of a d-c field along the z axis and an rf field in the x-y plane. The T_2 is commonly called the spin-spin or transverse relaxation time which contains damping due to phase interruption ($\Gamma_{12}{}^{ph}$). The T_1 is called the spin-lattice or longitudinal relaxation time when the spin interacts with phonons. These equations have been discussed very thoroughly in many places. The main point that we wish to emphasize is that they are a special case of our present treatment of damping and we have obtained the relaxation constants explicitly.

6.3 THE MARKOFF APPROXIMATION IN THE HEISENBERG PICTURE [7]

Equation of Motion for System Operator

We have shown that the mean value of a system operator M may be evaluated in either the SP

$$\langle M(t)\rangle = \mathrm{Tr}_S\, MS(t), \tag{6.3.1}$$

where we used (6.2.3) and where $S(t)$ is the reduced density operator

$$S(t) = \mathrm{Tr}_R\, \rho(t), \tag{6.3.2}$$

or in the HP

$$\langle M(t)\rangle = \mathrm{Tr}_{R,S}\, \rho(t_0)M(t)$$
$$= \mathrm{Tr}_S\, S(t_0)\langle M(t)\rangle_R \tag{6.3.3}$$

where $M(t)$ is in the HP, and we have let

$$\rho(t_0) = S(t_0)f_0(R) \tag{6.3.4}$$

and

$$\langle M(t)\rangle_R \equiv \mathrm{Tr}_R\, f_0(R)M(t). \tag{6.3.5}$$

In the prior section we obtained an equation of motion for $S(t)$ under the Markoff approximation. We may use this as follows to obtain the mean equation of motion for $\langle M(t)\rangle$ under this approximation. From (6.3.1) and (6.2.34b) we have

$$\frac{d\langle M(t)\rangle}{dt} = \mathrm{Tr}_S\, M\frac{\partial S}{\partial t}$$
$$\cong \mathrm{Tr}_S\, M\biggl\{\frac{1}{i\hbar}[H, S(t)] - \sum_{i,j}\delta(\omega_i, -\omega_j)$$
$$\times\, ([Q_iQ_jS(t) - Q_jS(t)Q_i]w_{ij}^{+} - [Q_iS(t)Q_j - S(t)Q_jQ_i]w_{ji}^{-})\biggr\}, \tag{6.3.6}$$

where all operators are in the SMP. We may next use the cyclic properties of traces to move $S(t)$ to the left in each term. This gives after slight rearrangement

$$\frac{d\langle M(t)\rangle}{dt} = \mathrm{Tr}\, M\frac{\partial S}{\partial t} \cong \mathrm{Tr}_S\, S(t)\biggl\{\frac{1}{i\hbar}[M, H] - \sum_{i,j}\delta(\omega_i, -\omega_j)$$
$$\times\, ([M, Q_i]Q_jw_{ij}^{+} - Q_j[M, Q_i]w_{ji}^{-})\biggr\}, \tag{6.3.7}$$

which is the equation of motion for $\langle M(t)\rangle$ under the Markoff approximation in the SP.

Let us next obtain the mean equation of motion for $\langle M(t)\rangle$ directly from (6.3.3) or (6.3.5) where we make the Markoff approximation in the HP. In the HP, $M^H(t)$ obeys the equation of motion

$$\frac{dM^H(t)}{dt} = \frac{1}{i\hbar}[M^H(t), H_T{}^H], \tag{6.3.8}$$

where by (6.1.2)

$$H_T = H + R + V \equiv H_0 + V. \tag{6.3.9}$$

If H_T is time-independent, it follows that

$$H_T{}^H = H_T{}^S. \tag{6.3.10}$$

That is, the total hamiltonian in the HP is equal to the total hamiltonian in the SP. Therefore, (6.3.8) becomes

$$\frac{dM^H}{dt} = \frac{1}{i\hbar}\,[M^H(t), H_0{}^S + V^S]. \tag{6.3.11}$$

If we next make the transformation

$$M^H(t) = e^{+(i/\hbar)H_0{}^S(t-t_0)}m(t)e^{-(i/\hbar)H_0{}^S(t-t_0)}, \tag{6.3.12}$$

we see that

$$\begin{aligned}\frac{dM^H}{dt} &= e^{+(i/\hbar)H_0{}^S(t-t_0)}\left\{\frac{1}{i\hbar}\,[m(t), H_0{}^S] + \frac{dm}{dt}\right\}e^{-(i/\hbar)H_0{}^S(t-t_0)}\\ &= \frac{1}{i\hbar}\,[M^H(t), H_0{}^S] + e^{(i/\hbar)H_0{}^S(t-t_0)}\,\frac{dm}{dt}\,e^{-(i/\hbar)H_0{}^S(t-t_0)}.\end{aligned} \tag{6.3.13}$$

If we use (6.3.12) and (6.3.13), we see that (6.3.11) reduces to

$$\frac{dm}{dt} = \frac{1}{i\hbar}\,[m(t), V(t_0 - t)] \tag{6.3.14}$$

where

$$V(t_0 - t) = e^{-(i/\hbar)H_0{}^S(t-t_0)}V^S e^{(i/\hbar)H_0(t-t_0)}. \tag{6.3.15}$$

We therefore see that the transformation (6.3.12) is analogous to transforming to the IP in that it has removed the high-frequency free (unperturbed) motion $[M^H, H_0{}^S]$ term from (6.3.11). We would therefore like to make the Markoff approximation on the exact equation (6.3.14) in the same way we did for the reduced density operator $s(t)$ in the last section.

We begin by writing the iterated solution of (6.3.14) up to second order in V as

$$\begin{aligned}m(t) = m(t_0) &+ \frac{1}{i\hbar}\int_{t_0}^{t}[m(t_0), V(t_0 - t')]\,dt'\\ &+ \left(\frac{1}{i\hbar}\right)^2\int_{t_0}^{t}dt'\int_{t_0}^{t'}dt''\;[[m(t_0), V(t_0 - t'')], V(t_0 - t')].\end{aligned} \tag{6.3.16}$$

As in the prior section, we assume that

$$\begin{gathered}\gamma^{-1} \gg t - t_0 \gg \tau_c\\ t - t_0 \gg \omega_i^{-1},\end{gathered} \tag{6.3.17}$$

as far as the behavior of $m(t)$ is concerned. At time t_0, we have that

$$m(t_0) = M^S, \tag{6.3.18}$$

which is the system operator in the SP. Therefore, in (6.3.16) we have

$$\frac{m(t)-m(t_0)}{t-t_0} \equiv \frac{\Delta m}{\Delta t} = \frac{1}{i\hbar\,\Delta t}\int_{t_0}^{t}[M^S, V(t_0-t')]\,dt' + \left(\frac{1}{i\hbar}\right)^2 \frac{1}{\Delta t}\int_{t}^{t} dt' \int_{t_0}^{t'} dt''\,[[M^S, V(t_0-t'')], V(t_0-t')], \quad (6.3.19)$$

where as far as the system is concerned

$$\Delta t = t - t_0 \quad (6.3.20)$$

is a small time by virtue of (6.3.17). Under this approximation we may re-write the first form of (6.3.13) as

$$\frac{dM^H}{dt} \cong e^{(i/\hbar)H_0{}^S(t-t_0)}\left\{\frac{1}{i\hbar}[M^S, H_0{}^S] + \frac{\Delta m}{\Delta t}\right\}e^{-(i/\hbar)H_0{}^S(t-t_0)}, \quad (6.3.21)$$

since $t \to t_0$ as far as the system behavior is concerned and in the $[m(t), H_0{}^S]$ term $m(t) \cong m(t_0) = M^S$ [the corrections are give by (6.3.19)].

For use in the next chapter, let us define $g_m(t)$ by means of

$$g_m(t) = \frac{1}{i\hbar}[M^S, V(t_0-t)] = \frac{1}{i\hbar}[M^S, e^{-(i/\hbar)H_0{}^S(t-t_0)}V^S e^{(i/\hbar)H_0{}^S(t-t_0)}], \quad (6.3.22)$$

which we shall tentatively call the Langevin force. With this notation (6.3.18) may be written as

$$\frac{\Delta m}{\Delta t} = \frac{1}{\Delta t}\int_{t_0}^{t} g_m(t')\,dt' + \left(\frac{1}{i\hbar}\right)^2 \frac{1}{\Delta t}\int_{t_0}^{t} dt' \int_{t_0}^{t'} dt''\,[[M^S, V(t_0-t'')], V(t_0-t')]. \quad (6.3.23)$$

Since $H_0 = H + R$ and $[H, R] = 0$, we see that if we multiply both sides of (6.3.21) by the reservoir equilibrium density operator $f_0(R)$, trace over the reservoir and use the cyclic property of traces for reservoir operators, we obtain

$$\frac{d\langle M^H\rangle_R}{dt} \cong e^{+(i/\hbar)H^S(t-t_0)}\left\{\frac{1}{i\hbar}[M^S, H^S] + \left\langle\frac{\Delta m}{\Delta t}\right\rangle_R\right\}e^{-(i/\hbar)H^S(t-t_0)}, \quad (6.3.24)$$

since

$$\mathrm{Tr}_R\, e^{(i/\hbar)R(t-t_0)}\{\quad\}e^{-(i/\hbar)R(t-t_0)}f_0(R) = \mathrm{Tr}_R\{\quad\}f_0(R), \quad (6.3.25)$$

and

$$\mathrm{Tr}_R\,[M^S, R]f_0(R) = [M^S, \mathrm{Tr}_R\, Rf_0(R)] = 0. \quad (6.3.26)$$

We therefore need the thermal average $\langle \Delta m/\Delta t\rangle_R$. If we use (6.2.15), (6.2.17), and (6.2.23) we see that

$$V(t_0 - t) = \sum_i Q_i{}^S e^{+i\omega_i(t_0-t)}F_i(t_0-t), \quad (6.3.27)$$

so that by (6.3.23) we have

$$\left\langle\frac{\Delta m}{\Delta t}\right\rangle_R = \frac{1}{\Delta t}\int_{t_0}^{t}\langle g_m(t')\rangle_R\,dt' - \sum_{i,j}\frac{1}{\Delta t}\int_{t_0}^{t}dt'\int_{t_0}^{t'}dt''$$
$$\times\,\langle[[M^S, Q_i^S F_i''], Q_j^S F_j']\rangle_R \exp i[\omega_i(t_0 - t'') + \omega_j(t_0 - t')], \quad (6.3.28)$$

and we have let $F_i'' \equiv F_i(t_0 - t'')$ and $F_j' \equiv F_j(t_0 - t')$. Also we have

$$g_m(t) = -i\sum_i [M^S, Q_i^S]F_i(t_0 - t)e^{i\omega_i(t_0-t)}, \quad (6.3.29)$$

so that on using the cyclic property of traces, we have

$$\langle g_m(t)\rangle_R = -i\sum_i [M^S, Q_i^S]\langle F_i^S\rangle_R e^{i\omega_i(t_0-t)} \quad (6.3.30)$$

since

$$\mathrm{Tr}_R\, f_0(R)F_i(t_0 - t) = \mathrm{Tr}_R\, f_0(R)e^{(i/\hbar)R(t_0-t)}F_i^S e^{-(i/\hbar)R(t_0-t)}$$
$$= \mathrm{Tr}_R\, f_0(R)F_i^S, \quad (6.3.31)$$

If we use this, expand the double commutator in (6.3.28) and use arguments as we did in (6.2.20), we readily obtain

$$\left\langle\frac{\Delta m}{\Delta t}\right\rangle_R = -\frac{i}{\Delta t}\sum_i [M^S, Q_i^S]\,\langle F_i^S\rangle_R\int_{t_0}^{t} e^{i\omega_i(t_0-t')}\,dt'$$
$$- \sum_{i,j}\frac{1}{\Delta t}\int_{t_0}^{t}dt'\int_{t_0}^{t'}dt''\,\{[M^S, Q_i^S]Q_j^S\langle F_i(t' - t'')F_j^S\rangle_R$$
$$- Q_j^S[M^S, Q_i^S]\langle F_j^S F_i(t' - t'')\rangle_R\}$$
$$\times \exp i[\omega_i(t_0 - t'') + \omega_j(t_0 - t')]. \quad (6.3.32)$$

If we use (6.2.21), we obtain

$$\left\langle\frac{\Delta m}{\Delta t}\right\rangle_R = -i\sum_i [M^S, Q_i^S]\langle F_i^S\rangle_R\frac{1}{\Delta t}\int_0^{\Delta t} e^{-i\omega_i x}\,dx$$
$$- \sum_{i,j}\frac{1}{\Delta t}\int_0^{t-t_0}d\xi\int_0^{t-t_0-\xi}d\tau\{[M^S, Q_i^S]Q_j^S\langle F_i(\tau)F_j^S\rangle_R$$
$$- Q_j^S[M^S, Q_i^S]\langle F_j F_i(\tau)\rangle_R\} \exp -i[(\omega_i + \omega_j)\xi + \omega_j\tau].$$
$$(6.3.33)$$

If we now use (6.3.17), we see as in the argument leading to (6.2.32) that (6.3.33) gives

$$\left\langle\frac{\Delta m}{\Delta t}\right\rangle_R = -\sum_{i,j}\delta(\omega_i, -\omega_j)\{[M, Q_i]Q_j w_{ij}^{+} - Q_j[M, Q_i]w_{ji}^{-}\}, \quad (6.3.34)$$

where all operators are system operators in the SMP and $w^{\pm}$ are defined by (6.2.27), since

$$\frac{1}{\Delta t}\int_t^{t+\Delta t}\langle g_m(t')\rangle_R\,dt' = 0. \tag{6.3.35}$$

This follows since

$$\frac{1}{\Delta t}\int_t^{t+\Delta t} e^{-i\omega_i x}\,dx = \begin{cases}1 & \text{if} \quad \omega_i = 0\\ 0 & \text{if} \quad \omega_i \neq 0,\end{cases}$$

when $\Delta t \gg \omega_i^{-1}$. Since no $\omega_i = 0$ for system operators in the IP, (6.3.35) follows.

If we next substitute (6.3.34) into (6.3.24) and insert

$$e^{(i/\hbar)H(t-t_0)}e^{-(i/\hbar)H(t-t_0)} = 1$$

between appropriate factors, we obtain

$$\frac{d}{dt}\langle M^H(t)\rangle_R \cong \Big\langle \frac{1}{i\hbar}[M(t), H] - \sum_{i,j}\delta(\omega_i, -\omega_j)$$
$$\times \{[M(t), Q_i(t)]Q_j(t)w_{ij}^{+} - Q_j(t)[M(t), Q_i(t)]w_{ji}^{-}\}\Big\rangle_R, \tag{6.3.36}$$

where we have let

$$\mathscr{G}(t) \equiv e^{(i/\hbar)H(t-t_0)}\mathscr{G}^S e^{-(i/\hbar)H(t-t_0)} \tag{6.3.37}$$

be any time-dependent operator appearing in (6.3.36). These operators are the Heisenberg operators under the Markoff approximation and we could, for convenience, say they are in the Heisenberg-Markoff picture (HMP). They obey the same commutation relations as the original Heisenberg operators.

If we multiply both sides of (6.3.36) by $S(t_0)$ and trace over the system, we obtain

$$\frac{d\langle M(t)\rangle}{dt} \cong \operatorname{Tr} S(t_0)\frac{dM(t)}{dt} = \operatorname{Tr} S(t_0)\Big\langle\Big\{\frac{1}{i\hbar}[M(t), H] - \sum_{i,j}\delta(\omega_i, -\omega_j)$$
$$\times ([M(t), Q_i(t)]Q_j(t)w_{ij}^{+} - Q_j(t)[M(t), Q_i(t)]w_{ji}^{-})\Big\}\Big\rangle_R, \tag{6.3.38}$$

where all operators are in the HMP.

Just as we may evaluate means exactly in the SP or HP

$$\langle M(t)\rangle = \operatorname{Tr}_S S(t)M = \operatorname{Tr}_S S(t_0)\langle M^H(t)\rangle_R, \tag{6.3.39a}$$

we may also evaluate under the Markoff approximation

$$\langle M(t)\rangle \cong \operatorname{Tr}_S S(t)M = \operatorname{Tr} S(t_0)M(t), \tag{6.3.39b}$$

where $S(t)$ satisfies (6.2.34b) and $M(t)$ here satisfies (6.3.36). Also the reader should compare (6.3.38) with (6.3.7) and note that the functional forms of these two equations are identical just as they would be under the exact transformation from the SP to the HP. This is the analog of thinking of a fixed coordinate system and rotating vectors as being equivalent to a rotating coordinate system and fixed vectors. This invariance of form will be important in our later work.

Driven Damped Oscillator

As a first example, let us consider again the damped driven oscillator. The total hamiltonian by (6.2.39)–(6.2.42) is

$$H_T = H_0 + V + W. \tag{6.3.40}$$

In this case (6.3.36) reduces to

$$\frac{d\langle M\rangle_R}{dt} = \langle -i\omega_c[M, a^\dagger a] - iv(t)[M, a^\dagger] - iv^*(t)[M, a] - [M, a^\dagger]aw_{12}^+ + a[M, a^\dagger]w_{21}^- - [M, a]a^\dagger w_{21}^+ + a^\dagger[M, a]w_{12}^-\rangle_R, \tag{6.3.41}$$

where *all* operators are in the HMP (Heisenberg-Markoff picture) and where

$$M \cong \langle M^H(t)\rangle_R. \tag{6.3.42}$$

The w's are given by (6.2.58). If we use this and the identity

$$[M, a^\dagger a] = [M, a^\dagger]a + a^\dagger[M, a], \tag{6.3.43}$$

then (6.3.41) reduces to

$$\left\langle\frac{dM}{dt}\right\rangle_R = \left\langle -iv(t)[M, a^\dagger] - iv^*(t)[M, a] - \left[i(\omega_c + \Delta\omega) + \frac{\gamma}{2}\right][M, a^\dagger]a - \left[i(\omega_c + \Delta\omega) - \frac{\gamma}{2}\right]a^\dagger[M, a] + \gamma\bar{n}[a, [M, a^\dagger]]\right\rangle_R, \tag{6.3.44}$$

and we have used the fact that

$$[a, [M, a^\dagger]] = [a^\dagger, [M, a]]. \tag{6.3.45}$$

If we make use of the commutation relations

$$\begin{aligned} [a, F(a, a^\dagger)] &= \frac{\partial F}{\partial a^\dagger} \\ [a^\dagger, F(a, a^\dagger)] &= -\frac{\partial F}{\partial a}, \end{aligned} \tag{6.3.46}$$

we may rewrite (6.3.44) as

$$\left\langle\frac{dM}{dt}\right\rangle_R = \left\langle -iv(t)\frac{\partial M}{\partial a} + iv^*(t)\frac{\partial M}{\partial a^\dagger} - \left[i(\omega_c + \Delta\omega) + \frac{\gamma}{2}\right]\frac{\partial M}{\partial a}a + \left[i(\omega_c + \Delta\omega) - \frac{\gamma}{2}\right]a^\dagger\frac{\partial M}{\partial a^\dagger} + \gamma\bar{n}\frac{\partial^2 M}{\partial a\, \partial a^\dagger}\right\rangle_R. \tag{6.3.47}$$

We emphasize *again* that all system operators are in the HMP which represents a reservoir thermal average under the Markoff approximation. If confusion is likely, we use a superscript M to represent the HMP or $\langle\ \rangle_R$.

In case $M = a$, we see that

$$\frac{d}{dt}\langle a\rangle_R = -iv(t) - \left[i(\omega_c + \Delta\omega) + \frac{\gamma}{2}\right]\langle a\rangle_R, \tag{6.3.48}$$

where we have emphasized that these are Markoff operators. Similarly, if we let $M = a^\dagger a$, we see that

$$\frac{d}{dt}\langle a^\dagger a\rangle_R = -iv(t)\langle a^\dagger\rangle_R + iv^*(t)\langle a\rangle_R - \gamma[\langle a^\dagger a\rangle_R - \bar{n}]. \tag{6.3.49}$$

The solution of (6.3.48) is easily seen to be

$$\langle a(t)\rangle_R = \langle a(t_0)\rangle_R e^{-[i\omega' + (\gamma/2)](t-t_0)} - ig(t), \tag{6.3.50}$$

where we have let

$$\omega' = \omega_c + \Delta\omega$$

$$g(t) = \int_{t_0}^{t} v(t')e^{[i\omega' + (\gamma/2)](t'-t)}\,dt'. \tag{6.3.51}$$

If we substitute (6.3.50) and its adjoint into (6.3.49) and solve, we obtain

$$\begin{aligned}
&\langle a^\dagger(t)a(t)\rangle_R \\
&\quad = \langle a^\dagger(t_0)a(t_0)\rangle_R e^{-\gamma(t-t_0)} + \bar{n}[1 - e^{-\gamma(t-t_0)}] \\
&\qquad + i\langle a(t_0)\rangle_R \int_{t_0}^{t} dt'\, v^*(t') \exp\left\{\gamma(t'-t) - \left(i\omega' + \frac{\gamma}{2}\right)(t'-t_0)\right\} \\
&\qquad - i\langle a^\dagger(t_0)\rangle_R \int_{t_0}^{t} dt'\, v(t') \exp\left\{\gamma(t'-t) + \left(i\omega' - \frac{\gamma}{2}\right)(t'-t_0)\right\} \\
&\qquad + \int_{t_0}^{t} dt'\, e^{\gamma(t'-t)}[v(t')g^*(t') + v^*(t')g(t')],
\end{aligned} \tag{6.3.52}$$

for the thermal average of the mean number of photons in the cavity mode.

Atom with Linewidth

Let us next apply (6.3.36) to the damped atom considered in the last section. If we again use (6.2.73), (6.2.62), and (6.2.75), we obtain

$$\begin{aligned}
\frac{d}{dt}\langle M(t)\rangle_R &= \frac{1}{i\hbar}\sum_l \epsilon_l\langle[M, |l\rangle\langle l|]\rangle_R - \sum_{klmn}\delta(\omega_{kl}, -\omega_{mn}) \\
&\quad \times \langle\{[M, |k\rangle\langle l|]|m\rangle\langle n|w^+_{klmn} - |m\rangle\langle n|[M, |k\rangle\langle l|]w^-_{mnkl}\rangle_R.
\end{aligned} \tag{6.3.53}$$

If we retain only the secular terms (6.2.77) for which (a) $k = n$, $l = m$ ($k \neq l$), (b) $k = l$, $m = n$ ($k \neq m$), and (c) $k = l = m = n$, this reduces to

$$\frac{d\langle M(t)\rangle_R}{dt} = \frac{1}{i\hbar}\sum_l \epsilon_l\langle[M, |l\rangle\langle l|]\rangle_R + \sum_{k,l}' w_{lk}\langle|k\rangle\langle l|M|l\rangle\langle k|\rangle_R$$
$$+ \sum_{k,l}\{(w_{llkk}^{+} + w_{llkk}^{-})\langle|l\rangle\langle l|M|k\rangle\langle k|\rangle_R$$
$$- w_{kllk}^{+}\langle M|k\rangle\langle k|\rangle_R - w_{kllk}^{-}\langle|k\rangle\langle k|M\rangle_R\}, \quad (6.3.54)$$

where we used (6.2.84).

If $M \equiv |i\rangle\langle j|$ and we use the orthogonality relations (6.2.63), we obtain

$$\frac{d}{dt}\langle|i\rangle\langle j|\rangle_R = i\omega_{ij}\langle|i\rangle\langle j|\rangle_R + \delta_{ij}\sum_k' w_{ik}\langle|k\rangle\langle k|\rangle_R$$
$$+ \left[(w_{iijj}^{+} + w_{iijj}^{-}) - \sum_l (w_{jllj}^{+} + w_{illi}^{-})\right]\langle|i\rangle\langle j|\rangle_R, \quad (6.3.55)$$

or when we use (6.2.87a)

$$\frac{d}{dt}\langle|i\rangle\langle j|\rangle_R = [i\omega_{ij} - \Gamma_{ij}{}^c]\langle|i\rangle\langle j|\rangle_R + \delta_{ij}\sum_k' w_{ik}\langle|k\rangle\langle k|\rangle_R. \quad (6.3.56)$$

These results are useful when we present the Langevin method in the next chapter. All operators are in the HMP.

6.4 ONE-TIME AVERAGES USING ASSOCIATED DISTRIBUTION FUNCTIONS [8–10]

In the prior section, we have shown that under the Markoff approximation we may evaluate one-time averages of system operators in either the SP or HP. In the former case, we must solve the master equation (6.2.34b):

$$\frac{\partial S}{\partial t} = \frac{1}{i\hbar}[H, S] - \sum_{i,j}\delta(\omega_i, -\omega_j)$$
$$\times \{(Q_iQ_jS - Q_jSQ_i)w_{ij}^{+} - (Q_iSQ_j - SQ_jQ_i)w_{ji}^{-}\}, \quad (6.4.1)$$

where all operators are in the SMP. That is, the H, Q_i, and Q_j are system operators at t_0 and S is at time t.

A formal solution of this operator equation may be written as

$$S(t) = X_{tt_0}[S(t_0)], \quad (6.4.2)$$

where X_{tt_0} is a linear functional of $S(t_0)$ which plays the role of the exact solution

$$\rho(t) = U(t, t_0)\rho(t_0)U^{-1}(t, t_0), \quad (6.4.3a)$$

where

$$U(t, t_0) \equiv U_{tt_0} = \exp - \frac{i}{\hbar} H_T(t - t_0), \tag{6.4.3b}$$

when H_T is time-independent.

The operators in the SMP obey the same commutation relations they did before we made the Markoff approximation.

Once we have a solution for $S(t)$ in the form (6.4.2), the mean of a function of system operators $a_1, a_2, \ldots, a_f \equiv \underset{\sim}{a}$ evaluated at t_0 in the SMP is given by (6.3.39b)

$$\langle M(\underset{\sim}{a}, t)\rangle = \mathrm{Tr}_S\, S(t)M(\underset{\sim}{a}) = \mathrm{Tr}_S\, X_{tt_0}[S(t_0)]M(\underset{\sim}{a}) \tag{6.4.4}$$

In the HMP, we must solve the operator equation (6.3.36)

$$\frac{dM}{dt} = \frac{1}{i\hbar}[M, H] - \sum_{i,j} \delta(\omega_i, -\omega_j)\{[M, Q_i]Q_j w_{ij}^{+} - Q_j[M, Q_i]w_{ji}^{-}\}, \tag{6.4.5}$$

where all operators are in the HMP at time t and

$$M[\underset{\sim}{a}(t)] \cong \langle M^H[\underset{\sim}{a}(t)]\rangle_R. \tag{6.4.6}$$

The mean of M by (6.3.39b) is then given by

$$\langle M(t)\rangle = \mathrm{Tr}\, S(t_0)M[\underset{\sim}{a}(t)]. \tag{6.4.7}$$

Just as we may use the cyclic property of traces to show in the exact case that

$$\begin{aligned}\langle M(t)\rangle &= \mathrm{Tr}\, \rho\ (t)M^S(t_0) = \mathrm{Tr}\, U_{tt_0}\rho^H(t_0)U_{tt_0}^{-1}M^S(t_0)\\ &= \mathrm{Tr}\, \rho^H(t_0)U_{tt_0}^{-1}M^S(t_0)U_{tt_0} = \mathrm{Tr}\, \rho^H(t_0)M^H(t),\end{aligned} \tag{6.4.8}$$

we may use the cyclic property of traces to rewrite (6.4.4) as

$$\begin{aligned}\langle M(\mathbf{a}, t)\rangle &= \mathrm{Tr}\, X_{tt_0}[S^H(t_0)][M^S[\underset{\sim}{a}(t_0)]\\ &= \mathrm{Tr}\, S^H(t_0)\tilde{X}_{tt_0}\{M^S[\underset{\sim}{a}(t_0)]\}\\ &= \mathrm{Tr}\, S^H(t_0)M^H[\underset{\sim}{a}(t)],\end{aligned} \tag{6.4.9}$$

where the last step follows from (6.4.7). Therefore, a formal solution of (6.4.5) may be written as

$$M^H[\underset{\sim}{a}(t)] = \tilde{X}_{tt_0}\{M^S[\underset{\sim}{a}(t_0)]\}, \tag{6.4.10}$$

where $\tilde{X}_{tt_0}$ is the "transpose" of X_{tt_0}. Actually the solution of (6.4.5) defines $\tilde{X}_{tt_0}$.

In both the SMP and the HMP we must solve operator equations. In this section we would like to take advantage of the quantum-"classical" correspondence that we developed in Chapter 3, Section 3.9 to transform the operator equations above to equivalent c-number equations so that classical mathematical methods of solution become available. We begin in the SMP with the density operator equation.

Equation of Motion for Associated Distribution Function

If $a_1, a_2, \ldots, a_f \equiv \underset{\sim}{a}$ are a complete set of system operators in the SMP and we choose the ordering $1, 2, \ldots, f$ as in Section 3.9, then by (3.9.3), (3.9.4), (3.9.22), and (3.9.23), (6.4.4) becomes

$$\langle M^c[\underset{\sim}{a}(t_0), t]\rangle = \operatorname{Tr} S(t)M^c[\underset{\sim}{a}(t_0)] = \langle \mathscr{C}\{\bar{M}^c[\underset{\sim}{\alpha}(t_0)]\}\rangle$$
$$= \int d\underset{\sim}{\alpha}_0\, \bar{M}^c(\underset{\sim}{\alpha}_0)P_c(\underset{\sim}{\alpha}_0, t), \tag{6.4.11}$$

where we have let $\underset{\sim}{\alpha}_0 \equiv \underset{\sim}{\alpha}(t_0)$, $d\underset{\sim}{\alpha}_0 = \prod_{i=1}^{f} d\alpha_i(t_0)$ and

$$P_c(\underset{\sim}{\alpha}_0, t) = \operatorname{Tr} S(t)\delta^c(\underset{\sim}{\alpha}_0 - \underset{\sim}{a}_0) \equiv \langle \delta^c(\underset{\sim}{\alpha}_0 - \underset{\sim}{a}_0)\rangle, \tag{6.4.12}$$

and $\underset{\sim}{a}_0 \equiv \underset{\sim}{a}(t_0)$. The $\delta^c(\underset{\sim}{\alpha}_0 - \underset{\sim}{a}_0)$ is the product of δ-functions in c-order. As we have noted, $P_c(\underset{\sim}{\alpha}_0, t)$ is the associated distribution function when system operators are in c-order. If M is in c-order, the quantum one-time average in (6.4.11) looks like a classical average if P_c were a classical probability distribution function. We would therefore like to obtain directly an equation of motion for P_c.

If we differentiate (6.4.12), we obtain

$$\frac{\partial P_c}{\partial t}(\underset{\sim}{\alpha}_0, t) = \operatorname{Tr} \frac{\partial S}{\partial t}\, \delta^c(\underset{\sim}{\alpha}_0 - \underset{\sim}{a}_0) = \frac{d}{dt}\langle \delta^c(\underset{\sim}{\alpha}_0 - \underset{\sim}{a}_0)\rangle. \tag{6.4.13}$$

If we use (6.3.7) and let $M^c(\underset{\sim}{a}_0) \equiv \delta^c(\underset{\sim}{\alpha}_0 - \underset{\sim}{a}_0)$, we obtain

$$\frac{\partial P_c}{\partial t}(\underset{\sim}{\alpha}_0, t) = \frac{d}{dt}\langle \delta^c(\underset{\sim}{\alpha}_0 - \underset{\sim}{a}_0)\rangle$$
$$= \operatorname{Tr} S(t)\left\{\frac{1}{i\hbar}[\delta^c(\underset{\sim}{\alpha}_0 - \underset{\sim}{a}_0), H] - \sum_{i,j} \delta(\omega_i, -\omega_j)\right.$$
$$\left. \times ([\delta^c(\underset{\sim}{\alpha}_0 - \underset{\sim}{a}_0), Q_i]Q_j w_{ij}^{+} - Q_j[\delta^c(\underset{\sim}{\alpha}_0 - \underset{\sim}{a}_0), Q_i]w_{ji}^{-})\right\}. \tag{6.4.14}$$

We easily see that we may express the δ-function as

$$\delta(\alpha - a) = \frac{1}{2\pi}\int_{-\infty}^{\infty} e^{-i\xi(\alpha - a)}\, d\xi = \frac{1}{2\pi}\int_{-\infty}^{\infty} e^{-i\xi\alpha}\sum_{n=0}^{\infty}\frac{(i\xi a)^n}{n!}\, d\xi$$
$$= \frac{1}{2\pi}\int_{-\infty}^{\infty} d\xi \sum_{n=0}^{\infty}\frac{(-a)^n}{n!}\frac{\partial^n}{\partial\alpha^n}e^{-i\xi\alpha} = e^{-a(\partial/\partial\alpha)}\frac{1}{2\pi}\int_{-\infty}^{\infty} e^{-i\xi\alpha}\, d\xi$$
$$= e^{-a(\partial/\partial\alpha)}\,\delta(\alpha), \tag{6.4.15a}$$

where the steps are self-explanatory. Then since

$$\delta^c(\underset{\sim}{\alpha}_0 - \underset{\sim}{a}_0) = \prod_{i=1}^{f} e^{-a_{i0}(\partial/\partial\alpha_{i0})}\, \delta(\alpha_{i0}) \equiv e^{-\underset{\sim}{a}_0(\partial/\partial\underset{\sim}{\alpha}_0)}\, \delta(\underset{\sim}{\alpha}_0), \tag{6.4.15b}$$

(6.4.14) becomes

$$\begin{aligned}\frac{\partial P_c}{\partial t}(\underset{\sim}{\alpha}_0, t) &= \operatorname{Tr} S(t)\Big\{\frac{1}{i\hbar}\,[e^{-\underset{\sim}{a}_0(\partial/\partial\underset{\sim}{\alpha}_0)}, H] \\ &\qquad - \sum_{i,j} \delta(\omega_i - \omega_j)([e^{-\underset{\sim}{a}_0(\partial/\partial\underset{\sim}{\alpha}_0)}, Q_i]Q_j w_{ij}^{+} \\ &\qquad - Q_j[e^{-\underset{\sim}{a}_0(\partial/\partial\underset{\sim}{\alpha}_0)}, Q_i]w_{ji}^{-})\Big\}\, \delta(\underset{\sim}{\alpha}_0) \\ &\equiv \operatorname{Tr} S(t)L'\Big\{\frac{\partial}{\partial\underset{\sim}{\alpha}_0}, \underset{\sim}{a}_0\Big\}\, \delta(\underset{\sim}{\alpha}_0).\end{aligned} \tag{6.4.16}$$

We show by several examples below that we may use the commutation relations to put all operators in L' into chosen order so that $L' = L'^c$. Then we may proceed to rewrite (6.4.16) as follows.

$$\begin{aligned}\frac{\partial P_c}{\partial t} &= \operatorname{Tr} S(t)L'^c\Big(\frac{\partial}{\partial\underset{\sim}{\alpha}_0}, \underset{\sim}{a}_0\Big)\, \delta(\underset{\sim}{\alpha}_0) \\ &= \operatorname{Tr} S(t)\int \bar{L}'^c\Big(\frac{\partial}{\partial\underset{\sim}{\alpha}_0}, \underset{\sim}{\beta}_0\Big)\, \delta^c(\underset{\sim}{\beta}_0 - \underset{\sim}{a}_0)\, d\underset{\sim}{\beta}_0\, \delta(\underset{\sim}{\alpha}_0).\end{aligned}$$

From the form of L' in (6.4.16), we see that we may always rewrite L'^c as

$$\bar{L}'^c\Big(\frac{\partial}{\partial\underset{\sim}{\alpha}_0}, \underset{\sim}{\beta}_0\Big) = \bar{L}^c\Big(\frac{\partial}{\partial\underset{\sim}{\alpha}_0}, \underset{\sim}{\beta}_0\Big)e^{-\underset{\sim}{\beta}_0(\partial/\partial\underset{\sim}{\alpha}_0)}. \tag{6.4.17a}$$

Thus

$$\frac{\partial P_c}{\partial t} = \operatorname{Tr} S(t)\int \bar{L}^c\Big(\frac{\partial}{\partial\underset{\sim}{\alpha}_0}, \underset{\sim}{\beta}_0\Big)e^{-\underset{\sim}{\beta}_0(\partial/\partial\underset{\sim}{\alpha}_0)}\, \delta^c(\underset{\sim}{\beta}_0 - \underset{\sim}{a}_0)\, d\underset{\sim}{\beta}_0\, \delta(\underset{\sim}{\alpha}_0).$$

But

$$e^{-\underset{\sim}{\beta}_0(\partial/\partial\underset{\sim}{\alpha}_0)}\, \delta(\underset{\sim}{\alpha}_0) = \delta(\underset{\sim}{\beta}_0 - \underset{\sim}{\alpha}_0),$$

so that

$$\begin{aligned}\frac{\partial P_c}{\partial t} &= \operatorname{Tr} S(t)\int \bar{L}^c\Big(\frac{\partial}{\partial\underset{\sim}{\alpha}_0}, \underset{\sim}{\beta}_0\Big)\, \delta(\underset{\sim}{\beta}_0 - \underset{\sim}{\alpha}_0)\, \delta(\underset{\sim}{\beta}_0 - \underset{\sim}{a}_0)\, d\underset{\sim}{\beta}_0 \\ &= \operatorname{Tr} S(t)\bar{L}^c\Big(\frac{\partial}{\partial\underset{\sim}{\alpha}_0}, \underset{\sim}{\alpha}_0\Big)\, \delta(\underset{\sim}{\alpha}_0 - \underset{\sim}{a}_0).\end{aligned} \tag{6.4.17b}$$

We illustrate below how the intermediate steps above may be taken. If we use the definition (6.4.12), this reduces to

$$\frac{\partial P_c}{\partial t}(\underset{\sim}{\alpha}_0, t) = \bar{L}^c\Big\{\frac{\partial}{\partial\underset{\sim}{\alpha}_0}, \underset{\sim}{\alpha}_0\Big\}P_c(\underset{\sim}{\alpha}_0, t), \tag{6.4.18}$$

which is the desired c-number equation for the distribution function. In general, all derivatives with respect to $\alpha_i(t_0)$ occur. In many cases of interest those derivatives higher than the second are missing or are small and (6.4.18) then reduces to the Fokker–Planck equation

$$\begin{aligned}\frac{\partial P_c}{\partial t}(\underset{\sim}{\alpha}_0, t) &= \bar{L}^c\left\{\frac{\partial}{\partial \underset{\sim}{\alpha}_0}, \underset{\sim}{\alpha}_0\right\}P_c(\underset{\sim}{\alpha}_0, t)\\ &\cong -\frac{\partial}{\partial \alpha_i}[A_i(\underset{\sim}{\alpha})P_c] + \frac{\partial^2}{\partial \alpha_i\,\partial \alpha_j}[D_{ij}(\underset{\sim}{\alpha})P_c], \qquad (6.4.19)\end{aligned}$$

where the α's refer to t_0 and we sum over repeated indices from 1 to f. This is the c-number equivalent equation of the master equation for $S(t)$.

Damped Oscillator

To illustrate this procedure we consider the first term on the right side of (6.4.16) for a damped oscillator where chosen order is normal order. Then

$$\begin{aligned}I &= -i\omega \operatorname{Tr} S(t)[e^{-a^\dagger(\partial/\partial\alpha^*)}e^{-a(\partial/\partial\alpha)}, a^\dagger a]\,\delta(\alpha)\,\delta(\alpha^*)\\ &= -i\omega \operatorname{Tr} S(t)\{e^{-a^\dagger(\partial/\partial\alpha^*)}e^{-a(\partial/\partial\alpha)}a^\dagger a - a^\dagger a e^{-a^\dagger(\partial/\partial\alpha^*)}e^{-a(\partial/\partial\alpha)}\}\,\delta(\alpha)\,\delta(\alpha^*).\end{aligned}$$
(6.4.20)

We put the terms in the curly bracket into normal order. Since $[a, F] = \partial F/\partial a^\dagger$ and $[a^\dagger, F] = -\partial F/\partial a$, we see that

$$\begin{aligned}e^{-a(\partial/\partial\alpha)}a^\dagger &= [a^\dagger - (\partial/\partial\alpha)]e^{-a(\partial/\partial\alpha)}\\ a e^{-a^\dagger(\partial/\partial\alpha^*)} &= e^{-a^\dagger(\partial/\partial\alpha^*)}\left(a - \frac{\partial}{\partial\alpha^*}\right).\end{aligned} \qquad (6.4.21)$$

If we use these in (6.4.20), we obtain

$$I = +i\omega \operatorname{Tr} S(t)e^{-a^\dagger(\partial/\partial\alpha^*)}\left[\frac{\partial}{\partial\alpha}a - \frac{\partial}{\partial\alpha^*}a^\dagger\right]e^{-a(\partial/\partial\alpha)}\,\delta(\alpha)\,\delta(\alpha^*), \qquad (6.4.22)$$

which is now in normal order. We may therefore use (3.9.3) and (3.9.4) to write this as

$$\begin{aligned}I = +i\omega \operatorname{Tr} S(t)\int\left(\frac{\partial}{\partial\alpha}\beta - \frac{\partial}{\partial\alpha^*}\beta^*\right)e^{-\beta^*(\partial/\partial\alpha^*)}\\ \times\,\delta(\alpha^*)e^{-\beta(\partial/\partial\alpha)}\,\delta(\alpha)\,\delta(\beta^* - a^\dagger)\,\delta(\beta - a)\,d^2\beta. \qquad (6.4.23)\end{aligned}$$

By arguments like those in (6.4.15a) it follows that

$$e^{-\lambda(\partial/\partial\mu)}\,\delta(\mu) = \delta(\lambda - \mu), \qquad (6.4.24)$$

and we have

$$I = +i\omega \operatorname{Tr} S(t) \int \left(\frac{\partial}{\partial \alpha} \beta - \frac{\partial}{\partial \alpha^*} \beta^* \right) \delta(\beta - \alpha)\, \delta(\beta^* - \alpha^*)\, \delta(\beta^* - a^\dagger)\, \delta(\beta - a)\, d^2\beta$$

$$= +i\omega \operatorname{Tr} S(t) \left[\frac{\partial}{\partial \alpha} \alpha - \frac{\partial}{\partial \alpha^*} \alpha^* \right] \delta(\alpha^* - a^\dagger)\, \delta(\alpha - a)$$

$$= +i\omega \left[\frac{\partial}{\partial \alpha} \alpha - \frac{\partial}{\partial \alpha^*} \alpha^* \right] P_c(\alpha, \alpha^*, t), \tag{6.4.25}$$

which is in the required form.

In the case of a damped oscillator, there is a close connection between the associated distribution function P_c when the chosen order is *normal* order and the *antinormally* ordered function associated with the density operator. This connection may be obtained as follows. The δ-function in *normal* order is

$$\delta^{(n)} \equiv \delta(\alpha^* - a^\dagger)\delta(\alpha - a). \tag{6.4.26}$$

If $|\beta\rangle$ is a coherent state where $F(a)|\beta\rangle = F(\beta)|\beta\rangle$, then the normally ordered function associated with $\delta^{(n)}$ is by (3.2.12)

$$\langle \beta | \delta(\alpha^* - a^\dagger)\, \delta(\alpha - a) | \beta \rangle = \delta(\alpha^* - \beta^*)\, \delta(\alpha - \beta). \tag{6.4.27}$$

If we assume the density operator is in *antinormal* order, we have by (6.4.12)

$$P_{(n)}(\alpha, \alpha^*, t) = \operatorname{Tr} S^{(a)}(a, a^\dagger, t)\, \delta(\alpha^* - a^\dagger)\, \delta(\alpha - a)$$

$$= \iint \frac{d^2\beta}{\pi} \bar{S}^{(a)}(\beta, \beta^*, t)\, \delta(\alpha^* - \beta^*)\, \delta(\alpha - \beta), \tag{6.4.28}$$

where we used (3.2.34) to evaluate the trace. The $\bar{S}^{(a)}(\beta, \beta^*, t)$ is the function associated with the reduced system density operator after it has been put in antinormal order. We also used the normally ordered function (6.4.27) associated with the δ-functions in normal order. If we carry out the β integrals in (6.4.28) we see that

$$P_{(n)}(\alpha, \alpha^*, t) = \frac{1}{\pi} \bar{S}^{(a)}(\alpha, \alpha^*, t), \tag{6.4.29}$$

which shows the desired connection between the associated distribution function for normal ordering and the function associated with the density operator in antinormal order. We therefore see that

$$S^{(a)}(a, a^\dagger, t) = \mathscr{A}\{\bar{S}^{(a)}(\alpha, \alpha^*, t)\} = \pi \mathscr{A}\{P_{(n)}(\alpha, \alpha^*, t)\}. \tag{6.4.30}$$

By (3.2.23) we see that $P_{(n)}(\alpha, \alpha^*, t)$ is identical to the P-representation of the reduced density operator.

Rather than go through the formalism above, we obtain the equation of motion of $P_{(n)}(\alpha, \alpha^*, t)$ directly as follows. By (6.2.59), the reduced density operator obeys the equation of motion

$$\frac{\partial S}{\partial t} = -i(\omega_c + \Delta\omega)\{[a^\dagger, S]a + a^\dagger[a, S]\} - iv(t)[a^\dagger, S] - iv^*(t)[a, S]$$
$$+ \frac{\gamma}{2}[2aSa^\dagger - a^\dagger aS - Sa^\dagger a] + \gamma\bar{n}[a^\dagger Sa + aSa^\dagger - a^\dagger aS - Saa^\dagger], \tag{6.4.31}$$

where we used the identity

$$[a^\dagger a, S] = [a^\dagger, S]a + a^\dagger[a, S]. \tag{6.4.32}$$

Let us now assume we have somehow managed to put $S(a, a^\dagger, t)$ into antinormal order and proceed by means of (3.3.9)

$$S^{(a)}a = aS^{(a)} - \frac{\partial S^{(a)}}{\partial a^\dagger}$$
$$a^\dagger S = Sa^\dagger - \frac{\partial S^{(a)}}{\partial a} \tag{6.4.33}$$

to put all terms in (6.4.31) into antinormal order. We have for example that

$$a^\dagger aS^{(a)} = (aa^\dagger - 1)S^{(a)} = a\left[S^{(a)}a^\dagger - \frac{\partial S^{(a)}}{\partial a}\right] - S^{(a)}$$
$$= aS^{(a)}a^\dagger - a\frac{\partial S^{(a)}}{\partial a} - S^{(a)}, \tag{6.4.34}$$

and it follows that all terms on the right of (6.4.34) are in antinormal order. If we proceed in this way, (6.4.31) may be rewritten in the equivalent form

$$\frac{\partial S^{(a)}}{\partial t} = -i\omega_c'\left\{-a\frac{\partial S^{(a)}}{\partial a} + \frac{\partial S^{(a)}}{\partial a^\dagger}a^\dagger\right\} + \frac{\gamma}{2}\left\{\frac{\partial}{\partial a}(aS^{(a)}) + \frac{\partial}{\partial a^\dagger}(S^{(a)}a^\dagger)\right\}$$
$$+ \gamma\bar{n}\frac{\partial^2 S^{(a)}}{\partial a\,\partial a^\dagger} + iv(t)\frac{\partial S^{(a)}}{\partial a} - iv^*(t)\frac{\partial S^{(a)}}{\partial a^\dagger}, \tag{6.4.35}$$

where we have let

$$\omega_c' = \omega_c + \Delta\omega. \tag{6.4.36}$$

Since now every term is in antinormal order, we may apply the $\mathscr{A}^{-1}$ operator to both sides and use (3.2.10) which gives

$$\frac{\partial \bar{S}^{(a)}}{\partial t}(\alpha, \alpha^*, t) = \left(\frac{\gamma}{2} + i\omega_c'\right)\frac{\partial}{\partial \alpha}(\alpha\bar{S}^{(a)}) + \left(\frac{\gamma}{2} - i\omega_c'\right)\frac{\partial}{\partial \alpha^*}(\alpha^*\bar{S}^{(a)})$$
$$+ \gamma\bar{n}\frac{\partial^2 \bar{S}^{(a)}}{\partial \alpha\,\partial \alpha^*} + iv(t)\frac{\partial \bar{S}^{(a)}}{\partial \alpha} - iv^*(t)\frac{\partial \bar{S}^{(a)}}{\partial \alpha^*}, \tag{6.4.37}$$

where we used (6.4.30). An equation of this type is known as a Fokker–Planck equation in agreement with (6.4.19). In Appendix K we discuss a few of the properties of such equations and we solve (6.4.37) in the following section. It is of course identical to (6.4.19) for the damped driven oscillator by virtue of (6.4.30).

We again remind the reader that if $M(a, a^\dagger)$ is an arbitrary function of a and $a^\dagger$ in the SP, then by (3.2.34), the average under the Markoff approximation is

$$\langle M(a, a^\dagger, t)\rangle = \iint \bar{M}^{(n)}(\alpha, \alpha^*)\bar{S}^{(a)}(\alpha, \alpha^*, t)\frac{d^2\alpha}{\pi}, \tag{6.4.38}$$

where $\bar{M}^{(n)}$ is the function associated with M in the chosen (here *normal*) order.

Homogeneously Broadened Three Level Atoms. We wish to obtain the equation of motion for the associated distribution function for homogeneously broadened atoms.

The equation of motion for the reduced density operator for an atom under the Markoff approximation in the IP is given by (6.2.85). If we use (6.2.36) and (6.2.74), this equation in the SP becomes

$$\frac{\partial S}{\partial t} = \sum_l \epsilon_l \frac{1}{i\hbar}[|l\rangle\langle l|, S] + \sum_{k,l}{}' w_{lk}|l\rangle\langle k|S|k\rangle\langle l|$$
$$+ \sum_{k,l}\{|k\rangle\langle k|S|l\rangle\langle l|(\overset{+}{w}_{llkk} + \overset{-}{w}_{llkk}) - \overset{+}{w}_{kllk}|k\rangle\langle k|S - \overset{-}{w}_{kllk}S|k\rangle\langle k|\}. \tag{6.4.39}$$

The imaginary parts of the $w^{\pm}$'s cause small shifts in the energy levels of the atom which we may neglect for simplicity and retain only their real parts. If we use (6.2.75b) and (6.2.84) we see that

$$\operatorname{Re} \overset{+}{w}_{kllk} = \operatorname{Re} \overset{-}{w}_{kllk} = \tfrac{1}{2}w_{lk}. \tag{6.4.40}$$

By (6.2.87a) we see that

$$\operatorname{Re}(\overset{+}{w}_{llkk} + \overset{-}{w}_{llkk}) = -\operatorname{Re}\Gamma^c_{lk} + \operatorname{Re}\sum_m(\overset{+}{w}_{kmmk} + \overset{-}{w}_{lmml}). \tag{6.4.41}$$

If we use (6.2.107) and (6.4.40), this becomes

$$\operatorname{Re}(\overset{+}{w}_{llkk} + \overset{-}{w}_{llkk}) = -\Gamma_{lk} + \tfrac{1}{2}\sum_m(w_{mk} + w_{ml}). \tag{6.4.42}$$

If we next use (6.2.115) and (6.2.116), we see that (6.4.42) reduces to

$$\operatorname{Re}(\overset{+}{w}_{llkk} + \overset{-}{w}_{llkk}) = -\Gamma_{lk}{}^{\text{ph}} + \tfrac{1}{2}w_{kk} + \tfrac{1}{2}w_{ll}. \tag{6.4.43}$$

If we use (6.4.40) and (6.4.43), (6.4.39) becomes

$$\frac{\partial S}{\partial t} = \sum_l \epsilon_l \frac{1}{i\hbar} [|l\rangle\langle l|, S] + \sum_{k,l}{}' w_{lk}|l\rangle\langle k|S|k\rangle\langle l|$$
$$+ \sum_{k,l} \{|k\rangle\langle k|S|l\rangle\langle l|(-\Gamma_{lk}{}^{\text{ph}} + \tfrac{1}{2}w_{kk} + \tfrac{1}{2}w_{ll}) - \tfrac{1}{2}w_{lk}(|k\rangle\langle k|S + S|k\rangle\langle k|)\}. \tag{6.4.44}$$

Since

$$\sum |l\rangle\langle l| = 1$$
$$\sum |k\rangle\langle k| = 1 \tag{6.4.45}$$

we may simplify the w_{kk} and w_{ll} term. Also we have that

$$\Gamma_{ll}{}^{\text{ph}} = \Gamma_{kk}{}^{\text{ph}} = 0 \tag{6.4.46}$$

by (6.2.117). Accordingly, we may write (6.4.44) as

$$\frac{\partial S}{\partial t} = \sum_l \epsilon_l \frac{1}{i\hbar} [|l\rangle\langle l|, S] + \sum_{k,l}{}' w_{lk}|l\rangle\langle k|S|k\rangle\langle l|$$
$$- \sum_{k,l}{}' \Gamma_{lk}{}^{\text{ph}}|k\rangle\langle k|S|l\rangle\langle l| + \sum_k \tfrac{1}{2}w_{kk}|k\rangle\langle k|S + \sum_l \tfrac{1}{2}w_{ll}S|l\rangle\langle l|$$
$$- \sum_{k,l} \tfrac{1}{2}w_{lk}[|k\rangle\langle k|S + S|k\rangle\langle k|]. \tag{6.4.47}$$

In the term $\frac{1}{2}\sum_l w_{ll}S|l\rangle\langle l|$ we may change the dummy summation index to k and combine the last four terms. This leaves

$$\frac{\partial S}{\partial t} = \sum_l \epsilon_l \frac{1}{i\hbar} [|l\rangle\langle l|, S]$$
$$- \sum_{k,l}{}' \{\Gamma_{lk}{}^{\text{ph}}|k\rangle\langle k|S|l\rangle\langle l| + w_{lk}[|l\rangle\langle k|S|k\rangle\langle l| - \tfrac{1}{2}S|k\rangle\langle k| - \tfrac{1}{2}|k\rangle\langle k|S]\}. \tag{6.4.48}$$

If we use (6.2.115) this may be written in the alternative equivalent form

$$\frac{\partial S}{\partial t} = \frac{1}{i\hbar}\sum_l \epsilon_l[|l\rangle\langle l|, S] - \sum_{k,l}{}' \Gamma_{lk}{}^{\text{ph}}|k\rangle\langle k|S|l\rangle\langle l|$$
$$+ \sum_{k,l}{}' w_{lk}|l\rangle\langle k|S|k\rangle\langle l| - \sum_k \frac{\Gamma_k}{2} [|k\rangle\langle k|S + S|k\rangle\langle k|]. \tag{6.4.49}$$

Suppose next that we have a large number N of identical independent atoms, each coupled to its own reservoir. We assume the damping of all atoms is the same and we say the atoms are homogeneously broadened. In this case if we let $(|k\rangle\langle l|)_\lambda$ be an operator for the λth atom, (6.4.49) becomes

$$\frac{\partial S}{\partial t} = \sum_{\lambda=1}^{N}\left\{\sum_l \left[\left(\frac{\epsilon_l}{i\hbar} - \frac{\Gamma_l}{2}\right)(|l\rangle\langle l|)_\lambda S - \left(\frac{\epsilon_l}{i\hbar} + \frac{\Gamma_l}{2}\right) S(|l\rangle\langle l|)_\lambda\right]\right.$$
$$\left. + \sum_{k,l}{}' [w_{lk}(|l\rangle\langle k|)_\lambda S(|k\rangle\langle l|)_\lambda - \Gamma_{lk}{}^{\text{ph}}(|k\rangle\langle k|)_\lambda S(|l\rangle\langle l|)_\lambda].\right\} \tag{6.4.50}$$

The S is of course in a Hilbert space corresponding to N atoms so that a state vector would be written as $|n_1\rangle\,|n_2\rangle \cdots |n_N\rangle$, a product of state vectors for each atom.

The distribution function by (6.4.13) and (6.4.50) for the N atoms obeys the equation of motion

$$\frac{\partial P_c}{\partial t} = \operatorname{Tr} S(t) \sum_{\lambda=1}^{N} \left\{ \sum_l \left[\left(\frac{\epsilon_l}{i\hbar} - \frac{\Gamma_l}{2} \right) \delta^c(|l\rangle\langle l|)_\lambda - \left(\frac{\epsilon_l}{i\hbar} + \frac{\Gamma_l}{2} \right) (|l\rangle\langle l|)_\lambda \delta^c \right] \right.$$
$$\left. + \sum_{k,l}{}' \left[w_{lk}(|k\rangle\langle l|)_\lambda \delta^c (|l\rangle\langle k|)_\lambda - \Gamma_{lk}{}^{\mathrm{ph}}(|l\rangle\langle l|)_\lambda \delta^c (|k\rangle\langle k|)_\lambda \right] \right\}, \quad (6.4.51)$$

where we used the cyclic property of traces and δ^c is the δ-function product in a chosen order which we must yet specify.

Let us now specialize to the case of N three level atoms which we later need for the theory of a laser. In this case the sums above will run from the ground level which we label 0 to the upper level which we label 2. In the applications we use the ground level only as a source of atoms which can somehow be pumped up to levels 1 and 2 as well as a sink into which atoms may decay. We are primarily concerned with the interaction of these atoms with radiation of frequency $\omega \simeq (\epsilon_2 - \epsilon_1)/\hbar$ which induces radiative transitions between levels 1 and 2. Accordingly, we are interested in the populations of levels 0, 1, and 2 and the dipole moment of the atoms induced by the radiation field between levels 1 and 2. The operators of physical interest are therefore the level populations

$$N_l = \sum_{\lambda=1}^{N} (|l\rangle\langle l|)_\lambda \qquad l = 0, 1, 2, \quad (6.4.52)$$

where it is observed that

$$\sum_{l=0}^{2} N_l = \sum_{\lambda=1}^{N} \sum_{l=0}^{2} (|l\rangle\langle l|)_\lambda = N, \quad (6.4.53)$$

by virtue of the completeness relation and the dipole moment operators

$$M = \sum_{\lambda=1}^{N} (|1\rangle\langle 2|)_\lambda$$
$$M^\dagger = \sum_{\lambda=1}^{N} (|2\rangle\langle 1|)_\lambda. \quad (6.4.54)$$

These operators have the property effectively of "annihilating" an atom in level 2 and "creating" one simultaneously in level 1 for M and vice versa for $M^\dagger$ and summing for all N atoms. These transitions are physically necessary to induce a dipole moment. For our problem a complete set of operators is therefore $M^\dagger$, N_1, N_2, M since $N_0 = N - N_1 - N_2$ and we choose the ordering defined by

$$\delta^c \equiv \delta(\mathscr{M}^* - M^\dagger)\delta(\mathscr{N}_1 - N_1)\delta(\mathscr{N}_2 - N_2)\delta(\mathscr{M} - M). \quad (6.4.55)$$

The script variables play the role of the various α's in the formal theory above.

By (6.4.15) we have

$$\delta(\alpha - a) = e^{-a(\partial/\partial\alpha)}\,\delta(\alpha), \tag{6.4.56}$$

where α is a c-number and a is an operator. Accordingly, we may write (6.4.55) in the very useful form

$$\delta^c \equiv e^{-M\dagger\partial/\partial\mathscr{M}^*}\, e^{-N_1\partial/\partial\mathscr{N}_1}\, e^{-N_2\partial/\partial\mathscr{N}_2}\, e^{-M\partial/\partial\mathscr{M}}\,\delta(\mathscr{M}^*)\,\delta(\mathscr{N}_1)\,\delta(\mathscr{N}_2)\,\delta(\mathscr{M}). \tag{6.4.57}$$

Note that $\mathscr{M}^*$, $\mathscr{M}$, $\mathscr{N}_1$, and $\mathscr{N}_2$ are now considered as independent c-number variables.

We next write out (6.4.50) explicitly for three level atoms. If we measure energy from the ground state, then $\epsilon_0 = 0$. Also we use the completeness relation (6.4.53). This gives on using (6.4.52)–(6.4.54) after minor algebra

$$\begin{aligned}\frac{\partial P_c}{\partial t} = \operatorname{Tr} S(t)\Big\{&-N\Gamma_0\,\delta^c + \left(\frac{\epsilon_1}{i\hbar} - \frac{\Gamma_1}{2}\right)\delta^c N_1 - \left(\frac{\epsilon_1}{i\hbar} + \frac{\Gamma_1}{2}\right) N_1\,\delta^c \\ &+ \left(\frac{\epsilon_2}{i\hbar} - \frac{\Gamma_2}{2}\right)\delta^c N_2 - \left(\frac{\epsilon_2}{i\hbar} + \frac{\Gamma_2}{2}\right) N_2\,\delta^c \\ &+ \sum_\lambda [w_{12}(|2\rangle\langle 1|)_\lambda\,\delta^c\,(|1\rangle\langle 2|)_\lambda + w_{21}(|1\rangle\langle 2|)_\lambda\,\delta^c\,(|2\rangle\langle 1|)_\lambda \\ &+ w_{01}(|1\rangle\langle 0|)_\lambda\,\delta^c\,(|0\rangle\langle 1|)_\lambda + w_{10}(|0\rangle\langle 1|)_\lambda\,\delta^c\,(|1\rangle\langle 0|)_\lambda \\ &+ w_{02}(|2\rangle\langle 0|)_\lambda\,\delta^c\,(|0\rangle\langle 2|)_\lambda + w_{20}(|0\rangle\langle 2|)_\lambda\,\delta^c\,(|2\rangle\langle 0|)_\lambda \\ &- \Gamma_{12}{}^{\mathrm{ph}}(|1\rangle\langle 1|)_\lambda\,\delta^c\,(|2\rangle\langle 2|)_\lambda - \Gamma_{12}{}^{\mathrm{ph}}(|2\rangle\langle 2|)_\lambda\,\delta^c\,(|1\rangle\langle 1_\lambda)|]\Big\}. \end{aligned} \tag{6.4.58}$$

We have used the result (6.2.117) that $\Gamma_{12}{}^{\mathrm{ph}} = \Gamma_{21}{}^{\mathrm{ph}}$. Also the terms proportional to $\Gamma_{10}{}^{\mathrm{ph}} = \Gamma_{20}{}^{\mathrm{ph}}$ vanish since for example

$$(|1\rangle\langle 1|)_\lambda\,\delta^c(|0\rangle\langle 0|)_\lambda = (|1\rangle\langle 1|0\rangle\langle 0|)_\lambda\,\delta^c = 0.$$

This follows since from the definition of the operators M, $M^\dagger$, N_1, and N_2 and the orthogonality relations

$$_\lambda\langle k|l\rangle_\lambda = \delta_{kl} \tag{6.4.59}$$

for the λth atom, we see that $(|0\rangle\langle 0|)_\lambda$ commutes with all operators in δ^c and $\langle 1|0\rangle = 0$.

In the first term in (6.4.58) we should have

$$-\frac{\Gamma_0}{2}\{\delta^c N_0 + N_0\,\delta^c\} = -\frac{\Gamma_0}{2}\{\delta^c(N - N_1 - N_2) + (N - N_1 - N_2)\,\delta^c\}.$$

In our problem it is reasonable to assume that $N_1 \ll N$ and $N_2 \ll N$ without requiring that N_1 and N_2 be small. We have made this approximation. Physically we are assuming that there is such a large supply of atoms in the ground state that we may neglect depletion of the ground state when atoms are pumped into levels 1 and 2.

The next step in the procedure is to put all terms on the right side of (6.4.58) into the chosen order. We must first bring terms like $(|2\rangle\langle 1|)_\lambda$ and $(|1\rangle\langle 2|)_\lambda$ in the w_{12} terms together and reduce them to bilinear form by means of the above orthogonality relation (6.4.59). This process is quite tedious and we carry out the details for two terms and refer the reader to Ref. 8 for further details.

Consider the term

$$\begin{aligned} I &= \operatorname{Tr} S(t)\, \delta^c\, N_1 \\ &= \operatorname{Tr} S(t) e^{-M^\dagger(\partial/\partial \mathscr{M}^*)} e^{-N_1(\partial/\partial \mathscr{N}_1)} e^{-N_2(\partial/\partial \mathscr{N}_2)} e^{-M(\partial/\partial \mathscr{M})} N_1 \\ &\qquad \times \delta(\mathscr{M}^*)\, \delta(\mathscr{N}_1)\, \delta(\mathscr{N}_2)\, \delta(\mathscr{M}), \end{aligned} \tag{6.4.60}$$

where we used the representation (6.4.57) for δ^c and the fact that N_1 commutes with the c-number δ-functions. We wish to commute N_1 through the M and N_2 terms to its left. Since

$$e^{M(\partial/\partial \mathscr{M})} e^{-M(\partial/\partial \mathscr{M})} \equiv 1, \tag{6.4.61}$$

we may write (6.4.60) as

$$\begin{aligned} I = \operatorname{Tr} S(t) &e^{-M^\dagger(\partial/\partial \mathscr{M}^*)} e^{-N_1(\partial/\partial \mathscr{N}_1)} e^{-N_2(\partial/\partial \mathscr{N}_2)} e^{-M(\partial/\partial \mathscr{M})} N_1 \\ &\times e^{M(\partial/\partial \mathscr{M})} e^{-M(\partial/\partial \mathscr{M})}\, \delta(\mathscr{M}^*)\, \delta(\mathscr{M})\, \delta(\mathscr{N}_1)\, \delta(\mathscr{N}_2). \end{aligned} \tag{6.4.62}$$

Consider the term

$$e^{-M(\partial/\partial \mathscr{M})} N_1 e^{M(\partial/\partial \mathscr{M})} = N_1 + \frac{\partial}{\partial \mathscr{M}} M. \tag{6.4.63}$$

To prove this, we first note that since operators for different atoms commute, we have the commutation relation

$$\begin{aligned} [N_1, M] &= \sum_\lambda \sum_{\lambda'} [(|1\rangle\langle 1|)_\lambda, (|1\rangle\langle 2|)_{\lambda'}] \\ &= \sum_\lambda [(|1\rangle\langle 1|)_\lambda, (|1\rangle\langle 2|)_\lambda] \\ &= \sum_\lambda (|1\rangle\langle 2|)_\lambda = M. \end{aligned} \tag{6.4.64}$$

Consider the function

$$f(\xi) = e^{-\xi M} N_1 e^{\xi M}, \tag{6.4.65}$$

where

$$f(0) = N_1. \tag{6.4.66}$$

If we differentiate $f(\xi)$ with respect to ξ, we have

$$\begin{aligned} \frac{df}{d\xi} &= e^{-\xi M}[N_1, M] e^{\xi M} \\ &= e^{-\xi M} M e^{\xi M} = M, \end{aligned} \tag{6.4.67}$$

where we used (6.4.64). If we integrate this and use (6.4.66), we obtain

$$f(\xi) = N_1 + \xi M. \tag{6.4.68}$$

If we identity ξ with $\partial/\partial\mathscr{M}$, (6.4.63) follows. Accordingly, (6.4.62) reduces to

$$I = \text{Tr } S(t)e^{-M^\dagger(\partial/\partial\mathscr{M}^*)}e^{-N_1(\partial/\partial\mathscr{N}_1)}e^{-N_2(\partial/\partial\mathscr{N}_2)}\left[N_1 + M\frac{\partial}{\partial\mathscr{M}}\right]e^{-M(\partial/\partial\mathscr{M})}$$
$$\times\ \delta(\mathscr{M}^*)\ \delta(\mathscr{N}_1)\ \delta(\mathscr{N}_2)\ \delta(\mathscr{M}). \quad (6.4.69)$$

The $M(\partial/\partial\mathscr{M})$ term is now in chosen order. Also it is easy to show that

$$[N_1, N_2] = 0. \quad (6.4.70)$$

Therefore,

$$\begin{aligned} I = {} & \text{Tr } S(t)e^{-M^\dagger(\partial/\partial\mathscr{M}^*)}e^{-N_1(\partial/\partial\mathscr{N}_1)}N_1e^{-N_2(\partial/\partial\mathscr{N}_2)}e^{-M(\partial/\partial\mathscr{M})} \\ & \times\ \delta(\mathscr{M}^*)\ \delta(\mathscr{N}_1)\ \delta(\mathscr{N}_2)\ \delta(\mathscr{M}) \\ & + \frac{\partial}{\partial\mathscr{M}}\text{Tr } S(t)e^{-M^\dagger(\partial/\partial\mathscr{M}^*)}e^{-N_1(\partial/\partial\mathscr{N}_1)}e^{-N_2(\partial/\partial\mathscr{N}_2)}Me^{-M(\partial/\partial\mathscr{M})} \\ & \times\ \delta(\mathscr{M}^*)\ \delta(\mathscr{N}_1)\ \delta(\mathscr{N}_2)\ \delta(\mathscr{M}), \end{aligned} \quad (6.4.71)$$

and we have put all operators into chosen order. We may therefore use (3.9.3) and (3.9.4) to write this as

$$\begin{aligned} I = {} & \text{Tr } S(t)\int\cdots\int[e^{-\mathscr{M}^{*\prime}(\partial/\partial\mathscr{M}^*_1)}e^{-\mathscr{N}_1'(\partial/\partial\mathscr{N}_1)}\mathscr{N}_1'e^{-\mathscr{N}_2'(\partial/\partial\mathscr{N}_2)}e^{-\mathscr{M}'(\partial/\partial\mathscr{M})} \\ & \times\ \delta(\mathscr{M}^*)\ \delta(\mathscr{N}_1)\ \delta(\mathscr{N}_2)\ \delta(\mathscr{M})]\ \delta(\mathscr{M}^{*\prime} - M^\dagger)\ \delta(\mathscr{N}_1' - N_1)\ \delta(\mathscr{N}_2' - N_2) \\ & \times\ \delta(\mathscr{M}' - M)\ d^2\mathscr{M}'\ d\mathscr{N}_1'\ d\mathscr{N}_2' + \frac{\partial}{\partial\mathscr{M}}\text{Tr } S(t)\int\cdots\int[e^{-\mathscr{M}^{*\prime}(\partial/\partial\mathscr{M}^*_1)} \\ & \times\ e^{-\mathscr{N}_1'(\partial/\partial\mathscr{N}_1)}e^{-(\mathscr{N}_2'\partial/\partial\mathscr{N}_2)}e^{-\mathscr{M}'(\partial/\partial\mathscr{M})}\ \mathscr{M}'\ \delta(\mathscr{M}^*)\ \delta(\mathscr{N}_1)\ \delta(\mathscr{N}_2)\ \delta(\mathscr{M})] \\ & \times\ \delta(\mathscr{M}^{*\prime} - M^\dagger)\ \delta(\mathscr{N}_1' - N_1)\ \delta(\mathscr{N}_2' - N_2)\ \delta(\mathscr{M}' - M)\ d^2\mathscr{M}'\ d\mathscr{N}_1'\ d\mathscr{N}_2', \end{aligned}$$
$$(6.4.72)$$

where we replaced the operators by c-numbers designated by primes and multiplied by δ-functions in chosen order. Now by (6.4.15), it follows that

$$e^{-\alpha'(\partial/\partial\alpha)}\ \delta(\alpha) = \delta(\alpha' - \alpha). \quad (6.4.73)$$

If we also use the definition (6.4.12), we see that (6.4.72) reduces to

$$\begin{aligned} I = {} & \int\cdots\int\delta(\mathscr{M}^{*\prime} - \mathscr{M}^*)\ \delta(\mathscr{N}_1' - \mathscr{N}_1)\ \delta(\mathscr{N}_2' - \mathscr{N}_2) \\ & \times\ \delta(\mathscr{M}' - \mathscr{M})\mathscr{N}_1'P_c(\mathscr{M}^{*\prime}, \mathscr{N}_1', \mathscr{N}_2', \mathscr{M}', t)\ d^2\mathscr{M}'\ d\mathscr{N}_1'\ d\mathscr{N}_2' \\ & + \frac{\partial}{\partial\mathscr{M}}\int\cdots\int\delta(\mathscr{M}^{*\prime} - \mathscr{M}^*)\ \delta(\mathscr{N}_1' - \mathscr{N}_1)\ \delta(\mathscr{N}_2' - \mathscr{N}_2) \\ & \times\ \delta(\mathscr{M}' - \mathscr{M})\mathscr{M}'P_c(\mathscr{M}^{*\prime}, \ldots, \mathscr{M}', t)\ d^2\mathscr{M}'\ d\mathscr{N}'\ d\mathscr{N}_2' \\ = {} & \left[\mathscr{N}_1 + \frac{\partial}{\partial\mathscr{M}}\mathscr{M}\right]P_c(\mathscr{M}^*, \mathscr{N}_1, \mathscr{N}_2, \mathscr{M}, t), \end{aligned} \quad (6.4.74)$$

which is the desired result for this term.

Consider next the term

$$I = \mathrm{Tr}\, S(t) \sum_{\lambda} (|2\rangle\langle 1|)_\lambda \delta^c(|1\rangle\langle 2|)_\lambda$$
$$= \mathrm{Tr}\, S(t) \sum_{\lambda} (|2\rangle\langle 1|)_\lambda e^{-M^\dagger(\partial/\partial\mathscr{M}^*)} e^{-N_1(\partial/\partial\mathscr{N}_1)} e^{-N_2(\partial/\partial\mathscr{N}_2)} e^{-M(\partial/\partial\mathscr{M})}$$
$$\times (|1\rangle\langle 2|)_\lambda\, \delta(\mathscr{M}^*)\, \delta(\mathscr{N}_1)\, \delta(\mathscr{N}_2)\, \delta(\mathscr{M}). \tag{6.4.75}$$

Since operators for different atoms commute, we may use (6.4.52)–(6.4.54) to write this as

$$I = \mathrm{Tr}\, S(t) \sum_{\lambda} (|2\rangle\langle 1|)_\lambda$$
$$\times \prod_{\lambda'} [e^{-|2\rangle\langle 1|(\partial/\partial\mathscr{M}^*)} e^{-|1\rangle\langle 1|(\partial/\partial\mathscr{N}_1)} e^{-|2\rangle\langle 2|(\partial/\partial\mathscr{N}_2)} e^{-|1\rangle\langle 2|(\partial/\partial\mathscr{M})}]_{\lambda'}$$
$$\times (|1\rangle\langle 2|)_\lambda\, \delta(\mathscr{M}^*) \cdots \delta(\mathscr{M}) \tag{6.4.76}$$
$$= \mathrm{Tr}\, S(t) \prod_{\lambda' \neq \lambda}{}' T_{\lambda'}$$
$$\times \sum_{\lambda} [|2\rangle\langle 1| e^{-|2\rangle\langle 1|(\partial/\partial\mathscr{M}^*)} e^{-|1\rangle\langle 1|(\partial/\partial\mathscr{N}_1)} e^{-|2\rangle\langle 2|(\partial/\partial\mathscr{N}_2)} e^{-|1\rangle\langle 2|(\partial/\partial\mathscr{M})} |1\rangle\langle 2|]_\lambda$$
$$\times \delta(\mathscr{M}^*)\, \delta(\mathscr{N}_1)\, \delta(\mathscr{N}_2)\, \delta(\mathscr{M}), \tag{6.4.77}$$

where we have let

$$T_{\lambda'} = [e^{-|2\rangle\langle 1|(\partial/\partial\mathscr{M}^*)} e^{-|1\rangle\langle 1|(\partial/\partial\mathscr{N}_1)} e^{-|2\rangle\langle 2|(\partial/\partial\mathscr{N}_2)} e^{-|1\rangle\langle 2|(\partial/\partial\mathscr{M})}]_{\lambda'}. \tag{6.4.78}$$

Our first task is to commute $|1\rangle\langle 2|$ and $|2\rangle\langle 1|$ through the various terms until they come together at the middle. Now I may be obviously rewritten as

$$I = \mathrm{Tr}\, S(t) \prod_{\lambda'}{}' T_{\lambda'} \sum_{\lambda} [e^{-|2\rangle\langle 1|(\partial/\partial\mathscr{M}^*)} e^{-|1\rangle\langle 1|(\partial/\partial\mathscr{N}_1)} e^{+|1\rangle\langle 1|(\partial/\partial\mathscr{N}_1)} |2\rangle\langle 1|$$
$$\times e^{-|1\rangle\langle 1|(\partial/\partial\mathscr{N}_1)} e^{-|2\rangle\langle 2|(\partial/\partial\mathscr{N}_2)} |1\rangle\langle 2| e^{|2\rangle\langle 2|(\partial/\partial\mathscr{N}_2)}$$
$$\times e^{-|2\rangle\langle 2|(\partial/\partial\mathscr{N}_2)} e^{-|1\rangle\langle 2|(\partial/\partial\mathscr{M})}]_\lambda\, \delta(\mathscr{M}^*) \cdots \delta(\mathscr{M}), \tag{6.4.79}$$

since, for example,

$$[|2\rangle\langle 1|, e^{-|2\rangle\langle 1|(\partial/\partial\mathscr{M}^*)}] = 0,$$

and

$$e^{-|1\rangle\langle 1|(\partial/\partial\mathscr{N}_1)} e^{|1\rangle\langle 1|(\partial/\partial\mathscr{N}_1)} \equiv 1.$$

Consider the term

$$e^{|1\rangle\langle 1|(\partial/\partial\mathscr{N}_1)} |2\rangle\langle 1| e^{-|1\rangle\langle 1|(\partial/\partial\mathscr{N}_1)}$$
$$= [1 + (e^{\partial/\partial\mathscr{N}_1} - 1)|1\rangle\langle 1|]|2\rangle\langle 1|[1 + (e^{-(\partial/\partial\mathscr{N}_1)} - 1)|1\rangle\langle 1|]$$
$$= |2\rangle\langle 1| + (e^{-(\partial/\partial\mathscr{N}_1)} - 1)|2\rangle\langle 1|$$
$$= e^{-(\partial/\partial\mathscr{N}_1)}|2\rangle\langle 1|, \tag{6.4.80}$$

where we expanded the exponential operators and used the orthogonality relations. Similarly, we see that

$$e^{-|2\rangle\langle 2|(\partial/\partial\mathcal{N}_2)}|1\rangle\langle 2|e^{|2\rangle\langle 2|(\partial/\partial\mathcal{N}_2)} = e^{\partial/\partial\mathcal{N}_2}|1\rangle\langle 2|. \tag{6.4.81}$$

If we use these (6.4.79) reduces to

$$\begin{aligned}
I &= \mathrm{Tr}\, S(t) \prod_{\lambda'}{}' T_{\lambda'} \sum_{\lambda} [e^{-|2\rangle\langle 1|(\partial/\partial\mathcal{M}^*)} e^{-|1\rangle\langle 1|(\partial/\partial\mathcal{N}_1)} e^{-(\partial/\partial\mathcal{N}_1)}|2\rangle\langle 1|1\rangle\langle 2| \\
&\quad \times e^{\partial/\partial\mathcal{N}_2} e^{-|2\rangle\langle 2|(\partial/\partial\mathcal{N}_2)} e^{-|1\rangle\langle 2|(\partial/\partial\mathcal{M})}]_{\lambda}\, \delta(\mathcal{M}^*)\, \delta(\mathcal{M})\, \delta(\mathcal{N}_1)\, \delta(\mathcal{N}_2) \\
&= e^{(\partial/\partial\mathcal{N}_2)-(\partial/\partial\mathcal{N}_1)}\, \mathrm{Tr}\, S(t) \prod_{\lambda'}{}' T_{\lambda'} \sum_{\lambda} [e^{-|2\rangle\langle 1|(\partial/\partial\mathcal{M}^*)} e^{-|1\rangle\langle 1|(\partial/\partial\mathcal{N}_1)}|2\rangle\langle 2| \\
&\quad \times e^{-|2\rangle\langle 2|(\partial/\partial\mathcal{N}_2)} e^{-|1\rangle\langle 2|(\partial/\partial\mathcal{M})}]_{\lambda}\, \delta(\mathcal{M}^*) \cdots \delta(\mathcal{M}) \\
&= e^{(\partial/\partial\mathcal{N}_2)-(\partial/\partial\mathcal{N}_1)}\, \mathrm{Tr}\, S(t) \sum_{\lambda} e^{-M^\dagger(\partial/\partial\mathcal{M}^*)} e^{-N_1(\partial/\partial\mathcal{N}_1)}(|2\rangle\langle 2|)_{\lambda} \\
&\quad \times e^{-N_2(\partial/\partial\mathcal{N}_2)} e^{-M(\partial/\partial\mathcal{M})}\, \delta(\mathcal{M}^*) \cdots \delta(\mathcal{M}) \\
&= e^{(\partial/\partial\mathcal{N}_2)-(\partial/\partial\mathcal{N}_1)}\, \mathrm{Tr}\, S(t) e^{-M^\dagger(\partial/\partial\mathcal{M}^*)} e^{-N_1(\partial/\partial\mathcal{N}_1)} N_2 e^{-N_2(\partial/\partial\mathcal{N}_2)} \\
&\quad \times e^{-M(\partial/\partial\mathcal{M})}\, \delta(\mathcal{M}^*) \cdots \delta(\mathcal{M}),
\end{aligned} \tag{6.4.82}$$

where the intermediate steps should be self-explanatory. Since this is now in chosen order, we see by arguments like those used in (6.4.72)–(6.4.74) that

$$I = e^{(\partial/\partial\mathcal{N}_2)-(\partial/\partial\mathcal{N}_1)}\mathcal{N}_2 P_c(\mathcal{M}^*, \mathcal{N}_1, \mathcal{N}_2, \mathcal{M}, t), \tag{6.4.83}$$

which is the desired form of the term.

If we continue in this way with all terms, the result is

$$\begin{aligned}
\frac{\partial P_c}{\partial t} &= \left\{-(R_1 + R_2) + \left(\frac{\epsilon_1}{i\hbar} - \frac{\Gamma_1}{2}\right)\left(\mathcal{N}_1 + \frac{\partial}{\partial\mathcal{M}}\mathcal{M}\right)\right. \\
&\quad + \left(-\frac{\epsilon_1}{i\hbar} - \frac{\Gamma_1}{2}\right)\left(\mathcal{N}_1 + \frac{\partial}{\partial\mathcal{M}^*}\mathcal{M}^*\right) + \left(\frac{\epsilon_2}{i\hbar} - \frac{\Gamma_2}{2}\right)\left(\mathcal{N}_2 - \frac{\partial}{\partial\mathcal{M}}\mathcal{M}\right) \\
&\quad + \left(-\frac{\epsilon_2}{i\hbar} - \frac{\Gamma_2}{2}\right)\left(\mathcal{N}_2 - \frac{\partial}{\partial\mathcal{M}^*}\mathcal{M}^*\right) + w_{12}\left[\exp\left(\frac{\partial}{\partial\mathcal{N}_2} - \frac{\partial}{\partial\mathcal{N}_1}\right)\right]\mathcal{N}_2 \\
&\quad + w_{21}\left[\left(\exp\left(\frac{\partial}{\partial\mathcal{N}_1} - \frac{\partial}{\partial\mathcal{N}_2}\right) + \frac{\partial^2}{\partial\mathcal{M}\,\partial\mathcal{M}^*}\right)\mathcal{N}_1\right. \\
&\quad + \left(\frac{\partial^2}{\partial\mathcal{M}\,\partial\mathcal{M}^*} + \exp\left(\frac{\partial}{\partial\mathcal{N}_2} - \frac{\partial}{\partial\mathcal{N}_1}\right)\frac{\partial^4}{\partial\mathcal{M}^2\,\partial\mathcal{M}^{*2}}\right)\mathcal{N}_2 \\
&\quad + \left(\exp\left(\frac{\partial}{\partial\mathcal{N}_1} - \frac{\partial}{\partial\mathcal{N}_2}\right)\frac{\partial}{\partial\mathcal{M}} + \frac{\partial^3}{\partial\mathcal{M}^2\,\partial\mathcal{M}^*}\right)\mathcal{M} \\
&\quad \left. + \left(\exp\left(\frac{\partial}{\partial\mathcal{N}_1} - \frac{\partial}{\partial\mathcal{N}_2}\right)\frac{\partial}{\partial\mathcal{M}^*} + \frac{\partial^3}{\partial\mathcal{M}^{*2}\,\partial\mathcal{M}}\right)\mathcal{M}^*\right]
\end{aligned}$$

$$+ w_{01}\left[\left(\exp\frac{\partial}{\partial\mathcal{N}_1}\right)\left(\mathcal{N}_1 + \frac{\partial}{\partial\mathcal{M}}\mathcal{M} + \frac{\partial}{\partial\mathcal{M}^*}\mathcal{M}^*\right)\right.$$

$$\left. + \left(\exp\frac{\partial}{\partial\mathcal{N}_2}\right)\frac{\partial^2}{\partial\mathcal{M}\,\partial\mathcal{M}^*}\mathcal{N}_2\right]$$

$$+ w_{10}\left[\left(\exp -\frac{\partial}{\partial\mathcal{N}_1}\right)(N - \mathcal{N}_1 - \mathcal{N}_2)\right] + w_{02}\left(\exp\frac{\partial}{\partial\mathcal{N}_2}\right)\mathcal{N}_2$$

$$+ w_{20}\left[\left(\exp -\frac{\partial}{\partial\mathcal{N}_2} + \frac{\partial^2}{\partial\mathcal{M}\,\partial\mathcal{M}^*}\exp -\frac{\partial}{\partial\mathcal{N}_1}\right)(N - \mathcal{N}_1 - \mathcal{N}_2)\right]$$

$$+ \Gamma_{12}{}^{\mathrm{ph}}\left[\exp\left(\frac{\partial}{\partial\mathcal{N}_2} - \frac{\partial}{\partial\mathcal{N}_1}\right)\frac{\partial^2}{\partial\mathcal{M}\,\partial\mathcal{M}^*}2\mathcal{N}_2\right.$$

$$\left.\left. + \frac{\partial}{\partial\mathcal{M}}\mathcal{M} + \frac{\partial}{\partial\mathcal{M}^*}\mathcal{M}^*\right]\right\}P_c, \tag{6.4.84}$$

where we have let

$$R_i = Nw_{i0} \qquad i = 1, 2 \tag{6.4.85}$$

be the rate at which atoms are pumped by the reservoir from level 0 to level i. If we note by (6.2.116) that

$$\Gamma_{12} = \Gamma_{12}{}^{ph} + \tfrac{1}{2}(\Gamma_1 + \Gamma_2), \tag{6.4.86}$$

and by (6.2.115) that

$$\begin{aligned}\Gamma_1 &= w_{01} + w_{21}\\ \Gamma_2 &= w_{02} + w_{12},\\ \Gamma_0 &= w_{10} + w_{20}\end{aligned} \tag{6.4.87}$$

and if we add and subtract the terms

$$(w_{01} + w_{21})\left(\frac{\partial}{\partial\mathcal{M}}\mathcal{M}P_c + \frac{\partial}{\partial\mathcal{M}^*}\mathcal{M}^*P_c\right)$$

to (6.4.84), we find after minor rearrangement that the equation for the associated distribution function becomes

$$\frac{\partial P_c}{\partial t} = \left\{\frac{\partial}{\partial\mathcal{M}}(\Gamma_{12} + i\omega_{21})\mathcal{M} + \frac{\partial}{\partial\mathcal{M}^*}(\Gamma_{12} - i\omega_{21})\mathcal{M}^*\right.$$

$$+ \left[\left(\exp -\frac{\partial}{\partial\mathcal{N}_1}\right) - 1\right]R_1 + \left[\left(\exp\frac{\partial}{\partial\mathcal{N}_1}\right) - 1\right]w_{01}\mathcal{N}_1$$

$$+ \left[\left(\exp -\frac{\partial}{\partial\mathcal{N}_2}\right) - 1\right]R_2 + \left[\left(\exp\frac{\partial}{\partial\mathcal{N}_2}\right) - 1\right]w_{02}\mathcal{N}_2$$

$$+ \left[\exp\left(\frac{\partial}{\partial \mathcal{N}_1} - \frac{\partial}{\partial \mathcal{N}_2}\right) - 1\right] w_{21}\mathcal{N}_1$$

$$+ \left[\exp\left(\frac{\partial}{\partial \mathcal{N}_2} - \frac{\partial}{\partial \mathcal{N}_1}\right) - 1\right] w_{12}\mathcal{N}_2$$

$$+ \frac{\partial^2}{\partial \mathcal{M}\,\partial \mathcal{M}^*}\left[\left(\exp - \frac{\partial}{\partial \mathcal{N}_1}\right) R_2 + \left(\exp \frac{\partial}{\partial \mathcal{N}_2}\right) w_{01}\mathcal{N}_2\right.$$

$$\left. + \exp\left(\frac{\partial}{\partial \mathcal{N}_2} - \frac{\partial}{\partial \mathcal{N}_1}\right) 2\Gamma_{12}{}^{\mathrm{ph}}\mathcal{N}_2 + w_{21}(\mathcal{N}_1 + \mathcal{N}_2)\right]$$

$$+ \frac{\partial}{\partial \mathcal{M}}\left[\exp\left(\frac{\partial}{\partial \mathcal{N}_1} - \frac{\partial}{\partial \mathcal{N}_2}\right) - 1\right] w_{21}\mathcal{M}$$

$$+ \frac{\partial}{\partial \mathcal{M}}\left[\left(\exp \frac{\partial}{\partial \mathcal{N}_1}\right) - 1\right] w_{01}\mathcal{M}$$

$$+ \frac{\partial}{\partial \mathcal{M}^*}\left[\exp\left(\frac{\partial}{\partial \mathcal{N}_1} - \frac{\partial}{\partial \mathcal{N}_2}\right) - 1\right] w_{21}\mathcal{M}^*$$

$$+ \frac{\partial}{\partial \mathcal{M}^*}\left[\left(\exp \frac{\partial}{\partial \mathcal{N}_1}\right) - 1\right] w_{01}\mathcal{M}^*$$

$$+ \frac{\partial^3}{\partial \mathcal{M}^2\,\partial \mathcal{M}^*} w_{21}\mathcal{M} + \frac{\partial^3}{\partial \mathcal{M}\,\partial \mathcal{M}^{*2}} w_{21}\mathcal{M}^*$$

$$+ \frac{\partial^4}{\partial \mathcal{M}^2\,\partial \mathcal{M}^{*2}} \exp\left(\frac{\partial}{\partial \mathcal{N}_2} - \frac{\partial}{\partial \mathcal{N}_1}\right) w_{21}\mathcal{N}_2\Bigg\} P_c, \tag{6.4.88}$$

where we let

$$\hbar\omega_{21} = \epsilon_2 - \epsilon_1, \tag{6.4.89}$$

and we have again neglected $\mathcal{N}_1$ and $\mathcal{N}_2$ compared with N which means we have neglected depletion of the ground state. The $\mathcal{N}_1$ and $\mathcal{N}_2$ may still be large compared with unity. Since derivatives of all orders are involved, we say this is a generalized Fokker–Planck equation. Since $\mathcal{N}_1$, $\mathcal{N}_2$, $\mathcal{M}$, and $\mathcal{M}^*$ are large, we see, for example, that

$$\exp\left(-\frac{\partial}{\partial \mathcal{N}_1}\right) - 1 \cong -\frac{\partial}{\partial \mathcal{N}_1} + \frac{1}{2}\frac{\partial^2}{\partial \mathcal{N}_1{}^2} - \frac{1}{6}\frac{\partial^3}{\partial \mathcal{N}_1{}^3} + \ldots,$$

and the higher derivative terms may be neglected since they are down by order $\mathcal{N}_1{}^{-1}$. If we retain only terms up to second order, we obtain the Fokker–Planck equation [see (6.4.19)].

$$\frac{\partial P_c}{\partial t} = \left\{\frac{\partial}{\partial \mathcal{M}}(\Gamma_{12} + i\omega_{21})\mathcal{M} + \frac{\partial}{\partial \mathcal{M}^*}(\Gamma_{12} - i\omega_{21})\mathcal{M}^*\right.$$

$$
\begin{aligned}
&- \frac{\partial}{\partial \mathscr{N}_1}[R_1 - \Gamma_1 \mathscr{N}_1 + w_{12}\mathscr{N}_2] - \frac{\partial}{\partial \mathscr{N}_2}[R_2 + w_{21}\mathscr{N}_1 - \Gamma_2 \mathscr{N}_2] \\
&+ \frac{1}{2}\frac{\partial^2}{\partial \mathscr{N}_1^{\,2}}[R_1 + w_{12}\mathscr{N}_2 + \Gamma_1 \mathscr{N}_1] \\
&+ \frac{1}{2}\frac{\partial^2}{\partial \mathscr{N}_2^{\,2}}[R_2 + w_{21}\mathscr{N}_1 + \Gamma_2 \mathscr{N}_2] \\
&+ \frac{\partial^2}{\partial \mathscr{N}_1 \, \partial \mathscr{N}_2}[-w_{12}\mathscr{N}_2 - w_{21}\mathscr{N}_1] \\
&+ \frac{\partial^2}{\partial \mathscr{N}_1 \, \partial \mathscr{M}} \Gamma_1 \mathscr{M} + \frac{\partial^2}{\partial \mathscr{N}_1 \, \partial \mathscr{M}^*} \Gamma_1 \mathscr{M}^* \\
&+ \frac{\partial^2}{\partial \mathscr{N}_2 \, \partial \mathscr{M}} (-w_{21}\mathscr{M}) + \frac{\partial^2}{\partial \mathscr{N}_2 \, \partial \mathscr{M}^*} (-w_{21}\mathscr{M}^*) \\
&+ \frac{\partial^2}{\partial \mathscr{M} \, \partial \mathscr{M}^*}[R_2 + (\Gamma_1 + 2\Gamma_{12}^{\mathrm{ph}})\mathscr{N}_2 + w_{21}\mathscr{N}_1]\Big\} P_c. \qquad (6.4.90)
\end{aligned}
$$

The first derivative terms give the drift or mean motion while the second derivatives give the diffusion as shown in Appendix K and below.

We have therefore shown that the equation of motion for the c-number distribution function obeys a Fokker–Planck equation for a damped harmonic oscillator and for N three level atoms when we neglect depletion of atoms in the ground state.

Equation of Motion for System Operator

We may also use our ordering techniques to convert the equation of motion for system operators in the HP under the Markoff approximation to c-number equations.

We begin by considering again (6.3.7) and (6.3.38):

$$
\begin{aligned}
\frac{d}{dt}\langle M(t)\rangle &\cong \operatorname{Tr} M \frac{\partial S}{\partial t} \\
&= \operatorname{Tr}_S S(t)\Big\{\frac{1}{i\hbar}[M, H] \\
&\quad - \sum_{i,j} \delta(\omega_i, -\omega_j)([M, Q_i]Q_j w_{ij}^{+} - Q_j[M, Q_i]w_{ji}^{-})\Big\}, \qquad (6.4.91)
\end{aligned}
$$

where all system operators are in the SMP at t_0, and

$$
\begin{aligned}
\frac{d}{dt}\langle M(t)\rangle &\approx \operatorname{Tr} S(t_0)\left\langle \frac{dM}{dt} \right\rangle_R \\
&= \operatorname{Tr} S(t_0)\Big\langle \Big\{\frac{1}{i\hbar}[M, H] \\
&\quad - \sum_{i,j} \delta(\omega_i, -\omega_j)([M, Q_i]Q_j w_{ij}^{+} - Q_j[M, Q_i]w_{ji}^{-})\Big\}\Big\rangle, \qquad (6.4.92)
\end{aligned}
$$

where all system operators are in the HMP at time t. That the *form* of these two equations is the same is of the utmost importance. In evaluating these means in the SMP, we may visualize the operators as fixed and the coordinates (state vectors) rotating while in the HMP we visualize the system operators as rotating and the coordinates as fixed just as we do in the case of a true SP or HP.

In the HMP, because the forms above are the same, we may order the operators at time t, rather than at t_0, as we did in the SMP. Thus we have $a_1(t)a_2(t)\cdots a_f(t) \equiv \underset{\sim}{a}(t)$ or $\underset{\sim}{a}_t$. The functional form of M^c ordered at t and t_0 will obviously be the same. Therefore, we see that

$$\begin{aligned}\langle M^c(\underset{\sim}{a}, t)\rangle &= \mathrm{Tr}\,\{S(t_0)\langle M^c(\underset{\sim}{a}_t, t)\rangle_R\} \\ &= \mathrm{Tr}\, S(t_0)\langle \mathscr{C}\{\bar{M}^c(\underset{\sim}{\alpha}_t, t)\rangle_R\} \\ &= \mathrm{Tr}\, S(t_0)\int \bar{M}^c(\underset{\sim}{\alpha}_t, t)\langle \delta^c(\underset{\sim}{\alpha}_t - \underset{\sim}{a}_t)\rangle_R\, d\underset{\sim}{\alpha}_t,\end{aligned} \tag{6.4.93}$$

where we let $\underset{\sim}{a}_t \equiv \underset{\sim}{\alpha}(t)$ be our independent variables. If we use (6.4.92), we have

$$\begin{aligned}\frac{d\langle M^c(\underset{\sim}{a}, t)\rangle}{dt} &= \mathrm{Tr}\, S(t_0)\frac{d}{dt}\langle \mathscr{C}\{\bar{M}^c(\underset{\sim}{\alpha}_t, t)\}\rangle_R \\ &= \mathrm{Tr}\, S(t_0)\int \bar{M}^c(\underset{\sim}{\alpha}_t, t)\, d\underset{\sim}{\alpha}_t\Big\langle\Big\{\frac{1}{i\hbar}[\delta^c(\underset{\sim}{\alpha}_t - \underset{\sim}{a}_t), H] \\ &\quad - \sum_{i,j}([\delta^c(\underset{\sim}{\alpha}_t - \underset{\sim}{a}_t), Q_i]Q_j w_{ij}^+ - Q_j[\delta^c(\underset{\sim}{\alpha}_t - \underset{\sim}{a}_t), Q_i]w_{ji}^-\Big\}\Big\rangle_R.\end{aligned} \tag{6.4.94}$$

The expression in the curly brackets now has the identical form as in (6.4.14) except the variables here are $\underset{\sim}{\alpha}_t$ and $\underset{\sim}{a}_t$ instead of $\underset{\sim}{\alpha}_0$ and $\underset{\sim}{a}_0$. If we therefore proceed exactly as in (6.4.15)–(6.4.17) we see that

$$\begin{aligned}\frac{d\langle M(t)\rangle}{dt} &= \mathrm{Tr}\, S(t_0)\int \bar{M}^c(\underset{\sim}{\alpha}_t, t)\, d\underset{\sim}{\alpha}_t\, \bar{L}^c\Big(\frac{\partial}{\partial\underset{\sim}{\alpha}_t}, \underset{\sim}{\alpha}_t\Big)\langle \delta^c(\underset{\sim}{\alpha}_t - \underset{\sim}{a}_t)\rangle_R \\ &= \mathrm{Tr}\, S(t_0)\Big\langle\frac{d}{dt}\mathscr{C}\{\bar{M}^c(\underset{\sim}{\alpha}_t, t)\}\Big\rangle_R.\end{aligned} \tag{6.4.95}$$

In case $\bar{L}^c$ contains derivatives only up to second order, it follows from (6.4.19) that

$$\begin{aligned}\frac{d\langle M(t)\rangle}{dt} &= \mathrm{Tr}\, S(t_0)\frac{d}{dt}\langle \mathscr{C}\{\bar{M}^c(\underset{\sim}{\alpha}_t, t)\}\rangle_R \\ &= \mathrm{Tr}\, S(t_0)\int \bar{M}^c(\underset{\sim}{\alpha}_t, t)\, d\underset{\sim}{\alpha}_t\Big\{-\frac{\partial}{\partial\alpha_{it}}\mathscr{A}_i(\underset{\sim}{\alpha}_t) + \frac{\partial^2}{\partial\alpha_{it}\,\partial\alpha_{jt}}\mathscr{D}_{ij}(\underset{\sim}{\alpha}_t)\Big\} \\ &\quad \times \langle \delta^c(\underset{\sim}{\alpha}_t - \underset{\sim}{a}_t)\rangle_R,\end{aligned} \tag{6.4.96}$$

where we again sum from 1 to f over repeated indices and $\alpha_{it} \equiv \alpha_i(t)$. If we now integrate the first derivative terms by parts once and the second derivative terms by parts twice [the integrated parts vanish because the δ-functions are zero at the limits], we obtain

$$\frac{d\langle M(t)\rangle}{dt} = \mathrm{Tr}\, S(t_0)\frac{d}{dt}\langle \mathscr{C}\{\bar{M}^c(\underset{\sim}{\alpha}_t)\}\rangle_R$$
$$= \mathrm{Tr}\, S(t_0)\int d\underset{\sim}{\alpha}_t\langle \delta^c(\underset{\sim}{\alpha}_t - \underset{\sim}{a}_t)\rangle_R\left\{\mathscr{A}_i(\underset{\sim}{\alpha}_t)\frac{\partial \bar{M}^c(\underset{\sim}{\alpha}_t)}{\partial \alpha_{it}} + \mathscr{D}_{ij}(\underset{\sim}{\alpha}_t)\frac{\partial^2 \bar{M}^c(\underset{\sim}{\alpha}_t)}{\partial \alpha_{it}\,\partial \alpha_{jt}}\right\}. \tag{6.4.97}$$

We should next observe that the integration over the δ-function has the effect of putting the quantity in the curly bracket into c-order; that is,

$$\int \langle \delta^c(\underset{\sim}{\alpha}_t - \underset{\sim}{a}_t)\rangle_R F(\underset{\sim}{\alpha}_t)\, d\alpha_t \equiv \langle \mathscr{C}\{F(\underset{\sim}{\alpha}_t)\}\rangle_R = \langle F^c(\underset{\sim}{a}_t)\rangle_R. \tag{6.4.98}$$

Therefore, (6.4.97) may be written as

$$\frac{d\langle M(t)\rangle}{dt} = \mathrm{Tr}\, S(t_0)\frac{d}{dt}\langle \mathscr{C}\{\bar{M}^c(\underset{\sim}{\alpha}_t)\}\rangle_R$$
$$= \mathrm{Tr}\, S(t_0)\left\langle \mathscr{C}\left\{\mathscr{A}_i(\underset{\sim}{\alpha}_t)\frac{\partial \bar{M}^c(\underset{\sim}{\alpha}_t)}{\partial \alpha_{it}} + \mathscr{D}_{ij}(\underset{\sim}{\alpha}_t)\frac{\partial^2 \bar{M}^c(\underset{\sim}{\alpha}_t)}{\partial \alpha_{it}\,\partial \alpha_{jt}}\right\}\right\rangle_R. \tag{6.4.99}$$

Next it follows directly since $S(t_0)$ is completely arbitrary that the arguments of the traces must be equal so that we have

$$\frac{d}{dt}\langle \mathscr{C}\{\bar{M}^c(\underset{\sim}{\alpha}_t)\}\rangle_R = \left\langle \mathscr{C}\left\{\mathscr{A}_i(\underset{\sim}{\alpha}_t)\frac{\partial \bar{M}^c(\underset{\sim}{\alpha}_t)}{\partial \alpha_{it}} + \mathscr{D}_{ij}(\underset{\sim}{\alpha}_t)\frac{\partial^2 \bar{M}^c(\underset{\sim}{\alpha}_t)}{\partial \alpha_{it}\,\partial \alpha_{jt}}\right\}\right\rangle_R. \tag{6.4.100}$$

Under the Markoff approximation we may interchange the ordering and differentiation operation

$$\frac{d}{dt}\langle \mathscr{C}\{\bar{M}^c(\underset{\sim}{\alpha}_t)\}\rangle_R = \mathscr{C}\left\{\left\langle \frac{d}{dt}\bar{M}^c(\underset{\sim}{\alpha}_t)\right\rangle_R\right\}. \tag{6.4.101}$$

As a result, since both sides of (6.4.100) are in chosen order, we may operate on both sides with $\mathscr{C}^{-1}$ and obtain the c-number equations

$$\frac{d\langle \bar{M}^c(\underset{\sim}{\alpha}_t)\rangle_R}{dt} = \left\langle \mathscr{A}_i(\underset{\sim}{\alpha}_t)\frac{\partial \bar{M}^c(\underset{\sim}{\alpha}_t)}{\partial \alpha_{it}}\right\rangle_R + \left\langle \mathscr{D}_{ij}(\underset{\sim}{\alpha}_t)\frac{\partial^2 \bar{M}^c(\underset{\sim}{\alpha}_t)}{\partial \alpha_{it}\,\partial \alpha_{jt}}\right\rangle_R, \tag{6.4.102}$$

where we sum i and j from 1 to f. Note that $\bar{M}^c(\underset{\sim}{\alpha}_t)$ may still be an operator as far as the reservoir variables are concerned. This will be discussed in more detail in Chapter 7.

If we let $M = a_k$ we have $\bar{M}^c = \alpha_k(t)$ then it follows from (6.4.102) that

$$\frac{d}{dt}\langle\alpha_k(t)\rangle_R = \langle\mathscr{A}_k(\underset{\sim}{\alpha}_t)\rangle_R. \tag{6.4.103}$$

We therefore see that the $\mathscr{A}_k$ first derivative terms in the Fokker–Planck equation give the mean drift motion of the corresponding operator a_k.

If we let $M = a_k a_l$ and these operators are in chosen order, then $\bar{M}^c = \alpha_k(t)\alpha_l(t)$ and by (6.4.102) we obtain

$$\frac{d}{dt}\langle\alpha_k(t)\alpha_l(t)\rangle_R = \langle\alpha_k(t)\mathscr{A}_l(\underset{\sim}{\alpha}_t) + \alpha_l(t)\mathscr{A}_k(\underset{\sim}{\alpha}_t) + 2\mathscr{D}_{kl}(\underset{\sim}{\alpha}_t)\rangle_R. \tag{6.4.104}$$

So we see that if we know the mean equation of motion for α_k and α_l, we know $\mathscr{A}_k$ and $\mathscr{A}_l$ and if in addition we know the mean equation of motion for $\alpha_k\alpha_l$, we can obtain the diffusion coefficients from (6.4.104). If we apply the $\mathscr{C}$ operator to both sides of (6.4.104) we obtain

$$\mathscr{C}\langle\mathscr{D}_{kl}\rangle_R = \left\langle\frac{d}{dt}\mathscr{C}(\alpha_k\alpha_l)\right\rangle_R - \langle\mathscr{C}\{\alpha_k\mathscr{A}_l + \alpha_l\mathscr{A}_k\}\rangle_R. \tag{6.4.105}$$

Damped Oscillator. As a simple example, consider the Fokker–Planck equation for the damped oscillator for normal order given in (6.4.37). We have by inspection

$$\begin{aligned} \mathscr{A}_\alpha &= -\left(\frac{\gamma}{2} + i\omega_c'\right)\alpha - iv(t) \\ \mathscr{A}_{\alpha^*} &= \mathscr{A}_\alpha^* \\ 2\mathscr{D}_{\alpha^*\alpha} &= \gamma\bar{n}; \qquad \mathscr{D}_{\alpha\alpha} = \mathscr{D}_{\alpha^*\alpha^*} = 0. \end{aligned} \tag{6.4.106}$$

Then (6.4.103) becomes

$$\frac{d\langle\alpha\rangle_R}{dt} = -\left[\frac{\gamma}{2} + i\omega_c'\right]\langle\alpha\rangle_R + iv(t), \tag{6.4.107}$$

and the solution is

$$\langle\alpha(t)\rangle_R = \alpha(0)e^{-[(\gamma/2)+i\omega_c']t} + i\int_0^t v(t')e^{[(\gamma/2)+i\omega_c'](t'-t)}\,dt', \tag{6.4.108a}$$

or on applying the $\mathscr{C}$ operator

$$\langle a(t)\rangle_R = a(0)e^{-[(\gamma/2)+i\omega_c']t} + i\int_0^t v(t')e^{[(\gamma/2)+i\omega_c'](t'-t)}\,dt'. \tag{6.4.108b}$$

Also if $M = a^\dagger a$, we have that

$$\frac{d}{dt}\langle\mathscr{C}\{\alpha^*\alpha\}\rangle_R = \langle\mathscr{C}\{\alpha^*\mathscr{A}_\alpha + \alpha\mathscr{A}_\alpha^* + \gamma\bar{n}\}\rangle_R, \tag{6.4.109a}$$

or

$$\frac{d}{dt}\langle a^\dagger(t)a(t)\rangle_R = -\gamma\langle a^\dagger(t)a(t)\rangle_R + \gamma\bar{n} - iv(t)\langle a^\dagger(t)\rangle_R + iv^*(t)\langle a(t)\rangle_R. \tag{6.4.109b}$$

The associated c-number equation is

$$\frac{d}{dt}\langle \alpha^*\alpha\rangle_R = -\gamma\langle \alpha^*\alpha\rangle_R + \gamma\bar{n} - iv(t)\langle \alpha^*\rangle_R + iv^*(t)\langle \alpha\rangle_R. \tag{6.4.109c}$$

Three-Level Atoms. As a second example, the Fokker–Planck equation for a set of N-damped three-level atoms (6.4.90) gives by inspection the following drift vectors and diffusion coefficients for the ordering specified:

$$\begin{aligned}
\mathscr{A}_{\mathscr{M}} &= -(\Gamma_{12} + i\omega_{21})\mathscr{M} = \mathscr{A}^*_{\mathscr{M}^*} \\
\mathscr{A}_{\mathscr{N}_1} &= +(R_1 - \Gamma_1\mathscr{N}_1 + w_{12}\mathscr{N}_2) \\
\mathscr{A}_{\mathscr{N}_2} &= +(R_2 + w_{21}\mathscr{N}_1 - \Gamma_2\mathscr{N}_2) \\
2\mathscr{D}_{\mathscr{N}_1\mathscr{N}_1} &= [R_1 + w_{12}\mathscr{N}_2 - \Gamma_1\mathscr{N}_1] \\
2\mathscr{D}_{\mathscr{N}_2\mathscr{N}_2} &= R_2 + w_{21}\mathscr{N}_1 - \Gamma_2\mathscr{N}_2 \\
2\mathscr{D}_{\mathscr{N}_1\mathscr{N}_2} &= -w_{12}\mathscr{N}_2 - w_{21}\mathscr{N}_1 \\
2\mathscr{D}_{\mathscr{N}_1\mathscr{M}} &= \Gamma_1\mathscr{M} = 2\mathscr{D}^*_{\mathscr{N}_1\mathscr{M}^*} \\
2\mathscr{D}_{\mathscr{N}_2\mathscr{M}} &= -w_{12}\mathscr{M} = 2\mathscr{D}^*_{\mathscr{N}_2\mathscr{M}^*} \\
2\mathscr{D}_{\mathscr{M}^*\mathscr{M}} &= R_2 + (\Gamma_1 + 2\Gamma_{12}{}^{\text{ph}})\mathscr{N}_2 + w_{21}\mathscr{N}_1.
\end{aligned} \tag{6.4.110}$$

We therefore see by (6.4.103) that

$$\begin{aligned}
\frac{d\langle\mathscr{N}_1\rangle_R}{dt} &= R_1 - \Gamma_1\langle\mathscr{N}_1\rangle_R + w_{12}\langle\mathscr{N}_2\rangle_R \\
\frac{d\langle\mathscr{N}_2\rangle_R}{dt} &= R_2 + w_{21}\langle\mathscr{N}_1\rangle_R - \Gamma_2\langle\mathscr{N}_2\rangle_R \\
\frac{d\langle\mathscr{M}\rangle_R}{dt} &= -(\Gamma_{12} + i\omega_{21})\langle\mathscr{M}\rangle_R.
\end{aligned} \tag{6.4.111}$$

The first two are simple classical rate equations for the populations in level 1 and 2. Since

$$\begin{aligned}
\Gamma_1 &= w_{01} + w_{21} \\
\Gamma_2 &= w_{02} + w_{12},
\end{aligned} \tag{6.4.112}$$

we see that $\langle d\mathscr{N}_1/dt\rangle_R$ gives the net change in the average population of level 1. The $R_1 = Nw_{10}$ is the rate at which the reservoir "pumps" atoms from the ground state to level 1. The $\Gamma_1\langle\mathscr{N}_1\rangle_R$ term gives the rate of loss of atoms from level 1 to levels 0 and 2 while the $w_{12}\langle\mathscr{N}_2\rangle_R$ term gives the rate of atoms entering level 1 from level 2. The w_{ij} give the transition probability per second for atoms to go from level j to level i as we have shown earlier. A similar interpretation can be given to the equation for $\langle d\mathcal{N}_2/dt\rangle_R$.

The solution of the $\mathcal{M}$ equation is readily seen to be

$$\langle \mathcal{M}(t)\rangle_R = \mathcal{M}(0)e^{-(\Gamma_{12}+i\omega_a)t}, \tag{6.4.113}$$

which shows that the dipole moment decays in a time of order Γ_{12}^{-1}.

The populations approach steady state ($d/dt = 0$) values given by

$$\begin{aligned}\langle \mathcal{N}_1(t)\rangle_{R,\text{ss}} &= \frac{R_1\Gamma_2 + w_{12}R_2}{\Gamma_1\Gamma_2 - w_{12}w_{21}}\\ \langle \mathcal{N}_2(t)\rangle_{R,\text{ss}} &= \frac{R_1 w_{21} + \Gamma_1 R_2}{\Gamma_1\Gamma_2 - w_{12}w_{21}}.\end{aligned} \tag{6.4.114}$$

The populations will be inverted when the population difference satisfies the inequality

$$\langle(\mathcal{N}_2 - \mathcal{N}_1)\rangle_{R,\text{ss}} = \frac{R_1(w_{21} - \Gamma_2) + R_2(\Gamma_1 - w_{12})}{\Gamma_1\Gamma_2 - w_{12}w_{21}} > 0. \tag{6.4.115}$$

When $\langle\mathcal{N}_2\rangle_{R,\text{ss}} = \langle\mathcal{N}_1\rangle_{R,\text{ss}}$, the average populations in the two levels are equal and we say the transition is saturated.

We obtain the transient solution of the rate equations later.

6.5 SOLUTION OF THE FOKKER–PLANCK EQUATION

Damped Oscillator [5]

In this section we solve the Fokker–Planck equation for the damped harmonic oscillator in two different ways and obtain characteristic functions.

Eigenvalues and Eigenfunctions. The Fokker–Planck equation (6.4.37) for a damped driven oscillator under the Markoff approximation is

$$\begin{aligned}\frac{\partial P}{\partial t} = \left(\frac{\gamma}{2} + i\omega_c\right)\frac{\partial}{\partial \alpha}(\alpha P) + \left(\frac{\gamma}{2} - i\omega_c'\right)\frac{\partial}{\partial \alpha^*}(\alpha^* P) + \gamma\bar{n}\frac{\partial^2 P}{\partial\alpha\,\partial\alpha^*}\\ + iv(t)\frac{\partial P}{\partial \alpha} - iv^*(t)\frac{\partial P}{\partial \alpha^*},\end{aligned} \tag{6.5.1a}$$

where we have let

$$P(\alpha, \alpha^*, t) = \frac{1}{\pi}\bar{S}^{(a)}(\alpha, \alpha^*, t), \tag{6.5.1b}$$

and P is normalized so that

$$\begin{aligned}\text{Tr}\, S^{(a)}(a, a^\dagger, t) &= \text{Tr}\iint \bar{S}^{(a)}(\alpha, \alpha^*, t)|\alpha\rangle\langle\alpha|\,\frac{d^2\alpha}{\pi}\\ &= \iint P(\alpha, \alpha^*, t)\, d^2\alpha = 1,\end{aligned} \tag{6.5.1c}$$

since $\text{Tr}\,|\alpha\rangle\langle\alpha| = \langle\alpha|\alpha\rangle = 1$.

We begin by removing the high-frequency behavior from (6.5.1a) which is equivalent to going to the IP. For this purpose, we let

$$\begin{aligned} \alpha &= \beta e^{-i\omega_c' t} \\ \alpha^* &= \beta^* e^{i\omega_c' t} \\ P(\alpha, \alpha^*, t) &= p(\beta, \beta^*, t). \end{aligned} \tag{6.5.2}$$

Then

$$\begin{aligned} \frac{\partial}{\partial \alpha}\alpha &= \frac{\partial}{\partial \beta}\beta = \left(\frac{\partial}{\partial \beta^*}\beta^*\right)^* \\ \frac{\partial^2}{\partial \alpha\,\partial \alpha^*} &= \frac{\partial^2}{\partial \beta\,\partial \beta^*} \\ \frac{\partial}{\partial \alpha} &= e^{i\omega_c' t}\frac{\partial}{\partial \beta} = \left(e^{-i\omega_c' t}\frac{\partial}{\partial \beta^*}\right)^* \\ \frac{\partial P}{\partial t} &= \frac{\partial p}{\partial \beta}\frac{\partial \beta}{\partial t} + \frac{\partial p}{\partial \beta^*}\frac{\partial \beta^*}{\partial t} + \frac{\partial p}{\partial t} \\ &= i\omega_c'\left(\beta\frac{\partial p}{\partial \beta} - \beta^*\frac{\partial p}{\partial \beta}\right) + \frac{\partial p}{\partial t} \\ &= i\omega_c'\left[\frac{\partial}{\partial \beta}(\beta p) - \frac{\partial}{\partial \beta^*}(\beta^* p)\right] + \frac{\partial p}{\partial t}. \end{aligned} \tag{6.5.3}$$

If we use these we see that (6.5.1a) becomes

$$\frac{\partial p}{\partial t}(\beta, \beta^*, t) = \frac{\gamma}{2}\left[\frac{\partial}{\partial \beta}(\beta p) + \frac{\partial}{\partial \beta^*}(\beta^* p)\right] + \gamma\bar{n}\frac{\partial^2 p}{\partial \beta\,\partial \beta^*} + iv(t)e^{i\omega_c t}\frac{\partial p}{\partial \beta} - iv^*(t)e^{-i\omega_c t}\frac{\partial p}{\partial \beta^*}. \tag{6.5.4}$$

If we restrict our driving term to

$$v(t) = v_0 e^{-i\omega_c' t}, \tag{6.5.5}$$

then (6.5.4) becomes separable in the time. If we let

$$p(\beta, \beta^*, t) = e^{-\lambda t}Q(\beta, \beta^*), \tag{6.5.6}$$

then (6.5.4) reduces to the eigenvalue equation

$$LQ = -\lambda Q, \tag{6.5.7}$$

where the operator L is

$$L \equiv \frac{\gamma}{2}\left[\frac{\partial}{\partial \beta}\left(\beta + \frac{2i}{\gamma}v_0\right) + \frac{\partial}{\partial \beta^*}\left(\beta^* - \frac{2i}{\gamma}v_0\right)\right] + \gamma\bar{n}\frac{\partial^2}{\partial \beta\,\partial \beta^*}. \tag{6.5.8}$$

To simplify we may introduce the real variables x and y defined by

$$\begin{aligned} \beta + \frac{2iv_0}{\gamma} &= \sqrt{\bar{n}}(x + iy) \\ \beta^* - \frac{2iv_0^*}{\gamma} &= \sqrt{\bar{n}}(x - iy). \end{aligned} \tag{6.5.9}$$

It is easy to show that

$$\begin{aligned} \frac{\partial}{\partial\beta}\left(\beta + \frac{2iv_0}{\gamma}\right) &= \frac{1}{2}\left[\frac{\partial}{\partial x} - i\frac{\partial}{\partial y}\right](x + iy) \\ \frac{\partial}{\partial\beta^*}\left(\beta^* - \frac{2iv_0}{\gamma}\right) &= \frac{1}{2}\left[\frac{\partial}{\partial x} + i\frac{\partial}{\partial y}\right](x - iy) \\ \bar{n}\frac{\partial^2}{\partial\beta\,\partial\beta^*} &= \frac{\partial}{\partial\alpha}\left[\frac{1}{2}\left(\frac{\partial}{\partial x} + i\frac{\partial}{\partial y}\right)\right] \\ &= \frac{1}{4}\left(\frac{\partial}{\partial x} - i\frac{\partial}{\partial y}\right)\left(\frac{\partial}{\partial x} + i\frac{\partial}{\partial y}\right) \\ &= \frac{1}{4}\left[\frac{\partial^2}{\partial x^2} + \frac{\partial^2}{\partial y^2}\right], \end{aligned} \tag{6.5.10}$$

so that (6.5.7) becomes

$$\left\{\frac{\gamma}{2}\left(x\frac{\partial}{\partial x} + y\frac{\partial}{\partial y} + 2\right) + \frac{\gamma}{4}\left(\frac{\partial^2}{\partial x^2} + \frac{\partial^2}{\partial y^2}\right)\right\}Q(x, y) = -\lambda Q(x, y). \tag{6.5.11}$$

We would like to change variables to eliminate the first derivative terms. When this is done, the equation is said to be self-adjoint. To attempt this elimination the standard procedure is to let

$$Q(x, y) = e^{-\chi(x,y)}R(x, y). \tag{6.5.12}$$

If we substitute this into (6.5.11), we obtain

$$\begin{aligned} &\frac{\partial^2 R}{\partial x^2} + \frac{\partial^2 R}{\partial y^2} + 2\left(x - \frac{\partial\chi}{\partial x}\right)\frac{\partial R}{\partial x} + 2\left(y - \frac{\partial\chi}{\partial y}\right)\frac{\partial R}{\partial y} \\ &\quad + \left[\left(\frac{\partial\chi}{\partial x}\right)^2 + \left(\frac{\partial\chi}{\partial y}\right)^2 - \frac{\partial^2\chi}{\partial x^2} - \frac{\partial^2\chi}{\partial y^2} - 2x\frac{\partial\chi}{\partial x} - 2y\frac{\partial\chi}{\partial y}\right]R \\ &\qquad = -\left(\frac{4\lambda}{\gamma} + 2\right)R. \end{aligned} \tag{6.5.13}$$

We choose χ to eliminate the first derivatives $\partial R/\partial x$ and $\partial R/\partial y$:

$$\frac{\partial \chi}{\partial x} = x$$
$$\frac{\partial \chi}{\partial y} = y, \tag{6.5.14}$$

so that

$$\chi = \frac{x^2}{2} + \frac{y^2}{2}. \tag{6.5.15}$$

With this choice, (6.5.13) reduces to

$$\frac{\partial^2 R}{\partial x^2} + \frac{\partial^2 R}{\partial y^2} + [\epsilon - x^2 - y^2]R = 0, \tag{6.5.16}$$

where we have let

$$\epsilon = \frac{4\lambda}{\gamma} + 2, \tag{6.5.17}$$

and

$$p(\beta, \beta^*, t) = e^{-\lambda t} e^{-\frac{1}{2}(x^2+y^2)} R(x, y). \tag{6.5.18}$$

Equation (6.5.16) is now self-adjoint (see Appendix K) and must be solved subject to the boundary conditions that $R(x, y) \to 0$ as $x \to \pm\infty$ and $y \to \pm\infty$. This equation with these boundary conditions is well-known since it is just the Schrödinger equation for a two-dimensional isotropic harmonic oscillator. The eigenvalues are given by

$$\epsilon_{n_x, n_y} = \frac{4}{\gamma} \lambda_{n_x, n_y} + 2 = (2n_x + 1) + (2n_y + 1), \tag{6.5.19}$$

where n_x and $n_y = 0, 1, 2, 3, \ldots$. The eigenfunctions are just the oscillator eigenfunctions given by

$$R_{n_x, n_y}(x, y) = N_{n_x, n_y} e^{-\frac{1}{2}(x^2+y^2)} H_{n_x}(x) H_{n_y}(y), \tag{6.5.20}$$

where $H_n(x)$ is the hermite polynomial and

$$N_{n_x, n_y} = \left[\frac{1}{\sqrt{\pi}\, 2^{n_x} n_x!} \frac{1}{\sqrt{\pi}\, 2^{n_y} n_y!}\right]^{\frac{1}{2}} \tag{6.5.21}$$

is a normalizing constant such that

$$\iint_{-\infty}^{\infty} R^2_{n_x, n_y}(x, y)\, dx\, dy = 1. \tag{6.5.22}$$

Because (6.5.16) is self-adjoint, the eigenfunctions are orthogonal so that

$$\iint_{-\infty}^{\infty} R_{n_x,n_y}(x, y) R_{m_x,m_y}(x, y)\, dx\, dy = \delta_{n_x,m_x} \delta_{n_y,m_y}, \tag{6.5.23}$$

since

$$\int_{-\infty}^{\infty} e^{-u^2} H_n(u) H_m(u)\, du = \delta_{nm} 2^n n!\, \sqrt{\pi}. \tag{6.5.24}$$

The eigenfunctions $R_{n_x,n_y}(x, y)$ form a complete set so that

$$\sum_{n_x=0}^{\infty} \sum_{n_y=0}^{\infty} R_{n_x,n_y}(x, y) R_{n_x,n_y}(x', y') = \delta(x - x')\, \delta(y - y'). \tag{6.5.25}$$

To state this another way, if the eigenfunctions form a complete set, we may expand $f(x, y)$ as

$$f(x, y) = \sum_{n_x=0}^{\infty} \sum_{n_y=0}^{\infty} c_{n_x,n_y} R_{n_x,n_y}(x, y). \tag{6.5.26}$$

If we multiply both sides by $R_{m_x,m_y}(x, y)$, integrate over all x and y and use the orthogonality relations (6.5.23), we see that the expansion coefficients are given by

$$c_{n_x,n_y} = \iint_{-\infty}^{\infty} dx'\, dy'\, f(x', y') R_{n_x,n_y}(x', y'). \tag{6.5.27}$$

If we substitute this into (6.5.26), we have

$$f(x, y) = \iint_{-\infty}^{\infty} dx'\, dy'\, f(x', y') \sum_{n_x=0}^{\infty} \sum_{n_y=0}^{\infty} R_{n_x,n_y}(x, y) R_{n_x,n_y}(x', y'), \tag{6.5.28}$$

where we interchanged the order of integration and summation. We see that (6.5.25) must follow in order that $f(x, y) = f(x, y)$.

If we use (6.5.18), (6.5.19), and (6.5.20), we see that the eigenfunctions are given by

$$\begin{aligned} p_{n_x,n_y}(x, y, t) &= A N_{n_x,n_y} e^{-(\gamma/2)(n_x+n_y)t} e^{-(x^2+y^2)} H_{n_x}(x) H_{n_y}(y) \\ &= A e^{-(\gamma/2)(n_x+n_y)t} e^{-\frac{1}{2}(x^2+y^2)} R_{n_x,n_y}(x, y). \end{aligned} \tag{6.5.29}$$

We choose A so that each eigenfunction is normalized to unity. By (6.5.1c) and (6.5.2), we have

$$\iint p(\beta, \beta^*, t)\, d^2\beta = \iint_{-\infty}^{\infty} p_{n_x,n_y}(x, y, t) \bar{n}\, dx\, dy = 1, \tag{6.5.30}$$

since $d^2\beta = \bar{n}\, dx\, dy$ by (6.5.9). We have by (6.5.29) that

$$\iint_{-\infty}^{\infty} dx\, dy\, \bar{n} A N_{n_x,n_y} e^{-(\gamma/2)(n_x+n_y)t} e^{-x^2} H_{n_x}(x) e^{-y^2} H_{n_y}(y) \equiv X. \quad (6.5.31)$$

If we note that $H_0(u) = 1$ and use (6.5.24), we see that

$$\int_{-\infty}^{\infty} e^{-x^2} H_n(x)\, dx = \delta_{n0}\sqrt{\pi}. \quad (6.5.32)$$

Therefore,

$$X = A\sqrt{\pi}\,\bar{n} = 1,$$

where by (6.5.21) $N_{00} = \pi^{-1/2}$. With this choice of A, the eigenfunctions become

$$p_{n_x,n_y}(x, y, t) = \frac{1}{\bar{n}\sqrt{\pi}} N_{n_x,n_y} e^{-(\gamma/2)(n_x+n_y)t} e^{-(x^2+y^2)} H_{n_x}(x) H_{n_y}(y)$$

$$= \frac{1}{\bar{n}\sqrt{\pi}} e^{-(\gamma/2)(n_x+n_y)t} e^{-1/2(x^2+y^2)} R_{n_x,n_y}(x, y). \quad (6.5.34)$$

The steady-state solution is the lowest order eigenfunction

$$p_{00}(x, y) = \frac{1}{\pi\bar{n}} \exp -(x^2 + y^2) = \frac{1}{\pi\bar{n}} \exp -\frac{1}{\bar{n}}\left|\beta + \frac{2iv_0}{\gamma}\right|^2. \quad (6.5.35)$$

All other eigenfunctions vanish as $t \to \infty$.

Next we obtain the conditional probability or Green's function solution for (6.5.7). This is the solution which satisfies the conditions at $t = 0$ that $x = x'$ and $y = y'$. We therefore take a linear superposition of eigenfunctions

$$p(x, y, t) = \sum_{n_x,n_y} c_{n_x,n_y} p_{n_x,n_y}(x, y, t). \quad (6.5.36)$$

At $t = 0$, we require that

$$p(x, y, 0) = \delta(x - x')\,\delta(y - y') = \sum_{n_x,n_y} c_{n_x,n_y} A R_{n_x,n_y}(x, y) e^{-1/2(x^2+y^2)}, \quad (6.5.37)$$

where we used (6.5.34) at $t = 0$. If we multiply both sides by $R_{m_x,m_y}(x, y) \times \exp +\frac{1}{2}(x^2 + y^2)$, integrate over all x and y and use (6.5.25) we have

$$\iint_{-\infty}^{\infty} e^{1/2(x^2+y^2)} R_{m_x,m_y}(x, y)\,\delta(x - x')\,\delta(y - y')\, dx\, dy$$

$$= A \sum_{n_x,n_y} c_{n_x,n_y} \delta_{n_x,m_x} \delta_{n_y,m_y} \quad (6.5.38)$$

or

$$A c_{m_x,m_y} = e^{1/2(x'^2+y'^2)} R_{m_x,m_y}(x', y'). \quad (6.5.39)$$

Therefore, (6.5.36) becomes when we use (6.5.34) the Green's function $p(x, y, t/x', y', 0) =$

$$\sum_{n_x, n_y} e^{-(\gamma/2)(n_x+n_y)t} \frac{R_{n_x,n_y}(x, y)R_{n_x,n_y}(x', y')}{e^{-(x'^2+y'^2)}} e^{-\frac{1}{2}(x^2+y^2)} e^{-\frac{1}{2}(x'^2+y'^2)}. \tag{6.5.40}$$

Clearly, at $t = 0$ we see by (6.5.25) that (6.5.37) is satisfied. This solution is the conditional probability of having x and y at time t given they had the values x' and y' at $t = 0$. If we use (6.5.34) and the steady-state solution (6.5.35), this may be written as

$$p(x, y, t/x', y', 0) = \bar{n} \sum_{n_x, n_y=0}^{\infty} \frac{p_{n_x,n_y}(x, y, t)p_{n_x,n_y}(x', y', 0)}{p_{00}(x', y')}. \tag{6.5.41}$$

If $p(x', y')$ is the probability the system has the values x' and y' at $t = 0$, then we have

$$p(x, y, t) = \iint p(x, y, t/x', y', 0)p(x', y')\, dx'\, dy', \tag{6.5.42}$$

as the total probability of finding the system at x and y at time t. If the system is initially in the steady-state, then $p(x', y') = p_{00}(x', y')$ so that by (6.5.41) we have

$$\begin{aligned} p(x, y, t) &= \iint_{-\infty}^{\infty} dx'\, dy'\, \bar{n} \sum_{n_x, n_y=0}^{\infty} p_{n_x,n_y}(x, y, t)p_{n_x,n_y}(x', y', 0) \\ &= \sum_{n_x, n_y=0}^{\infty} N_{n_x,n_y} p_{n_x,n_y}(x, y, t) \iint_{-\infty}^{\infty} \frac{1}{\sqrt{\pi}} e^{-(x'^2+y'^2)} H_{n_x}(x') H_{n_y}(y')\, dx'\, dy' \\ &= p_{00}(x, y), \end{aligned} \tag{6.5.43}$$

where we used (6.5.24). Therefore, the system will always remain in the steady-state.

General Solution for Arbitrary Driving Term. The eigenvalue technique used above is not applicable when the driving term $v(t)$ in (6.5.4) is arbitrary since we cannot then separate out the time dependence. In this section we derive a general solution when $v(t)$ is arbitrary.

We begin by finding the Green's function or conditional probability of finding the system at β and β^* at time t given that it had the values β' and β'^* at $t = 0$ when the system-reservoir interaction was turned on. This initial condition may be represented by the δ-functions

$$\delta(\beta - \beta')\, \delta(\beta^* - \beta'^*) = \lim_{\epsilon \to \infty} \frac{\epsilon}{\pi} e^{-\epsilon(\beta-\beta')(\beta^*-\beta'^*)} \tag{6.5.44}$$

[see (3.5.6)]. The representation of the δ-functions by a gaussian is particularly appropriate for the method of solution that we use.

We now look for a solution of the equation (6.5.4)

$$\frac{\partial p}{\partial t} = \frac{\gamma}{2}\left[\frac{\partial}{\partial \beta}(\beta p) + \frac{\partial}{\partial \beta^*}(\beta^* p)\right] + \gamma \bar{n} \frac{\partial^2 p}{\partial \beta\, \partial \beta^*} + iv(t)e^{i\omega_c t}\frac{\partial p}{\partial \beta} - iv^*(t)e^{-i\omega_c t}\frac{\partial p}{\partial \beta^*} \quad (6.5.45)$$

of the form

$$p(\beta, \beta^*, t/\beta', \beta'^*, 0) = e^{G(t)}, \quad (6.5.46a)$$

where

$$G(t) = -\frac{1}{\zeta(t)}[\beta - \eta(t)][\beta^* - \eta^*(t)] + \ln \nu(t). \quad (6.5.46b)$$

We could include higher powers of β and β^* but, as we shall see, we can, by suitable choice of η, η^*, ζ, and ν, solve (6.5.45) subject to the initial condition

$$p(\beta, \beta^*, 0/\beta', \beta'^*, 0) = \lim_{\epsilon \to \infty} \frac{\epsilon}{\pi} e^{-\epsilon(\beta-\beta')(\beta^*-\beta'^*)}, \quad (6.5.47)$$

with only the terms given. As we have shown in (3.5.5), since $\mathscr{A}\{p(\beta, \beta^*, 0/\beta', \beta'^*, 0\} = 1/\pi\, |\beta'\rangle\langle\beta'|$, this is a coherent state. Comparison of this with $G(0)$ shows that

$$\zeta(0) = \frac{1}{\epsilon}; \qquad \eta(0) = \beta'; \qquad \eta^*(0) = \beta'^*; \qquad \nu(0) = \frac{\epsilon}{\pi}. \quad (6.5.48)$$

If we put (6.5.46) into (6.5.45) and equate the coefficients of equal powers of β and β^*, we obtain after minor algebraic simplification the set of equations

$$\frac{d\zeta}{dt} = -\gamma\zeta + \gamma\bar{n} \quad (6.5.49a)$$

$$\frac{d\eta}{dt} = -\frac{\gamma}{2}\eta - iv(t)e^{i\omega_c t} \quad (6.5.49b)$$

$$\frac{1}{\nu}\frac{d\nu}{dt} = -\frac{1}{\zeta}\frac{d\zeta}{dt} \quad (6.5.49c)$$

together with the conjugate of the $d\eta/dt$ equation. The solutions of these which satisfy the initial conditions above are easily seen to be

$$\zeta(t) = \frac{1}{\epsilon}e^{-\gamma t} + \bar{n}(1 - e^{-\gamma t}) \xrightarrow[\epsilon \to \infty]{} \bar{n}(1 - e^{-\gamma t})$$
$$\eta(t) = \beta' e^{-(\gamma/2)t} + w(t)e^{i\omega_c t} \quad (6.5.50)$$
$$\nu(t) = \frac{1}{\pi\zeta(t)} \xrightarrow[\epsilon \to \infty]{} \frac{1}{\pi\bar{n}(1 - e^{-\gamma t})},$$

where

$$w(t) = -i\int_0^t v(t - t')e^{-[i\omega_c+(\gamma/2)]t'}\,dt'. \tag{6.5.51}$$

Therefore, the conditional probability is given by

$$p(\beta, \beta^*, t/\beta', \beta'^*, 0) = \frac{1}{\pi\bar{n}(1 - e^{-\gamma t})}\exp\left\{-\frac{|\beta - \beta' e^{-(\gamma/2)t} + w(t)e^{i\omega_c t}|^2}{\bar{n}(1 - e^{-\gamma t})}\right\}. \tag{6.5.52}$$

Clearly, as $t \to 0$, $w(t) \to 0$, and $\bar{n}(1 - e^{-t\gamma}) \equiv 1/\epsilon \to 0$ so that $p \xrightarrow[t\to 0]{} \delta(\beta - \beta')\,\delta(\beta^* - \beta'^*)$. Note that

$$\iint d^2\beta\, p\,(\beta, \beta^*, t/\beta', \beta'^*, 0) = 1, \tag{6.5.53}$$

so that the Green's function is normalized to unity. Furthermore, the effect of diffusion is clearly illustrated by this solution. At $t = 0$, the distribution function is a δ-function but as time goes on it spreads to a gaussian centered at

$$\beta = \beta' e^{-\gamma t} - w(t)e^{i\omega_c t}, \tag{6.5.54a}$$

and variance

$$\zeta(t) = \bar{n}(1 - e^{-\gamma t}). \tag{6.5.54b}$$

In the special case

$$v(t) = v_0 e^{-i\omega_c t}, \tag{6.5.55}$$

we see that

$$e^{iw_c t}w(t) = i\frac{2}{\gamma}v_0[e^{-(\gamma/2)t} - 1], \tag{6.5.56}$$

and

$$p(\beta, \beta^*, t/\beta', \beta'^*, 0)$$
$$= \frac{1}{\pi\bar{n}(1 - e^{-\gamma t})}\exp\left\{-\frac{\left|\beta - \beta' e^{-(\gamma/2)t} + \dfrac{2iv_0}{\gamma}(e^{-(\gamma/2)t} - 1)\right|^2}{\bar{n}(1 - e^{-\gamma t})}\right\}, \tag{6.5.57}$$

which is another representation of the Green's function given in (6.5.40).

For the case above, the steady-state solution is

$$p_{ss}(\beta, \beta^*) \xrightarrow[t\to\infty]{} \frac{1}{\pi\bar{n}}\exp -\left\{\frac{1}{\bar{n}}\left|\beta - \frac{2iv_0}{\gamma}\right|^2\right\}, \tag{6.5.58}$$

which agrees with (6.5.35).

The general solution of the Fokker-Planck equation (6.5.45) is

$$p(\beta, \beta^*, t) = \iint p\,(\beta, \beta^*, t/\beta', \beta'^*, 0)\, p(\beta', \beta'^*)\, d^2\beta', \tag{6.5.59}$$

where $p(\beta, \beta'^*)$ is the probability the system has the values β', β'^* at $t = 0$.

Another interesting limiting case occurs when the reservoir is at absolute zero. Then

$$\bar{n} = \frac{1}{e^{\hbar\omega_c/kT} - 1} \xrightarrow[T\to 0]{} 0 \tag{6.5.60}$$

and the Green's function (6.5.52) approaches

$$p(\beta, \beta^*, t/\beta', \beta'^*, 0) \xrightarrow[T\to 0]{} \delta[\beta - \beta' e^{-(\gamma/2)t} + w(t)e^{i\omega_c t}]$$
$$\times\ \delta[\beta^* - \beta'^* e^{-(\gamma/2)t} + w^*(t)e^{-i\omega_c t}]. \tag{6.5.61}$$

Therefore, if the system is initially in a coherent state, it will always remain in a coherent state whose center of gravity moves according to (6.5.54a). In other words, when the reservoir is at absolute zero, there is no diffusion to cause the distribution function to spread. This may also be seen directly from (6.5.45) since when $\bar{n} \to 0$, the diffusion term vanishes.

As $t \to \infty$, (6.5.52) approaches

$$p(\alpha, \alpha^*, t/\alpha', \alpha'^*, 0) \xrightarrow[t\to\infty]{} \frac{1}{\pi\bar{n}} \exp - \frac{|\alpha + u|^2}{\bar{n}} \tag{6.5.62a}$$

where

$$u = \lim_{t\to\infty} e^{i\omega_c t} \int_0^t v(t - t')e^{-[i\omega_c + (\gamma/2)]t'}\, dt'. \tag{6.5.62b}$$

This corresponds to a coherent signal in the presence of gaussian noise. To show this we argue as follows. At $t = 0$ the system is in a pure coherent state and this corresponds to a coherent signal. As time goes on, the reservoir introduces thermal noise and the resultant distribution corresponds to a coherent signal plus thermal noise. If the signal is zero ($v = 0$), then (6.5.62) corresponds to pure thermal noise.

To obtain the distribution function in the SP, we use (6.5.2) so that the Green's function (6.5.52) becomes ($\beta' \equiv \alpha'$, $\beta'^* \equiv \alpha'^*$ at $t = 0$)

$$P(\alpha, \alpha^*, t/\alpha', \alpha'^*, 0) = \frac{1}{\pi\bar{n}(1 - e^{-\gamma t})} \exp\left\{-\frac{|\alpha e^{i\omega_c t} - \alpha' e^{-(\gamma/2)t} + w(t)e^{i\omega_c t}|^2}{\bar{n}(1 - e^{-\gamma t})}\right\}$$
$$= \frac{1}{\pi\bar{n}(1 - e^{-\gamma t})} \exp\left\{-\frac{|\alpha - \alpha' e^{-[i\omega_c + (\gamma/2)]t} + w(t)|^2}{\bar{n}(1 - e^{-\gamma t})}\right\}. \tag{6.5.63}$$

Characteristic Function. Let us calculate the characteristic function defined by

$$C^n(\xi, \xi^*, t) = \mathrm{Tr}_{R,S}\, \rho(t)e^{i\xi^* a^\dagger} e^{i\xi a}$$
$$= \mathrm{Tr}_S\, S(a, a^\dagger, t)e^{i\xi^* a^\dagger} e^{i\xi a}, \tag{6.5.64}$$

where we traced over system and reservoir and used (6.2.3). The $S(t)$ is the reduced density operator in the Schrödinger picture. Since the exponentials are in normal order, we may evaluate the trace by means of (3.2.34):

$$C^{(n)}(\xi, t) = \iint \frac{d^2\alpha}{\pi} \bar{S}^{(a)}(\alpha, \alpha^*, t) \exp [i(\xi\alpha + \xi^*\alpha^*)]. \tag{6.5.65}$$

But

$$\begin{aligned} P_n(\alpha, \alpha^*, t) &= \frac{1}{\pi} \bar{S}^{(a)}(\alpha, \alpha^*, t) \\ &= \iint P(\alpha, \alpha^*, t/\alpha', \alpha'^*, 0)\, P(\alpha', \alpha'^*)\, d^2\alpha'. \end{aligned} \tag{6.5.66}$$

If we use this and (6.5.63), the characteristic function (CF) becomes

$$\begin{aligned} C^{(n)}(\xi, t) = \int P(\alpha', \alpha'^*)\, d^2\alpha' \int \frac{d^2\alpha}{\pi\bar{n}(1 - e^{-\gamma t})} \\ \times \exp \left\{ -\frac{|\alpha - \alpha' e^{-[i\omega_c + (\gamma/2)]t} + w(t)|^2}{\bar{n}(1 - e^{-\gamma t})} \right\} \exp i(\xi\alpha + \xi^*\alpha^*) \end{aligned} \tag{6.5.67}$$

If we let

$$\begin{aligned} z = \alpha - U; \qquad d^2z = d^2\alpha \\ U = \alpha' e^{-[i\omega_c + (\gamma/2)]t} - w(t), \end{aligned} \tag{6.5.68}$$

then when we complete the square, (6.5.67) becomes

$$\begin{aligned} C^{(n)}(\xi, t) = \int P(\alpha', \alpha'^*)\, d^2\alpha' \int \frac{d^2z}{\pi\zeta} \exp \left\{ -\frac{1}{\zeta} [z^* - i\xi\zeta][z - i\xi^*\zeta] \right\} \\ \times \exp \{ -|\xi|^2\zeta + i[\xi U + \xi^* U^*] \} \end{aligned} \tag{6.5.69}$$

where

$$\zeta \equiv \bar{n}(1 - e^{-\gamma t}). \tag{6.5.70}$$

Since

$$\int \frac{d^2z}{\pi\zeta} \exp -\frac{1}{\zeta} |z - i\xi^*\zeta|^2 = 1, \tag{6.5.71}$$

the CF reduces to

$$\begin{aligned} C^{(n)}(\xi, \xi^*, t) = \int P(\alpha', \alpha'^*)\, d^2\alpha' \exp \Big\{ -|\xi|^2\zeta + i\xi[\alpha' e^{-[i\omega_c + (\gamma/2)]t} - w(t)] \\ + i\xi^*[\alpha'^* e^{[i\omega_c - (\gamma/2)]t} - w^*(t)] \Big\}, \end{aligned} \tag{6.5.72}$$

where we used (6.5.68). From (6.5.64) we see that

$$\langle a^{\dagger r} a^s \rangle = \frac{\partial^{r+s} C^{(n)}}{\partial(i\xi^*)^r\, \partial(i\xi)^s} \bigg|_{\xi = \xi^* = 0} \tag{6.5.73}$$

As a special case, let the initial distribution correspond to a signal plus gaussian noise. Then

$$P(\alpha', \alpha'^*) = \frac{1}{\pi \bar{n}} \exp -\left(\frac{|\alpha' - u|^2}{\bar{n}}\right). \tag{6.5.74}$$

If we use this in (6.5.72), let $z' = \alpha' - u$, complete the square and integrate, we obtain

$$C^{(n)}(\xi, \xi^*, t) = \exp\{-|\xi|^2\zeta - i[\xi^*(e^{[i\omega_c-(\gamma/2)]t}u^* - w^*(t)) + \xi(e^{-[i\omega_c+(\gamma/2)]t}u - w(t))]\}. \tag{6.5.75}$$

The mean number of photons in this case by (6.5.73) is

$$\langle a^\dagger(t)a(t)\rangle = |u|^2 e^{-\gamma t} + |w(t)|^2 + [ue^{-[i\omega_c+(\gamma/2)]t}w^*(t) + \text{cc}]. \tag{6.5.76}$$

If no signal is initially present ($u = 0$), then

$$\langle a^\dagger(t)a(t)\rangle = |w(t)|^2, \tag{6.5.77}$$

while if there is no driving term ($w = 0$)

$$\langle a^\dagger(t)a(t)\rangle = |u|^2 e^{-\gamma t}, \tag{6.5.78}$$

so that the initial coherent signal decays to zero.

Damped Atoms

So far the Fokker-Planck equation (6.4.90) for N homogeneously broadened atoms has not been solved. However, it is of some interest to solve it for the case in which we may neglect all diffusion (all second derivative terms). In this case it reduces to

$$\frac{\partial P_c}{\partial t} - (\Gamma_{12} + i\omega_a)\mathscr{M}\frac{\partial P_c}{\partial \mathscr{M}} - (\Gamma_{12} - i\omega_a)\mathscr{M}^*\frac{\partial P_c}{\partial \mathscr{M}^*} + (R_1 - \Gamma_1\mathscr{N}_1 + w_{12}\mathscr{N}_2)\frac{\partial P_c}{\partial \mathscr{N}_1} + (R_2 - \Gamma_2\mathscr{N}_2 + w_{21}\mathscr{N}_1)\frac{\partial P_c}{\partial \mathscr{N}_2} - (2\Gamma_{12} + \Gamma_1 + \Gamma_2)P_c = 0. \tag{6.5.79}$$

We may solve this equation by the method of characteristics (see Appendix A). The characteristic equations are

$$\frac{dP_c}{\Lambda P_c} = \frac{dt}{1} = \frac{d\mathscr{M}}{-(\Gamma_{12} + i\omega_a)\mathscr{M}} = \frac{d\mathscr{M}^*}{-(\Gamma_{12} - i\omega_a)\mathscr{M}^*} = \frac{d\mathscr{N}_1}{R_1 + w_{12}\mathscr{N}_2 - \Gamma_1\mathscr{N}_1} = \frac{d\mathscr{N}_2}{R_2 + w_{21}\mathscr{N}_1 - \Gamma_2\mathscr{N}_2} \tag{6.5.80}$$

where

$$\Lambda \equiv 2\Gamma_{12} + \Gamma_1 + \Gamma_2. \tag{6.5.81}$$

The reader will quickly note that these characteristic equations are identical in form with the mean equations of motion (6.4.111). In other words when we neglect diffusion, the "motion" of the associated c-number variables is just the "classical" motion of the atoms.

The solution of the first three characteristics equations are easily seen to be

$$\begin{aligned} P_c e^{\Lambda t} &= P_c(0) \\ \mathscr{M} e^{(\Gamma_{12}+i\omega_a)t} &= \mathscr{M}_0 \\ \mathscr{M}^* e^{(\Gamma_{12}-i\omega_a)t} &= \mathscr{M}_0^*, \end{aligned} \tag{6.5.82}$$

where $P_c(0)$, $\mathscr{M}_0$, and $\mathscr{M}_0^*$ are constants of integration. The solution of the $d\mathscr{N}_1/dt$ and $d\mathscr{N}_2/dt$ equations may be obtained in a form useful for present purposes as follows. We multiply both sides of the $d\mathscr{N}_1/dt$ equation by α and the $d\mathscr{N}_2/dt$ equation by β and add. This gives

$$\frac{d}{dt}(\alpha\mathscr{N}_1 + \beta\mathscr{N}_2) + (\Gamma_1\alpha - w_{21}\beta)\mathscr{N}_1 + (\Gamma_2\beta - w_{12}\alpha)\mathscr{N}_2 = R_1\alpha + R_2\beta. \tag{6.5.83}$$

We choose α and β by requiring that

$$\begin{aligned} \Gamma_1\alpha - w_{21}\beta &= \lambda\alpha \\ \Gamma_2\beta - w_{12}\alpha &= \lambda\beta \end{aligned} \tag{6.5.84a}$$

or

$$\begin{aligned} (\Gamma_1 - \lambda)\alpha - w_{21}\beta &= 0 \\ -w_{12}\alpha + (\Gamma_2 - \lambda)\beta &= 0. \end{aligned} \tag{6.5.84b}$$

In order for these two equations to have a nontrivial solution, the determinant of the coefficients of α and β must vanish. This will be true if λ takes on either of the two values λ_1 or λ_2 given by

$$\begin{aligned} \lambda_1 &= \frac{\Gamma_1 + \Gamma_2}{2} + \Delta \\ \lambda_2 &= \frac{\Gamma_1 + \Gamma_2}{2} - \Delta, \end{aligned} \tag{6.5.85}$$

where we define

$$\Delta = +\sqrt{\left(\frac{\Gamma_1 - \Gamma_2}{2}\right)^2 + w_{12}w_{21}}. \tag{6.5.86}$$

In this case, we see from the first of (6.5.84) that

$$\begin{aligned} \beta_1 &= \frac{(\Gamma_1 - \lambda_1)}{w_{21}}\alpha_1 = \left[\frac{\Gamma_1 - \Gamma_2}{2} - \Delta\right]\frac{\alpha_1}{w_{21}} \\ \beta_2 &= \frac{(\Gamma_1 - \lambda_2)}{w_{21}}\alpha_1 = \left[\frac{\Gamma_1 - \Gamma_2}{2} + \Delta\right]\frac{\alpha_2}{w_{21}}. \end{aligned} \tag{6.5.87}$$

We are at liberty to choose α_1 and α_2 in any convenient way. We therefore let

$$\alpha_1 = \alpha_2 = w_{21} \tag{6.5.88}$$

so that

$$\begin{aligned} \beta_1 &= \Omega - \Delta \\ \beta_2 &= \Omega + \Delta, \end{aligned} \tag{6.5.89}$$

where we have let

$$\Omega = \frac{\Gamma_1 - \Gamma_2}{2}. \tag{6.5.90}$$

If we next substitute (6.5.84a) into (6.5.83), we obtain

$$\frac{d}{dt}(\alpha\mathcal{N}_1 + \beta\mathcal{N}_2) + \lambda(\alpha\mathcal{N}_1 + \beta\mathcal{N}_2) = R_1\alpha + R_2\beta. \tag{6.5.91}$$

The solution is easily seen to be

$$\alpha\mathcal{N}_1(t) + \beta\mathcal{N}_2(t) = e^{-\lambda t}\left[\alpha\mathcal{N}_{10} + \beta\mathcal{N}_{20} + \frac{(R_1\alpha + R_2\beta)}{\lambda}(e^{\lambda t} - 1)\right], \tag{6.5.92}$$

where $\mathcal{N}_{10}$ and $\mathcal{N}_{20}$ are constants of integration. This may be rewritten as the two equations

$$\xi_1 \equiv \mathcal{N}_{10} + \frac{\beta_1}{\alpha}\mathcal{N}_{20} = e^{\lambda_1 t}\left(\mathcal{N}_1 + \frac{\beta_1}{\alpha}\mathcal{N}_2\right) - \frac{1}{\lambda_1}\left(R_1 + R_2\frac{\beta_1}{\alpha}\right)(e^{\lambda_1 t} - 1) \tag{6.5.93}$$

$$\xi_2 \equiv \mathcal{N}_{10} + \frac{\beta_2}{\alpha}\mathcal{N}_{20} = e^{\lambda_2 t}\left(\mathcal{N}_1 + \frac{\beta_2}{\alpha}\mathcal{N}_2\right) - \frac{1}{\lambda_2}\left(R_1 + R_2\frac{\beta_2}{\alpha}\right)(e^{\lambda_2 t} - 1)$$

where we solved for $\mathcal{N}_{10} + \beta_j\mathcal{N}_{20}/\alpha$ and we used the two solutions for β_1, β_2 corresponding to the eigenvalues λ_1 and λ_2. These equations may be solved easily for $\mathcal{N}_{10}$ and $\mathcal{N}_{20}$

$$\begin{aligned} \mathcal{N}_{10} &= \frac{\beta_2\xi_1 - \beta_1\xi_2}{\beta_2 - \beta_1} \equiv u(\mathcal{N}_1, \mathcal{N}_2, t) \\ \mathcal{N}_{20} &= \frac{\alpha(\xi_2 - \xi_1)}{\beta_2 - \beta_1} \equiv v(\mathcal{N}_1, \mathcal{N}_2, t); \end{aligned} \tag{6.5.94}$$

ξ_1 and ξ_2 are defined by (6.5.93). These give the two remaining integrals which are required. The general solution of the distribution function equation (6.5.79) is therefore

$$P_c(\mathcal{N}_1, \mathcal{N}_2, \mathcal{M}, \mathcal{M}^*, t) = e^{-\Lambda t} g[\mathcal{M} e^{(\Gamma_{12}+i\omega_a)t}, \mathcal{M}^* e^{(\Gamma_{12}-i\omega_a)t}, u, v], \quad (6.5.95)$$

where g is an arbitrary function. If at $t = 0$, we know the initial values $\mathcal{M}_0$, $\mathcal{M}^*$, $\mathcal{N}_{10}$, and $\mathcal{N}_{20}$, we obtain the Green's function solution

$$\begin{aligned} P_c(\mathcal{N}_1, \mathcal{N}_2, \mathcal{M}, \mathcal{M}^*, t/\mathcal{N}_{10}, \mathcal{N}_{20}, \mathcal{M}_0, \mathcal{M}_0^*, 0) \\ = e^{-\Lambda t}\, \delta[\mathcal{M} e^{(\Gamma_{12}+i\omega_2)t} - \mathcal{M}_0]\, \delta[\mathcal{M}^* e^{(\Gamma_{12}-i\omega_a t} - \mathcal{M}_0^*] \\ \times \delta[u(\mathcal{N}_1, \mathcal{N}_2, t) - \mathcal{N}_{10}]\, \delta[v(\mathcal{N}_1, \mathcal{N}_2, t) - \mathcal{N}_{20}], \quad (6.5.96) \end{aligned}$$

where we have determined the functional form to agree with the initial conditions. This shows that the variables follow their "classical" trajectories for all time. This is the course to be expected since we have neglected all diffusion.

If we use the definition of Λ (6.5.81) and use the well known property of δ-functions

$$\delta(ax) = \frac{1}{|a|}\, \delta(x) \quad (6.5.97)$$

then the Green's function solution may be written at

$$\begin{aligned} P_c(\mathcal{N}_1, \mathcal{N}_2, \mathcal{M}, \mathcal{M}^*, t/\mathcal{N}_{20}, \mathcal{N}_{20}, \mathcal{M}_0, \mathcal{M}_0^*, 0) \\ = \delta[\mathcal{M} - \mathcal{M}_0 e^{-(\Gamma_{12}+i\omega_a)t}]\, \delta[\mathcal{M}^* - \mathcal{M}_0^* e^{-(\Gamma_{12}-i\omega_a)t}] \\ \times \delta[e^{-\Gamma_1 t} u(\mathcal{N}_1, \mathcal{N}_2, t) - \mathcal{N}_{10} e^{-\Gamma_1 t}]\, \delta[e^{-\Gamma_2 t} v(\mathcal{N}_1, \mathcal{N}_2, t) - \mathcal{N}_{20} e^{-\Gamma_2 t}]. \\ (6.5.98) \end{aligned}$$

6.6 TWO-TIME AVERAGES, SPECTRA [10]

Up to this point we have developed the theory in such a way that we may calculate the average of a function of system operators which are all evaluated at the same time under the Markoff approximation. However, a complete statistical description requires mean values of operators at different times. For example, in the case of a mode of the radiation field, the fluctuation spectrum is defined by

$$\int_{-\infty}^{\infty} e^{-i\omega t} \langle a^\dagger(t) a(0) \rangle\, dt. \quad (6.6.1)$$

The intensity spectrum is given by

$$\int_{-\infty}^{\infty} e^{-i\omega t} \langle a^\dagger(0) a^\dagger(t) a(t) a(0) \rangle\, dt, \quad (6.6.2)$$

while the photon number spectrum is given by

$$\int_{-\infty}^{\infty} e^{-i\omega t}\langle a^\dagger(t)a(t)a^\dagger(0)a(0)\rangle\, dt \tag{6.6.3}$$

(see Appendix I). These are related directly to measured experimental quantities which the theory must be capable of predicting. In this section we show that under the Markoff approximation, such two time averages may be converted to one time averages.

We have shown in several cases of interest that the distribution function P_c obeys the Fokker–Planck equation (6.4.19)

$$\frac{\partial P_c}{\partial t}(\underset{\sim}{\alpha}, t) = -\frac{\partial}{\partial \alpha_i}[\mathscr{A}_i(\underset{\sim}{\alpha})P_c(\underset{\sim}{\alpha}, t)] + \frac{\partial^2}{\partial \alpha_i\, \partial \alpha_j}[\mathscr{D}_{ij}(\underset{\sim}{\alpha})P_c(\underset{\sim}{\alpha}, t)]. \tag{6.6.4}$$

We may write the solution of this equation subject to the initial distribution $P_c(\underset{\sim}{\beta}, t_0)$ as

$$P_c(\underset{\sim}{\alpha}, t) = \int d\underset{\sim}{\beta}\, P_c(\underset{\sim}{\alpha}, t/\underset{\sim}{\beta}, t_0)P_c(\underset{\sim}{\beta}, t_0), \tag{6.6.5}$$

where $P_c(\underset{\sim}{\alpha}, t/\underset{\sim}{\beta}, t_0)$ is the conditional "probability" or Green's function which obeys (6.6.4) and has the property that

$$P_c(\underset{\sim}{\alpha}, t/\underset{\sim}{\beta}_0, t_0) \xrightarrow[t\to t_0]{} \delta(\underset{\sim}{\alpha} - \underset{\sim}{\beta}). \tag{6.6.6}$$

We then see by (6.6.5) and (6.6.6) that

$$P_c(\underset{\sim}{\alpha}, t) \xrightarrow[t\to t_0]{} P_c(\underset{\sim}{\alpha}, t_0) \tag{6.6.7}$$

as required.

By (6.4.11), we have that the one-time average of $M(\underset{\sim}{a})$ evaluated at t_0 under the Markoff approximation is

$$\langle M(t)\rangle \cong \int \bar{M}^c(\underset{\sim}{\alpha})P_c(\underset{\sim}{\alpha}, t)\, d\underset{\sim}{\alpha}. \tag{6.6.8}$$

If we use (6.6.5), this becomes

$$\langle M(t)\rangle \cong \int \bar{M}^c(\underset{\sim}{\alpha})\, d\underset{\sim}{\alpha} \int P_c(\underset{\sim}{\alpha}, t/\underset{\sim}{\beta}, t_0)P_c(\underset{\sim}{\beta}, t_0)\, d\underset{\sim}{\beta}. \tag{6.6.9}$$

From the definition (6.4.12)

$$P_c(\underset{\sim}{\beta}, t_0) = \operatorname{Tr} S(t_0)\, \delta^c(\underset{\sim}{\beta} - \underset{\sim}{a}), \tag{6.6.10}$$

we may write (6.6.9) as

$$\langle M(t)\rangle \cong \operatorname{Tr} S(t_0)\left\{\int d\underset{\sim}{\alpha} \int d\underset{\sim}{\beta}\, \bar{M}^c(\underset{\sim}{\alpha})P_c(\underset{\sim}{\alpha}, t/\underset{\sim}{\beta}, t_0)\, \delta^c(\underset{\sim}{\beta} - \underset{\sim}{a})\right\}. \tag{6.6.11}$$

If we compare this with (6.4.7) we see that the solution of (6.4.5) for $M(t)$ under the Markoff approximation is

$$\langle M^H(t)\rangle_R \equiv \mathrm{Tr}_R f_0(R) M^H(t)$$
$$\cong M(t) = \int d\underset{\sim}{\alpha} \int d\underset{\sim}{\beta}\, \bar{M}^c(\underset{\sim}{\alpha}) P_c(\underset{\sim}{\alpha}, t/\underset{\sim}{\beta}, t_0)\, \delta^c(\underset{\sim}{\beta} - \underset{\sim}{a}). \quad (6.6.11)$$

That is, if we know the Green's function solution of the Fokker–Planck equation, we have a solution for (6.4.5).

Consider next the exact two time average

$$\langle M(t)N(t_0)\rangle = \mathrm{Tr}_{R,S}\, \rho(t_0) M^H(t) N(t_0), \quad (6.6.12)$$

where $N(t_0)$ is a system operator in the SP at time t_0 and $M^H(t)$ is the full system operator in the HP. The time t_0 is any time at which the density operator factors as $S(t_0)f_0(R)$ when the system has lost its prior memory. Then, we have

$$\langle M(t)N(t_0)\rangle = \mathrm{Tr}_S\, S(t_0)\langle M^H(t)\rangle_R N(t_0). \quad (6.6.13)$$

Under the Markoff approximation, we may replace $\langle M^H(t)\rangle_R$ by (6.6.11) so that

$$\langle M(t)N(t_0)\rangle \cong \mathrm{Tr}_S\, S(t_0) M(t) N(t_0)$$
$$= \int d\underset{\sim}{\alpha} \int d\underset{\sim}{\beta}\, \bar{M}^c(\underset{\sim}{\alpha}) P_c(\underset{\sim}{\alpha}, t/\underset{\sim}{\beta}, t_0)\, \mathrm{Tr}\,[S(t_0)\, \delta^c(\underset{\sim}{\beta} - \underset{\sim}{a}) N(t_0)]. \quad (6.6.14)$$

We have therefore reduced the average to a trace of operators all evaluated at the same time t_0. All that is needed is the same Green's function solution. If $N = 1$, this reduces to (6.6.9).

Let us illustrate this result for the damped oscillator. The Green's function is given by (6.5.63) with $t_0 = 0$

$$P_{(n)}(\alpha, \alpha^*, t/\beta, \beta^*, 0) = \frac{1}{\pi\zeta(t)} \exp - \frac{|\alpha - U(t)|^2}{\zeta(t)}, \quad (6.6.15)$$

where

$$\zeta(t) = \bar{n}(1 - e^{-\gamma t})$$
$$U(t) = \beta e^{-[i\omega_c + (\gamma/2)]t} - w(t) \quad (6.6.16)$$
$$w(t) = -i \int_0^t v(t - \tau) e^{-[i\omega_c + (\gamma/2)]\tau}\, d\tau.$$

If we let $M = a^\dagger$ and $N = a$, we have by (6.6.14) [using normal order]

$$\langle a^\dagger(t) a(0)\rangle = \int d^2\alpha \int d^2\beta\, \alpha^* P_{(n)}(\alpha, \alpha^*, t/\beta, \beta^*, 0)\, \mathrm{Tr}\,[S(0)\, \delta(\beta^* - a^\dagger)\, \delta(\beta - a) a]. \quad (6.6.17)$$

If $S(0)$ is in antinormal order, we see that since the other terms are in normal order that

$$\operatorname{Tr} S^{(a)}(0)\,\delta(\beta^* - a^\dagger)\,\delta(\beta - a)a = \int \bar{S}^{(a)}(\alpha', \alpha'^*, 0)\alpha'\,\delta(\beta^* - \alpha'^*)\,\delta(\beta - \alpha')\frac{d^2\alpha'}{\pi}$$

$$= \frac{\bar{S}^{(a)}}{\pi}(\beta, \beta^*, 0)\beta. \tag{6.6.18}$$

Therefore,

$$\langle a^\dagger(t)a(0)\rangle = \int d^2\alpha \int d^2\beta\, \alpha^*\beta P_{(n)}(\alpha, \alpha^*, t/\beta\,, \beta^*, 0)P_{(n)}(\beta, \beta^*, 0). \tag{6.6.19}$$

We have shown earlier in (6.4.30) that

$$\frac{\bar{S}^{(a)}}{\pi}(\beta, \beta^*) = P_{(n)}(\beta, \beta^*). \tag{6.6.20}$$

If we use (6.6.15) and carry out the integration over α, we obtain

$$\langle a^\dagger(t)a(0)\rangle = \iint \beta U^*(\beta, \beta^*, t)P_{(n)}(\beta, \beta^*, 0)\, d^2\beta$$

$$= \iint d^2\beta\{|\beta|^2\, e^{[i\omega_c - (\gamma/2)]t} - \beta w^*(t)\}P_{(n)}(\beta, \beta^*, 0)$$

$$\equiv \langle a^\dagger(0)a(0)\rangle e^{[i\omega_c - (\gamma/2)]t} - \langle a(0)\rangle w^*(t). \tag{6.6.21}$$

So long as t is greater than a reservoir correlation time, we may take the initial distribution to be the steady-state distribution for thermal noise, namely,

$$P_{(n)}(\beta, \beta^*, 0) = \frac{1}{\pi\bar{n}} \exp - \frac{|\beta|^2}{\bar{n}}. \tag{6.6.22}$$

When we carry out the integrals in (6.6.21), we obtain

$$\langle a^\dagger(t)a(0)\rangle = \bar{n}e^{i\omega_c t}e^{-(\gamma/2)|t|}. \tag{6.6.23}$$

Since the process must be stationary, it must also decay if $t \to -t$. Therefore,

$$\langle a^\dagger(t)a(0)\rangle = \bar{n}e^{i\omega_c t}e^{-(\gamma/2)|t|}. \tag{6.6.24}$$

When we take the Fourier transform, we obtain the fluctuation spectrum

$$\int_{-\infty}^{\infty} \langle a^\dagger(t)a(0)\rangle e^{i\omega t}\, dt = \frac{\gamma\bar{n}}{(\omega_c - \omega)^2 + (\gamma/2)^2}, \tag{6.6.25}$$

which is seen to be Lorentzian with half-width given by $\gamma/2$ centered at $\omega = \omega_c$.

We leave as an exercise the evaluation of the intensity and photon number fluctuation spectra.

6.7 ROTATING WAVE VAN DER POL OSCILLATOR

It has been shown [11] that the equation of motion for the associated distribution function for the reduced density operator in antinormal order for the radiation field of a single mode laser is the Fokker–Planck equation

$$\frac{\partial p}{\partial t}(\alpha, \alpha^*, t) = -\frac{\partial}{\partial \alpha}[(g - |\alpha|^2)\alpha p] - \frac{\partial}{\partial \alpha^*}[(g - |\alpha|^2)\alpha^* p] + 4\frac{\partial^2 p}{\partial \alpha\, \partial \alpha^*}(\alpha, \alpha^*, t), \tag{6.7.1}$$

where g is a numerical pumping parameter and $p(\alpha, t)$ refers to the interaction picture. It describes a radiation field mode at frequency ω_c which is damped for $g < 0$ and starts to grow exponentially when $g > 0$; $g = 0$ is called the oscillation threshold. For $g > 0$ the oscillation builds up but because of the nonlinear term $|\alpha|^2$, the oscillations will stabilize at a fixed amplitude since $g - |\alpha|^2$ eventually goes to zero as $|\alpha|^2$ increases. An equation such as (6.7.1) describes a classical Van der Pol oscillator which is the associated "classical" description of a laser. The quantum density operator is given by

$$s(a, a^\dagger, t) = \pi\mathscr{A}\{p(\alpha, \alpha^*, t)\}, \tag{6.7.2}$$

where $\mathscr{A}$ is the antinormal ordering operator.

Because of the $|\alpha|^2$ terms in (6.7.1), we cannot solve it analytically. However, there are two interesting regimes where it may be solved, namely, far below threshold ($|\alpha|^2 \ll |g|$) and far above threshold ($|\alpha|^2 \gg |g|$). We solve it in these two regions and obtain the spectra of the laser.

Far Below Threshold ($|\alpha|^2 \ll |g|$)

In this region, (6.7.1) reduces to

$$\frac{\partial p}{\partial t} = -\frac{\partial}{\partial \alpha}(g\alpha p) - \frac{\partial}{\partial \alpha^*}(g\alpha^* p) + 4\frac{\partial^2 p}{\partial \alpha\, \partial \alpha^*}, \tag{6.7.3}$$

which by (6.5.4) is the equation for a damped oscillator if we make the identification $v = 0$

$$\gamma\bar{n} = 4$$
$$g = -\frac{\gamma}{2}. \tag{6.7.4}$$

The Green's function associated with the Schrödinger picture can be obtained directly from (6.5.63), namely,

$$P(\alpha, \alpha^*, t/\alpha', \alpha'^*, 0) = \frac{1}{\pi\zeta(t)} \exp - \frac{|\alpha - \alpha' e^{-(i\omega_c - g)t}|^2}{\zeta(t)}, \tag{6.7.5}$$

where

$$\zeta(t) = \frac{2}{g}(e^{2gt} - 1) \qquad g < 0. \tag{6.7.6}$$

From (6.6.25), we see that the fluctuation spectrum below threshold is Lorentzian with half width $|g|$ and is given by

$$\int_{-\infty}^{\infty} \langle a^\dagger(t)a(0)\rangle e^{+i\omega t}\, dt = \frac{4}{(\omega_c - \omega)^2 + g^2}. \tag{6.7.7}$$

Far Above Threshold $|\alpha|^2 \gg |g|$

Before we attempt an analytic solution in this regime, it is advantageous to convert (6.7.1) to polar coordinates. We let

$$\alpha = re^{i\varphi} \qquad d^2\alpha = r\, dr\, d\varphi. \tag{6.7.8}$$

By the usual rules of differentiation, we have

$$\begin{aligned} \frac{\partial}{\partial\alpha}\alpha &= \frac{1}{2}\left[\frac{\partial}{\partial r}r + 1 - \frac{i}{2}\frac{\partial}{\partial\varphi}\right] \\ \frac{\partial}{\partial\alpha^*}\alpha^* &= \frac{1}{2}\left[\frac{\partial}{\partial r}r + 1 + \frac{i}{2}\frac{\partial}{\partial\varphi}\right] \end{aligned} \tag{6.7.9}$$

$$4\frac{\partial^2}{\partial\alpha\,\partial\alpha^*} = \frac{1}{r}\frac{\partial}{\partial r}\left(r\frac{\partial}{\partial r}\right) + \frac{1}{r^2}\frac{\partial^2}{\partial\varphi^2}. \tag{6.7.10}$$

If we use these, (6.7.1) becomes after minor algebra

$$\frac{\partial p}{\partial t} = \frac{1}{r}\frac{\partial}{\partial r}\left(r\frac{\partial p}{\partial r}\right) + \frac{1}{r^2}\frac{\partial^2 p}{\partial\varphi^2} - \frac{1}{r}\frac{\partial}{\partial r}[(g - r^2)r^2 p]. \tag{6.7.11}$$

We normalize $p(r, \varphi, t)$ so that

$$\int_0^{2\pi} d\varphi \int_0^{\infty} p(r, \varphi, t) r\, dr = 1. \tag{6.7.12}$$

Equation (6.7.11) is still exactly equivalent to (6.7.1). We may obtain the steady-state solution which is independent of φ very easily. Since

$$\frac{\partial}{\partial t} = 0 \qquad \frac{\partial}{\partial\varphi} = 0, \tag{6.7.13}$$

it becomes

$$\frac{1}{r}\frac{\partial}{\partial r}\left\{r\frac{\partial p}{\partial r} - (g - r^2)r^2 p\right\} = 0, \tag{6.7.14}$$

or if we integrate once and take the constant of integration to be zero we have

$$\frac{dp}{dr} = (g - r^2)rp \tag{6.7.15}$$

or

$$\begin{aligned} p_s(r) &= N e^{(g/2)r^2 - \frac{1}{4}r^4} \\ &= \frac{e^{-\frac{1}{4}(r^2-g)^2}}{\displaystyle\int_0^\infty e^{-\frac{1}{4}(r^2-g)^2} r\,dr} = \frac{e^{-\frac{1}{4}(r^2-g)^2}}{\displaystyle\int_{-g/2}^\infty e^{-x^2}\,dx}, \end{aligned} \tag{6.7.16}$$

where we normalized the steady-state solution so that

$$\int_0^\infty r\,dr\,p_s(r) = 1 \tag{6.7.17}$$

We sketch this solution in Figure 6.4 for $g = -5$ (far below threshold), at $g = 0$ (threshold) and $g = 5$ (far above threshold).

The mean number of photons in the steady-state is given by

$$\begin{aligned} \langle a^\dagger a \rangle &= \int_0^\infty r^2 p_s(r) r\,dr \\ &= \frac{\displaystyle\int_{-g/2}^\infty (g + 2x)e^{-x^2}\,dx}{\displaystyle\int_{-g/2}^\infty e^{-x^2}\,dx}. \end{aligned} \tag{6.7.18}$$

This is sketched in Figure 6.5 as a function of g. Far above threshold, we may take the lower limit to $-\infty$ so that

$$\langle a^\dagger a \rangle_s \approx g. \tag{6.7.19}$$

Far above threshold, the mean number of photons will have small fluctuations above this steady-state value. Let us expand r as

$$r \cong \sqrt{g} + r_1(t); \qquad r\,dr \cong \sqrt{p}\,dr_1 \tag{6.7.20}$$

where $r_1(t) \ll \sqrt{g}$. If we put this into (6.7.11) and neglect quantities of order $r_1{}^2$, we have

$$\frac{\partial p}{\partial t} \cong \frac{\partial^2 p}{\partial r_1{}^2} + 2g\frac{\partial}{\partial r_1}(r_1 p) + \frac{1}{g}\frac{\partial^2 p}{\partial \varphi^2}. \tag{6.7.21}$$

This equation is now separable and may be solved. We therefore let

$$p(r, \varphi, t) = \frac{R(r_1, t)\Phi(\varphi, t)}{\sqrt{g}}, \tag{6.7.22}$$

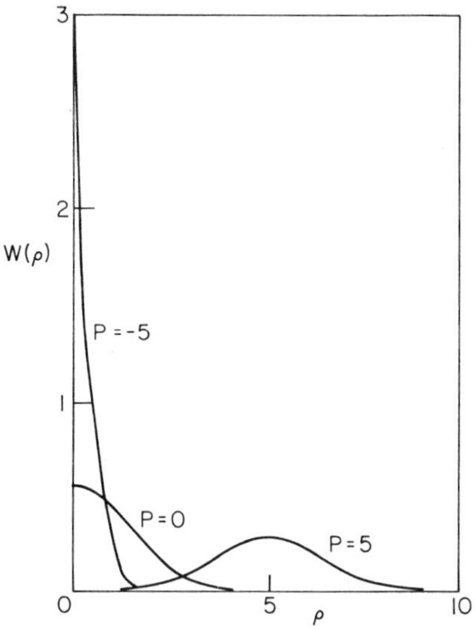

Figure 6.4 The steady-state photon probability distribution function for three values of the net pump rate g. $g = 0$ corresponds to threshold. (Reprinted from reference 11.)

where we normalize so that

$$\int_0^{2\pi} d\varphi \int_0^{\infty} p(r, \varphi, t) r \, dr = 1 = \int_0^{2\pi} \Phi(\varphi, t) \, d\varphi \int_{-\sqrt{g}}^{\infty} R(r_1, t) \, dr_1, \quad (6.7.23)$$

or since g is large, we require

$$\begin{aligned} &\int_0^{2\pi} \Phi(\varphi, t) \, d\varphi = 1 \\ &\int_{-\infty}^{\infty} R(r_1, t) \, dr_1 = 1. \end{aligned} \quad (6.7.24)$$

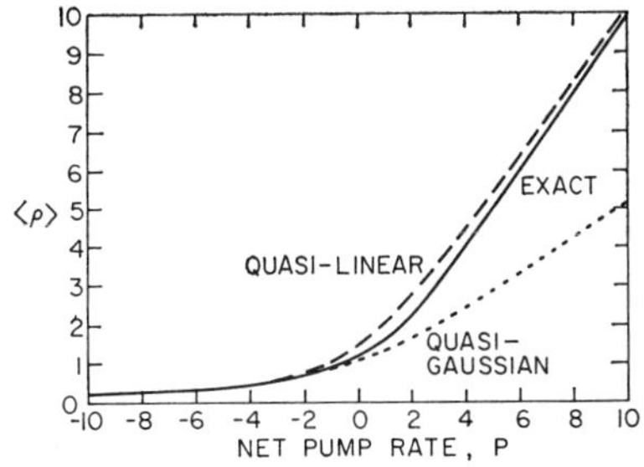

Figure 6.5 Mean number of photons in the steady-state as a function of the pump parameter g. (Reprinted from reference 11.)

When we put (6.7.22) into (6.7.21) and separate, we obtain

$$\frac{\partial R}{\partial t} = \frac{\partial^2 R}{\partial r_1{}^2} + \frac{\partial}{\partial r_1}(2gr_1R)$$

$$\frac{\partial \Phi}{\partial t} = \frac{1}{g}\frac{\partial^2 \Phi}{\partial \varphi^2}.$$

The steady-state solutions of these ($\partial/\partial t = 0$) such that $\Phi(\varphi + 2\pi) = \Phi(\varphi)$ are

$$R_s(r_1) = \sqrt{\frac{g}{\pi}}\, e^{-gr_1{}^2}$$
$$\Phi_s(\varphi) = \frac{1}{2\pi}, \tag{6.7.26}$$

so that the amplitude is a gaussian and the phase is uniformly distributed. Thus we see that in the steady-state

$$\langle r\rangle_s = \sqrt{g}$$
$$\langle r^2\rangle_s = g + \langle r_1{}^2\rangle = g + \frac{1}{2g}$$
$$\langle r^3\rangle_s = g^{3/2} + 3\sqrt{g}\,\langle r_1{}^2\rangle = g^{3/2} + \frac{3}{2g^{1/2}} \tag{6.7.27}$$
$$\langle r^4\rangle_s = g^2 + 6g\langle r_1{}^2\rangle + \langle r_1{}^4\rangle = g^2 + 3 + \frac{3}{4g^2}.$$

We next look for the conditional probability solutions of (6.7.25) such that

$$\lim_{t\to 0} R(r_1, t/r_{10}, 0) = \delta(r_1 - r_{10}) = \lim_{\epsilon\to\infty}\sqrt{\frac{\varepsilon}{\pi}}\, e^{-\varepsilon(r_1 - r_{10})^2}$$
$$\lim_{t\to 0} \Phi(\varphi, t/\varphi_0, 0) = \delta(\varphi - \varphi_0) = \frac{1}{2\pi}\sum_{-\infty}^{\infty} e^{in(\varphi-\varphi_0)}. \tag{6.7.28}$$

In the radial equation we look for a solution of the form

$$R(r_1, t/r_{10}, 0) = \sqrt{\frac{\nu(t)}{\pi}}\exp\left\{-\frac{[r_1 - \eta(t)]^2}{\zeta(t)}\right\}, \tag{6.7.29}$$

where

$$\nu(0) = \epsilon; \qquad \eta(0) = r_{10}; \qquad \zeta(0) = \frac{1}{\epsilon}. \tag{6.7.30}$$

If we substitute (6.7.29) in the radial equation (6.7.25) and equate equal powers of r_1, we obtain

$$\frac{d\zeta}{dt} = -4g\zeta + 4$$
$$\frac{d\eta}{dt} = -2g\eta \tag{6.7.31}$$
$$\frac{1}{\nu}\frac{d\nu}{t} = -\frac{1}{\zeta}\frac{d\zeta}{dt}.$$

When we solve these subject to the initial conditions (6.7.30) and let $\epsilon \to \infty$, we obtain

$$\zeta(t) = \frac{1}{g}(1 - e^{-4gt}),$$
$$\eta(t) = r_{10}e^{-2gt} \tag{6.7.32}$$
$$\nu(t) = \frac{1}{\zeta(t)}.$$

To obtain the Green's function solution for the phase equation, we let

$$\Phi(\varphi, t/\varphi_0, 0) = \sum_{-\infty}^{\infty} c_n(t)e^{in\varphi}. \tag{6.7.33}$$

If we put this in (6.7.25), we obtain

$$\sum_{-\infty}^{\infty}\left(\frac{dc_n}{dt} + \frac{n^2}{g}c_n\right)e^{in\varphi} = 0 \tag{6.7.34}$$

or

$$c_n(t) = c_n(0)\exp\left(-\frac{n^2}{g}t\right). \tag{6.7.35}$$

If we use the initial condition (6.7.28), we have

$$\Phi(\varphi, t/\varphi_0, 0) = \sum_{-\infty}^{\infty} c_n(0)e^{in\varphi}\exp\left(-\frac{n^2}{g}t\right) \xrightarrow[t\to 0]{} \frac{1}{2\pi}\sum_{-\infty}^{\infty} e^{in(\varphi-\varphi_0)}. \tag{6.7.36}$$

Therefore,

$$\Phi(\varphi, t/\varphi_0, 0) = \frac{1}{2\pi}\sum_{-\infty}^{\infty} e^{in(\varphi-\varphi_0)}\exp\left(-\frac{n^2}{g}t\right). \tag{6.7.37}$$

The field correlation function is given by (6.6.19) as

$$\langle a^\dagger(t)a(0)\rangle = \iint \beta P(\beta, \beta^*)\, d^2\beta \iint \alpha^* P(\alpha, \alpha^*, t/\beta, \beta^*, 0)\, d^2\alpha, \tag{6.7.38}$$

where by (6.5.2), we must replace α by $\alpha e^{i\omega_c t}$ and α^* by $\alpha^* e^{-i\omega_c t}$ in our distribution functions or

$$\begin{aligned} re^{i\varphi} &\to re^{i(\varphi+\omega_c t)} \\ re^{-i\varphi} &\to re^{-i(\varphi+\omega_c t)}. \end{aligned} \tag{6.7.39}$$

Therefore,

$$P(\alpha, \alpha^*, t/\beta, \beta^*, 0) \to \sqrt{\frac{\nu(t)}{\pi g}} \exp\left\{-\frac{[r_1 - \eta(t)]^2}{\zeta(t)}\right\} \frac{1}{2\pi} \sum_{-\infty}^{\infty} e^{in(\varphi-\varphi_0+\omega_c t)} \exp\left(-\frac{n^2}{g} t\right) \tag{6.7.40}$$

$$\begin{aligned} d^2\alpha &\to \sqrt{g}\, dr_1\, d\varphi \\ \alpha^* &\to (\sqrt{g} + r_1) e^{-i\varphi}. \end{aligned} \tag{6.7.41}$$

In the steady-state, by (6.7.32) we see that

$$\zeta \to \frac{1}{g}; \qquad \eta \to 0; \qquad \nu \to g \tag{6.7.42}$$

so the steady-state distribution is

$$P_s(\beta, \beta^*) \to \frac{1}{\sqrt{\pi}} e^{-g r_{10}^2} \frac{1}{2\pi}, \tag{6.7.43}$$

since only the $n = 0$ term contributes as $t \to \infty$. Also

$$d^2\beta = \sqrt{g}\, dr_{10}\, d\varphi_0 \tag{6.7.44}$$

while in the integrand

$$\beta = (\sqrt{g} + r_{10}) e^{i\varphi_0}. \tag{6.7.45}$$

When we use these, (6.7.38) becomes for large g

$$\begin{aligned} \langle a^\dagger(t) a(0)\rangle &= \int_0^{2\pi}\int_{-\infty}^{\infty} \sqrt{g}\, dr_{10}\, d\varphi_0 (\sqrt{g} + r_{10}) e^{i\varphi_0} \frac{1}{\sqrt{\pi}} e^{-g r_{10}^2} \frac{1}{2\pi} \\ &\quad \times \sum_{-\infty}^{\infty} \int_0^{2\pi}\int_{-\infty}^{\infty} \sqrt{g}\, dr_1\, d\varphi (\sqrt{g} + r_1) \frac{e^{-i\varphi}}{2\pi} \sqrt{\frac{\nu}{\pi g}} \\ &\quad \times \exp\left\{-\frac{(r_1 - \eta)^2}{\zeta} + in(\varphi - \varphi_0 + \omega_0 t) - \frac{n^2 t}{g}\right\} \\ &= e^{i\omega_c t - (1/g)t} \sqrt{\frac{g}{\pi}} \int_{-\infty}^{\infty} dr_{10} (\sqrt{g} + r_{10}) [\sqrt{g} + r_{10} e^{-2gt}] e^{-g r_{10}^2} \\ &= e^{i\omega_c t - (1/g)t} \left[g + \frac{1}{2g} e^{-2gt}\right]. \end{aligned} \tag{6.7.46}$$

Since the process is stationary, we have

$$\langle a^\dagger(t)a(0)\rangle = e^{i\omega_c t}\left[g e^{g^{-1}|t|} + \frac{1}{2g}\, e^{(g^{-1}+2g)|t|}\right]. \tag{6.7.47}$$

The fluctuation spectrum is given by

$$\int_{-\infty}^{\infty} e^{-i\omega t}\langle a^\dagger(t)a(0)\rangle\, dt = \frac{2}{(\omega_c - \omega)^2 + (1/g)^2} + \frac{[2 + (1/g^2)]}{(\omega_c - \omega)^2 + \left(\dfrac{1}{g} + 2g\right)^2}. \tag{6.7.48}$$

The spectrum far above threshold consists of two Lorentzians. The first has half-width g^{-1} which is extremely narrow. The second has half-width $2g$ approximately and is very broad.

We leave as an exercise the calculation of the intensity and photon number spectra for this oscillator.

For a more rigorous discussion of some of the topics treated in this chapter the reader may consult Ref. 12. Reference 13 will be of interest for presenting a different approach to a quantum-classical correspondence. A few other recent examples of the use Fokker–Planck equations are given in Ref. 14.

REFERENCES

[1] W. H. Louisell, *Coupled Mode and Parametric Electronics*, New York: Wiley, 1960.

[2] W. H. Louisell and L. R. Walker, *Phys. Rev.*, **137**, B204 (1965).

[3] M. Lax, *J. Phys. Chem. Solid*, **25**, 487 (1964).

[4] W. H. Louisell, "Quantum Theory of Noise", in *International School of Physics "Enrico Fermi" XLII Course in Quantum Optics*, 1967, Varenna, Italy.

[5] W. H. Louisell and J. H. Marburger, *J. Quantum Electron.*, **QE-3**, 348 (1967).

[6] F. Reif, *Fundamentals of Statistical and Thermal Physics*, New York: McGraw-Hill, 1965, Chapt. 15.

[7] M. Lax, *Phys. Rev.*, **145**, 110 (1966).

[8] M. Lax and H. Yuen, *Phys. Rev.*, **172**, 362 (1968).

[9] J. P. Gordon, *Phys. Rev.*, **161**, 367 (1967).

[10] M. Lax, *Phys. Rev.*, **172**, 350 (1968); W. H. Louisell and J. H. Marburger, *Phys. Rev.*, **186**, 174 (1969).

[11] M. Lax and W. H. Louisell, *J. Quantum Electronics*, **QE-3**, 47 (1967).

[12] J. R. Klauder and E. E. G. Sudarshan, *Fundamentals of Quantum Optics*, New York: W. A. Benjamin, 1968.

[13] G. S. Agarwal, *Phys. Rev.*, **A2**, 2038 (1970) and **3**, 1783 (1971).

[14] T. von Foerster and R. J. Glauber, *Phys. Rev.* **A3**, 1481 (1971); R. Brambilla and M. Gronchi, *Lett. Nuovo Cimento*, **2**, 511 (1971); H. Hubner, *Z. Phys.*, **239**, 103 (1970); R. Graham and H. Haken, *Z. Phys.*, **235**, 166 (1970); Y. M. Golubev, *Optics Spectrosc.*, **28**, 528 (1970).

PROBLEMS

6.1 Derive (6.2.37) from (6.2.35) using the transformation from the SP to the IP (6.2.36).

6.2 If the reservoir is at absolute zero and there is no driving term, solve (6.2.61) as an operator equation.

6.3 Evaluate the diagonal matrix elements of both sides of (6.2.61) in the coherent state representation ($a|\alpha\rangle = \alpha|\alpha\rangle$).

6.4 Evaluate the m, n matrix elements of both sides of (6.2.61) in the number representation ($a^\dagger a|n\rangle = n|n\rangle$).

6.5 In Problem 6.3 above, the density operator is in the IP. Transform it to the SP.

6.6 Find the eigenvalues and eigenfunctions in Problem 6.2 in the coherent state representation when $v(t) = v_0 e^{-iw_c t}$.

6.7 Find the conditional probability solution of Problem 6.3 when $\alpha = \alpha'$ and $\alpha^* = \alpha'^*$ at $t = 0$. Note that

$$|\alpha'\rangle\langle\alpha'| = \lim_{\epsilon\to 1} \mathcal{N}\{\exp - \epsilon\,|\alpha - \alpha'|^2\}.$$

6.8 Verify (6.2.75b).

6.9 Can we use the master equation (6.2.85) when the "atom" is a harmonic oscillator? Why?

6.10 Verify (6.2.87b).

6.11 Solve the Pauli equation (6.2.126) for a two-level atom.

6.12 Show that σ_e in (6.2.151) is given by

$$\sigma_e = \tanh \frac{1}{kT}(\epsilon_2 - \epsilon_1).$$

6.13 Solve the Bloch equations (6.2.148) when $\hbar\omega = \epsilon_2 - \epsilon_1 = \hbar\omega_{21}$ and $v(t)$ is given by (6.2.140). Find the steady-state solutions also.

6.14 Solve (6.3.48) and (6.3.49). Obtain $\langle a(t)\rangle$ and $\langle a^\dagger(t)a(t)\rangle$.

6.15 Solve (6.5.11) in polar coordinates $x + iy = re^{i\varphi}$. Show that the energy eigenvalues are given by $\epsilon = 2(|m| + 2n - 1)$ where $m = \pm 1, \pm 2, \ldots,$ and $n = 1, 2, 3, \ldots$ and that the eigenfunctions are

$$R_{nm}(r, \varphi) = Ne^{im\varphi}e^{-\frac{1}{2}r^2}r^{|m|}L^{|m|}_{n+|m|-1}(r^2)$$

where $L_q^{\,p}(u)$ is a generalized Laguerre polynomial and N is a normalizing constant.

6.16 Solve (6.5.45) directly when $\bar{n} = 0$ subject to the initial condition that $p(0) = \delta(\alpha - \alpha')\,\delta(\alpha^* - \alpha'^*)$. *Hint:* Use the method of characteristics.

6.17 Obtain the density operator from (6.5.57) by applying the antinormal ordering operator to α and α^*.

6.18 If there is no driving term [$v(t) = 0$] in problem 6.17 show that the diagonal elements of the density operator are (as $t \to \infty$)

$$\langle n|s|n\rangle = \frac{\bar{n}^n}{(1 + \bar{n})^{n+1}} \equiv (1 - e^{-\lambda})e^{-\lambda n},$$

where $\lambda = \hbar\omega/kT$. This is just a Boltzmann distribution corresponding to gaussian thermal noise which is approached when the system started in a coherent state.

6.19 Evaluate the intensity and photon number spectra defined by (6.6.2) and (6.6.3) for a damped driven mode of the radiation field.

6.20 Obtain the steady-state solution of the Fokker–Planck equation (6.4.90). *Hint:* Use the method of Appendix K.

7

Quantum Theory of Damping—Langevin Approach

In this chapter we present the quantum theory of damping from a Langevin viewpoint [1]. Essentially the method consists of replacing the reservoir by damping terms in the Heisenberg equations of motion for a lossless system and adding random forces as driving terms which add fluctuations to the system. The forces must be chosen in such a way that the system has the correct statistical properties to agree with those obtained in Chapter 6.

We begin by considering a damped harmonic oscillator in Section 7.1 in the Heisenberg picture. The results we obtain here will allow us to visualize better the general quantum Langevin theory of noise sources which we present in Section 7.2. We illustrate this formulation in Section 7.3 and 7.4 for a single multilevel atom and N homogeneously broadened three level atoms, respectively. In the final section we give the c-number formulation and relate the drift vectors and diffusion matrix of the prior chapter to our c-number Langevin equations.

7.1 LANGEVIN EQUATIONS OF MOTION FOR DAMPED OSCILLATOR

In this section we obtain solutions of the Heisenberg equations of motion for a damped oscillator under the Wigner–Weisskopff approximation. These allow us to obtain the Langevin equations of motion for a damped oscillator and are useful when we present the general Langevin approach to damping in the following section.

The damped oscillator is described by the hamiltonian

$$H_T = \hbar\omega_c a^\dagger a + \hbar \sum_j \omega_j b_j^\dagger b_j + \hbar \sum_j (\kappa_j b_j a^\dagger + \kappa_j^* b_j^\dagger a). \qquad (7.1.1)$$

Let $M(a, a^\dagger)$ be an arbitrary function of a and $a^\dagger$. It satisfies the Heisenberg equation of motion

$$\frac{dM}{dt} = \frac{1}{i\hbar}[M, H_T] = -i\omega_c[M, a^\dagger a] - i\sum_j \kappa_j^* b_j^\dagger[M, a] - i[M, a^\dagger]\sum_j \kappa_j b_j, \tag{7.1.2}$$

where all operators are in the HP. If we use the identity

$$[M, a^\dagger a] = [M, a^\dagger]a + a^\dagger[M, a], \tag{7.1.3}$$

and the commutation relations

$$\begin{aligned}[M, a] &= -\frac{\partial M}{\partial a^\dagger}\\ [M, a^\dagger] &= \frac{\partial M}{\partial a},\end{aligned} \tag{7.1.4}$$

(7.1.2) reduces to

$$\frac{dM}{dt} = -i\omega_c \frac{\partial M}{\partial a} a + i\omega_c a^\dagger \frac{\partial M}{\partial a^\dagger} + i\sum_j \kappa_j^* b_j^\dagger \frac{\partial M}{\partial a^\dagger} - i\frac{\partial M}{\partial a}\sum_j \kappa_j b_j, \tag{7.1.5}$$

where we again emphasize that all operators are in the HP. Similarly, we see that

$$\frac{db_j}{dt} = -i\omega_j b_j - i\kappa_j^* a. \tag{7.1.6a}$$

We may write this as the integral equation

$$b_j(t) = e^{-i\omega_j t} b_j(0) - i\kappa_j^* \int_0^t a(t')e^{i\omega_j(t'-t)}\, dt', \tag{7.1.6b}$$

where $b_j(0)$ is in the SP and $a(t')$ is in the HP. Although $b_j(t)$ and $b_j^\dagger(t)$ [obtained by taking the adjoint of (7.1.6)] commute with all functions of $a(t)$ and $a^\dagger(t)$ in the HP, the two separate parts of $b_j(t)$ and $b_j^\dagger(t)$ in (7.1.6b) do not. We must therefore be careful of order when we put $b_j(t)$ and $b_j^\dagger(t)$ back into (7.1.5). When we do this, we obtain after minor algebra

$$\begin{aligned}\frac{dM}{dt} &= -i\omega_c \frac{\partial M}{\partial a} a + i\omega_c a^\dagger \frac{\partial M}{\partial a^\dagger} - \sum_j |\kappa_j|^2 \\ &\quad \times \int_0^t \left\{ a^\dagger(t')e^{-i\omega_j(t'-t)} \frac{\partial M}{\partial a^\dagger} + \frac{\partial M}{\partial a} a(t')e^{i\omega_j(t'-t)} \right\} dt' + G_M,\end{aligned} \tag{7.1.7a}$$

where we have let

$$G_M = i\sum_j \left\{ \kappa_j^* b_j^\dagger(0)e^{i\omega_j t} \frac{\partial M}{\partial a^\dagger} - \frac{\partial M}{\partial a} \kappa_j b_j(0)e^{-i\omega_j t} \right\}. \tag{7.1.7b}$$

Again all operators are in the HP except $b_j(0)$ and $b_j^\dagger(0)$ which are in the SP. So far (7.1.7) is exact. Let us consider two special cases. First, let $M = a$. Then we have

$$\frac{da}{dt} = -i\omega_c a - \sum_j |\kappa_j|^2 \int_0^t a(t')e^{+i\omega_j(t'-t)}\,dt' + G_a \tag{7.1.8a}$$

$$G_a = -i \sum_j \kappa_j b_j(0) e^{-i\omega_j t}. \tag{7.1.8b}$$

If we let

$$a(t) = A(t)e^{-i\omega_c t} \tag{7.1.9}$$

to remove the high-frequency behavior from (7.1.8), we see that

$$[a(t), a^\dagger(t)] = [A(t), A^\dagger(t)] = 1, \tag{7.1.10}$$

and (7.1.8) reduces to

$$\frac{dA}{dt} = -\sum_j |\kappa_j|^2 \int_0^t dt'\, A(t') \exp i(\omega_j - \omega_c)(t' - t) + G_A, \tag{7.1.11a}$$

where we have let

$$G_A = -i \sum_j \kappa_j b_j(0) \exp -i(\omega_j - \omega_c)t. \tag{7.1.11b}$$

Unfortunately, this exact integrodifferential equation cannot be solved and we must resort to the Wigner-Weisskopff approximation. If we take the Laplace transform of (7.1.11) and use the results of Section 5.3, we have after minor algebra

$$\bar{A}(s) = \frac{a(0) + \bar{G}_A(s)}{s + \sum_j \dfrac{|\kappa_j|^2}{s + i(\omega_j - \omega_c)}}, \tag{7.1.12a}$$

where

$$\bar{A}(s) = \int_0^\infty e^{-st} A(t)\,dt, \tag{7.1.12b}$$

and

$$\bar{G}_A(s) = -i \sum_j \frac{\kappa_j b_j(0)}{s + i(\omega_j - \omega_c)}. \tag{7.1.12c}$$

Under the Wigner-Weisskopff approximation (see Section 5.3) we let

$$-i \sum_j \frac{|\kappa_j|^2}{(\omega_j - \omega_c) - is} \to \lim_{s\to 0} -i \int \frac{g(\omega_j)\,|\kappa(\omega_j)|^2\,d\omega_j}{(\omega_j - \omega_c) - is}$$

$$= -i \int d\omega_j\, g\,|\kappa|^2 \left\{\frac{1}{\omega_j - \omega_c} + i\pi\,\delta(\omega_j - \omega_c)\right\}$$

$$\equiv \frac{\gamma}{2} + i\,\Delta\omega, \tag{7.1.13}$$

where by (6.2.56)

$$\gamma = 2\pi g(\omega_c)\,|\kappa(\omega_c)|^2$$
$$\Delta\omega = -\int \frac{g(\omega_j)\,|\kappa(\omega_j)|^2\,d\omega_j}{\omega_j - \omega_c}. \tag{7.1.14}$$

Thus (7.1.12a) becomes

$$\bar{A}(s) = \frac{a(0)}{s + \frac{1}{2}\gamma + i\,\Delta\omega} - i\sum_j \frac{\kappa_j b_j(0)}{[s + i(\omega_j - \omega_c)][s + \frac{1}{2}\gamma + i\,\Delta\omega]}. \tag{7.1.15}$$

The inverse transform gives

$$a(t) = u(t)a(0) + \sum_j v_j(t)b_j(0) = e^{-i\omega_c t}A(t), \tag{7.1.16a}$$

where

$$u(t) = \exp -[\tfrac{1}{2}\gamma + i(\omega_c + \Delta\omega)]t$$
$$v_j(t) = \frac{-\kappa_j e^{-i\omega_j t}[1 - \exp i(\omega_j - \omega_c - \Delta\omega)te^{-\gamma t/2}]}{\omega_c - \omega_j + \Delta\omega - i\gamma/2}. \tag{7.1.16b}$$

If we neglect the small frequency shifts, we see that the effect of the Wigner-Weisskopff approximation is to replace the exact (7.1.11) by the Langevin equation

$$\frac{dA}{dt} = -\tfrac{1}{2}\gamma A + G_A(t), \tag{7.1.17a}$$

where

$$G_A(t) = -i\sum_j \kappa_j b_j(0)e^{-i(\omega_j - \omega_c)t}, \tag{7.1.17b}$$

since the solution of this equation is given by (7.1.16) as the reader may readily verify. In this case G_A is the random operator Langevin noise source and the $-(\gamma/2)A$ gives the drift motion. The noise sources are always chosen so their reservoir average is zero. If the reservoir is in thermal equilibrium, we see that

$$\langle G_A(t)\rangle_R = \mathrm{Tr}_R f_0(R)G_A(t) = 0, \tag{7.1.18}$$

when we use (7.1.17b). Therefore, we see from (7.1.17) that

$$\frac{d}{dt}\langle A(t)\rangle_R = -\tfrac{1}{2}\gamma\langle A(t)\rangle_R \tag{7.1.19}$$

so that

$$\langle A(t)\rangle_R = e^{-\gamma t/2}a(0), \tag{7.1.20}$$

since $\langle A(0)\rangle_R = \langle a(0)\rangle_R \equiv a(0)$. This result also follows from (7.1.16) directly.

As a second example, let $M = a^\dagger a \equiv A^\dagger A$. In this case (7.1.7) becomes

$$\frac{d}{dt} a^\dagger a = -\sum_j |\kappa_j|^2 \int_0^t [a^\dagger(t')e^{-i\omega_j(t'-t)}a(t) + a^\dagger(t)a(t')e^{i\omega_j(t'-t)}]\, dt' + G_{a^\dagger a}$$

$$= -\gamma a^\dagger(t)a(t) + G_{a^\dagger a}, \tag{7.1.21}$$

where we have used the Wigner-Weisskopff approximation as we did in going from (7.1.11) to (7.1.17) and where

$$G_{a^\dagger a}(t) = i \sum_j \{\kappa_j^* b_j^\dagger(0)e^{i\omega_j t}a(t) - \kappa_j a^\dagger(t)b_j(0)e^{-i\omega_j t}\}$$

$$= i \sum_j \kappa_j^* b_j(0)e^{i(\omega_j-\omega_c)t}A(t) + \text{ha}, \tag{7.1.22}$$

where ha means hermitian adjoint. If we use the solution (7.1.16), $G_{a^\dagger a}$ becomes

$$G_{a^\dagger a}(t) = +i \sum_j \kappa_j^* b_j^\dagger(0)e^{i(\omega_j-\omega_c)t}e^{-(\gamma/2)t}a(0)$$

$$+ \sum_{j,k} \kappa_j^* b_j^\dagger(0)e^{i(\omega_j-\omega_c)t}\kappa_k b_k(0)\frac{[e^{-i(\omega_k-\omega_c)t} - e^{-(\gamma/2)t}]}{(\gamma/2) - i(\omega_k - \omega_c)} + \text{ha}. \tag{7.1.23}$$

Since for the reservoir in thermal equilibrium

$$\langle b_j^\dagger(0)\rangle_R = \langle b_j(0)\rangle_R = 0$$

$$\langle b_j^\dagger(0)b_k(0)\rangle_R = \delta_{jk}\bar{n}_j \equiv \delta_{jk}\frac{1}{e^{\hbar\omega_j/kT} - 1}, \tag{7.1.24}$$

we see that

$$\langle G_{a^\dagger a}(t)\rangle_R$$

$$= \sum_j |\kappa_j|^2 \bar{n}_j \left\{\frac{1 - e^{[i(\omega_j-\omega_c)-(\gamma/2)]t}}{(\gamma/2) - i(\omega_j - \omega_c)} + \text{cc}\right\}$$

$$= \sum_j |\kappa_j|^2 \bar{n}_j \frac{\{\gamma - \gamma e^{-(\gamma/2)t}\cos(\omega_j - \omega_c)t + 2(\omega_j - \omega_c)e^{-(\gamma/2)t}\sin(\omega_j - \omega_c)t\}}{(\gamma/2)^2 + (\omega_j - \omega_c)^2}. \tag{7.1.25}$$

Since $|\kappa_j|^2\bar{n}_j$ is slowly varying and the summand is so strongly peaked at $\omega_j = \omega_c$, we may convert this sum to an integral and remove the slowly varying factors. This gives

$$\langle G_{a^\dagger a}(t)\rangle_R \cong |\kappa(\omega_c)|^2 g(\omega_c)\bar{n}(\omega_c)\int_{-\infty}^{\infty} dx \frac{\{\gamma - \gamma e^{-(\gamma/2)t}\cos xt + 2xe^{-(\gamma/2)t}\sin xt\}}{(\gamma/2)^2 + x^2} \tag{7.1.26}$$

where we let $x = \omega_j - \omega_c$ and $dx = d\omega_j$. We extended the limits from $-\infty$ to $+\infty$ since the main contribution comes at $x = 0$. Since

$$\int_{-\infty}^{\infty} \frac{dx}{(\gamma/2)^2 + x^2} = \frac{2\pi}{\gamma}$$

$$\int_{-\infty}^{\infty} \frac{\cos xt}{(\gamma/2)^2 + x^2}\, dx = \frac{2\pi}{\gamma} e^{-(\gamma/2)|t|} \tag{7.1.27}$$

$$\int_{-\infty}^{\infty} \frac{x \sin x\, |t|}{(\gamma/2)^2 + x^2}\, dx = +\begin{cases} \pi e^{-(\gamma/2)|t|} & t \neq 0 \\ 0 & t = 0, \end{cases}$$

we see that (7.1.26) reduces to

$$\langle \mathscr{G}_{a^\dagger a}(t) \rangle_R = \gamma \bar{n}, \tag{7.1.28}$$

where we used (6.2.56). If we therefore take the thermal reservoir average of both sides of (7.1.21), we obtain

$$\frac{d}{dt} \langle a^\dagger a \rangle_R = -\gamma \langle a^\dagger a \rangle_R + \gamma \bar{n}, \tag{7.1.29}$$

which has the solution

$$\langle a^\dagger(t) a(t) \rangle_R = e^{-\gamma t} a^\dagger(0) a(0) + \bar{n}[1 - e^{-\gamma t}]. \tag{7.1.30}$$

Let us now rewrite (7.1.21) and include the reservoir average of $G_{a^\dagger a}$ so that (since $A^\dagger A = a^\dagger a$)

$$\frac{d}{dt} A^\dagger A = -\gamma A^\dagger A + \gamma \bar{n} + G_{A^\dagger A}, \tag{7.1.31}$$

where we have let

$$\begin{aligned} G_{A^\dagger A} &= G_{a^\dagger a} - \langle G_{a^\dagger a} \rangle_R \\ &= G_{a^\dagger a} - \gamma \bar{n}. \end{aligned} \tag{7.1.32}$$

Since now $\langle G_{A^\dagger A}(t) \rangle_R = 0$, we see that the thermal average over the reservoir of (7.1.31) gives the identical result as the thermal average of (7.1.21). The Langevin force $G_{A^\dagger A}$ is chosen so that its reservoir thermal average is zero and the remaining terms in (7.1.31) give the thermally averaged drift motion of the operator on the left side. It is the Langevin equation of motion for $a^\dagger a = A^\dagger A$, the photon number. The Langevin force and the damping term in these equations replace the reservoir in an equivalent circuit representation. The damping term plays the role of a resistance and the *operator* force plays the role of a noise generator which puts fluctuations into the circuit. Since it is a quantum noise generator, it maintains the commutation relations and therefore insures that the uncertainty principle is not violated under the approximations (see Problem 7.1).

Spectra

Once we have the approximate solutions of the Heisenberg equations, we may calculate the various spectra. The fluctuation spectrum defined by (6.6.1) is

$$\int_{-\infty}^{\infty} e^{-i\omega t}\langle a^\dagger(t)a(0)\rangle\, dt = \int_{-\infty}^{\infty} e^{-i(\omega-\omega_c)t}\langle A^\dagger(t)A(0)\rangle\, dt, \tag{7.1.33}$$

where we used (7.1.9) and its adjoint. This spectrum is just the Fourier transform of the correlation function

$$K_{A^\dagger A}(t) \equiv \langle A^\dagger(t)A(0)\rangle = \mathrm{Tr}_{R,S} A^\dagger(t)a(0)\rho(0), \tag{7.1.34}$$

where the initial density operator is

$$\rho(0) = S(0)f_0(R) = \frac{S(0)e^{-\beta R}}{\mathrm{Tr}_R\, e^{-\beta R}}, \tag{7.1.35}$$

and we trace over both system and reservoir. If use the adjoint of (7.1.16) we have for the two-time correlation function

$$\begin{aligned} K_{A^\dagger A}(t) &= \mathrm{Tr}_{R,S} f_0(R)S(0)\left\{\left[e^{-(\gamma/2)t}a^\dagger(0) + \int_0^t e^{(\gamma/2)(t'-t)}G_A^\dagger(t')\, dt'\right]a(0)\right\} \\ &= \mathrm{Tr}_S\, S(0)a^\dagger(0)a(0)e^{-(\gamma/2)t} + \mathrm{Tr}_S\left\{S(0)\int_0^t e^{(\gamma/2)(t'-t)}\langle G_A^\dagger(t')\rangle_R\, dt' a(0)\right\} \end{aligned} \tag{7.1.36}$$

or

$$K_{A^\dagger A}(t) = e^{-(\gamma/2)|t|}\langle a^\dagger(0)a(0)\rangle \tag{7.1.37}$$

where we used the adjoint of (7.1.18) and have let

$$\langle a^\dagger(0)a(0)\rangle = \mathrm{Tr}_S\, S(0)a^\dagger(0)a(0). \tag{7.1.38}$$

We have used the absolute value of t since for a stationary process [2]

$$K(t) = K(-t). \tag{7.1.39}$$

The fluctuation spectrum (7.1.33) reduces to

$$\frac{\gamma\langle a^\dagger(0)a(0)\rangle}{(\omega-\omega_c)^2 + (\gamma/2)^2}, \tag{7.1.40}$$

which is Lorentzian centered at $\omega = \omega_c$ with half-width $\gamma/2$. This result agrees with (6.6.25) if at $t = 0$ the cavity is in thermal equilibrium with the reservoir so that

$$\bar{n} = \langle a^\dagger(0)a(0)\rangle. \tag{7.1.41}$$

Note that (6.6.25) was calculated under the Markoff approximation whereas here we have used the Wigner-Weisskopff approximation which to our present accuracy is equivalent.

Consider next the intensity spectrum (6.6.2)

$$I(\omega) = \int_{-\infty}^{\infty} dt\, e^{-i\omega t} \langle a^\dagger(0) a^\dagger(t) a(t) a(0)\rangle$$

$$= \int_{-\infty}^{\infty} dt\, e^{-i\omega t} \langle a^\dagger(0) A^\dagger(t) A(t) a(0)\rangle. \tag{7.1.42}$$

The correlation function is

$$\begin{aligned} K(t) &= \mathrm{Tr}_{R,S}\, \rho(0) a^\dagger(0) A^\dagger(t) A(t) a(0)\rangle \\ &= \mathrm{Tr}_S\, S(0) a^\dagger(0) \langle A^\dagger(t) A(t)\rangle_R a(0). \end{aligned} \tag{7.1.43}$$

From (7.1.30) we see that

$$\begin{aligned} K(t) &= \mathrm{Tr}_S\, S(0) a^\dagger(0) \{a^\dagger(0) a(0) e^{-\gamma t} + \bar{n}[1 - e^{-\gamma t}]\} a(0) \\ &= \langle a^{\dagger 2}(0) a^2(0)\rangle e^{-\gamma|t|} + \bar{n}\langle a^\dagger(0) a(0)\rangle [1 - e^{-\gamma|t|}]. \end{aligned} \tag{7.1.44}$$

The intensity spectrum is

$$I(\omega) = \frac{2\gamma}{\gamma^2 + \omega^2} \{\langle a^{\dagger 2}(0) a^2(0)\rangle - \bar{n}\langle a^\dagger(0) a(0)\rangle\} + 2\pi\bar{n}\langle a^\dagger(0) a(0)\rangle\, \delta(\omega), \tag{7.1.45}$$

which consists of a Lorentzian of half-width γ centered at $\omega = 0$ plus a δ-function at $\omega = 0$.

Diffusion Coefficients. Fluctuation-Dissipation Theorem [2]

In this section we obtain two time correlation functions for the Langevin forces and derive the fluctuation-dissipation theorem.

We have shown in (7.1.18) that

$$\langle G_A(t)\rangle_R = 0. \tag{7.1.46a}$$

It also follows since $G_A^\dagger(t) = G_{A^\dagger}(t)$ that

$$\langle G_{A^\dagger}(t)\rangle_R = 0. \tag{7.1.46b}$$

Similarly, one may easily show that the autocorrelation functions are zero:

$$\begin{aligned} K_{AA}(t_1 - t_2) &\equiv \langle G_A(t_1) G_A(t_2)\rangle_R = \mathrm{Tr}_R\, f_0(R) G_A(t_1) G_A(t_2) = 0 \\ K_{A^\dagger A^\dagger}(t_1 - t_2) &\equiv \langle G_{A^\dagger}) t_1) G_{A^\dagger}(t_2)\rangle_R = 0 \end{aligned} \tag{7.1.47}$$

(see Problem 7.4).

Consider next the cross-correlation function defined by

$$\begin{aligned} K_{A^\dagger A}(t_1 - t_2) &= \langle G_A^\dagger(t_1) G_A(t_2) \rangle_R \\ &= \sum_{j,k} \kappa_j^* \kappa_k \langle b_j^\dagger(0) b_k(0) \rangle_R e^{i[(\omega_j-\omega_c)]t_1 - (\omega_k - \omega_c)t_2]} \\ &= \sum_j |\kappa_j|^2 \, \bar{n}_j \, e^{i(\omega_j-\omega_c)(t_1-t_2)} \\ &= \int_0^\infty g(\omega_j) \, |\kappa(\omega_j)|^2 \, \bar{n}(\omega_j) e^{i(\omega_j-\omega_c)(t_1-t_2)} \, d\omega_j, \qquad (7.1.48) \end{aligned}$$

where we used (7.1.24). Note that this is a function of the time difference $t_1 - t_2$ only which is characteristic of stationary random processes. In a similar way we see that

$$\begin{aligned} K_{AA^\dagger}(t_1 - t_2) = \langle G_A(t_1) G_A^\dagger(t_2) \rangle_R &= \sum_j |\kappa_j|^2 \, (1 + \bar{n}_j) e^{-i(\omega_j-\omega_c)(t_1-t_2)} \\ &= \int_0^\infty d\omega_j \, g(\omega_j) \, |\kappa(\omega_j)|^2 \, [1 + \bar{n}(\omega_j)] e^{-i(\omega_j-\omega_c)(t_1-t_2)}. \qquad (7.1.49) \end{aligned}$$

We therefore see that

$$K_{A^\dagger A}(t_1 - t_2) \neq K_{AA^\dagger}(t_1 - t_2), \qquad (7.1.50)$$

as would be the case with a classical cross-correlation function.

Rather than pull out the slowly varying terms in (7.1.48) and (7.1.49) and replacing the integral of the exponential by a δ-function, let us proceed a little more carefully to show that this approximation corresponds to the Markoff approximation made in the last chapter.

We expect that as $|t_1 - t_2| \to \infty$, the rapid oscillations in (7.1.48) and (7.1.49) will cause the correlations of the Langevin forces to vanish so that for a sufficiently long time difference, they will become uncorrelated. Let τ_c be the correlation time of these forces. If we multiply both sides of $K_{A^\dagger A}(t_1 - t_2)$ by dt_1 and dt_2 and integrate both from t to $t + \Delta t$ where $\Delta t \gg \tau_c$ but short compared with the system damping time γ^{-1} and divide by Δt we obtain the diffusion coefficient *defined* by

$$\begin{aligned} 2\langle D_{A^\dagger A} \rangle_R &\equiv \frac{1}{\Delta t} \int_t^{t+\Delta t} dt_1 \int_t^{t+\Delta t} dt_2 \langle G_A^\dagger(t_1) G_A(t_2) \rangle_R \\ &= \frac{1}{\Delta t} \int_0^\infty d\omega_j \, g(\omega_j) \, |\kappa(\omega_j)|^2 \, \bar{n}(\omega_j) \int_t^{t+\Delta t} dt_1 \int_t^{t+\Delta t} dt_2 \, e^{+i(\omega_j-\omega_c(t_1-t_2)} \\ &= \Delta t \int_0^\infty d\omega_j \, g(\omega_j) \, |\kappa(\omega_j)|^2 \, \bar{n}(\omega_j) \, \frac{4 \sin^2 \frac{1}{2}(\omega_j - \omega_c) \Delta t}{(\Delta t)^2 (\omega_j - \omega_c)^2} . \end{aligned}$$

$$(7.1.51)$$

The integrand is now highly peaked at $\omega_j = \omega_c$ so that we may without serious error let the lower limit extend to $-\infty$ and remove the slowly varying terms $g|\kappa|^2\bar{n}$ from the integral. If we use (6.2.56) again

$$\gamma = 2\pi g(\omega_c)|\kappa(\omega_c)|^2, \tag{7.1.52}$$

then

$$2\langle D_{A^\dagger A}\rangle_R = \frac{\gamma\bar{n}}{2\pi}\frac{2}{\Delta t}\int_{-\infty}^{\infty}\frac{\sin^2 x\,\Delta t}{x^2}\,dx, \tag{7.1.53}$$

where we let $2x = \omega_j - \omega_c$. Then since

$$\int_{-\infty}^{\infty}\frac{\sin^2 x\,\Delta t}{x^2}\,dx = \pi\,\Delta t, \tag{7.1.54}$$

the diffusion coefficient becomes

$$\begin{aligned} 2\langle D_{A^\dagger A}\rangle_R &= \gamma\bar{n} \\ &= \frac{1}{\Delta t}\int_t^{t+\Delta t} dt_1\int_t^{t+\Delta t} dt_2\langle G_A{}^\dagger(t_1)G_A(t_2)\rangle_R, \end{aligned} \tag{7.1.55}$$

provided

$$\gamma^{-1} \gg \Delta t \gg \tau_c. \tag{7.1.56}$$

We see that we could have obtained this identical result if we replaced (7.1.48) by

$$\begin{aligned} K_{A^\dagger A}(t_1 - t_2) &= \langle G_A{}^\dagger(t_1)G_A(t_2)\rangle_R \\ &= g\,|\kappa|^2\,\bar{n}\int_{-\infty}^{\infty} e^{i(\omega_j-\omega_c(t_1-t_2)}\,d\omega_j \\ &= \gamma\bar{n}\,\delta(t_1 - t_2). \end{aligned} \tag{7.1.57}$$

Therefore the Markoff and Wigner-Weisskopff approximations to the present accuracy give the same diffusion coefficients, since

$$\frac{1}{\Delta t}\int_t^{t+\Delta t} dt_1\int_t^{t+\Delta t} dt_2\,\gamma\bar{n}\,\delta(t_1 - t_2) = \gamma\bar{n} = 2\langle D_{A^\dagger A}\rangle. \tag{7.1.58}$$

We always assume that the random forces are δ correlated provided that the system is Markoffian—its future is determined by the present and not the past. The reservoir becomes uncorrelated long before the system has changed very much and the system therefore cannot develop any memory of the past through its interaction with the reservoir.

Similarly, we see that

$$\begin{aligned} K_{AA^\dagger}(t_1 - t_2) &= \langle G_A(t_1)G_A{}^\dagger(t_2)\rangle_R = \gamma(\bar{n} + 1)\,\delta(t_1 - t_2) \\ 2\langle D_{AA^\dagger}\rangle_R &= \gamma(\bar{n} + 1). \end{aligned} \tag{7.1.59}$$

From (7.1.57) we see that on integrating both sides we obtain

$$\gamma = \frac{1}{\bar{n}} \int_{-\infty}^{\infty} \langle G_A{}^\dagger(\tau) G_A(0) \rangle_R \, d\tau. \tag{7.1.60}$$

This states that the system damping γ is determined by the reservoir fluctuating forces which introduce fluctuations into the system. This is one formulation of the fluctuation-dissipation theorem. Similarly, by (7.1.59) we obtain another form of the theorem

$$\gamma = \frac{1}{\bar{n}+1} \int_{-\infty}^{\infty} \langle G_A(\tau) G_A{}^\dagger(0) \rangle_R \, d\tau. \tag{7.1.61}$$

The diffusion coefficients $\langle D_{AA} \rangle_R$ and $\langle D_{A^\dagger A^\dagger} \rangle$ are easily seen to be zero.

Langevin Equation for Photon Number

We have already obtained the Langevin equation of motion for the photon number (7.1.31). To obtain more insight into the nature of the Markoff approximation and to aid us in formulating the Langevin method for more general systems, let us obtain the Langevin equation for the photon number by another method.

By the usual rules of differentiation, we have that

$$\frac{d}{dt} A^\dagger A = A^\dagger \frac{dA}{dt} + \frac{dA^\dagger}{dt} A. \tag{7.1.62}$$

If we use the Langevin equation (7.1.17) and its adjoint, we obtain

$$\frac{d}{dt} A^\dagger A = -\gamma A^\dagger A + A^\dagger G_A + A G_A^\dagger,$$

where we have been careful about order. The reader should convince himself that this equation is identical to (7.1.21) and (7.1.22). We need the reservoir average to give us the drift motion

$$\frac{d}{dt} \langle A^\dagger(t) A\ t) \rangle_R = -\gamma \langle A^\dagger(t) A(t) \rangle_R + \langle A^\dagger(t) G_A(t) \rangle_R + \langle G_{A^\dagger}(t) A(t) \rangle_R. \tag{7.1.64}$$

We have already evaluated $\langle [A^\dagger(t) G_A(t) + G_A{}^\dagger(t) A(t)] \rangle_R = \gamma \bar{n}$. We used the solution for $A(t)$ and $A^\dagger(t)$. The method we now use relies more directly on the Markoff approximation and does not require knowledge of the solution for $A(t)$ and $A^\dagger(t)$. Consequently, it is more general.

We begin by writing the identity

$$\begin{aligned} \langle G_{a^\dagger a}(t) \rangle_R &= \langle [A^\dagger(t) G_A(t) + G_{A^\dagger}(t) A(t)] \rangle_R \\ &\equiv \left\langle \left[A^\dagger(t_c) + \int_{t_c}^{t} \frac{dA^\dagger}{ds} ds \right] G_A(t) \right\rangle_R + \left\langle G_{A^\dagger}(t) \left[A(t_c) + \int_{t_c}^{t} ds \frac{dA}{ds} \right] \right\rangle_R \end{aligned} \tag{7.1.65}$$

where $t > t_c$ and $\gamma^{-1} \gg t - t_c \gg \tau_c$. Clearly,

$$\begin{aligned} \langle A^\dagger(t_c) G_A(t) \rangle_R &= 0 \\ \langle G_{A^\dagger}(t) A(t_c) \rangle_R &= 0. \end{aligned} \tag{7.1.66}$$

This result must be true under the Markoff approximation for if the system operator and reservoir Langevin force were correlated over this time interval, the system would develop memory. Since we happen to know the solution for $A^\dagger(t_c)$, let us verify that (7.1.66) is indeed true before proceeding. We have by (7.1.11b) and the adjoint of (7.1.16) that for $t > t_c > 0$ with $\Delta\omega = 0$

$$\begin{aligned} &\langle A^\dagger(t_c) G_A(t) \rangle_R \\ &= -i \left\langle \left\{ e^{-\gamma t_c} a(0) + i \sum_j \kappa_j^* b_j^\dagger(0) \frac{[e^{i(\omega_j - \omega_c)t_c} - e^{-(\gamma/2)t_c}]}{(\gamma/2) + i(\omega_j - \omega_c)} \right\} \sum_k \kappa_k b_k(0) e^{-i(\omega_k - \omega_c)t} \right\rangle_R \\ &= \sum_j |\kappa_j|^2 \bar{n}_j \frac{[e^{i(\omega_j - \omega_c)t_c} - e^{-(\gamma/2)t_c}]}{(\gamma/2) + i(\omega_j - \omega_c)} e^{-i(\omega_j - \omega_c)t} \\ &= \frac{\gamma \bar{n}}{2\pi} \int_{-\infty}^{\infty} \frac{[e^{-ix(t - t_c)} - e^{-(\gamma/2)t_c} e^{-ixt}]}{(\gamma/2) + ix} dx, \end{aligned} \tag{7.1.67}$$

where we used (7.1.52), (7.1.24), and let $x = \omega_j - \omega_c$. Since for $\alpha > 0$

$$\int_{-\infty}^{\infty} \frac{e^{-i\alpha x}}{(\gamma/2) + ix} dx = 0 \tag{7.1.68}$$

(7.1.66) follows directly. A similar argument verifies the second relation of (7.1.66). Again (7.1.66) is a direct consequence of the Markoff approximation and is not peculiar to the damped oscillator. Accordingly, (7.1.65) reduces to

$$\langle G_{a^\dagger a}(t) \rangle_R = \int_{t_c}^{t} \left\{ \left\langle \frac{dA^\dagger}{ds} G_A(t) \right\rangle_R + \left\langle G_A^\dagger(t) \frac{dA}{ds} \right\rangle_R \right\} ds. \tag{7.1.69}$$

If we next use (7.1.17) and its adjoint, we obtain

$$\begin{aligned} &\langle G_{a^\dagger a}(t) \rangle_R \\ &= \int_{t_c}^{t} ds \left\langle \left\{ \left[-\frac{\gamma}{2} A^\dagger(s) + G_{A^\dagger}(s) \right] G_A(t) + G_A^\dagger(t) \left[-\frac{\gamma}{2} A(s) + G_A(s) \right] \right\} \right\rangle_R. \end{aligned} \tag{7.1.70}$$

By the Markoff approximation again since $t > s$, we have that

$$\begin{aligned} \langle A^\dagger(s) G_A(t) \rangle_R &= 0 \\ \langle G_{A^\dagger}(t) A(s) \rangle_R &= 0. \end{aligned} \tag{7.1.71}$$

Also we have shown that over the interval $t - t_c$ that

$$\langle G_{A^\dagger}(s) G_A(t)\rangle_R = \langle G_{A^\dagger}(t) G_A(s)\rangle_R = 2\langle D_{A^\dagger A}\rangle_R\, \delta(t - s) = \gamma\bar{n}\, \delta(t - s). \tag{7.1.72}$$

Therefore, (7.1.70) reduces to

$$\langle G_a{}^\dagger{}_a(t)\rangle_R = \int_{t_0}^{t} ds\, 2\gamma\bar{n}\, \delta(t - s) = \gamma\bar{n}, \tag{7.1.73}$$

since

$$\int_{t_c}^{t} F(s)\, \delta(t - s)\, ds = \tfrac{1}{2}F(t). \tag{7.1.74}$$

This agrees of course with our prior calculations. Therefore, (7.1.64) becomes

$$\frac{d}{dt}\langle A^\dagger A\rangle_R = -\gamma\langle A^\dagger A\rangle_R + \gamma\bar{n}. \tag{7.1.75}$$

We may take as our Langevin equation

$$\frac{d}{dt} A^\dagger A = -\gamma A^\dagger A + \gamma\bar{n} + G_{A^\dagger A}, \tag{7.1.76}$$

where the Langevin force is

$$G_{A^\dagger A}(t) = A^\dagger(t_c) G_A(t) + G_A^\dagger(t) A(t_c), \tag{7.1.77}$$

and has the property that

$$\langle G_{A^\dagger A}(t)\rangle_R = 0. \tag{7.1.78}$$

Again the drift motion is explicitly included and a random force is added to retain the correct quantum fluctuations. The first term in (7.1.76) is the rate of loss of photons into the reservoir while the second gives the rate at which photons enter from the reservoir. The force causes fluctuations from the mean photon number.

Einstein Relation

Suppose we know the thermal average equations

$$\frac{d\langle A\rangle_R}{dt} = -\frac{\gamma}{2}\langle A\rangle_R$$

$$\frac{d\langle A^\dagger\rangle_R}{dt} = -\frac{\gamma}{2}\langle A^\dagger\rangle_R \tag{7.1.79}$$

$$\frac{d}{dt}\langle A^\dagger A\rangle_R = -\gamma\langle A^\dagger A\rangle_R + \gamma\bar{n}.$$

We may then write the Langevin equations

$$\begin{aligned}\frac{dA}{dt} &= -\frac{\gamma}{2}A + G_A \\ \frac{dA^\dagger}{dt} &= -\frac{\gamma}{2}A^\dagger + G_{A^\dagger} \\ \frac{dA^\dagger A}{dt} &= -\gamma A^\dagger A + \gamma\bar{n} + G_{A^\dagger A}.\end{aligned} \tag{7.1.80}$$

In a time interval t to $t + \Delta t$ where $\gamma^{-1} \gg \Delta t \gg \tau_c$, we may integrate these as

$$\begin{aligned}\frac{\Delta A}{\Delta t} &= -\frac{\gamma}{2}A + \frac{1}{\Delta t}\int_t^{t+\Delta t} G_A(s)\,ds \\ \frac{\Delta A^\dagger}{\Delta t} &= -\frac{\gamma}{2}A^\dagger + \frac{1}{\Delta t}\int_t^{t+\Delta t} G_{A^\dagger}(s')\,ds' \\ \frac{\Delta A^\dagger A}{\Delta t} &= -\gamma A^\dagger A + \gamma\bar{n} + \frac{1}{\Delta t}\int_t^{t+\Delta t} G_{A^\dagger A}(s)\,ds,\end{aligned} \tag{7.1.81}$$

where

$$\begin{aligned}\Delta A &= A(t+\Delta t) - A(t) \\ \Delta A^\dagger A &= A^\dagger(t+\Delta t)A(t+\Delta t) - A^\dagger(t)A(t).\end{aligned} \tag{7.1.82}$$

That is, during Δt, A, $A^\dagger$, and $A^\dagger A$ do not change whereas the forces do.

We see that

$$\begin{aligned}\left\langle\frac{\Delta A^\dagger A}{\Delta t}\right\rangle_R &= \left\langle\frac{[A^\dagger(t) + \Delta A^\dagger][A(t) + \Delta A] - A^\dagger(t)A(t)}{\Delta t}\right\rangle_R \\ &= \left\langle A^\dagger(t)\frac{\Delta A}{\Delta t} + \frac{\Delta A^\dagger}{\Delta t}A + \frac{\Delta A^\dagger\,\Delta A}{\Delta t}\right\rangle_R.\end{aligned} \tag{7.1.83}$$

If we use the first two of (7.1.81) on the right above, we obtain

$$\begin{aligned}\left\langle\frac{\Delta A^\dagger A}{\Delta t}\right\rangle_R &= -\gamma\langle A^\dagger(t)A(t)\rangle_R - \frac{\gamma}{2}\cdot\frac{2}{\Delta t} \\ &\quad\times\int_t^{t+\Delta t}\left\langle A^\dagger(t)G_A(s) + G_{A^\dagger}(s)A(t)\right\rangle_R ds + \left(\frac{\gamma}{2}\right)^2\langle A^\dagger(t)A(t)\rangle_R\Delta t \\ &\quad+\frac{1}{\Delta t}\int_t^{t+\Delta t} ds\int_t^{t+\Delta t} ds'\langle G_{A^\dagger}(s)G_A(s')\rangle_R\end{aligned} \tag{7.1.84}$$

By the Markoff approximation, we know that

$$\begin{aligned}\langle A^\dagger(t)G_A(s)\rangle_R &= 0 \qquad s > t \\ \langle G_{A^\dagger}(s)A(t)\rangle_R &= 0 \qquad s > t,\end{aligned} \tag{7.1.85}$$

so that as $\Delta t \to 0$ as far as system operators are concerned the $(\gamma/2)^2 \Delta t \langle A^\dagger A \rangle_R$ term vanishes and we have

$$\frac{d}{dt} \langle A^\dagger A \rangle_R = -\gamma \langle A^\dagger A \rangle + 2\langle D_{A^\dagger A} \rangle, \tag{7.1.86}$$

since

$$\langle G_{A^\dagger}(s) G_A(s') \rangle_R = 2\langle D_{A^\dagger A} \rangle_R \, \delta(s - s'). \tag{7.1.87}$$

W th the aid of (7.1.17) and its adjoint, we may rewrite (7.1.86) as

$$2\langle D_{A^\dagger A} \rangle = \frac{d}{dt} \langle A^\dagger A \rangle_R - \left\langle A^\dagger \left[\frac{dA}{dt} - G_A \right] \right\rangle_R - \left\langle \left[\frac{dA^\dagger}{dt} - G_A^\dagger \right] A \right\rangle_R, \tag{7.1.88}$$

which is called the Einstein relation to determine the diffusion coefficient. If we use (7.1.80), we see that (7.1.88) reduces to

$$2\langle D_{A^\dagger A} \rangle_R = \gamma \bar{n} \tag{7.1.89}$$

so that the diffusion constants may be obtained from a knowledge of the mean equations of motion according to the Einstein relation (7.1.88).

7.2 QUANTUM THEORY OF LANGEVIN NOISE SOURCES [1]

A quantum system experiences damping and fluctuations when it interacts with a reservoir as we have seen in the previous chapter. The Langevin approach adopts the philosophy that the reservoir may be completely eliminated provided that the frequency shifts and damping caused by the system-reservoir interaction are incorporated into the equations of motion and provided suitable quantum (operator) noise sources are added as driving terms to the equations of motion. The moments or correlation functions of the Langevin noise sources must give the correct statistical behavior to the system (correct diffusion coefficients).

According to (6.3.36), the thermal reservoir average of a system operator is given by

$$\frac{d\langle M \rangle_R}{dt} = -\frac{i}{\hbar} \langle [M, H] \rangle_R - \sum_{i,j} \delta(\omega_i, -\omega_j) \times \{ \langle [M, Q_i] Q_j \rangle_R w_{ij}^+ - \langle Q_j [M, Q_i] \rangle_R w_{ji}^- \}. \tag{7.2.1}$$

We have taken into account the frequency shifts and damping caused by the reservoir in this equation. We may remove the reservoir averages if we add a random Langevin force G_M which will be a function of the system operators as well as reservoir operators which has the correct statistical properties. We therefore write the Langevin equation

$$\frac{dM}{dt} = -\frac{i}{h}[M, H] - \sum_{i,j} \delta(\omega_i, -\omega_j)\{[M, Q_i]Q_j w_{ij}^+ - Q_j[M, Q_i]w_{ji}^-\} + G_M(t)$$

$$\equiv A_M + G_M. \tag{7.2.2}$$

In order for (7.2.2) to be consistent with (7.2.1), we must first require that

$$\langle G_M(t)\rangle_R = 0. \tag{7.2.3}$$

Furthermore, since we have retained terms up to second order in the system-reservoir in (7.2.1), we see from the derivation of (7.2.1) given in Section 6.3 that G_M be only first order in the interaction. We therefore let G_M be given by

$$G_M(t) = -i \sum_i \langle [M, Q_i]\rangle_R \{F_i(t) - \langle F_i\rangle_R\}, \tag{7.2.4}$$

where

$$F_i(t) = e^{(i/\hbar)R(t-t_0)} F_i^S e^{-(i/\hbar)R(t-t_0)}. \tag{7.2.5}$$

We have obviously not "derived" (7.2.4). We show that it gives the correct results for a damped oscillator and that its two time correlation functions give the correct diffusion coefficients in special cases.

We first note that the choice (7.2.4) satisfies (7.2.3). Consider again the special case of the damped oscillator. If we let $M = a$, and use (6.2.41) and (6.2.58), we see that

$$\begin{aligned} \frac{da}{dt} &= -\left[i\omega_c' + \frac{\gamma}{2}\right]a + G_a \\ G_a &= -i \sum_j \kappa_j b_j(t_0) e^{-i\omega_j(t-t_0)}. \end{aligned} \tag{7.2.6}$$

If we use (7.1.9), we see that this agrees with (7.1.17).

If $M = a^\dagger a$, we have on algebraic simplification of (7.2.2)

$$\frac{d}{dt} a^\dagger a = -\gamma a^\dagger a + \gamma \bar{n} + G_{A^\dagger A} \tag{7.2.7a}$$

$$\begin{aligned} G_{A^\dagger A} &= -i\langle [a^\dagger a, a^\dagger]\rangle_R F_1(t) - i\langle [a^\dagger a, a]\rangle_R F_2(t) \\ &= -i\langle a(t)\rangle_R \sum_j \kappa_j^* b_j^\dagger(t) e^{i\omega_j(t-t_0)} + i\langle a^\dagger(t)\rangle_R \sum_j \kappa_j b_j(t_0) e^{-i\omega_j(t-t_0)}, \end{aligned} \tag{7.2.7b}$$

which clearly agrees with (7.1.31) if we use the solution of the reservoir average of (7.2.6) and its adjoint.

Consider next a set of system operators $\underset{\sim}{a} = \{a_1, a_2, \ldots\}$. Then the Langevin equation for a_μ is

$$\frac{da_\mu}{dt} = -\frac{i}{\hbar}[a_\mu, H] - \sum_{i,j} \delta(\omega_i, -\omega_j)\{[a_\mu, Q_i]Q_j w_{ij}^+ - Q_j[a_\mu, Q_i]w_{ji}^-\} + G_\mu$$
$$\equiv A_\mu + G_\mu. \tag{7.2.8}$$

For simplicity assume

$$-\frac{i}{\hbar}[a_\mu, H] = i\omega_\mu a_\mu. \tag{7.2.9}$$

It is always imperative that this high-frequency free motion be removed. We therefore let

$$a_\mu(t) = e^{i\omega_\mu(t-t_0)}a'_\mu(t). \tag{7.2.10}$$

Then (7.2.8) becomes

$$\frac{da'_\mu}{dt} = -\sum_{i,j} \delta(\omega_i, -\omega_j)\{[a'_\mu, Q_i]Q_j w_{ij}^+ - Q_j[a'_\mu, Q_i]w_{ij}^-\} + G'_\mu$$
$$\equiv A'_\mu + G'_\mu, \tag{7.2.11}$$

where

$$G'_\mu = -i\sum_i \langle [a'_\mu, Q_i]\rangle_R[F_i(t) - \langle F_i\rangle_R]. \tag{7.2.12}$$

Of course, the Q_i and Q_j will be functions of the a's and they should also be transformed according to (7.2.10). It then follows that

$$\frac{d\langle a'_\mu\rangle_R}{dt} = \langle A'_\mu\rangle_R. \tag{7.2.13}$$

After removing the high-frequency behavior, we may repeat the analysis of Section 7.1 to derive the Einstein relation

$$2\langle D'_{\mu\nu}\rangle_R = \frac{1}{\Delta t}\int_t^{t+\Delta t} ds \int_t^{t+\Delta t} ds' \langle G'_\mu(s)G'_\nu(s')\rangle_R$$
$$= \frac{d}{dt}\langle a'_\mu a'_\nu\rangle_R - \langle a'_\mu A'_\nu\rangle_R - \langle A'_\mu a'_\nu\rangle_R, \tag{7.2.14}$$

while the random forces are δ-correlated and given by

$$\langle G'_\mu(s)G'_\nu(s')\rangle_R = 2\langle D'_{\mu\nu}\rangle_R\,\delta(s - s'). \tag{7.2.15a}$$

Also, we have a statement of the Markoff approximation that

$$\langle F[\underset{\sim}{a}'(t)]G'_\mu(s)\rangle_R = 0 \qquad \text{if} \quad s > t, \tag{7.2.15b}$$

where F is an arbitrary function of the system operators. The reader may verify these results for the damped oscillator.

7.3 LANGEVIN EQUATIONS FOR A MULTILEVEL ATOM

Let us apply the Langevin Theory to a multilevel atom coupled to a reservoir. For simplicity let us adopt the notation of Equations (3.10.38) and (3.10.39) in the following equations. According to (6.3.56) and (7.2.8) if

$$M = a_i^\dagger a_j \equiv |i\rangle\langle j|, \tag{7.3.1}$$

(7.2.8) becomes

$$\frac{d}{dt} a_i^\dagger a_j = (i\omega_{ij} - \Gamma_{ij}{}^c) a_i^\dagger a_j + \delta_{ij} \sum_p{}' w_{ip} a_p^\dagger a_p + G_{ij}, \tag{7.3.2}$$

where by (7.2.4)

$$G_{ij}(t) = -i \sum_{mn} \langle [a_i^\dagger a_j, a_m^\dagger a_n] \rangle_R [f_{mn}(t) - \langle f_{mn} \rangle_R], \tag{7.3.3}$$

since

$$V = \hbar \sum_i F_i Q_i = \hbar \sum_{mn} f_{mn} a_m^\dagger a_n.$$

We must first remove the high-frequency terms from (7.3.2). We let

$$a_i^\dagger(t) a_j(t) = e^{i\omega'_{ij}(t-t_0)} b_i^\dagger(t) b_j(t), \tag{7.3.4}$$

where ω_{ij} includes the imaginary part of $\Gamma_{ij}{}^c$. Then (7.3.2) reduces to

$$\frac{db_i^\dagger b_j}{dt} = -\Gamma_{ij} b_i^\dagger b_j + \delta_{ij} \sum_p{}' w_{ip} b_p^\dagger b_p + g_{ij}, \tag{7.3.5a}$$

where Γ_{ij} is the real part of $\Gamma_{ij}{}^c$ and

$$g_{ij}(t) = -i \sum_{mn} \langle [b_i^\dagger b_j, b_m^\dagger b_n] \rangle_R e^{i\omega'_{mn}(t-t_0)} \{f_{mn}(t) - \langle f_{mn} \rangle_R\}. \tag{7.3.5b}$$

The reader should note that the b's obey the same orthogonality relations as the a's. For example,

$$a_i^\dagger a_j a_\kappa^\dagger a_l = \delta_{jk} a_i^\dagger a_l = \delta_{jk} e^{i\omega'_{il}(t-t_0)} b_i^\dagger b_l \tag{7.3.6}$$

whereas

$$\begin{aligned} a_i^\dagger a_j a_k^\dagger a_l &= e^{i(\omega'_{ij}+\omega'_{kl})(t-t_0)} b_i^\dagger b_j b_k^\dagger b_l \\ &= \delta_{jk} b_i^\dagger b_l e^{\omega'_{il}(t-t_0)}, \end{aligned} \tag{7.3.7}$$

in order for these two to give the same result.

To convince the reader the random forces g_{ij} give the correct diffusion coefficients, we calculate them first from the Einstein relation

$$2\langle D'_{ijkl} \rangle_R = \frac{d}{dt} \langle b_i^\dagger b_j b_k^\dagger b_l \rangle_R - \langle b_i^\dagger b_j A_{kl} \rangle_R - \langle A_{ij} b_k^\dagger b_l \rangle_R \tag{7.3.8}$$

where

$$A_{ij} = -\Gamma_{ij} b_i^\dagger b_j + \delta_{ij} \sum_p{}' w_{ip} b_p^\dagger b_p. \tag{7.3.9}$$

As an alternative we use the random forces to calculate

$$2\langle D'_{ijkl}\rangle_R = \frac{1}{\Delta t}\int_t^{t+\Delta t} ds \int_t^{t+\Delta t} ds' \langle g_{ij}(s) g_{kl}(s')\rangle_R, \tag{7.3.10a}$$

and show that the results agree. Note that the Einstein relation does not involve the random forces directly. Equation (7.3.10) implies the forces are δ-correlated in this time interval so that

$$\langle g_{ij}(s) g_{kl}(s')\rangle_R = 2\langle D_{ijkl}\rangle_R\, \delta(s - s'). \tag{7.3.10b}$$

By the orthogonality relations, (7.3.8) becomes

$$\begin{aligned} 2\langle D'_{ijkl}\rangle_R &= \delta_{jk}\frac{d}{dt}\langle b_i^\dagger b_l\rangle_R - \left\langle b_i^\dagger b_j\left\{-\Gamma_{kl} b_k^\dagger b_l + \delta_{kl}\sum_p{}' w_{kp} b_p^\dagger b_p\right\}\right\rangle_R \\ &\quad - \left\langle\left\{-\Gamma_{ij} b_i^\dagger b_j + \delta_{ij}\sum_p{}' w_{ip} b_p^\dagger b_p\right\} b_k^\dagger b_l\right\rangle_R \\ &= \delta_{jk}\left\{\frac{d}{dt}\langle b_i^\dagger b_l\rangle_R + (\Gamma_{kl} + \Gamma_{ij})\langle b_i^\dagger b_l\rangle_R\right\} \\ &\quad - \delta_{kl}\sum_{p\neq k}{}' w_{kp}\,\delta_{jp}\langle b_i^\dagger b_p\rangle_R - \delta_{ij}\sum_{p\neq i}{}' w_{ip}\,\delta_{kp}\langle b_p^\dagger b_l\rangle_R. \end{aligned} \tag{7.3.11}$$

If we use the reservoir average of (7.3.5) and let $j = l$, we obtain

$$\begin{aligned} 2\langle D'_{ijkl}\rangle_R &= \delta_{jk}(-\Gamma_{il} + \Gamma_{kl} + \Gamma_{ij})\langle b_i^\dagger b_l\rangle_R \\ &\quad + \delta_{il}\delta_{jk}\sum_p{}' w_{ip}\langle b_p^\dagger b_p\rangle_R - \delta_{kl} w_{kj}\langle b_i^\dagger b_j\rangle_R - \delta_{ij} w_{ik}\langle b_k^\dagger b_l\rangle_R, \end{aligned} \tag{7.3.12}$$

where the w_{jj} and w_{ii} terms are missing.

Let us next use (7.3.10). First we must obtain $g_{ik}(t)$ in a more useful form. From (7.3.5), we see that

$$\frac{d}{dt}\langle b_i^\dagger b_j\rangle_R = -\Gamma_{ij}\langle b_i^\dagger b_j\rangle_R + \delta_{ij}\sum_p{}' w_{ip}\langle b_p^\dagger b_p\rangle_R. \tag{7.3.13}$$

If $i \neq j$, the solution is

$$\langle b_i^\dagger(t) b_j(t)\rangle_R = e^{-\Gamma_{ij}(t-t_0)} b_i^\dagger(t_0) b_j(t_0). \tag{7.3.14}$$

When $i = j$, we have an infinite set of coupled equations which we cannot solve exactly. However, $g_{ij}(t)$ is already of first order in the system reservoir interaction and the correction to (7.3.14) for $i = j$ due to the w_{ip} terms would be of second order. We therefore may use (7.3.14) in g_{ij} when $i = j$.

By (7.3.10) and (7.3.5) we have

$$2\langle D'_{ijkl}\rangle_R = -\sum_{mnpq} \frac{1}{t-t_0} \int_{t_0}^{t} ds \int_{t_0}^{t} ds' \langle [b_i^\dagger(s)b_j(s), b_m^\dagger(s)b_n(s)]\rangle_R$$
$$\times \langle [b_k^\dagger(s')b_l(s'), b_p^\dagger(s')b_q(s')]\rangle_R e^{i\omega'_{mn}(s-t_0)+i\omega'_{pq}(s'-t_0)}$$
$$\times \langle [f_{mn}(s) - \langle f_{mn}\rangle_R][f_{pq}(s') - \langle f_{pq}\rangle_R]\rangle_R. \tag{7.3.15}$$

Now

$$\Gamma_{ij}^{-1} \gg \Delta t \gg \tau_c \tag{7.3.16}$$
$$\langle f_{mn}(s)\rangle_R = \langle f_{mn}\rangle_R.$$

We may therefore remove the slowly varying system terms from (7.3.15) evaluated at t_0. This gives

$$2\langle D'_{ijkl}\rangle_R = -\sum_{mnpq} [b_i^\dagger b_j, b_m^\dagger b_n][b_k^\dagger b_l, b_p^\dagger b_q] \frac{1}{\Delta t} \int_{t_0}^{t} ds \int_{t_0}^{t} ds'\, e^{i\omega'_{mn}(s-t_0)+i\omega'_{pq}(s-t_0)}$$
$$\times \{\langle f_{mn}(s)f_{pq}(s')\rangle_R - \langle f_{mn}\rangle_R\langle f_{pq}\rangle_R\}. \tag{7.3.17}$$

However, as we saw in Chapter 6,

$$\langle f_{mn}(s)f_{pq}(s')\rangle_R = \langle f_{mn}(s-s')f_{pq}(0)\rangle_R. \tag{7.3.18}$$

If we therefore let

$$s - s' = \tau \tag{7.3.19}$$
$$s' - t_0 = \xi,$$

(7.3.17) reduces to

$$2\langle D'_{ijkl}\rangle_R$$
$$= -\sum_{mnpq} [b_i^\dagger b_j, b_m^\dagger b_n][b_k^\dagger b_l, b_p^\dagger b_q] \frac{1}{\Delta t} \int_0^{\Delta t} d\xi\, e^{i(\omega'_{mn}+\omega'_{pq})\xi} \int_{-\xi}^{-\xi+\Delta t} d\tau\, e^{i\omega'_{mn}\tau}$$
$$\times \{\langle f_{mn}(\tau)f_{pq}(0)\rangle_R - \langle f_{mn}\rangle_R\langle f_{pq}\rangle_R\}. \tag{7.3.20}$$

Again because $\langle f_{mn}(\tau)f_{pq}(0)\rangle_R = 0$ for $\tau > \tau_c$ we may replace the upper limit on the τ integral by $+\infty$ and the lower by $-\infty$. Also since

$$\Delta t \gg (\omega'_{mn} + \omega'_{pq})^{-1}, \tag{7.3.21}$$

we see that we may let $\Delta t \to \infty$ in

$$\frac{1}{\Delta t} \int_0^{\Delta t} d\xi\, e^{i(\omega'_{mn}+\omega'_{pq})\tau} = \delta(\omega'_{mn}, -\omega'_{pq}). \tag{7.3.22}$$

Since $\omega'_{mn} = -\omega'_{pq}$, we see that (7.3.20) reduces to

$$2\langle D'_{ijkl}\rangle_R = -\sum_{mnpq} [b_i^\dagger b_j, b_m^\dagger b_n][b_k^\dagger b_l, b_m^\dagger b_n]\, \delta(\omega'_{mn}, -\omega'_{pq}) w_{mnpq}, \tag{7.3.23}$$

where

$$w_{mnpq} = \int_{-\infty}^{\infty} e^{-i\omega'_{pq}\tau} \langle f_{mn}(\tau)f_{pq}(0)\rangle_R\, d\tau. \tag{7.3.24}$$

(See Problem 7.6 to show why the last term in (7.3.20) does not contribute.) If we break this up from $-\infty$ to 0 plus 0 to $+\infty$, use the fact that $\omega'_{pq} = -\omega'_{mn}$ and let $\tau \to -\tau$ in the integral from $-\infty$ to 0, we may easily show that

$$w_{mnpq} = w^{+}_{mnpq} + w^{-}_{mnpq}, \tag{7.3.25}$$

where $w^{\pm}$ are given by (6.2.75a).

Next we may take the limit as $\Delta t \to 0 (t_0 \to t)$ in (7.3.23) which then becomes

$$2\langle D_{ijkl}\rangle_R = -\sum_{mnpq} \langle [b_i^\dagger b_j, b_m^\dagger b_n][b_k^\dagger b_l, b_p^\dagger b_q]\rangle_R\, \delta(\omega'_{mn}, -\omega'_{pq})[w^{+}_{mnpq} + w^{-}_{mnpq}], \tag{7.3.26}$$

where the b's are now evaluated at time t. Finally, if we use the orthogonality relations and pick out the secular terms for which (a) $m = q, n = p$ $(m \neq n)$, (b) $m = n, p = q$ $(m \neq p)$, and (c) $m = n = p = q$ and use the definition of the Γ's in terms of the $w^{\pm}$, we see that (7.3.26) reduces to (7.3.12).

Although the replacement of $[b_i^\dagger b_j, b_m^\dagger b_n][b_k^\dagger b_l, b_p^\dagger b_q]$ evaluated at t_0 by its reservoir average at time t as $t_0 \to t$ may seem strange, the reader should note that this is *exactly* the same procedure we used in obtaining the equation for $\langle dM/dt\rangle_R$ in Section 6.3.

7.4 LANGEVIN EQUATIONS FOR N HOMOGENEOUSLY BROADENED THREE-LEVEL ATOMS

In this section we obtain the Langevin equations of motion for N identical three-level atoms which are independent, each coupled to its own independent reservoir. The atoms ultimately will be coupled to a radiation field of frequency $\omega_c \cong (\epsilon_2 - \epsilon_1)/\hbar \equiv \omega_{21}$ which coincides approximately with the energy separation of the upper two levels of the atoms. This field will induce the atoms to make radiative transitions between these two levels. The ground state, 0, will act only as a source of atoms which may be pumped incoherently from the ground state to the upper two levels and a sink into which atoms in the upper levels may decay. Therefore, we shall be interested in the operators $(a_i^\dagger a_i)_\lambda$ where $i = 0, 1, 2$ which give the probability of finding the λth atom in level i and the dipole moment or radiative transition operators $(a_1^\dagger a_2)_\lambda$ and $(a_2^\dagger a_1)_\lambda$ for the λth atom between the upper two levels.

The equations of motion for the λth atom for these operators follow from (7.3.2):

$$\frac{d}{dt}(a_1^\dagger a_2)_\lambda = (-i\omega_{21}{}^\lambda - \Gamma_{12}{}^\lambda)(a_1^\dagger a_2)_\lambda + G_{12}{}^\lambda(t) \tag{7.4.1}$$

$$\frac{d}{dt}(a_2^\dagger a_2)_\lambda = w_{20}{}^\lambda(a_0^\dagger a_0)_\lambda + w_{21}{}^\lambda(a_1^\dagger a_1)_\lambda - \Gamma_2{}^\lambda(a_2^\dagger a_2)_\lambda + G_{22}{}^\lambda(t) \tag{7.4.2}$$

$$\frac{d}{dt}(a_1^\dagger a_1)_\lambda = w_{10}{}^\lambda(a_0^\dagger a_0)_\lambda - \Gamma_1{}^\lambda(a_1^\dagger a_1)_\lambda + w_{12}{}^\lambda(a_2^\dagger a_2)_\lambda + G_{11}{}^\lambda(t), \tag{7.4.3}$$

together with the adjoint of (7.4.1) and the completeness relation

$$(a_0^\dagger a_0)_\lambda + (a_1^\dagger a_1)_\lambda + (a_2^\dagger a_2)_\lambda = 1. \tag{7.4.4}$$

The Langevin forces are given by

$$G_{ij}{}^\lambda(t) = -i\sum_{m=1}^{2}\sum_{n=1}^{2}\langle[(a_i^\dagger a_j)_\lambda, (a_m^\dagger a_n)_\lambda]\rangle_R[f_{mn}{}^\lambda(t) - \langle f_{mn}{}^\lambda\rangle_R], \tag{7.4.5}$$

for $i, j = 1$ or 2. The $m = n = 0$ terms are not included since these would introduce operators other than those of interest. Again we are actually considering two level atoms which have another level as a source or sink.

Again we wish to remove the high-frequency behavior from (7.4.1). We, therefore, let

$$(a_1^\dagger a_2)_\lambda = e^{i\omega_0(t-t_0)}(b_1^\dagger b_2)_\lambda \tag{7.4.6}$$
$$(a_i^\dagger a_i)_\lambda \equiv (b_i^\dagger b_i)_\lambda,$$

where

$$\omega_0 = \omega_{21}{}^\lambda \equiv \omega_a{}^\lambda. \tag{7.4.7}$$

When we couple the atom to the radiation field, ω_0 will only be approximately equal to ω_a as we shall see.

If we now use (7.4.6) in (7.4.1)–(7.4.3) and assume that all atoms are identical so that the transition probabilities, $w_{ij}{}^\lambda$, damping constants, $\Gamma_{ij}{}^\lambda$, and frequencies are the same for all atoms we may omit the λ and we see that

$$\frac{d}{dt}(b_1^\dagger b_2)_\lambda = +[i(\omega_0 - \omega_a) - \Gamma_{12}](b_1^\dagger b_2)_\lambda + g_{12}{}^\lambda(t) \tag{7.4.8}$$

$$\frac{d}{dt}(b_2^\dagger b_2)_\lambda = w_{20} + w'_{21}(b_1^\dagger b_1)_\lambda - \Gamma'_2(b_2^\dagger b_2)_\lambda + g_{22}{}^\lambda(t) \tag{7.4.9}$$

$$\frac{d}{dt}(b_1^\dagger b_1)_\lambda = w_{10} - \Gamma'_1(b_1^\dagger b_1)_\lambda + w'_{12}(b_2^\dagger b_2)_\lambda + g_{11}{}^\lambda(t). \tag{7.4.10}$$

In this case we say the atoms are homogeneously broadened. Here we have used (7.4.4) to eliminate $(b_0{}^\dagger b_0)_\lambda$ and have let

$$\Gamma'_2 = \Gamma_2 + w_{20} \qquad w'_{21} = w_{21} - w_{20}$$
$$\Gamma'_1 = \Gamma_1 + w_{10} \qquad w'_{12} = w_{12} - w_{10}. \tag{7.4.11}$$

Also

$$g_{12}{}^\lambda(t) = e^{i\omega_0(t-t_0)}G_{12}{}^\lambda(t) = [g_{21}{}^\lambda(t)]^\dagger \qquad g_{ii}{}^\lambda(t) = G_{ii}{}^\lambda(t). \tag{7.4.12}$$

For homogeneously broadened atoms, the $f_{kl}{}^\lambda \equiv f_{kl}$ in (7.4.5).

Since the atoms are independent and the reservoirs are independent, the Langevin forces are independent. If we define

$$M = \sum_{\lambda=1}^{N}(b_1^\dagger b_2)_\lambda = \sum_{\lambda=1}^{N}(a_1^\dagger a_2)_\lambda e^{i\omega_0(t-t_0)} \tag{7.4.13}$$
$$N_j = \sum_{\lambda=1}^{N}(b_j^\dagger b_j)_\lambda = \sum_{\lambda=1}^{N}(a_j^\dagger a_j)_\lambda \qquad j = 0, 1, 2$$
$$g_{ij}(t) = \sum_{\lambda=1}^{N} g_{ij}{}^\lambda(t),$$

then for N atoms, (7.4.8)–(7.4.10) become

$$\begin{aligned}\frac{dM}{dt} &= [i(\omega_0 - \omega_a) - \Gamma_{12}]M + g_{12}(t) \\ \frac{dN_2}{dt} &= w_{20} + w'_{21}N_1 - \Gamma'_2 N_2 + g_{22}(t) \\ \frac{dN_1}{dt} &= w_{10} - \Gamma'_1 N_1 + w'_{12}N_2 + g_{11}(t).\end{aligned} \tag{7.4.14}$$

From the orthogonality relations

$$(b_i b_j^\dagger)_\lambda \equiv (\langle i | j \rangle)_\lambda = \delta_{ij}, \tag{7.4.15}$$

and the commutation relations

$$[(b_i^\dagger b_j)_\lambda, (b_k^\dagger b_l)_{\lambda'}] = \delta_{\lambda\lambda'}[b_i^\dagger b_j, b_k^\dagger b_l]_\lambda, \tag{7.4.16}$$

we see that

$$\begin{aligned} &[N_1, N_2] = 0 \qquad && [M^\dagger, M] = N_2 - N_1 \\ &[M, N_1] = -M \qquad && [M^\dagger, N_1] = M^\dagger \\ &[M, N_1] = M \qquad && [M^\dagger, N_2] = -M^\dagger. \end{aligned} \tag{7.4.17}$$

The random forces for different atoms are uncorrelated so that $\langle g_{ij}{}^\lambda(s) g_{kl}{}^{\lambda'}(s')\rangle_R = \delta_{\lambda\lambda'}\langle g_{ij}{}^\lambda(s) g_{kl}{}^\lambda(s')\rangle_R$. It therefore follows that

$$\begin{aligned}\langle g_{ij}(s) g_{kl}(s')\rangle_R &= \sum_{\lambda=1}^{N}\sum_{\lambda'=1}^{N}\langle g_{ij}{}^\lambda(s) g_{kl}{}^{\lambda'}(s')\rangle_R \\ &= \sum_\lambda \sum_{\lambda'} \delta_{\lambda\lambda'}\langle g_{ij}{}^\lambda(s) g_{kl}{}^\lambda(s')\rangle_R \\ &= N\sum_\lambda \langle g_{ij}{}^\lambda(s) g_{kl}{}^\lambda(s')\rangle_R \\ &= N\sum_\lambda 2\langle D'^{\lambda}_{ijkl}\rangle_R\,\delta(s - s') \\ &\equiv 2\langle D'_{ijkl}\rangle_R\,\delta(s - s'),\end{aligned} \tag{7.4.18}$$

or

$$2\langle D'_{ijkl}\rangle_R = \frac{1}{\Delta t}\int_t^{t+\Delta t} ds \int_t^{t+\Delta t} ds' \langle g_{ij}(s) g_{kl}(s')\rangle_R. \tag{7.4.19}$$

From the above, we may write the explicit forces as

$$\begin{aligned} g_{12}(t) &= -i\langle M(t)\rangle_R[f_{22}(t) - f_{11}(t) - (\langle f_{22}\rangle_R - \langle f_{11}\rangle_R)] \\ &\quad + i\langle (N_2(t) - N_1(t))\rangle_R[f_{21}(t) - \langle f_{21}\rangle_R]e^{i\omega_0(t-t_0)} \\ g_{22}(t) &= i\langle M(t)\rangle_R[f_{12}(t) - \langle f_{12}\rangle_R] + \text{ha} \\ g_{11}(t) &= -g_{22}(t). \end{aligned} \tag{7.4.20}$$

From (7.4.14) the reader may obtain the solutions

$$\langle M(t)\rangle_R = M(t_0)\exp\,[i(\omega_0 - \omega_a) - \Gamma_{12}](t - t_0) \qquad (7.4.21)$$

$$\begin{aligned}\langle N_2(t)\rangle_R = \Big\{ & w_{20} + (\Gamma_1' - \alpha)N_2(t_0) + w_{21}'N_1(t_0) \\ & - \frac{\alpha(w_{20}\Gamma_1' + w_{21}'w_{10})}{\beta(\alpha^2 - \beta^2)}\Big\} e^{-\alpha(t-t_0)}\, sh\, \beta(t - t_0) \\ & + \Big\{N_2(t_0) - \frac{w_{20}\Gamma_1' + w_{21}'w_{10}}{\alpha^2 - \beta^2}\Big\} e^{-\alpha(t-t_0)}\, ch\, \beta(t - t_0) \\ & + \frac{w_{20}\Gamma_1' + w_{21}'w_{10}}{\alpha^2 - \beta^2},\end{aligned} \qquad (7.4.22)$$

where

$$\alpha = \frac{\Gamma_1' + \Gamma_2'}{2}$$
$$\beta^2 = \left(\frac{\Gamma_1' - \Gamma_2'}{2}\right)^2 + w_{12}'w_{21}'. \qquad (7.4.23)$$

If 1 and 2 subscripts are interchanged everywhere in (7.4.22), we obtain $\langle N_1(t)\rangle_R$. These solutions may be needed to obtain equations of motion for operators such as MN_2.

7.5 LANGEVIN THEORY OF NOISE SOURCES; ASSOCIATED FUNCTION FORMULATION

We may also convert the operator Langevin theory to an associated c-number theory. If we write A_μ and G_μ in c-order in (7.2.8) as far as the system operators are concerned at time t, we have

$$\frac{da_\mu}{dt} = A_\mu{}^c + G_\mu{}^c. \qquad (7.5.1)$$

Then since $A_\mu{}^c$ is in chosen order, we may write

$$\frac{d\langle a_\mu\rangle_R}{dt} = \langle A_\mu{}^c\rangle_R \equiv \langle \mathscr{C}\{\bar{A}_\mu{}^c(\underset{\sim}{\alpha}_t)\}\rangle_R. \qquad (7.5.2)$$

If we compare this with (6.4.100) when $M = a_\mu$, we see that

$$\mathscr{A}_\mu(\underset{\sim}{\alpha}_t) \equiv \bar{A}_\mu{}^c(\underset{\sim}{\alpha}_t). \qquad (7.5.3)$$

We may therefore obtain the drift vectors for the c-number Fokker–Planck equation by ordering A_μ (7.2.8) and replacing a_t by α_t. We may therefore write a c-number Langevin equation for α_μ

$$\frac{d\alpha_\mu}{dt} = \mathscr{A}_\mu + \mathscr{G}_\mu^{\,c}, \tag{7.5.4}$$

where $\mathscr{G}_\mu$ has also had its system operators put into chosen order in G_μ and a_t is replaced by α_t.

Similarly from (7.2.2) we may write

$$\frac{d}{dt} a_\mu a_\nu = A_{\mu\nu} + G_{\mu\nu}, \tag{7.5.5}$$

where we assume $a_\mu a_\nu$ is in chosen order. If we put $A_{\mu\nu}$ and $G_{\mu\nu}$ system operators in chosen order we see that

$$\frac{d}{dt}\langle \mathscr{C}\{\alpha_\mu\alpha_\nu\}\rangle_R = \langle \mathscr{C}\{\bar{A}_{\mu\nu}^{\,c}(\underset{\sim}{\alpha}_t)\}\rangle_R \equiv \langle \mathscr{C}\{\mathscr{A}_{\mu\nu}\}\rangle_R. \tag{7.5.6}$$

We may write the c-number equation

$$\frac{d}{dt}\alpha_\mu\alpha_\nu = \mathscr{A}_{\mu\nu} + \mathscr{G}_{\mu\nu}$$

$$= \alpha_\mu \frac{d\alpha_\nu}{dt} + \alpha_\nu \frac{d\alpha_\mu}{dt}. \tag{7.5.7}$$

If we use (7.5.4) we see that

$$\frac{d}{dt}\alpha_\mu\alpha_\nu = \mathscr{A}_{\mu\nu} + \mathscr{G}_{\mu\nu}$$

$$= \alpha_\mu\mathscr{A}_\nu + \alpha_\nu\mathscr{A}_\mu + \alpha_\mu\mathscr{G}_\nu + \alpha_\nu\mathscr{G}_\mu. \tag{7.5.8}$$

Under the Markoff approximation we clearly see by the techniques used in Section 7.1 in the operator case that

$$\langle \alpha_\mu\mathscr{G}_\nu + \alpha_\nu\mathscr{G}_\mu\rangle_R = \langle 2\mathscr{D}_{\mu\nu}\rangle_R, \tag{7.5.9}$$

so that

$$\frac{d}{dt}\langle \alpha_\mu\alpha_\nu\rangle_R = \langle \mathscr{A}_{\mu\nu}\rangle_R$$

$$= \langle \alpha_\mu\mathscr{A}_\nu + \alpha_\nu\mathscr{A}_\mu + 2\mathscr{D}_{\mu\nu}\rangle_R. \tag{7.5.10}$$

If we apply the $\mathscr{C}$ operator to each side, this agrees with (6.4.105). Therefore,

$$\frac{d\alpha_\mu\alpha_\nu}{dl} = \alpha_\mu\mathscr{A}_\nu + \alpha_\nu\mathscr{A}_\mu + 2\mathscr{D}_{\mu\nu} + \mathscr{G}_{\mu\nu}. \tag{7.5.11}$$

We have for simplicity neglected removing the high-frequency free motion in this section.

The reader should show that in general

$$2\langle \mathscr{D}_{\mu\nu}(\underset{\sim}{a})\rangle_R \neq 2\langle \mathscr{C}\{\mathscr{D}_{\mu\nu}(\underset{\sim}{\alpha})\}\rangle_R. \tag{7.5.12}$$

Since no ordering is involved in the operator Langevin equations for the N damped atoms, we may replace the operators directly by their associated c-numbers.

$$M \rightarrow \mathscr{M}; \qquad M^\dagger \rightarrow \mathscr{M}^*; \qquad N_i \rightarrow \mathscr{N}_i. \tag{7.5.13}$$

This also applies to the random forces as may be seen by inspection of (7.4.20). Thus quantum and c-number Langevin equations in this case are identical in form. If we also use (7.3.12) for atom λ and the result (7.4.18),

$$2\langle \mathscr{D}'_{ijkl}\rangle = N \sum_\lambda 2\langle \mathscr{D}'^{\lambda}_{ijkl}\rangle_R, \tag{7.5.14}$$

we see that again no ordering is involved since only M, $M^\dagger$, and N_i occur singly so that the classical and quantum diffusion coefficients also have identical forms in the cases in which i, j, k, and l equal 1 or 2.

PROBLEMS

7.1 Verify that the solution (7.1.16) and its adjoint satisfies the commutation relation $[a(t), a^\dagger(t)] = 1$.
7.2 If $S(0) = (1 - e^{-\lambda})e^{-\lambda a^\dagger a}$, evaluate the means in (7.1.45).
7.3 Calculate the photon number spectrum for a damped driven oscillator using the techniques of Section 7.1.
7.4 Verify (7.1.47).
7.5 Show that (7.3.26) reduces to (7.3.12).
7.6 Show that the last term involving $\langle f_{mn}\rangle_R\langle f_{pq}\rangle_R$ in (7.3.20) vanishes identically when $\Delta t \rightarrow 0$. *Hint:* Because $\langle f_{mn}(0)\rangle_R\langle f_{pq}(0)\rangle_R$ is not peaked, we cannot extend the limits on the τ integral in this term to ∞; $\Delta t \gg \omega_{mn}{}^{-1}$.
7.7 Verify the commutation relations (7.4.17).
7.8 Give a physical interpretation to each of the terms in (7.4.14).
7.9 Obtain the diffusion coefficient $2\langle D_{2222}\rangle_R$ directly from the random force $g_{22}(t)$ in (7.4.19).
7.10 Verify (7.5.9) under the Markoff approximation.

REFERENCES

[1] M. Lax, *Phys. Rev.*, **145,** 110 (1966).
[2] F. Reif, *Fundamentals of Statistical and Thermal Physics,* New York: McGraw-Hill, 1965, Chapt. 15.

8

Lamb's Semiclassical Theory of a Laser [1]

In Chapter 5 we considered stimulated and spontaneous emission and stimulated absorption of radiation by an atom. We showed if the light of frequency ω was approximately equal to the energy separation between two atomic energy levels ($\hbar\omega \cong \epsilon_b - \epsilon_a \equiv \hbar\omega_{ba}$) that absorption occurred if the atom was in lower level $|a\rangle$ and stimulated emission occurred if the atom was in the upper level $|b\rangle$. Furthermore, we showed that the transition probability per second for induced emission w_{ab} and absorption w_{ba} were the same. If we have a large number of atoms in a cavity consisting of two mirrors and the gas is in thermal equilibrium at temperature T, there will be more atoms in the lower state, N_a, then in the upper state, N_b, given by the Boltzmann distribution

$$\frac{N_b}{N_a} = \exp\frac{-(\epsilon_b - \epsilon_a)}{kT},$$

where k is Boltzmann's constant. If therefore light of frequency $\omega \cong (\epsilon_b - \epsilon_a)/kT$ passes through the gas there will be a net absorption of radiation since $N_b w_{ab} > N_a w_{ba}$.

If, on the other hand, by some means we can invert the populations so that the number of atoms is greater in the upper level than in the lower level, a net amount of energy will be supplied to the radiation field by the atoms. If the amount supplied is greater than the losses, the field in the cavity will grow. The process by which we achieve population inversion is called pumping and is the source of energy for the field. As the field builds up to a large value, nonlinear effects come into play which tend to equalize the populations. When the populations are equalized, the atoms are said to be saturated and no further gain occurs. Then we must supply additional atoms with inverted populations to maintain the laser oscillation.

In Figure 8.1 we show a diagram of the first He–Ne gas laser [2]. The tube which is approximately 100 cm long is filled with a He–Ne gas mixture (1 torr pressure of He and 0.1 mm of Ne). The "cavity" is a Fabry–Perot

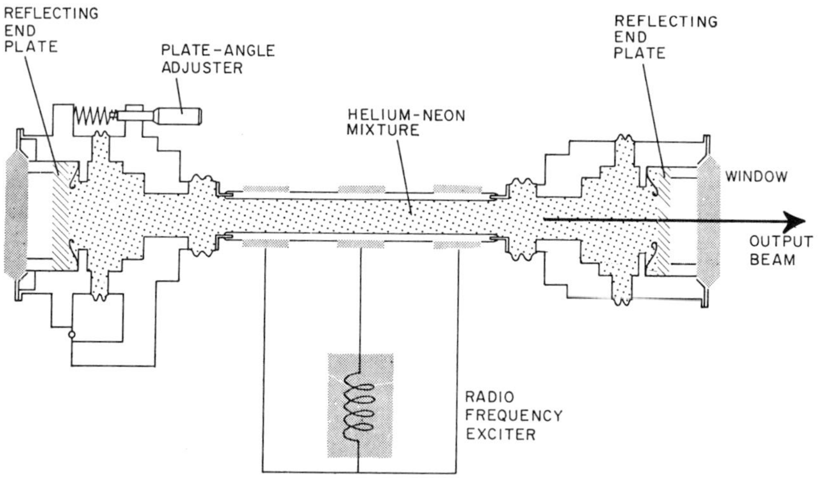

Figure 8.1 Sketch of He–Ne gas laser. [Reprinted from A. L. Schowlow, Sci. Amer., June, 1961.]

interferometer consisting of two parallel plane semi-transparent mirrors [3]. The pumping which we describe briefly is provided by an rf discharge as shown in the figure.

In Figure 8.2 we see a rough energy level diagram of He and Ne. The first excited level of He is a metastable level which means that the atom cannot be excited in first approximation by an electromagnetic wave as we have just described. The transition is forbidden by selection rules which means the dipole moment vanishes for these two states. However, an electron moving through the tube because of the discharge can collide inelastically with the He atom and excite it to the metastable level. A metastable level also cannot radiatively decay spontaneously to the lower state because the dipole moment matrix element is zero. However, the He ion can collide with a Ne ion in the discharge and transfer its energy to the Ne by an inelastic collision. This happens because the Ne $2s$-level is in approximate synchronism with the 2^3S metastable He level. When energies are approximately the same, a resonance exchange of energy on collision is very likely. Since the $2p$-levels of the Ne are practically empty, we obtain a population inversion with more atoms in the $2s$-levels than in the $2p$-level of Ne.

The oscillation starts with a spontaneous decay of a Ne atom from a $2s$- to a $2p$-level. If the emitted radiation is emitted in a nonaxial direction in the discharge tube, it escapes and is lost. If it is approximately axial, there is a large probability that it will collide with another Ne atom and *induce* it to radiate. Since the probability for induced emission is proportional to the radiation present, it is greater than for spontaneous emission and the induced

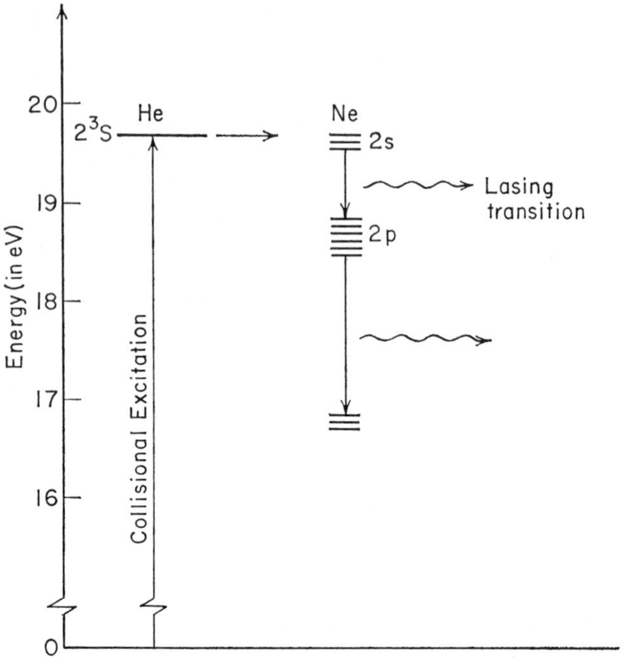

Figure 8.2 Energy level diagrams of He and Ne.

field is in phase with the incident field. We therefore obtain a cascade buildup of the wave. At the mirror most of the light is reflected so that the mirrors provide a feedback mechanism for the buildup of energy. Light going in the wrong direction is lost by diffraction, some (say 5%) is emitted through the semi-transparent ends (which act like an antenna) and some is lost in the mirrors. The oscillations will build up from spontaneous emission if we have pumped enough atoms into the upper state to overcome these losses.

It is easy to see that the emerging laser light will be very well collimated, since it would escape by diffraction otherwise. Also the intensity will be greater than with conventional sources of light, since each atom emits in phase. In an ordinary source, the atoms emit their radiation by spontaneous decay (rather than by stimulated emission) which is a random process. Consequently, the fields sometime add in phase and sometime cancel. In fact the laser intensity is orders of magnitude greater than ordinary sources. Furthermore, since the frequency is determined by the atomic energy level separation, it is a resonant phenomenon and will be very monochromatic. Laser light is also coherent.

It is of interest to discuss the resonator briefly [4–6]. We consider it in more detail in the next section. Since the mirrors are assumed to be perfect conductors, each mode will have an integral number l of half-wave lengths

$$d = \frac{\lambda}{2} l,$$

where d is the mirror separation. For $d = 100$ cm and $\lambda = 6000$ Å, this corresponds to approximately 10^6 half-wavelengths. To obtain these normal modes, one would have to solve Maxwell's equations subject to boundary conditions on the mirrors and in free space, since there are no mirrors on the side. Fox and Li [4] circumvented this difficult boundary value problem by using Huygen's principle. They assumed a given field distribution on one mirror and used Huygen's principle to calculate the field on the other mirror. After allowing for reflection loss, they used the calculated field on mirror 2 to recalculate that on mirror 1. Each time some energy was lost out the sides by diffraction. After several hundred such calculations, the field energy distribution did not change (except for diffraction losses), and this distribution was taken as a normal mode.

With this brief introduction, we present the semiclassical theory of a laser, neglecting all unessential complications [1]. We first discuss cavity modes for spherical mirrors. We then proceed in two steps. First, we assume the atoms in the cavity act as radiating dipoles which act as a source of radiation in the cavity. Next, we obtain the dipole moments of the atoms which are induced by the electric field in the cavity which is treated classically. We then solve for the field in a self-consistent way. Pumping and losses are taken into account in a phenomenological way.

8.1 MODES IN "COLD" SPHERICAL RESONATOR

Instead of using Huygen's principle as Fox and Li as well as Boyd and Gordon did to obtain the resonator modes, we present a simpler treatment due to Kogelnik and Li [5]. Although the latter is not as general, it has the advantage of being analytically tractable.

We would like to solve the wave equation

$$\nabla^2 \mathbf{E} - \frac{1}{c^2} \frac{\partial^2 \mathbf{E}}{\partial t^2} = 0 \qquad (8.1.1)$$

inside a spherical resonator as shown in Figure 8.3. Each mirror has a radius of curvature b and the mirrors are separated by a distance d. It is not necessary for both mirrors to have the same radius. When $b = d$, the resonators are said to be confocal. The equation of the right mirror is

$$r^2 + \left[z + b - \frac{d}{2}\right]^2 = b^2, \qquad (8.1.2)$$

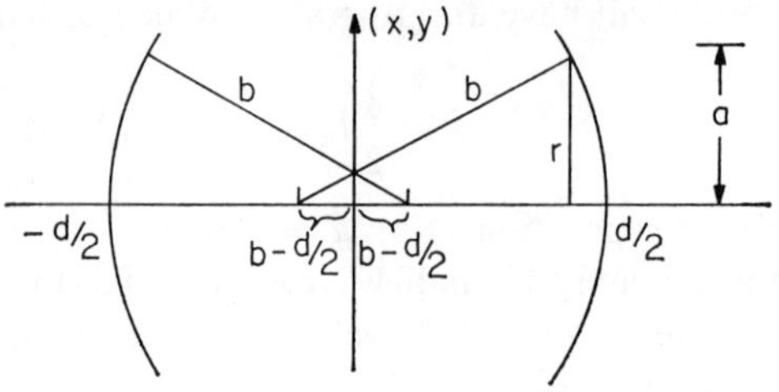

Figure 8.3 Diagram of symmetric spherical resonator. The radius of curvature is b and the separation is d.

where

$$r^2 = x^2 + y^2. \tag{8.1.3}$$

In the event $r \ll b$, a standard approximation, (8.1.2) reduces to

$$z \cong \frac{d}{2} - \frac{r^2}{2b}. \tag{8.1.4}$$

In this approximation, we therefore cannot distinguish between a spherical and paraboloidal mirror. The equation for the left mirror is under the same approximation

$$z \cong -\frac{d}{2} + \frac{r^2}{2b}. \tag{8.1.5}$$

We expect the laser beam to propagate approximately as a plane wave in the z-direction but to have a nonuniform radial intensity distribution concentrated near the axis of propagation. We also expect the phase fronts to be slightly curved. We would like to obtain solutions of the wave equation in which the mirrors coincide with a surface of constant phase.

As a zeroth approximation, let us look for a solution of the form

$$E_x = \psi(x, y, z)e^{i(kz-\omega t)}, \tag{8.1.6}$$

where $k = 2\pi/\lambda$ and

$$\frac{\partial \psi}{\partial z} \ll ik\psi. \tag{8.1.7}$$

Then the wave equation becomes approximately

$$\frac{\partial^2 \psi}{\partial x^2} + \frac{\partial^2 \psi}{\partial y^2} + 2ik\psi = 0, \tag{8.1.8}$$

where we neglected $\partial^2\psi/\partial z^2$ compared with $\partial\psi/\partial z$.

In an attempt to find solutions in which the surfaces of constant phase coincide with the mirrors, we look for solutions of the form

$$\psi(x, y, z) = g\left[\frac{\sqrt{2}\,x}{w(z)}\right]h\left[\frac{\sqrt{2}\,y}{w(z)}\right] \exp i\left[P(z) + \frac{kr^2}{2q(z)}\right], \tag{8.1.9}$$

where for convenience we let

$$\frac{1}{q(z)} = \frac{1}{R(z)} + \frac{2i}{kw^2(z)}, \tag{8.1.10}$$

and R and w are real. If we combine this with (8.1.6), we see that the constant phase surfaces are determined by

$$k\left[z + \frac{r^2}{2R(z)}\right] + R_e P(z) = \text{constant}, \tag{8.1.11}$$

where R_e is the real part and we used (8.1.10). This is very close in form to (8.1.4). The $w(z)$ is a scaling factor and $R(z)$ is effectively the radius of curvature of the wave front for a given z. The $R_e P(z)$ will not alter the phase front very much as we shall see. The $\sqrt{2}$ is used for convenience.

If we substitute (8.1.9) and its derivatives into (8.1.8), we obtain after some algebra

$$\frac{1}{g}\left\{\frac{d^2 g}{d\xi^2} - ik\left[w\frac{dw}{dz} - \frac{w^2}{q}\right]\xi\frac{dg}{d\xi}\right\} + \frac{1}{h}\left\{\frac{d^2 h}{d\eta^2} - ik\left[w\frac{dw}{dz} - \frac{w^2}{q}\right]\eta\frac{dh}{d\eta}\right\}$$
$$+ kw^2\left(\frac{i}{q} - \frac{dP}{dz}\right) + \frac{k^2 w^4}{4q^2}\left(\frac{dq}{dz} - 1\right)(\xi^2 \eta^2) = 0, \tag{8.1.12}$$

where we have let

$$\xi = \frac{\sqrt{2}\,x}{w(z)}; \qquad \eta = \frac{\sqrt{2}\,y}{w(z)}. \tag{8.1.13}$$

If we are to have any luck in solving this equation, we must choose P, w, and q or equivalently P, w, and R so that

$$w\frac{dw}{dz} - \frac{w^2}{q} = c_1 \tag{8.1.14a}$$

$$kw^2\left(\frac{i}{q} - \frac{dP}{dz}\right) = c_2 \tag{8.1.14b}$$

$$\frac{w^4}{q^2}\left(\frac{dq}{dz} - 1\right) = c_3. \tag{8.1.14c}$$

Then the equation will be separable in ξ and η at least. Unfortunately, these equations are so nonlinear that the solutions are extremely awkward. If we let $c_3 = 0$, the solutions are then quite easy. In this case

$$\frac{dq}{dz} = 1$$

or (8.1.15)

$$q(z) = z + q_0.$$

By (8.1.11) we see that $R(z)$ plays the role of the radius of curvature of the wave front at a given z. In the present case the symmetry would suggest that at $z = 0$, the wave front is a plane so that $R(0) = \infty$. By (8.1.10) and (8.1.15) we see that

$$q = z + q_0 = z - i\,\frac{kw_0^2}{2}, \tag{8.1.16}$$

where $w_0 \equiv w(0)$, a parameter still at our disposal. If we then use this and (8.1.10), we find after minor algebra that

$$\begin{aligned} w^2(z) &= w_0^2\left[1 + \left(\frac{2z}{kw_0^2}\right)^2\right] \\ R(z) &= \left(\frac{kw_0^2}{2}\right)^2 \frac{[1 + (2z/kw_0^2)^2]}{z}. \end{aligned} \tag{8.1.17}$$

Thus by choosing $c_3 = 0$, q, w, and therefore R are completely determined. By happy coincidence when we use (8.1.16) and (8.1.17) we see that

$$w\,\frac{dw}{dz} - \frac{w^2}{q} = -\frac{2i}{k} = c_1 \tag{8.1.18}$$

which is a constant. Therefore, (8.1.12) becomes

$$\frac{1}{g}\left\{\frac{d^2g}{d\xi^2} - 2\xi\,\frac{dg}{d\xi}\right\} + \frac{1}{h}\left\{\frac{d^2h}{d\eta^2} - 2\eta\,\frac{dh}{d\eta}\right\} + \epsilon_\xi + \epsilon_\eta = 0, \tag{8.1.19}$$

where we have let

$$kw^2\left(\frac{i}{q} - \frac{dP}{dz}\right) = c_2 \equiv \epsilon_\xi + \epsilon_\eta, \tag{8.1.20}$$

or when we use (8.1.16) and (8.1.17) we obtain on integrating

$$iP(z) = -\log\sqrt{1 + \left(\frac{2z}{kw_0^2}\right)^2} - i[1 + \tfrac{1}{2}(\epsilon_\xi + \epsilon_\eta)]\tan^{-1}\left(\frac{2z}{kw_0^2}\right). \tag{8.1.21}$$

The solution then is of the form

$$\begin{aligned} &E_x(x, y, z, t) \\ &\quad = \frac{Ag(\xi)h(\eta)}{\sqrt{1+\zeta^2}} \\ &\quad\times \exp\left\{ik\left[z + \frac{r^2}{2R(\zeta)}\right] - (1 + \epsilon_\xi + \epsilon_\eta)\tan^{-1}\zeta - \omega t\right\}\exp -\tfrac{1}{2}(\xi^2 + \eta^2), \end{aligned} \tag{8.1.22}$$

where we have let

$$\zeta = \frac{2z}{k{w_0}^2} \tag{8.1.23}$$

and also

$$R(\zeta) = \frac{k{w_0}^2}{2}\frac{(1 + \zeta^2)}{\zeta} \tag{8.1.24}$$

$$w^2(\zeta) = w_0^2(1 + \zeta^2). \tag{8.1.25}$$

The constant $c_2 = \epsilon_\xi + \epsilon_\eta$ is still at our disposal. One can show that the solution of

$$\frac{d^2g}{d\xi^2} - 2\xi\frac{dg}{d\xi} + \epsilon_\xi g = 0 \tag{8.1.26}$$

blows up faster than $\exp(\frac{1}{2}\xi^2)$ for large ξ unless we let

$$\epsilon_\xi = 2m \tag{8.1.27}$$

where $m = 0, 1, 2, \ldots$. Since E_x must be finite as $\xi \to \infty$, this is the only physical solution. Then the solutions are

$$g = H_m(\xi), \tag{8.1.28}$$

where $H_m(\xi)$ are hermite polynomials. Thus we have

$$\begin{aligned} E(x, y, z, t) = {}& \frac{AH_m}{\sqrt{1 + \zeta^2}}\left[\sqrt{\frac{2}{1 + \zeta^2}}\frac{x}{w_0}\right]H_n\left[\sqrt{\frac{2}{1 + \zeta^2}}\frac{y}{w_0}\right]\exp\left\{-\frac{1}{1 + \zeta^2}\frac{r^2}{{w_0}^2}\right\} \\ & \times \exp i\left\{kz + \frac{\zeta}{1 + \zeta^2}\frac{r^2}{{w_0}^2} - (m + n + 1)\tan^{-1}\zeta\right\}. \end{aligned} \tag{8.1.29}$$

If we next require that at $z = d/2$, $R(d/2) = b$, the mirror radius of curvature, we see by (8.1.24) that

$$k{w_0}^2 = \sqrt{\frac{2d(b - d)}{2}}. \tag{8.1.30}$$

Then both mirrors coincide approximately to surfaces of constant phase aside from the phase term

$$(m + n + 1)\tan^{-1} \pm \frac{d}{\sqrt{2d(b - d/2)}} \approx \pm\frac{\pi}{4}(m + n + 1), \tag{8.1.31}$$

which for moderate m and n is small compared with the $\pm kd/2 = \pm\pi d/\lambda$ term.

Finally, the phases at $z = \pm d/2$ and $r = 0$

$$\begin{aligned} \varphi\left(\frac{-d}{2}\right) &= -\frac{kd}{2} + (m + n + 1)\tan^{-1}\frac{d}{\sqrt{2d(b - d/2)}} \\ \varphi\left(\frac{+d}{2}\right) &= \frac{kd}{2} - (m + n + 1)\tan^{-1}\frac{d}{\sqrt{2d(b - d/2)}}. \end{aligned} \tag{8.1.32}$$

To determine k and therefore the resonant frequencies we require that the round trip phase change

$$\Delta\varphi = 2\left[\varphi\left(\frac{d}{2}\right) - \varphi\left(\frac{-d}{2}\right)\right]$$

$$= 2kd - 4(m + n + 1)\tan^{-1}\frac{d}{\sqrt{2d(b - d/2)}} = 2\pi l, \quad (8.1.33)$$

where l is an integer. Thus the normal mode frequencies are given by

$$kd = \frac{2\pi d}{\lambda_{mnl}} = \frac{d}{c}\,\omega_{mnl} = \pi l + 2(m + n + 1)\tan^{-1}\frac{d}{\sqrt{2d(b - d/2)}}, \quad (8.1.34)$$

where c is the velocity of light. In the confocal case $b = d$ and

$$\omega_{mnl} = \frac{c\pi}{d}\,[l + \tfrac{1}{2}(m + n + 1)]. \quad (8.1.35)$$

For a CO_2 laser resonant at 10.6 μ and 100 cm long we see that

$$l + \tfrac{1}{2}(m + n + 1) \cong 2 \times 10^5. \quad (8.1.36)$$

Since $m = n = 0$ is the lowest order transverse mode

$$l \sim 2 \times 10^5 \quad (8.1.37)$$

and we see that the resonant frequencies are primarily determined by the longitudinal quantum number l in agreement with the statement made in the introduction. Since $H_0(x) = 1$, the fundamental TEM_{00l} mode is given by

$$E_{00l}(x, y, z, t)$$
$$= \frac{A}{\sqrt{1 + \zeta^2}} \exp - \frac{1}{1 + \zeta^2}\frac{r^2}{w_0^2} \exp i\left\{kz + \frac{\zeta}{1 + \zeta^2}\frac{r^2}{w_0^2} - \tan^{-1}\zeta - \omega t\right\} \quad (8.1.38)$$

where

$$\frac{\omega_{00l}}{c} = k_{00l} \cong \frac{\pi l}{d} \quad (8.1.39)$$

and

$$w_0^2 = \frac{d\sqrt{2d(b - d/2)}}{\pi l} \quad (8.1.40)$$

$$\zeta = \frac{2z}{\sqrt{2d(b - d/2)}}. \quad (8.1.41)$$

It should be remembered that we have assumed the mirror radii are small compared with their radius of curvature so that these modes are approximately correct for low order modes in which the beam intensity is concentrated near the centers of the mirrors so that diffraction losses are quite small.

8.2 THE CAVITY FIELD DRIVEN BY ATOMS

In this semiclassical theory of a laser oscillator we treat the field classically. The field obeys the Maxwell equations (4.8.1)–(4.8.4). Since there are no free charges, we let $\rho = 0$. In the laser medium we take the constitutive relations

$$\mathbf{B} = \mu_0 \mathbf{H}; \qquad \mathbf{D} = \epsilon_0 \mathbf{E} + \mathbf{P}; \qquad \mathbf{J} = \sigma \mathbf{E}; \tag{8.2.1}$$

$\mathbf{P}$ is the polarization of the medium and will act as the source of radiation to drive the field in the cavity. The third relation above accounts phenomenologically for the cavity losses including diffraction. If we take the curl of both sides of (4.8.2) and use (4.8.4) and (8.2.1), we obtain the wave equation

$$\text{curl curl } \mathbf{E} + \frac{1}{c^2}\frac{\partial^2 \mathbf{E}}{\partial t^2} + \mu_0 \sigma \frac{\partial \mathbf{E}}{\partial t} = -\mu_0 \frac{\partial^2 \mathbf{P}}{\partial t^2}, \tag{8.2.2}$$

where we see that the polarization drives the electric field. We are interested only in the solutions to the inhomogeneous equation, since the loss term will cause all solutions of the homogeneous equation to damp out. As we have noted, the driving term will be provided by the dipole moments of the atoms which radiate into the cavity mode. In this part of the analysis, we assume that the polarization is known and calculate the field it produces in the cavity.

We assume the field is plane polarized in the x-direction and we neglect the transverse field variation for simplicity. Then (8.2.2) reduces to

$$\frac{\partial^2 E_x(z, t)}{\partial t^2} + \frac{\sigma}{\epsilon_0}\frac{\partial E_x}{\partial t} - c^2 \frac{\partial^2 E_x}{\partial z^2} = -\frac{1}{\epsilon_0}\frac{\partial^2 P_x(z, t)}{\partial t^2}. \tag{8.2.3}$$

When we neglect the losses and driving term, we obtain the wave equation

$$\frac{\partial^2 E_x}{\partial t^2} = c^2 \frac{\partial^2 E_x}{\partial z^2}. \tag{8.2.4}$$

The solutions which satisfy the boundary conditions $E_x(z, t) = 0$ at $z = 0$ and $z = d$ for all time are

$$(E_n e^{-i\Omega_n t} + E_n^* e^{i\Omega_n t}) \sin K_n z, \tag{8.2.5}$$

where the eigenfrequencies are

$$\Omega_n = cK_n = \frac{n\pi}{d} c = \frac{2\pi c}{\lambda_n}, \tag{8.2.6}$$

n is an integer of order 10^6 when $d = 100$ cm in a typical laser. The modes are separated by

$$\frac{\Delta\Omega_n}{2\pi} = \frac{c}{2d}\Delta n \sim 150 \text{ MHz}, \tag{8.2.7}$$

for $d = 100$ cm and $\Delta n = 1$.

To attempt to solve (8.2.3) let us expand the field in normal modes

$$E_x(z, t) = \sum_n A_n(t) \sin K_n z, \tag{8.2.8}$$

since E_x must still vanish at the perfectly conducting mirrors. We assume the losses are due to diffraction. If we substitute (8.2.8) into (8.2.3), multiply both sides by $\sin K_m z$ and integrate from $z = 0$ to $z = d$, we obtain

$$\frac{d^2 A_m}{dt^2} + \frac{\sigma}{\epsilon_0}\frac{dA_m}{dt} + c^2 K_m^{\ 2} A_m = -\frac{1}{\epsilon_0}\frac{d^2 P_m}{dt^2}, \tag{8.2.9}$$

where we have let

$$P_m(t) = \frac{2}{d}\int_0^d P_x(z, t) \sin K_m z \, dz. \tag{8.2.10}$$

If the losses are small, the resonant cavity frequencies will not change very much from the loss free frequencies given by (8.2.6) and the driving term will excite these frequencies. We therefore look for solutions of the form

$$A_m(t) = \mathscr{E}(t)e^{-i\omega t} + \mathscr{E}^*(t)e^{i\omega t}, \tag{8.2.11}$$

where we expect ω to be approximately one of the loss free frequencies Ω_m and that $\mathscr{E}(t)$ will be slowly varying:

$$\left|\frac{d\mathscr{E}}{dt}\right| \ll |\omega\mathscr{E}|. \tag{8.2.12}$$

The polarization is expected to oscillate at frequency

$$\omega_a = \frac{\epsilon_b - \epsilon_a}{\hbar} \approx \omega; \tag{8.2.13}$$

we therefore assume we may expand the polarization as

$$P_m(t) = P(t)e^{-i\omega t} + P^*(t)e^{i\omega t}, \tag{8.2.14}$$

where if the atomic lifetimes in states $|a\rangle$ and $|b\rangle$ are long compared with the atomic period we expect that

$$\left|\frac{dP}{dt}\right| \ll \omega\,|P|. \tag{8.2.15}$$

If we substitute (8.2.11) and (8.2.14) into (8.2.9) and use the approximations (8.2.12) and (8.2.15), we obtain

$$\frac{d\mathscr{E}}{dt} + \left[\frac{\gamma}{2} + \frac{i(\Omega_m^{\ 2} - \omega^2)}{2\omega}\right]\mathscr{E} \cong \frac{i\omega P}{2\epsilon_0} \tag{8.2.16}$$

together with its complex conjugate. Here we have assumed small loss so that

$$\left|\frac{\sigma}{2\epsilon_0\omega}\right| \ll 1, \tag{8.2.17}$$

and have let

$$\gamma \equiv \frac{\sigma}{\epsilon_0}. \tag{8.2.18}$$

Since we expect $\omega \approx \Omega_m \approx \omega_a$, we have

$$\frac{\Omega_m{}^2 - \omega^2}{2\omega} = \frac{(\Omega_m - \omega)(\Omega_m + \omega)}{2\omega} \approx \Omega_m - \omega, \tag{8.2.19}$$

so that (8.2.16) becomes

$$\frac{d\mathscr{E}}{dt} + \left[\frac{\gamma}{2} + i(\Omega_m - \omega)\right]\mathscr{E} \cong \frac{i\omega_a}{2\epsilon_0} P. \tag{8.2.20}$$

Under the approximations above (essentially small damping and long atomic lifetimes) this is the equation for the field amplitude driven by an atomic polarization whose frequency is approximately in resonance with a natural cavity mode frequency. We have removed the high-frequency field and polarization oscillations by the expansions (8.2.11) and (8.2.14) so that $\mathscr{E}$ and P are expected to change very little during a period of the high-frequency oscillation. It is the first fundamental equation of a laser oscillator. In the next section we study the effect the field has in inducing the atomic polarization.

8.3 THE INDUCED ATOMIC DIPOLE MOMENT

In the previous section we derived an equation for the electric field in a cavity which was radiated by an atomic polarization. In this section we derive the polarization for a large number of atoms which is induced by a given electric field in the cavity.

Consider a two-level atom in a cavity in which a polarized electric field exists. The Hamiltonian is

$$H = H_A + V, \tag{8.3.1}$$

where H_A is the free atom hamiltonian and

$$\begin{aligned} V &= -e\mathbf{r}\cdot\mathbf{E}(\mathbf{r}, t) = -exE_x(z, t) \\ &= -ex[\mathscr{E}(t)e^{-i\omega t} + \mathscr{E}^*(t)e^{i\omega t}] \sin K_n z, \end{aligned} \tag{8.3.2}$$

where we have assumed that only one mode of the field is excited and that it is polarized in the x-direction. The field is evaluated at the position of the "atom."

We now look for solutions of the Schrödinger equation

$$i\hbar \frac{\partial|\psi(z, t)\rangle}{\partial t} = (H_A + V)|\psi(z, t)\rangle \tag{8.3.3}$$

of the form

$$|\psi(z, t)\rangle = c_a(z, t)e^{i\omega t/2}|a\rangle + c_b(z, t)e^{-i\omega t/2}|b\rangle, \tag{8.3.4}$$

where we shall find later that $\hbar\omega \approx \epsilon_b - \epsilon_a$. The two atomic states are orthogonal and assumed complete in the present context. The exponential time factors were introduced for convenience and cause no loss of generality. If we insert (8.3.4) into (8.3.3) and use the eigenvalue equations

$$H_A|i\rangle = \epsilon_i|i\rangle \qquad i = a, b, \tag{8.3.5}$$

we obtain

$$i\hbar\left[\frac{dc_a}{dt} + \frac{i\omega}{2}c_a\right]e^{i\omega t/2}|a\rangle + i\hbar\left[\frac{dc_b}{dt} - \frac{i\omega}{2}c_b\right]e^{-i\omega t/2}|b\rangle$$

$$= \epsilon_a c_a e^{i\omega t/2}|a\rangle + \epsilon_b c_b e^{-i\omega t/2}|b\rangle + V(z, t)[c_a e^{i\omega t/2}|a\rangle + c_b e^{-i\omega t/2}|b\rangle]. \tag{8.3.6}$$

If we next take the scalar product of both sides of this equation with $\langle a|$ and use the orthogonality relations $\langle a|b\rangle = \langle b|a\rangle = 0$, we obtain

$$\frac{dc_a}{dt} = -i\left(\frac{\omega}{2} + \frac{\epsilon_a}{\hbar}\right)c_a + \frac{1}{i\hbar}\langle a|V(z, t)|b\rangle e^{-i\omega t}c_b. \tag{8.3.7}$$

We have assumed as is usually the case for most lasers that the diagonal matrix elements $\langle a|V|a\rangle$ and $\langle b|V|b\rangle$ are zero. In the integration

$$\langle a|V(z, t)|b\rangle = -e\langle a|x \sin K_n z|b\rangle[\mathscr{E}e^{-i\omega t} + \text{cc}], \tag{8.3.8}$$

the atomic wave functions are strongly peaked near the center of mass of the atom, and we may remove the $\sin K_n z$ evaluated at the atomic center of mass.† If we let

$$\wp \equiv e\langle a|x|b\rangle = e\langle b|x|a\rangle^*, \tag{8.3.9}$$

then (8.3.7) becomes

$$\frac{dc_a}{dt} = -i\left(\frac{\omega}{2} + \frac{\epsilon_a}{\hbar}\right)c_a + i\frac{\wp}{\hbar}\sin K_n z[\mathscr{E}(t)e^{-2i\omega t} + \mathscr{E}^*(t)]c_b. \tag{8.3.10}$$

Under the rotating wave approximation, $\mathscr{E}(t)\exp(-2i\omega t)$ is very rapidly varying compared with $\mathscr{E}^*(t)$ and will average approximately to zero. We therefore neglect it and obtain

$$\frac{dc_a}{dt} = -i\left(\frac{\omega}{2} + \frac{\epsilon_a}{\hbar}\right)c_a + i\frac{\wp}{\hbar}\sin K_n z\mathscr{E}^*(t)c_b. \tag{8.3.11}$$

† In this case we do not make the "dipole approximation," since atoms are spatially distributed throughout the cavity.

In a similar way we obtain

$$\frac{dc_b}{dt} = i\left(\frac{\omega}{2} - \frac{\epsilon_b}{\hbar}\right)c_b + i\frac{\wp^*}{\hbar}\sin K_n z\mathscr{E}(t)c_a. \tag{8.3.12}$$

These equations give the time development of the probability amplitudes for finding the atom in levels $|a\rangle$ and $|b\rangle$ in the presence of a driving electric field.

So far we have not taken into account the finite lifetimes of the atom in levels $|a\rangle$ and $|b\rangle$. We have assumed the energy levels are infinitely sharp. Based on our prior work, we assume there is a lower energy level to which atoms in the two levels may decay whose energy separation is large compared with $\epsilon_b - \epsilon_a$ so that the driving field will be far out of resonance and will not induce transitions to the ground state. We then introduce the atomic lifetimes Γ_a^{-1} and Γ_b^{-1} so that (8.3.11) and (8.3.12) become

$$\frac{dc_a}{dt} = -\left[\tfrac{1}{2}\Gamma_a + i\left(\frac{\omega}{2} + \frac{\epsilon_a}{\hbar}\right)\right]c_a + i\frac{\wp}{\hbar}\sin K_n z\mathscr{E}^*(t)c_b \tag{8.3.13}$$

$$\frac{dc_b}{dt} = -\left[\tfrac{1}{2}\Gamma_b - i\left(\frac{\omega}{2} - \frac{\epsilon_b}{\hbar}\right)\right]c_b + i\frac{\wp^*}{\hbar}\sin K_n z\mathscr{E}(t)c_a. \tag{8.3.14}$$

We are interested in obtaining the polarization induced by the field in the active medium. This is defined by

$$P(z, t) = N\langle\psi(z, t)|ex|\psi(z, t)\rangle, \tag{8.3.15}$$

where there are N atoms per unit volume in the cavity. If we use (8.3.4) and let $\langle a|x|a\rangle = \langle b|x|b\rangle = 0$, we have

$$\begin{aligned} P(z, t) &= N\{c_a^* e^{-i\omega t/2}\langle a| + c_b^* e^{i\omega t/2}\langle b|\}ex\{c_a e^{i\omega t/2}|a\rangle + c_b e^{-i\omega t/2}|b\rangle\} \\ &= P_n(z, t)e^{-i\omega t} + \text{cc}, \end{aligned} \tag{8.3.16}$$

where we have let

$$P_n(z, t) \equiv N\wp c_a^*(z, t)c_b(z, t). \tag{8.3.17}$$

We therefore see that we need the bilinear combinations $c_a^* c_b$ and $c_a c_b^*$ rather than c_a and c_b separately. Also we need to know the mean number of atoms per unit volume in the two levels given by $N|c_a(z, t)|^2$ and $N|c_b(z, t)|^2$.

If we multiply both sides of (8.3.13) by c_a^* and add it to its complex conjugate, we obtain

$$\frac{d}{dt}c_a^* c_a = -\Gamma_a c_a^* c_a + i\left[\frac{\wp}{\hbar}\mathscr{E}^* c_a^* c_b - \frac{\wp^*}{\hbar}\mathscr{E} c_a c_b^*\right]\sin K_n z, \tag{8.3.18}$$

which gives the time dependence of the probability of finding an atom in level $|a\rangle$. The first term represents the loss of atoms out of level $|a\rangle$ due to such incoherent processes as spontaneous decay to lower levels, collisions, and such. The second term causes the population to change by the induced action of the radiation field present. To have laser action, we must in addition provide some pumping mechanism to pump atoms incoherently into levels $|a\rangle$ and $|b\rangle$; thus we add a phenomenological pumping term and we obtain

$$\frac{dN_a}{dt}(z, t) = R_a(z, t) - \Gamma_a N_a(z, t) + i[P_n(z, t)\mathscr{E}^* - P_n^*(z, t)\mathscr{E}] \sin K_n z\, \hbar^{-1}, \tag{8.3.19}$$

where we multiplied both sides of (8.3.18) by N, used (8.3.17) and have let

$$N_a(z, t) = N|c_a(z, t)|^2. \tag{8.3.20}$$

Also $R_a(z, t) = Nw_{a0}(z, t)$ is the rate at which atoms are pumped from the ground state to state $|a\rangle$ seconds per unit volume.

If we next multiply both sides of (8.3.19) by $(2/d)$ and integrate over the cavity from $z = 0$ to $z = d$, we obtain

$$\frac{dN_a(t)}{dt} = R_a(t) - \Gamma_a N_a(t) + \frac{i}{\hbar}[P_n(t)\mathscr{E}^*(t) - P_n^*(t)\mathscr{E}(t)], \tag{8.3.21}$$

where we have let

$$\begin{aligned} N_a(t) &= \frac{2}{d}\int_0^d N_a(z, t)\, dz \\ R_a(t) &= \frac{2}{d}\int_0^d R_a(z, t)\, dz \\ P_n(t) &= \frac{2}{d}\int_0^d P_n(z, t) \sin K_n z\, dz. \end{aligned} \tag{8.3.22}$$

The reader may proceed in a similar fashion to obtain

$$\frac{dN_b(t)}{dt} = R_b(t) - \Gamma_b N_b(t) - \frac{i}{\hbar}[P_n^*(t)\mathscr{E}^*(t) - P_n(t)\mathscr{E}(t)] \tag{8.3.23}$$

$$\frac{dP_n}{dt} = -[\Gamma_{ab} + i(\omega_a - \omega)]P_n - i\frac{|\wp|^2}{\hbar}(N_b - N_a)\mathscr{E}, \tag{8.3.24}$$

together with the conjugate to (8.3.24). We have let

$$\begin{aligned} \Gamma_{ab} &= \tfrac{1}{2}(\Gamma_a + \Gamma_b) \\ N_b(t) &= \frac{2}{d}\int_0^d N_b(z, t)\, dz \\ R_b(t) &= \frac{2}{d}\int_0^d R(z, t)\, dz \\ \hbar\omega_a &= \epsilon_b - \epsilon_a. \end{aligned} \tag{8.3.25}$$

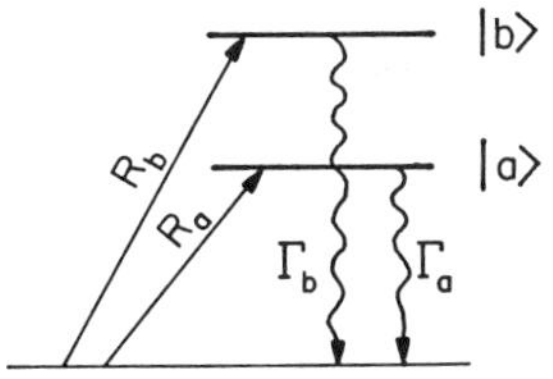

Figure 8.4 Pumping and decay rates for three-level atom. The ground state acts as a reservoir for atoms and the laser action is between levels a and b.

The physical meaning of these equations should be quite clear. Equation (8.3.21) gives the net rate of change at which atoms are entering and leaving state $|a\rangle$. The R_a term gives the rate at which atoms are being "pumped" into level $|a\rangle$. The $-\Gamma_a N_a$ term represents the incoherent decay of atoms from level $|a\rangle$ to lower levels. We could also add a term $+w_{ab}N_b$ to represent incoherent transitions from $|b\rangle$ to $|a\rangle$, but we omit this for simplicity. The Γ_a^{-1} is the lifetime of the atom in level $|a\rangle$ in the absence of driving field. These first terms are incoherent since they contain no phase information [see Figure 8.4].

The last term $i(P\mathscr{E}^* - P^*\mathscr{E})$ represents the net induced population change in level $|a\rangle$ due to the presence of a driving field. A similar interpretation applies to (8.3.23).

Equation (8.3.24) gives the net rate of change of the induced polarization due to the presence of a driving field. The first term allows for the decay of the induced polarization at the average rate Γ_{ab}. This decay can arise from spontaneous decay, collisions, and such. We could redefine Γ_{ab} as

$$\Gamma_{ab} = \tfrac{1}{2}(\Gamma_a + \Gamma_b) + \Gamma_{ab}^{\text{phase}} \tag{8.3.26}$$

to allow for phase interrupting elastic collisions as we have seen from our previous work.

The term $-i(\omega_a - \omega)P$ shows that we have removed the high-frequency behavior from P since it will turn out that $\omega \approx \omega_a$ and $P(t)$ is slowly varying in a time of order ω_a^{-1}. The deviation of ω from ω_a represents a detuning effect due to the coupling of the field to the atoms.

The last term $|\wp|^2(N_b - N_a)\mathscr{E}$ represents the induced polarization of the atoms by the field. The field causes the atom to oscillate between levels $|a\rangle$ and $|b\rangle$ which induces a dipole moment.

Equations (8.3.21), (8.3.23), (8.3.24), and its conjugate together with (8.3.15) and its conjugate represent the complete set of coupled equations which describe a laser oscillator. To solve these nonlinear equations exactly is impossible. We therefore proceed to make some approximations which are valid for a gas laser. We should note that we have neglected such things as the inhomogeneous broadening effect of Doppler motion. The reader should refer to Lamb's classic paper for such a discussion.

8.4 ADIABATIC ELIMINATION OF THE ATOMIC VARIABLES; PROPERTIES OF THE OSCILLATOR

In a gas laser the "cold" atomic linewidths Γ_a and Γ_b are large compared with the "cold" cavity linewidth γ:

$$\Gamma_a \gg \gamma; \qquad \Gamma_b \gg \gamma. \tag{8.4.1}$$

If we neglect the weak coupling between the atoms and the field in (8.3.21), (8.3.23), and (8.3.24), the polarization P_n decays to zero and the populations approach the steady-state values R_b/Γ_b and R_a/Γ_a in a time of order $\Gamma_{ab}{}^{-1}$. From (8.2.20) the field $\mathscr{E}$ approaches zero in a time of order γ^{-1} in the absence of the atomic driving term. Since $\Gamma \gg \gamma$ we see that the atomic variables will decay so quickly that the field has not had time to change appreciably. The atomic variables will therefore follow the electric field adiabatically when the coupling is weak. This allows us to eliminate the atomic variables as we shall show.

Consider (8.3.24). The polarization and populations are changing very rapidly compared with the field. Let us therefore integrate both sides over a time t_1 such that

$$\frac{1}{\gamma} \gg t_1 \gg \frac{1}{\Gamma_{ab}}. \tag{8.4.2}$$

This gives

$$\frac{P(t_1) - P(0)}{t_1} + [\Gamma_{ab} + i(\omega_a - \omega)]\frac{1}{t_1}\int_0^{t_1} P(t')\,dt' = \frac{|\wp|^2}{i\hbar}\mathscr{E}(t_1)\frac{1}{t_1}\int_0^{t_1}[N_b(t') - N_a(t')]\,dt'. \tag{8.4.3}$$

We removed $\mathscr{E}(t')$ from the integral since its value does not change much between $t' = 0$ and $t' = t_1 \ll \gamma^{-1}$. Since $t_1 \gg \Gamma_{ab}{}^{-1}$, the first term will be very small and we neglect it. The second term gives us a "coarse grained" average of $P(t')$. If we let

$$\bar{P}(t_1) = \frac{1}{t_1}\int_0^{t_1} P(t')\,dt', \tag{8.4.4}$$

and

$$\bar{N}_j(t_1) = \frac{1}{t_1}\int_0^{t_1} N_j(t')\,dt \qquad j = a, b \tag{8.4.5}$$

then we have

$$\bar{P}(t_1) \cong \frac{|\wp|^2}{i\hbar}\,\frac{\mathscr{E}(t_1)[\bar{N}_b(t_1) - \bar{N}_a(t_1)]}{\Gamma_{ab} + i(\omega_a - \omega)}. \tag{8.4.6}$$

A similar "coarse grained" average of (8.3.21) and (8.3.23) gives upon using (8.4.6)

$$0 = R_b - \Gamma_b \bar{N}_b - \frac{2|\wp|^2}{\hbar^2} \frac{\Gamma_{ab}}{\Gamma_{ab}^{\ 2} + (\omega_a - \omega)^2} |\mathscr{E}|^2 (\bar{N}_b - \bar{N}_a) \quad (8.4.7)$$

$$0 = R_a - \Gamma_a \bar{N}_a + \frac{2|\wp|^2}{\hbar^2} \frac{\Gamma_{ab}}{\Gamma_{ab}^{\ 2} + (\omega_a - \omega)^2} |\mathscr{E}|^2 (\bar{N}_b - \bar{N}_a). \quad (8.4.8)$$

We here assume R_a and R_b are time independent. Since $N_j = Nc_j^* c_j$ $(j = a, b)$, we see that the quantity

$$w \equiv \frac{|\wp|^2}{\hbar^2} \frac{2\Gamma_{ab}}{\Gamma_{ab}^{\ 2} + (\omega_a - \omega)^2} |\mathscr{E}|^2 \quad (8.4.9)$$

is the probability per second that the field induces an atom in level b to go to level a as well as an atom in level a to go to level b. When detuning is neglected, $\omega = \omega_a$, and we have

$$w_{\max} = \frac{|\wp|^2 |\mathscr{E}|^2}{\hbar^2} \frac{2}{\Gamma_{ab}}. \quad (8.4.10)$$

which is the maximum transition probability.

We may now solve (8.4.7) and (8.4.8) for the "coarse grained" average populations where we averaged over times large compared with the lifetimes of the excited states of the atoms. We easily find that the population difference is

$$\bar{N}_b(t) - \bar{N}_a(t) = \frac{(N_{b0} - N_{a0})}{1 + (w 2\Gamma_{ab}/\Gamma_a \Gamma_b)} = \frac{N_{b0} - N_{a0}}{1 + [\mu\, |\mathscr{E}(t)|^2 / \Gamma_{ab}^{\ 2} + (\omega_a - \omega)^2]}, \quad (8.4.11)$$

where we define

$$N_{j0} = \frac{R_j}{\Gamma_j} \qquad j = a, b, \quad (8.4.12)$$

and

$$\mu = \frac{|\wp|^2 4\Gamma_{ab}^{\ 2}}{\hbar^2 \Gamma_a \Gamma_b}. \quad (8.4.13)$$

The population difference adiabatically follows the instantaneous field intensity $|\mathscr{E}(t)|^2$. As we know, the field will build up when $\bar{N}_b(t) - \bar{N}_a(t) > 0$. For fixed pump rates, R_b and R_a, we see that as the intensity builds up, the population difference will start to decrease. This nonlinear saturation effect causes the laser to come to a steady state intensity. Physically, we start with more atoms in level b than level a. They decay spontaneously and are induced to go from level b to a and a to b. Since the induced probabilities are equal, more will come down than will go up with a net emission of radiation. This process tends to equalize the populations with a reduction in the build up of

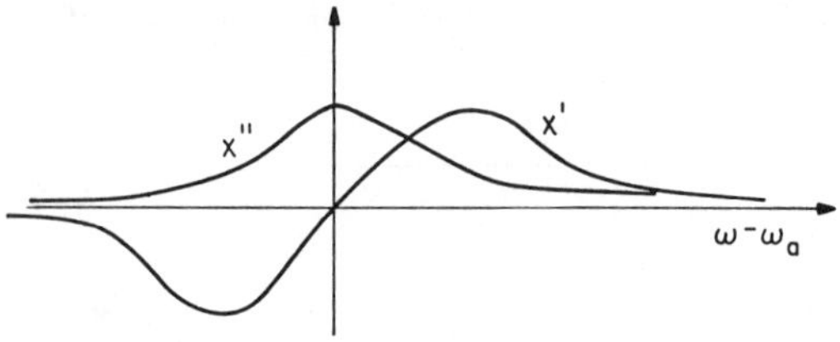

Figure 8.5 Real and imaginary parts of nonlinear susceptibility of atoms induced by a resonant radiation field in the steady state.

the field. A dynamic equilibrium is reached between the pumping and the losses and gives a steady-state oscillation.

The polarization (8.4.6) which is induced by the field with the help of (8.4.11) reduces to

$$\bar{P}(t) = \frac{(N_{b0} - N_{a0})[\omega - \omega_a - i\Gamma_{ab}]}{[\Gamma_{ab}{}^2 + (\omega - \omega_a)^2 + \mu\,|\mathscr{E}(t)|^2]}\frac{|\wp|^2}{\hbar^2}\mathscr{E}(t)$$
$$\equiv \epsilon_0 \chi \mathscr{E}(t) \equiv \epsilon_0(\chi' - i\chi'')\mathscr{E}(t), \tag{8.4.14}$$

where we have defined a complex nonlinear susceptibility of the atoms which depends on the field intensity $I(t) \equiv |\mathscr{E}(t)|^2$. In Figure 8.5 we sketch χ' and χ'' versus the detuning when the intensity has its steady-state value and $N_{b0} > N_{a0}$. The polarization increases as the field first builds up but will also saturate when the nonlinear term becomes dominant.

We have therefore succeeded in obtaining the polarization in terms of the driving electric field alone under the assumptions that Γ_a and Γ_b (the atomic linewidths) are large compared with γ (the cavity linewidth). Part of the induced polarization is in phase with the inducing field and will add to it while part is out of phase due to losses (decay of atoms to the ground state). The field radiated by the dipole must also be the field causing the dipole for self-consistency. We therefore substitute $\bar{P}(t)$ for $P(t)$ in (8.2.20). This gives the equation for the field

$$\frac{d\mathscr{E}}{dt} = \left\{\left[-\frac{\gamma}{2} + \frac{\mathscr{G}_0}{2}\frac{\Gamma_{ab}}{\Gamma_{ab}{}^2 + (\omega - \omega_a)^2 + \mu I}\right]\right.$$
$$\left. + i\left[(\omega - \Omega_n) + \frac{\mathscr{G}_0}{2}\frac{(\omega - \omega_a)}{\Gamma_{ab}{}^2 + (\omega - \omega_a)^2 + \mu I}\right]\right\}\mathscr{E}, \tag{8.4.15}$$

where

$$I \equiv |\mathscr{E}(t)|^2$$
$$\mathscr{G}_0 = \frac{\Omega_n\,|\wp|^2}{\epsilon_0 \hbar}(N_{b0} - N_{a0}) > 0. \tag{8.4.16}$$

The $\gamma/2$ term is the cavity loss while the second term with $\mathscr{G}_0$ gives the gain.

Let us first study the steady-state behavior of the oscillator. In this case $d\mathscr{E}/dt = 0$. Since $\mathscr{E} \neq 0$, we see from the above that each square bracket must vanish separately. That is,

$$\frac{\mathscr{G}_0}{2} \frac{\Gamma_{ab}}{\Gamma_{ab}^2 + (\omega_s - \omega_a)^2 + \mu I_s} = \frac{\gamma}{2} \tag{8.4.17}$$

$$\omega_s - \Omega_n = -\frac{\mathscr{G}_0}{2} \frac{(\omega_s - \omega_a)}{\Gamma_{ab}^2 + (\omega_s - \omega_a)^2 + \mu I_s}. \tag{8.4.18}$$

These two equations determine the intensity in the steady-state, I_s, and the oscillation frequency, ω_s. If we use (8.4.17) in (8.4.18), we see that

$$\omega_s - \Omega_n = (\omega_a - \omega_s)\frac{\gamma}{2\Gamma_{ab}}, \tag{8.4.19}$$

or the steady-state frequency is given by

$$\omega_s = \frac{\mathscr{S}\omega_a + \Omega_n}{\mathscr{S} + 1}, \tag{8.4.20}$$

where we have let

$$\mathscr{S} \equiv \frac{\gamma}{2\Gamma_{ab}} = \frac{\gamma}{\Gamma_a + \Gamma_b}, \tag{8.4.21}$$

which is the ratio of the cavity linewidth to the average atomic linewidth. We have assumed that $\Gamma_a + \Gamma_b \gg \gamma$ so that $\mathscr{S} \ll 1$. Therefore, to first order in $\mathscr{S}$, the steady-state frequency is

$$\begin{aligned} \omega_s &\cong (1 - \mathscr{S})(\Omega_n + \mathscr{S}\omega_a) \\ &= \Omega_n + \mathscr{S}(\omega_a - \Omega_n) + \cdots. \end{aligned} \tag{8.4.22}$$

The cavity tends to oscillate at the cavity resonant frequency rather than ω_a because the 'Q' of the cavity mode is high.

The steady-state intensity from (8.3.17) is

$$\mu I_s + (\omega_s - \omega_a)^2 = \Gamma_{ab}\left(\frac{\mathscr{G}_0}{\gamma} - \Gamma_{ab}\right). \tag{8.4.23}$$

We sketch this detuning curve in Figure 8.6. Since $I_s = |\mathscr{E}_s|^2$, I_s cannot be negative. There will be no steady-state oscillation unless

$$\mathscr{G}_0 \geq \gamma\Gamma_{ao} \tag{8.4.24}$$

or by (8.4.16).

$$N_{b0} - N_{a0} \geq \frac{\epsilon_0 \hbar}{\Omega_n |\wp|^2}\gamma\Gamma_{ab}, \tag{8.4.25}$$

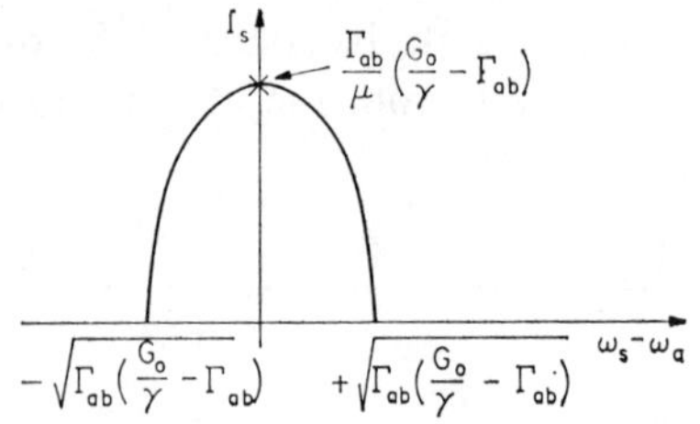

Figure 8.6 Detuning curve of oscillator which is the steady intensity versus the detuning.

as we see from Figure 8.6. Therefore, the population in level b must exceed the population in level a by this amount to overcome the losses and build to a steady-state value. The equality is the "start oscillation" condition for the laser. We obtain maximum steady-state intensity if $\omega_s \approx \Omega_n \approx \omega_a$. Less pump power is needed if $N_{a0} = 0$. It is therefore desirable to have Γ_a large and Γ_b small so that the lower state decays much more rapidly than the upper state. Then since $\Gamma_{ab} \approx \Gamma_a/2$ the pump threshold reduces to

$$R_b \geq \frac{\hbar\epsilon_0}{\Omega_n\,|\wp|^2}\,\frac{\gamma\Gamma_a\Gamma_b}{2}. \tag{8.4.26}$$

When $\omega_s = \omega_a$, the maximum steady-state intensity by (8.4.23), (8.4.13), and (8.4.16) becomes

$$(|\mathscr{E}_s|^2)_{\max} = (I_s)_{\max} = \frac{\Gamma_a\Gamma_b}{4}\left[\frac{\hbar\Omega_n}{\epsilon_0\gamma\Gamma_a}(N_{b0}-N_{a0}) - \frac{\hbar^2}{|\wp|^2}\right]. \tag{8.4.27}$$

Far above threshold, the second term here may be neglected.

We may obtain the steady-state population difference from (8.4.11), (8.4.13), (8.4.16), and (8.4.23):

$$\begin{aligned}(\bar{N}_b - \bar{N}_a)_s &= \frac{(N_{b0}-N_{a0})}{\mathscr{G}_0}\,\frac{\gamma}{\Gamma_{ab}}\,[\Gamma_{ab}^{\;2} + (\omega_a-\omega_s)^2] \\ &= \frac{\epsilon_0\hbar}{\Omega_n\,|\wp|^2}\,\frac{\gamma}{\Gamma_{ab}}\,[\Gamma_{ab}^{\;2} + (\omega_a-\omega_s)^2].\end{aligned} \tag{8.4.28}$$

If we let the threshold value by (8.4.25) be

$$(N_{b0}-N_{a0})_{th} = \frac{\epsilon_0\hbar}{\Omega_n\,|\wp|^2}\,\gamma\Gamma_{ab}, \tag{8.4.29}$$

then the steady-state population is

$$(\bar{N}_b - \bar{N}_a)_s = (N_{b0}-N_{a0})_{th}\left[1 + \left(\frac{\omega_a-\omega_s}{\Gamma_{ab}}\right)^2\right]. \tag{8.4.30}$$

This varies between the limits $(N_{b0} - N_{a0})_{th}$ when $\omega_s = \omega_a$ to $N_{b0} - N_{a0} = (R_b/\Gamma)$ when $\omega_s = \omega_a \pm \sqrt{\Gamma_{ab}(\mathscr{G}_0 - \gamma\Gamma_{ab})/\gamma}$.

If we use (8.4.29) and (8.4.16), we may rewrite (8.4.23) as

$$\mu I_s + (\omega_s - \omega_a)^2 = \Gamma_{ab}^{\ 2}\left\{\frac{(N_{b0} - N_{a0})}{(N_{b0} - N_{a0})_{th}} - 1\right\}. \tag{8.4.31}$$

Let us next consider the transient behavior of the oscillator. For this purpose let us write

$$\mathscr{E}(t) = \sqrt{I(t)}\, e^{i\varphi(t)}, \tag{8.4.32}$$

where $I(t) = |\mathscr{E}(t)|^2$ is the intensity and φ the phase. Then since

$$\frac{d\mathscr{E}}{dt} = \left[\frac{1}{2I}\frac{dI}{dt} + i\frac{d\varphi}{dt}\right] I^{1/2} e^{i\varphi}, \tag{8.4.33}$$

if we put this into (8.4.15) and equate real and imaginary parts, we obtain the two equations for the intensity and the phase

$$\frac{1}{I}\frac{dI}{dt} = -\gamma + \mathscr{G}_0 \frac{\Gamma_{ab}}{\Gamma_{ab}^{\ 2} + (\omega - \omega_a)^2 + \mu I} \tag{8.4.34}$$

$$\frac{d\varphi}{dt} = \omega - \Omega_n + \frac{\mathscr{G}_0}{2}\frac{(\omega - \omega_a)}{\Gamma_{ab}^{\ 2} + (\omega - \omega_a)^2 + \mu I}. \tag{8.4.35}$$

We may integrate the first equation exactly. Let

$$\beta = \Gamma_{ab}^{\ 2} + (\omega - \omega_a)^2$$

$$\alpha = \frac{\Gamma_{ab}\mathscr{G}_0}{\gamma}. \tag{8.4.36}$$

Then it may be written as

$$\frac{dI}{I}\frac{(\beta + \mu I)}{(\alpha - \beta - \mu I)} = \gamma\, dt,$$

or as partial fractions

$$\frac{\mu\, dI}{(\alpha - \beta)}\left[\frac{\beta}{\mu I} + \frac{\alpha}{\alpha - \beta - \mu I}\right] = \gamma\, dt,$$

so that on integrating we obtain

$$\beta \ln \frac{I}{I_0} - \alpha \log \frac{(\alpha - \beta - \mu I)}{(\alpha - \beta - \mu I_0)} = \gamma (\alpha - \beta) t \tag{8.4.37}$$

where I_0 is the initial intensity. Unfortunately, this is a transcendental equation so we cannot obtain $I(t)$ explicitly. It may be rewritten as

$$\left[\frac{I(t)}{I_0}\right]^\beta \left[\frac{\alpha - \beta - \mu I_0}{\alpha - \beta - \mu I(t)}\right]^\alpha = e^{\gamma(\alpha - \beta)t}. \tag{8.4.38}$$

It is actually easier to visualize the behavior of the intensity in two limiting cases.

CASE 1

$$\mu I \ll \beta \equiv \Gamma_{ab}^2 + (\omega - \omega_a)^2 \qquad \text{(far below threshold)}$$

In this case the solution of (8.4.34) reduces to

$$I(t) = I_0 \exp \left[\frac{\alpha}{\beta} - 1\right] \gamma t, \tag{8.4.39}$$

which shows that if

$$\alpha \geq \beta$$

or

$$\mathscr{G}_0 \geq \frac{\gamma}{\Gamma_{ab}} [\Gamma_{ab}^2 + (\omega - \omega_a)^2], \tag{8.4.40}$$

the intensity will grow exponentially as the oscillations start to build up from the initial intensity (see Figure 8.7). Note that the intensity will build up only if radiation is initially present. This arises from our semiclassical treatment in which fluctuations due to the spontaneous emission of the atom are neglected when the field is treated classically. We correct this in the next chapter.

The phase equation (8.4.35) in this case reduces to

$$\varphi(t) = \varphi(0) + \left[\omega - \Omega_n + \frac{\mathscr{G}_0(\omega - \omega_a)}{2[\Gamma_{ab}^2 + (\omega - \omega_a)^2]}\right] t. \tag{8.4.41}$$

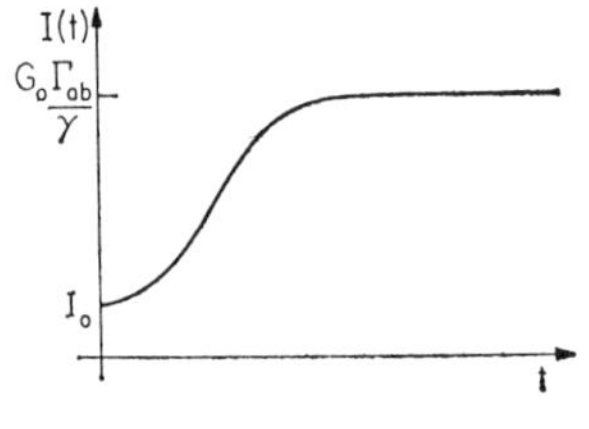

Figure 8.7 Buildup of laser intensity from below threshold to its steady value.

CASE 2

$$\mu I \gg \beta = \Gamma_{ab}^2 + (\omega - \omega_a)^2 \qquad \text{(for above threshold)}$$

In this case, by (8.4.34) we have

$$\frac{1}{I}\frac{dI}{dt} = -\gamma + \frac{\mathscr{G}_0\Gamma_{ab}}{\mu I}$$

or

$$I(t) = e^{-\gamma(t-t_1)}I_i + \frac{\mathscr{G}_0\Gamma_{ab}}{\mu\gamma}[1 - e^{-\gamma(t-t_1)}]$$
$$\xrightarrow[t\to\infty]{} \frac{\mathscr{G}_0\Gamma_{ab}}{\gamma},$$

which is the approximate steady-state value by (8.4.23) when $\mu I_s \gg \Gamma_{ab}^2 + (\omega - \omega_a)^2$ (see Figure 8.7).

We therefore see that the intensity starts to grow exponentially and as it builds up, it saturates and approaches a steady-state value.

The phase approaches a steady-state value also under these conditions.

REFERENCES

[1] W. E. Lamb, Jr., *Phys. Rev.*, **134,** A1429 (1964); 1963 Varenna Summer School Lectures, Course XXXI, p. 78; M. Scully, 1967 Varenna Summer School Lectures, Course XLI.
[2] A. Javan, W. R. Bennet, Jr., and D. R. Herriott, *Phys. Rev. Lett.*, **6,** 106 (1961); A. L. Schawlow, *Sci. Amer.*, June 1961.
[3] A. L. Schawlow and C. H. Townes, *Phys. Rev.*, **112,** 1940 (1958).
[4] A. G. Fox and T. Li, *Bell Syst. Tech. J.*, **40,** 453 (1961).
[5] H. Kogelnik and T. Li, *Proc. IEEE,* **54,** 1312 (1966).
[6] G. Boyd and J. P. Gordon, *Bell Syst. Tech. J.*, **40,** 489 (1961).

PROBLEMS

8.1 If we define a density matrix by

$$\rho = |\psi(t)\rangle\langle\psi(t)|$$

where $|\psi(t)\rangle$ is given by (8.3.4) and where its matrix elements are given by

$$\rho_{ij}(t) = \langle i|\rho(t)|j\rangle \qquad i, j = a \text{ or } b$$

show by using (8.3.13) and (8.3.14) with $\sin K_n z = 1$ that ρ satisfies the equation

$$\frac{\partial \rho}{\partial t} = \frac{1}{i\hbar}[K, \rho] + \tfrac{1}{2}\{\Gamma, \rho\}_+,$$

where we have let

$$K \equiv \begin{pmatrix} 0 & \wp^* \mathscr{E} e^{-i\omega t} \\ \wp \mathscr{E} e^{i\omega t} & 0 \end{pmatrix}; \qquad \Gamma \equiv \begin{pmatrix} \Gamma_a & 0 \\ 0 & \Gamma_b \end{pmatrix},$$

and $[A, B] \equiv AB - BA$ and $\{A, B\}_+ \equiv AB + BA$. This is just a model for a two-level atom with linewidth.

8.2 Expand the nonlinear terms in (8.4.15) in powers of $I - I_s$ where I_s is the steady-state value. To first order in $(I - I_s)$ show that the resulting equation is just that for a classical rotating wave van der Pol oscillator.

8.3 From (8.4.34) and (8.4.35) solve for the relation between φ and I.

9

Statistical Properties of a Laser

9.1 THE LASER MODEL [1–4]

In the prior chapter we discussed the semiclassical theory of a single mode homogeneously broadened laser in which we treated the electromagnetic field classically. To discuss the statistical properties of the laser oscillator output, we must repeat the analysis and quantize the radition field. In addition we must account for the damping of the radiation field as well as the quantum statistical fluctuations of the reservoir associated with the damping as we did in Chapters 6 and 7. Finally, we must account for the fluctuations in the atomic system due to various damping mechanisms.

In Figure 9.1 we show a block diagram of the laser model. The reservoir coupled to a mode of the field is shown which accounts for the mode damping and the quantum fluctuations which the damping introduces into the field. We have discussed such a reservoir in detail in Chapters 6 and 7 since the field mode is equivalent to a damped harmonic oscillator. Below we show N three-level atoms per unit volume coupled to a pumping and damping reservoir. Each atom is thought of as coupled to its own reservoir and they are coupled to each other only through the atom-field interaction. Again we discussed the damped atoms in detail in Chapters 6 and 7. Finally, the atoms and field are coupled by a dipole interaction as discussed in the prior chapter. However, we treat the field quantum mechanically.

The causal unperturbed hamiltonian of the atom and field is

$$H_0 = \sum_{\lambda=1}^{N} \sum_{l=0}^{2} \epsilon_l (a_l^\dagger a_l)_\lambda + \hbar\omega_c a^\dagger a, \tag{9.1.1}$$

where we are using the notation introduced in Section 3.10. Again we let

$$N_l = \sum_{\lambda=1}^{N} (a_l^\dagger a_l)_\lambda \tag{9.1.2}$$

be the "number" operator for the number of atoms in level $|l\rangle$.

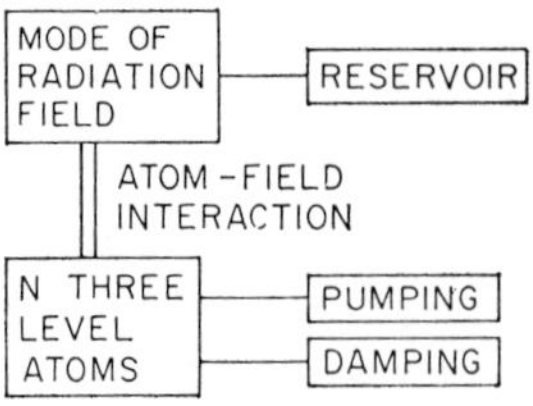

Figure 9.1 Block diagram of laser model.

Under the dipole and rotating wave approximations, the causal atom-field interaction energy is

$$W_{\text{atom field}} = i\hbar\mu \sum_{\lambda=1}^{N} [a^\dagger(a_1^\dagger a_2)_\lambda - (a_2^\dagger a_1)_\lambda a]$$
$$\equiv i\hbar\mu[a^\dagger M - M^\dagger a], \tag{9.1.3}$$

where

$$M = \sum_{\lambda=1}^{N} (a_1^\dagger a_2)_\lambda. \tag{9.1.4}$$

The μ is proportional to the dipole moment matrix element of the atom in levels $|1\rangle$ and $|2\rangle$; $\hbar\mu$ has the dimension of energy since the operators are dimensionless. We treat the interaction exactly instead of using a perturbation approach since we want to allow for nonlinear effects.

The "system" operators are a, $a^\dagger$, M, $M^\dagger$, and N_l. The reduced density operator which describes a homogeneously broadened single mode laser shown schematically in Figure 9.1 obeys the equation of motion

$$\frac{\partial S}{\partial t} = \frac{1}{i\hbar}[S, H_0 + W] + \left(\frac{\partial S}{\partial t}\right)_F + \left(\frac{\partial S}{\partial t}\right)_A, \tag{9.1.5}$$

where the first terms describe the causal behavior. The second term describes the interaction of the field mode with its reservoir and is given in (6.2.59) with $v(t) = 0$. The last term describes the interaction of the atoms with their pumping and damping reservoirs and is given by (6.4.50) where we restrict l to 0, 1, 2. The reduced density operator describes the behavior of both the atoms and the field mode.

In the semiclassical theory we neglected the last two terms in (9.1.5).

9.2 THE FOKKER–PLANCK EQUATION FOR A LASER

In this section we convert the operator equation (9.1.5) into its c-number associated distribution equivalent to free ourselves of the operators. We have by definition that

$$P_c(\underset{\sim}{\alpha}, t) = \operatorname{Tr} S(t)\, \delta^c(\underset{\sim}{\alpha} - \underset{\sim}{a}), \tag{9.2.1}$$

where we shall take as our chosen ordering $a^\dagger$, $M^\dagger$, N_1, N_2, M, a. The corresponding c-numbers are α^*, $\mathcal{M}^*$, $\mathcal{N}_1$, $\mathcal{N}_2$, $\mathcal{M}$, α. Then by (9.2.1) and

(9.1.1)–(9.1.5) we have

$$\frac{\partial P_c}{\partial t} = \mathrm{Tr}\left\{\frac{1}{i\hbar}[S, H_0 + W] + \left(\frac{\partial S}{\partial t}\right)_F + \left(\frac{\partial S}{\partial t}\right)_A\right\}\delta^c(\underset{\sim}{\alpha} - \underset{\sim}{a}), \qquad (9.2.2)$$

where

$$\delta^c(\underset{\sim}{\alpha} - \underset{\sim}{a}) \equiv \delta(\alpha^* - a^\dagger)\,\delta(\mathscr{M}^* - M^\dagger)\,\delta(\mathscr{N}_1 - N_1) \times \delta(\mathscr{N}_2 - N_2)\,\delta(\mathscr{M} - M)\,\delta(\alpha - a). \qquad (9.2.3)$$

We have already evaluated the last two terms in (9.2.2), since the atom and field operators commute. The field term for normal ordering is given by (6.4.37) with $v(t) = 0$ and the atomic term is given by (6.4.90) under the assumption that the number of atoms in levels 1 and 2 are small compared with those in the ground state, although we still may have a large number in these upper levels. The H_0 term has also been included in (6.4.37) and (6.4.90). We therefore need only to evaluate the Tr $[S, W]$ term. We have

$$I = \frac{1}{i\hbar}\mathrm{Tr}_S\,[S(t), W_{A-F}]\,\delta^c(\underset{\sim}{\alpha} - \underset{\sim}{a}) = \frac{1}{i\hbar}\mathrm{Tr}_S\,S(t)[\delta^c(\underset{\sim}{\alpha} - \underset{\sim}{a}), W_{A-F}], \qquad (9.2.4)$$

where we used the cyclic property of the trace. If we use (6.4.15), (9.2.3), and (9.1.3), we have

$$\begin{aligned} I = \mu\,\mathrm{Tr}_S\,S(t)\{[&e^{-a^\dagger(\partial/\partial\alpha^*)}e^{-M^\dagger(\partial/\partial\mathscr{M}^*)}e^{-N_1(\partial/\partial\mathscr{N}_1)}e^{-N_2(\partial/\partial\mathscr{N}_2)}e^{-a(\partial/\partial\alpha)}(a^\dagger M - M^\dagger a) \\ &- (a^\dagger M - M^\dagger a)e^{-a^\dagger(\partial/\partial\alpha^*)}e^{-M^\dagger(\partial/\partial\mathscr{M}^*)}e^{-N_1(\partial/\partial\mathscr{N}_1)}e^{-N_2(\partial/\partial\mathscr{N}_2)}e^{-a(\partial/\partial\alpha)}] \\ &\times \delta(\alpha)\,\delta(\alpha^*)\,\delta(\mathscr{M})\,\delta(\mathscr{M})\,\delta(\mathscr{N}_1)\,\delta(\mathscr{N}_2). \end{aligned} \qquad (9.2.5)$$

We must next use the commutation relations to put all terms into chosen order. Since these techniques have already been developed, we leave the details as an exercise. The result is

$$\begin{aligned} I = \mu\Big\{&-\frac{\partial}{\partial\alpha}\mathscr{M} - \frac{\partial}{\partial\alpha^*}\mathscr{M}^* - \frac{\partial}{\partial\mathscr{M}}(\mathscr{N}_2 - \mathscr{N}_1)\alpha - \frac{\partial}{\partial\mathscr{M}^*}(\mathscr{N}_2 - \mathscr{N}_1)\alpha^* \\ &+ (\alpha^*\mathscr{M} + \alpha\mathscr{M}^*)[1 - e^{(\partial/\partial\mathscr{N}_1)-(\partial/\partial\mathscr{N}_2)}] + \frac{\partial^2}{\partial\mathscr{M}^2}\mathscr{M}\alpha + \frac{\partial^2}{\partial\mathscr{M}^{*2}}\mathscr{M}^*\alpha^*\Big\}P_c. \end{aligned} \qquad (9.2.6)$$

When $\mathscr{N}_1$ and $\mathscr{N}_2$ are small compared with N, we may expand the exponential to second order as we have done in our previous work. This yields

$$e^{(\partial/\partial\mathscr{N}_1)-(\partial/\partial\mathscr{N}_2)} \cong 1 + \frac{\partial}{\partial\mathscr{N}_1} - \frac{\partial}{\partial\mathscr{N}_2} + \frac{1}{2}\frac{\partial^2}{\partial\mathscr{N}_1^2} + \frac{1}{2}\frac{\partial^2}{\partial\mathscr{N}_2^2} - \frac{\partial^2}{\partial\mathscr{N}_1\,\partial\mathscr{N}_2} \qquad (9.2.7)$$

so that the interaction term reduces to

$$I = \mu\left\{-\frac{\partial}{\partial\alpha}\mathcal{M} - \frac{\partial}{\partial\alpha^*}\mathcal{M}^* - \frac{\partial}{\partial\mathcal{M}}(\mathcal{N}_2 - \mathcal{N}_1)\alpha - \frac{\partial}{\partial\mathcal{M}^*}(\mathcal{N}_2 - \mathcal{N}_1)\alpha^* \right.$$
$$+ \left[\frac{\partial}{\partial\mathcal{N}_2} - \frac{\partial}{\partial\mathcal{N}_1} - \frac{1}{2}\frac{\partial^2}{\partial\mathcal{N}_2^{\,2}} - \frac{1}{2}\frac{\partial^2}{\partial\mathcal{N}_1^{\,2}} + \frac{\partial^2}{\partial\mathcal{N}_1\partial\mathcal{N}_2}\right](\alpha^*\mathcal{M} + \alpha\mathcal{M}^*)$$
$$\left. + \frac{\partial^2}{\partial\mathcal{M}^2}\alpha\mathcal{M} + \frac{\partial^2}{\partial\mathcal{M}^{*2}}\alpha^*\mathcal{M}^*\right\}P_c. \tag{9.2.8}$$

If we combine this with (6.4.37) and (6.4.90), we obtain the desired Fokker–Planck equation

$$\frac{\partial P_c}{\partial t} = \left\{-\frac{\partial}{\partial\alpha}\left[\mu\mathcal{M} - \left(\frac{\gamma}{2} + i\omega_c\right)\alpha\right] - \frac{\partial}{\partial\alpha^*}\left[\mu\mathcal{M}^* - \left(\frac{\gamma}{2} - i\omega_c\right)\alpha^*\right]\right.$$
$$- \frac{\partial}{\partial\mathcal{M}}[\mu(\mathcal{N}_2 - \mathcal{N}_1)\alpha - (\Gamma_{12} + i\omega_a)\mathcal{M}]$$
$$- \frac{\partial}{\partial\mathcal{M}^*}[\mu(\mathcal{N}_2 - \mathcal{N}_1)\alpha^* - (\Gamma_{12} - i\omega_a)\mathcal{M}^*]$$
$$- \frac{\partial}{\partial\mathcal{N}_2}[R_2 + w_{21}\mathcal{N}_1 - \Gamma_2\mathcal{N}_2 - \mu(\alpha^*\mathcal{M} + \alpha\mathcal{M}^*)]$$
$$- \frac{\partial}{\partial\mathcal{N}_1}[R_1 - \Gamma_1\mathcal{N}_1 + w_{12}\mathcal{N}_2 + \mu(\alpha^*\mathcal{M} + \alpha\mathcal{M}^*)]$$
$$+ \frac{1}{2}\frac{\partial^2}{\partial\mathcal{N}_2^{\,2}}[R_2 + w_{21}\mathcal{N}_1 + \Gamma_2\mathcal{N}_2 - \mu(\alpha^*\mathcal{M} + \alpha\mathcal{M}^*)]$$
$$+ \frac{1}{2}\frac{\partial^2}{\partial\mathcal{N}_1^{\,2}}[R_1 + w_{12}\mathcal{N}_2 + \Gamma_1\mathcal{N}_1 - \mu(\alpha^*\mathcal{M} + \alpha\mathcal{M}^*)]$$
$$+ \frac{\partial^2}{\partial\mathcal{N}_1\,\partial\mathcal{N}_2}[-w_{12}\mathcal{N}_2 - w_{21}\mathcal{N}_1 + \mu(\alpha^*\mathcal{M} + \alpha\mathcal{M}^*)]$$
$$+ \frac{\partial^2}{\partial\mathcal{N}_1\,\partial\mathcal{M}}\Gamma_1\mathcal{M} + \frac{\partial^2}{\partial\mathcal{N}_1\,\partial\mathcal{M}^*}\Gamma_1\mathcal{M}^*$$
$$+ \frac{\partial^2}{\partial\mathcal{N}_2\,\partial\mathcal{M}}(-w_{21}\mathcal{M}) + \frac{\partial^2}{\partial\mathcal{N}_2\,\partial\mathcal{M}^*}(-w_{21}\mathcal{M}^*)$$
$$+ \frac{\partial^2}{\partial\mathcal{M}\,\partial\mathcal{M}^*}[R_2 + (\Gamma_1 + 2\Gamma_{12}^{\,ph})\mathcal{N}_2 + w_{21}\mathcal{N}_1]$$
$$\left. + \frac{\partial^2}{\partial\mathcal{M}^2}\mu\alpha\mathcal{M} + \frac{\partial^2}{\partial\mathcal{M}^{*2}}\mu\alpha^*\mathcal{M}^* + \frac{\partial^2}{\partial\alpha\,\partial\alpha^*}\gamma\bar{n}\right\}P_c. \tag{9.2.9}$$

We have let $\omega_{21} \equiv \omega_a$ above. The first derivative terms give the mean drift motion while the second derivative terms give the diffusion.

9.3 THE LASER ASSOCIATED LANGEVIN EQUATIONS

We may immediately write down the Langevin c-number equations of motion by inspection from the Fokker–Planck equation (9.2.9). According to the theory given in Chapter 7, we have

$$\frac{d\alpha}{dt} = -\left(\frac{\gamma}{2} + i\omega_c\right)\alpha + \mu\mathcal{M} + \mathcal{G}_\alpha \tag{9.3.1}$$

$$\frac{d\mathcal{M}}{dt} = -(\Gamma_{12} + i\omega_a)\mathcal{M} + \mu(\mathcal{N}_2 - \mathcal{N}_1)\alpha + \mathcal{G}_{\mathcal{M}} \tag{9.3.2}$$

$$\frac{d\mathcal{N}_2}{dt} = R_2 + w_{21}\mathcal{N}_1 - \Gamma_2\mathcal{N}_2 - \mu(\alpha^*\mathcal{M} + \alpha\mathcal{M}^*) + \mathcal{G}_{\mathcal{N}_2} \tag{9.3.3}$$

$$\frac{d\mathcal{N}_1}{dt} = R_1 - \Gamma_1\mathcal{N}_1 + w_{12}\mathcal{N}_2 + \mu(\alpha^*\mathcal{M} + \alpha\mathcal{M}^*) + \mathcal{G}_{\mathcal{N}_1} \tag{9.3.4}$$

together with the adjoints of (9.3.1) and (9.3.2).

The frequency $\omega_c \simeq \omega_a$ and the laser will oscillate at a frequency close to this, say ω_0. We may remove this high frequency from the equations by letting

$$\begin{aligned} \alpha(t) &= \alpha'(t)e^{-i\omega_0 t} \\ \mathcal{M}(t) &= \mathcal{M}'(t)e^{-i\omega_0 t}. \end{aligned} \tag{9.3.5}$$

Then (9.3.1)–(9.3.4) become

$$\frac{d\alpha'}{dt} = -\left[\frac{\gamma}{2} + i(\omega_c - \omega_0)\right]\alpha' + \mu\mathcal{M}' + g_\alpha \tag{9.3.6}$$

$$\frac{d\mathcal{M}'}{dt} = -[\Gamma_{12} + i(\omega_a - \omega_0)]\mathcal{M}' + \mu(\mathcal{N}_2 - \mathcal{N}_1)\alpha' + g_{\mathcal{M}} \tag{9.3.7}$$

$$\frac{d\mathcal{N}_2}{dt} = R_2 + w_{21}\mathcal{N}_1 - \Gamma_2\mathcal{N}_2 - \mu(\alpha'^*\mathcal{M}' + \alpha'\mathcal{M}'^*) + g_{\mathcal{N}_2} \tag{9.3.8}$$

$$\frac{d\mathcal{N}_1}{dt} = R_1 - \Gamma_1\mathcal{N}_1 + w_{12}\mathcal{N}_2 + \mu(\alpha'^*\mathcal{M}' + \alpha'\mathcal{M}'^*) + g_{\mathcal{N}_1}, \tag{9.3.9}$$

where the new random forces are related to the old by

$$\begin{aligned} g_\alpha &= e^{i\omega_0 t}\mathcal{G}_\alpha \qquad & g_{\mathcal{N}_1} &= \mathcal{G}_{\mathcal{N}_1} \\ g_{\mathcal{M}} &= e^{i\omega_0 t}\mathcal{G}_{\mathcal{M}} \qquad & g_{\mathcal{N}_2} &= \mathcal{G}_{\mathcal{N}_2}. \end{aligned} \tag{9.3.10}$$

If we neglect the random forces, these equations reduce to the semiclassical laser equations of Chapter 8 as we shall show.

We have that

$$\hbar\mu = \sqrt{\frac{\hbar\omega_c}{2\epsilon_0 L^3}}\,\wp, \tag{9.3.11}$$

where $\wp = \langle 1|ex|2\rangle$. The polarization is

$$P = \frac{\wp}{V}\sum_{\lambda=1}^{N}(a_1^\dagger a_2)_\lambda = \frac{\wp}{V}M \rightarrow \frac{\wp\mathscr{M}'}{V}, \tag{9.3.12}$$

since we removed the high frequency in (8.3.24) also. Furthermore, we make the identification

$$\mathscr{E}(t) = -i\sqrt{\frac{\hbar\omega_c}{2\epsilon_0 V}}\,\alpha'(t). \tag{9.3.13}$$

With these identifications, (9.3.6) reduces to (8.2.20) and the others reduce to (8.3.21), (8.3.23), and (8.3.24) aside from the nonradiative transition terms $w_{21}\mathscr{N}_1$ and $w_{12}\mathscr{N}_2$ which were neglected in the semiclassical approach. We also assumed here that $\wp = \wp^*$ for simplicity.

The moments of the Langevin forces may be read off by inspection from the Fokker–Planck equation (9.2.9), since from the general theory of Chapter 7 we have

$$\langle g_i(t)g_j(u)\rangle_R = 2\langle\mathscr{D}_{ij}\rangle\,\delta(t-u) \tag{9.3.14}$$

As an example

$$\langle g_\alpha^*(t)g_\alpha(u)\rangle_R = \gamma\bar{n}\,\delta(t-u) \tag{9.3.15}$$

$$\langle g_{\mathscr{M}^*}(t)g_{\mathscr{M}}(u)\rangle_R = [R_2 + (\Gamma_1 + 2\Gamma_{12}{}^{ph})\mathscr{N}_2 + w_{21}\mathscr{N}_1]\,\delta(t-u). \tag{9.3.16}$$

9.4 ADIABATIC ELIMINATION OF ATOMIC VARIABLES

In a typical gas laser the atomic decay rates (linewidths) may be of order 10^8 sec^{-1} while the radiative decay rate γ (cavity linewidth) may be of order 10^7 sec^{-1}. Under these conditions we may adiabatically eliminate the atomic variables as we did in the semiclassical theory in Chapter 8. To simplify the model for tutorial purposes without losing its essential physical features, we first assume that Γ_1 is very large compared with other decay rates. This means that atoms in level 1 decay to the ground state so rapidly that we may set

$$\mathscr{N}_1 = 0, \tag{9.4.1}$$

and neglect the $\mathscr{N}_1$ equation 9.3.9 completely. Also the distribution function which satisfies the Fokker–Planck equation (9.2.9) becomes independent of

$\mathcal{N}_1$ and we may therefore neglect all derivatives with respect to $\mathcal{N}_1$:

$$\frac{\partial P_c}{\partial \mathcal{N}_1} = 0$$
$$\frac{\partial^2 P_c}{\partial \mathcal{N}_1^{\,2}} = 0. \tag{9.4.2}$$

Under this approximation the Langevin equations 9.3.6–9.3.8 reduce to

$$\frac{d\alpha'}{dt} = -\left[\frac{\gamma}{2} + i(\omega_c - \omega_0)\right]\alpha' + \mu\mathcal{M}' + g_\alpha \tag{9.4.3}$$

$$\frac{d\mathcal{M}'}{dt} = -[\Gamma_{12} - i(\omega_0 - \omega_a)]\mathcal{M}' + \mu\mathcal{N}_2\alpha' + g_{\mathcal{M}} \tag{9.4.4}$$

$$\frac{d\mathcal{N}_2}{dt} = R_2 - \Gamma_2\mathcal{N}_2 - B + g_{\mathcal{N}_2}, \tag{9.4.5}$$

where

$$B = \mu(\alpha'^*\mathcal{M}' + \alpha'\mathcal{M}'^*). \tag{9.4.6}$$

In addition we assume

$$\Gamma_{12} = \frac{\Gamma_1 + \Gamma_2}{2} + \Gamma_{12}^{\;ph} \approx \frac{\Gamma_1}{2} \tag{9.4.7}$$

is very large compared with γ and Γ_2 so that we may neglect $d\mathcal{M}'/dt$ compared with $\Gamma_{12}\mathcal{M}'$. We may therefore solve (9.4.4) for the adiabatic value of $\mathcal{M}'$

$$\mathcal{M}' = \frac{\mu\mathcal{N}_2\alpha' + g_{\mathcal{M}}}{\Gamma_{12} - i(\omega_0 - \omega_a)}. \tag{9.4.8}$$

If we use this and its conjugate, (9.4.3) and (9.4.5) become

$$\frac{d\alpha'}{dt} = \frac{1}{2}\left\{\pi\mathcal{N}_2 - \gamma + i\left[\frac{\pi\mathcal{N}_2}{\Gamma_{12}}(\omega_0 - \omega_a) - 2(\omega_c - \omega_0)\right]\right\}\alpha' + f_\alpha \tag{9.4.9}$$

$$\frac{d\mathcal{N}_2}{dt} = R_2 - (\Gamma_2 + \pi\,|\alpha'|^2)\mathcal{N}_2 - \mu\left[\frac{\alpha'^* g_{\mathcal{M}}}{\Gamma_{12} - i(\omega_0 - \omega_a)} + \frac{\alpha' g_{\mathcal{M}^*}}{\Gamma_{12} + i(\omega_0 - \omega_a)}\right] + g_{\mathcal{N}_2}, \tag{9.4.10}$$

where we have let

$$\pi \equiv \frac{2\Gamma_{12}\mu^2}{\Gamma_{12}^{\;2} + (\omega_0 - \omega_a)^2} \tag{9.4.11}$$

$$f_\alpha(t) = g_\alpha(t) + \frac{\mu g_{\mathcal{M}}(t)}{\Gamma_{12} - i(\omega_0 - \omega_a)}. \tag{9.4.12}$$

In order to be self-consistent, as $\Gamma_1 \to \infty$, $\Gamma_{12} \to \infty$, we must also assume $\mu \to \infty$ in such a way that π defined above remains finite to retain the essential physical features of the model.

Equation (9.4.9) is a Langevin equation in which $\langle f_\alpha(t)\rangle_R = 0$ and the first curly bracket term in (9.4.9) gives the entire drift motion. However, the square bracket term involving $g_{\mathscr{M}}$ and $g_{\mathscr{M}^*}$ in (9.4.10) has a finite reservoir average which we must evaluate and add to the drift motion before it is a correct Langevin equation in which the reservoir average of the random force is zero. Consider therefore the quantity

$$I = \langle \alpha'^*(t) g_{\mathscr{M}}(t)\rangle_R. \tag{9.4.13}$$

We may rewrite this as

$$I = \left\langle \left\{ \left[\alpha'^*(t_c) + \int_{t_c}^{t} \frac{d\alpha'^*}{ds}\, ds \right] g_{\mathscr{M}}(t) \right\rangle_R, \tag{9.4.14}$$

where $t_c = t - \epsilon$ and ultimately we let $\epsilon \to 0$. Under the Markoff approximation

$$\langle \alpha'^*(t_c) g_{\mathscr{M}}(t)\rangle_R = 0 \qquad t > t_c. \tag{9.4.15}$$

If we use this and (9.4.9), we obtain

$$I = \frac{1}{2}\int_{t_c}^{t} \left\langle \left\{ \pi\mathscr{N}_2(s) - \gamma - i\left[\frac{\pi\mathscr{N}_2(s)}{\Gamma_{12}}(\omega_0 - \omega_a) - 2(\omega_c - \omega_0)\right]\right\} \right.$$
$$\left. \times\, \alpha'^*(s) g_{\mathscr{M}}(t) \right\rangle_R ds + \int_{t_c}^{t} \left\langle \left[g_{\alpha^*}(s) + \frac{\mu g_{\mathscr{M}^*}(s)}{\Gamma_{12} + i(\omega_0 - \omega_a)} \right] g_{\mathscr{M}}(t) \right\rangle_R ds, \tag{9.4.16}$$

where we used (9.4.12). Again under the Markoff approximation since $t > s$, the first integral vanishes. Next since the forces are δ-correlated we have

$$\begin{aligned} \langle g_{\alpha^*}(s) g_{\mathscr{M}}(t)\rangle_R &= 2\langle \mathscr{D}_{\alpha^*\mathscr{M}}\rangle\, \delta(s - t) \\ \langle g_{\mathscr{M}^*}(s) g_{\mathscr{M}}(t)\rangle_R &= 2\langle \mathscr{D}_{\mathscr{M}^*\mathscr{M}}\rangle \delta(s - t), \end{aligned} \tag{9.4.17}$$

so that I becomes on carrying out the integrals

$$I = \langle \alpha'^*(t) g_{\mathscr{M}}(t)\rangle_R = \langle \mathscr{D}_{\alpha^*\mathscr{M}}\rangle + \langle \mathscr{D}_{\mathscr{M}^*\mathscr{M}}\rangle \frac{\mu}{\Gamma_{12} + i(\omega_0 - \omega_a)}. \tag{9.4.18}$$

In a similar way we see that

$$\langle \alpha'(t) g_{\mathscr{M}^*}(t)\rangle_R = \langle \mathscr{D}_{\alpha\mathscr{M}^*}\rangle + \frac{\langle \mathscr{D}_{\mathscr{M}^*\mathscr{M}}\rangle \mu}{\Gamma_{12} - i(\omega_0 - \omega_a)}. \tag{9.4.19}$$

Therefore, we see that the reservoir average of the $g_{\mathscr{M}}$ and $g_{\mathscr{M}^*}$ terms in (9.4.10) become

$$-\frac{\pi}{\Gamma_{12}}\langle\mathscr{D}_{\mathscr{M}^*\mathscr{M}}\rangle_R - \frac{\mu}{\Gamma_{12} - i(\omega_0 - \omega_a)}\langle\mathscr{D}_{\alpha^*\mathscr{M}}\rangle - \frac{\mu}{\Gamma_{12} + i(\omega_0 - \omega_a)}\langle\mathscr{D}_{\alpha\mathscr{M}^*}\rangle \tag{9.4.20}$$

These coefficients may be read off directly from the Fokker–Planck equation (9.2.9):

$$\begin{aligned} 2\langle\mathscr{D}_{\mathscr{M}^*\mathscr{M}}\rangle_R &= R_2 + (\Gamma_1 + 2\Gamma_{12}{}^{ph})\mathscr{N}_2 \\ &\equiv R_2 + 2\Gamma_{12}\mathscr{N}_2 \\ \langle\mathscr{D}_{\alpha^*\mathscr{M}}\rangle &= \langle\mathscr{D}_{\alpha\mathscr{M}^*}\rangle = 0 \end{aligned} \tag{9.4.21}$$

We used (9.4.7) and the fact that $\Gamma_{12} \gg \Gamma_2$. If we therefore include the contribution (9.4.20) in the drift motion and define a new random force

$$f_{\mathscr{N}_2}(t) = g_{\mathscr{N}_2}(t) - \mu\left[\frac{\alpha_c'^* g_{\mathscr{M}}(t)}{\Gamma_{12} - i(\omega_0 - \omega_a)} + \frac{\alpha_c' g_{\mathscr{M}^*}(t)}{\Gamma_{12} + i(\omega_0 - \omega_a)}\right], \tag{9.4.22}$$

where $\alpha_c' \equiv \alpha'(t_c)$ whose reservoir average vanishes we may replace (9.4.10) with

$$\frac{d\mathscr{N}_2}{dt} = R_2\left[1 - \frac{\pi}{2\Gamma_{12}}\right] - [\Gamma_2 + \pi + \pi\,|\alpha'|^2]\mathscr{N}_2 + f_{\mathscr{N}_2}. \tag{9.4.23}$$

However,

$$\frac{\pi}{\Gamma_{12}} \sim 10^{-8} \tag{9.4.24}$$

for a gas laser so that the entire correction to the drift motion of $\mathscr{N}_2$ is negligible and we have for the equation of motion for the population difference

$$\frac{d\mathscr{N}_2}{dt} = R_2 - [\Gamma_2 + \pi\,|\alpha'|^2]\mathscr{N}_2 + f_{\mathscr{N}_2}, \tag{9.4.25}$$

where $f_{\mathscr{N}_2}$ is given by (9.4.22). The reservoir average of the $g_{\mathscr{M}}$ and $g_{\mathscr{M}^*}$ in (9.4.10) make negligible contribution to the drift motion.

If we take the reservoir average of both sides of (9.4.9), we obtain

$$\frac{d\langle\alpha'\rangle_R}{dt} = \tfrac{1}{2}\langle[\pi\mathscr{N}_2 - \gamma]\alpha'\rangle_R + i\left\langle\left[\frac{\pi\mathscr{N}_2}{\Gamma_{12}}(\omega_0 - \omega_a) - 2(\omega_c - \omega_0)\right]\alpha'\right\rangle_R.$$

In the steady-state, we see that the real and imaginary parts must vanish so that we obtain the steady state frequency of the laser when

$$\pi\langle\mathscr{N}_2\alpha'\rangle_R = \gamma\langle\alpha'\rangle_R, \tag{9.4.26a}$$

so that

$$\frac{\gamma}{2\Gamma_{12}}(\omega_0 - \omega_a) = (\omega_c - \omega_0), \tag{9.4.26b}$$

which agrees with the classical result (8.4.19). Since $\Gamma_{12} \gg \gamma$, we see that $\omega_0 \cong \omega_c$. If we tune the cavity so that $\omega_c \cong \omega_a$, we may neglect detuning without serious error. The laser equation for the field and populations therefore reduce to

$$\frac{d\alpha'}{dt} = \tfrac{1}{2}[\pi\mathcal{N}_2 - \gamma]\alpha' + f_\alpha \tag{9.4.27}$$

$$\frac{d\mathcal{N}_2}{dt} = R_2 - (\Gamma_2 + \pi\,|\alpha'|^2)\mathcal{N}_2 + f_{\mathcal{N}_2} \tag{9.4.28}$$

$$\pi = \frac{2\mu^2}{\Gamma_{12}}, \tag{9.4.29}$$

where

$$f_\alpha(t) = g_\alpha(t) + \frac{\mu}{\Gamma_{12}}\, g_{\mathcal{M}}(t) \tag{9.4.30}$$

$$f_{\mathcal{N}_2}(t) = g_{\mathcal{N}_2}(t) - \frac{\mu}{\Gamma_{12}}\, [\alpha_c'^{*} g_{\mathcal{M}}(t) + \alpha_c' g_{\mathcal{M}^*}(t)]. \tag{9.4.31}$$

We next need the two time correlation functions of these forces. From (9.2.9), we have by inspection (neglecting the nonradiative transition probabilities $w_{12} = w_{21} = 0$)

$$\begin{aligned}\langle g_{\mathcal{N}_2}(t) g_{\mathcal{N}_2}(u)\rangle_R &= 2\langle \mathscr{D}_{\mathcal{N}_2\mathcal{N}_2}\rangle_R\, \delta(t-u)\\ &= \langle [R_2 + \Gamma_2\mathcal{N}_2 - B]\rangle_R \end{aligned}\tag{9.4.32}$$

$$\begin{aligned}\langle g_{\mathcal{M}^*}(t) g_{\mathcal{M}}(u)\rangle_R &= 2\langle \mathscr{D}_{\mathcal{M}^*\mathcal{M}}\rangle_R\, \delta(t-u)\\ &\cong \langle [R_2 + 2\Gamma_{12}\mathcal{N}_2]\rangle_R\, \delta(t-u)\end{aligned}\tag{9.4.33}$$

$$\begin{aligned}\langle g_{\mathcal{M}}(t) g_{\mathcal{M}}(u)\rangle_R &= 2\langle \mathscr{D}_{\mathcal{M}\mathcal{M}}\rangle_R\, \delta(t-u) e^{2i\omega_0 t}\\ &= \mu\langle \alpha'\mathcal{M}'\rangle_R\, \delta(t-u) = \langle g_{\mathcal{M}^*}(t) g_{\mathcal{M}^*}(u)\rangle_R^*\end{aligned}\tag{9.4.34}$$

$$\begin{aligned}\langle g_{\alpha^*}(t)\, g_\alpha(u)\rangle_R &= 2\langle D_{\alpha^*\alpha}\rangle_R\, \delta(t-u)\\ &= \gamma\bar{n}\, \delta(t-u)\end{aligned}\tag{9.4.35}$$

We have used (9.4.7) to simplify (9.4.34) where $\Gamma_{12} \gg \Gamma_2$ and let $\mathcal{N}_1 = 0$ All others vanish. The B is given by (9.4.6).

We next replace $\mathcal{M}'$ in the correlation functions above by its adiabatic value. Therefore, we have on using (9.4.6), (9.4.8), (9.4.18), (9.4.19), and (9.4.21)

$$\langle B\rangle_R = \pi\langle|\alpha'|^2 \mathscr{N}_2\rangle_R + \frac{\mu}{\Gamma_{12}}\langle[g_{\mathscr{M}}\alpha'^* + g_{\mathscr{M}^*}\alpha']\rangle_R$$

$$= \pi\langle|\alpha'|^2 \mathscr{N}_2\rangle_R + \frac{\pi}{\Gamma_{12}}\langle\mathscr{D}_{\mathscr{M}^*\mathscr{M}}\rangle_R$$

$$= \pi\langle|\alpha'|^2 \mathscr{N}_2\rangle_R + \frac{\pi}{2\Gamma_{12}}[R_2 + 2\Gamma_{12}\langle\mathscr{N}_2\rangle]. \qquad (9.4.36)$$

Since $\pi/\Gamma_{12} \ll 1$, (9.4.32) reduces to

$$\langle g_{\mathscr{N}_2}(t)g_{\mathscr{N}_2}(u)\rangle_R = \langle[R_2 + (\Gamma_2 - \pi\,|\alpha'|^2)\mathscr{N}_2]\rangle_R\,\delta(t-u). \qquad (9.4.37)$$

Equations (9.4.33) and (9.4.35) are unaltered, while (9.4.34) becomes

$$\langle g_{\mathscr{M}}(t)g_{\mathscr{M}}(u)\rangle_R \cong \frac{\pi}{2}\langle\mathscr{N}_2\alpha'^2\rangle_R\,\delta(t-u) \cong \langle g_{\mathscr{M}^*}(t)g_{\mathscr{M}^*}(u)\rangle_R^*. \qquad (9.4.38)$$

We may next use (9.4.37), (9.4.38), (9.4.33), and (9.4.35) to evaluate the two time correlation functions of f_α and $f_{\mathscr{N}_2}$ given by (9.4.30) and (9.4.31). We therefore have that

$$\langle f_\alpha(t)f_\alpha(u)\rangle_R = \frac{\pi}{2\Gamma_{12}}\langle g_{\mathscr{M}}(t)g_{\mathscr{M}}(u)\rangle_R = \frac{\pi^2}{4\Gamma_{12}^2}\langle\alpha'^2\mathscr{N}_2\rangle_R\,\delta(t-u)$$
$$= \langle f_{\alpha^*}(t)f_{\alpha^*}(u)\rangle_R^*. \qquad (9.4.39)$$

Also, we have

$$\langle f_{\alpha^*}(t)f_\alpha(u)\rangle_R = \langle g_{\alpha^*}(t)g_\alpha(u)\rangle_R + \frac{\pi}{2\Gamma_{12}}\langle g_{\mathscr{M}^*}(t)g_{\mathscr{M}}(u)\rangle_R$$

$$= \left\langle\left\{\gamma\bar{n} + \frac{\pi}{2\Gamma_{12}}[R_2 + 2\Gamma_{12}\mathscr{N}_2]\right\}\right\rangle_R\delta(t-u). \qquad (9.4.40)$$

Similarly, we have

$$\langle f_\alpha(t)f_{\mathscr{N}_2}(u)\rangle = -\frac{\pi}{2\Gamma_{12}}\left\langle\left\{R_2 + \mathscr{N}_2\left(2\Gamma_{12} + \frac{\pi}{2}|\alpha'|^2\right)\right\}\alpha'\right\rangle_R\delta(t-u)$$
$$= \langle f_{\alpha^*}(t)f_{\mathscr{N}_2}(u)\rangle_R^*, \qquad (9.4.41)$$

and

$$\langle f_{\mathscr{N}_2}(t)f_{\mathscr{N}_2}(u)\rangle_R$$
$$= \left\langle\left\{R_2\left[1 + \frac{\pi\,|\alpha'|^2}{\Gamma_{12}}\right] + \left[\Gamma_2 + \pi\,|\alpha'|^2 + \frac{\pi^2\,|\alpha'|^4}{2\Gamma_{12}}\right]\mathscr{N}_2\right\}\right\rangle_R\delta(t-u). \qquad (9.4.42)$$

We may use these correlation functions to write down a new Fokker–Planck equation which describes the field and population. See Problem 9.3.

We next assume we may neglect $d\mathcal{N}_2/dt$ compared with $\Gamma_2\mathcal{N}_2$ since $\Gamma_2 \gg \gamma$. From (9.4.28) we then obtain the adiabatic population difference

$$\mathcal{N}_2 \cong \frac{R_2 + f_{\mathcal{N}_2}}{\Gamma_2 + \pi\,|\alpha'|^2}. \tag{9.4.43}$$

When we put this in (9.4.27), we obtain

$$\frac{d\alpha'}{dt} = \frac{1}{2}\left[\frac{\pi R_2}{\Gamma_2 + \pi\,|\alpha'|^2} - \gamma\right]\alpha' + f_\alpha + \frac{\pi f_{\mathcal{N}_2}\alpha'}{2(\Gamma_2 + \pi\,|\alpha'|^2)}. \tag{9.4.44}$$

We leave it as an exercise for the reader to show that the reservoir average of the last term in (9.4.44) is negligible compared with the drift term [5] (see Problem 9.4). The Langevin equation for the field alone may therefore be written as

$$\frac{d\alpha'}{dt} = \mathscr{A}_\alpha + \mathscr{G}_\alpha \tag{9.4.45}$$

where

$$\begin{aligned} \mathscr{A}_\alpha &= \frac{1}{2}\left[\frac{\pi R_2}{\Gamma_2 + \pi\,|\alpha'|^2} - \gamma\right]\alpha' \\ \mathscr{G}_\alpha(t) &= \frac{\pi}{2}\,\frac{\alpha_c'}{\Gamma_2 + \pi\,|\alpha_c'|^2}\,f_{\mathcal{N}_2}(t) + f_\alpha(t). \end{aligned} \tag{9.4.46}$$

The two-time correlation functions are found to be

$$\begin{aligned} \langle \mathscr{G}_\alpha(t)\mathscr{G}_\alpha(u)\rangle_R &= \frac{\pi^2 R_2}{4\Gamma_{12}}\,\delta(t-u) \\ &\quad\times \left\langle \frac{\alpha'^2}{(\Gamma_2 + \pi\,|\alpha'|^2)}\left\{1 + \frac{2\Gamma_{12}}{(\Gamma_2 + \pi\,|\alpha'|^2)} - \frac{\pi^2\,|\alpha'|^4}{2(\Gamma_2 + \pi\,|\alpha'|^2)^2}\right\}\right\rangle \\ &\equiv 2\langle D_{\alpha\alpha}{}^F\rangle\,\delta(t-u) \end{aligned} \tag{9.4.47}$$

$$\begin{aligned} \langle \mathscr{G}_{\alpha^*}(t)\mathscr{G}_\alpha(u)\rangle_R &= \delta(t-u)\left\langle\left\{\gamma\bar{n} + \frac{\pi R_2}{(\Gamma_2 + \pi\,|\alpha'|^2)}\left[1 - \frac{\pi\,|\alpha'|^2}{2(\Gamma_2 + \pi\,|\alpha'|^2)}\right]\right\}\right\rangle \\ &\equiv 2\langle D_{\alpha^*\alpha}{}^F\rangle\,\delta(t-u). \end{aligned} \tag{9.4.48}$$

In the limit in which $2\Gamma_{12} \approx \Gamma_1 \to \infty$, we see that

$$\begin{aligned} \langle \mathscr{G}_\alpha(t)\mathscr{G}_\alpha(u)\rangle_R &\to -\frac{\pi^2 R_2}{2}\left\langle\frac{\alpha'^2}{(\Gamma_2 + \pi\,|\alpha'|^2)^2}\right\rangle\delta(t-u) \\ &= 2\langle D_{\alpha\alpha}{}^F\rangle\,\delta(t-u). \end{aligned} \tag{9.4.49}$$

We may also obtain the Langevin equation for the photon number. We have by (9.4.45) and (9.4.46)

$$\frac{d}{dt}\alpha'^*\alpha' = \alpha'^*(\mathscr{A}_\alpha + \mathscr{G}_\alpha) + \alpha'(\mathscr{A}_\alpha^* + \mathscr{G}_\alpha^*)$$
$$= \left[\frac{\pi R_2}{\Gamma_2 + \pi|\alpha'|^2} - \gamma\right]|\alpha'|^2 + \alpha'^*\mathscr{G}_\alpha + \alpha'\mathscr{G}_\alpha^*. \qquad (9.4.50)$$

The reservoir average of the $\mathscr{G}_\alpha$ and $\mathscr{G}_\alpha^*$ terms are easily seen to be

$$\langle\alpha'^*\mathscr{G}_\alpha + \alpha'\mathscr{G}_\alpha^*\rangle_R = \gamma\bar{n} + \frac{\pi R_2}{(\Gamma_2 + \pi|\alpha'|^2)}\left[1 - \frac{\pi|\alpha'|^2}{2(\Gamma_2 + \pi|\alpha'|^2)}\right]. \qquad (9.4.51)$$

Therefore, we rewrite (9.4.50) as

$$\frac{d}{dt}|\alpha'|^2 = \gamma(\bar{n} - |\alpha'|^2) + \frac{\pi R_2(|\alpha'|^2 + 1)}{\Gamma_2 + \pi|\alpha'|^2} - \frac{\pi R_2 \pi|\alpha'|^2}{2(\Gamma_2 + \pi|\alpha'|^2)^2} + \mathscr{G}_{\alpha^*\alpha}, \qquad (9.4.52)$$

where $\mathscr{G}_{\alpha^*\alpha}$ is a new Langevin force with zero reservoir average. The ratio of the second to first term in πR_2 is

$$\rho_{21} \equiv \frac{\pi|\alpha'|^2}{2(|\alpha'|^2 + 1)(\Gamma_2 + \pi|\alpha'|^2)}.$$

At very low operating levels

$$|\alpha'|^2 \ll \frac{\Gamma_2}{\pi} \sim 10^8$$

$$\rho_{21} \sim \frac{\pi}{\Gamma_2}\frac{|\alpha'|^2}{|\alpha'|^2 + 1} \ll 1.$$

At high operating levels

$$|\alpha'|^2 \gg \frac{\Gamma_2}{\pi}$$

$$\rho_{21} \cong \frac{1}{2|\alpha'|^2} \ll 1.$$

At intermediate levels

$$\pi|\alpha'|^2 \sim \Gamma_2$$

$$\rho_{21} \sim \frac{\pi}{2\Gamma_2} \ll 1.$$

We therefore may always neglect the second term in R_2 so that

$$\frac{d}{dt}|\alpha'|^2 = \gamma(\bar{n} - |\alpha'|^2) + \frac{\pi R_2(|\alpha'|^2 + 1)}{\Gamma_2 + \pi|\alpha'|^2} + \mathscr{G}_{\alpha^*\alpha}$$
$$\equiv \mathscr{A}_{\alpha^*\alpha} + \mathscr{G}_{\alpha^*\alpha}. \qquad (9.4.53)$$

If we neglect fluctuations, we may set $\mathscr{A}_{\alpha^*\alpha} = 0$ to obtain the steady-state operating intensity:

$$|\alpha'|_0^2 \equiv I_0 = \frac{1}{2}\left(\frac{R_2}{\gamma} - \frac{\Gamma_2}{\pi} + \bar{n}\right) + \sqrt{\frac{1}{4}\left(\frac{R_2}{\gamma} - \frac{\Gamma_2}{\pi} + \bar{n}\right)^2 + \bar{n}\frac{\Gamma_2}{\pi} + \frac{R_2}{\gamma}}. \tag{9.4.54}$$

We may use the drift vector (9.4.46) and the diffusion coefficients (9.4.48) and (9.4.49) to write the Fokker–Planck equation

$$\frac{\partial P_{(a)}}{\partial t} = -\frac{\partial}{\partial \alpha'}\mathscr{A}_\alpha P_{(a)} - \frac{\partial}{\partial \alpha'^*}\mathscr{A}_\alpha^* P_{(a)}$$
$$+ \frac{\partial^2}{\partial \alpha'^2} D_{\alpha\alpha}{}^F P_{(a)} + \frac{\partial^2}{\partial \alpha'^{*2}} D^F_{\alpha^*\alpha^*} P_{(a)} + \frac{\partial^2}{\partial \alpha'\, \partial \alpha'^*} 2D^F_{\alpha^*\alpha} P_{(a)}, \tag{9.4.55}$$

which describes the statistical properties of the radiation field of a single mode homogeneously broadened laser when the atomic linewidths are large compared with the cavity linewidth.

9.5 THE LASER AS A ROTATING WAVE VAN DER POL OSCILLATOR

Consider the Langevin equation for the field (9.4.45)

$$\frac{d\alpha'}{dt} = \mathscr{A}_\alpha + \mathscr{G}_\alpha \tag{9.5.1}$$

where

$$\mathscr{A}_\alpha = \frac{1}{2}\left[\frac{\pi R_2}{\Gamma_2 + \pi I} - \gamma\right]\alpha', \tag{9.5.2}$$

where we have let

$$I = |\alpha'|^2. \tag{9.5.3}$$

Let us linearize this equation about the operating intensity given in (9.4.54). That is, we write

$$\Gamma_2 + \pi I \equiv \Gamma_2 + \pi I_0 + \pi(I - I_0) \tag{9.5.4}$$

and

$$\mathscr{A}_\alpha \simeq \frac{1}{2}\left\{\frac{\pi R_2}{\Gamma_2 + \pi I_0}\left[1 + \frac{\pi I_0}{\Gamma_2 + \pi I_0}\right] - \gamma - \frac{\pi^2 R_2 I}{(\Gamma_2 + \pi I_0)^2}\right\}\alpha'. \tag{9.5.5}$$

If we let

$$\Pi = \frac{\pi R_2}{\gamma(\Gamma_2 + \pi I_0)}\left[1 + \frac{\pi I_0}{\Gamma_2 + \pi I_0}\right] - 1$$
$$S = \frac{\pi^2 R_2}{\gamma(\Gamma_2 + \pi I_0)^2}, \tag{9.5.6}$$

then

$$\mathscr{A}_\alpha = \frac{\gamma}{2}\{\Pi - SI\}\alpha'. \tag{9.5.7}$$

This expansion is very good in the range of operating intensities

$$1 \leq I \leq \frac{\Gamma_2}{\pi}, \tag{9.5.8}$$

which includes the threshold region and far above since

$$I_{th} \sim \sqrt{\frac{\Gamma_2}{\pi}} \sim 10^4, \tag{9.5.9}$$

as we shall show. In this range we see from (9.4.48) that the diffusion coefficients reduce to

$$2\langle D^F_{\alpha^*\alpha}\rangle \cong \gamma\bar{n} + \frac{\pi R_2}{\Gamma_2}$$
$$2\langle D^F_{\alpha\alpha}\rangle \cong 0 \cong 2\langle D^F_{\alpha\alpha}\rangle^*. \tag{9.5.10}$$

Under this approximation, (9.5.1) is seen to be the equation for a rotating wave van der Pol oscillator.

We may express the Langevin equation (9.5.1) in terms of dimensionless (scaled) variables as follows. Let

$$t = T\tau \qquad \alpha' = \xi\beta. \tag{9.5.11}$$

Then (9.5.1) becomes on using (9.5.7)

$$\frac{d\beta}{d\tau} = \frac{\gamma T}{2}\{\Pi - S\xi^2|\beta|^2\}\beta + F_\beta, \tag{9.5.12}$$

where

$$F_\beta = \frac{T}{\xi}\mathscr{G}_\alpha. \tag{9.5.13}$$

Also

$$\langle \mathscr{G}^*_\alpha(t)\mathscr{G}_\alpha(u)\rangle = \frac{\xi^2}{T^2}\langle F^*_\beta(t)F_\beta(u)\rangle = 2\langle \mathscr{D}^F_{\alpha^*\alpha}\rangle\,\delta(t-u)\rangle$$

$$= \left[\gamma\bar{n} + \frac{\pi R_2}{\Gamma_2}\right]\delta(t-u). \tag{9.5.14}$$

We may now choose ξ and T as follows. Define a pump parameter g as

$$g = \frac{\gamma}{2}T\Pi = T\left\{\frac{\pi R_2}{2\Gamma_2} - \frac{\gamma}{2}\right\}. \tag{9.5.15}$$

Here we neglected πI compared with Γ_2 as well as π-compared with Γ_2. At threshold $g = 0$ while $g > 0$ above threshold and $g < 0$ below threshold.

We next require that the coefficient of $|\beta|^2$ be unity in (9.5.11) as one constraint to determine ξ and T. Thus

$$\frac{\gamma TS}{2}\xi^2 = 1 \tag{9.5.16}$$

so that (9.5.12) becomes

$$\frac{d\beta}{d\tau} = (g - |\beta|^2)\beta + F_\beta. \tag{9.5.17}$$

By (9.5.14) we have that

$$\langle F_{\beta^*}(t_1)F_\beta(t_2)\rangle = \frac{T^2}{\xi^2}\left[\gamma\bar{n} + \frac{\pi R_2}{\Gamma_2}\right]\delta(t_1 - t_2). \tag{9.5.18}$$

But since

$$\delta(ax) = \frac{1}{|a|}\delta(x), \tag{9.5.19}$$

we have

$$\langle F_{\beta^*}(t_1)F_\beta(t_2)\rangle = \frac{T}{\xi^2}\left[\gamma\bar{n} + \frac{\pi R_2}{\Gamma_2}\right]\delta(\tau_1 - \tau_2), \tag{9.5.20}$$

since $t = T\tau$. We require this to be

$$\langle F_{\beta*}(t_1)F_\beta(t_2)\rangle = 4\delta(\tau_1 - \tau_2) \tag{9.5.21}$$

as our second constraint. From these two constraints, we easily see that the intensity scaling is

$$\xi^2 = \left\{\frac{\bar{n} + (\pi R_2/\gamma\Gamma_2)}{(2\pi/\Gamma_2)(\pi R_2/\gamma\Gamma_2)}\right\}^{1/2}, \tag{9.5.22}$$

while the time scaling is

$$\gamma T = \left\{\frac{\pi}{8\Gamma_2}\left(\frac{\pi R_2}{\gamma\Gamma_2}\right)\left[\bar{n} + \frac{\pi R_2}{\gamma\Gamma_2}\right]\right\}^{-1/2}. \tag{9.5.23}$$

At threshold, $p = 0$, so that

$$(R_2)_{th} = \frac{\gamma}{\pi}\Gamma_2. \tag{9.5.24}$$

Accordingly, we have

$$\xi^2 = \left\{\frac{\bar{n} + [R_2/(R_2)_{th}]}{(2\pi/\Gamma_2)[R_2/(R_2)_{th}]}\right\}^{1/2}. \tag{9.5.25}$$

If we neglect $\bar{n}$, we see that at threshold

$$\xi_{th}^2 = \left\{\frac{\Gamma_2}{2\pi}\right\}^{1/2} \approx 10^4, \tag{9.5.26}$$

which is the threshold intensity.

Similarly, we see that at threshold

$$\gamma T_{th} \sim \left\{\frac{8\Gamma_2}{\pi}\right\}^{1/2}, \tag{9.5.27}$$

or

$$T_{th} \cong \frac{1}{\gamma} 10^4. \tag{9.5.28}$$

We may therefore write g as

$$g = \frac{\sqrt{2}\,(R_2/R_{2th}) - 1}{\{(\pi/\Gamma_2)(R_2/R_{2th})[\bar{n} + (R_2/R_{2th})]\}^{1/2}}. \tag{9.5.29}$$

We therefore conclude that the laser is equivalent to a rotating wave van der Pol oscillator over a wide range of operating intensities.

Under the approximation that $1 \leq I \leq \Gamma_2/\pi$, the Fokker–Planck equation (9.4.55) becomes

$$\frac{\partial P}{\partial t} = -\frac{\partial}{\partial \alpha'}\frac{\gamma}{2}[\Pi - SI]\alpha' P - \frac{\partial}{\partial \alpha'^*}\frac{\gamma}{2}[\Pi - SI]\alpha'^* P + \frac{\partial^2}{\partial \alpha'\,\partial \alpha'^*}\left(\gamma\bar{n} + \frac{\pi R_2}{\gamma}\right)P. \tag{9.5.30}$$

When we use our scaled variables, this becomes

$$\frac{\partial P}{\partial \tau} = -\frac{\partial}{\partial \beta}[g - |\beta|^2]\beta P - \frac{\partial}{\partial \beta^*}[g - |\beta|^2]\beta^* P + \frac{\partial^2}{\partial \beta\,\partial \beta^*}4P, \tag{9.5.31}$$

which is the form we have already studied in great detail in Chapter 6.

9.6 PHASE AND AMPLITUDE FLUCTUATIONS, STEADY-STATE SOLUTION, LASER LINEWIDTH

Since we are interested in phase and amplitude fluctuations let us transform the Fokker–Planck equation (9.4.45) into these variables. We let

$$\begin{aligned} \alpha' &= I^{1/2}e^{-i\varphi} \\ \alpha'^* &= I^{1/2}e^{i\varphi}, \end{aligned} \tag{9.6.1}$$

so that

$$I = \alpha'\alpha'^*; \qquad \varphi = \frac{1}{2i}\ln\frac{\alpha'^*}{\alpha'}. \tag{9.6.2}$$

By the usual rules of differentiation, we see that

$$d^2\alpha' = \tfrac{1}{2} dI\, d\varphi$$

$$\frac{\partial}{\partial \alpha'}\alpha' = \frac{\partial}{\partial I} I + \frac{i}{2}\frac{\partial}{\partial \varphi} = \left(\frac{\partial}{\partial \alpha'^*}\alpha'^*\right)^*$$

$$\frac{\partial^2}{\partial \alpha'\,\partial \alpha'^*} = \frac{\partial^2}{\partial I^2} I - \frac{\partial}{\partial I} + \frac{\partial^2}{\partial \varphi^2}\frac{1}{4I} \tag{9.6.3}$$

$$\frac{\partial^2}{\partial \alpha'^2} = \frac{\partial^2}{\partial I^2} I^2 - \frac{1}{4}\frac{\partial^2}{\partial \varphi^2} + \frac{i}{2}\frac{\partial}{\partial \varphi} + i\frac{\partial^2}{\partial \varphi\,\partial I} I$$

$$= \left(\frac{\partial^2}{\partial \alpha'^{*2}}\right)^*.$$

After minor algebra we see that (9.4.55) becomes

$$\frac{\partial P_1}{\partial t} = -\frac{\partial}{\partial I} A_I P_1 + \frac{\partial^2}{\partial I^2} D_{II} P_1 + \frac{\partial^2}{\partial \varphi^2} D_{\varphi\varphi} P_1, \tag{9.6.4}$$

where we used (9.4.46), (9.4.48), and (9.4.49), and we have let

$$A_I = \frac{\pi R_2}{\Gamma_2 + \pi I}\left\{I + 1 - \frac{\pi I}{2(\Gamma_2 + \pi I)}\right\} + \gamma(\bar{n} - I)$$

$$= I\left\{\gamma \bar{n} + \frac{\pi R_2 \Gamma_2}{(\Gamma_2 + \pi I)^2}\right\} \tag{9.6.5}$$

$$D_{\varphi\varphi} = \frac{1}{4I}\left\{\gamma \bar{n} + \frac{\pi R_2}{(\Gamma_2 + \pi I)}\right\}.$$

Also we have let

$$P(\alpha', \alpha'^*, t) = 2P_1(I, \varphi, t), \tag{9.6.6}$$

so that

$$\int P(\alpha', \alpha'^*, t)\, d^2\alpha' = \int P_1(I, \varphi, t)\, dI\, d\varphi = 1. \tag{9.6.7}$$

The corresponding Langevin equations are

$$\frac{dI}{dt} = A_I + F_I \tag{9.6.8}$$

$$\frac{d\varphi}{dt} = F_\varphi \tag{9.6.9}$$

where the first moments are

$$\begin{aligned} \langle F_I(s)F_I(u)\rangle &= 2D_{II}\,\delta(s-u) \\ \langle F_\varphi(s)F_\varphi(u)\rangle &= 2D_{\varphi\varphi}\,\delta(s-u) \\ \langle F_I(s)F_\varphi(u)\rangle &= 0. \end{aligned} \tag{9.6.10}$$

We should note that only at extremely high operating levels do we need retain the diffusion correction to A_I in (9.6.8). This follows, since if $I \gg 1$

$$I + 1 - \frac{\pi I}{\Gamma_2 + \pi I} \cong I\left[1 - \frac{\pi}{\Gamma_2 + \pi I}\right] = \frac{I[\Gamma_2 + \pi I - \pi]}{(\Gamma_2 + \pi I)} \cong I$$

since $\Gamma_2 \gg \pi$. Therefore,

$$A_I \cong \left(\frac{\pi R_2}{\Gamma_2 + \pi I} - \gamma\right) I + \gamma \bar{n} \qquad I \gg 1. \tag{9.6.11}$$

Because A_I and the diffusion coefficients depend on I only, the Fokker–Planck equation (9.6.4) is separable, and we may look for solutions of the form

$$P_1(I, \varphi, t) = e^{-\lambda t} e^{il\varphi} Q(I), \tag{9.6.12}$$

where l is an integer or zero in order that P_1 be single-valued. Then (9.6.4) becomes

$$\frac{d}{dI}\left\{\frac{d}{dI}(D_{II}Q) - A_I Q\right\} + (\lambda - l^2 D_{\varphi\varphi})Q = 0, \tag{9.6.13}$$

which is an eigenvalue equation. The steady-state solutions correspond to $\lambda = 0$. The steady-state solution which is independent of φ ($l = 0$) is given by

$$\frac{d}{dI}(D_{II}Q_S) - A_I Q_S = 0, \tag{9.6.14}$$

or

$$Q_S = e^U, \tag{9.6.15}$$

where

$$U = \int \frac{A_I - (dD_{II}/dI)}{D_{II}}\, dI. \tag{9.6.16}$$

As long as $I \gg 1$, the reader may easily show that

$$\frac{dD_{II}}{dI} \ll A_I. \tag{9.6.17}$$

Therefore, if we let $\bar{n} = 0$ for simplicity, we see after minor algebra that

$$\frac{dU}{dI} \cong \frac{A_I}{D_{II}} = \frac{R_{2th}}{R_2}\left(\frac{R_2}{R_{2th}} - 1 - \frac{\pi I}{\Gamma_2}\right)\left(1 + \frac{\pi I}{\Gamma_2}\right) \tag{9.6.18}$$

We used (9.5.24). In the vicinity of threshold, we may neglect the quadratic term in $\pi I/\Gamma_2$ so that the steady-state solution becomes

$$Q_S = \exp \frac{R_{2th}}{R_2}\left\{\left(\frac{R_2}{R_{2th}} - 1\right)I + \left(\frac{R_2}{R_{2th}} - 2\right)\frac{\pi I^2}{2\Gamma_2}\right\}, \tag{9.6.19}$$

which is a gaussian for

$$0 < R_2 < 2R_{2th}. \tag{9.6.20}$$

As a next application let us obtain the laser linewidth due to phase fluctuations above threshold where the amplitude is stabilized sufficiently that we may adiabatically eliminate the intensity.

From (9.6.5) and (9.6.10) we have that

$$\begin{aligned}\langle F_\varphi(t)F_\varphi(u)\rangle &= 2D_{\varphi\varphi}\,\delta(t-u)\\ &= \frac{1}{4I}\{\gamma\bar{n} + \pi\langle N_2\rangle\},\end{aligned} \tag{9.6.21}$$

where the mean population difference is

$$\langle N_2\rangle = \frac{R_2}{\Gamma_2 + \pi I} \tag{9.6.22}$$

We used (9.4.43).

In the region for above threshold (a factor of 10 in photon number), the fluctuations in I become relatively small. This stabilization of the oscillator permits us to replace I in (9.6.21) by its operating value given by

$$I_0 = \frac{R_2}{\gamma} - \frac{\Gamma_2}{\pi} \tag{9.6.23}$$

(compare 9.4.54). Since I is stable, the Fokker–Planck equation (9.6.4) reduces to

$$\frac{\partial P_1}{\partial t} = D_{\varphi\varphi}\frac{\partial^2 P_1}{\partial \varphi^2}, \tag{9.6.24}$$

which is a diffusion equation which has the Green's function solution (as we have seen earlier)

$$P_1(\varphi, t/\varphi_0, 0) = \frac{1}{2\pi}\sum_{-\infty}^{\infty} e^{in(\varphi-\varphi_0)}e^{-n^2 D_{\varphi\varphi}t}. \tag{9.6.25}$$

The steady-state distribution is

$$P_{SS}(\varphi_0) = \frac{1}{2\pi}. \tag{9.6.26}$$

The laser spectrum when amplitude fluctuations are suppressed is given by the Fourier transform of

$$\langle a^\dagger(t)a(0)\rangle = I_0\langle e^{i[\varphi(t)-\varphi_0]}\rangle. \tag{9.6.27}$$

But by (9.6.25) and (9.6.26) we have

$$\begin{aligned}\langle e^{i(\varphi-\varphi_0)}\rangle &= \int P_{SS}(\varphi_0)\, d\varphi_0 \int e^{i(\varphi-\varphi_0)} P(\varphi, t/\varphi_0, 0)\, d\varphi \\ &= \int_0^{2\pi} \frac{1}{2\pi}\, d\varphi_0 \int_0^{2\pi} \frac{1}{2\pi} \sum_{-\infty}^{\infty} e^{i(n+1)(\varphi-\varphi_0)-n^2 D_{\varphi\varphi} t}\, d\varphi \\ &= e^{-D_{\varphi\varphi}|t|}. \end{aligned} \tag{9.6.28}$$

The spectrum is

$$\int_{-\infty}^{\infty} e^{i\omega t} \exp\left[-D_{\varphi\varphi}|t|\right] dt = \frac{2D_{\varphi\varphi}}{\omega^2 + D_{\varphi\varphi}^{\ 2}}, \tag{9.6.29}$$

which is Lorentzian whose full width at half power is

$$2D_{\varphi\varphi} = \frac{\gamma\bar{n} + (\pi R_2/\Gamma_2 + \pi I_0)}{4I_0}, \tag{9.6.30}$$

and is due to phase fluctuations at this operating level.

Lax and Zwanziger [6] have recently applied these techniques to study laser photon statistics near threshold. Stephens [7] has also used these techniques to study the noise properties of Josephson junction oscillators.

REFERENCES

[1] M. Lax, *Phys. Rev.*, **145,** 110 (1966).
[2] M. Lax and W. H. Louisell, *IEEE J. Quantum Electron.*, **QE-3,** 47 (1967).
[3] M. Lax, *Phys. Rev.*, **172,** 350 (1968).
[4] M. Lax, *Phys. Rev.*, **172,** 362 (1968).
[5] M. Lax and W. H. Louisell, *Phys. Rev.*, **185,** 568 (1969).
[6] M. Lax and M. Zwanziger, presented at VII International Quantum Electronics Conference, Montreal, 1972, and to be published.
[7] M. J. Stephens, *Phys. Rev.*, **182,** 531 (1969); M. O. Scully and P. A. Lee, *Ann. N.Y. Acad. Sci.*, **168,** 387 (1970).

PROBLEMS

9.1 Derive (9.2.6).

9.2 Give a physical interpretation to each term in (9.4.27) and (9.4.28).

9.3 From the two-time correlation functions (9.4.39)–(9.4.42) and the drift vectors given by (9.4.27) and (9.4.28) write out the new Fokker–Planck equation in the variables α', α'^*, and $\mathcal{N}_2$. The new distribution function depends only on these variables. How is it related to the old distribution function of (9.2.29)?

9.4 Evaluate the reservoir average of the last term in (9.4.44) and show that it is negligible compared with the drift term.

9.5 Derive (9.4.51).

9.6 Calculate the mean number of photons in steady state using (9.6.19).

9.7 Justify (9.6.27).

9.8 Calculate the laser spectrum far below threshold.

Appendix A

Method of Characteristics

Consider the two equations

$$\frac{dx}{P} = \frac{dy}{Q} = \frac{dz}{R}, \tag{A.1}$$

where P, Q, and R are functions of x, y, z. Assume we can find a solution of these equations of the form

$$u(x, y, z) = a, \tag{A.2}$$

where a is a constant. Then, we have that

$$\begin{aligned} 0 = du &= \frac{\partial u}{\partial x}\,dx + \frac{\partial u}{\partial y}\,dy + \frac{\partial u}{\partial z}\,dz \\ &= \left[\frac{\partial u}{\partial x} + \frac{\partial u}{\partial y}\frac{dy}{dx} + \frac{\partial u}{\partial z}\frac{dz}{dx}\right] dx \\ &= \left[\frac{\partial u}{\partial x} + \frac{\partial u}{\partial y}\frac{Q}{P} + \frac{\partial u}{\partial z}\frac{R}{P}\right] dx \\ &= \left[P\frac{\partial u}{\partial x} + Q\frac{\partial u}{\partial y} + R\frac{\partial u}{\partial z}\right]\frac{dx}{P}, \end{aligned} \tag{A.3}$$

where we used (A.1). From (A.3) it therefore follows that u satisfies the partial differential equation

$$P\frac{\partial u}{\partial x} + Q\frac{\partial u}{\partial y} + R\frac{\partial u}{\partial z} = 0, \tag{A.4}$$

as well as the system (A.1).

From (A.2) we also have when we assume x and y are independent variables

$$\begin{aligned} \frac{\partial u}{\partial x} + \frac{\partial u}{\partial z}\frac{\partial z}{\partial x} &= 0 \\ \frac{\partial u}{\partial y} + \frac{\partial u}{\partial z}\frac{\partial z}{\partial y} &= 0. \end{aligned} \tag{A.5}$$

When we solve for $\partial u/\partial x$ and $\partial u/\partial y$ and substitute into (A.4), we have

$$\left[-P\frac{\partial z}{\partial x} - Q\frac{\partial z}{\partial y} + R\right]\frac{\partial u}{\partial z} = 0 \tag{A.6}$$

or

$$P\frac{\partial z}{\partial x} + Q\frac{\partial z}{\partial y} = R. \tag{A.7}$$

That is, $u(x, y, z) = a$ also solves (A.7). Therefore, to obtain solutions for (A.4) or (A.7), we need only find a solution of (A.1). If l, m, and n are any functions of x, y, z, we also easily see that

$$\begin{aligned} l\,dx + m\,dy + n\,dz &= \left[l + m\frac{dy}{dx} + n\frac{dz}{dx}\right]dx \\ &= \left[l + m\frac{Q}{P} + n\frac{R}{P}\right]dx \\ &= (lP + mQ + nR)\frac{dx}{P}, \end{aligned} \tag{A.8}$$

where we used (A.1). Therefore, we see that

$$\frac{dx}{P} = \frac{dy}{Q} = \frac{dz}{R} = \frac{l\,dx + m\,dy + n\,dz}{lP + mQ + nR}. \tag{A.9}$$

If we can solve the original set, we may be able to find functions l, m, and n which we may solve. In fact if we can find l, m, n such that

$$l\,dx + m\,dy + n\,dz = 0, \tag{A.10}$$

then this is an exact differential and may be integrated so that

$$du = 0 = l\,dx + m\,dy + n\,dz, \tag{A.11}$$

and it follows from (A.8) that

$$lP + mQ + nR = 0 \tag{A.12}$$

also. But this is just (A.4), since

$$\frac{\partial u}{\partial x} = l; \qquad \frac{\partial u}{\partial y} = m; \qquad \frac{\partial u}{\partial z} = n \tag{A.13}$$

if $l\,dx + m\,dy + n\,dz = du$ is exact.

It is straightforward to generalize these results to more variables: solutions of the set

$$\frac{dx_1}{P_1} = \frac{dx_2}{P_2} = \dots = \frac{dx_n}{P_n} \tag{A.14}$$

also satisfy

$$\sum_{i=1}^{n} P_i \frac{\partial u}{\partial x_i} = 0. \tag{A.15}$$

Equation (A.1) are the so-called characteristic equations associated with (A.4) or (A.7). We next show that if $u(x, y, z) = a$ and $v(x, y, z) = b$ are two independent solutions of the characteristic equations [which also satisfy (A.4) or (A.7)], then any arbitrary function

$$\phi(u, v) = 0 \tag{A.16}$$

or equivalently,

$$u = g(v) \tag{A.17}$$

also satisfies (A.4) or (A.7). This follows, since we have

$$\frac{\partial \varphi}{\partial x} = \frac{\partial \varphi}{\partial u}\left[\frac{\partial u}{\partial x} + \frac{\partial u}{\partial z}\frac{\partial z}{\partial x}\right] + \frac{\partial \varphi}{\partial v}\left[\frac{\partial v}{\partial x} + \frac{\partial v}{\partial z}\frac{\partial z}{\partial x}\right] = 0 \tag{A.18}$$

$$\frac{\partial \varphi}{\partial y} = \frac{\partial \varphi}{\partial u}\left[\frac{\partial u}{\partial y} + \frac{\partial u}{\partial z}\frac{\partial z}{\partial y}\right] + \frac{\partial \varphi}{\partial v}\left[\frac{\partial v}{\partial x} + \frac{\partial v}{\partial z}\frac{\partial z}{\partial y}\right] = 0. \tag{A.19}$$

However, since u and v both satisfy (A.5), both relations are automatically satisfied identically.

Appendix B

Hamiltonian for Radiation Field in Plane-Wave Representation

We derive the hamiltonian (4.3.45) for the energy in the radiation field in a cubical cavity in the plane-wave representation.

We substitute for **E** and **H** from (4.3.44) into (4.3.45) and obtain, after minor algebra,

$$H = -\frac{\hbar}{4\tau}\sum_{l,\sigma}\sum_{l',\sigma'}\sqrt{\omega_l\omega_{l'}}\,[\hat{\mathbf{e}}_{l\sigma}\cdot\hat{\mathbf{e}}_{l'\sigma'} + (\hat{\mathbf{e}}_{l\sigma}\times\hat{\mathbf{k}}_l)\cdot(\hat{\mathbf{e}}_{l'\sigma'}\times\hat{\mathbf{k}}_{l'})]$$

$$\times\int_{\text{cavity}} d\tau\,[a_{l\sigma}(t)\exp(i\mathbf{k}_l\cdot\mathbf{r}) - a_{l\sigma}^{\dagger}(t)\exp(-i\mathbf{k}_l\cdot\mathbf{r})]$$

$$\times\,[a_{l'\sigma'}(t)\exp(i\mathbf{k}_{l'}\cdot\mathbf{r}) - a_{l'\sigma'}^{\dagger}(\mathbf{t})\exp(-i\mathbf{k}_{l'}\cdot\mathbf{r})]. \quad \text{(B.1)}$$

Since, by (4.3.36),

$$\mathbf{k}_{\pm l} = \pm\frac{2\pi}{L}(l_1\hat{\mathbf{i}} + l_2\hat{\mathbf{j}} + l_3\hat{\mathbf{k}}), \quad \text{(B.2)}$$

we see that

$$\frac{1}{\tau}\int_{\text{cavity}} d\tau\,\exp[\pm i(\mathbf{k}_l + \mathbf{k}_{l'})\cdot\mathbf{r}] = \delta_{l',-l}$$

$$\frac{1}{\tau}\int_{\text{cavity}} d\tau\,\exp[\pm i(\mathbf{k}_l - \mathbf{k}_{l'})\cdot\mathbf{r}] = \delta_{l',l}. \quad \text{(B.3)}$$

If we substitute these into (B.1) and carry out the sum over l', we have (since $\omega_l \equiv \omega_{-l}$)

$$H = \sum_{l,\sigma,\sigma'}\frac{\hbar\omega_l}{4}\{(a_{l\sigma}a_{l\sigma}^{\dagger} + a_{l\sigma'}^{\dagger}a_{l\sigma'})[(\hat{\mathbf{e}}_{l\sigma}\cdot\hat{\mathbf{e}}_{l\sigma'}) + (\hat{\mathbf{e}}_{l\sigma}\times\hat{\mathbf{k}}_l)\cdot(\hat{\mathbf{e}}_{l\sigma'}\times\hat{\mathbf{k}}_l)]$$

$$-\,(a_{l\sigma}a_{-l\sigma'} + a_{l\sigma}^{\dagger}a_{-l\sigma'}^{\dagger})[(\hat{\mathbf{e}}_{l\sigma}\cdot\hat{\mathbf{e}}_{-l\sigma'}) + (\hat{\mathbf{e}}_{l\sigma}\times\hat{\mathbf{k}}_l)\cdot(\hat{\mathbf{e}}_{-l\sigma'}\times\hat{\mathbf{k}}_{-l})]\}. \quad \text{(B.4)}$$

By the ordinary rules of vector analysis,

$$
\begin{aligned}
(\hat{\mathbf{e}}_{l\sigma} \times \hat{\mathbf{k}}_l) \cdot (\hat{\mathbf{e}}_{l\sigma'} \times \hat{\mathbf{k}}_l) &= [\hat{\mathbf{e}}_{l\sigma} \times (\hat{\mathbf{k}}_l \times \hat{\mathbf{e}}_{l\sigma'})] \cdot \hat{\mathbf{k}}_l \\
&= [(\hat{\mathbf{e}}_{l\sigma} \cdot \hat{\mathbf{e}}_{l\sigma'})\hat{\mathbf{k}}_l - (\hat{\mathbf{e}}_{l\sigma} \cdot \hat{\mathbf{k}}_l)\hat{\mathbf{e}}_{l\sigma'}] \cdot \hat{\mathbf{k}}_l \\
&= \hat{\mathbf{e}}_{l\sigma} \cdot \hat{\mathbf{e}}_{l\sigma'} = \delta_{\sigma\sigma'}, \qquad \text{(B.5)}
\end{aligned}
$$

since $\hat{\mathbf{e}}_{l\sigma} \cdot \hat{\mathbf{k}}_l = 0$ and $\hat{\mathbf{k}}_l \cdot \hat{\mathbf{k}}_l = 1$. We also have by the same argument

$$(\hat{\mathbf{e}}_{l\sigma} \times \hat{\mathbf{k}}_l) \cdot (\hat{\mathbf{e}}_{-l\sigma'} \times \hat{\mathbf{k}}_{-l}) = (\hat{\mathbf{e}}_{l\sigma} \cdot \hat{\mathbf{e}}_{-l\sigma'})\hat{\mathbf{k}}_l \cdot \hat{\mathbf{k}}_{-l} = -(\hat{\mathbf{e}}_{l\sigma} \cdot \hat{\mathbf{e}}_{-l\sigma'}), \qquad \text{(B.6)}$$

since $\hat{\mathbf{k}}_{-l} = -\hat{\mathbf{k}}_l$. From (B.6), the last term in (B.4) vanishes identically, while by (B.5) the first term reduces to $2\delta_{\sigma\sigma'}$ and so

$$H = \sum_{l,\sigma} \frac{\hbar\omega_l}{2}(a_{l\sigma}a_{l\sigma}^\dagger + a_{l\sigma}^\dagger a_{l\sigma})$$

in agreement with (4.3.45).

Appendix C

Momentum of Field in Cavity

In this appendix we evaluate the classical momentum associated with the radiation field in a cavity.

If we substitute **E** and **H** from (4.3.44) into (4.3.47) we obtain

$$\mathbf{G} = -\frac{\hbar}{2c}\sum_{l,\sigma}\sum_{l'\sigma'}\sqrt{\omega_l\omega_{l'}}\,[\hat{\mathbf{e}}_{l\sigma}\times(\hat{\mathbf{e}}_{l'\sigma'}\times\hat{\mathbf{k}}_{l'})]$$
$$\times\frac{1}{\tau}\int_{\text{cavity}} d\tau[a_{l\sigma}(t)\exp(i\mathbf{k}_l\cdot\mathbf{r}) - a_{l\sigma}{}^\dagger(t)\exp(-i\mathbf{k}_l\cdot\mathbf{r})]$$
$$\times[a_{l'\sigma'}(t)\exp(i\mathbf{k}_{l'}\cdot\mathbf{r}) - a_{l'\sigma'}{}^\dagger(t)\exp(-i\mathbf{k}_{l'}\cdot\mathbf{r})]. \qquad (C.1)$$

The integral is identical with the one evaluated in Appendix B for the field energy. If we use this result, we see that **G** becomes

$$\mathbf{G} = \sum_{l,\sigma,\sigma'}\frac{\hbar\omega_l}{2c}\{(a_{l\sigma}a_{l\sigma'}{}^\dagger + a_{l\sigma}{}^\dagger a_{l\sigma'})[\hat{\mathbf{e}}_{l\sigma}\times(\hat{\mathbf{e}}_{l\sigma'}\times\hat{\mathbf{k}}_l)]$$
$$-(a_{l\sigma}a_{-l\sigma'} + a_{l\sigma}{}^\dagger a_{-l\sigma'}{}^\dagger)[\hat{\mathbf{e}}_{l\sigma}\times(\hat{\mathbf{e}}_{-l\sigma'}\times\hat{\mathbf{k}}_{-l})]\}. \qquad (C.2)$$

We have omitted the explicit time dependence for simplicity.

By a well-known vector identity, we see that

$$\begin{aligned}\hat{\mathbf{e}}_{l\sigma}\times(\hat{\mathbf{e}}_{l\sigma'}\times\hat{\mathbf{k}}_l) &= (\hat{\mathbf{e}}_{l\sigma}\cdot\hat{\mathbf{e}}_{l\sigma'})\hat{\mathbf{k}}_l = \delta_{\sigma\sigma'}\hat{\mathbf{k}}_l\\ \hat{\mathbf{e}}_{l\sigma}\times(\hat{\mathbf{e}}_{-l\sigma'}\times\hat{\mathbf{k}}_{-l}) &= -(\hat{\mathbf{e}}_{l\sigma}\cdot\hat{\mathbf{e}}_{-l\sigma'})\hat{\mathbf{k}}_l,\end{aligned} \qquad (C.3)$$

since $\hat{\mathbf{e}}_{l\sigma}\cdot\hat{\mathbf{k}}_l = 0$ and $\hat{\mathbf{k}}_{-l} = -\hat{\mathbf{k}}_l$. If we use these, (C.2) reduces to

$$\mathbf{G} = \tfrac{1}{2}\sum_{l,\sigma}\hbar\mathbf{k}_l(a_{l\sigma}a_{l\sigma}{}^\dagger + a_{l\sigma}{}^\dagger a_{l\sigma})$$
$$+\tfrac{1}{2}\sum_{l,\sigma,\sigma'}\hbar\mathbf{k}_l(a_{l\sigma}a_{-l\sigma'} + a_{l\sigma}{}^\dagger a_{-l\sigma'}{}^\dagger)(\hat{\mathbf{e}}_{l\sigma}\cdot\hat{\mathbf{e}}_{-l\sigma'}), \qquad (C.4)$$

since $\mathbf{k}_l = \omega_l \hat{\mathbf{k}}_l / c$. We now show that the last sums over l, σ, and σ' vanish identically. The $a_{l\sigma}$ and $a_{-l\sigma'}$ commute classically as well as quantum-mechanically. Since $\mathbf{k}_l = -\mathbf{k}_{-l}$ and since σ' and σ are dummy indices, we may write for the first sum over l, σ, and σ'

$$\tfrac{1}{2} \sum \hbar \mathbf{k}_l a_{l\sigma} a_{-l\sigma'} (\hat{\mathbf{e}}_{l\sigma} \cdot \hat{\mathbf{e}}_{-l\sigma'}) = \tfrac{1}{4} \sum \hbar \mathbf{k}_l a_{l\sigma} a_{-l\sigma'} (\hat{\mathbf{e}}_{l\sigma} \cdot \hat{\mathbf{e}}_{-l\sigma'} - \hat{\mathbf{e}}_{l\sigma} \cdot \hat{\mathbf{e}}_{-l\sigma'}) \equiv 0.$$

A similar argument shows that the $a^\dagger a^\dagger$ term also vanishes. Therefore, (C.4) reduces to the value given in (4.3.48) of the text.

Appendix D

Properties of Transverse Delta Function*

The transverse δ function is defined by

$$\delta_{ij}{}^{T}(\boldsymbol{\rho}) = \frac{1}{L^3}\sum_{l}[\delta_{ij} - (\hat{\mathbf{k}}_l)_i(\hat{\mathbf{k}}_l)_j]\exp(i\mathbf{k}_l\cdot\boldsymbol{\rho}) \tag{D.1}$$

or

$$\delta_{ij}{}^{T}(\boldsymbol{\rho}) \xrightarrow[L\to\infty]{} \frac{1}{(2\pi)^3}\int_{-\infty}^{\infty} d\mathbf{k}\left(\delta_{ij} - \frac{k_ik_j}{k^2}\right)\exp(i\mathbf{k}\cdot\boldsymbol{\rho}), \tag{D.2}$$

where $\mathbf{k} = k\hat{\mathbf{k}}$, $k^2 = |\mathbf{k}|^2$, and $d\mathbf{k} \equiv dk_x\, dk_y\, dk_z$.

We shall now derive some useful properties of $\delta_{ij}{}^{T}(\boldsymbol{\rho})$.

$$1^\circ \qquad \delta_{ij}{}^{T}(\boldsymbol{\rho}) = \delta_{ji}{}^{T}(\boldsymbol{\rho}). \tag{D.3}$$

This property follows by simple inspection of (D.2) since $\delta_{ij} = \delta_{ji}$ and $k_ik_j = k_jk_i$.

$$2^\circ \qquad \delta_{ij}{}^{T}(\boldsymbol{\rho}) = \delta_{ij}{}^{T}(-\boldsymbol{\rho}) \tag{D.4}$$

If in (D.2) we let $\mathbf{k} \to -\mathbf{k}$, then $d\mathbf{k} \to -d\mathbf{k}$ and we have from (D.2)

$$\begin{aligned}\delta_{ij}{}^{T}(\boldsymbol{\rho}) &= -\iiint_{+\infty}^{-\infty} d\mathbf{k}\left(\delta_{ij} - \frac{k_ik_j}{k^2}\right)\exp(-i\mathbf{k}\cdot\boldsymbol{\rho}) \\ &= +\iiint_{-\infty}^{\infty} d\mathbf{k}\left(\delta_{ij} - \frac{k_ik_j}{k^2}\right)\exp(-i\mathbf{k}\cdot\boldsymbol{\rho}) \\ &\equiv \delta_{ij}{}^{T}(-\boldsymbol{\rho}).\end{aligned}$$

Q.E.D.

$$3^\circ \qquad \sum_j \frac{\partial \delta_{ij}{}^{T}}{\partial x_j} = 0. \tag{D.5}$$

* These properties of the transverse δ function were presented in a course at Stanford University by Dr. D Walecka.

From (D.2), when we differentiate with respect to x_j and sum over j,

$$\sum_j \frac{\partial \delta_{ij}{}^T(\boldsymbol{\rho})}{\partial x_j} = \frac{i}{(2\pi)^3} \sum_j \iiint_{-\infty}^{\infty} k_j \, d\mathbf{k} \exp(i\mathbf{k} \cdot \boldsymbol{\rho}) \left(\delta_{ij} - \frac{k_i k_j}{k^2} \right).$$

However,

$$\sum_j k_j \left(\delta_{ij} - \frac{k_i k_j}{k^2} \right) = k_i - \frac{k_i \sum_j k_j k_j}{k^2} \equiv 0,$$

since

$$\sum_j k_j k_j = k^2. \qquad \text{Q.E.D}$$

$$4^\circ \qquad \sum_i \frac{\partial \delta_{ij}{}^T(\boldsymbol{\rho})}{\partial x_i} = 0. \tag{D.6}$$

This follows from (D.3) and (D.5) directly.

$$5^\circ \qquad \delta_{ij}{}^T(\boldsymbol{\rho}) = \delta_{ij}\, \delta(\boldsymbol{\rho}) + \frac{1}{4\pi} \frac{\partial^2}{\partial x_i \, \partial x_j} \frac{1}{|\boldsymbol{\rho}|}, \tag{D.7}$$

where

$$\delta(\mathbf{r}) = \frac{1}{(2\pi)^3} \iiint_{-\infty}^{\infty} d\mathbf{k} \exp(i\mathbf{k} \cdot \mathbf{r}) \equiv \delta(x)\, \delta(y)\, \delta(z) \tag{D.8}$$

is the ordinary three-dimensional Dirac δ function since

$$\delta(x) = \frac{1}{2\pi} \int_{-\infty}^{\infty} dk_x e^{ik_x x}. \tag{D.9}$$

To prove (D.7), we have from (D.2) and (D.8)

$$\delta_{ij}{}^T(\boldsymbol{\rho}) = \delta_{ij}\, \delta(\boldsymbol{\rho}) - \frac{1}{(2\pi)^3} \int_{-\infty}^{\infty} d\mathbf{k} \frac{k_i k_j}{k^2} \exp(i\mathbf{k} \cdot \boldsymbol{\rho}). \tag{D.10}$$

However,

$$\frac{\partial}{\partial x_i} \frac{\partial}{\partial x_j} \int_{-\infty}^{\infty} \frac{d\mathbf{k} \exp(i\mathbf{k} \cdot \boldsymbol{\rho})}{k^2} = -\int_{-\infty}^{\infty} \frac{k_i k_j \exp(i\mathbf{k} \cdot \boldsymbol{\rho})}{k^2} d\mathbf{k}. \tag{D.11}$$

Therefore (D.10) may be written

$$\delta_{ij}{}^T(\boldsymbol{\rho}) = \delta_{ij}\, \delta(\boldsymbol{\rho}) + \frac{\partial^2}{\partial x_i \, \partial x_j} \frac{1}{(2\pi)^3} \int_{-\infty}^{\infty} \frac{d\mathbf{k} \exp(i\mathbf{k} \cdot \boldsymbol{\rho})}{k^2}. \tag{D.12}$$

The last integral may be indirectly evaluated from electrostatic theory. The potential of a charge e is

$$V = \frac{e}{4\pi r}. \tag{D.13}$$

This satisfies the Poisson equation

$$\nabla^2 V = -e\,\delta(\mathbf{r}). \tag{D.14}$$

We therefore find that

$$\nabla^2\left(\frac{1}{r}\right) = -4\pi\,\delta(\mathbf{r}) = -\frac{1}{2\pi^2}\int_{-\infty}^{\infty} d\mathbf{k}\,\exp\,(i\mathbf{k}\cdot\mathbf{r}), \tag{D.15}$$

where we used (D.8). It is easily verified that

$$\frac{1}{4\pi r} = \frac{1}{(2\pi)^3}\int \frac{d\mathbf{k}\,\exp\,(i\mathbf{k}\cdot\mathbf{r})}{k^2} \tag{D.16}$$

that is, if we operate with ∇^2 on both sides of (D.16), we see that (D.15) is satisfied. If we then use (D.16) in (D.12), (D.7) follows:

$$\frac{\partial \delta_{ij}{}^T(\boldsymbol{\rho})}{\partial x_k} = -\frac{\partial \delta_{ij}{}^T(\boldsymbol{\rho})}{\partial x'_k} \tag{D.17}$$

if $\boldsymbol{\rho} = \mathbf{r} - \mathbf{r}'$.

This property follows very easily by differentiating (D.2) alternately with respect to x_k and x'_k.

Appendix E

Commutation Relations for **D** and **B**

In this appendix we derive the commutation relations (4.6.17) and (4.6.18) of the text.

Since $\mathbf{B} = \text{curl } \mathbf{A}$, we see that in the Schrödinger picture

$$[D_1(\mathbf{r}), B_1(\mathbf{r}')] = \left[D_1(\mathbf{r}), \frac{\partial A_3(\mathbf{r}')}{\partial y'} - \frac{\partial A_2(\mathbf{r}')}{\partial z'} \right]$$

$$\equiv \frac{\partial}{\partial y'} [D_1(\mathbf{r}), A_3(\mathbf{r}')] - \frac{\partial}{\partial z'} [D_1(\mathbf{r}), A_2(\mathbf{r}')]. \tag{E.1}$$

From (4.6.5) of the text and the symmetry properties (D.3) and (D.4) of Appendix D, we may write (E.1) as

$$[D_1(\mathbf{r}), B_1(\mathbf{r}')] = i\hbar \frac{\partial}{\partial y'} \delta_{13}{}^T(\boldsymbol{\rho}) - i\hbar \frac{\partial}{\partial z'} \delta_{12}{}^T(\boldsymbol{\rho}), \tag{E.2}$$

where $\boldsymbol{\rho} = \mathbf{r} - \mathbf{r}'$. From (D.2) of Appendix D,

$$\begin{aligned} \frac{\partial \delta_{13}{}^T}{\partial y'} &= + \frac{i}{(2\pi)^3} \int_{-\infty}^{\infty} d\mathbf{k} \, \frac{k_1 k_3 k_2}{k^2} \exp(i\mathbf{k} \cdot \boldsymbol{\rho}) \\ \frac{\partial \delta_{12}{}^T}{\partial x'} &= + \frac{i}{(2\pi)^3} \int_{-\infty}^{\infty} d\mathbf{k} \, \frac{k_1 k_2 k_3}{k^3} \exp(i\mathbf{k} \cdot \boldsymbol{\rho}). \end{aligned} \tag{E.3}$$

If we substitute these equations into (E.2), we see that $[D_1(\mathbf{r}), B_1(\mathbf{r}')] = 0$ in the Schrödinger picture so that (4.6.17) of the text follows in the Heisenberg picture. (A similar proof follows for $[D_2, B_2]$ and $[D_3, B_3]$.)

We next consider the commutator for two perpendicular components, say D_1 and B_2. We therefore have

$$
\begin{aligned}
[D_1(\mathbf{r}), B_2(\mathbf{r}')] &= \left[D_1(\mathbf{r}), \frac{\partial A_1(\mathbf{r}')}{\partial z'} - \frac{\partial A_3(\mathbf{r}')}{\partial x'} \right] \\
&= \frac{\partial}{\partial z'} [D_1(\mathbf{r}), A_1(\mathbf{r}')] - \frac{\partial}{\partial x'} [D_1(\mathbf{r}), A_3(\mathbf{r}')] \\
&= i\hbar \frac{\partial}{\partial z'} \delta_{11}{}^T(\boldsymbol{\rho}) - i\hbar \frac{\partial}{\partial x'} \delta_{13}{}^T(\boldsymbol{\rho}),
\end{aligned} \tag{E.4}
$$

where we used the commutation relations (4.6.5) of the text.

Again from (D.2) and (D.8) of Appendix D, we see that

$$
\begin{aligned}
\frac{\partial}{\partial z'} \delta_{11}{}^T(\boldsymbol{\rho}) &= \frac{\partial}{\partial z'} \delta(\boldsymbol{\rho}) + \frac{i}{(2\pi)^3} \int_{-\infty}^{\infty} \frac{k_1{}^2 k_3}{k^2} \exp(i\mathbf{k} \cdot \boldsymbol{\rho})\, d\mathbf{k} \\
\frac{\partial}{\partial x'} \delta_{13}{}^T(\boldsymbol{\rho}) &= \frac{i}{(2\pi)^3} \int_{-\infty}^{\infty} \frac{k_1 k_3 k_1}{k^2} \exp(i\mathbf{k} \cdot \boldsymbol{\rho})\, d\mathbf{k}.
\end{aligned} \tag{E.5}
$$

If we put these into (E.4), we have

$$
\begin{aligned}
[D_1(\mathbf{r}), \mathbf{B}_2(\mathbf{r}')] &= +i\hbar \frac{\partial}{\partial z'} \delta(\boldsymbol{\rho}) \\
&\equiv -i\hbar \frac{\partial}{\partial z} \delta(\boldsymbol{\rho}),
\end{aligned} \tag{E.6}
$$

since $\boldsymbol{\rho} = \mathbf{r} - \mathbf{r}'$. Similar proofs follow for $[D_2, B_3]$ and $[D_3, B_1]$ so that (4.6.18a) is proved. The proof of (4.6.18b) is similar.

Appendix F

Heisenberg Equations of Motion for **D** and **B**

We derive the Heisenberg equations of motion given in (4.6.20) of the text. The hamiltonian is given by (4.6.19). Since by (4.6.13a) all components of **D** commute, we see by (4.6.19) that for the x component of **D**

$$i\hbar \frac{dD_1(\mathbf{r}, t)}{dt} = \frac{1}{2\mu_0} \int d\tau' \, [D_1(\mathbf{r}, t), \mathbf{B}^2(\mathbf{r}', t)], \tag{F.1}$$

where $\mathbf{B}^2 = B_1^2 + B_2^2 + B_3^2$. From (4.6.17), $[D_1, B_1^2] = 0$ so that (F.1) reduces to

$$i\hbar \frac{dD_1(\mathbf{r}, t)}{dt} = \frac{1}{2\mu_0} \int d\tau' \, \{[D_1(\mathbf{r}, t), B_2^2(\mathbf{r}', t)] + [D_1(\mathbf{r}, t), B_3^2(\mathbf{r}', t)]\}. \tag{F.2}$$

From Problem 1.8(f), we see that

$$[D_1, B_2'^2] = [D_1, B_2']B_2' + B_2'[D_1, B_2'], \tag{F.3}$$

where we use the notation $\mathbf{B}' \equiv \mathbf{B}(\mathbf{r}', t)$. From (4.6.18), Eq. F.3 is given by

$$[D_1, B_2'^2] = -2i\hbar B_2' \frac{\partial}{\partial z} \delta(\boldsymbol{\rho}), \tag{F.4}$$

since $(\partial/\partial z)\delta(\boldsymbol{\rho})$ is a c number and commutes with B_2'.

Similarly,

$$[D_1, B_3'^2] = +2i\hbar B_3' \frac{\partial}{\partial y} \delta(\boldsymbol{\rho}). \tag{F.5}$$

If we substitute (F.4) and (F.5) into (F.2), we have

$$i\hbar \frac{dD_1}{dt} = \frac{i\hbar}{\mu_0}\left[\frac{\partial}{\partial y} \int d\tau' \, B_3(\mathbf{r}', t)\, \delta(\boldsymbol{\rho}) - \frac{\partial}{\partial z} \int d\tau' \, B_2(\mathbf{r}', t)\, \delta(\boldsymbol{\rho})\right]. \tag{F.6}$$

But from the definition of the δ function, we know that

$$\int_{-\infty}^{\infty} d\tau' \, f(\mathbf{r}') \, \delta(\mathbf{r} - \mathbf{r}') = f(\mathbf{r}), \tag{F.7}$$

so that (F.6) reduces to

$$i\hbar \frac{dD_1(\mathbf{r}, t)}{dt} = \frac{i\hbar}{\mu_0}\left[\frac{\partial B_3(\mathbf{r}, t)}{\partial y} - \frac{\partial B_2(\mathbf{r}, t)}{\partial z}\right]. \tag{F.8}$$

But $\mathbf{B} = \mu_0 \mathbf{H}$, and the expression in brackets is just the x component of the curl $\mathbf{B}$.

Similar proofs follow for the y and z components of $\mathbf{D}$, and we have proved (4.6.20a) of the text. We leave as an exercise the proof of (4.6.20b).

Appendix G

Evaluation of Field Commutation Relations

We show here that

$$\left[A_k(\mathbf{r}, t), \int \frac{\epsilon_0}{2}(\dot{\mathbf{A}}(\mathbf{r}', t))^2\, d\tau'\right] = i\hbar \dot{A}_k^{\,T}(\mathbf{r}, t). \tag{G.1}$$

Since $\mathbf{r}'$ is a variable of integration, we may write the left side of (G.1) as

$$\sum_l \int \frac{\epsilon_0}{2}\, d\tau'\, [A_k(\mathbf{r}), \dot{A}_l(\mathbf{r}')\dot{A}_l(\mathbf{r}')] = i\hbar \sum_l \int d\tau'\, \dot{A}_l(\mathbf{r}')\, \delta_{kl}^{\,T}(\mathbf{r} - \mathbf{r}'), \tag{G.2}$$

where we used the identity

$$[A, BC] = [A, B]C + B[A, C], \tag{G.3}$$

and the commutation relations (4.9.2).

To proceed, we need another property of the transverse δ function.

We consider the expansion of an arbitrary vector

$$\mathbf{B}(\mathbf{r}) = \frac{1}{L^{3/2}} \sum_l \sum_{\sigma=1}^{3} B_{l\sigma} \hat{\mathbf{e}}_{l\sigma} \exp(-i\mathbf{k}_l \cdot \mathbf{r}), \tag{G.4}$$

where $\hat{\mathbf{e}}_{l\sigma}$ ($\sigma = 1, 2, 3$) are three mutually perpendicular unit vectors and $B_{l\sigma}$ are expansion coefficients.

We may write $\mathbf{B}$ as the sum of its component parallel to $\mathbf{k}_l$ (longitudinal) and its two components perpendicular to $\mathbf{k}_l$ (transverse). We have

$$\mathbf{B}(\mathbf{r}) = \mathbf{B}^T + \mathbf{B}^L. \tag{G.5}$$

If we let $\hat{\mathbf{e}}_{l3} = \mathbf{k}_l/|\mathbf{k}_l|$, then from (G.4)

$$\begin{aligned} \mathbf{B}^T(\mathbf{r}) &= \frac{1}{L^{3/2}} \sum_l \sum_{\sigma=1}^{2} B_{l\sigma} \hat{\mathbf{e}}_{l\sigma} \exp(-i\mathbf{k}_l \cdot \mathbf{r}) \\ \mathbf{B}^L(\mathbf{r}) &= \frac{1}{L^{3/2}} \sum_l B_{l3} \hat{\mathbf{e}}_{l3} \exp(-i\mathbf{k}_l \cdot \mathbf{r}), \end{aligned} \tag{G.6}$$

where div $\mathbf{B}^T = 0$ and curl $\mathbf{B}^L = 0$.

We consider now the integral

$$I = \sum_{i=1}^{3} \int d\mathbf{r}\, B_i(\mathbf{r})\, \delta_{ij}{}^T(\mathbf{r} - \mathbf{r}'). \tag{G.7}$$

We shall show that

$$I = B_j{}^T(\mathbf{r}'), \tag{G.8}$$

that is, the transverse δ function projects out the transverse component of the vector.

If we put (G.4) and (D.1) into (G.7), we have

$$I = \frac{1}{L^{3/2}} \sum_{i=1}^{3} \sum_{l} \sum_{\sigma=1}^{3} B_{l\sigma} \int d\mathbf{r} \exp(-i\mathbf{k}_l \cdot \mathbf{r})(\hat{\mathbf{e}}_{l\sigma})_i \exp(-i\mathbf{k}_l \cdot \mathbf{r})$$
$$\times \sum_{l'} \frac{1}{L^3} \exp[i\mathbf{k}_{l'} \cdot (\mathbf{r} - \mathbf{r}')](\delta_{ij} - \hat{\mathbf{k}}_{l'i}\hat{\mathbf{k}}_{l'j}). \tag{G.9}$$

If we note that

$$\frac{1}{L^3} \int_0^L d\mathbf{r} \exp[i(\mathbf{k}_{l'} - \mathbf{k}_l) \cdot \mathbf{r}] = \delta_{ll'} \tag{G.10}$$

recall that

$$\mathbf{k}_l = \frac{2\pi}{L}(l_1\hat{\mathbf{i}} + l_2\hat{\mathbf{j}} + l_3\hat{\mathbf{k}}) \tag{G.11}$$

and carry out the sum over l' in (G.9), then

$$I = \frac{1}{L^{3/2}} \sum_{l} \sum_{\sigma=1}^{3} B_{l\sigma} \exp(-i\mathbf{k}_l \cdot \mathbf{r}') \sum_{i=1}^{3} (\hat{\mathbf{e}}_{l\sigma})_i(\delta_{ij} - \hat{\mathbf{k}}_{li}\hat{\mathbf{k}}_{lj}). \tag{G.12}$$

It is easily shown that

$$\begin{aligned} \sum_{i=1}^{3} (\hat{\mathbf{e}}_{l\sigma})_i(\delta_{ij} - \hat{\mathbf{k}}_{li}\hat{\mathbf{k}}_{lj}) &= (\hat{\mathbf{e}}_{l\sigma})_j \qquad \text{if} \quad \sigma = 1, 2 \\ &= 0 \qquad\qquad \text{if} \quad \sigma = 3. \end{aligned} \tag{G.13}$$

When we use this, (G.12) becomes

$$I = \frac{1}{L^{3/2}} \sum_{l} \sum_{\sigma=1}^{2} B_{l\sigma} \exp(-i\mathbf{k}_l \cdot \mathbf{r}')(\hat{\mathbf{e}}_{l\sigma})_j.$$

From (G.6), this reduces to (G.8), the desired result. If we recall that $\delta_{ij}{}^T = \delta_{ji}{}^T$, we see that (G.1) follows from (G.2) when (G.7) and (G.8) are used.

Appendix H

Evaluation of Sums in Equation (5.10.17.)*

In this appendix we evaluate the sum in (5.10.17) of the text.

We consider the sum over the polarizations σ. Figure H.1 shows $\hat{\mathbf{k}}_l$, θ_l, φ_l, $\mathbf{x}$, and $\mathbf{x}'$. For simplicity, we have assumed that $\mathbf{x}$ and $\mathbf{x}'$ have only x components. The propagation vector has components

$$\hat{k}_l = (\sin\theta_l \cos\varphi_l, \sin\theta_l \sin\varphi_l, \cos\theta_l). \tag{H.1}$$

The unit orthogonal polarization vectors $\hat{\mathbf{e}}_{l1}$ and $\hat{\mathbf{e}}_{l2}$ may be taken as

$$\begin{aligned} \hat{\mathbf{e}}_{l1} &= (-\cos\theta_l \cos\varphi_l, -\cos\theta_l \sin\varphi_l, \sin\theta_l) \\ \hat{\mathbf{e}}_{l2} &= (\sin\varphi_l, -\cos\varphi_l, 0), \end{aligned} \tag{H.2}$$

and the atom dipole moments have components

$$\begin{aligned} \mathbf{x} &= |\mathbf{x}|[1, 0, 0] \\ \mathbf{x}' &= |\mathbf{x}'|[1, 0, 0]. \end{aligned} \tag{H.3}$$

With this choice of polarization vectors, the sum over σ in (5.10.17) becomes

$$\sum_{\sigma=1}^{2} (\hat{\mathbf{e}}_{l\sigma} \cdot \mathbf{x})(\hat{\mathbf{e}}_{l\sigma} \cdot \mathbf{x}') = |\mathbf{x}|\,|\mathbf{x}'|\,(\cos^2\theta_l \cos^2\varphi_l + \sin^2\varphi_l). \tag{H.4}$$

We may change the sum over l in (5.10.17) by means of (4.5.4) and (4.5.10) to an integral

$$\sum_l \to \frac{L^3}{(2\pi c)^3} \int_0^\infty \omega_l^2\, d\omega_l \int_0^\pi \sin\theta_l\, d\theta_l \int_0^{2\pi} d\varphi_l, \tag{H.5}$$

* E. Fermi, *Rev. Mod. Phys.*, **4**, 87 (1932).

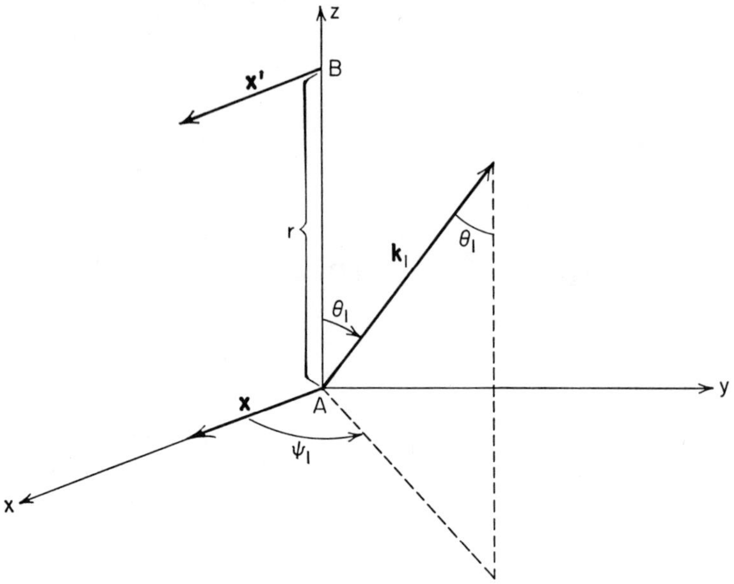

Figure H.1 Geometry for propagation vector and atomic dipole moments.

so that (5.10.17) becomes

$$c(1;2;0;t) = -\frac{e^2\,|\mathbf{x}|\,|\mathbf{x}'|\,\omega\omega'}{2\hbar\epsilon_0(2\pi c)^3 i}\int_0^\infty \frac{\omega_l\,d\omega_l[1 - e^{-i(\omega_l-\omega')t}]}{[i(\omega-\omega_l)+\gamma_A/2](\omega_l-\omega')}$$
$$\times \int_0^\pi \sin\theta_l\,d\theta_l \int_0^\pi d\varphi_l\,(\sin^2\varphi_l + \cos^2\theta_l\cos^2\varphi_l)e^{ik_lr\cos\theta_l}. \tag{H.6}$$

The integrals over φ_l and θ_l are easily evaluated:

$$\int_0^{2\pi}\sin^2\varphi_l\,d\varphi_l = \int_0^{2\pi}\cos^2\varphi_l\,d\varphi_l = \pi. \tag{H.7}$$

If we let $u = \cos\theta_l$, we have

$$\pi\int_{-1}^{1} du\,(1+u^2)e^{ik_lru} = 4\pi\left[\frac{\sin k_lr}{k_lr} + \frac{\cos k_lr}{(k_lr)^2} - \frac{\sin k_lr}{(k_lr)^3}\right]. \tag{H.8}$$

We make the assumption that the atoms are located many wavelengths apart so that $k_lr \gg 1$. In this case, we need retain only the first term in (H.8) so that (H.6) reduces to

$$c(1;2;0;t) = -\frac{1}{r}\frac{e^2\,|\mathbf{x}|\,|\mathbf{x}'|\,\omega\omega'}{4\pi^2\epsilon_0\hbar c^2 i}\int_0^\infty \frac{\sin(\omega_l/c)r[1 - e^{-i(\omega_l-\omega')t}]}{[i(\omega-\omega_l)+\gamma_A/2](\omega_l-\omega')}\,d\omega_l, \tag{H.9}$$

since $\omega_l = ck_l$.

It has been assumed that $1/\gamma_A$, the half life of atom A, is very short so that the emission may be considered as part of a continuous spectrum. Since the integral in (H.9) is very strongly peaked at $\omega_l = \omega'$, we may remove the factor $[i(\omega - \omega_l) + \gamma/2]^{-1}$ from under the integral sign and replace ω_l by ω'. For the same reason, with little error, we may let the lower limit on ω_l go to $-\infty$. In this case (H.9) reduces to

$$c(1; 2; 0; t) \cong -\frac{1}{r}\frac{e^2\,|\mathbf{x}|\,|\mathbf{x}'|\,\omega\omega'\mu_0}{4\pi^2\hbar i[i(\omega - \omega') + \gamma_A/2]} \times \int_{-\infty}^{\infty} \frac{d\omega_l \sin(\omega_l/c)r[1 - e^{-i(\omega_l - \omega')t}]}{\omega_l - \omega'}. \quad \text{(H.10)}$$

To evaluate the integral, we let

$$2\pi\zeta = \omega_l - \omega', \quad \text{(H.11)}$$

and the integral may be written as

$$\begin{aligned} I &= \int_{-\infty}^{\infty} \frac{d\zeta}{\zeta} \sin\frac{r}{c}(2\pi\zeta + \omega')(1 - e^{-i2\pi t\zeta}) \\ &= \sin\frac{\omega' r}{c}\int_{-\infty}^{\infty}\frac{d\zeta}{\zeta}\cos\frac{2\pi r\zeta}{c}(1 - \cos 2\pi t\zeta) \\ &\quad - i\cos\frac{\omega' r}{c}\int_{-\infty}^{\infty}\frac{d\zeta}{\zeta}\sin\frac{2\pi r\zeta}{c}\sin 2\pi t\zeta \\ &\quad + \cos\frac{\omega' r}{c}\int_{-\infty}^{\infty}\frac{d\zeta}{\zeta}\sin\frac{2\pi r\zeta}{c}(1 - \cos 2\pi t\zeta) \\ &\quad - i\sin\frac{\omega' r}{c}\int_{-\infty}^{\infty}\frac{d\zeta}{\zeta}\sin 2\pi t\zeta\cos\frac{2\pi r\zeta}{c}. \end{aligned} \quad \text{(H.12)}$$

The first two integrals vanish since the integrands are odd functions of ζ. The last two are easily evaluated by means of the following integrals:

$$\text{(1)} \qquad \int_{-\infty}^{\infty} \frac{\sin \tau x}{x}\,dx = \pi. \quad \text{(H.13)}$$

This may be verified by taking the Laplace transform with respect to τ of both sides.

$$\text{(2)} \qquad \int_{-\infty}^{\infty} \frac{\sin px \cos qx}{x}\,dx = \begin{cases} \pi & p > q \\ 0 & p < q \end{cases} \quad \text{(H.14)}$$

This is proved by writing the integrand as

$$\sin px \cos qx \equiv \tfrac{1}{2}[\sin(p + q)x + \sin(p - q)x],$$

and using (H.13). When we use (H.13) and (H.14), (H.12) reduces to

$$I = \begin{cases} 0 & t < \dfrac{r}{c} \\ +\pi e^{i\omega' r/c} & t > \dfrac{r}{c}. \end{cases} \tag{H.15}$$

Therefore, (H.10) becomes

$$c(1;2;0;t) = \begin{cases} 0 & t < \dfrac{r}{c} \\ -\dfrac{1}{r}\dfrac{e^2|\mathbf{x}||\mathbf{x}'|\,\omega\omega'\mu_0 e^{+i\omega' r/c}}{4\pi\hbar i[i(\omega-\omega') + \gamma_A/2]} & t > \dfrac{r}{c}. \end{cases} \tag{H.16}$$

This is the result we wished to prove.

Appendix I

Wiener–Khintchine Relations*

Let $v(t)$ be a random variable. Define

$$v_T(t) = \begin{cases} v(t) & \text{for} \quad -T < t < T \\ 0 & \text{otherwise.} \end{cases} \tag{I.1}$$

Let $V(\omega)$ be the Fourier transform of $v_T(t)$:

$$V(\omega) = \frac{1}{2\pi} \int_{-\infty}^{\infty} v_T(t') e^{-i\omega t'}\, dt'. \tag{I.2}$$

If we multiply by $e^{i\omega t}$ and integrate over $d\omega$ from $-\infty$ to $+\infty$, we obtain

$$v_T(t) = \int_{-\infty}^{\infty} V(\omega) e^{i\omega t}\, d\omega, \tag{I.3}$$

where we used the result

$$\delta(t - t') = \frac{1}{2\pi} \int_{-\infty}^{\infty} d\omega\, e^{i\omega(t-t')}. \tag{I.4}$$

If v is real, it follows that

$$\begin{aligned} v^*(t) &= v(t) \\ V^*(\omega) &= V(-\omega). \end{aligned} \tag{I.5}$$

The ensemble average of $v(t)$ is defined by

$$\langle v(t) \rangle = \frac{1}{N} \sum_{k=1}^{N} v^{(k)}(t), \tag{I.6}$$

where $v^{(k)}(t)$ is the value of $v(t)$ for the kth system of the ensemble and N is the number of systems in the ensemble (N is very large).

The time average of $v(t)$ for a given system of the ensemble is defined by

$$\overline{v^{(k)}(t)} = \frac{1}{2T} \int_{-T}^{T} v^{(k)}(t + t')\, dt' \tag{I.7}$$

where T is very long.

* See Ref. 6, Chap. 4.

An ensemble is stationary with respect to v if there is no preferred origin in time so that the ensembles $\{v^{(k)}(t)\}$ and $\{v^{(k)}(t + t')\}$ have the same statistical properties. If in the course of time $v^{(k)}(t)$ for each system passes through all values accessible to it, the ensemble is ergodic. For an ensemble which is ergodic, we have that

$$\overline{v^{(k)}(t)} = \overline{v(t)}, \tag{I.8}$$

which means that the time average will be the same for all systems in the ensemble. Moreover if the ensemble is stationary

$$\overline{v(t)} = \bar{v} \tag{I.9}$$

will be independent of t. Similarly, for a stationary ensemble,

$$\langle v(t) \rangle = \langle v \rangle. \tag{I.10}$$

Therefore, the time and ensemble averages are equal for stationary ergodic ensembles.

The correlation function of $v(t)$ for a stationary ensemble is

$$K(s) = \langle v(t)\, v(t + s) \rangle. \tag{I.11}$$

It is independent of t since v is stationary. Note that

$$K(0) = \langle v^2 \rangle. \tag{I.12}$$

If we take the Fourier transform of $K(s)$, we have

$$\begin{aligned} J(\omega) &= \frac{1}{2\pi} \int_{-\infty}^{\infty} e^{-i\omega s} K(s)\, ds \\ &= \frac{1}{2\pi} \int_{-\infty}^{\infty} e^{-i\omega s} \langle v(t) v(t + s) \rangle\, ds, \end{aligned} \tag{I.13}$$

which is defined as the spectral density of $v(t)$. If we multiply both sides of $e^{i\omega t'}$ and integrate over all ω and use (I.4), we obtain

$$K(s) = \int_{-\infty}^{\infty} J(\omega) e^{+i\omega s}\, d\omega. \tag{I.14}$$

Equations (I.13) and (I.14) are the Wiener–Khintchine relations.

We next relate these to $V(\omega)$. If $v(t)$ is ergodic and stationary, we have

$$\begin{aligned} K(s) &= \langle v(0) v(s) \rangle = \overline{v(0) v(s)} \\ &= \frac{1}{2T} \int_{-T}^{T} dt'\, v(t') v(s + t'). \end{aligned} \tag{I.15}$$

If we use (I.1), we have (on using (I.3) twice)

$$\begin{aligned}
K(s) &= \frac{1}{2T}\int_{-\infty}^{\infty} dt'\, v_T(t')v_T(s+t') \\
&= \frac{1}{2T}\int_{-\infty}^{\infty} dt' \int_{-\infty}^{\infty} V(\omega')e^{i\omega' t'}\, d\omega' \int_{-\infty}^{\infty} V(\omega)e^{i\omega(t'+s)}\, d\omega \\
&= \frac{1}{2T}\int_{-\infty}^{\infty} d\omega' \int_{-\infty}^{\infty} d\omega\, V(\omega')V(\omega)e^{i\omega s}\int_{-\infty}^{\infty} dt'\, e^{i(\omega+\omega')t'} \\
&= \frac{\pi}{T}\int_{-\infty}^{\infty} d\omega \int_{-\infty}^{\infty} d\omega'\, V(\omega)V(\omega')e^{i\omega s}\delta(\omega+\omega') \\
&= \frac{\pi}{T}\int_{-\infty}^{\infty} d\omega\, V(\omega)V(-\omega)e^{i\omega s} \\
&= \frac{\pi}{T}\int_{-\infty}^{\infty} d\omega\, |V(\omega)|^2\, e^{i\omega s}, \qquad \text{(I.16)}
\end{aligned}$$

where we used (I.5). Comparison with (I.14) shows that

$$J(\omega) = \frac{\pi}{T}\,|V(\omega)|^2, \qquad \text{(I.17)}$$

which is the spectral density of $v(t)$. Note also that

$$K(0) = \langle v^2 \rangle = \frac{\pi}{T}\int_{-\infty}^{\infty} |V(\omega)|^2\, d\omega. \qquad \text{(I.18)}$$

Appendix J

Atom-Field Hamiltonian under Dipole Approximation*

In this appendix we show that under the dipole approximation, we may approximate the hamiltonian for an atom interacting with an electromagnetic field

$$H = \frac{1}{2m}[\mathbf{p} - e\mathbf{A}(\mathbf{r}, t)]^2 + eV(\mathbf{r}), \tag{J.1}$$

by

$$H = \frac{1}{2m}\mathbf{p}^2 + eV(\mathbf{r}) - \boldsymbol{\mu} \cdot \mathbf{E}(\mathbf{r}, t), \tag{J.2}$$

where $\boldsymbol{\mu} = e\mathbf{r}$ is the atomic dipole moment and $\mathbf{E}$ is the electric field.

If we make the unitary transformation

$$|\psi(t)\rangle = \exp\left[i\frac{e\mathbf{r}}{\hbar} \cdot \mathbf{A}(\mathbf{r}, t)\right]|\chi(t)\rangle \equiv U|\chi(t)\rangle, \tag{J.3}$$

the Schrödinger equation

$$i\hbar\frac{\partial|\psi\rangle}{\partial t} = H|\psi\rangle \tag{J.4}$$

becomes

$$i\hbar U\frac{\partial|\chi\rangle}{\partial t} + i\hbar\frac{\partial U}{\partial t}|\chi\rangle = HU|\chi\rangle, \tag{J.5}$$

where U is defined by (J.3) and H is given by (J.1). Since

$$U^\dagger = U^{-1},$$

we have on multiplying (J.5) from the left by $U^\dagger$

$$i\hbar\frac{\partial|\chi\rangle}{\partial t} = K|\chi\rangle, \tag{J.6a}$$

* Paul I. Richards, *Phys. Rev.*, **73,** 254 (1948); see E. A. Power and S. Zineau, *Phil. Trans. A.*, **251,** 427 (1959) and references contained therein.

where

$$K \equiv U^\dagger H U - i\hbar U^\dagger \frac{\partial U}{\partial t} \tag{J.6b}$$

From (J.3) we see first that

$$-i\hbar U^\dagger \frac{\partial U}{\partial t} = e\mathbf{r} \cdot \frac{\partial \mathbf{A}}{\partial t}. \tag{J.7}$$

In the Coulomb gauge

$$\text{div } \mathbf{A} = 0, \tag{J.8}$$

and when we neglect the field source [$\rho = 0 = \mathbf{J}$], we have [cf. Section 4.8]

$$\begin{aligned} \mathbf{E}(\mathbf{r}, t) &= -\frac{\partial \mathbf{A}}{\partial t} \\ \mathbf{B}(\mathbf{r}, t) &= \text{curl } \mathbf{A}. \end{aligned} \tag{J.9}$$

If we use (J.7)–(J.9), the transformed hamiltonian (J.6b) becomes

$$K = -e\mathbf{r} \cdot \mathbf{E}(\mathbf{r}, t) + U^\dagger H U. \tag{J.10}$$

We next proceed to evaluate $U^\dagger H U$. We have

$$H' \equiv U^\dagger H U = \frac{1}{2m} U^\dagger\{\mathbf{p}^2 - e(\mathbf{A} \cdot \mathbf{p} + \mathbf{p} \cdot \mathbf{A})\}U + \frac{e^2}{2m}\mathbf{A}^2 + eV(\mathbf{r}), \tag{J.11}$$

where we used (J.1), (J.3), and the fact that U commutes through $V(\mathbf{r})$ and $\mathbf{A}^2(\mathbf{r}, t)$. Also

$$\sum_{j=1}^{3} [p_j, A_j] = \frac{\hbar}{i} \sum_j \frac{\partial A_j}{\partial x_j} = 0 \tag{J.12}$$

by (J.8) so that $\mathbf{p} \cdot \mathbf{A} = \mathbf{A} \cdot \mathbf{p}$ in the Coulomb gauge. Therefore, we have

$$H' = U^\dagger \frac{\mathbf{p}^2}{2m} U - \frac{e}{m}\mathbf{A} \cdot (U^\dagger \mathbf{p} U) + \frac{e^2}{2m}\mathbf{A}^2 + eV. \tag{J.13}$$

Our next task is to commute U through $\mathbf{p}^2$ and $\mathbf{p}$. We have

$$[p_i, U] = \frac{\hbar}{i}\frac{\partial U}{\partial x_i} = eU \frac{\partial(\mathbf{r} \cdot \mathbf{A})}{\partial x_i}. \tag{J.14}$$

Therefore,

$$\begin{aligned} -\frac{e}{m}\mathbf{A} \cdot (U^\dagger \mathbf{p} U) &= -\frac{e}{m}\mathbf{A} \cdot U^\dagger[U\mathbf{p} + eU \text{ grad } (\mathbf{r} \cdot \mathbf{A})] \\ &= -\frac{e}{m}\mathbf{A} \cdot \mathbf{p} - \frac{e^2}{m}\mathbf{A} \cdot [\text{grad } (\mathbf{r} \cdot \mathbf{A})], \end{aligned} \tag{J.15}$$

since $U^\dagger U = 1$

Next, since

$$[AB, C] \equiv A[B, C] + [A, C]B, \tag{J.16}$$

we have

$$\begin{aligned}\sum_{i=1}^{3} [p_i^2, U] &= \sum_i \{p_i[p_i, U] + [p_i, U]p_i\} \\ &\equiv \sum \{[p_i, [p_i, U]] + 2[p_i, U]p_i\}.\end{aligned} \tag{J.17}$$

By (J.14), we have

$$\begin{aligned}[p_i, [p_i, U]] &= e\left[p_i, U\frac{\partial(\mathbf{r}\cdot\mathbf{A})}{\partial x_i}\right] = \frac{e\hbar}{i}\frac{\partial}{\partial x_i}\left[U\frac{\partial(\mathbf{r}\cdot\mathbf{A})}{\partial x_i}\right] \\ &= U\left\{\frac{e\hbar}{i}\frac{\partial^2(\mathbf{r}\cdot\mathbf{A})}{\partial x_i^2} + e^2\left[\frac{\partial(\mathbf{r}\cdot\mathbf{A})}{\partial x_i}\right]^2\right\}.\end{aligned} \tag{J.18}$$

Thus (J.17) becomes when we use (J.14)

$$[\mathbf{p}^2, U] = U\left\{\frac{e\hbar}{i}\nabla^2(\mathbf{r}\cdot\mathbf{A}) + e^2[\text{grad}\,(\mathbf{r}\cdot\mathbf{A})]^2 + 2e\,\text{grad}\,(\mathbf{r}\cdot\mathbf{A})\cdot\mathbf{p}\right\}. \tag{J.19}$$

If we use (J.15) and (J.19), (J.13) becomes

$$\begin{aligned}H' = \frac{\mathbf{p}^2}{2m} &+ \frac{e\hbar}{2im}\nabla^2(\mathbf{r}\cdot\mathbf{A}) + \frac{e^2}{2m}[\nabla(\mathbf{r}\cdot\mathbf{A})]^2 + \frac{e}{m}\nabla(\mathbf{r}\cdot\mathbf{A})\cdot\mathbf{p} \\ &- \frac{e}{m}\mathbf{A}\cdot\mathbf{p} - \frac{e^2}{m}\mathbf{A}\cdot\nabla(\mathbf{r}\cdot\mathbf{A}) + \frac{e^2}{2m}\mathbf{A}^2 + eV.\end{aligned} \tag{J.20}$$

Now

$$\begin{aligned}\frac{\partial}{\partial x_i}(\mathbf{r}\cdot\mathbf{A}) &= A_i + \mathbf{r}\cdot\frac{\partial\mathbf{A}}{\partial x_i} \\ \sum_i\left[\frac{\partial}{\partial x_i}(\mathbf{r}\cdot\mathbf{A})\right]^2 &= \mathbf{A}^2 + \sum_{i,j} 2A_i x_j\frac{\partial A_j}{\partial x_i} + \sum_{i,j,k} x_j x_k\frac{\partial A_j}{\partial x_i}\frac{\partial A_k}{\partial x_i} \\ \nabla^2(\mathbf{r}\cdot\mathbf{A}) &= \sum_{i,j} x_i\frac{{}^2A}{\partial x_j^2}.\end{aligned} \tag{J.21}$$

In the last term we used (J.8).

When we use these, (J.20) reduces to

$$\begin{aligned}H' &= \frac{\mathbf{p}^2}{2m} + \frac{e}{m}\sum_{i,j} x_i\frac{\partial A_i}{\partial x_j}p_j + \frac{e\hbar}{2im}\sum_{i,j} x_i\frac{\partial^2 A_i}{\partial x_j^2} \\ &\quad + \frac{e^2}{2m}\sum_{i,j,k} x_i x_j\frac{\partial A_i}{\partial x_k}\frac{\partial A_j}{\partial x_k} + eV \\ &= K + e\mathbf{r}\cdot\mathbf{E},\end{aligned} \tag{J.22}$$

where we used (J.10).

As an example, consider a plane wave

$$\mathbf{A} = \mathbf{A}_0 e^{-i\omega t + i\mathbf{k}\cdot\mathbf{r}}, \tag{J.23}$$

where

$$\omega^2 = c^2 k^2. \tag{J.24}$$

Then (J.22) becomes

$$K = \frac{\mathbf{p}^2}{2m} + eV(\mathbf{r}) - i\omega e\mathbf{r}\cdot\mathbf{A}(\mathbf{r}, t) + i\frac{e}{m}(\mathbf{r}\cdot\mathbf{A})(\mathbf{k}\cdot\mathbf{p})$$

$$+ i\frac{e\hbar}{2m}k^2(\mathbf{r}\cdot\mathbf{A}) - \frac{e^2}{2m}k^2(\mathbf{r}\cdot\mathbf{A})^2, \tag{J.25}$$

where we used (J.9).

In doing perturbation theory, we take the unperturbed hamiltonian as

$$K_0 = \frac{\mathbf{p}^2}{2m} + eV(\mathbf{r}), \tag{J.26}$$

and we would expect that the first order terms would be those linear in e. However, we see that these terms are in the ratio

$$\omega : \frac{\mathbf{k}\cdot\mathbf{p}}{m} : \frac{\hbar k^2}{2m} = ck : kv : \frac{\hbar k^2}{2m}$$

$$= c : v : \frac{\hbar k}{2m} = 1 : \frac{v}{c} : \frac{\hbar k}{2mc},$$

where we have let $v = p/m$, the electron velocity and used (J.24). Since $v \ll c$ we may neglect the $\mathbf{k}\cdot\mathbf{p}$ term in (J.25). Also at optical frequencies $\hbar k/2mc \ll 1$ and we may neglect the k^2 term linear in e in (J.25). Thus for plane waves under the dipole approximation the hamiltonian is

$$K \cong \frac{\mathbf{p}^2}{2m} + eV(\mathbf{r}) - e\mathbf{r}\cdot\mathbf{E}(\mathbf{r}, t), \tag{J.27}$$

so that it is legitimate to treat $-e\mathbf{r}\cdot\mathbf{E}$ as the perturbing energy rather than $-e\mathbf{A}\cdot\mathbf{p}/m$.

Appendix K

Properties of Fokker–Planck Equations

Let $\alpha_1, \alpha_2, \ldots, \alpha_f \equiv \underset{\sim}{\alpha}$ be a set of c-number independent variables. Let $P(\underset{\sim}{\alpha}, t)$ is a probability distribution function which obeys an equation of the form

$$\frac{\partial P}{\partial t} = -\frac{\partial}{\partial \alpha_i}(\mathscr{A}_i P) + \frac{\partial^2}{\partial \alpha_i \, \partial \alpha_j}(\mathscr{D}_{ij} P) \equiv -\frac{\partial}{\partial \alpha_i} J_i, \tag{K.1}$$

where the repeated indices are summed over from 1 to f and the $\mathscr{A}_i$ and $\mathscr{D}_{ij}$ are functions of the $\underset{\sim}{\alpha}$'s. This equation is called a Fokker–Planck equation. The $\mathscr{A}_i$ are called the components of a drift vector and the $\mathscr{D}_{ij}$ are called the components of a diffusion matrix

To obtain insight into the meaning of the drift vector, let us obtain the equation for the mean value of α_k defined by

$$\langle \alpha_k \rangle = \int P(\alpha, t) \alpha_k \, d\underset{\sim}{\alpha}, \tag{K.2}$$

where $d\underset{\sim}{\alpha} \equiv d\alpha_1, d\alpha_2, \ldots, d\alpha_f$ If we take the time derivative of both sides of (K.2) and use (K.1), we obtain

$$\begin{aligned} \frac{d}{dt}\langle \alpha_k \rangle &= \int \frac{\partial P}{\partial t}(\underset{\sim}{\alpha}, t)\alpha_k \, d\underset{\sim}{\alpha} \\ &= \int \alpha_k \, d\underset{\sim}{\alpha} \left\{ -\frac{\partial}{\partial \alpha_i}(\mathscr{A}_i P) + \frac{\partial^2}{\partial \alpha_j \, \partial \alpha_i}(\mathscr{D}_{ij} P) \right\}. \end{aligned} \tag{K.3}$$

If we integrate the first term by parts once and the second by parts twice, note that

$$\begin{aligned} \frac{\partial \alpha_k}{\partial \alpha_i} &= \delta_{ki} \\ \frac{\partial^2}{\partial \alpha_i \, \partial \alpha_j} \alpha_k &= 0, \end{aligned} \tag{K.4}$$

and assume the integrated parts vanish since $P \to 0$ as $\alpha_j \to \pm\infty$, we obtain

$$\frac{d\langle\alpha_k\rangle}{dt} = \int \mathscr{A}_k P\, d\alpha \equiv \langle\mathscr{A}_k\rangle. \tag{K.5}$$

Therefore, the mean equation of motion for α_k is determined by the kth component of the drift vector.

Consider next the equation of motion

$$\begin{aligned}\frac{d}{dt}\langle\alpha_k\alpha_l\rangle &= \int d\underset{\sim}{\alpha}\, \frac{\partial P}{\partial t}\, \alpha_k\alpha_l \\ &= \int d\underset{\sim}{\alpha}\, \alpha_k\alpha_l \left\{ -\frac{\partial}{\partial\alpha_i}(\mathscr{A}_i P) + \frac{\partial^2}{\partial\alpha_i\,\partial\alpha_j}(\mathscr{D}_{ij}P) \right\}. \end{aligned} \tag{K.6}$$

If we again integrate the first term by parts once and the second twice, we obtain

$$\begin{aligned}\frac{d}{dt}\langle\alpha_k\alpha_l\rangle &= \int d\underset{\sim}{\alpha}\{\alpha_k\mathscr{A}_l P + \alpha_l\mathscr{A}_k P + (\mathscr{D}_{kl} + \mathscr{D}_{lk})P\} \\ &= \langle\alpha_k\mathscr{A}_l\rangle + \langle\alpha_l\mathscr{A}_k\rangle + 2\langle\mathscr{D}_{lk}\rangle \end{aligned} \tag{K.7}$$

since

$$\mathscr{D}_{kl} = \mathscr{D}_{lk}. \tag{K.8}$$

As a simple example consider the Fokker–Planck equation

$$\frac{\partial P}{\partial t} = -\frac{\partial}{\partial x}[(A' - B'x)P] + \mathscr{D}\frac{\partial^2 P}{\partial x^2} = -\frac{\partial}{\partial x}\left[(A' - B'x)P - \mathscr{D}\frac{\partial P}{\partial x}\right], \tag{K.9}$$

where A', B', and $\mathscr{D}$ are constants. By (K.5), we have

$$\frac{d}{dt}\langle x\rangle = A' - B'\langle x\rangle \tag{K.10}$$

since

$$\int P\, dx = 1. \tag{K.11}$$

From (K.10) it follows that

$$\langle x(t)\rangle = e^{-B't}\langle x(0)\rangle + \frac{A'}{B'}(1 - e^{-B't}) \xrightarrow[t\to\infty]{} \frac{A'}{B'}. \tag{K.12}$$

Thus $\langle x(t)\rangle$ approaches a steady-state value when $B' > 0$. The mean motion is unaffected by the diffusion constant $\mathscr{D}$.

By (K.7) and (K.9) we see that

$$\begin{aligned}\frac{d}{dt}\langle x\,(t)\rangle &= 2\langle x(A' - B'x)\rangle + 2\mathscr{D} \\ &= 2A'\langle x\rangle - 2B'\langle x^2\rangle + 2\mathscr{D}. \end{aligned} \tag{K.13}$$

If we use (K.12) and integrate, we find that

$$\langle x^2(t)\rangle - \langle x(t)\rangle^2 = \frac{\mathscr{D}}{B'}(1 - e^{-2B't}) + [\langle x^2(0)\rangle - \langle x(0)\rangle^2]e^{-2B't}, \quad \text{(K.14)}$$

which gives the variance or the mean square deviation. The initial fluctuations die out and in the steady-state

$$(\Delta x)^2 \equiv \langle x^2\rangle - \langle x\rangle^2 \to \frac{\mathscr{D}}{B'}, \quad \text{(K.15)}$$

which shows that residual fluctuations are due to the presence of the diffusion term. This result could have been obtained more directly from (K.10), since

$$\frac{d}{dt}\langle x\rangle^2 = 2\langle x\rangle \frac{d\langle x\rangle}{dt} = 2A'\langle x\rangle - 2B'\langle x\rangle^2. \quad \text{(K.16)}$$

If we subtract this from (K.13), we obtain

$$\frac{d}{dt}\{\langle x^2\rangle - \langle x\rangle^2\} = 2\mathscr{D} - 2B'\{\langle x^2\rangle - \langle x\rangle^2\}, \quad \text{(K.17)}$$

whose solution is given by (K.14).

We next return to the general equation (K.1). To solve this equation we begin by separating out the time dependence. We let

$$P(\underset{\sim}{\alpha}, t) = e^{-\Lambda_l t}P_l(\underset{\sim}{\alpha}). \quad \text{(K.18)}$$

If we put this in (K.1) P_l satisfies the eigenvalue equation

$$LP_l(\underset{\sim}{\alpha}) = \Lambda_l P_l(\underset{\sim}{\alpha}), \quad \text{(K.19)}$$

where L is the operator

$$-L = \frac{\partial}{\partial \alpha_i}\mathscr{A}_i + \frac{\partial^2}{\partial \alpha_i\, \partial \alpha_j}\mathscr{D}_{ij}. \quad \text{(K.20)}$$

The steady-state solution is the eigenfunction associated with the eigenvalue $\Lambda_l = 0$.

We next look for a transformation of the type

$$P_l(\underset{\sim}{\alpha}) = e^{\chi(\alpha)}Q_l(\underset{\sim}{\alpha}), \quad \text{(K.21)}$$

which will remove the first derivative terms of P_l with respect to $\underset{\sim}{\alpha}$ in (K.19). If we can do this by some choice of χ the equation for Q_l is called self-adjoint.

If we note that

$$e^{-\chi}\frac{\partial}{\partial \alpha_i}e^{\chi} = \frac{\partial}{\partial \alpha_i} + \frac{\partial \chi}{\partial \alpha_i} \quad \text{(K.22)}$$

and

$$e^{-\chi}\frac{\partial^2}{\partial\alpha_i\,\partial\alpha_j}e^{\chi} = e^{-\chi}\frac{\partial}{\partial\alpha_i}e^{\chi}\left[\frac{\partial}{\partial\alpha_j}+\frac{\partial\chi}{\partial\alpha_j}\right]$$

$$= \left[\frac{\partial}{\partial\alpha_i}+\frac{\partial\chi}{\partial\alpha_i}\right]\left[\frac{\partial}{\partial\alpha_j}+\frac{\partial\chi}{\partial\alpha_j}\right] \tag{K.23}$$

we see that

$$-e^{-\chi}Le^{\chi} \equiv -L' = \frac{\partial}{\partial\alpha_i}\mathscr{D}_{ij}\frac{\partial}{\partial\alpha_j} + F(\underset{\sim}{\alpha}) + R, \tag{K.24}$$

where

$$F(\underset{\sim}{\alpha}) = \frac{\partial^2\mathscr{D}_{ij}}{\partial\alpha_i\,\partial\alpha_j} + \left(\frac{\partial\chi}{\partial\alpha_i}\frac{\partial\chi}{\partial\alpha_j} + \frac{\partial^2\chi}{\partial\alpha_i\,\partial\alpha_j}\right)\mathscr{D}_{ij} + 2\frac{\partial\chi}{\partial\alpha_i}\frac{\partial\mathscr{D}_{ij}}{\partial\alpha_j} - \frac{\partial\mathscr{A}_i}{\partial\alpha_i} - \mathscr{A}_i\frac{\partial\chi}{\partial\alpha_i} \tag{K.25}$$

and

$$R = \left[-\mathscr{A}_i + 2\mathscr{D}_{ij}\frac{\partial\chi}{\partial\alpha_j} + \frac{\partial\mathscr{D}_{ij}}{\partial\alpha_j}\right]\frac{\partial}{\partial\alpha_i}. \tag{K.26}$$

We therefore see that if we can choose χ so that R is zero, then all first derivative terms of Q_l with respect to α will vanish. We therefore require that

$$2\mathscr{D}_{ij}\frac{\partial\chi}{\partial\alpha_j} = \mathscr{A}_i - \frac{\partial\mathscr{D}_{ij}}{\partial\alpha_j}. \tag{K.27}$$

If the determinant of the matrix $\mathscr{D}_{ij}$ is not zero, its inverse exists such that

$$\mathscr{D}_{ki}^{-1}\mathscr{D}_{ij} = \delta_{kj}. \tag{K.28}$$

If we multiply both sides of (K.27) by $\mathscr{D}_{ki}^{-1}$ and sum on i, we obtain the equations for χ

$$\frac{\partial\chi}{\partial\alpha_k} = \tfrac{1}{2}\mathscr{D}_{ki}^{-1}\left[\mathscr{A}_i - \frac{\partial\mathscr{D}_{ij}}{\partial\alpha_j}\right] \equiv U_k. \tag{K.29}$$

To integrate these equations, it is sufficient that

$$\frac{\partial U_k}{\partial\alpha_l} = \frac{\partial U_l}{\partial\alpha_k}. \tag{K.30}$$

Then χ is independent of the path of integration and is given by

$$\chi(\underset{\sim}{\alpha}) = \int_{\underset{\sim}{\alpha}^0}^{\underset{\sim}{\alpha}} d\alpha_m\,U_m$$

$$= -\tfrac{1}{2}\int_{\underset{\sim}{\alpha}^0}^{\underset{\sim}{\alpha}}\left\{\mathscr{D}_{mi}^{-1}\left[\frac{\partial\mathscr{D}_{ij}}{\partial\alpha_j} - \mathscr{A}_i\right]\right\}d\alpha_m. \tag{K.31}$$

The relations (K.30) are called the integrability conditions.

If we assume the $\mathscr{A}_i$ and $\mathscr{D}_{ij}$ are such that (K.30) is satisfied then $R = 0$ and when we put (K.21) into (K.19) we have

$$Le^{\chi}Q_l = \Lambda_l e^{\chi} Q_l \tag{K.32a}$$

or

$$e^{-\chi}Le^{\chi}Q_l \equiv L'Q_l = \Lambda_l Q_l\,, \tag{K.32b}$$

where

$$-L' = \frac{\partial}{\partial\alpha_i}\mathscr{D}_{ij}\frac{\partial}{\partial\alpha_j} + F(\underset{\sim}{\alpha}). \tag{K.33}$$

Before proceeding to discuss (K.32), let us consider the steady state solution of (K.1) for which $\partial P/\partial t = 0$. In this case we have

$$0 = -\frac{\partial}{\partial\alpha_i}\left[\mathscr{A}_i P_s - \frac{\partial}{\partial\alpha_j}(\mathscr{D}_{ij}P_s)\right] \equiv -\frac{\partial J_i}{\partial\alpha_i}. \tag{K.34}$$

A possible solution corresponds to $J_i = 0$ or

$$\mathscr{D}_{ij}\frac{\partial P_s}{\partial\alpha_j} = \left(\mathscr{A}_i - \frac{\partial\mathscr{D}_{ij}}{\partial\alpha_j}\right)P_s. \tag{K.35}$$

If we multiply by $\mathscr{D}_{ki}^{-1}$ and sum over i, we have

$$\frac{\partial \ln P_s}{\partial\alpha_k} = -\mathscr{D}_{ki}^{-1}\left[\frac{\partial\mathscr{D}_{ij}}{\partial\alpha_j} - \mathscr{A}_i\right] \equiv 2U_k \tag{K.36}$$

by (K.29). Again the relations (K.30) are satisfied if $\ln P$ is given by the line integral

$$\ln P_s = 2\int_{\underset{\sim}{\alpha}}^{\underset{\sim}{\alpha}} U_k\, d\alpha_k \equiv 2\chi(\underset{\sim}{\alpha}) \tag{K.37}$$

and is independent of the path. Therefore, the steady-state distribution is

$$P_s = e^{2\chi(\underset{\sim}{\alpha})}, \tag{K.38}$$

which is the eigenfunction for $\Lambda_l = 0$.

Let us return to (K.32). We have

$$\left[\frac{\partial}{\partial\alpha_i}\mathscr{D}_{ij}\frac{\partial}{\partial\alpha_j} + F(\underset{\sim}{\alpha})\right]Q_l(\underset{\sim}{\alpha}) = -\Lambda_l Q_l(\underset{\sim}{\alpha}). \tag{K.39}$$

To show the eigenfunctions are orthogonal we write this for Λ_m and have

$$\left[\frac{\partial}{\partial\alpha_i}\mathscr{D}_{ij}\frac{\partial}{\partial\alpha_j} + F(\underset{\sim}{\alpha})\right]Q_m(\underset{\sim}{\alpha}) = -\Lambda_m Q_m(\underset{\sim}{\alpha}). \tag{K.40}$$

Then we multiply (K.39) by Q_m from the left, (K.40) by Q_l from the left and subtract. This gives on integrating

$$\int\left\{Q_l\frac{\partial}{\partial\alpha_i}\mathscr{D}_{ij}\frac{\partial}{\partial\alpha_j}Q_m - Q_m\frac{\partial}{\partial\alpha_i}\mathscr{D}_{ij}\frac{\partial}{\partial\alpha_i}Q_l\right\}d\underset{\sim}{\alpha} = (\Lambda_l - \Lambda_m)\int Q_l Q_m\, d\underset{\sim}{\alpha}\,. \tag{K.41}$$

If we integrate the first terms by parts twice and note that the integrated parts vanish, we see it cancels with the second term on the left (i and j are summed over so we may interchange them). Therefore, if $\Lambda_l \neq \Lambda_m$, we have

$$\int Q_l(\underset{\sim}{\alpha})Q_m(\underset{\sim}{\alpha})\, d\underset{\sim}{\alpha} = \delta_{lm}, \tag{K.42}$$

where we normalize $l = m$ to unity. We could not have carried through this proof if we had not eliminated the first derivative terms and made L' self-adjoint.

It can be proved that if the eigenvalues have a lower bound, then the Q_l form a complete set:

$$\sum_l Q_l(\underset{\sim}{\alpha})Q_l(\underset{\sim}{\alpha}') = \delta(\underset{\sim}{\alpha} - \underset{\sim}{\alpha}'). \tag{K.43}$$

In terms of the P_l's, we have

$$\begin{aligned} \int \frac{P_l(\underset{\sim}{\alpha})P_m(\underset{\sim}{\alpha})}{P_s(\underset{\sim}{\alpha})}\, d\underset{\sim}{\alpha} &= \delta_{lm} \\ \sum_l \frac{P_l(\underset{\sim}{\alpha})P_l(\underset{\sim}{\alpha}')}{P_s(\underset{\sim}{\alpha})} &= \delta(\underset{\sim}{\alpha} - \underset{\sim}{\alpha}'), \end{aligned} \tag{K.44}$$

where P_s is the steady-state weighting function.

To illustrate these techniques, consider (K.9). For the steady-state, we have

$$\frac{dP_s}{dx} = \frac{(A' - B'x)}{\mathscr{D}} P_s, \tag{K.45}$$

or

$$P_s = Ce^{A'x-(B'x^2/2)} = \sqrt{\frac{2\pi\mathscr{D}}{B'}} \exp - \frac{B'}{2\mathscr{D}}\left(x - \frac{A'}{B'}\right)^2, \tag{K.46}$$

where we have normalized so that

$$\int_{-\infty}^{\infty} P_s(x)\, dx = 1. \tag{K.47}$$

In this case (K.29) reduces to

$$U = \frac{1}{2\mathscr{D}}[A' - B'x] = \frac{d\chi}{dx}, \tag{K.48}$$

so that

$$\begin{aligned} \chi &= \frac{1}{2\mathscr{D}}\left[A'x - \frac{B'x^2}{2}\right] + \text{constant} \\ &= -\frac{B'}{4\mathscr{D}}\left(x - \frac{A'}{B}\right)^2 + \frac{1}{4}\ln\frac{2\pi\mathscr{D}}{B'} \end{aligned} \tag{K.49}$$

in agreement with the general relation between P_s and $\exp 2\chi$.

If we use (K.48) and (K.25), (K.32b) becomes

$$\left\{\mathscr{D}\frac{d^2}{dx^2} - \frac{1}{4\mathscr{D}}(A' - B'x) + \frac{B'}{2}\right\}Q_l = -\Lambda_l Q_l \tag{K.50}$$

or

$$\frac{d^2Q_l}{dx^2} + \left[\frac{B'}{2\mathscr{D}} + \frac{\Lambda_l}{\mathscr{D}} - \frac{1}{4\mathscr{D}^2}(A' - B'x)^2\right]Q_l = 0. \tag{K.51}$$

If we let

$$\alpha = \frac{(A' - B'x)}{\sqrt{2\mathscr{D}B'}}, \tag{K.52}$$

this reduces to

$$\frac{d^2Q_l}{d\alpha^2} + [2\Lambda_l + 1 - \alpha^2]Q_l = 0. \tag{K.53}$$

The eigenvalues of this are

$$\Lambda_l = l \qquad l = 0, 1, 2, \ldots \tag{K.54}$$

and

$$Q_l(\alpha) = N_l e^{-\frac{1}{2}\alpha^2} H_l(\alpha), \tag{K.55}$$

where the $H_l(\alpha)$ are hermite polynomials.

Finally, the Green's function solution may be written as

$$(P\underset{\sim}{\alpha}, t/\underset{\sim}{\alpha}_0, 0) = \sum_l e^{-\Lambda_l t}\frac{P_l(\underset{\sim}{\alpha})P_l(\underset{\sim}{\alpha}^0)}{P_s(\underset{\sim}{\alpha})}, \tag{K.56}$$

since as $t \to 0$, by (K.44) this solution approaches the required product of δ-functions:

$$\delta(\underset{\sim}{\alpha} - \underset{\sim}{\alpha}^0) \equiv \prod_{j=1}^{f} \delta(\alpha_j - \alpha_j^0). \tag{K.57}$$

The reader should try to solve (K.9) with a function of the form exp ($\mathscr{G}$) where

$$\mathscr{G} = -\frac{1}{\zeta(t)}[x - \eta(t)]^2 + \ln \nu(t), \tag{K.58}$$

which reduces at $t = 0$ to

$$P(x, t/x_0, 0) \underset{t\to 0}{\longrightarrow} \delta(x - x_0). \tag{K.59}$$

Index